D1251070

Handbook of Industrial Gas Utilization

Handbook of Industrial Gas Utilization

Engineering Principles and Practice

R. Pritchard, J. J. Guy, N. E. Connor

VNR VAN NOSTRAND REINHOLD COMPANY
NEW YORK CINCINNATI ATLANTA DALLAS SAN FRANCISCO
LONDON TORONTO MELBOURNE

First published 1977 in Great Britain by Bowker Publishing Company Limited, Epping, Essex

Van Nostrand Reinhold Company Regional Offices:
New York Cincinnati Atlanta Dallas San Francisco

Van Nostrand Reinhold Company International Offices:
London Toronto Melbourne

Library of Congress Catalog Card Number: 77-22278
ISBN: 0-442-26635-9

Published in 1977 by Van Nostrand Reinhold Company 450 West 33rd Street, New York, N.Y. 10001
Published simultaneously in Canada by Van Nostrand Reinhold Ltd.

15 14 13 12 11 10 9 8 7 6 5 4 3 2 1

Library of Congress Cataloging in Publication Data
Pritchard, R
Handbook of industrial gas utilization.
Includes bibliographies and index.
1. Combustion engineering. 2. Gas, Natural.
3. Gas as fuel. I. Guy, J.J., joint author.
II. Connor, N.E., joint author. III. Title.
TJ254.5.P67 1977 621.4'023 77-22278
ISBN 0-442-26635-9

Important notice
It is stressed that all legal requirements, particularly those relating to safety, must be strictly complied with. The information in this book is believed to represent good practice at the time of writing but is given without responsibility and no liability is accepted for any loss or damage arising out of such information.

Typeset in 10 on 11 point Times New Roman by
Amos Typesetters, Hockley, Essex
Illustrations drawn by Colin Taylor
Printed in Great Britain by Thomson Litho Ltd,
East Kilbride, Scotland

Contents

Foreword

Gas has a special place as an industrial fuel. This has been recognized for many years and a wide range of equipment has been developed to take advantage of its unique qualities — flexibility, cleanliness and ease of control. These made it advantageous to use in a wide variety of industrial processes from the very small to the very large, from the working flame burner used in the jewellery trade to large high-temperature steel-heating furnaces.

Natural gas has become available to industry in Great Britain in increasing quantities in recent years. The annual usage of gas by industry immediately prior to the introduction of natural gas was 96.5 PJ (915 million therms) but by 1974/75 the quantity used had risen to 633 PJ (6000 million therms). New discoveries have been made offshore and natural gas is destined to play an even greater part in the industrial scene in Great Britain. The number of nations using natural gas as an industrial fuel has also grown significantly in recent years.

With the need for energy conservation it is important that the best possible use is made of the supplies of this premium fuel. To this end, it is vital that those concerned with gas utilization, both suppliers and users, should have suitable training or guidance. This should include an understanding of the theoretical aspects of gas utilization and their application to practical engineering and heating processes. It should also embrace knowledge of the most recent advances in technology as well as of the experience of the past and that gained in the massive conversion operation that has taken place over recent years.

The authors, lecturers in gas engineering at the University of Salford, have within the one volume covered the fundamental, technological and engineering aspects of industrial gas utilization. In addition they have provided an extensive list of literature references which will enable the reader to study in further detail any particular aspect of the subject matter.

This book is an ambitious work, well prepared and executed and I commend it to the student and the practising engineer both within the gas industry and in industry at large.

Sir Denis Rooke
Chairman, British Gas Corporation

Preface

Since the introduction of natural gas in Britain industrial gas usage has increased more than sixfold and by the end of the present decade industry could be using over 844 PJ (8000 million therms) per annum representing a roughly tenfold increase over a twelve-year period. In terms of the total supply of energy to industry the share held by the gas industry has risen from 4 per cent towards 20 per cent and could rise above this. In terms of useful output, because of greater efficiency in use gas may well provide up to 30 per cent of the needs of the British industrial fuel market by the end of the 1970s.

It is against this background of rapid growth that this book has been written. The authors have been conscious of the absence of up-to-date and comprehensive textbooks on industrial gas utilization and it is hoped that this book will meet the needs of students of gas engineering and fuel technology and be useful to practising engineers in this field.

It should be made clear that coverage is restricted to natural gas and manufactured town gas. The technology of liquefied petroleum gases and their industrial utilization are outside the scope of this book.

The British gas industry has now virtually completed its immense conversion programme from manufactured town gas to natural gas. Consequently the reader may ask why we have dealt with the combustion properties of town gas and made reference to burners and equipment designed for its combustion. The answer is twofold. Firstly, much natural-gas practice has evolved out of long-established techniques developed for town gas and some background knowledge of these methods is necessary to appreciate fully the logic behind many developments. Secondly, it is hoped that this book will be read outside its country of origin and will be of value in countries which have gas industries based on manufactured town gas.

The use of heat produced by burning fuel for the purpose of carrying out industrial processes has a very long history. The design of early equipment based on the use of solid fuels resulted from the application of 'common sense' modified by experience and when new liquid and gaseous fuels became available modifications of the same basic designs were used.

In modern times a much wider range of process heating applications has emerged with a resulting diversification of furnace and appliance design but the same basic requirements of combustion space, heat-transfer area and flue system remain. Whatever the application, the fuel and oxidant have to be brought into contact in a suitable way, a controlled flame has to be produced, the heat transferred to the load at the right temperature, the combustion products removed and the unused heat recovered if possible.

In terms of basic engineering science then, design must be the application of the principles of fluid dynamics, combustion and heat transfer, and however empirical the original design may have been it can best be improved and its full potential achieved if these basic principles are correctly applied.

A brief outline of the appropriate basic principles followed by their application to a wide range of industrial gas-using processes forms the basis of this book.

Chapters 2 to 5 give a concise account of the relevant aspects of fluid dynamics, heat transfer, combustion and acoustics. Chapters 6 to 12 deal with the technology of industrial gas utilization including, for example, burner and furnace design, safety, measurement and control. In these chapters the subject-matter of the earlier chapters is applied extensively. In the final chapters the utilization of gas in a wide variety of industrial and commercial processes is reviewed.

It is a pleasure to thank the many people and organizations who have provided information, advice and help.

We are grateful to the British Gas Corporation for their sponsorship and guidance, in particular to Mr W. A. Fitzsimons for initiating the idea of the book and sustaining its momentum, to Mr V. H. Miller for carrying the burden of liaison between ourselves, BGC and the publishers and to the Director and staff of the BGC Midland Research Station for their valuable criticism and advice.

Thanks are due to Professor F. A. Holland, Chairman of the Department of Chemical Engineering, University of Salford, for granting permission to write the book, for the provision of typing facilities and for the interest he has shown in its preparation.

We are indebted to Mr D. French of Bowker Publishing Company Ltd for carrying out so well the unenviable task of removing errors, inconsistencies and ambiguities from the text. We would however be surprised if the book contained no errors and would be glad to know of those detected by readers.

R. Pritchard
J. J. Guy
N. E. Connor
Salford, 1976

Acknowledgements

Many of the illustrations in this book have previously appeared in other publications of which the copyright is owned by the British Gas Corporation. The authors are grateful to the Corporation for permission to utilize material in this way. The sources of the illustrations are noted in captions.

Grateful acknowledgement is made to the Institution of Gas Engineers for permission to reproduce diagrams and tables that originally appeared in its publications. The sources of these illustrations are noted in captions.

Grateful acknowledgement is also made to the following for permission to use their material:

Academic Press Inc. and Dr Guenther von Elbe: Figures 4:29, 4:30 and 4:39

Adaptogas Ltd: Figure 7:2

Aeromatic Company Ltd: Figure 7:15(*b*), (*d*) & (*f*)

Amal Ltd: Figure 7:15(*a*)

Applied Science Publishers Ltd and Dr R. D. Ford: Figures 5:8 and 5:9

Benn Bros Ltd: Figure 15:13

Geo. Bray & Company Ltd: Figure 7:15(*e*)

British Standards Institution: Figures 8:14, 8:45, 8:56 and 8:59

The Combustion Institute: Figure 4:31

Hartmann & Braun (UK) Ltd: Figure 8:13

The Institute of Fuel: Figures 4:21, 4:22, 6:2, 7:23, 7:51(*b*), 13:9(*a*) & (*b*), 13:11, 13:12, 13:31, 13:32, 13:37(*a*), 13:37(*c*), 15:10, 15:11 and 15:23

Institution of Chemical Engineers: Figure 9:5

Institution of Metallurgists: Figures 13:27 and 13:29

International Flame Research Foundation: Figures 8:3, 8:4 and 8:35

McGraw-Hill Book Company: Figures 3:13, 3:14, 4:1, 4:18, 4:20 and 12:1

Maywick (Hanningfield) Ltd: Figure 7:36(*b*) & (*d*)

Mercury House Publications Ltd: Figures 8:33 and 8:61

Monometer Manufacturing Company Ltd: Figure 13:4(*c*)

Newnes-Butterworths: Figures 8:40, 8:41, 8:42, 8:43 and 8:60

P. C. Compteurs Ltd: Figure 8:54

Pergamon Press Ltd: Figures 3:17, 8:27 and 8:29

The Editor, *Process Engineering:* Figures 8:9, 8:10, 8:21, 8:22, 8:23 and 8:34

Teknigas Ltd: Figure 7:15(*c*)

Turret Press Ltd: Figure 13:46(*d*)

Van Nostrand Reinhold Co Ltd: Figure 13:7

John Wiley & Sons, Inc: Figure 13:2(*a*) & (*b*)

Notes on the authors

Dr R. Pritchard was born in 1935 and joined North Western Gas Board in 1952 as a student engineer, studying Gas Engineering at the Royal Technical College, Salford, between 1954 and 1957 when he was awarded the College Associateship (ARTCS). He remained with North Western Gas Board, working as an industrial gas engineer, until 1961 when he was appointed Lecturer in Gas Engineering at the Royal College of Advanced Technology, Salford which later became the University of Salford. Dr Pritchard was awarded a PhD of the University of Salford in 1972 for research on the burning velocity of methane–air flames and is the author of a number of papers on flame propagation and fuel-gas interchangeability. He is a Chartered Engineer and a Member of the Institution of Gas Engineers and of the Institute of Fuel and currently serves on the Education Committee of the Institute of Fuel.

Mr J. J. Guy was born in London in 1923 and graduated in Chemical Engineering at Imperial College in 1945. After working at the Fuel Research Station he joined the newly formed Gas Research Board and worked on the corrosion of water heaters due to sulphur in gas and on the development of vortex combustion. Mr Guy then moved to gas-appliance design with R. and A. Main and Ascot Water Heaters, and in 1951 became Technical Manager of a vitreous enamelling works. He became a lecturer at Thames Polytechnic in 1960 and at the University of Salford in 1966 where he has been concerned mainly with the teaching of gas engineers, specializing mainly in applications of combustion technology and process control.

Mr N. E. Connor joined the North Western Gas Board as a student engineer in 1950 and began studying Mechanical Engineering at Warrington and St Helens Technical Colleges. In 1953 he was awarded a Whitworth Society prize and a Technical State Scholarship to study Gas Engineering at the University of Leeds, graduating in 1957 with an Honours BSc Degree. He returned to the North Western Gas Board as a production engineer in the South Lancashire group and was appointed Chief Chemist at the Warrington Production Station in 1958. In 1964 he left to take up an appointment as a Lecturer in Gas Engineering at the Royal College of Advanced Technology, Salford which later became the University of Salford. He is a Chartered Engineer, a Fellow of the Institution of Gas Engineers, and of the Institute of Fuel. His Institution activities include membership of the Council of Manchester Junior Gas Association and of the Committee of the North Western Section of the Institute of Fuel. In recent years he has also been actively engaged in running Symposia on Gas Engineering, Fuel Utilization and Chemical Engineering at the University.

List of symbols

For symbols for SI units of measurement, see pages 2 and 3. The only meanings listed here are those used several times in the book.

Signs

—	average; (over lower-case letters) direct radiant exchange area in either direction; (over capital letters) total radiant exchange area in either direction
$\longrightarrow$	(over lower-case letters) direct radiant exchange area in direction shown; (over capital letters) total radiant exchange area in direction shown; vector
$\ominus$	standard state (most stable form of substance at room temperature) for measurement of thermodynamic quantities
°C	degree(s) Celsius
°F	degree(s) Fahrenheit
°R	degree(s) Rankine

Greek alphabet

α	absorptance; acoustic absorption; linear expansivity
α_t	power transmission coefficient
β	blade angle; fraction of waste heat not recovered
γ	c_p/c_V; cubic expansivity
δ	boundary-layer thickness
δ^*	displacement thickness
Δ	change
ε	eddy diffusivity; emissivity; expansibility factor
η	efficiency; thickness of preheat zone
η_p	pressure efficiency
κ	K/p where K is defined in Section 12.3.1 and p is partial pressure
λ	thermal conductivity; wavelength
μ	dynamic viscosity
ν	kinematic viscosity; refractive index
ρ	density; reflectance
σ	Stefan–Boltzmann constant ($\approx 5.670 \times 10^{-8}$ W m^{-2} K^{-4})
τ	characteristic time; shear stress; transmittance

θ	angle; Thring–Newby number
φ	angle
φ_X	volume fraction of substance X
Φ	heat flow rate
χ_m	magnetic susceptibility
ω	angular frequency; weight per unit volume
Ω	angular velocity

Subscripts

a	air; apparent; atmosphere
atm	atmospheric
b	burned gas; burner bar
c	cold; contraction; critical
d	detonation; discharge
e	entrained; external
f	bulk of a fluid; flame cone; formation; forward; friction; furnace
F	flashback
g	gas
h	hot
i	ignition point; incident; inert gases; internal
j	jet
L	laminar; load
m	mean value; mixture; mixture tube; model
M	manometer
n	nozzle
N	normal direction
o	orifice
p	burner port; plane
p	at constant pressure
r	reaction; real; reflected; refractory; resonator
s	shock wave; stagnation; static; surface of a fluid
t	throat; total; transmitted; tube; tunnel exit
T	turbulent
u	unburned gas
v	valve; vent
V	at constant volume
w	wall; wind

Latin alphabet

a	area; constant; velocity of sound
A	amplitude; area; frequency factor of encounters between reactant molecules
$\boldsymbol{A}$	complex amplitude
(Ar)	Archimedes number, $l\Delta Tg/u^2T_g$
atm	standard atmosphere (= 101 325 Pa)
b	breadth; constant
B	radiant power per unit surface area; radiosity
Btu	British thermal unit
c	concentration; specific heat capacity
c_g	gross calorific value per unit mass

c_n	net calorific value per unit mass
c_p	specific heat capacity at constant pressure
$\bar{c}_p$	mean value of c_p over a specified temperature range
c_V	specific heat capacity at constant volume
C	heat capacity; molar heat capacity
C_c	coefficient of contraction
C_d	discharge coefficient
C_D	drag coefficient, $\Delta p/\rho U^2$
C_{do}	discharge coefficient at orifice
C_{dp}	discharge coefficient at burner ports
C_F	ratio of friction head loss to kinetic head
C_g	gross calorific value
C_m	molar heat capacity
C_n	net calorific value per unit volume
(Ct)	Craya–Curtet parameter
d	relative density (specific gravity); thickness
$d_{\parallel}$	quenching distance
d_p	depth of penetration of quenching
D	density of sound energy; diameter; diffusivity
D_0	quenching diameter
D_{eff}	see equation 4.72
E	electromotive force; energy; $(1-r_a)^{-\frac{1}{2}}$ (see Section 8.5.6)
E_a	activation energy of a reaction
E_{min}	minimum ignition energy
f	correction factor in equation 4.86; Fanning friction factor; frequency; friction factor
F	correction factor in bar-burner design (see Section 10.1.6); factor defined in Section 12.3.1; Faraday constant ($\approx 9.649 \times 10^4$ C mol^{-1}); force; view factor
F_D	drag force
$\mathbf{F}$	force (vector)
(Fr)	Froude number, U^2/gL
ft	foot
g	gaseous
g	acceleration due to gravity; boundary velocity gradient
g_B	boundary velocity gradient at which blowoff occurs
g_F	boundary velocity gradient at which flashback (lightback) occurs
G	weight
(Gr)	Grashof number, $L^3\gamma g\Delta T\rho^2/\mu^2$ (where γ is cubic expansivity)
h	hour
h	heat-transfer coefficient
h_D	mass-transfer coefficient
h_f	head loss due to friction
h_s	specific stagnation enthalpy
H	enthalpy; radiant power per unit surface area; weight of water in unit volume of moist air
i	momentum
I	electric current; sound intensity; specific enthalpy
in	inch

inHg	inch of mercury (conventional)
inH_2O	inch of water (conventional)
j	complex number with property $j^2+1=0$
j_D	Colburn *j* factor for mass transfer, $(h_D/u)(\mu/\rho D)^{2/3}$ (where *D* is diffusivity)
j_h	Colburn *j* factor for heat transfer, $(St)(Pr)^{2/3}$
J	rate of a reaction
k	circular wavenumber; rate constant of a reaction; transport coefficient
K	bulk modulus; constant; equilibrium constant; Karlovitz flame-stretch factor; proportionality factor defined in Section 12.3.1; vent area coefficient
K_{av}	see Section 7.4.5
l	liquid
l	length; mean free path; specific latent heat
L	characteristic length; flammability limit; radiance; scale of turbulence
L_{co}	outer cone height
L_f	flame height; length of flame run
L_p	height of primary flame cone
lbf	pound-force
(Le)	Lewis number, $\rho c_p D_{eff}/\lambda$ (see equation 4.72)
m^3 (st)	standard cubic metre, i.e. 1 m^3 measured at 15 °C and 101 325 Pa (dry)
m	mass; mass rate of flow
M	molar mass; mass per unit area; relative molecular mass
(Ma)	Mach number
mmHg	millimetre of mercury (conventional)
n	excess-air factor; number density of molecules
N	frequency of rotation; noise rating number
N_s	specific speed
(Nu)	Nusselt number, hL/λ
p	partial pressure; pressure
p_g	gauge pressure
$\mathbf{p}$	complex pressure
P	absolute pressure; sound energy flux
(Pr)	Prandtl number, $c_p\mu/\lambda$
q	quantity of heat
Q	torque
r	radius; ratio; ratio of volume rates of flow
r_a	area ratio
r_p	pressure ratio
r_r	recirculation ratio
r_T	temperature ratio
$\mathbf{r}$	position (vector)
R	electrical resistance; molar gas constant ($\approx$ 8.314 J K^{-1} mol^{-1}); rangeability; thermal resistance
(Re)	Reynolds number, $U\rho L/\mu$
s	solid
s	distance along a path; loudness
S	cross-sectional area; humid heat (specific heat capacity of moist air); swirl number; Weaver's flame-speed factor
S_f	flame speed
S_u	burning velocity

(Sc)	Schmidt number, $\mu/\rho D$ (where D is diffusivity)
sft^3	standard cubic foot (i.e. a volume of 1 ft^3 when measured at a temperature of 60 °F and a pressure of 30 inHg and saturated with water vapour)
(Sh)	Sherwood number, $h_D L/D$ (where D is diffusivity)
(St)	Stanton number, $(Nu)/(Re)(Pr)$
t	tonne
t	time; Celsius temperature
t_r	residence time
T	period; thermodynamic (absolute) temperature
T_d	dew point
TR	ton of refrigeration
u	velocity
u_d	detonation velocity
u_{ext}	blowoff velocity
u_g	axial unburned-gas velocity
u_u	velocity of unburned gas
$\boldsymbol{u}$	velocity (vector)
U	average velocity; overall coefficient of heat transfer
v	specific volume; velocity
V	volume; volume rate of flow
V_b	breakdown voltage
V_m	molar volume
w	tangential component of velocity; work
W	power; Wobbe number, $C_g/\sqrt{d}$ (where d is relative density)
x	mole fraction; rectangular coordinate
y	rectangular coordinate
z	rectangular coordinate
Z	acoustic impedance
$\boldsymbol{Z}$	complex acoustic impedance

1

Dimensions, units and data

All physical quantities may be expressed in terms of fundamental dimensions. The number and choice of these is somewhat arbitrary, but the inclusion of mass, length and time has proved convenient. In addition fundamental entities such as temperature, electricity and light have usually been added.

Unit quantities of these basic dimensions can be selected arbitrarily and a coherent system of units can then be derived for all physical quantities, in which the product or quotient of any two unit quantities in the system is the unit value of the resulting quantity.

In the SI (International System of Units), seven fundamental dimensions are used, the basic units of which are the kilogram, kg (mass), the metre, m (length), the second, s (time), the ampere, A (electric current), the kelvin, K (temperature), the candela, cd (light intensity) and (since 1971) the mole, mol (amount of substance). In addition, there are two 'supplementary units', the radian, rad (plane angle) and the steradian, sr (solid angle).

Figure 1:1 lists some of the named derived units and Figure 1:2 lists the prefixes used in forming multiples of units in the SI.

Engineers have to take their data from many sources, often expressed in older systems of units. Figures 1:3 to 1:20 list some of the more common conversion factors (to four significant figures) between systems of units.

The enthalpies of combustion products at normal combustion temperatures from the combustion of lean mixtures of methane and air can be obtained from Figures 1:23 and 1:25. At temperatures higher than about 2000 K the result will be slightly in error because of the change in dissociation consequent upon composition change. For example, to calculate the enthalpy at 2000 K of products from the combustion of methane with 50 per cent excess air. From Figure 1:23, 1 volume of methane produces 10.811 volumes of combustion products. From Figure 1:21, 50 per cent excess air = 4.88 volumes of air per volume of methane. Hence:

$$H = \frac{(10.811 \times 2.72416) + (4.88 \times 2.44932)}{10.811 + 4.88}$$
$$= 2.64 \text{ MJ m}^{-3} \text{ (st)}$$

Quantity	Name of SI unit	Symbol	Expressed in terms of SI base units or derived units
Frequency	hertz	Hz	$1\ \mathrm{Hz} = 1\ \mathrm{s}^{-1}$
Force	newton	N	$1\ \mathrm{N} = 1\ \mathrm{kg\,m\,s}^{-2}$
Pressure, stress	pascal	Pa	$1\ \mathrm{Pa} = 1\ \mathrm{N\,m}^{-2}$
Work, energy, quantity of heat	joule	J	$1\ \mathrm{J} = 1\ \mathrm{N\,m}$
Power	watt	W	$1\ \mathrm{W} = 1\ \mathrm{J\,s}^{-1}$
Quantity of electricity	coulomb	C	$1\ \mathrm{C} = 1\ \mathrm{A\,s}$
Electric potential, potential difference, tension, electromotive force	volt	V	$1\ \mathrm{V} = 1\ \mathrm{W A}^{-1}$
Electric capacitance	farad	F	$1\ \mathrm{F} = 1\ \mathrm{A\,s\,V}^{-1}$
Electric resistance	ohm	Ω	$1\ \Omega = 1\ \mathrm{V\,A}^{-1}$
Conductance	siemens	S	$1\ \mathrm{S} = 1\ \mathrm{A\,V}^{-1}$
Flux of magnetic induction, magnetic flux	weber	Wb	$1\ \mathrm{Wb} = 1\ \mathrm{V\,s}$
Magnetic flux density, magnetic induction	tesla	T	$1\ \mathrm{T} = 1\ \mathrm{Wb\,m}^{-2}$
Inductance	henry	H	$1\ \mathrm{H} = 1\ \mathrm{V\,s\,A}^{-1}$
Luminous flux	lumen	lm	$1\ \mathrm{lm} = 1\ \mathrm{cd\,sr}$
Illumination	lux	lx	$1\ \mathrm{lx} = 1\ \mathrm{lm\,m}^{-2}$
Radioactivity	becquerel	Bq	$1\ \mathrm{Bq} = 1\ \mathrm{s}^{-1}$
Radiation dose	gray	Gy	$1\ \mathrm{Gy} = 1\ \mathrm{J\,kg}^{-1}$

The names pascal and siemens were adopted in 1971. The names becquerel and gray were adopted in 1975

Figure 1:1 SI derived units with special names

Factor by which the unit is multiplied	Prefix	Symbol
10^{18}	exa	E
10^{15}	peta	P
10^{12}	tera	T
10^{9}	giga	G
10^{6}	mega	M
10^{3}	kilo	k
10^{2}	hecto	h
10	deca	da
10^{-1}	deci	d
10^{-2}	centi	c
10^{-3}	milli	m
10^{-6}	micro	μ
10^{-9}	nano	n
10^{-12}	pico	p
10^{-15}	femto	f
10^{-18}	atto	a

The symbol of a prefix is considered to be combined with the unit symbol to which it is directly attached, forming with it a new unit symbol which can be raised to a positive or negative power and which can be combined with other unit symbols to form symbols for compound units

Figure 1:2 SI prefixes for multiples and submultiples of units

	mm	in	ft	yd	m	km	mile
1 mm =	1	0.03937	3.281×10^{-3}	1.094×10^{-3}	10^{-3}	10^{-6}	0.6214×10^{-6}
1 in =	25.4	1	0.08333	0.02778	0.0254	25.4×10^{-6}	15.78×10^{-6}
1 ft =	304.8	12	1	0.3333	0.3048	0.3048×10^{-3}	189.4×10^{-6}
1 yd =	914.4	36	3	1	0.9144	0.9144×10^{-3}	568.2×10^{-6}
1 m =	10^3	39.37	3.281	1.094	1	10^{-3}	621.4×10^{-6}
1 km =	10^6	39.37×10^3	3.281×10^3	1094	10^3	1	0.6214
1 mile =	1.609×10^6	63360	5280	1760	1609	1.609	1

1 mil = 1 thou = 10^{-3} in, 1 Å = 10^{-10} m, 1 μm = 10^{-6} m

Figure 1:3 Conversion factors: length

	mm²	in²	ft²	yd²	m²	acre	km²	mile²
1 mm² =	1	1.550×10^{-3}	10.76×10^{-6}	1.196×10^{-6}	10^{-6}	247.1×10^{-12}	10^{-12}	386.1×10^{-15}
1 in² =	645.2	1	6.944×10^{-3}	771.6×10^{-6}	645.2×10^{-6}	159.4×10^{-9}	645.2×10^{-12}	249.1×10^{-12}
1 ft² =	92.90×10^3	144	1	0.1111	0.09290	23.0×10^{-6}	92.90×10^{-9}	35.87×10^{-9}
1 yd² =	836.1×10^3	1296	9	1	0.8361	206.6×10^{-6}	836.1×10^{-9}	322.8×10^{-9}
1 m² =	10^6	1550	10.76	1.196	1	247.1×10^{-6}	10^{-6}	386.1×10^{-9}
1 acre =	4047×10^6	6.273×10^6	43560	4840	4047	1	4.047×10^{-3}	1.562×10^{-3}
1 km² =	10^{12}	1550×10^6	10.76×10^6	1.196×10^6	10^6	247.1	1	0.3861
1 mile² =	2.59×10^{12}	400.6×10^6	27.88×10^6	3.097×10^6	2.59×10^6	640	2.590	1

1 are = 100 m², 1 hectare = 10^4 m², 1 cm² = 10^{-4} m²

Figure 1:4 Conversion factors: area

		mm^3	in^3	dm^3*	gal	ft^3	yd^3	**m^3**
1 mm^3	=	1	61.02×10^{-6}	10^{-6}	0.220×10^{-6}	35.31×10^{-9}	1.308×10^{-9}	10^{-9}
1 in^3	=	16.39×10^3	1	0.01639	3.605×10^{-3}	578.7×10^{-6}	21.43×10^{-6}	16.39×10^{-6}
1 dm^3*	=	10^6	61.02	1	0.220	0.03531	1.308×10^{-3}	10^{-3}
1 gal	=	4.546×10^6	277.4	4.546	1	0.1605	5.946×10^{-3}	4.546×10^{-3}
1 ft^3	=	28.32×10^6	1728	28.32	6.229	1	0.03704	0.02832
1 yd^3	=	764.6×10^6	46656	764.6	168.2	27	1	0.7646
1 m^3	=	10^9	61024	10^3	220.0	35.31	1.308	1

* The litre has been defined as 1 dm^3 exactly, and this definition has been provisionally accepted in the UK. The name litre should, however, be avoided when referring to precise measurements

1 pt = 0.5682 dm^3, 1 cm^3 = 10^3 mm^3, 1 ml = 1 cm^3, 1 US gal = 231 in^3

1 standard cubic foot (sft^3) (i.e. 1 ft^3 of a gas saturated with water vapour measured at a temperature of 60 °F and a pressure of 30 in of mercury measured at a latitude of 53°N where $g = 9.81329$ m s^{-2}) is equivalent to 0.02778 standard cubic metre (m^3 (st)) (i.e. when dry and measured at 15 °C and 101.325 kPa)

Figure 1:5 Conversion factors: volume and capacity

		mm/s	in/min	ft/min	in/s	ft/s	**m/s**	**km/s**
1 mm/s	=	1	2.368	0.1969	0.03937	3.281×10^{-3}	10^{-3}	10^{-6}
1 in/min	=	0.4233	1	0.08333	0.01667	1.389×10^{-3}	0.4233×10^{-3}	0.4233×10^{-6}
1 ft/min	=	5.08	12	1	0.2	0.0166	5.08×10^{-3}	5.08×10^{-6}
1 in/s	=	25.4	60	5	1	0.08333	0.0254	25.4×10^{-6}
1 ft/s	=	304.8	720	60	12	1	0.3048	304.8×10^{-6}
1 m/s	=	10^3	2368	196.9	39.37	3.281	1	10^{-3}
1 km/s	=	10^6	2368×10^3	196.9×10^3	39.37×10^3	3281	10^3	1

1 km/h = 0.6214 mph (mile/h), 1 mph = 1.609 km/h

Figure 1:6 Conversion factors: velocity

	mm³/s	ft³/h	ft³/min	m³/s
1 mm³/s =	1	127.1×10^{-6}	2.119×10^{-6}	10^{-9}
1 ft³/h =	7865	1	0.0166	7.865×10^{-6}
1 ft³/min =	471.9×10^{3}	60	1	471.9×10^{-6}
1 m³/s =	10^{9}	127.1×10^{3}	2119	1

Figure 1:7 Conversion factors: volume rate of flow

$1 \text{ in}^4 = 0.4162 \times 10^{-6} \text{ m}^4$ $\qquad$ $1 \text{ m}^4 = 2.403 \times 10^{6} \text{ in}^4$

Figure 1:8 Conversion factors: second moment of area

	g	lb	kg	Mg	ton
1 g =	1	2.205×10^{-3}	10^{-3}	10^{-6}	984.2×10^{-9}
1 lb =	453.6	1	0.4536	453.6×10^{-6}	446.4×10^{-6}
1 kg =	10^{3}	2.205	1	10^{-3}	984.2×10^{-6}
1 Mg =	10^{6}	2205	10^{3}	1	0.9842
1 ton =	1.016×10^{6}	2240	1016	1.016	1

1 slug = 14.59 kg, 1 oz = 28.35 g, 1 tonne (t) = 1 Mg

Figure 1:9 Conversion factors: mass

	kg/m^3	lb/ft^3	kg/ft^3	lb/in^3	g/mm^3
1 kg/m^3 =	1	0.06243	0.02832	36.05×10^{-6}	10^{-6}
1 lb/ft^3 =	16.02	1	0.4536	578.7×10^{-6}	16.02×10^{-6}
1 kg/ft^3 =	35.31	2.205	1	1.278×10^{-3}	35.31×10^{-6}
1 lb/in^3 =	27.68×10^3	1728	783.8	1	27.68×10^{-3}
1 g/mm^3 =	10^6	62.43×10^3	28.32×10^6	36.05	1

1 mg/ml = 1 kg/m^3, 1 gm/cm^3 (gm/cc) = 1 g/ml = 10^{-3} g/mm^3

Figure 1:10 Conversion factors: density

Dynamic	1 poise (P) = 10^{-1} Pa s	1 Pa s = 10 P
	1 lbf s/ft^2 = 47.88 Pa s	1 Pa s = 0.02089 lbf s/ft^2
Kinematic	1 stokes (St) = 10^{-4} m^2/s	1 m^2/s = 10^4 St
	1 ft^2/s = 0.09290 m^2/s	1 m^2/s = 10.76 ft^2/s

Figure 1:11 Conversion factors: viscosity

	pdl	N	lbf	kgf	kN	tonf
1 pdl =	1	0.1383	0.03108	0.01410	138.3×10^{-6}	13.88×10^{-6}
1 N =	7.233	1	0.2248	0.1020	10^{-3}	100.4×10^{-6}
1 lbf =	32.17	4.448	1	0.4536	4.448×10^{-3}	446.4×10^{-6}
1 kgf =	70.93	9.807	2.205	1	9.807×10^{-3}	0.9842×10^{-3}
1 kN =	7233	10^3	224.8	102.0	1	0.1004
1 tonf =	72.07×10^3	9964	2240	1016	9.964	1

1 dyne = 10^{-5} N

Figure 1:12 Conversion factors: force

	Pa	lb/ft² (psf)	inH_2O	**kPa**	inHg	lbf/in² (psi)	bar	tonf/in² (Tsi)
1 Pa =	1	0.02089	4.015×10^{-3}	10^{-3}	295.3×10^{-6}	145.0×10^{-6}	10^{-5}	64.75×10^{-9}
1 lbf/ft² =	47.88	1	0.1922	0.04788	0.01414	6.944×10^{-3}	478.8×10^{-6}	3.100×10^{-6}
1 inH_2O =	249.1	5.202	1	0.2491	0.07356	0.03613	2.491×10^{-3}	
1 kPa =	10^3	20.89	4.015	1	0.2953	0.1450	10^{-2}	64.75×10^{-6}
1 inHg =	3386	70.73	13.60	3.386	1	0.4912	0.03386	
1 lbf/in² =	6895	144	27.68	6.895	2.036	1	0.06895	446.4×10^{-6}
1 bar =	10^5	2089	401.5	10^2	29.53	14.50	1	6.475×10^{-3}
1 tonf/in² =	15.44×10^6	322.6×10^3		15.44×10^3		2240	154.4	1

1 kgf/cm² = 98.07 kPa, 1 tonf/ft² = 2223 kPa, 1 ksi = 10^3 lbf/in² = 6895 kPa
1 torr = 133.3 Pa, 1 atm = 1013 mbar, 1 mbar = 10^2 Pa
The bar is *not* an SI unit and if quoted should always be accompanied by an equivalent in SI units
Where any ambiguity is likely, absolute and gauge pressure should be distinguished by writing 'abs.' or 'gauge' after the unit of pressure. No other abbreviations are at present recognized for use with metric units

Figure 1:13 Conversion factors: pressure, stress

	ft pdl	**J**	ft lbf	cal	Btu	**MJ**	kW h	therm
1 ft pdl =	1	0.04214	0.03109	0.01007	39.94×10^{-6}	42.14×10^{-9}	11.70×10^{-9}	399.4×10^{-12}
1 J =	23.73	1	0.7376	0.2388	947.8×10^{-6}	10^{-6}	277.7×10^{-9}	9.478×10^{-9}
1 ft lbf =	32.17	1.356	1	0.3238	1.285×10^{-3}	1.356×10^{-6}	376.6×10^{-9}	12.85×10^{-9}
1 cal =	99.36	4.187	3.088	1	3.968×10^{-3}	4.187×10^{-6}	1.163×10^{-6}	39.68×10^{-9}
1 Btu	25030	1055	778.2	252.0	1	1.055×10^{-3}	293.1×10^{-6}	10^{-5}
1 MJ =	23.73×10^6	10^6	737.6×10^3	238.8×10^3	947.8	1	0.2778	9.478×10^{-3}
1 kW h =	85.43×10^6	3.6×10^6	2.655×10^6	859.8×10^3	3412	3.6	1	0.03412
1 therm =	2503×10^6	105.5×10^6	77.82×10^6	25.20×10^6	10^5	105.5	29.31	1

1 erg = 10^{-7} J
The calorie used here is the international table calorie, defined as 4.1868 J exactly

Figure 1:14 Conversion factors: heat, work, energy

		Btu/h	W	ft lbf/s	h.p.	kW
1 Btu/h	=	1	0.2931	0.2162	0.003930	293.1×10^{-6}
1 W	=	3.412	1	0.7376	0.01341	10^{-3}
1 ft lbf/s	=	4.626	1.356	1	0.01819	1.356×10^{-3}
1 h.p.	=	2545	745.7	550.5	1	0.7457
1 kW	=	3412	10^3	737.6	13.41	1

Figure 1:15 Conversion factors, power, heat flow rate

		MJ/m³ (st)		Btu (international table)/sft³		Btu (15 °C)/sft³	
		Dry	Saturated	Dry	Saturated	Dry	Saturated
MJ/m³ (st)	Dry	1	0.9832	26.80	26.33	26.81	**26.34**
	Saturated	1.017	1	27.26	26.78	27.27	26.79
Btu (international table)/sft³	Dry	0.037 31	0.036 69	1	0.9826	1.000	0.9829
	Saturated	0.037 97	0.037 34	1.018	1	1.018	1.000
Btu (15 °C)/sft³	Dry	0.037 30	0.036 67	0.9997	0.9823	1	0.9826
	Saturated	**0.037 96**	0.037 32	1.017	0.9997	1.018	1

For practical purposes, there is no difference between the two definitions of the British thermal unit. In the gas industry, the 15 °C British thermal unit is the standard and is approximately equal to 1054.73 J
In the British system, measurements of calorific value are referred to conditions in which the gas is saturated and measured at 60 °F and 30 in mercury (measured at latitude 53°N). The 'metric standard conditions' are that the gas is dry and measured at 15 °C (= 59 °F) at a pressure of 101.325 kPa. Conversion factors between these two standards are bold in the above table. See also Figure 1:5

Figure 1:16 Conversion factors: calorific value

Let the same temperature be measured as t °F, θ °C and T K. Then:

$$\theta = \tfrac{5}{9}(t - 32)$$
$$T = \tfrac{5}{9}(t + 459.67)$$
$$T = \theta + 273.15$$

Temperature difference: 1 °C = 1 K = $\tfrac{9}{5}$ °F

Figure 1:17 Conversion factors: temperature

1 Btu/ft² h = 3.155 W/m²
1 W/m² = 0.3170 Btu/ft² h

Figure 1:18 Conversion factors: heat loss, intensity of heat flow rate

		W/m² °C	Btu/ft² h °F	cal/cm² s °C
1 W/m² °C	=	1	0.1761	23.88×10^{-6}
1 Btu/ft² h °F	=	5.678	1	135.6×10^{-6}
1 cal/cm² s °C	=	41 868	7373	1

The calorie used here is the international table calorie = 4.1868 J exactly

Figure 1:19 Conversion factors: thermal conductance, heat-transfer coefficient

		W/m °C	Btu/ft h °F	Btu in/ft² s °F
1 W/m °C	=	1	0.5778	1.926×10^{-3}
1 Btu/ft h °F	=	1.731	1	3.333×10^{-3}
1 Btu in/ft² s °F	=	519.2	300	1

Figure 1:20 Conversion factors: thermal conductivity

Constituent	N_2	CO_2	CH_4	C_2H_6	C_3H_8	C_4H_{10}	C_5H_{12}
Mole fraction (%)	1.46	0.05	94.4	3.22	0.6	0.2	0.07

Calorific values

	Gross	Net
MJ/m³(st)	38.65	34.88
MJ/kg	53.77	48.53
Btu/sft³	1018	919

Effective molecular weight 16.98
Specific gravity 0.586
Stoichiometric mixture 9.76 mol (volumes) air per mol (volume) gas

Figure 1:21 Average composition and properties of a natural gas (Bacton No. 2)

Temperature (K)	N_2	H_2O	H_2	O_2	O	OH	H	NO	N	CO	CO_2	Relative molecular mass (molecular weight)
4000	0.50731	0.00292	0.01744	0.00915	0.14139	0.02111	0.20948	0.02030	0.00125	0.06832	0.00132	20.10
3800	0.52525	0.00843	0.02668	0.01457	0.11873	0.03126	0.17968	0.02265	0.00059	0.06958	0.00259	20.83
3600	0.55136	0.02131	0.03609	0.02087	0.09040	0.04103	0.13920	0.02375	0.00026	0.07067	0.00506	21.86
3400	0.56390	0.04464	0.04117	0.02607	0.06095	0.04576	0.09445	0.02292	0.00010	0.07045	0.00958	23.10
3200	0.61803	0.07661	0.03887	0.02813	0.03587	0.04282	0.05514	0.02009	0.00000	0.06714	0.01728	24.37
3000	0.64867	0.11053	0.03074	0.02634	0.01825	0.03377	0.02756	0.01589	0.00000	0.05912	0.02914	25.48
2800	0.67325	0.13961	0.02093	0.02136	0.00788	0.02266	0.01179	0.01127	0.00000	0.04631	0.04494	26.34
2600	0.69135	0.16074	0.01258	0.01478	0.00281	0.01299	0.00429	0.00705	0.00000	0.03115	0.06225	26.90
2400	0.70344	0.17413	0.00675	0.00860	0.00080	0.00633	0.00130	0.00383	0.00000	0.01754	0.07728	27.37
2200	0.71034	0.18163	0.00321	0.00416	0.00017	0.00257	0.00032	0.00177	0.00000	0.00812	0.08751	27.61
2000	0.71409	0.18530	0.00132	0.00164	0.00003	0.00085	0.00006	0.00068	0.00000	0.00301	0.09302	27.72
1800	0.71556	0.18685	0.00044	0.00052	0.00000	0.00022	0.00001	0.00021	0.00000	0.00086	0.09534	27.77
1600	0.71604	0.18738	0.00011	0.00013	0.00000	0.00004	0.00000	0.00005	0.00000	0.00017	0.09609	27.79
1400	0.71616	0.18751	0.00001	0.00004	0.00000	0.00000	0.00000	0.00001	0.00000	0.00002	0.09625	27.79
1200	0.71617	0.18753	0.00000	0.00003	0.00000	0.00000	0.00000	0.00000	0.00000	0.00000	0.09627	27.79
1000	0.71617	0.18753	0.00000	0.00003	0.00000	0.00000	0.00000	0.00000	0.00000	0.00000	0.09627	27.79
800	0.71617	0.18753	0.00000	0.00003	0.00000	0.00000	0.00000	0.00000	0.00000	0.00000	0.09627	27.79
600	0.71617	0.18753	0.00000	0.00003	0.00000	0.00000	0.00000	0.00000	0.00000	0.00000	0.09627	27.79
400	0.71617	0.18753	0.00000	0.00003	0.00000	0.00000	0.00000	0.00000	0.00000	0.00000	0.09627	27.79
288	0.71617	0.18753	0.00000	0.00003	0.00000	0.00000	0.00000	0.00000	0.00000	0.00000	0.09627	27.79
Flame temperature conditions												
2200	0.70961	0.18066	0.00365	0.00477	0.00026	0.00308	0.00045	0.00200	0.00000	0.00942	0.08611	27.57

Figure 1:22 Composition of the products of stoichiometric combustion of methane and air at various temperatures

Temperature (K)	Moles of combustion products per mole of natural gas*	Enthalpy of undissociated products† ($MJ\ m^{-3}$(st))	Enthalpy of undissociated products† ($MJ\ kg^{-1}$)	Density ($kg\ m^{-3}$)	'Frozen' specific heat capacity‡ ($J\ kg^{-1}\ °C^{-1}$)	Equilibrium specific heat capacity§ ($J\ kg^{-1}\ °C^{-1}$)
4000	14.910	14.48364	12.32284	0.0612	1589.0	—
3800	14.388	13.13006	11.17120	0.0668	1580.3	—
3600	13.711	11.47513	9.76316	0.0740	1569.9	—
3400	12.973	9.68260	8.23806	0.0828	1558.7	7572.4
3200	12.298	7.98655	6.79505	0.0928	1548.1	6743.0
3000	11.764	6.54235	5.56630	0.1035	1538.9	5538.6
2800	11.378	5.37749	4.57523	0.1146	1530.6	4402.7
2600	11.116	4.45746	3.79246	0.1263	1522.1	3453.6
2400	10.950	3.73895	3.18114	0.1389	1511.9	2639.7
2200	10.857	3.17724	2.70323	0.1529	1498.6	2124.8
2000	10.811	2.72416	2.31774	0.1639	1481.8	1762.3
1800	10.793	2.33598	1.98748	0.1880	1460.6	1562.0
1600	10.787	1.98195	1.68627	0.2116	1434.1	1461.6
1400	10.785	1.64533	1.39987	0.2418	1401.1	1406.1
1200	10.785	1.32012	1.12317	0.2821	1360.1	1360.1
1000	10.785	1.00604	0.85595	0.3386	1309.8	1309.8
800	10.785	0.70502	0.59984	0.4232	1250.0	1250.0
600	10.785	0.41864	0.35618	0.5643	1185.6	1185.6
400	10.785	0.14698	0.12505	0.8464	1128.7	1128.7
288	10.785	0.00000	0.00000	1.1755	1104.7	1104.7

* Rises with temperature because of dissociation
† Including the dissociation heat of reaction
‡ Sum of the molar heat capacities of the constituent gases
§ Defined as $c_p(\text{eq}) = (\partial h/\partial T)_p$, which therefore includes the heat taken up or released in dissociation/recombination reactions

Figure 1:23 Properties of the products of stoichiometric combustion of methane and air at various temperatures

Temperature (K)	'Frozen' viscosity‖ (μPa s)	'Frozen' thermal conductivity¶ (W m^{-1} °C^{-1})	'Frozen' Prandtl number**	Equilibrium conductivity†† (W m^{-1} °C^{-1})	Equilibrium Prandtl number§§
4000	97.57	0.3130	0.4953	1.1873	—
3800	94.74	0.2931	0.5109	1.5486	—
3600	91.94	0.2699	0.5349	1.8288	—
3400	89.08	0.2451	0.5666	1.8458	0.3855
3200	86.03	0.2214	0.6017	1.5794	0.3873
3000	82.71	0.2010	0.6334	1.1810	0.3879
2800	79.16	0.1845	0.6568	1.8058	0.4525
2600	75.46	0.1710	0.6716	0.5257	0.4957
2400	71.61	0.1593	0.6797	0.3426	0.5622
2200	67.59	0.1481	0.6838	0.2329	0.6168
2000	63.51	0.1373	0.6855	0.1713	0.6534
1800	59.28	0.1252	0.6916	0.1365	0.6782
1600	54.86	0.1127	0.6982	0.1156	0.6956
1400	50.22	0.0997	0.7056	0.0997	0.7081
1200	45.43	0.0870	0.7101	0.0870	0.7101
1000	40.18	0.0738	0.7135	0.0738	0.7135
800	34.54	0.0599	0.7204	0.0599	0.7204
600	28.29	0.0465	0.7220	0.0465	0.7220
400	21.05	0.0326	0.7297	0.0326	0.7297
288	16.26	0.0243	0.7386	0.0243	0.7386

‖ Viscosity of the composition at each temperature

¶ Obtained by summing the conductivities of the individual species, representing energy transfer by molecular conduction only

** Using 'frozen' specific heat and conductivity

†† Includes the heat transferred by diffusion and recombination of dissociated species, and is appropriate when dealing with convective heat

§§ Uses equilibrium values of specific heat and thermal conductivity

Figure 1:23 (continued)

Temperature (K)	N_2	H_2O	H_2	O_2	O	OH	H	NO	N	CO	CO_2	Relative molecular mass (molecular weight)
4000	0.66453	0.00000	0.00000	0.03100	0.26027	0.00000	0.00000	0.04277	0.00143	0.00000	0.00000	25.18
3800	0.67739	0.00000	0.00000	0.05118	0.22255	0.00000	0.00000	0.04821	0.00067	0.00000	0.00000	25.74
3600	0.69479	0.00000	0.00000	0.07826	0.17504	0.00000	0.00000	0.05163	0.00029	0.00000	0.00000	26.43
3400	0.71482	0.00000	0.00000	0.10877	0.12450	0.00000	0.00000	0.05181	0.00011	0.00000	0.00000	27.17
3200	0.73441	0.00000	0.00000	0.13774	0.07938	0.00000	0.00000	0.04847	0.00000	0.00000	0.00000	27.82
3000	0.75098	0.00000	0.00000	0.16149	0.04518	0.00000	0.00000	0.04235	0.00000	0.00000	0.00000	28.32
2800	0.76363	0.00000	0.00000	0.17884	0.02281	0.00000	0.00000	0.03473	0.00000	0.00000	0.00000	28.64
2600	0.77264	0.00000	0.00000	0.19052	0.01009	0.00000	0.00000	0.02676	0.00000	0.00000	0.00000	28.83
2400	0.77832	0.00000	0.00000	0.19802	0.00383	0.00000	0.00000	0.01933	0.00000	0.00000	0.00000	28.92
2200	0.78303	0.00000	0.00000	0.20277	0.00121	0.00000	0.00000	0.01299	0.00000	0.00000	0.00000	28.96
2000	0.78539	0.00000	0.00000	0.20582	0.00030	0.00000	0.00000	0.00799	0.00000	0.00000	0.00000	28.97
1800	0.78778	0.00000	0.00000	0.20777	0.00006	0.00000	0.00000	0.00439	0.00000	0.00000	0.00000	28.97
1600	0.78897	0.00000	0.00000	0.20897	0.00000	0.00000	0.00000	0.00207	0.00000	0.00000	0.00000	28.97
1400	0.78981	0.00000	0.00000	0.20961	0.00000	0.00000	0.00000	0.00078	0.00000	0.00000	0.00000	28.97
1200	0.78989	0.00000	0.00000	0.20989	0.00000	0.00000	0.00000	0.00021	0.00000	0.00000	0.00000	28.97
1000	0.78989	0.00000	0.00000	0.20989	0.00000	0.00000	0.00000	0.00021	0.00000	0.00000	0.00000	28.97
800	0.78989	0.00000	0.00000	0.20989	0.00000	0.00000	0.00000	0.00021	0.00000	0.00000	0.00000	28.97
600	0.78989	0.00000	0.00000	0.20989	0.00000	0.00000	0.00000	0.00021	0.00000	0.00000	0.00000	28.97
400	0.78989	0.00000	0.00000	0.20989	0.00000	0.00000	0.00000	0.00021	0.00000	0.00000	0.00000	28.97
288	0.78989	0.00000	0.00000	0.20989	0.00000	0.00000	0.00000	0.00021	0.00000	0.00000	0.00000	28.97

Figure 1:24 Composition of dry air at various temperatures

Temperature (K)	Ratio of moles at given temperature to moles of undissociated air*	Enthalpy of undissociated products† (MJ m^{-3}(st))	Enthalpy of undissociated products† (MJ kg^{-1})	Density (kg m^{-3})	'Frozen' specific heat capacity‡ (J kg^{-1} °C^{-1})	Equilibrium specific heat capacity§ (J kg^{-1} °C^{-1})
4000	1.151	9.07368	7.40468	0.0767	1329.8	—
3800	1.126	8.21674	6.70537	0.0825	1324.8	—
3600	1.096	7.25449	5.92011	0.0894	1319.3	—
3400	1.066	6.28007	5.12492	0.0973	1313.6	3853.8
3200	1.041	5.39368	4.40158	0.1059	1307.4	3340.6
3000	1.023	4.65109	3.79558	0.1150	1300.8	2726.1
2800	1.011	4.05061	3.30554	0.1246	1293.7	2195.6
2600	1.005	3.56289	2.90754	0.1351	1285.9	1809.8
2400	1.002	3.15188	2.57213	0.1468	1277.2	1564.9
2200	1.000	2.78735	2.27465	0.1603	1267.2	1423.1
2000	1.000	2.44934	1.99879	0.1765	1255.8	1342.7
1800	1.000	2.12677	1.73557	0.1961	1242.3	1292.4
1600	1.000	1.81495	1.48111	0.2206	1225.7	1253.2
1400	1.000	1.51215	1.23401	0.2521	1205.2	1218.8
1200	1.000	1.21768	0.99370	0.2941	1179.0	1179.0
1000	1.000	0.93267	0.76112	0.3530	1145.1	1145.1
800	1.000	0.65717	0.53629	0.4412	1102.2	1102.2
600	1.000	0.39279	0.32054	0.5883	1054.2	1054.2
400	1.000	0.13960	0.11392	0.8824	1016.3	1016.3
288	1.000	0.00000	0.00000	1.2256	1012.2	1012.2

* Rises with temperature because of composition changes due to dissociation and no formation
† Including dissociation heat of reaction
‡ Sum of the molar heat capacities of the constituent gases
§ Defined as $c_p(\mathrm{eq}) = (\partial h/\partial T)_p$, which therefore includes the heat taken up or released in dissociation/recombination reactions

Figure 1:25 Properties of dry air at various temperatures

Temperature (K)	'Frozen' viscosity‖ (μPa s)	'Frozen' thermal conductivity¶ (W m⁻¹ °C⁻¹)	'Frozen' Prandtl number**	Equilibrium conductivity†† (W m⁻¹ °C⁻¹)	Equilibrium Prandtl number§§
4000	101.09	0.1943	0.6918	0.4696	—
3800	97.96	0.1872	0.6931	0.5438	—
3600	94.77	0.1799	0.6949	0.5842	—
3400	91.47	0.1724	0.6968	0.5571	0.6328
3200	88.03	0.1648	0.6985	0.4721	0.6229
3000	84.51	0.1571	0.6998	0.3690	0.6243
2800	80.89	0.1494	0.7004	0.2782	0.6384
2600	77.15	0.1416	0.7007	0.2119	0.6589
2400	73.19	0.1334	0.7006	0.1688	0.6786
2200	69.12	0.1251	0.7002	0.1422	0.6916
2000	65.03	0.1167	0.6997	0.1252	0.6977
1800	60.84	0.1081	0.6992	0.1124	0.6994
1600	56.28	0.0988	0.6984	0.1009	0.6992
1400	51.50	0.0890	0.6975	0.0890	0.7054
1200	46.59	0.0789	0.6961	0.0789	0.6961
1000	41.38	0.0682	0.6942	0.0682	0.6942
800	35.86	0.0571	0.6922	0.0571	0.6922
600	29.69	0.0454	0.6894	0.0454	0.6894
400	22.54	0.0333	0.6873	0.0333	0.6873
288	17.79	0.0261	0.6905	0.0261	0.6905

‖ Viscosity of the composition at each temperature

¶ Obtained by summing the conductivities of the individual species, representing energy transfer by molecular conduction only

** Using 'frozen' specific heat capacity and conductivity

†† Includes the heat transferred by diffusion and recombination of dissociated species, and is appropriate when dealing with convective heat

§§ Uses equilibrium values of specific heat capacity and thermal conductivity

Figure 1:25 (continued)

Metal	Thermal conductivity ($W\ m^{-1}\ °C^{-1}$)						Density ($kg\ m^{-3}$)
	0 °C	100 °C	200 °C	300 °C	400 °C	liquid	
0.2 C steel		52					7750
Cast iron		45					7340
Ni steel .15C		49					
Cr steel .6C		41					
Copper	378		372		363		8860
Brass 12 % Zn	100	108	113	115	117		8420
Aluminium	227		227		227	87	
Duralumin 4 % Cu	159	181	194				2770
Mg alloys	145		134		134		1800
Nickel, pure	78						
Nickel, cast	55		52		49		
Zinc	113		105		93	58	
Lead	35		33			14	
Nimonic 75		13					8310
Inconel		15					
Monel		26					8860

Figure 1:26 Thermal conductivities and densities of metals

Scaling temperature (°C)	900	1000	1100	1200
Conductivity ($W\ m^{-1}\ °C^{-1}$)	1.45	1.63	1.86	2.1

Figure 1:27 Thermal conductivity of scale on mild steels

Temperature (°C)	200	400	600	800	1000	1200
Conductivity as a percentage of value at 0 °C	95	85	75	68	68	73

Figure 1:28 Thermal conductivity of steels related to temperature

Substance	Mean specific heat 15 °C to melting point ($kJ\ kg^{-1}\ °C^{-1}$)	Melting point (K)	Heat in solid at melting point ($kJ\ kg^{-1}$)	Latent heat of fusion ($kJ\ kg^{-1}$)	Total heat in liquid at melting temp ($kJ\ kg^{-1}$)	Average pouring temp (K)	Total heat in liquid at pouring temp ($kJ\ kg^{-1}$)
Aluminium	1.038	930	665.2	393.1	1058.3	1022	1156.0
Antimony	0.226	903	138.9	162.8	301.7	989	321.0
Babbitt—lead base	0.163	512	36.8	60.9	97.7	602	111.6
Babbitt—tin base	0.297	513	52.6	79.3	157.5	764	211.7
Bismuth	0.138	543	35.1	43.0	78.2	600	86.5
Brass, Muntz metal	0.440	1161	383.8	160.5	544.3	1283	607.1
Brass, red	0.435	1340	458.2	201.2	695.4	1505	739.2
Brass, yellow (85 Cu–15 Zn)	0.435	1298	439.6	196.3	635.9	1450	709.4
Bronze, aluminium	0.528	1323	546.6	229.3	776.0	1477	856.0
Bronze, bearing	0.398	1273	391.5	185.8	577.3	1394	632.7
Bronze, bell metal	0.419	1163	366.1	177.5	543.6	1310	617.3
Bronze, gun metal	0.448	1283	445.4	195.8	611.3	1420	702.5
Bronze, Tobin	0.448	1158	389.6	171.0	560.6	1283	625.5
Cadmium	0.243	594	87.0	45.4	132.3	672	155.6
Chromium	0.657	1882	1114.2		1114.2		
Copper	0.435	1356	465.2	211.7	676.9	1477	732.7
Die casting, aluminium base	0.988	894	598.5	379.1	977.6	1033	1118.8
Die casting, lead base	0.159	589	47.7	40.7	88.4	710	339.6
Die casting, tin base	0.293	505	64.2	70.2	134.4	615	162.8
Die casting, zinc base	0.431	689	172.1	111.6	283.8	800	348.9

Figure 1:29 Thermal properties of metals

Substance	Mean specific heat 15 °C to melting point ($kJ\ kg^{-1}\ °C^{-1}$)	Melting point (K)	Heat in solid at melting point ($kJ\ kg^{-1}$)	Latent heat of fusion ($kJ\ kg^{-1}$)	Total heat in liquid at melting temp ($kJ\ kg^{-1}$)	Average pouring temp (K)	Total heat in liquid at pouring temp ($kJ\ kg^{-1}$)
German silver	0.456	1283	451.2	200.5	651.7	1420	723.4
Gold	0.138	1335	144.7	66.3	211.0	1450	227.3
Iron 15.5 °C to 1530 °C	0.691	1803	1039.7	207.0	1246.7	1873	1290.9
Iron 15.5 °C to 1404 °C	0.662		918.8*	8.1†	928.1‡		
Iron 15.5 °C to 927 °C	0.662		600.1*	29.1†	629.2‡		
Iron 15.5 °C to 724 °C	0.628		436.1*	26.7†	462.9‡		
Lead	0.134	600	41.9	23.3	65.1	655	72.1
Linotype	0.151	525	35.6	50.0	85.6	600	96.8
Magnesium	0.139	924	723.9	194.7	918.5	1020	1027.4
Manganese	0.716	1503	870.0	153.5	1023.4	1590	1090.9
Molybdenum	0.251	2894					
Monel metal	0.540	1597	707.1	273.1	980.2	1783	1088.6
Nickel 15.5 °C to 1451 °C	0.561	1724	804.8	305.9	1110.7	1840	1174.6
Platinum	0.134	2046		111.6			
Rhodium	0.251	2239					

* Heat in solid up to transformation point
† Latent heat associated with change of structure
‡ Heat in solid after allotropic change

Figure 1:29 (continued)

Substance	Mean specific heat 15 °C to melting point ($kJ\ kg^{-1}\ °C^{-1}$)	Melting point (K)	Heat in solid at melting point ($kJ\ kg^{-1}$)	Latent heat of fusion ($kJ\ kg^{-1}$)	Total heat in liquid at melting temp ($kJ\ kg^{-1}$)	Average pouring temp (K)	Total heat in liquid at pouring temp ($kJ\ kg^{-1}$)
Silver	0.264	1234	248.9	108.9	357.7	255	388.4
Solder, bismuth	0.167	384	21.6	38.1	59.8	439	68.6
Solder, plumber's	0.214	485	41.9	53.5	95.4	533	104.7
Steel	0.691		Not given because of variable characteristics of steels				
Stereotype	0.151	531	36.1	60.9	97.0	600	107.0
Tin	0.289	505	62.8	58.2	121.0	615	148.9
Tungsten	0.147	3655		183.3			
Vanadium	0.494	1983					
Zinc	0.448	692	181.0	111.6	292.6	755	330.3
Zirconium	0.276	1972					

Figure 1:29 (continued)

	Frequency in Hz					
	125	250	500	1000	2000	4000
Brick, unpainted	0.03	0.03	0.03	0.04	0.05	0.07
Concrete, unpainted	0.01	0.01	0.02	0.02	0.02	0.03
Tiled floor, solid backing	0.02	0.03	0.03	0.03	0.03	0.02
Parquet floor	0.04	0.04	0.07	0.06	0.06	0.07
Wood joist floor	0.15	0.11	0.10	0.07	0.06	0.07
Plate glass window	0.18	0.06	0.04	0.03	0.02	0.02
Normal glass window	0.35	0.25	0.18	0.12	0.07	0.04
Plasterboard, 12.5 mm on timber frame	0.29	0.10	0.05	0.04	0.07	0.09
Plastered brick	0.01	0.01	0.02	0.03	0.04	0.05
Acoustic plaster, 25 mm	0.25	0.45	0.78	0.92	0.89	0.87
Fibreboard, 12.5 mm on solid backing	0.05	0.10	0.15	0.25	0.30	0.30
Mineral, glass wool, 25 mm solid backing	0.15	0.35	0.70	0.85	0.90	0.90
Same faced with 5% perforated hardboard	0.10	0.35	0.85	0.85	0.35	0.15
Carpet on good underlay	0.08	0.24	0.57	0.69	0.71	0.73
Curtains, heavy draped	0.07	0.31	0.49	0.75	0.70	0.60
Water surface	0.01	0.01	0.01	0.01	0.02	0.02
Audience, m^2 per person	0.18	0.40	0.46	0.46	0.51	0.46
Hard seats, unoccupied, m^2 per seat	0.07	0.10	0.15	0.17	0.18	0.20
Upholstered seats, unoccupied, m^2 per seat	0.12	0.20	0.28	0.30	0.32	0.37

Figure 1:30 Acoustic absorption coefficients of some common materials

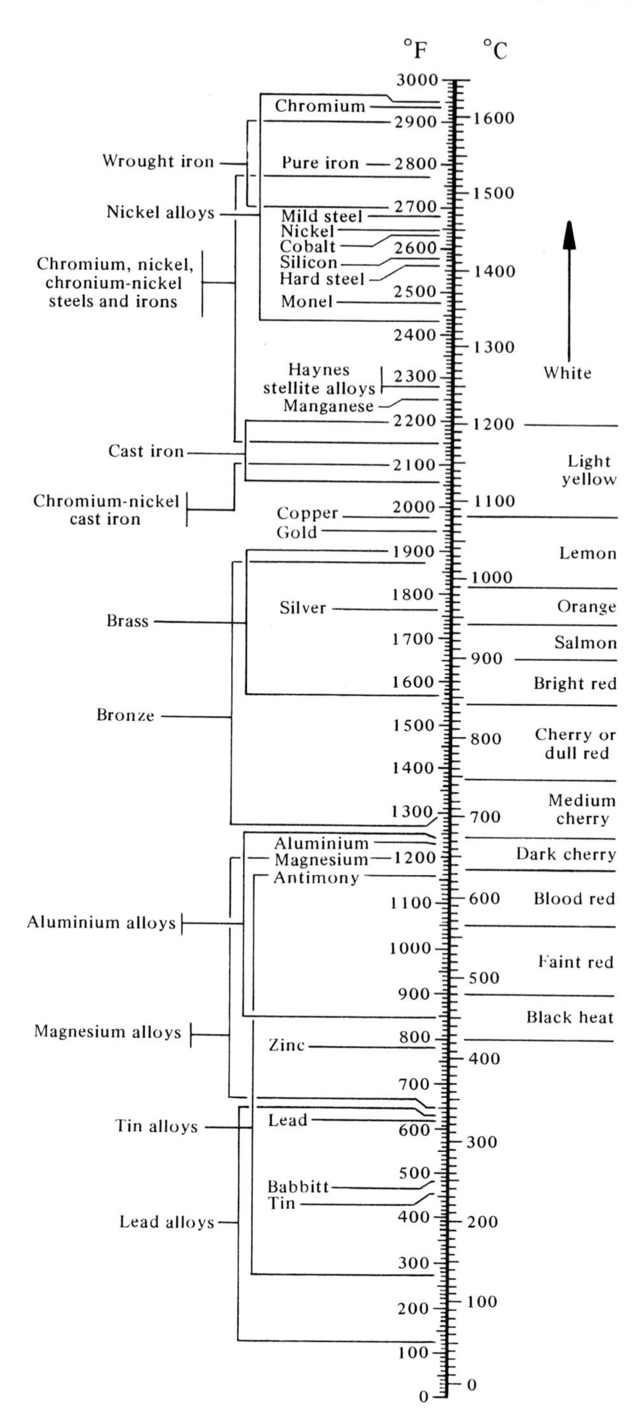

Figure 1:31 Melting points of metals and alloys of practical importance

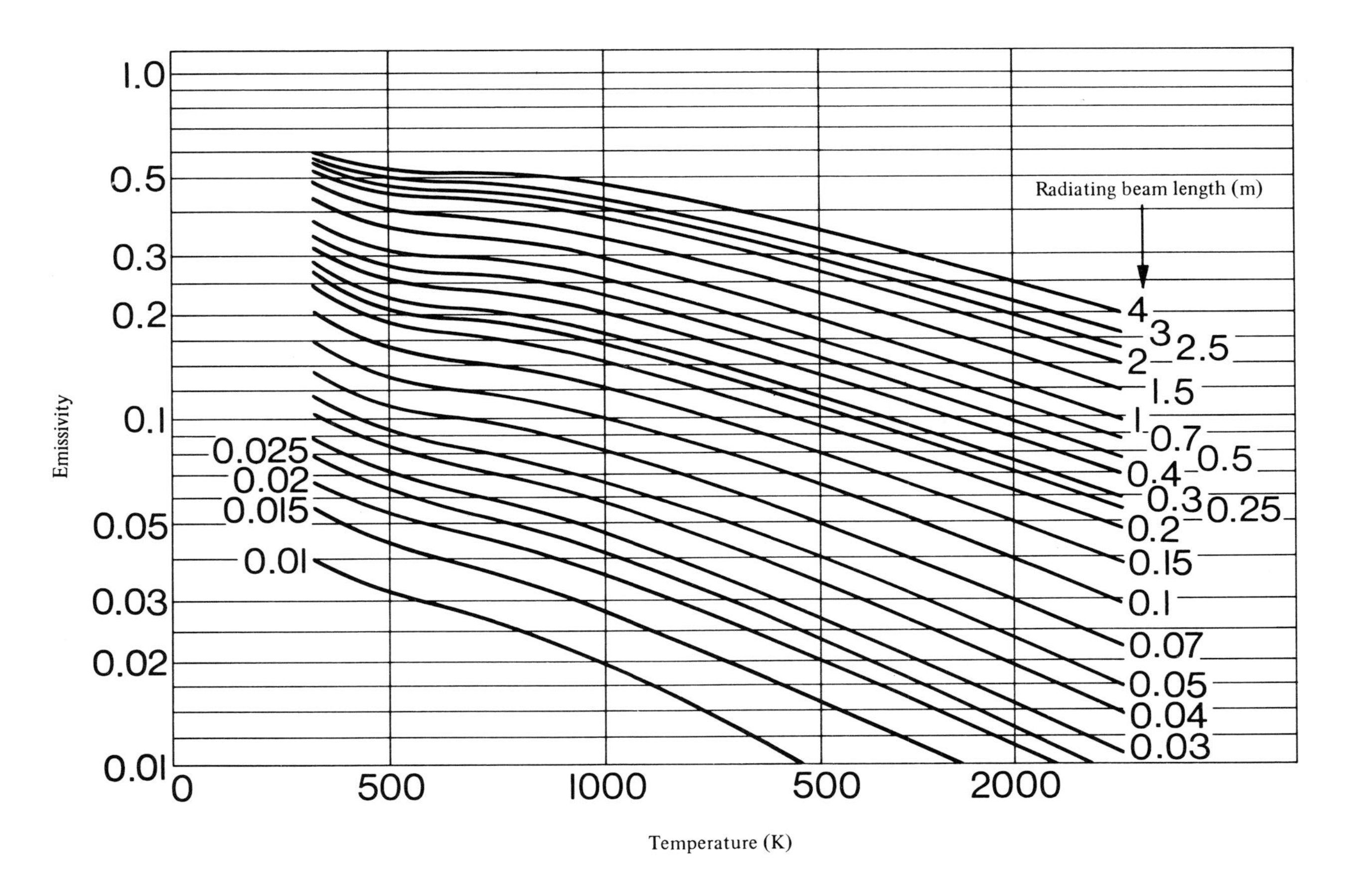

Figure 1:32 Emissivity of combustion products from stoichiometric combustion of methane

2

Fluid mechanics

2.1 FUNDAMENTAL EQUATIONS

This chapter is about the mechanics of Newtonian fluids such as gases and most common liquids like water and lubricating oils. Newtonian fluids have the property that they deform continuously when a shear stress is applied to them. The relationship between shear stress and deformation rate is shown in Figure 2:1 where a thin

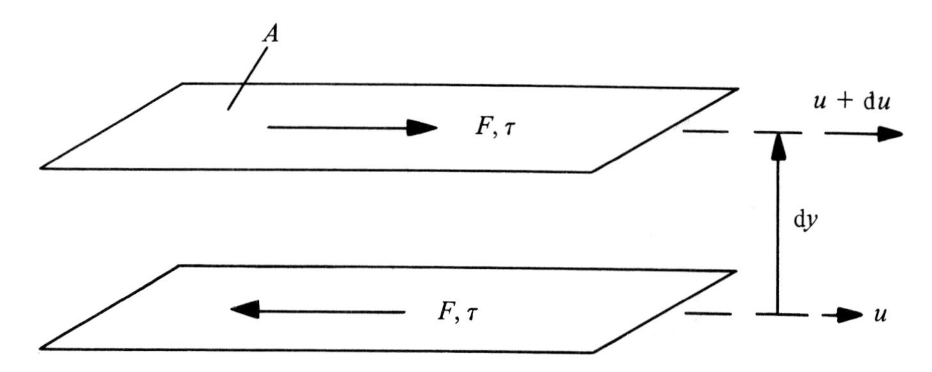

Figure 2:1 Shear stress in fluid due to velocity gradient

layer of fluid is subjected to a shear stress τ which causes a velocity gradient. For a Newtonian fluid:

$$\tau = F/A = \mu \mathrm{d}u/\mathrm{d}y \tag{2.1}$$

where μ is the coefficient of viscosity.

Equation 2.1 may be written:

$$\tau/\rho = \nu \, \mathrm{d}u/\mathrm{d}y \tag{2.2}$$

where ρ is the density of the fluid and $\nu = \mu/\rho$ is the coefficient of kinematic viscosity.

For liquids, μ decreases with temperature increase, while for gases μ increases with temperature increase.

2.1.1 Corollaries of the fundamental equations

The following statements follow immediately from the preceding section:

1 If a fluid is at rest or moving with uniform velocity without relative motion within the fluid then the stress acting on any surface element in the fluid must be normal to that surface. This normal stress is called the *pressure.*

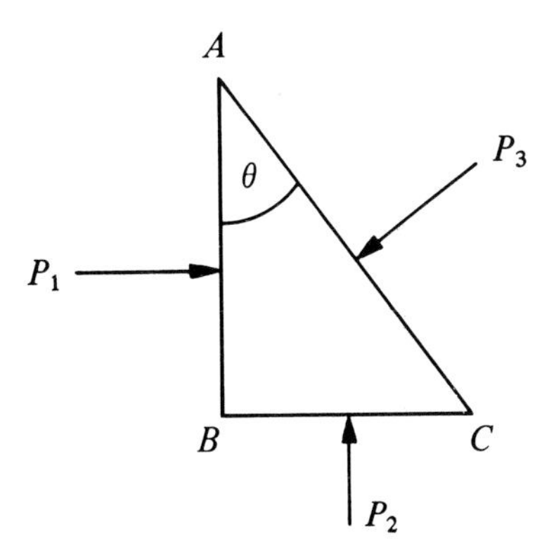

Figure 2:2 Equilibrium of fluid elements

2 The pressure at a point in a fluid which is at rest or moving without deformation is the same in all directions. Figure 2:2 shows a small element of such a fluid. Resolving horizontally gives:

$$P_1 \times AB = P_3 \times AC\cos\theta = P_3 \times AB$$

Therefore:

$$P_1 = P_3$$

Similarly, by resolving vertically, it follows that in the limit as the element size decreases so that its weight can be ignored then $P_2 = P_3$ also.

3 In such a fluid the pressure at a given horizontal level is everywhere the same. This is clear from Figure 2:3. If there are no horizontal shear stresses on the cylindrical element then, for equilibrium, $P_1 = P_2$.

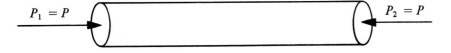

Figure 2:3 Equilibrium of fluid elements

The above corollaries are not true when relative motion is occurring in the fluid, since there are then shear stresses on the elements. Nevertheless, if the velocity gradients or viscosity coefficients are small, as in the case of gases, they may often be taken as true without significant error.

2.1.2 Equations of state for fluids

An equation of state gives the relationship between pressure, density and temperature changes in a fluid.

Gases

A gas can be liquefied by the application of a sufficiently high pressure provided its temperature is below a critical value which depends on the gas. For temperatures well above the critical value, and pressures well below the critical pressure, gases have the following equation of state:

$$PV_{\mathrm{m}} = RT \tag{2.3}$$

where P is the pressure, V_{m} is the molar volume of the gas, and T is its thermodynamic (absolute) temperature. R is called the molar gas constant and its value is approximately 8.314 J K^{-1} mol^{-1} (11.04 ft lbf/°R mol).

At pressure and temperature conditions nearer to the critical values a closer approximation to the state equation is the van der Waals equation:

$$\{P + (a/V_{\mathrm{m}}^2)\}\,(V_{\mathrm{m}} - b) = RT \tag{2.4}$$

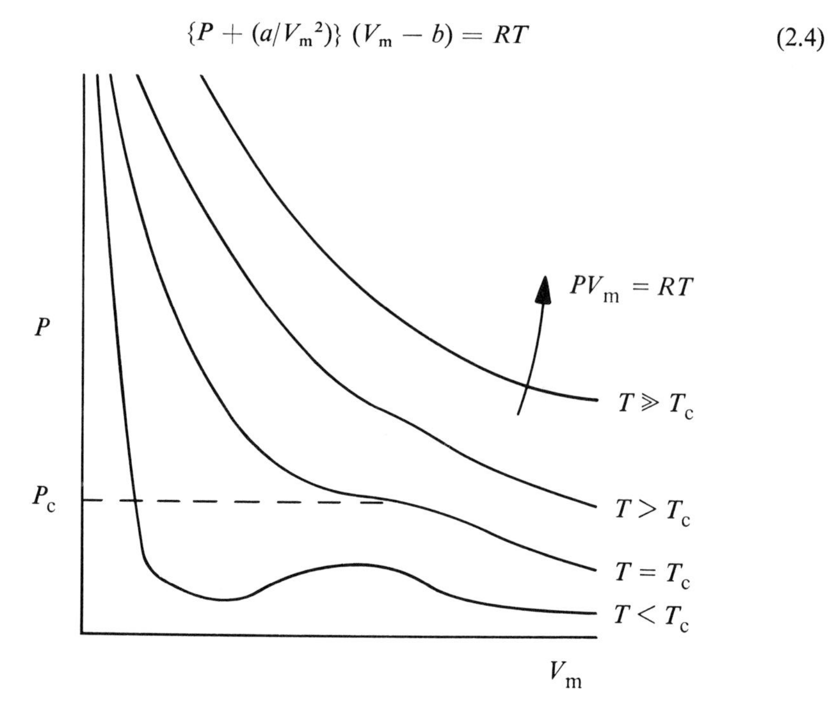

Figure 2:4 Equation of state for a gas

where a and b are constants that are characteristic of each gas. Figure 2:4 shows the form of the pressure–density relation for various temperatures.

Liquids

$$\mathrm{d}P/\mathrm{d}\rho = K/\rho \tag{2.5}$$

where K is the bulk modulus for the liquid, which varies with the temperature.

2.1.3 Variation of pressure with height in a static fluid

From the equilibrium of the fluid element shown in Figure 2:5:

$$\mathrm{d}P + \rho g \mathrm{d}z = 0$$

or

$$\mathrm{d}P/\mathrm{d}z = -\rho g \tag{2.6}$$

or

$$(\mathrm{d}P/\rho g) + \mathrm{d}z = 0$$

Any of equations 2.6 can be integrated to give the pressure change occurring with a change of height if ρ is given as a function of P or of z.

If the density of the fluid is constant then integration of 2.6 gives:

$$P_1 - P_2 = \rho g(z_2 - z_1) \tag{2.7}$$

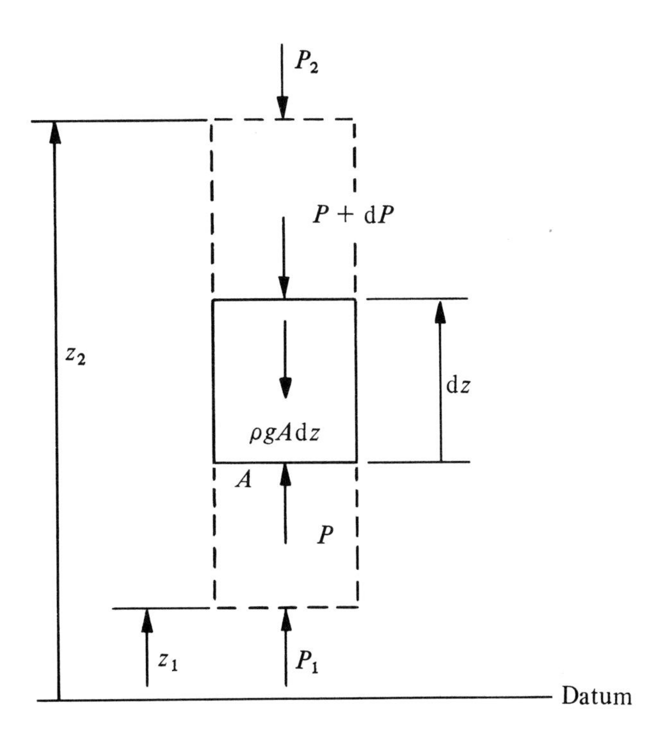

Figure 2:5 Change of pressure with height in a fluid

2.1.4 Manometers

A manometer is a device for measuring the pressure in a fluid and at its simplest takes the form of the U-tube shown in Figure 2:6. The densities of the three fluids are ρ_1, ρ_2 and ρ_M respectively and there must be a clear interface between the fluids.

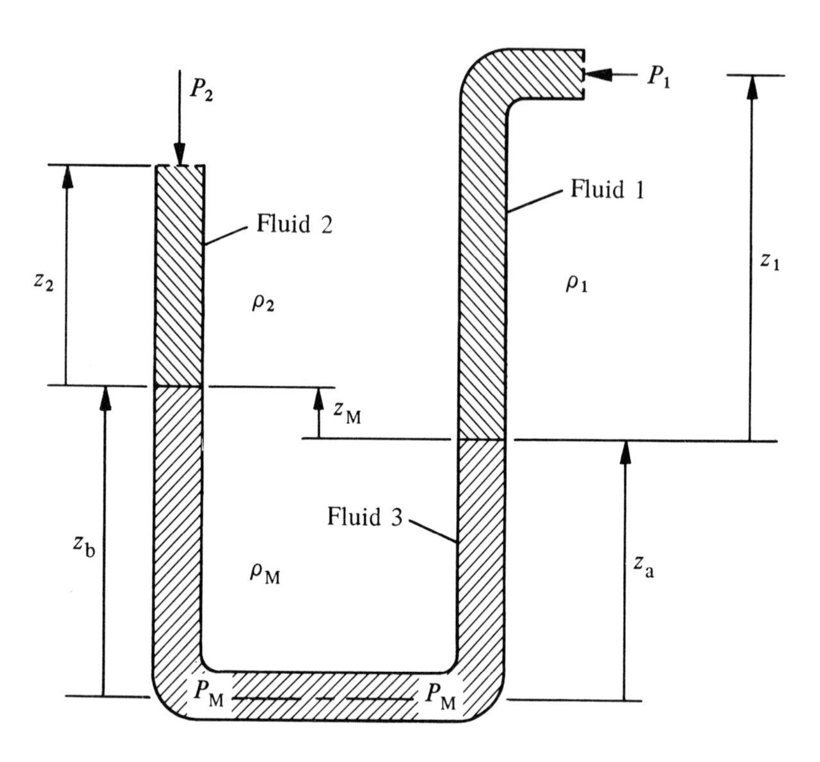

Figure 2:6 Manometer

This will be achieved if fluid 3 is a liquid and fluids 1 and 2 are gases or liquids immiscible with fluid 3. From corollary (3) of Section 2.1.1 there is a constant pressure P_M at a horizontal datum lying in fluid 3. Then on the right-hand limb, from equation 2.7:

$$P_1 = P_M - \rho_M g z_a - \rho_1 g z_1$$
$$P_2 = P_M - \rho_M g z_b - \rho_2 g z_2$$

Subtracting:

$$P_1 - P_2 = \rho_M g(z_b - z_a) + \rho_2 g z_2 - \rho_1 g z_1 \tag{2.8}$$

If $\rho_2 z_2$ and $\rho_1 z_1$ are small compared with $\rho_M z_M$, as is often the case when fluids 1 and 2 are gases, then:

$$P_1 - P_2 \approx \rho_M g z_M \tag{2.9}$$

The sensitivity of a manometer is clearly higher as ρ_M becomes smaller. A high sensitivity can be achieved by making ρ_1 and ρ_2 nearly equal to ρ_M. For example, suppose $\rho_1 = \rho_2 = \rho$, and suppose P_1 and P_2 are acting at the same height in an arrangement like that shown in Figure 2:7 where clearly $z_1 - z_2 = z_M$ provided

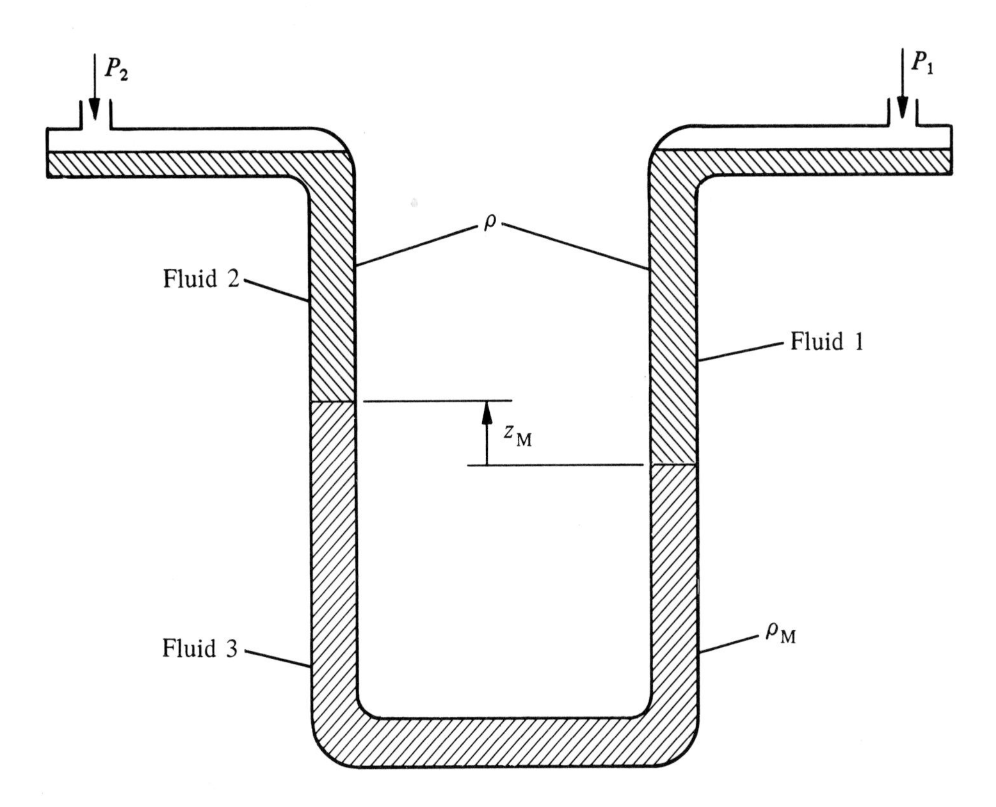

Figure 2:7 Manometer

that the upper reservoirs have a very much larger cross-sectional area than the manometer tube so that there is negligible change in the surface levels of liquids 1 and 2. Then equation 2.8 becomes:

$$P_1 - P_2 = (\rho_M - \rho)gz_M \tag{2.10}$$

and by choosing two liquids to give a small value of $(\rho_M - \rho)$ a high sensitivity is obtained.

Many other manometer arrangements are used, and all may be analysed by the method used above. The manometer clearly measures a pressure difference, and often one of the pressures, say P_2, is made a standard reference pressure, such as atmospheric pressure P_a. In this case what is measured by the instrument is $P_1 - P_a = p_{g1}$ which is called the *gauge pressure* at position 1. Very often the pressure in a gas is expressed as gauge pressure, but it must be remembered that whenever the equation of state 2.3 is involved in a calculation then the absolute pressure must be used.

2.1.5 The force on a surface immersed in a fluid

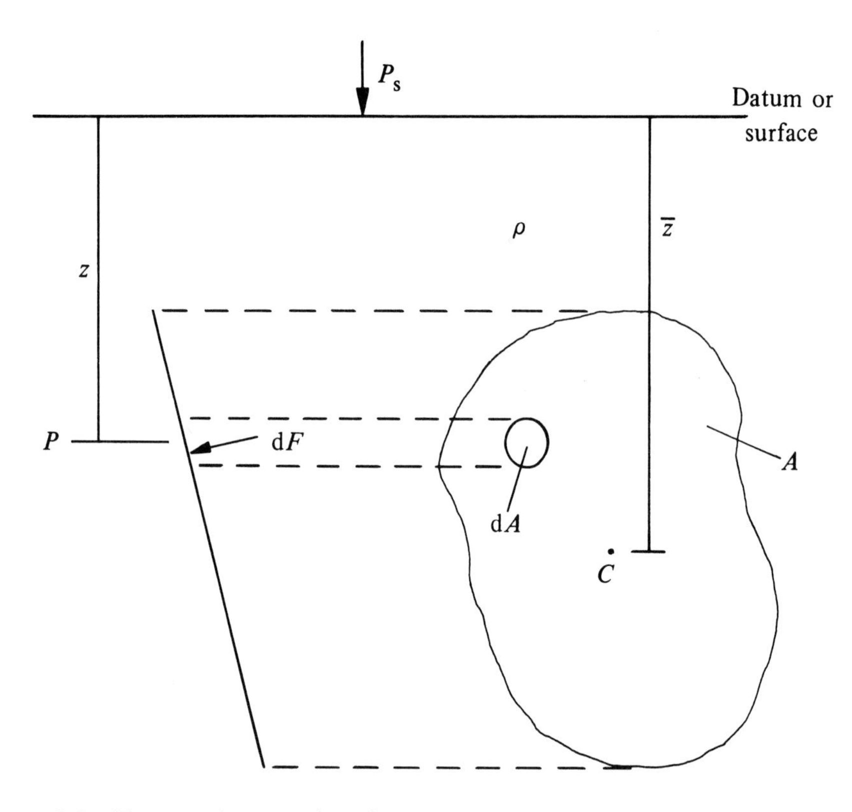

Figure 2:8 Force on immersed surface

Figure 2:8 shows a surface immersed in a fluid of density ρ which is assumed to be constant. The force dF on the surface element dA is PdA which, from equation 2.7, is $(P_s + \rho gz)\mathrm{d}A$. Then the total force on the surface is:

$$F = \sum(P_s\mathrm{d}A + \rho gz\mathrm{d}A)$$

i.e.

$$F = P_sA + \rho g\bar{z}A \tag{2.11}$$

where $\bar{z}$ is the depth of the centroid of A below the datum. The force F acts through a point called the centre of pressure whose position may be found by taking moments about a suitable axis.

2.2 FLUIDS IN MOTION: STREAMLINES

A flowing fluid may move in either a *laminar* or a *turbulent* fashion. In steady laminar flow the motion is regular so that the path of an individual particle is a smooth line, and all particles passing through a given point will follow the same path or *streamline.* In turbulent flow, which occurs when the fluid velocity exceeds some critical value, the fluid particles move in an irregular manner, but it is still possible to use the concept of a streamline as being the mean line along which fluid particles

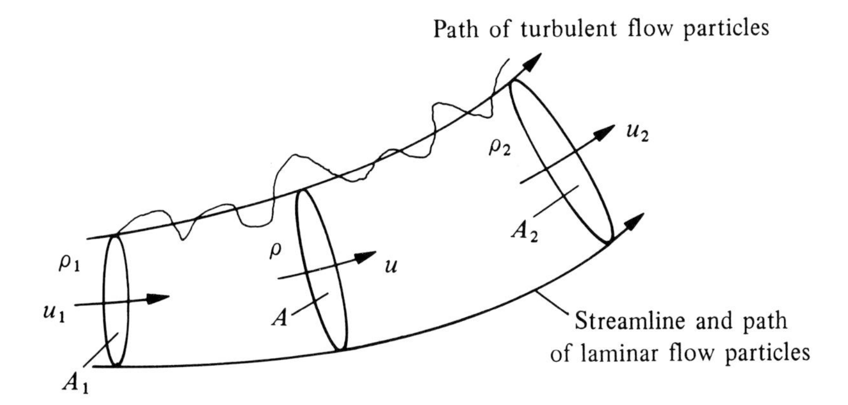

Figure 2:9 Streamtube

flow. Figure 2:9 illustrates this concept. It is clear from the above definition of a streamline that it is *a line in the fluid across which there is no net flow of fluid.* Consequently, a bundle of streamlines defines a *streamtube* in which, for steady flow, the mass flow rate is constant. Thus, referring to Figure 2:9,

$$\mathrm{d}(\rho u A) = 0 \tag{2.12}$$

or

$$\rho u A = m \tag{2.12a}$$

providing the streamtube area A is sufficiently small that the velocity is uniform over the section. m is the mass flow rate. Equations 2.12 and 2.12a are an expression of the law of conservation of matter, and either of these equations is usually called the *equation of continuity.*

Streamlines form a very convenient curvilinear coordinate system in which to describe fluid motion because velocity is always along the streamline.

It is clear that the walls of any vessel containing a fluid must be streamtubes, and in some cases, e.g. straight or slightly curved pipes, it may reasonably be assumed that all streamlines are parallel.

2.2.1 Rate of change: equation of motion along a streamline

It is often necessary to consider the rate of change of some quantity associated with a fluid particle as it moves. If X denotes the quantity, such as density, velocity, temperature, then in general X is a function of position s along the streamline and time t, i.e. $X = X(s, t)$. Hence:

$$\mathrm{d}X = \frac{\partial X}{\partial s}\mathrm{d}s + \frac{\partial X}{\partial t}\mathrm{d}t$$

Therefore:

$$\frac{\mathrm{d}X}{\mathrm{d}t} = \left(\frac{\partial X}{\partial s}\right)\left(\frac{\mathrm{d}s}{\mathrm{d}t}\right) + \left(\frac{\partial X}{\partial t}\right)\left(\frac{\mathrm{d}t}{\mathrm{d}t}\right) = u\frac{\partial X}{\partial s} + \frac{\partial X}{\partial t} \tag{2.13}$$

This process is called 'differentiation following the motion of the fluid' and the quantity $\mathrm{d}X/\mathrm{d}t$, often written as $\mathrm{D}X/\mathrm{D}t$, is called the total or substantive differential. In particular, if X is u, the velocity of the fluid, then the particle acceleration is:

$$\mathrm{D}u/\mathrm{D}t = u(\partial u/\partial s) + (\partial u/\partial t) \tag{2.14}$$

and $\partial u/\partial t = 0$ if the motion is steady.

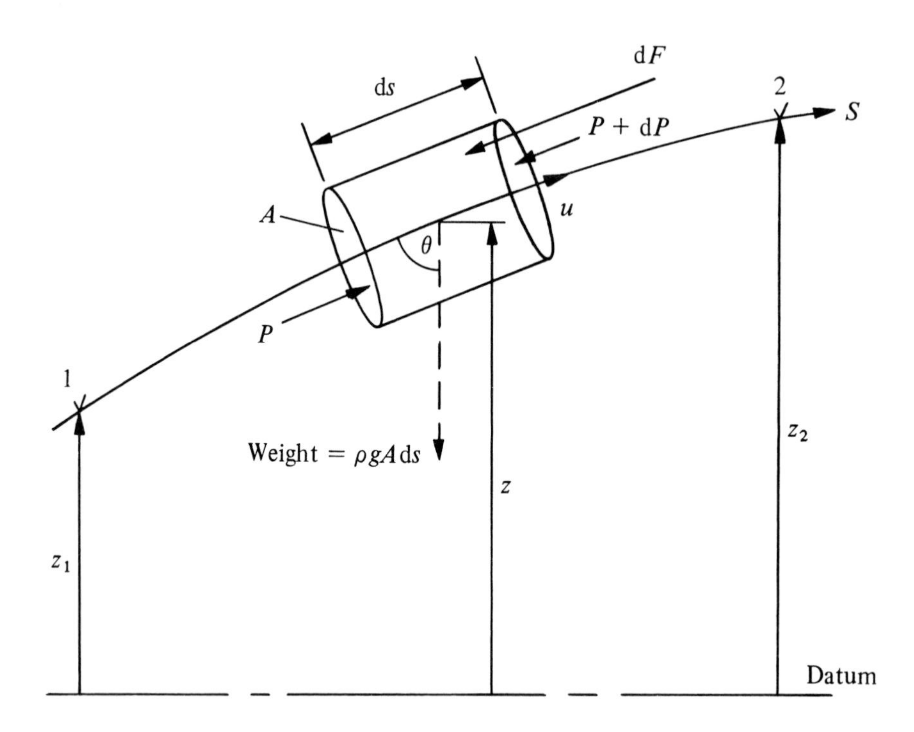

Figure 2:10 Equation of motion along a streamline

To derive the equation of motion for a fluid particle moving along a streamline, consider Figure 2:10 showing the forces acting on a fluid element. F is any external force impressed on the particle, expressed as force per unit weight of fluid, so that $\mathrm{d}F = F\rho gA\mathrm{d}s$. This force might arise as a viscous stress due to a velocity gradient perpendicular to the streamline, or it might be a force applied by, say, a pump impeller. Writing Newton's second law of motion for the fluid element gives:

$$-A\mathrm{d}P - \rho gA(\mathrm{d}s)\cos\theta - F\rho gA\mathrm{d}s = \rho A(\mathrm{d}s)u(\partial u/\partial s)$$

By dividing through by ρgA and writing $\mathrm{d}s\cos\theta = \mathrm{d}z$ and $\rho g = \omega$ this becomes:

$$(u/g)\mathrm{d}u + (1/\omega)\mathrm{d}P + \mathrm{d}z + F\mathrm{d}s = 0 \tag{2.15}$$

which is the general equation of motion for the steady flow of a fluid particle along a streamline. When $F = 0$, equation 2.15 is called *Bernoulli's equation*. Note that in the given form each term of the equation has the dimension of length. Provided ω is known as a function of P, and F is known as a function of s, it is possible to integrate equation 2.15 between any points 1 and 2 on the streamline. For example, if $F = 0$ and the fluid is incompressible, so that ω is constant, then:

$$(u_1{}^2/2g) + (P_1/\omega) + z_1 = (u_2{}^2/2g) + (P_2/\omega) + z_2 \tag{2.16}$$

Equation 2.16 will also apply to a gas provided there is only a small relative change in pressure and temperature between points 1 and 2 so that ω may be considered constant.

Note also that when $u = 0$ and $F = 0$, equation 2.15 becomes identical with the hydrostatic equation 2.6.

2.2.2 Equation of motion perpendicular to a streamline

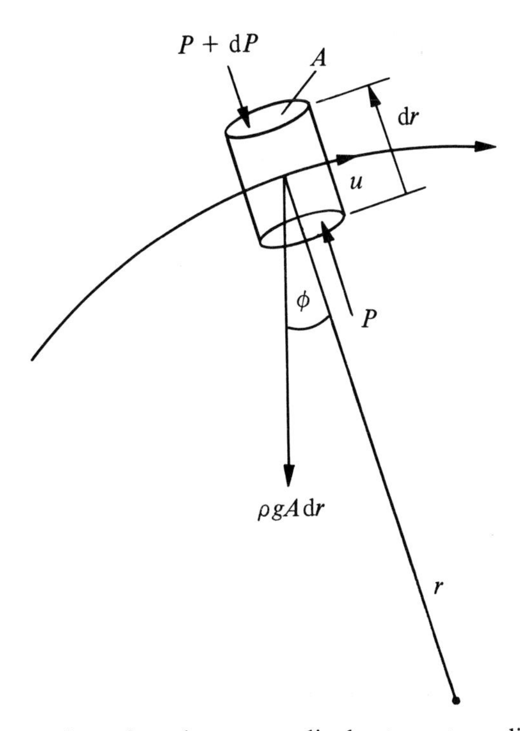

Figure 2:11 Equation of motion perpendicular to a streamline

Figure 2:11 shows a fluid element at a position on a streamline where its radius of curvature is r. From kinematics it is known that the particle has an inward acceleration of u^2/r. Hence the second law of motion applied in the direction normal to the streamline is:

$$A\mathrm{d}P - \omega A(\mathrm{d}r)\cos\varphi = (\omega/g)A(\mathrm{d}r)u^2/r$$

or

$$dP/\mathrm{d}r = \omega\cos\varphi + (\omega u^2/rg) \tag{2.17}$$

If ω and u are known as functions of r then equation 2.17 may be integrated between any two radii r_1 and r_2 to give the pressure change between streamlines.

2.2.3 Energy equation for steady flow along a streamline

This equation is a statement of the law of conservation of energy applied to a fluid particle moving along a streamline. The energy per unit weight of fluid comprises the following terms:

1 The enthalpy $c_pT/g = (c_VT/g) + (P/\omega)$. For a gas this may be written as $(c_VT/g) + (RT/Mg)$ where M is molar mass
2 The kinetic energy $u^2/2g$
3 The potential energy above a datum z

If, between sections 1 and 2 of a streamline, heat energy q per unit weight of fluid is added to the fluid, while each unit weight of fluid does w units of work on its surroundings, then energy conservation is expressed as:

$$\begin{aligned}&(c_VT_1/g) + (P_1/\omega_1) + (u_1{}^2/2g) + z_1 + q \\ &= (c_VT_2/g) + (P_2/\omega_2) + (u_2{}^2/2g) + z_2 + w\end{aligned} \tag{2.18}$$

where, of course, all terms must be expressed in the same units, which in equation 2.18 are units of length, i.e. energy per unit weight.

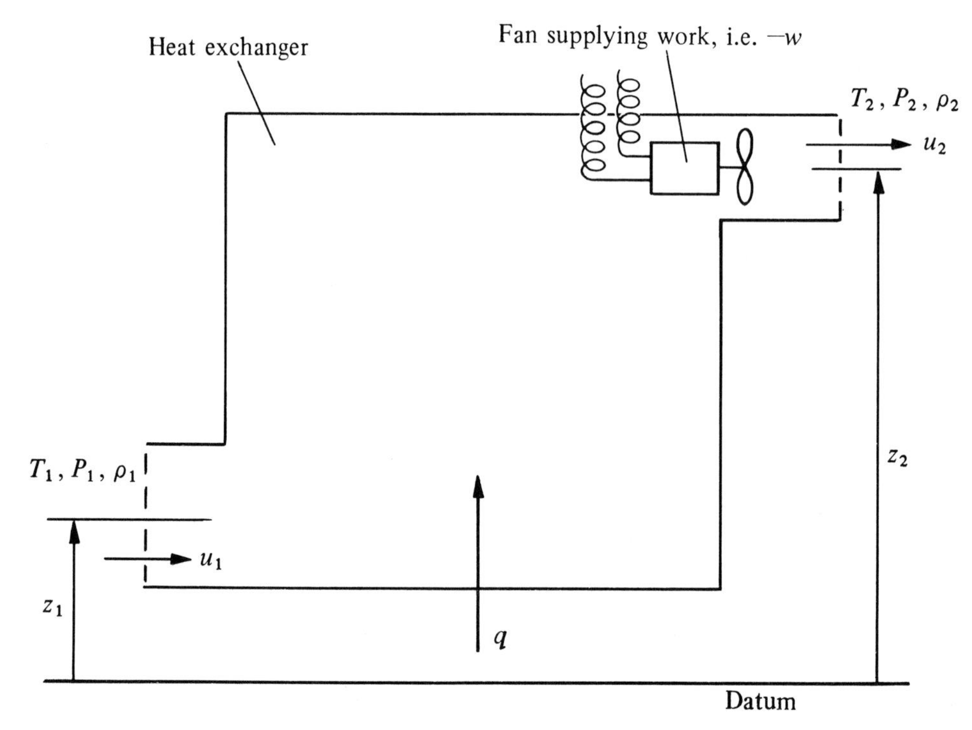

Figure 2:12 Conservation-of-energy equation

Equation 2.18 may be written in differential form and it then expresses energy conservation over an elemental streamline:

$$(c_V/g)dT + d(P/\omega) + (u/g)du + dz - dq + dw = 0 \tag{2.19}$$

It is clear that the energy-conservation equation must apply to any collection of streamtubes, i.e. to any system through which fluid flows. As an example, Figure 2:12 shows this as applied to a heat exchanger. In this kind of application it is necessary to be able to evaluate the enthalpy and kinetic energy at the inlet and outlet sections by some kind of averaging or integration over the cross-sections.

2.2.4 The momentum equation

This relates the forces on a region through which a fluid flows to the momentum changes in the fluid passing through the region. Referring to Figure 2:13, the fluid particle $dAds$ is acted on by forces dF_e from outside the region and also by internal forces dF_i from neighbouring particles. A particular direction, shown as the x-direction, is selected and u_x is the component in the x-direction of the fluid velocity along the streamline. Then, by the results of Section 2.2.1, for steady flow $du_x/dt = udu_x/ds$. Hence the equation of motion is:

$$dF_{ex} + dF_{ix} = \rho(dAds)u(du_x/ds)$$

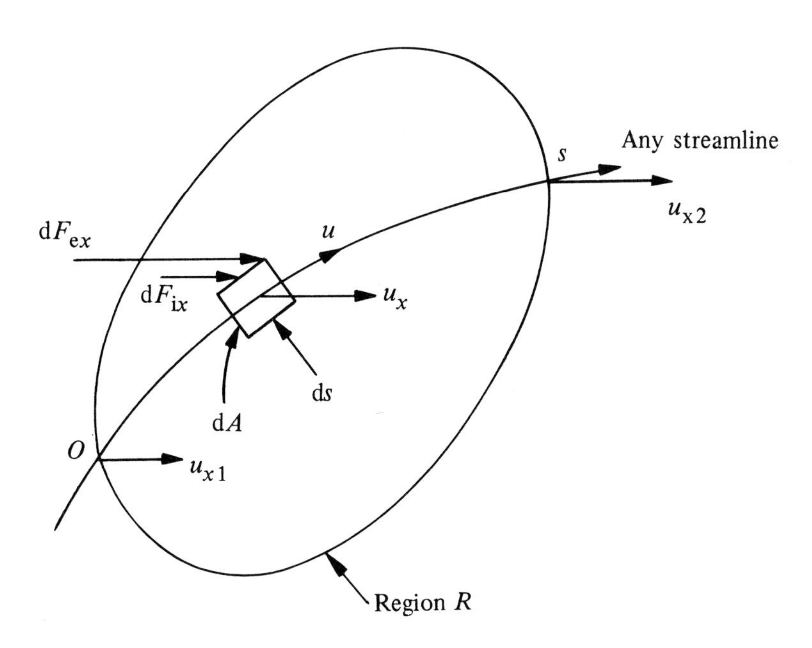

Figure 2:13 Conservation of momentum for a fluid passing through a defined region

Write $\rho(dA)u = m$ = the mass flow rate along the streamtube shown. From equation 2.12a, m is constant. Integrating along the streamtube gives:

$$\int_0^s \mathrm{d}F_{ex} + \int_0^s \mathrm{d}F_{ix} = \int_0^s m\mathrm{d}u_x = m(u_{x2} - u_{x1})$$

Then summing over the whole region for all streamlines:

$$\sum_R \left(\int_0^s \mathrm{d}F_{ex} + \int_0^s \mathrm{d}F_{ix}\right) = \sum_R m(u_{x2} - u_{x1})$$

Now from the third law of motion:

$$\sum \int_0^s \mathrm{d}F_{ix} = 0$$

hence:

$$F_{ex} = \sum m(u_{x2} - u_{x1}) \tag{2.20}$$

or, stated in words: the force on the region in any direction is equal to the momentum leaving per second minus the momentum entering per second, in the given direction. This may also be stated in vector form as:

$$\boldsymbol{F} = \sum m(\boldsymbol{u}_2 - \boldsymbol{u}_1) \tag{2.21}$$

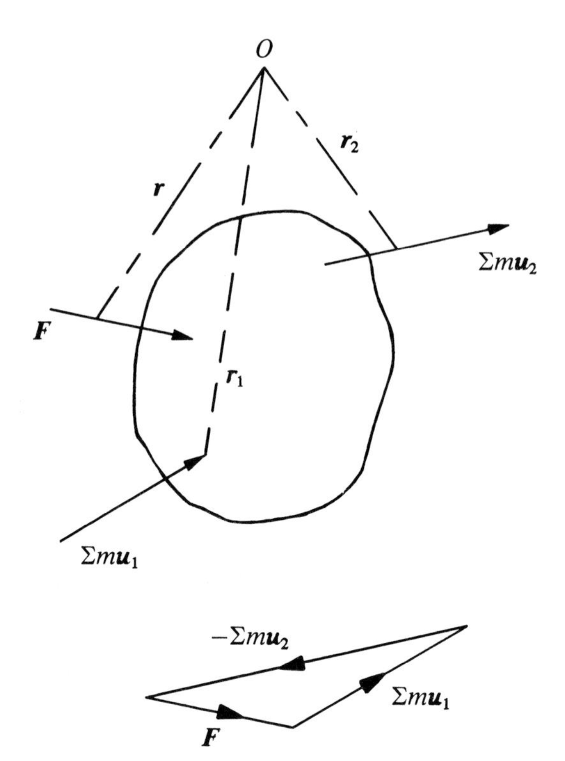

Figure 2:14 Conservation of momentum and of moment of momentum for a fluid passing through a defined region

Figure 2:14 shows the significance of the momentum equation in this form. Similarly, by taking moments about a point O as in Figure 2:14, the corresponding equation for moment of momentum (or angular momentum) can be derived:

$$\boldsymbol{F} \wedge \boldsymbol{r} = \sum m(\boldsymbol{u}_2 \wedge \boldsymbol{r}_2 - \boldsymbol{u}_1 \wedge \boldsymbol{r}_1) \qquad (2.22)$$

In applying the momentum equation it is most important to define clearly the region of space being considered so that there can be no doubt about which are internal forces and which external forces.

2.2.5 Summary and examples using the equations for steady fluid flow along a streamline

The equations derived in Sections 2.2 to 2.2.3 are usually sufficient, together with the equation of state, to solve most problems in fluid mechanics in which the stream-tube pattern is known. They will be summarized in this section.

For convenience, $1/\rho$ will be replaced by v, the specific volume of the fluid.

Equation of state for a gas

$$PV_m = RT \qquad (2.3)$$

Equation of continuity for any fluid

$$\mathrm{d}(\rho uA) = 0 \qquad (2.12)$$

or:

$$\rho uA = m = \text{constant mass flow rate} \qquad (2.12a)$$

Equation of motion (Bernoulli) for any fluid

$$(u/g)\mathrm{d}u + (v/g)\mathrm{d}P + \mathrm{d}z + F\mathrm{d}s = 0 \qquad (2.15a)$$

$$(u^2/2g) + (1/g)\int v\mathrm{d}P + z + \int F\mathrm{d}s = \text{constant} \qquad (2.15b)$$

Equation of energy for any fluid

$$(c_V/g)\mathrm{d}T + \mathrm{d}(Pv/g) + (u/g)\mathrm{d}u + \mathrm{d}z - \mathrm{d}q + \mathrm{d}w = 0 \qquad (2.19a)$$

or, for a gas:

$$[(c_V/g) + (R/Mg)]\mathrm{d}T + (u/g)\mathrm{d}u + \mathrm{d}z - \mathrm{d}q + \mathrm{d}w = 0 \qquad (2.19b)$$

or, writing $c_p/c_V = \gamma$ and $c_p - c_V = R/M$:

$$[\gamma/(\gamma - 1)g]\mathrm{d}(Pv) + (u/g)\mathrm{d}u + \mathrm{d}z - \mathrm{d}q + \mathrm{d}w = 0 \qquad (2.19c)$$

or, in the integrated form, for any fluid:

$$(c_pT_1/g) + (u_1{}^2/2g) + z_1 + q - w = (c_pT_2/g) + (u_2{}^2/2g) + z_2 \qquad (2.19d)$$

Equation of motion for any fluid

$$\boldsymbol{F} = \sum m(\boldsymbol{u}_2 - \boldsymbol{u}_1) \qquad (2.21)$$

and:

$$\boldsymbol{F} \wedge \boldsymbol{r} = \sum m(\boldsymbol{u}_2 \wedge \boldsymbol{r}_2 - \boldsymbol{u}_1 \wedge \boldsymbol{r}_1) \qquad (2.22)$$

Example 1

Show that, when gas flows with no heat or work addition (i.e. reversible adiabatic or isentropic flow), Pv^{γ} is constant and $(P_2/P_1)^{(\gamma-1)/\gamma} = T_2/T_1$.

From equation 2.19c:

$$[\gamma/(\gamma-1)]\,(P\mathrm{d}v + v\mathrm{d}P) + u\mathrm{d}u + g\mathrm{d}z = 0$$

From equation 2.15a:

$$v\mathrm{d}P + u\mathrm{d}u + g\mathrm{d}z = 0$$

Subtract:

$$\gamma P\mathrm{d}v + v\mathrm{d}P = 0$$

or
$$(\mathrm{d}P/P) + \gamma(\mathrm{d}v/v) = 0$$

Integrate:
$$\ln p + \gamma \ln v = \text{constant}$$

or
$$\ln(Pv^{\gamma}) = \text{constant}$$

or
$$Pv^{\gamma} = \text{constant} \tag{2.23}$$

Hence $P_2^{\,1/\gamma}v_2 = P_1^{\,1/\gamma}v_1$ or $(P_2/P_1)^{1/\gamma}(v_2/v_1) = 1$. Also, from 2.3, $P_2v_2/P_1v_1 = T_2/T_1$. Eliminating v_2/v_1 between these gives:

$$(P_2/P_1)^{(\gamma-1)/\gamma} = T_2/T_1 \tag{2.24}$$

Example 2

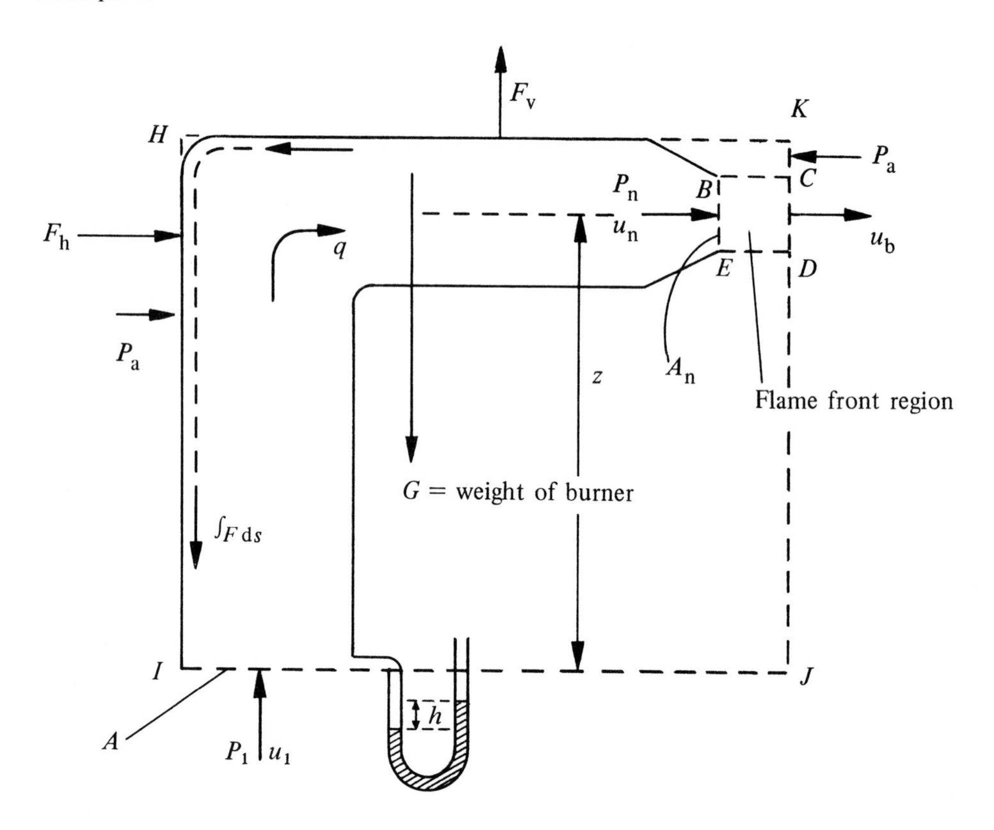

Figure 2:15 Burner; see Section 2.2.5, example 2

Figure 2:15 shows a burner supplied with a stoichiometric air–gas mixture which burns as a flat flame at the nozzle exit. Determine (i) the pressure at the nozzle exit, (ii) the average friction force per unit weight of gas between pipe and gas between inlet and outlet nozzle, (iii) the force required to hold the burner still, given the flow rate q m^3/s measured at atmospheric pressure P_a, the inlet gas pressure measured on a manometer as h m of water, the density ρ of the mixture at atmospheric pressure, and the rise in temperature T through the flame from an original mixture temperature of T_a absolute.

From equation 2.7, $P_1 = P_a + 9800h$, since the density of water is about 1000 kg/m^3 and g is about 9.8 m/s^2. In a practical situation of this kind, h would be small —of the order of 30 mm—so that p_{g1}, the gauge pressure at the inlet, would be of the order of 980 Pa. Since normal atmospheric pressure is 101 300 Pa, the absolute pressure at the inlet is only about 1 per cent higher than P_a. Hence it is quite justifiable to assume constant density up to the nozzle, providing the temperature of the gas is constant.

(i) The momentum equation over the flame front section *BCDE* is:

$$(P_n - P_a)A_n = m(u_b - u_n) \tag{2.25}$$

where m is given as ρq. The same result can be obtained by using the equation of motion 2.15:

$$\mathrm{d}P + \rho u \mathrm{d}u = 0$$

Since there is no change of cross-section in *BCDE*, $\rho u = m/A_n$, which can be integrated to give:

$$A_n \int_{BE}^{CD} \mathrm{d}P - m \int_{BE}^{CD} \mathrm{d}u = 0$$

which is identical with equation 2.25. Also, by continuity, $u_n = q/A_n$ and, by the equation of state and continuity, $u_b = u_n(T_a + T)/T$. Hence P_n can be found from equation 2.25. It will be found, if practical values are used, that P_n is only very slightly above P_a, but this difference in pressure has a bearing on flame structure and stability.

(ii) The friction force between wall and gas is, of course, tangential to the wall, i.e. along the streamline direction, and corresponds to the force F in the equation of motion 2.15 and this is the equation that must be used here. The momentum equation cannot be used here because it would be necessary to consider all the forces between wall and fluid, including pressure forces which are normal to the streamline and which are unknown. In applying the equation of motion the whole pipe is taken as a streamtube and the average velocity is assumed to be q/A at inlet and q/A_n at nozzle exit. If, as is usually the case in industrial gas equipment, the flow is turbulent, then the velocity over the cross-section is reasonably uniform and this assumption is justified.

Using the integrated form of 2.15b:

$$(u_1{}^2/2g) + (P_1/\omega) = (u_n{}^2/2g) + (P_n/\omega) + \int F\mathrm{d}s + z \tag{2.26}$$

in which $\int F\mathrm{d}s$ is the required total friction force per unit weight and is the only unknown quantity in 2.26. Then $\int F\mathrm{d}s = F_{av}L$, where L is the length of the pipe from inlet to nozzle outlet, and this is called the head loss due to friction h_f.

(iii) Since the problem is to find the force on the whole burner, any region of space which encloses the whole burner can be taken, and a convenient region is *HIJK*. It is convenient to use the momentum equation in the horizontal and vertical directions separately. Horizontally:

$$F_h + P_a(HI) - P_a(JK) = m(u_b - 0) \tag{2.27}$$

since the horizontal velocity at the inlet is 0. But $P_a(HI) - P_a(JK) = 0$ since $HI = JK$. Hence F_h can be evaluated.

Vertically:

$$F_v + P_a(IJ) - P_a(HK) - G = m(0 - u_1) \tag{2.28}$$

Hence F_v can be evaluated.

If necessary the line of action of the force could be found by using the angular-momentum equation.

Example 3

Determine the change of pressure which occurs when two parallel streams of liquid mix.

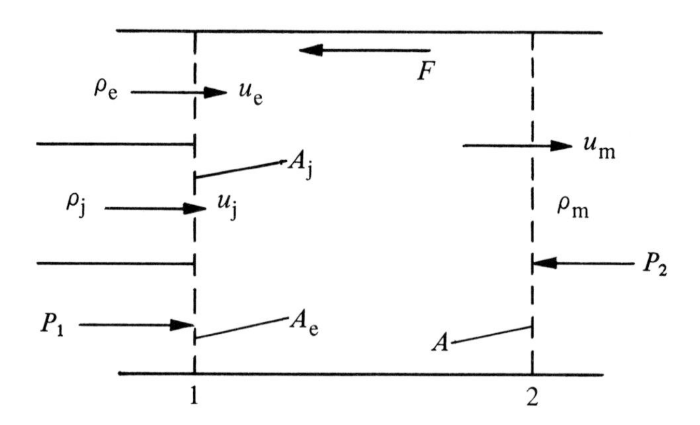

Figure 2:16 Momentum conservation applied to fluids during mixing

This is an important practical case since it occurs in injectors and jet pumps. It is shown in Figure 2:16, where the mixing is completed in the parallel section 1–2. For completeness a frictional force F has been shown acting on the fluid in the section.

Applying the momentum theorem 2.21 gives:

$$P_1 A - P_2 A - F = \rho_m u_m^2 - \rho_e u_e^2 - \rho_j u_j^2 \tag{2.29}$$

Also, the continuity equation applied between 1 and 2 is:

$$\rho_e u_e A_e + \rho_j u_j A_j = \rho_m u_m A \tag{2.30}$$

where A_e and A_j are the areas through which the fluids e and j flow respectively.

From these two equations $P_1 - P_2$ can be determined.

This application is dealt with more fully in Section 10.1.1. Further examples of the use of these equations are the orifice meter and venturi meter for flow measurement dealt with in Sections 8.5.6 and 10.5.

2.2.6 Flow accompanied by large density changes

This commonly occurs where a fluid at a relatively high pressure flows through a nozzle into a region of lower pressure to give a high-velocity jet, usually for the purpose of entraining another fluid. This is illustrated in Figure 2:17 showing two shapes of nozzle. The nozzles are fairly short and well designed to minimize friction losses, so it is usually assumed that there is a reversible adiabatic flow through the nozzle. If this assumption may be made then the equation of motion 2.15b, or the energy equation 2.19, can be applied between the tank and any other section such as 1, t or 2, using the adiabatic equation of state 2.23 to give v as a function of P.

For example, equation 2.18 gives:

$$I_0 = c_p T_0 = c_p T_1 + \tfrac{1}{2} u_1^2 \tag{2.31}$$

T_0 is called the *stagnation temperature* of the fluid, while T_1 is the temperature which would be recorded by a thermometer moving with the fluid. T_0 would be recorded by a fixed thermometer in a moving fluid since the fluid is brought to rest at a stagnation point on the leading surface of the thermometer, I_0 is thus the (specific) stagnation enthalpy.

If the equation of motion and the energy equation, or either of these with the adiabatic law, are applied between 0 and t then:

$$\tfrac{1}{2}u_t^2 = [\gamma/(\gamma - 1)]\, P_0 v_0 \,[1 - (P_t/P_0)^{(\gamma-1)/\gamma}] \tag{2.32}$$

and then continuity, equation 2.12, gives:

$$m^2/A_t^2 = 2[\gamma/(\gamma - 1)]\, P_0 \rho_0 \,[(P_t/P_0)^{2/\gamma} - (P_t/P_0)^{(\gamma+1)/\gamma}] \tag{2.33}$$

Similar expressions will, of course, obtain if section 2, or any other section is used instead of section t.

Equations 2.32 and 2.33 are normally sufficient for the applications found in gas utilization. However, a full treatment of compressible flow is rather complex, and the important points are summarized below, without proof. By differentiating equation 2.33 with respect to (P_t/P_0), the maximum mass flow occurs for a given P_0, when:

$$P_t/P_0 = [2/(\gamma + 1)]^{\gamma/(\gamma-1)} \tag{2.34}$$

which for air is 0.53.

It can be shown that the velocity at the throat u_t is then equal to the velocity of sound a_t at the throat. The ratio of fluid velocity to local sound velocity u_t/a_t is called the Mach number (Ma), so that another way of stating the above is that $(Ma) = 1$ at the throat when the critical pressure occurs, and under these conditions no further increase in mass flow can occur without increasing the throat area. The nozzle is then said to be in the 'choked' condition, and the flow is then given by:

$$m^2/A_t^2 = \gamma\, P_0\, \rho_0\, [2/(\gamma + 1)]^{(\gamma+1)/(\gamma-1)} \tag{2.33a}$$

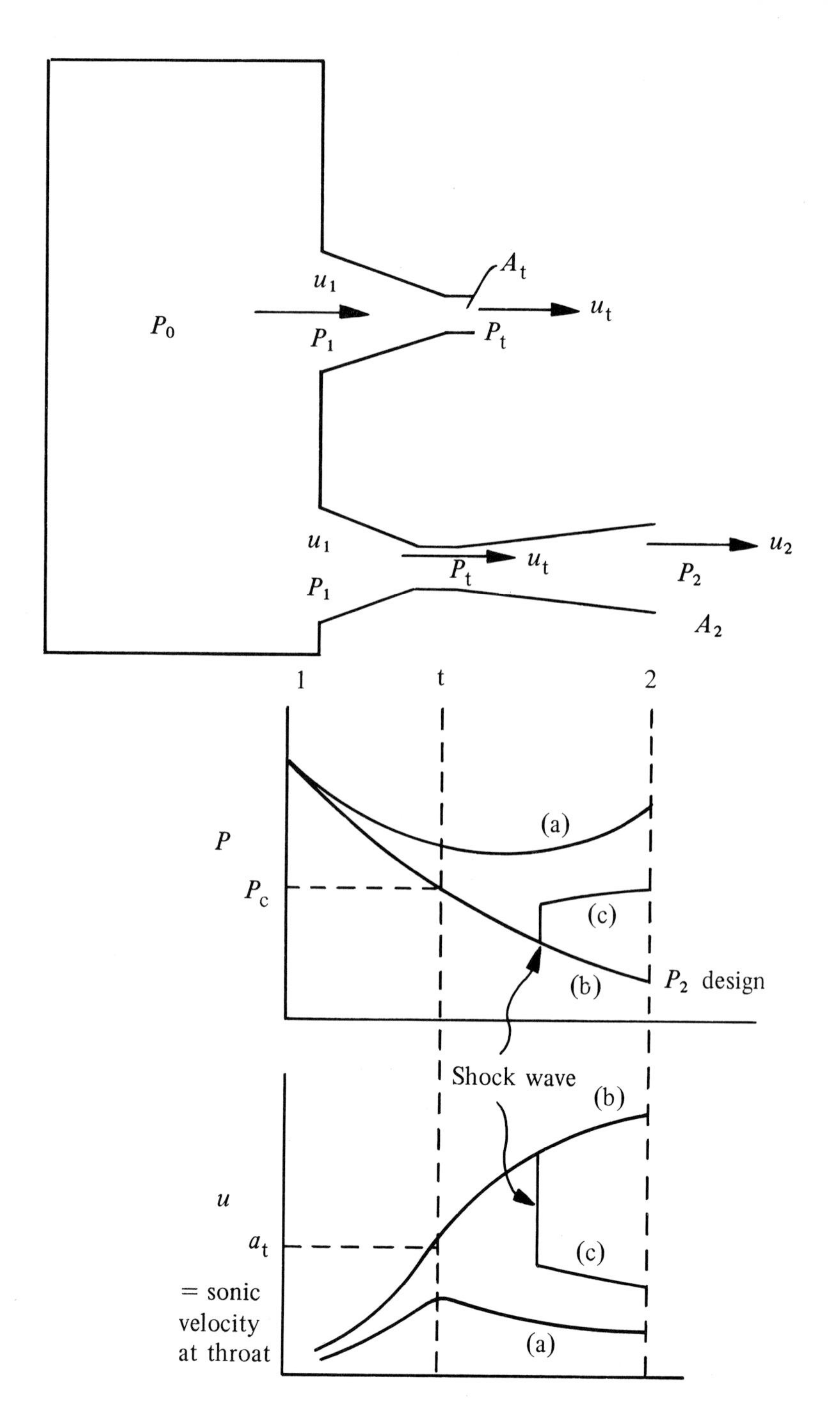

Figure 2:17 Gas flow through a nozzle

which indicates that as long as the nozzle is choked, the flow is dependent only upon upstream pressure, which makes it a useful method of flow control in situations where downstream pressure fluctuations occur.

If the throat is replaced by an orifice of area A_{or}, then the 'throat', i.e. the minimum cross-sectional area of the fluid stream, occurs just after the orifice at a section called the vena contracta where the stream cross-section is given by $C_d A_{or}$, C_d being an empirical discharge coefficient. For this case A_t is replaced by $C_d A_{or}$ in equation 2.33a.

If a gas velocity greater than a, i.e. supersonic velocity, is required, then the throat must be followed by a diverging section in which the pressure continues to fall. This may be proved by using the continuity equation together with the equation of motion to show that dA/A is positive for $(Ma) > 1$.

For a given exit area A_2 there is only one exit pressure P_2 which will permit reversible adiabatic flow through the nozzle. If the pressure P_2 rises above this design value of P_2 then a *shock wave* forms in the diverging portion of the nozzle, across which there is a sudden rise in pressure, and a velocity change from supersonic to subsonic. These changes in the shock wave may be found by applying the energy equation 2.19d, momentum, 2.21, and state 2.3 over the shock wave.

These various points are illustrated in the graph in Figure 2:17.

In gas-utilization practice, it is unusual for choked conditions to apply at the nozzle. Normally only convergent nozzles are used and the flow is subsonic throughout.

2.3 DIMENSIONAL ANALYSIS

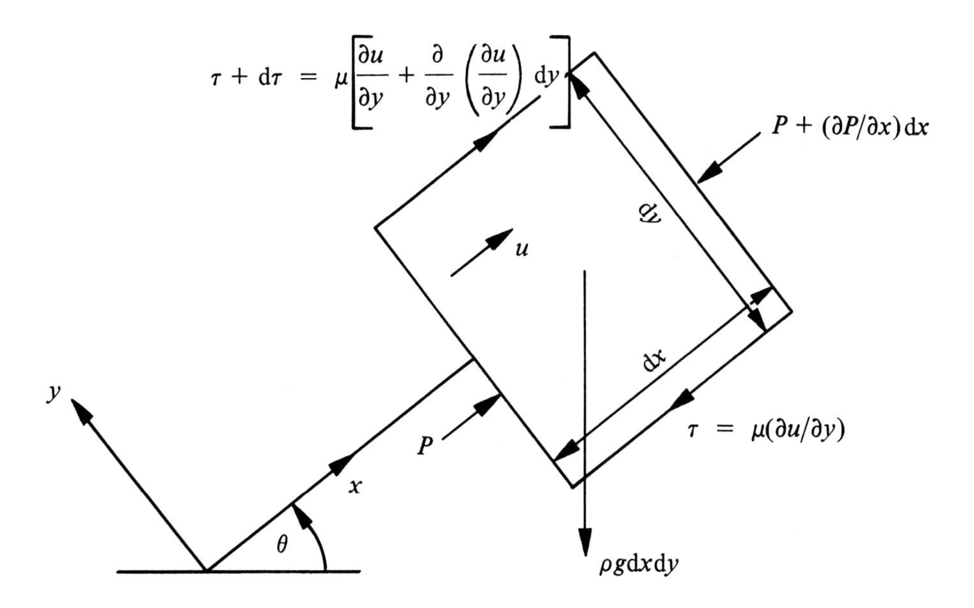

Figure 2:18 Equation of motion along a streamline in a viscous fluid

Consider the fluid particle shown in Figure 2:18 which is in steady motion along the streamline. The equation of motion in the x-direction is:

$$-(\partial P/\partial x)\mathrm{d}x\mathrm{d}y - \rho g\mathrm{d}x\mathrm{d}y\sin\theta - \mu(\partial^2 u/\partial y^2)\mathrm{d}x\mathrm{d}y = \rho\mathrm{d}x\mathrm{d}yu(\partial u/\partial x)$$

or

$$(\partial P/\partial x) + \rho g\sin\theta - \mu(\partial^2 u/\partial y^2) + \rho u(\partial u/\partial x) = 0 \quad (2.35)$$

This is, in fact, the same equation of motion as was given in Section 2.2.1, but with the force $\mathrm{d}F$ replaced by a viscous force due to a change in velocity gradient across the streamline, where, for simplicity, only two-dimensional motion is being considered. Suppose this particle is a typical particle in some fluid-flow system, such as the flow through a boiler. Then if L is some characteristic length of the system, ΔP is the pressure drop across the whole system and U is the average velocity in the system, it is possible to write $x = \alpha L$, $y = \beta L$, $P = \gamma \Delta P$, $u = \delta U$, where clearly, $\alpha,\beta,\gamma,\delta$ are dimensionless. Then 2.35 becomes:

$$(\Delta P/L)\frac{\partial\gamma}{\partial\alpha} + \rho g\sin\theta - (\mu U/L^2)\frac{\partial^2\delta}{\partial\beta^2} + (\rho U^2/L)\delta\frac{\partial\delta}{\partial\alpha} = 0$$

Now divide through by any one of the bracketed quantities, say $(\rho U^2/L)$ to give:

$$(\Delta P/\rho U^2)\frac{\partial\gamma}{\partial\alpha} + (gL/U^2)\sin\theta - (\mu/U\rho L)\frac{\partial^2\delta}{\partial\beta^2} + \delta\frac{\partial\delta}{\partial\alpha} = 0 \quad (2.36)$$

in which *all* terms are dimensionless. Clearly, the solution of equation 2.36 for any boundary conditions, i.e. for any shape of flow system, depends only on the values of the bracketed coefficients, regardless of the scale of the system, so that, for example, it would equally well represent the fluid motion in an exact scale model of the boiler as in the real boiler, provided the bracketed terms had the same values in the model as in the prototype. Note that this could be achieved using quite different fluids in the two systems.

In most complex systems it is not possible to solve an equation such as 2.36 (which of course would normally be three-dimensional) so that a useful method of investigation would be to perform experiments on the system, or on a model, and to correlate the results in terms of the dimensionless parameters in brackets. These particular groups are widely used in fluid dynamics.

$U\rho L/\mu$ is called the Reynolds number (*Re*).

U^2/gL is called the Froude number (*Fr*).

$\Delta P/\rho U^2$ is called the drag coefficient or friction factor (C_D or f) and a common method of correlating experimental results is to plot C_D against (*Re*) and (*Fr*).

The above discussion is intended to bring out the fundamental significance of the dimensionless groups, but clearly the dimensionless groups can easily be formed from the relevant quantities which characterize the system merely by inspection. Note also that quite different, but equally valid, dimensionless groups could have been produced by dividing through by a different bracketed quantity. The groups given above are usually the most convenient and they will be used in the following sections discussing fluid friction. Other groups may sometimes be more relevant—for example, the blade tip velocity would be more relevant than the fluid velocity in correlating results on pumps. These dimensionless groups may also be regarded as ratios of forces. For example, the viscous stresses in the system are proportional to $\mu U/L$, and the inertia stresses to ρU^2, so (Re) is a measure of the ratio of viscous forces to inertia forces.

Dimensionless similarity between model and prototype means keeping all these ratios the same in the two systems. If this is done then the patterns of motion in them will be identical. This topic will be discussed further in Section 12.8.

2.4 FLUID FRICTION DRAG ON SURFACES

When a fluid flows over a surface, the fluid in contact with the surface is at rest, i.e. there is no slip between fluid and surface. Hence there must be a velocity gradient extending from the surface into the fluid and, from equation 2.1, there must be a tangential stress between surface and moving fluid. In most cases it is not possible to determine the stress analytically, so experiments are carried out, for example in wind tunnels, on model systems, or on prototypes where possible, and the results plotted as C_D against (Re). In this case C_D would probably be expressed as $F_D/\frac{1}{2}A\rho U^2$ where F_D is the total drag force on the surface of area A. An example of the form of this relationship is shown in Figure 2:19.

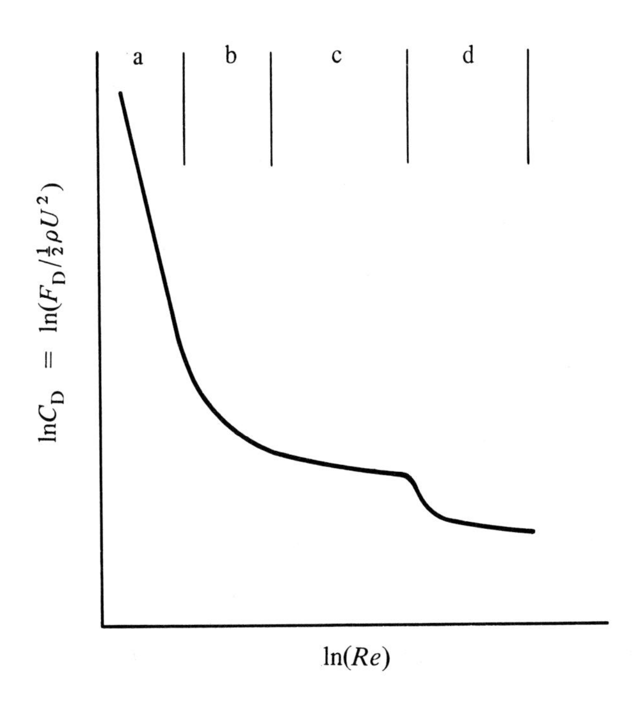

Figure 2:19 Drag coefficient for flow round cylinder or sphere

Region a laminar flow throughout, viscous friction only

Region b transition region, some turbulence

Region c breakaway has occurred, form drag mainly

Region d turbulent boundary layer, breakaway point moves downstream, form drag

2.4.1 Friction due to laminar flow in a pipe

This is one of the few cases which can be dealt with analytically. Consider the cylindrical fluid element in Figure 2:20. Since the streamlines are straight the pressure is

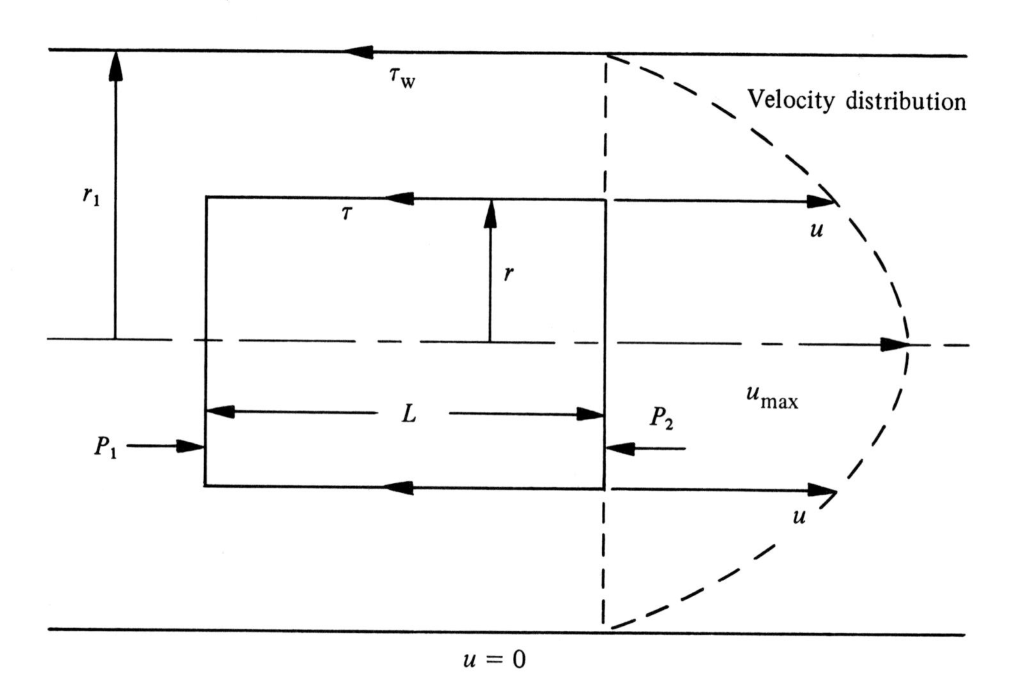

Figure 2:20 Equation of motion for cylindrical element in laminar flow in circular pipe

constant over any cross-section (equation 2.17) and, as there is no acceleration of the fluid, the equation of motion is:

$$\pi r^2(P_1 - P_2) = \pi r^2 \Delta P = 2\pi r L \tau = -2\pi r L \mu (\mathrm{d}u/\mathrm{d}r)$$

Therefore

$$-\int_{u_{max}}^{0} \mathrm{d}u = (\Delta P/2\mu L)\int_{0}^{r_1} r\mathrm{d}r$$

or

$$u_{max} = \Delta P r_1^2/4\mu L \tag{2.37}$$

and the velocity distribution is:

$$u = \Delta P(r_1^2 - r^2)/4\mu L \tag{2.38}$$

Since this is a parabolic distribution, the flow rate is given by $V = \frac{1}{2}\pi r_1^2 u_{max}$ and therefore equation 2.37 can be rewritten as:

$$\Delta P = 8\mu LV/\pi r_1{}^4 = 8\mu Lu_m/r_1{}^2 \tag{2.39}$$

where u_m is the mean velocity. This is the Poiseuille equation.

Also, from equation 2.38, the velocity gradient at the wall is:

$$(du/dr)_{r=r_1} = -\Delta Pr_1/2\mu L = -4u_m/r_1 \quad \text{(using 2.39)}$$

Therefore

$$\tau_w = 4\mu u_m/r_1 \tag{2.40}$$

and

$$\tau_w/\tfrac{1}{2}\rho u_m{}^2 = 16\mu/\rho Du_m = 16/(Re) \tag{2.41}$$

where the pipe diameter D is used in the Reynolds number.

2.4.2 Friction due to turbulent flow in a pipe

When the value of (Re) exceeds about 2000, the flow in a pipe usually becomes turbulent. This introduces an additional shear stress due to the random radial motion of fluid particles. This is shown qualitatively in Figure 2:21 where particle a moving

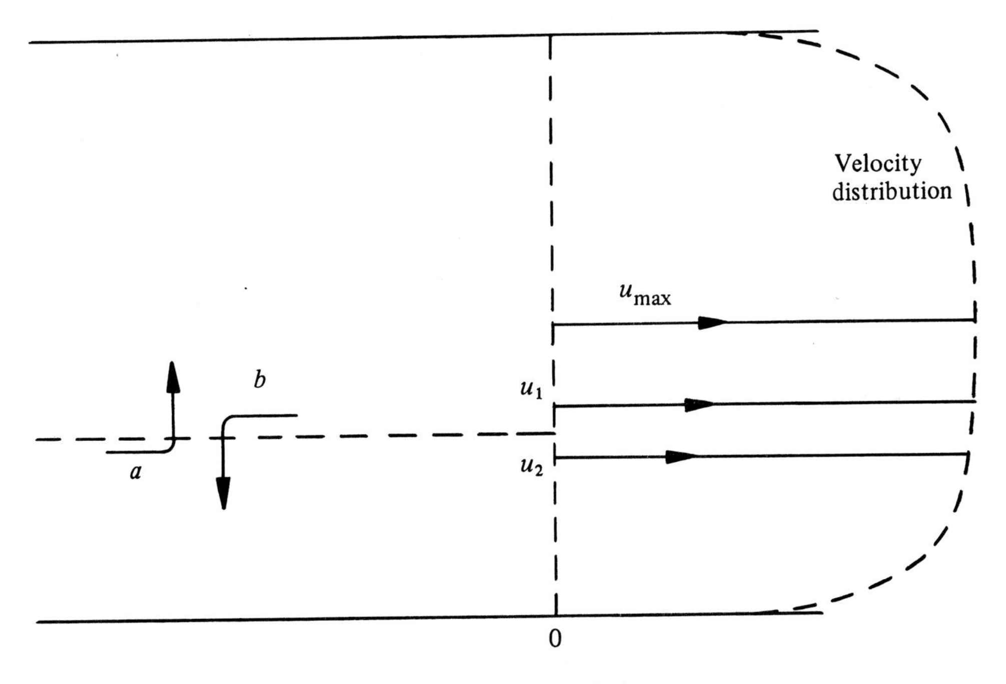

Figure 2:21 Change in velocity profile due to turbulence

at the velocity u_2 moves into the region of higher velocity u_1 and, therefore, by momentum interchange tends to reduce u_1. Similarly the motion of particle b tends to increase u_2. The result is a flattening of the velocity profile across the pipe compared with that in laminar flow, and therefore a greater velocity gradient at the pipe wall, and a greater value of τ_w. Also, the value of u_m is more nearly equal to u_{max}.

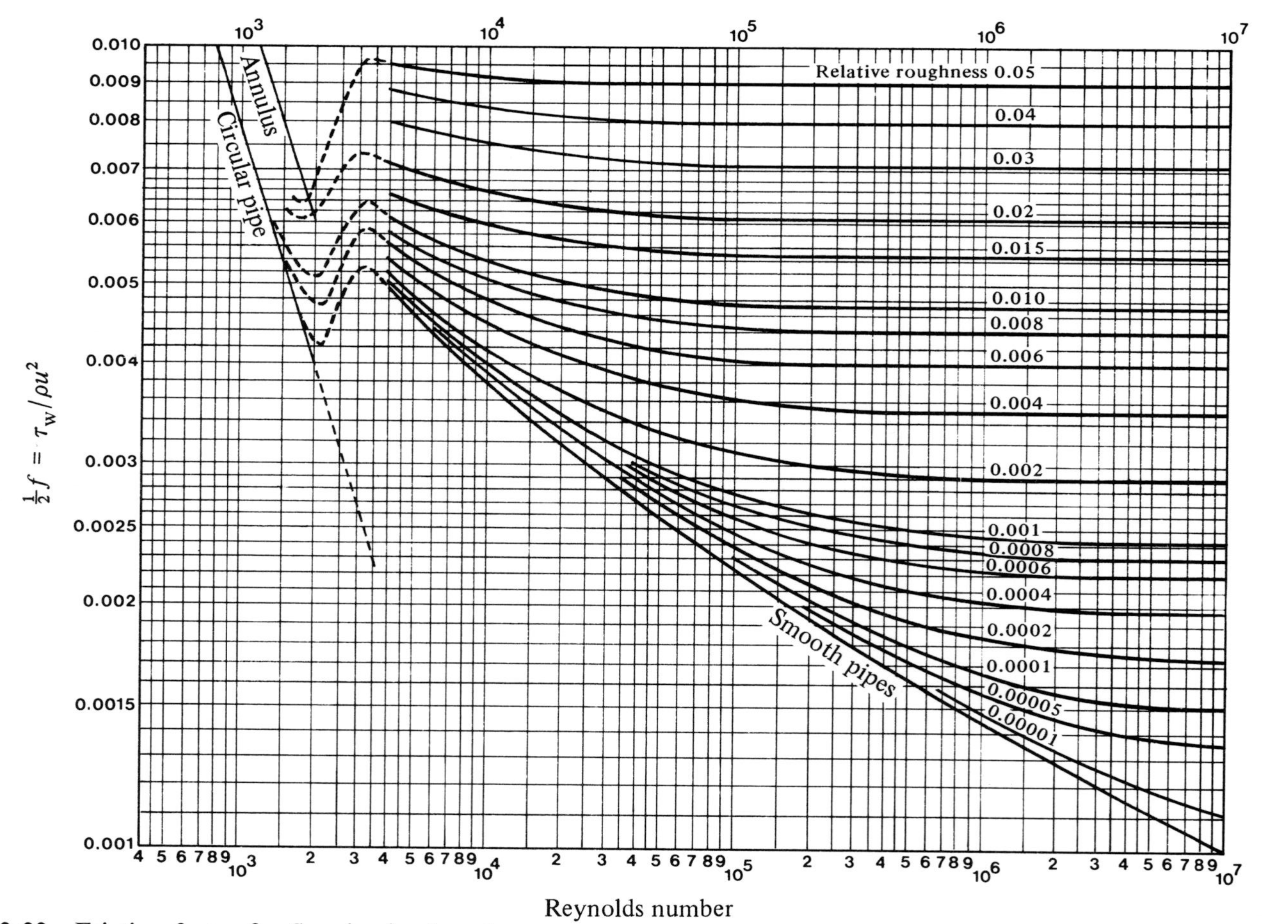

Figure 2:22 Friction factor for flow in circular pipes

The results of very extensive experimental work using various fluids in many different sizes of pipe have been correlated by plotting $\tau_w/\frac{1}{2}\rho u_m^2$ against (Re), and this is shown in Figure 2:22. $\tau_w/\frac{1}{2}\rho u_m^2$ is called the friction factor f and in the turbulent region this is dependent also upon pipe roughness which is expressed as pipe diameter D divided by mean height of surface projections k. It will be seen that for rough pipes, i.e. small D/k, the value of f tends to be constant at high values of (Re).

The value of F (i.e. force per unit weight of fluid) in the equation of motion 2.15 for turbulent friction is $4\tau_w \pi DL/\omega \pi D^2 L = 4\tau_w/\omega D$.

Since $\tau_w = \frac{1}{2}\rho u_m^2 f$ this becomes $F = 4fu_m^2/2gD$. Hence the integral

$$\int_0^L F ds$$

over a pipe length is:

$$4fLu_m^2/2gD = h_f \tag{2.42}$$

This is known as the D'Arcy formula.

If the flow is in a duct of non-circular cross-section, D is replaced in equation 2.42 by the hydraulic mean depth for the section defined as:

$$\frac{4 \times \text{cross-sectional area}}{\text{section perimeter}}$$

2.4.3 Secondary losses

Equations 2.41 and 2.42 refer to straight parallel pipes. In addition, friction forces are generated at pipe bends and at changes in cross-section. All such secondary friction forces are given by:

$$h_{fs} = c_f\, u_m^2/2g \tag{2.43}$$

Typical values of c_f are:

Pipe entrance	0.5
Sudden reduction in pipe diameter	0.5
Sudden enlargement	$[1 - (D_1/D_2)^2]^2$
Right-angle elbow or tee	1
Large-radius bend	0.25

2.5 BOUNDARY LAYER

This is the fluid layer next to the surface over which fluid is flowing, within which the velocity gradient occurs. Since the velocity of the undisturbed fluid u_0 is approached asymptotically the boundary-layer thickness δ can be arbitrarily defined as the distance at which the velocity is $0.99\, u_0$.

Momentum considerations show that the thickness of the boundary layer increases with distance from the leading edge. Referring to Figure 2:23, and applying the momentum equation 2.21 to the region 1122:

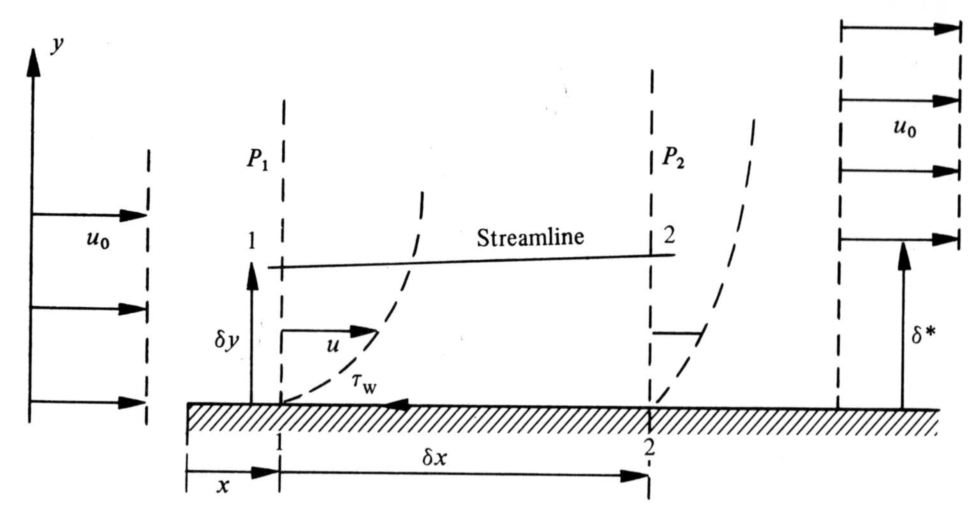

Figure 2:23 Thickening of boundary layer as fluid moves over surface

$$(P_1 - P_2)\delta y - \tau_w \mathrm{d}x = \underset{1}{\int_0^{\delta y} \rho u^2 \mathrm{d}y} - \underset{2}{\int_0^{\delta y} \rho u^2 \mathrm{d}y} \tag{2.44}$$

If the bulk fluid outside the boundary layer is moving in an unconstrained manner so that no velocity changes occur, and if the streamlines are not curved, then the pressure will be the same everywhere, so that the pressure term in equation 2.44 becomes zero, and

$$\int_2 < \int_1$$

i.e. the velocity gradient becomes less as the fluid moves over the surface.

Figure 2:23 also shows the definition of the *displacement thickness* where

$$u_0\,\delta^* = \int_0^\infty (u_0 - u)\,\mathrm{d}y$$

i.e. if the velocity were zero for a distance δ^* from the surface and then suddenly changed to u_0, the total flow would be unchanged.

The first part of the boundary layer is always laminar, but usually becomes turbulent downstream, depending on the Reynolds number $u_0 x/v$ but there is always a thin laminar sublayer next to the surface, and this is important in providing the main resistance to the flow of heat to a surface.

For a laminar boundary layer, by assuming a simple power-series velocity distribution in the boundary layer, Blasius found, essentially by solving equation 2.44, that $\delta = 5.48x/(Re_x)^{\frac{1}{2}}$, $\delta^* = 1.73x/(Re_x)^{\frac{1}{2}}$ and $C_D = 0.72\ (Re_x)^{-\frac{1}{2}}$.

For the turbulent boundary layer, experiment shows that $u/u_0 = (y/\delta)^{1/7}$, whence $\delta = 0.379x/(Re_x)^{0.2}$, $\delta^* = 0.047x/(Re_x)^{0.2}$ and $C_D = 0.072\ (Re_x)^{-0.2}$.

When the flow is confined, as within a duct or pipe, the boundary layers must eventually meet and then their thickness is constant, so that the right-hand side of equation 2.44 becomes zero and the pressure must fall in the direction of flow as discussed in Sections 2.4.1 and 2.4.2.

2.5.1 Curved boundary: separation of boundary layer

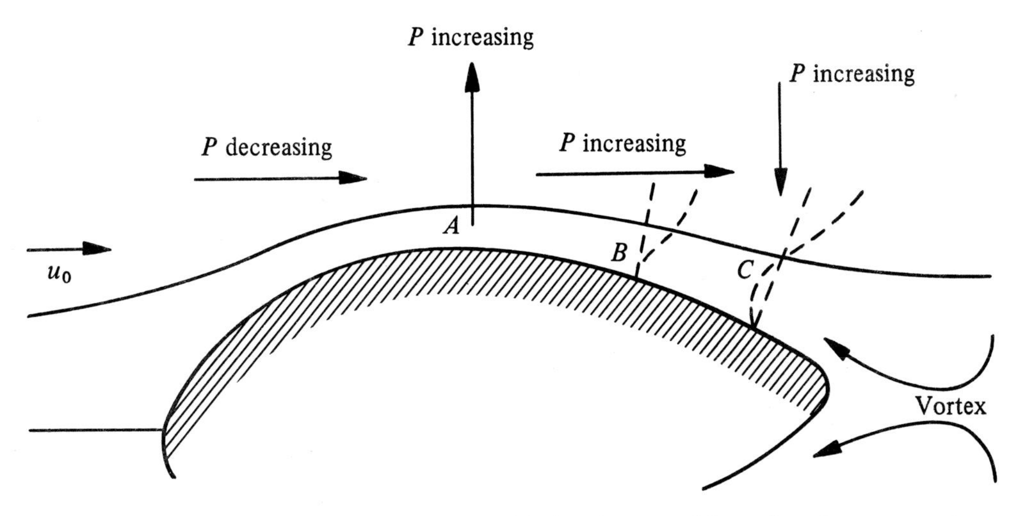

Figure 2:24 Flow over curved surface with breakaway of boundary layer occurring at B

Referring to Figure 2:24, the increase in velocity in the region A and the curvature of the streamlines give rise to the changes of pressure indicated. Under these conditions the momentum of the boundary layer falls until the velocity gradient at the wall may become zero as at B, and the boundary layer is said to separate from the surface. The reversal of flow which subsequently occurs as at C is supplied by the formation of a vortex behind the point of separation.

As the velocity u_0 increases the pairs of vortices are alternately shed from the back of the body and move downstream as a vortex wake, new vortices being continually formed.

The energy carried in the vortex requires an increased drag on the surface called *form drag* and the alternate shedding causes a periodic transverse force on the body which can cause oscillations.

Since a turbulent boundary layer carries a greater momentum than does a laminar layer, separation will occur later in a turbulent than in a laminar boundary layer.

3

Heat and mass transfer

3.1 HEAT TRANSFER

Heat energy is the energy of random atomic and molecular motion. Its intensity is described by the temperature and its capacity by the mass multiplied by the specific heat capacity. Heat flow takes place in the direction of the temperature gradient.

Three modes of heat transfer will be considered: conduction, convection and radiation.

Conduction occurs by direct exchange of energy from more energetic to less energetic electrons, atoms or molecules. This is the principal mode of heat transfer in solids, and the rate is given by:

$$Q = -\lambda A(\mathrm{d}T/dx) \tag{3.1}$$

where λ is the thermal conductivity.

Convection from a fluid to a surface takes place by a combination of conduction and actual transport of fluid particles between the bulk of the fluid and the surface. The rate of heat transfer is given by the Newton rate equation:

$$Q = hA(T_{\mathrm{f}} - T_{\mathrm{s}}) \tag{3.2}$$

where the heat-transfer coefficient h is dependent upon fluid properties and velocity and is therefore also a function of temperature.

Radiation is energy carried in electromagnetic waves which are emitted by atoms and travel across space with the velocity of light. The rate at which radiant heat is emitted from a surface is:

$$Q = \sigma\varepsilon AT^4 \tag{3.3}$$

where ε is emissivity and σ is the Stefan–Boltzmann constant.

3.2 CONDUCTION

Steady-state conduction through a plane block of material in which λ is constant

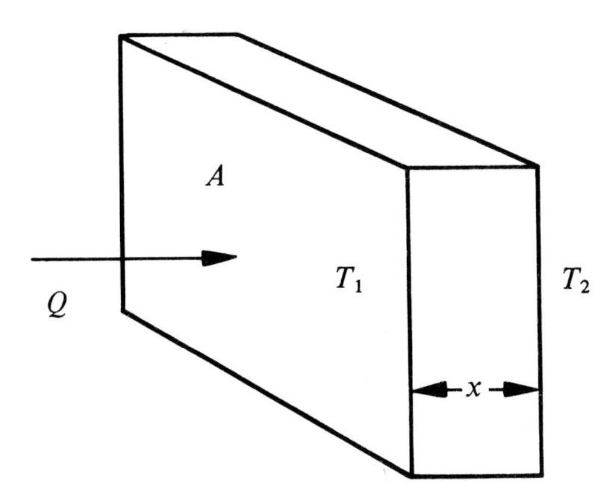

Figure 3:1 Heat conduction through a plane block with constant conductivity

is shown in Figure 3:1. From equation 3.1, $-\mathrm{d}T/\mathrm{d}x$ is constant and equal to $(T_1 - T_2)/x$. Hence in this case

$$Q = \lambda A(T_1 - T_2)/x \tag{3.4}$$

This is of the same form as equation 3.2. If λ is a function of temperature it may often be sufficiently accurate to evaluate λ at the mean temperature $\frac{1}{2}(T_1 + T_2)$. If λ varies greatly through the material it will be more accurate to divide the block into a number of sections as in Figure 3:2, and this method also applies to the case of a composite block, e.g. a typical furnace wall. Then:

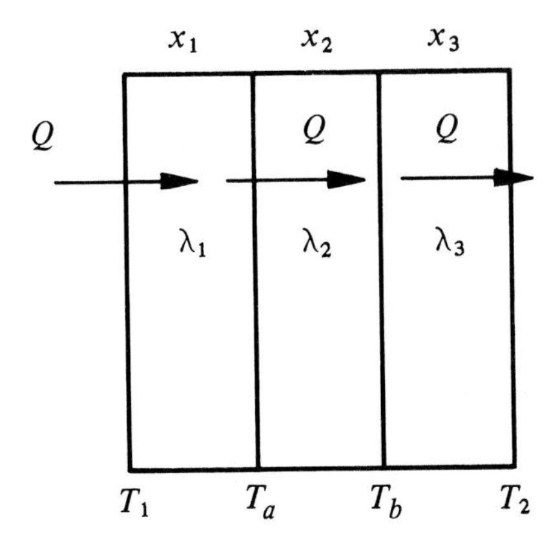

Figure 3:2 Heat conduction through a plane block with variable conductivity

$$Q = \lambda_1 A(T_1 - T_a)/x_1 = \lambda_2 A(T_a - T_b)/x_2$$

and so on, and upon bringing the temperature differences to one side of the equation and adding, the intermediate temperatures T_a, T_b vanish, giving

$$Q = A(T_1 - T_2)/(x_1 \lambda_1^{-1} + x_2 \lambda_2^{-1} + x_3 \lambda_3^{-1}) \tag{3.5}$$

For many solids λ is linearly related to temperature:

$$\lambda = \lambda_0 (1 + aT) \tag{3.6}$$

Using this in equation 3.1, separating the variables and integrating, gives:

$$\int_0^x Q/A\lambda_0 \, dx = -\int_{T_1}^{T_2} (1 + aT) \, dT$$

i.e.

$$Q = [1 + \tfrac{1}{2}a (T_2 + T_1)] (T_1 - T_2) \lambda_0 A/x \tag{3.7}$$

In the case of steady-state conduction through a curved block of material, A is a function of x. For many regular geometric shapes, integration of equation 3.1 gives an expression for the heat flow. For example, take the case of radial conduction

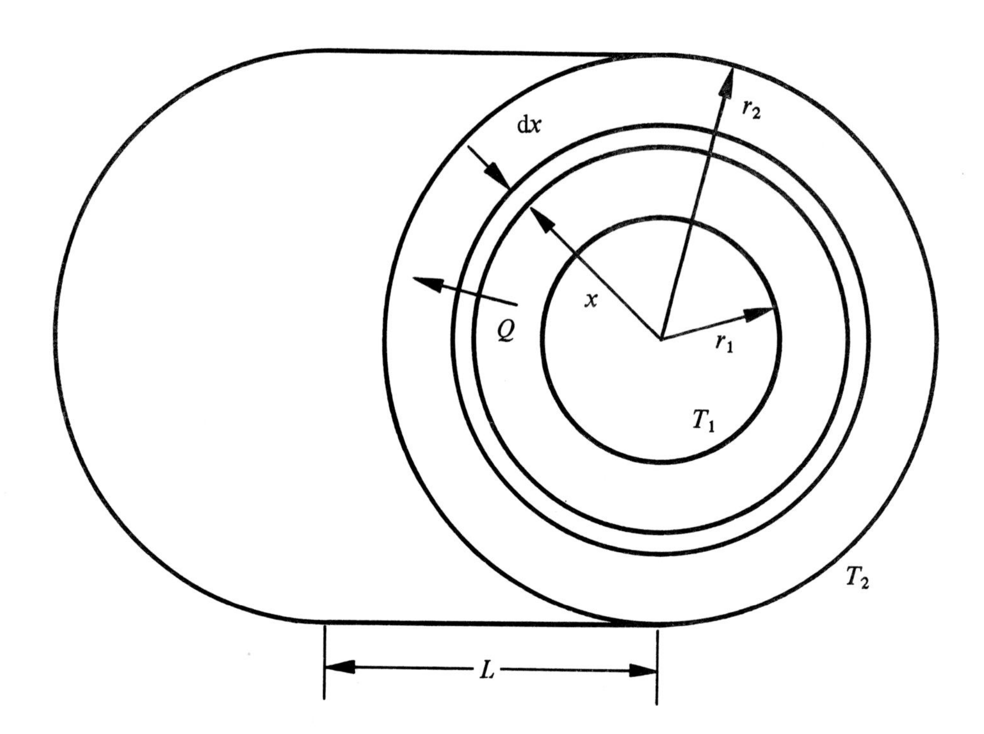

Figure 3:3 Heat conduction through a cylindrioal annulus

through a cylindrical annulus (Figure 3:3). For the element of radius x:

$$Q = -\lambda 2\pi x L\,(\mathrm{d}T/\mathrm{d}x)$$

and, on separating variables and integrating,

$$Q = 2\pi\lambda L(T_1 - T_2)/\ln(r_2/r_1) \tag{3.8}$$

Equations 3.4, 3.5 and 3.8 all have the form:

$$Q = (T_1 - T_2)/R \tag{3.9}$$

where the thermal resistance R is, respectively,

$$x/A\lambda, \qquad (x_1\lambda_1^{-1} + x_2\lambda_2^{-1} + \ldots)/A, \qquad [\ln(r_2/r_1)]/2\pi\lambda L$$

In like manner, the thermal resistance of a multiple annular cylinder, such as a lagged pipe, is:

$$R = \frac{\ln(r_{a2}/r_{a1})}{2\pi\lambda_a L} + \frac{\ln(r_{b2}/r_{b1})}{2\pi\lambda_b L} + \ldots \tag{3.10}$$

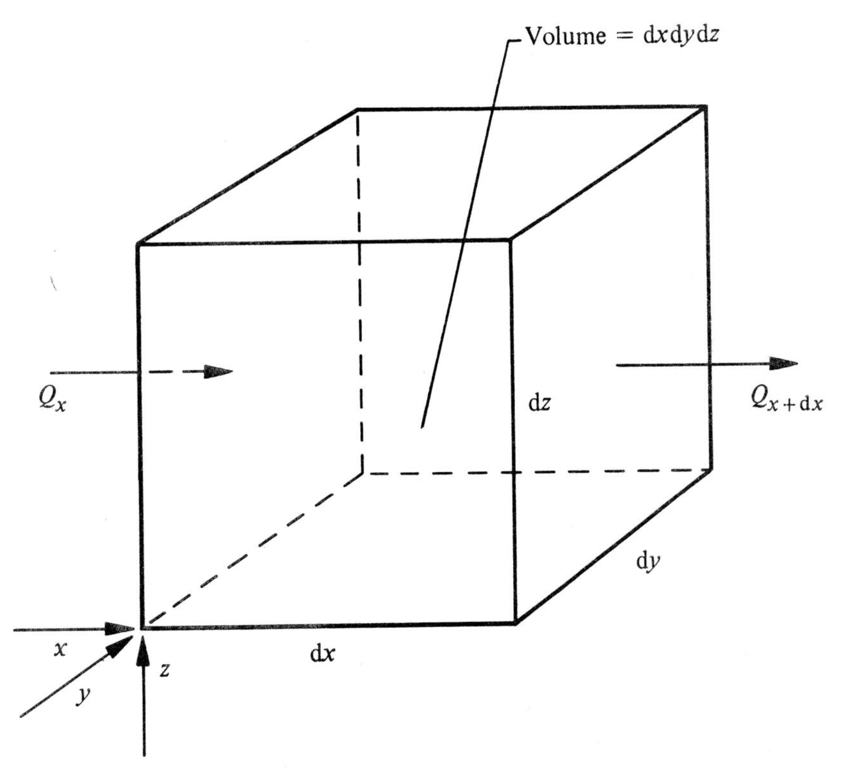

Figure 3:4 Three-dimensional heat flow

3.2.1 The general unsteady-state heat-conduction equation

The cases dealt with above are one-dimensional, that is the temperature gradient is in one coordinate direction only. Consider now the elemental cube of material in Figure 3:4. The temperature of the element is T. At the coordinate x, i.e. the left-hand face, the temperature gradient is $\partial T/\partial x$, and at $x + \mathrm{d}x$, it is:

$$\frac{\partial T}{\partial x} + \frac{\partial}{\partial x}\left(\frac{\partial T}{\partial x}\right) \mathrm{d}x$$

Then the net heat flow into the element in the x-direction is:

$$Q_x - Q_{x+\mathrm{d}x} = -\lambda \frac{\partial T}{\partial x}\mathrm{d}y\mathrm{d}z - \left[-\lambda \frac{\partial T}{\partial x} - \frac{\partial}{\partial x}\left(\lambda \frac{\partial T}{\partial x}\right) \mathrm{d}x\right] \mathrm{d}y\mathrm{d}z$$

$$= \frac{\partial}{\partial x}\left(\lambda \frac{\partial T}{\partial x}\right) \mathrm{d}x\mathrm{d}y\mathrm{d}z$$

The net flows into the element in the other directions are found similarly. Then the total net flow into the element is the rate of change of its heat content:

$$\frac{\partial}{\partial t}(\rho c_p T)\, \mathrm{d}x\mathrm{d}y\mathrm{d}z$$

where ρ is density. Hence a heat balance on the element is

$$\frac{\partial}{\partial x}\left(\lambda \frac{\partial T}{\partial x}\right) + \frac{\partial}{\partial y}\left(\lambda \frac{\partial T}{\partial y}\right) + \frac{\partial}{\partial z}\left(\lambda \frac{\partial T}{\partial z}\right) = \frac{\partial}{\partial t}(\rho c_p T) \tag{3.11}$$

This is perfectly general and may of course be simplified where appropriate. For example, if λ is constant and conditions are steady, then:

$$\frac{\partial^2 T}{\partial x^2} + \frac{\partial^2 T}{\partial y^2} + \frac{\partial^2 T}{\partial z^2} = 0 \tag{3.12}$$

These equations may, of course, be written in cylindrical or spherical coordinates if required. No general solutions are possible, but numerical finite-difference methods can always be used to solve any problem given its initial and boundary conditions. Electrical analogues may also be used to solve the Laplace equation 3.12, particularly in its two-dimensional form.

A particular problem which can often be solved analytically is the one-dimensional unsteady-state case, for example, the heating up of a wall or slab, which is expressed by:

$$\partial^2 T/\partial x^2 - (\rho c_p/\lambda)(\partial T/\partial t) = 0 \tag{3.11a}$$

This may be solved using Laplace transforms or, for certain boundary conditions, by separation of variables.

A very good account of various analogue and numerical methods for dealing with steady- and unsteady-state multi-dimensional conduction problems is given by Schenck [1], and such problems are further mentioned in Chapter 12.

3.3 CONVECTION

In convective heat transfer the rate-determining process is the conduction of heat across the laminar boundary layer of fluid adjacent to the surface.

3.3.1 Forced convection

Forced convection is the most important convective heat-transfer process. It occurs when fluid is moved past the surface by an externally applied pressure gradient such as that produced by a fan or a chimney. Two distinct cases occur according as the bulk fluid motion is laminar or turbulent. In the turbulent case there is a very narrow laminar sublayer next to the wall giving a large temperature gradient and a consequently high heat-transfer rate.

The detailed mechanism of convective heat transfer is complex and analytical methods of investigation are of limited use. However, it is instructive to consider a simplified analysis of the process. Figure 3:5 shows the situation close to the surface

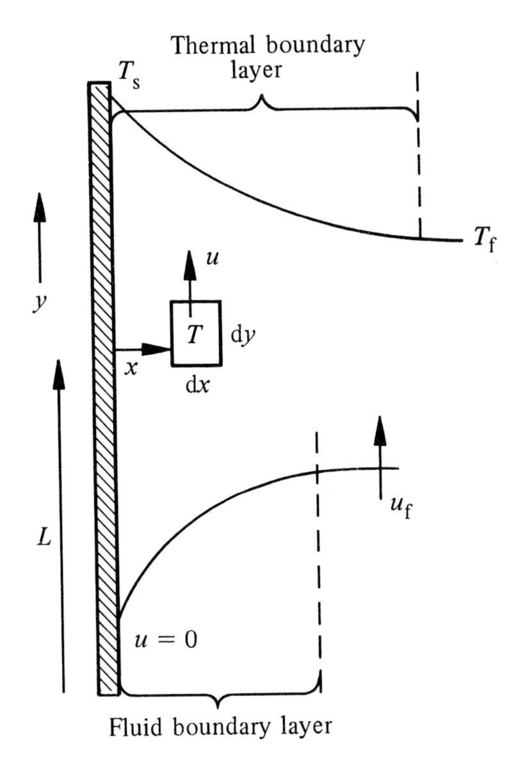

Figure 3:5 Convective heat transfer close to surface

where the fluid is cooler than the wall and moving laminarly. A heat balance on the small fluid element shown gives:

$$\partial^2 T/\partial x^2 = (\rho c_p u/\lambda)(\partial T/\partial y) \tag{3.13}$$

Also, the net viscous force on the element is equal to its mass $\times$ acceleration, that is:

$$\mu \partial^2 u/\partial x^2 = \rho u(\partial u/\partial y)$$

or

$$\partial^2 u/\partial x^2 = (\rho u/\mu)(\partial u/\partial y) \tag{3.14}$$

where μ is dynamic viscosity. Equations 3.13 and 3.14 are of the same form and, having similar boundary conditions, will give similar profiles for temperature and velocity if the coefficients in the equations have the same value, that is if $\rho c_p u/\lambda = \rho u/\mu$, i.e. if:

$$c_p \mu/\lambda = 1 \tag{3.15}$$

The dimensionless quantity $c_p\mu/\lambda$ is called the Prandtl number (Pr) and its value determines the relative thickness of fluid and thermal boundary layers. According to von Kármán, the thickness of the fluid boundary layer for flow over a flat plate is $1.73\ L/(Re)^{\frac{1}{2}}$, where the dimensionless number $(Re) = \rho L u_f/\mu$. Hence in the case where $(Pr) = 1$, the temperature gradient would be expected to be of the order $(T_s - T_f)(Re)^{\frac{1}{2}}/1.73L$ and the heat transfer therefore should be of the order:

$$Q/A = (T_s - T_f)(Re)^{\frac{1}{2}}\ \lambda/1.73L$$

Comparing this with the Newton rate equation 3.2 it can be seen that:

$$h = \lambda(Re)^{\frac{1}{2}}/1.73L$$

or

$$hL/\lambda = 0.58(Re)^{\frac{1}{2}} \tag{3.16}$$

The dimensionless group hL/λ is called the Nusselt number (Nu). From equation 3.18 it can be seen that equation 3.16 is of the right order of magnitude since if $(Pr) = 1$, equation 3.18 is $(Nu) = 0.66(Re)^{\frac{1}{2}}$. For gases (Pr) is typically between 0.7 and 1.

3.3.2 Natural convection

In natural convection, the fluid moves over the surface by reason of the buoyancy forces arising from the temperature gradients at the interface. The temperature and velocity profiles will be of the form shown in Figure 3:6. In this case the Reynolds number is inapplicable since the fluid velocity is not an independent variable. Instead, the dimensionless group which introduces the buoyancy forces is the Grashof number. $(Gr) = L^3\gamma g\, \Delta T \rho^2/\mu^2$ where γ is the cubic expansivity.

Particularly in the case of laminar convection in pipes, the further dimensionless group L/D may be involved. This is because the velocity distribution near the pipe entrance is not the same as that further along the pipe where the boundary layers have joined to give the final parabolic velocity profile.

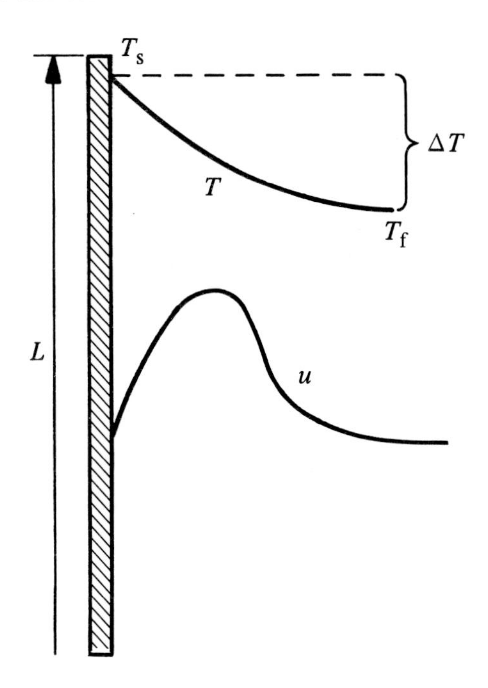

Figure 3:6 Temperature and velocity profiles in natural convection

3.3.3 Convected heat transfer: data correlations by dimensionless groups

Heat transfer by convection from various shapes of surface has been determined by extensive experiments, and is usually correlated in terms of the dimensionless numbers given above. Clearly, by combining dimensionless groups others may be formed and in studies of heat transfer the following additional groups are used:

$$\text{Stanton number } (St) = (Nu)/(Re)(Pr)$$
$$\text{Colburn } j \text{ factor } (j_h) = (St)(Pr)^{2/3}$$

For gases the value of (Pr) is close to unity so that in some expressions for convective heat transfer to gases (Pr) does not occur. An important matter is the temperature at which the various fluid properties are evaluated—usually the film or boundary-layer temperature. Since this is not known until the heat transfer has been calculated it may be necessary to carry out an initial calculation in which temperatures are assumed, and then to recalculate, using corrected temperatures.

Correlations for natural convection
These are of the form:

$$(Nu) = K\{(Gr)(Pr)\}^n \tag{3.17}$$

$$n = \begin{cases} 0.25 \text{ for laminar flow, i.e. when } (Gr)(Pr) < 10^8 \\ 0.33 \text{ for turbulent flow, i.e. when } (Gr)(Pr) > 10^8 \end{cases}$$

In the transition region around $(Gr)(Pr) = 10^8$ it is best to use the more conservative equation. The fluid properties in Figure 3:7 are evaluated at film temperature, $\frac{1}{2}(T_s + T_f)$.

System	K		Dimension taken as L in (Nu) and (Gr)
	Laminar	Turbulent	
Vertical plates or cylinders	0.56	0.12	Height of plate or cylinder
Horizontal plates from top face	0.54	0.14	Length of long side
Horizontal plates from bottom face	0.25	—	Length of long side
Horizontal cylinders	0.47	0.10	Diameter of cylinder

Figure 3:7 Correlations for natural convection

Correlations for forced convection
A very large number of expressions are quoted in the literature and only a few of the more generally applicable ones will be given here.

Laminar flow over flat plate:

$$j_h = \tfrac{2}{3}(Re)^{-\frac{1}{2}} \tag{3.18}$$

where L is the height of the plate. For this case also, the *local* heat-transfer coefficient at a height x above the bottom of the plate is given by

$$j_{hx} = \tfrac{1}{3}(Re)^{-\frac{1}{2}} \tag{3.18a}$$

where x is the relevant dimension in (Re).

Turbulent flow over flat plate:

$$j_h = 0.037(Re)^{-0.2} \tag{3.18b}$$

As in the case of equation 3.18 the linear dimension is the length of the plate and the velocity is that of the bulk fluid, but in the laminar case μ is evaluated at bulk-fluid temperature while in the turbulent case it is evaluated at film temperature.

Flow inside tubes for turbulent conditions of $(Re) = 10\,000$ to $120\,000$ and $(Pr) = 0.7$ to 120, and $L/D < 60$:

$$(St)(Pr) = 0.023\,(Re)^{-0.2}\,[1 + (D/L)^{0.7}] \tag{3.19}$$

The dimension used in evaluating (Re) is the tube diameter, and quantities are evaluated at film temperature.

For values of $L/D > 60$

$$(St)\,(Pr)^{0.6} = 0.023\,(Re)^{-0.2} \tag{3.19a}$$

and quantities are evaluated at bulk temperature.

For ducts of non-circular cross-section, the dimension to be used is the hydraulic mean diameter $= 4 \times$ cross-sectional area/perimeter.

Flow of gas over the outside of a tube and normal to the axis:

$$(Nu) = K(Re)^n \tag{3.20}$$

(Re)	n	K
1–4	0.33	0.89
4–40	0.385	0.82
40–4000	0.466	0.615
4000–40 000	0.62	0.174
40 000–250 000	0.81	0.24

Figure 3:8 Correlations for forced convection: flow over a circular tube normal to axis

The linear dimension is the outside diameter of the tube, and quantities are evaluated in Figure 3:8 at film temperatures. For shapes other than circular the values shown in Figure 3:9 apply.

Shape	(Re)	n	K
→ ◇	5000–100 000	0.59	0.22
→ ▭	5000–100 000	0.675	0.092
→ ▯	4000–15 000	0.73	0.205

Figure 3:9 Correlations for forced convection: flow over non-circular tubes normal to axis

For flow over *banks* of tubes, equation 3.20 also holds, but the velocity taken is that in the minimum clearance area between the tubes. The values of K and n vary somewhat according to tube spacing, but reasonable average values are:

For in-line arrangements $n = 0.6$, $K = 0.26$
For staggered arrangements $n = 0.57$, $K = 0.45$

3.3.4 Overall heat-transfer coefficients

Commonly heat is transferred between two fluids separated by a surface (Figure 3:10) and an overall heat-transfer coefficient U is defined by

$$Q = UA(T_h - T_c) \tag{3.21}$$

By using the same method as that used in deriving equation 3.5 it can be shown that:

$$U^{-1} = h_1^{-1} + x\lambda^{-1} + h_2^{-1} \tag{3.22}$$

In many cases one of the individual heat-transfer coefficients may be much smaller than the others, in which case only this single controlling coefficient needs to be considered. For example, if the hot fluid is a gas transferring heat to water through a thin metal wall then h_1 is much smaller than the other terms and the process is said to be gas-film controlled.

3.3.5 Mean temperature difference in heat exchangers

Where the fluids are flowing over the surface as in Figure 3:10 it is necessary to define a mean temperature difference for use in equation 3.21, and it is easily shown that the appropriate value to take is the logarithmic mean temperature difference, ΔT_{lm}, defined as

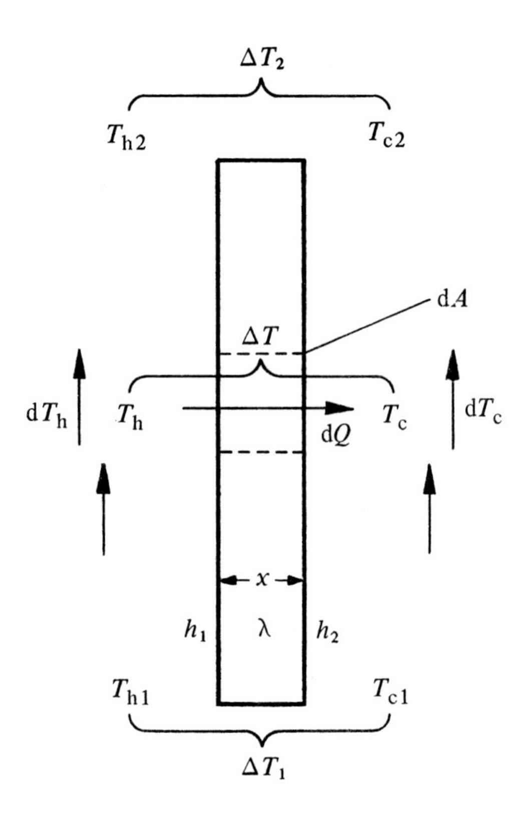

Figure 3:10 Transfer of heat between two fluids

$$\Delta T_{\mathrm{lm}} = (\Delta T_2 - \Delta T_1)/\ln(\Delta T_2/\Delta T_1) \tag{3.23}$$

for both counter- and co-current flows. For, considering an element of area $\mathrm{d}A$ as in Figure 3:10, $\mathrm{d}Q = U\mathrm{d}A\,\mathrm{d}T$, $\mathrm{d}T_{\mathrm{h}} = -\,\mathrm{d}Q/C_{\mathrm{h}}$ and $\mathrm{d}T_{\mathrm{c}} = \mathrm{d}Q/C_{\mathrm{c}}$, where C_{c} and C_{h} are the thermal capacities of the two flows.

Therefore $\quad \mathrm{d}(\Delta T) = \mathrm{d}T_{\mathrm{h}} - \mathrm{d}T_{\mathrm{c}} = -\,U\mathrm{d}A\,\mathrm{d}T\,(C_{\mathrm{h}}^{-1} + C_{\mathrm{c}}^{-1})$

Therefore $\quad \mathrm{d}(\Delta T)/\Delta T = -\,U\mathrm{d}A\,(C_{\mathrm{h}}^{-1} + C_{\mathrm{c}}^{-1})$ (3.24)

Therefore $\quad (\ln\Delta T)_1^2 = -\,UA\,(C_{\mathrm{h}}^{-1} - C_{\mathrm{c}}^{-1})$

The dimensionless quantity on the right side of 3.24 is called the number of transfer units (NTU). But also:

$$Q = C_{\mathrm{h}}\,(T_{\mathrm{h}1} - T_{\mathrm{h}2}) = C_{\mathrm{c}}\,(T_{\mathrm{c}2} - T_{\mathrm{c}1})$$

Therefore $\quad QC_{\mathrm{h}}^{-1} + QC_{\mathrm{c}}^{-1} = \Delta T_1 - \Delta T_2$

and using 3.24 to eliminate $C_{\mathrm{h}}^{-1} + C_{\mathrm{c}}^{-1}$ yields equation 3.23.

3.3.6 Analogies between convective heat transfer and friction

Equations 3.13 and 3.14 show the general similarity between the mechanism of heat transfer through the boundary layer and the drag on the surface which results in the fluid momentum being reduced to zero at the wall. Reynolds made the assumption that:

$$\frac{\text{heat transferred to surface}}{\text{heat available}} = \frac{\text{momentum lost by friction}}{\text{momentum available}}$$

i.e.

$$h\,(T_f - T_s)/\rho c_p u\,(T_f - T_s) = \tau/\rho u^2 = \tfrac{1}{2}f$$

Therefore $h = \frac{1}{2} f \rho c_p u$, or, in terms of dimensionless groups,

$$(Nu) = (Re)\,(Pr)\tfrac{1}{2}f \quad \text{(Reynolds analogy)} \tag{3.25}$$

This applies reasonably well when $(Pr) = 1$.

Other improved analogies having wider applicability have been derived:

$$(Nu) = \tfrac{1}{2}f\,(Re)\,(Pr)/[1 + 5\{(Pr) - 1\}\sqrt{(\tfrac{1}{2}f)}] \quad \text{(Prandtl–Taylor analogy)} \tag{3.26}$$

$$j_h = \tfrac{1}{2}f \quad \text{(Colburn analogy)} \tag{3.27}$$

These all reduce to the Reynolds analogy when $(Pr) = 1$ and are all applicable to simple cases such as heat transfer in conduits. Schenk [1] has empirically extended the use of the Colburn analogy to a wide variety of surfaces such as finned tubes of various sections and shapes and shows that for $(Re) > 5000$

$$j_h = 0.0298\,f^{0.475} \tag{3.28}$$

Separate equations are needed for different ranges of lower values of (Re). Equation 3.28 is useful in predicting heat transfer in compact heat exchangers from pressure-loss data.

3.4 RADIATION

All substances emit electromagnetic waves of various frequencies. Most solids emit a continuous spectrum of frequencies, the maximum intensity of radiation being emitted at a wavelength (in μm) of

$$2898/(T/\text{K}) \quad \text{(Wien's displacement law)} \tag{3.29}$$

Hence at low temperatures most of the radiation is in the infrared region, and as the temperature rises so a larger proportion falls in the visible range.

Many gases emit no radiation, but polyatomic gases with asymmetric molecules emit energy in certain discrete wavebands, usually in the infrared. In the case of luminous gases, the visible radiation is from hot particles of solid carbon in the flames.

When radiant energy falls on a substance, clearly

$$\alpha + \tau + \rho = 1 \tag{3.30}$$

where the symbols represent the respective fractions absorbed, transmitted and reflected. Most solid materials are practically opaque and therefore have $\tau = 0$.

The total radiation emitted from a body is given by Stefan's law:

$$Q = \sigma \varepsilon A T^4 \tag{3.31}$$

where T is the thermodynamic (absolute) temperature and:

$$\sigma = 5.67 \times 10^{-8}\ \mathrm{W/m^2\,K^4} = 0.171 \times 10^{-8}\ \mathrm{Btu/h\,ft^2\,{}^\circ R^4}$$

A *black body* is one for which the emissivity ε and the absorptance α are both 1. All actual materials have α and $\varepsilon < 1$, and their values vary with the wavelength of the radiation being absorbed or emitted, but α and ε are equal at any given frequency. Many radiation heat-transfer problems can be simplified if it is assumed that $\alpha = \varepsilon$ at all frequencies, i.e. at all temperatures of the radiation fluxes entering and leaving the surface, and a *grey body* is defined as one for which $\alpha = \varepsilon$ for all temperatures.

For most materials other than bright metallic surfaces the radiation is diffuse and the distribution is described by Lambert's law:

$$L = L_N \cos\theta \tag{3.32}$$

where L is the radiance of the surface in the direction making an angle θ with the normal, and L_N is the maximum value of the radiance, which occurs in the direction normal to the surface. L has the units radiant energy/unit time $\times$ unit area $\times$ unit solid angle.

In Figure 3:11 the solid angle subtended by the surface element dA_2 at the centre of dA_1 is, by definition, $(dA_2/r^2)\cos\phi$. Therefore the radiation falling on dA_2 from dA_1 is

$$dQ_{12} = L_{1N}\, dA_1 \cos\theta\, dA_2 \cos\phi / r^2 \tag{3.33}$$

In the particular case where dA_2 is an annular area on the surface of a hemisphere with dA_1 at the centre of the diametral plane, as in Figure 3:12, then $dA_2 = 2\pi r \sin\theta\, r d\theta$ and $\cos\phi = 1$, and in this case equation 3.33 can be integrated to give the total radiation from dA_1 to the whole hemisphere.
This is:

$$Q = \sigma \varepsilon_1\, dA_1\, T_1^{\,4} = \pi L_{1N}\, dA_1 \tag{3.34}$$

3.4.2 The view factor F

In order to find the total radiation from a surface A_1 to a surface A_2 as in Figure 3:11, equation 3.33 must be integrated over the areas A_1 and A_2 using L_{1N} from equation 3.34. This will give an expression of the form

$$Q_{12} = F_{12}\, \sigma \varepsilon_1\, A_1\, T_1^{\,4} \tag{3.35}$$

where

$$F_{12} = \pi^{-1} A_1^{-1} \int_{A_1} \int_{A_2} r^{-2} \cos\theta \cos\varphi\, dA_1\, dA_2 \tag{3.36}$$

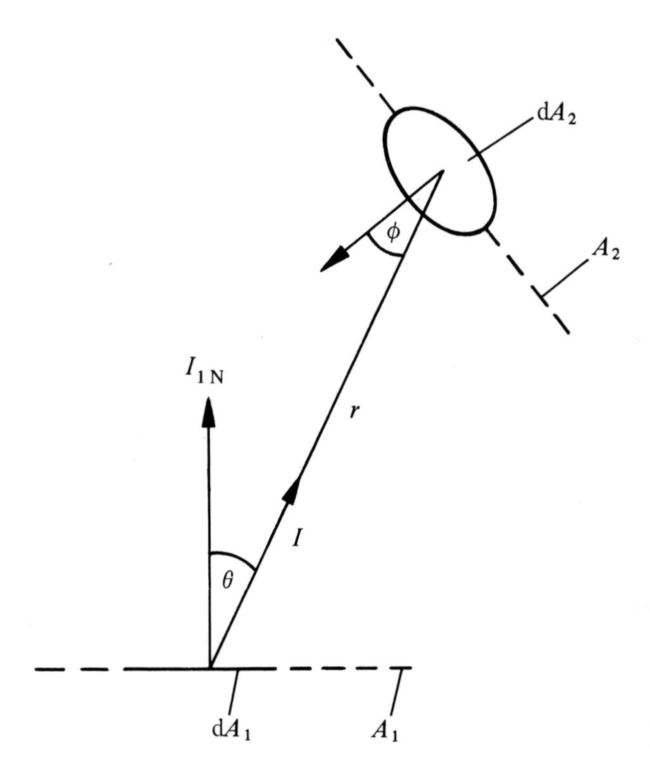

Figure 3:11 Luminous radiation from a surface

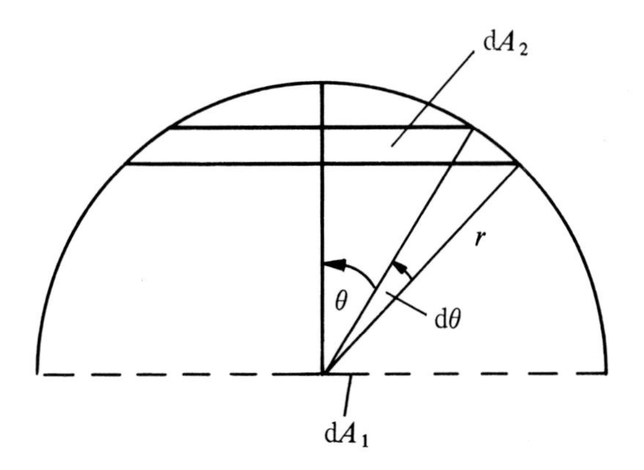

Figure 3:12 Luminous radiation on the interior of a hemisphere

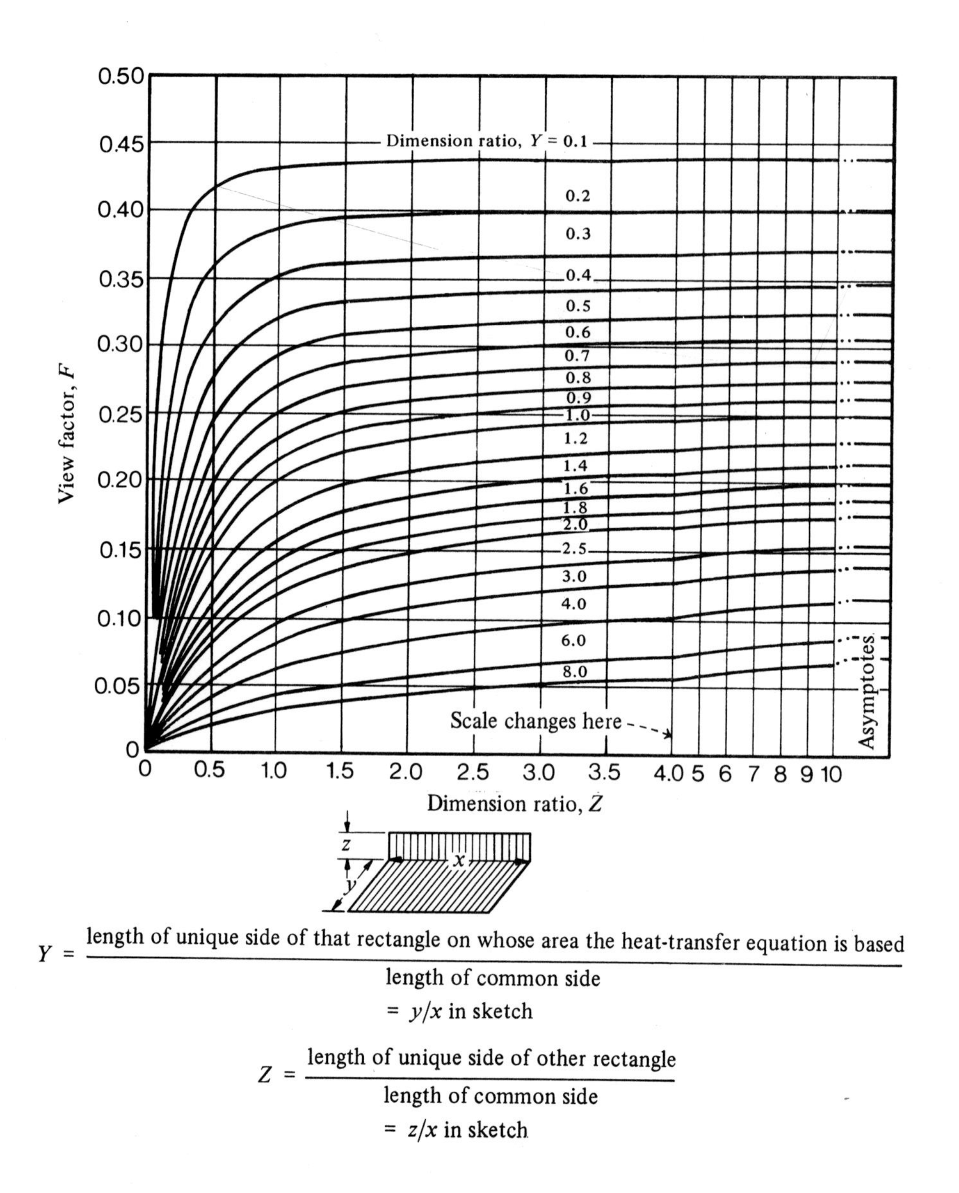

$$Y = \frac{\text{length of unique side of that rectangle on whose area the heat-transfer equation is based}}{\text{length of common side}}$$

$= y/x$ in sketch

$$Z = \frac{\text{length of unique side of other rectangle}}{\text{length of common side}}$$

$= z/x$ in sketch

Figure 3:13 View factor F for direct radiation between adjacent rectangles in perpendicular planes
(From *Heat Transmission* by William H. McAdams. Copyright 1933, 1942, 1954 William H. McAdams. Used with permission of McGraw-Hill Book Company)

and is the fraction of the total radiant energy leaving A_1 which is intercepted by A_2. It is clear from the symmetry of the integral that:

$$F_{12} A_1 = F_{21} A_2 \tag{3.37}$$

Values of F have been calculated for various common surface configurations and Love [2] gives comprehensive data on this. An example is shown in Figure 3:13.

Hottel [4] uses the notation

$$\overline{s_1 s_2} = \overline{s_2 s_1} \tag{3.37a}$$

3.4.3 Radiation interchange between a number of surfaces with a transparent medium between them

Several methods are given in the literature for solving this type of problem. A method well suited to computer solutions is given here. Defining B_i (the radiosity) as the rate at which energy leaves unit area of surface i, and H_i as the rate at which energy is incident upon unit area of i, we have:

$$B_i = \sigma \varepsilon_i T_i^4 + \rho_i H_i \tag{3.38}$$

assuming reflectance $= \rho_i = 1 - \alpha_i$, i.e. assuming opaque surfaces and

$$H_i = \sum_{k=1}^{n} B_k F_{ik} \tag{3.39}$$

because, for example, the energy leaving k is $B_k A_k$ of which $B_k A_k F_{ki}$ arrives at i, and from the reciprocity of equation 3.37 this is $B_k A_i F_{ik}$ which expressed as radiation per unit area is as in 3.39. From this set of simultaneous equations all the B_i can be found.

Also, the net heat transferred per unit time from a surface is:

$$Q_i = (\sigma \varepsilon_i T_i^4 - \alpha_i H_i) A_i = (B_i - H_i) A_i \tag{3.40}$$

Eliminating H_i between 3.38 and 3.40 gives

$$Q_i = \varepsilon_i (\sigma T_i^4 - B_i) A_i / (1 - \varepsilon_i) \tag{3.41}$$

for the case where $\alpha_i = \varepsilon_i$, i.e. for grey bodies.

3.4.4 Radiation from non-luminous gases

So far as combustion gases are concerned the triatomic gases CO_2 and H_2O are the most important, although CO and CH_4 and other fuel gases emit radiation. Since a gas is partly transparent the radiation emitted by a mass of gas is a function of the number of radiating molecules, and therefore the radiation per unit area of surface bounding the gas is a function of the partial pressure of the gas and the effective thickness or beam length, l, of the body of gas, as well as the temperature. The relation is not linear since partial reabsorption occurs within the gas, and is given in

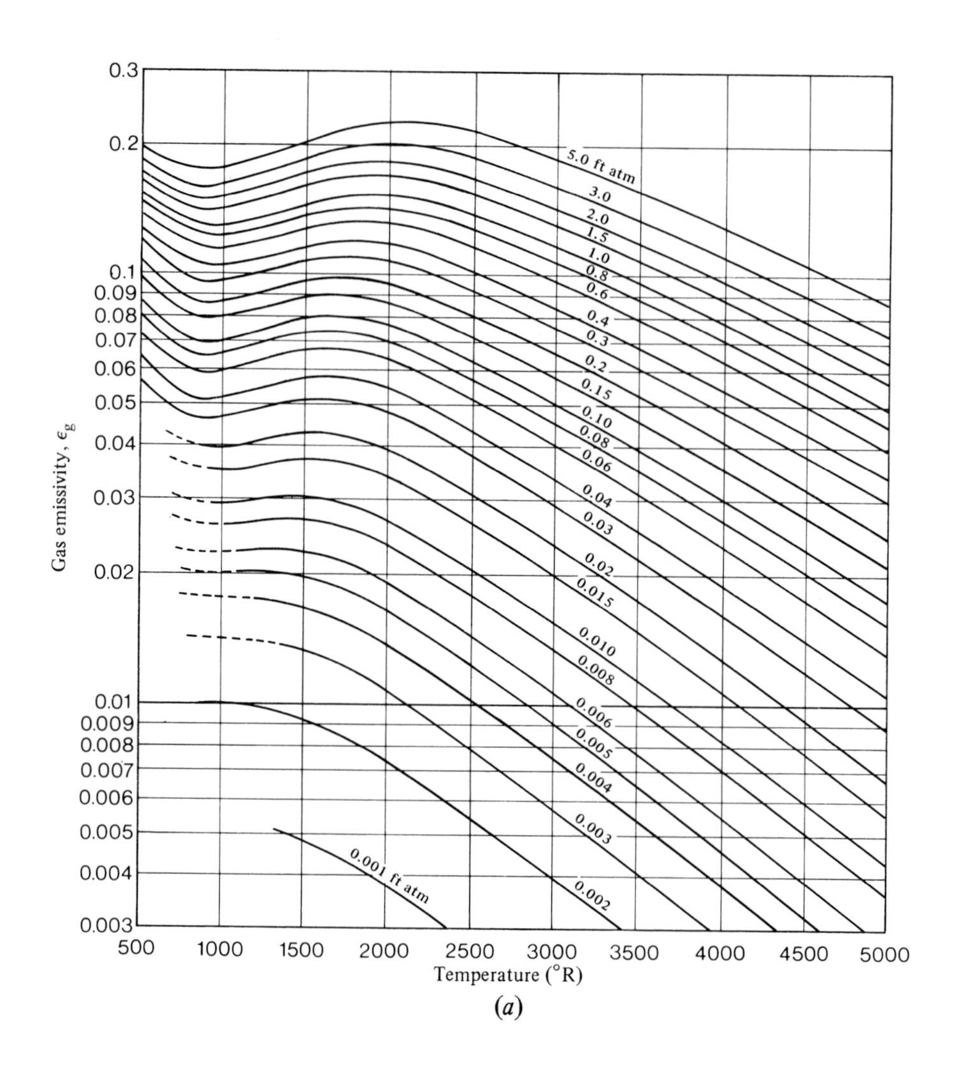
Gas emissivity, ϵ_g
0.3
0.2
0.1
0.09
0.08
0.07
0.06
0.05
0.04
0.03
0.02
0.01
0.009
0.008
0.007
0.006
0.005
0.004
0.003
500
1000
1500
2000
2500
3000
3500
4000
4500
5000
Temperature (°R)
5.0 ft atm
3.0
2.0
1.5
1.0
0.8
0.6
0.4
0.3
0.2
0.15
0.10
0.08
0.06
0.04
0.03
0.02
0.015
0.010
0.008
0.006
0.005
0.004
0.003
0.002
0.001 ft atm
(a)

Figure 3:14

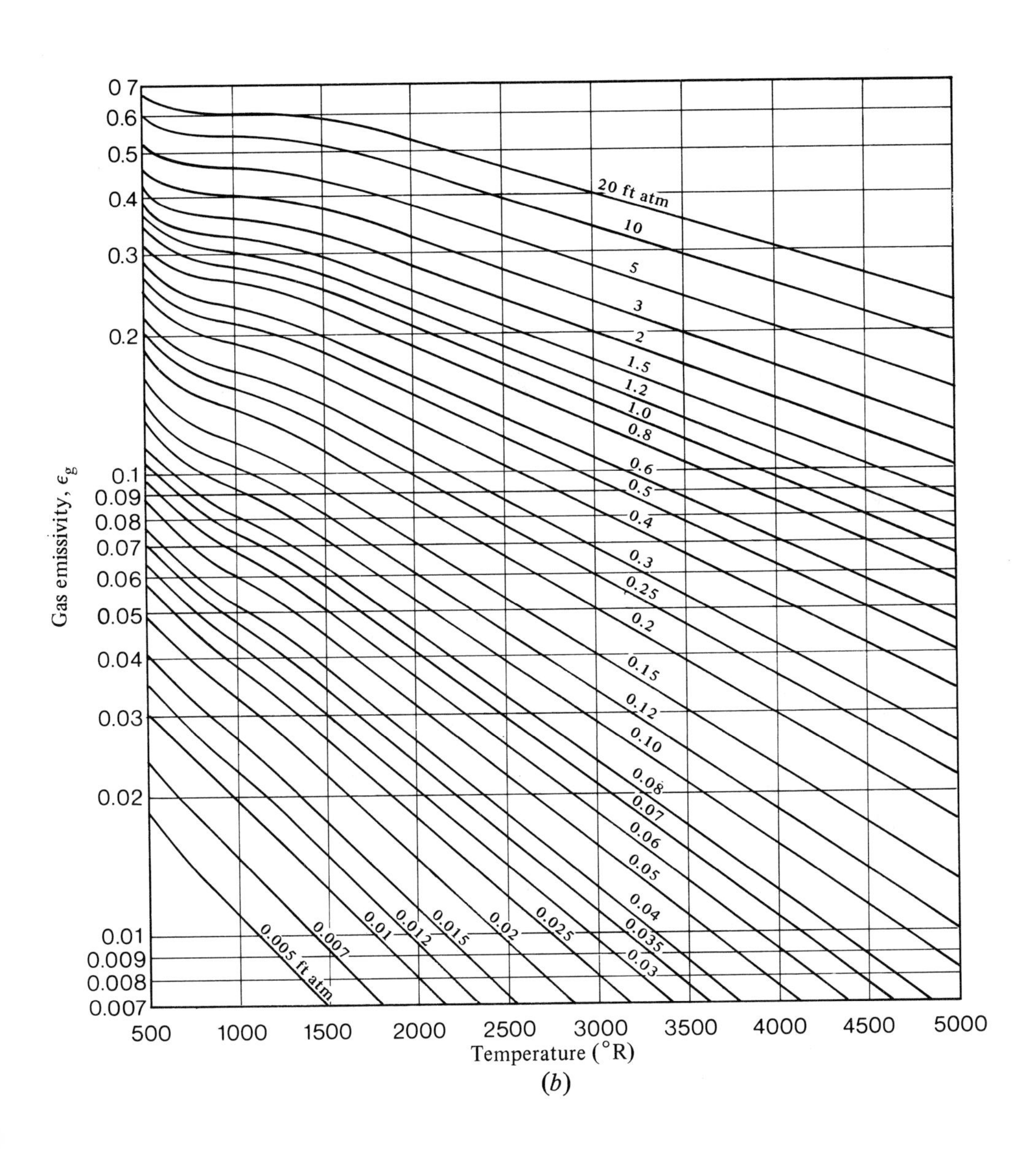
Gas emissivity, ϵ_g
0.7
0.6
0.5
0.4
0.3
0.2
0.1
0.09
0.08
0.07
0.06
0.05
0.04
0.03
0.02
0.01
0.009
0.008
0.007
500
1000
1500
2000
2500
3000
3500
4000
4500
5000
Temperature (°R)
20 ft atm
10
5
3
2
1.5
1.2
1.0
0.8
0.6
0.5
0.4
0.3
0.25
0.2
0.15
0.12
0.10
0.08
0.07
0.06
0.05
0.04
0.035
0.03
0.025
0.02
0.015
0.012
0.01
0.007
0.005 ft atm

(b)

Figure 3:14

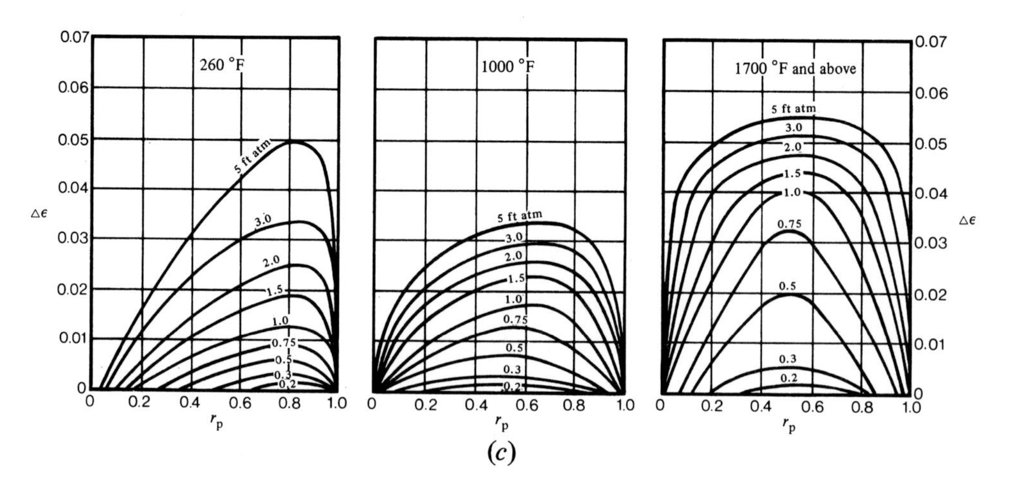

Figure 3:14 Gas emissivity

(*a*) Variation of emissivity of carbon dioxide with temperature for various values of $p_{CO_2}l$

(*b*) Variation of emissivity of water vapour with temperature for various values of $p_{H_2O}l$

(*c*) Correction of gas emissivity due to spectral overlap of water vapour and carbon dioxide: variation of correction with the value of $r_p = p_{H_2O}/(p_{CO_2} + p_{H_2O})$ for various values of $p_{CO_2}l + p_{H_2O}l$ at three temperatures

(From *Heat Transmission* by William H. McAdams. Copyright 1933, 1942, 1954 William H. McAdams. Used with permission of McGraw-Hill Book Company)

graphical form in Figure 3:14 as emissivity as a function of *pl*. When both CO_2 and H_2O are present the total emissivity is less than the sum of the individual emissivities since some mutual reabsorption occurs, hence the correction graph Figure 3:14(*c*).

Example

Gas composition 50% N_2, 20% CO_2, 30% H_2O at atmospheric pressure and 2000 °R. What is the emissivity for a 3 ft path length?

$$p_{CO_2} = 0.2 \text{ atm}$$

Therefore $$p_{CO_2}l = 0.6 \text{ ft atm}$$

Therefore $$\varepsilon_{CO_2} = 0.13$$

Similarly $p_{H_2O}l = 0.9$ ft atm
Therefore $\varepsilon_{H_2O} = 0.20$
Correction: $p_{H_2O}/(p_{CO_2} + p_{H_2O}) = 0.6$
and $p_{CO_2}l + p_{H_2O}l = 1.5$
Therefore $\Delta\varepsilon = 0.045$
Therefore $\varepsilon_g = 0.33 - 0.045 = 0.285$

The beam length depends on the geometry of the system but for low values of pl may be taken as 4 × gas volume/total surface area of gas, for all the surfaces forming an enclosure.

Hottel [4] shows that for a wide range of shapes the beam length may be taken as 3.5 × volume/surface.

3.4.5 Radiation interchange between gas and bounding surfaces

This may be dealt with by the method given above in equations 3.38 to 3.41, except that equation 3.39 is now replaced by

$$H_i = \sigma\varepsilon_{gi}T_g^4 + \sum_k (1 - \alpha_{gki})B_k F_{ik} \qquad (3.39a)$$

Figure 3:15 illustrates the meaning of these equations. ε_{gi} is the emissivity of the gas for the surface i. Clearly different path lengths may apply for different surfaces, but common engineering practice is to use a single value of 3.5 × gas volume/total surface area. α_{gki} is the absorptance of the gas for radiation at the temperature T_k going from surface k to surface i. Hottel gives the approximation:

$$\alpha_{gki} = \varepsilon_{gk}\,(T_g/T_k)^n$$

where $n = 0.65$ for CO_2 and 0.45 for H_2O vapour.

From Figure 3:15 it is clear that the net radiation leaving the gas to the n surrounding surfaces is made up of the heat radiated directly by the gas to each surface, minus the heat absorbed by the gas from radiation passing between every pair of surfaces. The latter term for radiation leaving any surface k is

$$\sum_{i=1}^{n} \alpha_{gki}\, B_k F_{ki} A_k$$

so the net radiation leaving the gas is

$$Q_g = \sigma \sum_{i=1}^{n} \varepsilon_{gi} A_i T_g^4 - \sum_{i=1}^{n}\sum_{k=1}^{n} \alpha_{gki} B_k F_{ki} A_k \qquad (3.42)$$

Using Hottel's notation this can be written as

$$Q_g = \sigma T_g^4 \sum_i \overrightarrow{gs_i} - \sum_k \overrightarrow{s_k g} B_k \qquad (3.42a)$$

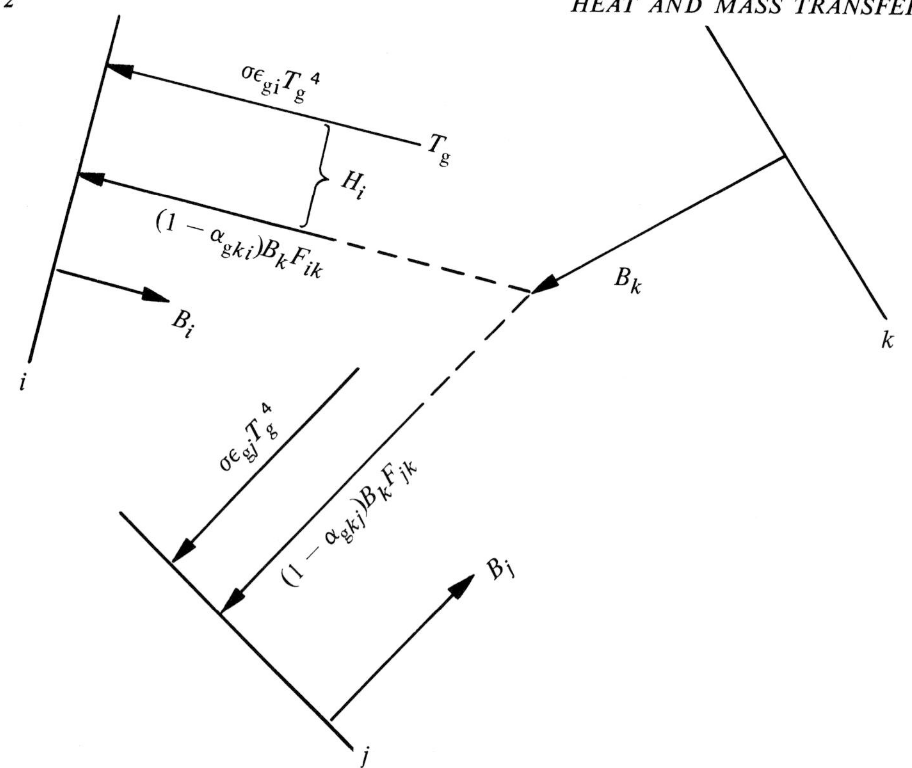

Figure 3:15 Radiation interchange between gas and bounding surfaces

A close approximation to the radiation from a gas to a surrounding grey surface is:

$$Q = \sigma A(\varepsilon_g T_g{}^4 - \alpha_{gs} T_s{}^4)\tfrac{1}{2}(\varepsilon_s + 1)$$

Kern [8] gives a number of approximate methods which have been shown to be sufficiently accurate in calculations on particular types of furnace. Examples of the use of radiation interchange are given in Section 12.3.

3.4.6 Radiation from luminous flames

The luminosity of flames is due to particles of carbon or heavy hydrocarbons. In the case of powdered-coal and oil flames the particles may be of the order of 100 μm in diameter and are substantially opaque, while soot particles in luminous gases are of the order 0.05 μm diameter and are partly transparent. The amount of soot in a luminous flame depends on the hydrocarbons present, the air–fuel ratio, and the homogeneity of mixing of fuel and air, which in turn may depend on momentum transfer between the streams of gas and air. Hence it is not possible to predict the luminosity in any particular case. The effective emissivity of a luminous flame is made up of the emissivity of the particles plus the emissivity of the gases minus the fraction of this intercepted by the particles, and this can be evaluated if the concentration and size of the particles are known (e.g. Thring *et al.* [5, 6]).

The emissivity of a luminous gas can be inferred from measurements of the apparent flame temperature at various wavelengths, using an optical pyrometer, or by measurements using a total radiation pyrometer.

For more detailed information on radiation generally reference may be made to McAdams [3] and to Hottel and Sarofim [4], and to Section 12.3 of this book.

3.4.7 The general heat balance at a surface

For steady-state conditions of a surface:

$$Q_{cond} + Q_{conv} + Q_{rad} = 0 \tag{3.44}$$

where these are the heat flows leaving the surface.

Similarly, for a gas which is flowing at a mass flow rate m:

$$Q_{g,\,cond} + Q_{g,\,conv} + Q_{g,\,rad} = mc_p\delta T \tag{3.45}$$

where the Qs are the heat flows leaving the gas and δT is the change in gas temperature. From this set of simultaneous equations 3.44 and 3.45, together with the appropriate equations for the various modes of heat transfer, the heat flows and temperatures can be calculated, although iterative procedures will usually be needed.

3.5 MASS TRANSFER

Molecular diffusion occurs in fluids as a result of random molecular motion, which means that there will be a net movement of molecules of a given species in the direction of the concentration gradient. For a mixture of two molecular species this is expressed as Fick's law:

$$N_A = -D_{AB}\,(dc_A/dx) = -(D_{AB}/RT)\,(dp_A/dx) \tag{3.46}$$

where N_A is the flux of constituent A in moles/unit area, unit time, c_A is the concentration of A in moles/unit volume, and p_A is the partial pressure of A in the mixture. D_{AB} is called the binary diffusion coefficient, and is a function of temperature.

Molecular diffusion is the only mass-transport process in a fluid at rest or in laminar motion, but in turbulent flow there is in addition an *eddy diffusion* due to eddies in the fluid, and eddy diffusivity is defined in a similar way:

$$N_{Ae} = -D_{ABe}\,(dc_A/dx) \tag{3.47}$$

If diffusion takes place in the gas phase then the pressure must remain constant throughout, so that $dp_A/dx = -dp_B/dx$ and $N_A = -N_B$, so:

$$D_{AB} = D_{BA} = D \tag{3.48}$$

Very close similarities exist between mass, heat and momentum transfer. In Figure 3:5 the curve representing T could be replaced by one representing c, the concentration of a vapour being transferred from the surface to the fluid. A mass balance on the element $dx\,dy$ gives:

$$(\partial^2 c/\partial x^2) - (u/D)(\partial c/\partial y) = 0 \qquad (3.49)$$

formed in the same way as were equations 3.13 and 3.14, and there will be the same concentration and velocity profiles if the ratio $(u/D)(\mu/\rho u) = 1$, i.e. if $\mu/\rho D = 1$. This ratio is called the Schmidt number (Sc) and is clearly analogous to (Pr).

A mass-transfer coefficient h_D is defined by:

$$N = h_D (c_s - c_f) \qquad (3.50)$$

where c_s and c_f are the concentrations of the diffusing species at the surface and in the bulk fluid respectively. Using the same argument as that used in deriving equation 3.16, if $(Sc) = 1$ then

$$h_D L/D \approx 0.58\,(Re)^{\frac{1}{2}} \qquad (3.51)$$

where $h_D L/D$ is the Sherwood number (Sh) which is clearly analogous to (Nu).

The general mass-transfer correlation will be

$$(Sh) = fn\,(Re)\,(Sc) \qquad (3.52)$$

In the same way as the Stanton number was formed it is possible to form the group $(Sh)/(Re)(Sc) = h_D/u$, and Colburn has derived a j factor for mass transfer which is

$$j_D = (h_D/u)(\mu/\rho D)^{2/3} \qquad (3.53)$$

which is clearly analogous to the j_h factor for heat transfer.

Experimental correlations for mass transfer from surfaces to moving gas show that:

$$j_h \approx j_D \qquad (3.54)$$

3.6 HUMIDITY, PSYCHROMETRY, DRYING

When the partial pressure of water vapour in air is equal to its vapour pressure p_{Wsa} at the air temperature the air is saturated. For lower values of the water vapour partial pressure p_W the percentage relative humidity of the unsaturated air is defined as:

$$(p_W/p_{Wsa}) \times 100 \qquad (3.55)$$

Alternatively, 'percentage humidity' is the ratio of the weight H of water contained in unit volume of moist air to the weight H_{Sa} of water in unit volume of saturated air:

$$(H/H_{Sa}) \times 100 \qquad (3.56)$$

Relative humidity and percentage humidity are equivalent at low values of p_W relative to the total air pressure.

For any value of p_W, if the air is cooled, a temperature T_d will be reached at which $p_W = p_{Wsa}$, i.e. the air becomes saturated, and T_d is called the dew point.

The specific heat capacity of moist air is called the humid heat:

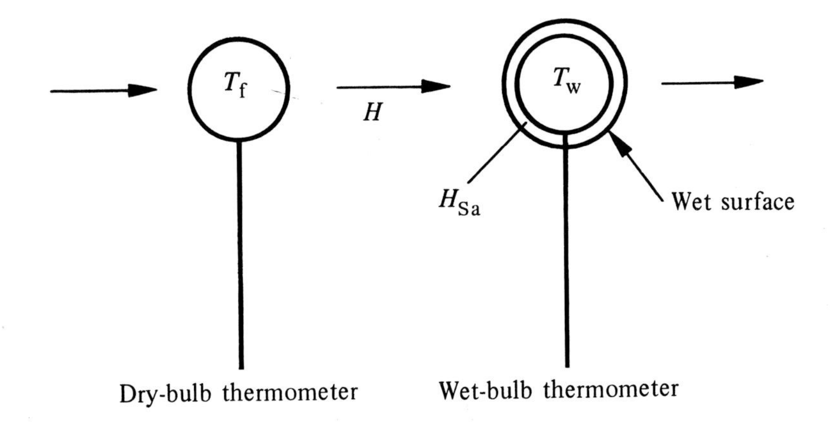

Figure 3:16 Measurement of humidity

$$S = (1 + 2H) \times 10^{-3} \text{ MJ kg}^{-1}\text{ °C}^{-1} = (0.24 + 0.48H) \text{ Btu/lb °F}$$

Figure 3:16 shows a stream of air passing over a wet surface. Water vapour will be transferred from the surface to the air and heat will have to flow from the air to the surface to supply the latent heat of evaporation. If the air flow rate is large so that its humidity does not change appreciably, then from equation 3.50 the rate of water transferred expressed as mass per unit time is:

$$m = h_D \rho_a A (H_{Sa} - H) \tag{3.57}$$

The heat which has to be supplied to the surface is:

$$Q = ml = h_D A \rho_a l(H_{Sa} - H) \tag{3.58}$$

where l is the latent heat of vaporization of water.

But also:
$$Q = hA(T_f - T_w) \tag{3.59}$$

Hence:
$$H_{Sa} - H = h(T_f - T_w)/lh_D \rho_a \tag{3.60}$$

The fact that $j_h = j_D$ implies that the ratio h/h_D is constant, so that from a measurement of the wet bulb temperature T_W and the dry bulb temperature T_f the value of H can be found since H_{Sa} is known at the temperature T_w.

Another temperature defined is the *adiabatic saturation temperature* T_{Sa}. Suppose a quantity of air at T_f and H is brought into contact with a wet surface under adiabatic conditions and equilibrium is reached so that H rises to H_{Sa}. Then the enthalpy of the system will remain unchanged so the air temperature will fall to T_{Sa}, where

$$S(T_f - T_{Sa}) = l(H_{Sa} - H)$$

or

$$H_{Sa} - H = S(T_f - T_{Sa})/l \tag{3.61}$$

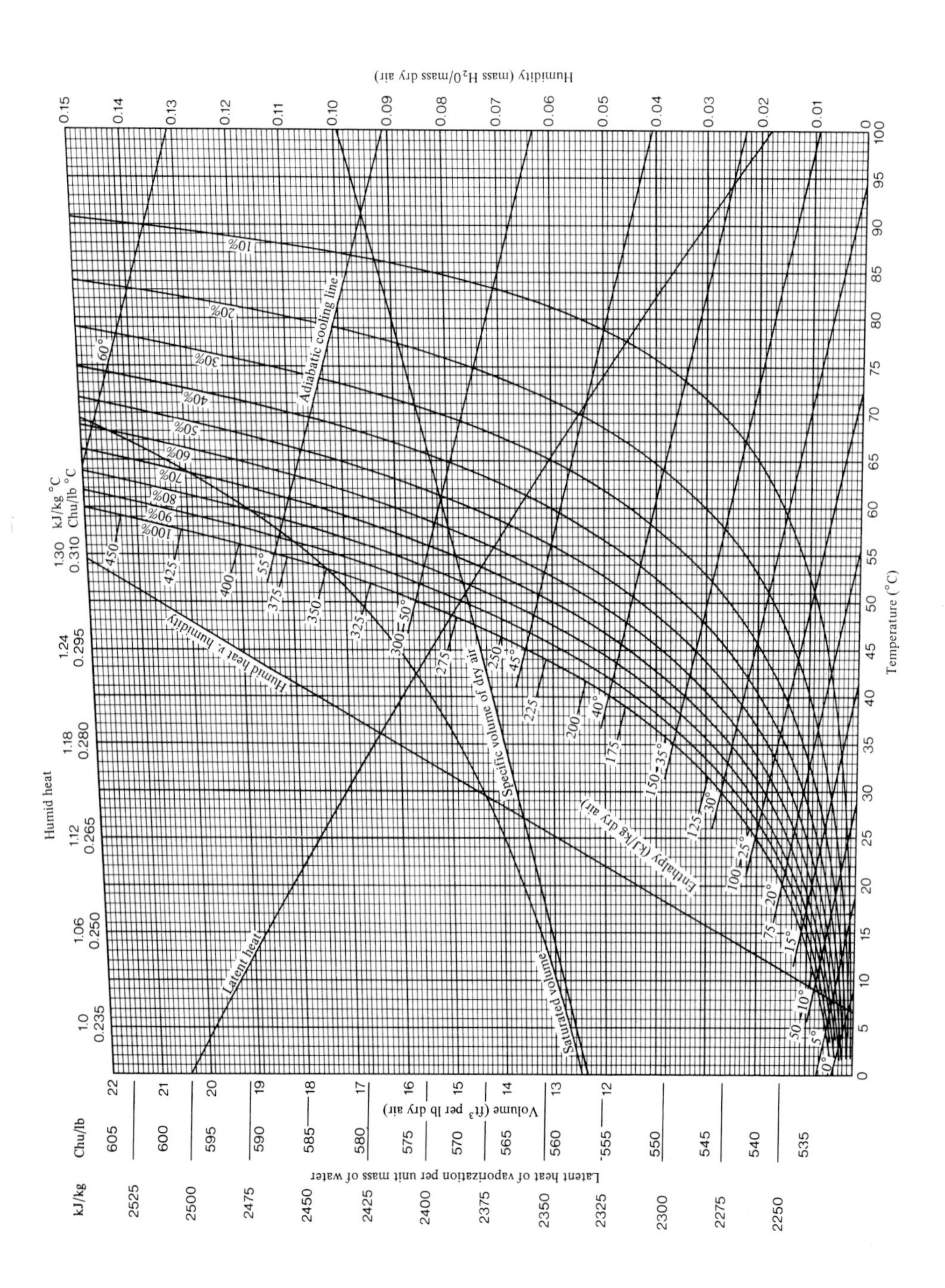

Figure 3.17 Psychrometric chart of air–water-vapour system

Comparison of equations 3.58 and 3.59 shows that T_{sa} and T_w will be equal if $S = h/h_D\rho_a$ and this is true for the water-vapour–air system. For other vapour–air systems T_w is larger than T_{sa}.

To facilitate calculations, various properties of the air–water-vapour system are plotted on the psychrometric chart in Figure 3:17.

3.6.1 Drying of materials

A common process is the drying of wet materials using a stream of air to carry away the moisture. Examples are the drying of granular solids and of steel sheets coated with wet vitreous enamel. Usually the air is heated to such a temperature that at a given flow rate it will leave the drier in an unsaturated condition.

Another method is to supply the heat to the surface to be dried by means of radiant heaters, using cool air to remove the moisture, and care has to be taken in this case lest the air becomes saturated when it will redeposit the water as droplets onto a cooler surface.

The psychrometric chart may be used to calculate the air flow or temperature required in such drying processes.

Example

100 kg/h of water is to be removed from wet articles. Air at 15 °C and 50 per cent relative humidity is heated to 90 °C and passed over the articles, and is to leave the drier at 80 per cent relative humidity. How much air is required, how much heat is to be added to it, and what is its exit temperature and dew point? Assume that the drying operation is adiabatic and neglect the heating up of the articles and conveyor.

The process is shown on the simplified chart of Figure 3:18 and the chart of Figure 3:17 gives:

$$H_1 = 0.0052, \qquad H_2 = 0.0290$$
$$I_1 = 48 \text{ kJ/kg dry air}, \qquad I_2 = 130 \text{ kJ/kg dry air}$$

Hence 0.0238 kg water is taken up per kg dry air so the air required is 100/0.0238 = 4200 kg/h.

The heat to be supplied to the air is $(I_2 - I_1)$ kJ/kg so the total heat required is 4200×82 kJ/h = 344 MJ/h (3.26 therm/h).

Dew point = 31 °C
Exit air temperature = 35 °C

3.6.2 Rate of drying of porous materials

Drying occurs in two stages. In stage 1 the surface is completely wet and the drying rate is therefore constant and can be calculated from correlations of the form of equation 3.51. In stage 2 the surface has partly dried and the rate-determining process is the transport of water from the interior of the material to its surface, either by capillarity or by molecular diffusion in the liquid phase. Figure 3:19 shows the general form of the drying rate curve. In stage 1 the drying rate is proportional to $(H_{sa} - H)u^{0.8}$ and empirical correlations are given by Coulson and Richardson [7] who also give a full account of the process generally.

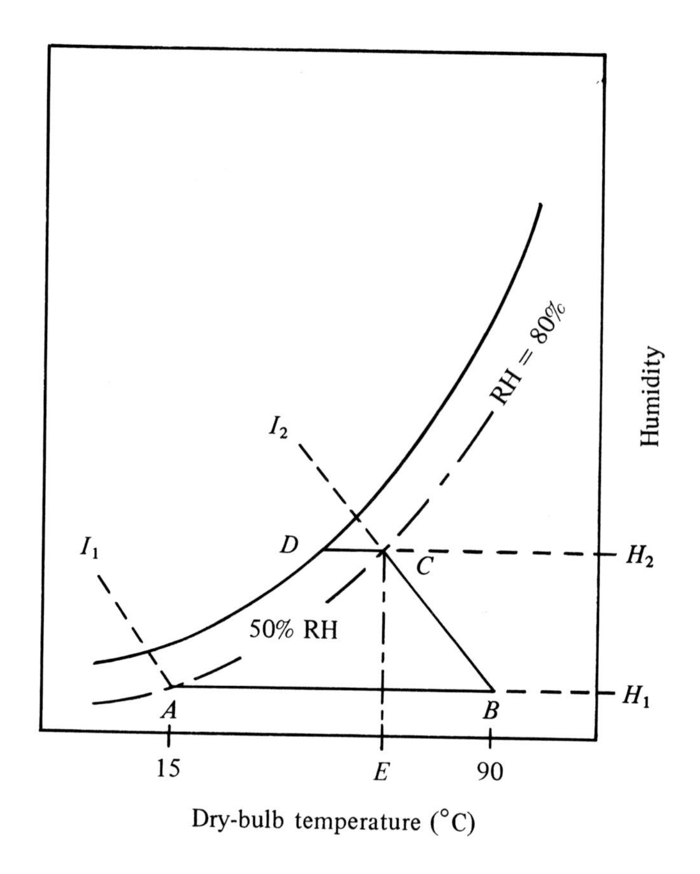

Figure 3:18 Drying process

3.7 FURTHER READING

Thring, M. W., Foster, P. J., McGrath, I. A. and Ashton, J. S., 'Prediction of the emissivity of hydrocarbon flames', *Papers Presented at the International Heat Transfer Conference 1961–2*, American Society of Mechanical Engineers (1961)

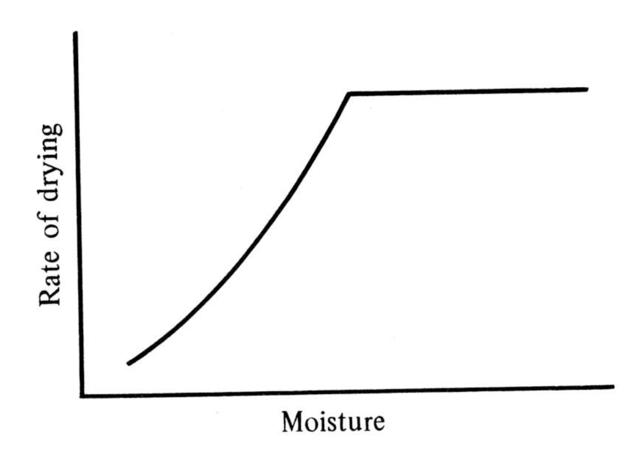

Figure 3:19 Rate of drying of porous materials

REFERENCES

1 Schenk, H., *Heat Transfer Engineering*, Englewood Cliffs, N.J., Prentice-Hall (1959)
2 Love, T. J., *Radiative Heat Transfer*, Columbus, Ohio, Merrill (1968)
3 McAdams, W. H., *Heat Transmission*, 3rd edition, New York, McGraw-Hill (1954)
4 Hottel, H. C. and Sarofim, A. F., *Radiative Transfer*, New York, McGraw-Hill (1967)
5 Thring, M. W., *The Science of Flames and Furnaces*, 2nd edition, London, Chapman & Hall (1962)
6 Thring, M. W., Beér, J. M. and Foster, P. J., 'The radiative properties of luminous flames', *Proceedings of 3rd International Heat Transfer Conference*, vol. 5, Chicago, American Institute of Chemical Engineers (1966)
7 Coulson, J. M. and Richardson, J. F., *Chemical Engineering*, 2nd edition, Oxford, Pergamon, 2 vols (1964–8)
8 Kern, D. Q., *Process Heat Transfer*, New York, McGraw-Hill (1950)

4

Combustion and flames

4.1 THE ELEMENTS OF CHEMICAL THERMODYNAMICS AND KINETICS

4.1.1 Introduction

In the combustion of fuels two aspects of the reactions involved are important:

1 What chemical changes occur
2 How fast they occur

The first question is difficult to answer for the period in which the changes actually take place since the mechanism of the reaction needs to be specified. Such mechanisms are complex and even now ill-understood for flame reactions. However, it is comparatively easy to determine the final chemical changes which have taken place when the reactions have been completed. Determining the final changes involves the use of *chemical thermodynamics.* How fast the reactions occur is equally important and makes up the subject of *chemical kinetics.*

In Sections 4.1.2 and 4.1.3 a brief elementary treatment of these two subjects is presented. The aim is simply to present the basic ideas and terminology so that the reader with no background in chemistry will find what follows intelligible. Oversimplification is inevitable and the reader is referred to the bibliography at the end of this chapter for detailed treatment of the subjects.

4.1.2 Chemical thermodynamics

The ideal-gas laws

An *ideal gas* may be defined as a gas to which the laws of Boyle, Charles (or Gay-Lussac) and Avogadro are applicable at all temperatures.

Boyle's law:

$$V \propto 1/P \text{ or } PV = \text{constant} \tag{4.1}$$

i.e. at constant temperature, the volume V of a definite mass of gas is inversely proportional to the pressure P.

Charles's law: at constant pressure, the volume of any gas expands by the same fraction of its volume at 0 °C for every 1 °C rise in temperature, i.e. if V_0 is the volume of a given mass of gas at 0 °C and V_t is that at t °C then:

$$V_t - V_0 = \gamma t V_0$$

or

$$V_t = V_0 (1 + \gamma t) \qquad (4.2)$$

where γ = the coefficient of cubical expansion which should be the same for all gases. In practice this equation is not obeyed exactly, but at low pressures and fairly high temperatures γ approaches a constant value. Measurements indicate that the limiting value is 0.0036609 and so this may be taken as γ for an ideal gas. For such a gas, equation 4.2 becomes:

$$V_t = V_0 (1 + 0.0036609t)$$

Rearranging:

$$V_t = \frac{V_0}{273.16}(273.16 + t) \qquad (4.3)$$

Clearly the volume of an ideal gas must become zero at a temperature of −273.16 °C, thus defining the *absolute zero* of the *ideal-gas temperature scale.*

The addition of 273.16 to the Celsius (centigrade) temperature gives the *absolute temperature* measured in kelvins (K). This is the temperature scale adopted for the International System of Units. Clearly, the temperature *interval* remains the same, i.e. 1 °C = 1 K.

Using absolute temperatures, equation 4.3 may be rewritten

$$V_T/T = V_0/T_0 \qquad (4.4)$$

i.e. V/T is constant for a given mass of ideal gas at constant pressure. Combining 4.1 and 4.4 for a given mass of ideal gas:

$$PV/T = \text{constant} \qquad (4.5)$$

or

$$P_1V_1/T_1 = P_2V_2/T_2 \qquad (4.6)$$

This relationship is widely used to convert gas volumes to any required temperature and pressure.

Avogadro's law: equal volumes of all gases at the same temperature and pressure contain equal numbers of molecules. Thus the constant in 4.5 should be independent of the type of gas provided that 1 mole (i.e. a quantity with a mass equal to its relative molecular mass in grams) is considered. The symbol V_m is used for the volume of 1 mole of a gas. Thus PV_m/T must be a universal constant, which is denoted by R, i.e.:

$$PV_m = RT \qquad (4.7)$$

where R is the *molar gas constant*. Equation 4.7 is the *ideal-gas equation* or *equation of state*.

For temperature and pressure conditions of 0 °C and 1 standard atmosphere (101 325 Pa or 760 mm Hg), 1 mole occupies 22.413 litres. In the SI (International System of Units) 1 gram mole of ideal gas occupies 23.645×10^{-3} m^3 (st) and at Imperial Standard Conditions (stp) 1 lb mole occupies 378 sft^3. Thus the numerical value of R is given by:

$$1 \times 22.413 = R \times 273.16$$

Therefore:

$$R = 0.0821 \text{ litre atm mol}^{-1} \text{ K}^{-1}$$

or in other units:

$$8.314 \text{ J mol}^{-1} \text{ K}^{-1}$$
$$0.073 \text{ ft}^3 \text{ atm lb mol}^{-1} \text{ °F}^{-1}$$
$$1.986 \text{ Btu lb mol}^{-1} \text{ °F}^{-1}$$

In virtually all combustion processes of interest to gas utilization engineers, the pressure remains very close to atmospheric and the gases may be regarded as ideal. For conditions closer to the *critical temperature and pressure*, intermolecular effects become important and a closer approximation to the equation of state is given by the van der Waals equation (equation 2.4).

A further gas law which may be involved in combustion calculations is *Dalton's law of partial pressures* which states that when two or more gases are mixed, the total pressure is equal to the sum of the partial pressures of the constituents or equal to the sum of the pressures which each constituent would exert if it alone occupied a volume equal to that of the mixture. Thus for a gas mixture, each constituent gas exerts that proportion of the total pressure corresponding to its proportion (by volume) in the mixture.

The first law of thermodynamics

The first law, which states that although energy may be converted from one form to another it cannot be created or destroyed, may be expressed as:

$$\Delta E = E_2 - E_1 = q_{to} + w_{on} \tag{4.8}$$

where ΔE is the change in *internal energy*, q_{to} is the heat added to the system and w_{on} is the work done on the system. In chemical thermodynamics the only work that needs to be considered is usually gaseous pressure–volume work.

Consider an infinitesimal expansion $\mathrm{d}V$ of a gas against an external pressure P_e. The work done on the system $\mathrm{d}w = -P_e\,\mathrm{d}V$ i.e.:

$$w = -\int P_e\,\mathrm{d}V$$

For a *reversible* process the external pressure P_e is virtually equal to the gas pressure P. Thus:

$$w = -\int P\mathrm{d}V$$

At *constant volume*, $\mathrm{d}V = 0$. Therefore:

$$-\int P\mathrm{d}V = 0$$

and equation 4.8 simplifies to:

$$\Delta E = q_V \qquad (4.9)$$

i.e. the heat energy added to a constant-volume system is equal to the increase in internal energy.

At *constant pressure* (which represents almost exclusively the conditions of interest in this book):

$$w = -\int P\mathrm{d}V = -P\Delta V$$

therefore

$$\Delta E = q_p - P\Delta V \qquad (4.10)$$

Enthalpy

Since virtually all the combustion processes of interest in this book occur at atmospheric pressure, the most important thermodynamic function is the *heat content at constant pressure* or *enthalpy H* defined as:

$$H = E + PV \qquad (4.11)$$

Since E, P and V are functions of state so also is H.

Only changes in H are accessible to measurement. Thus equation 4.11 may be more conveniently written:

$$\Delta H = \Delta E + P\Delta V \quad (P \text{ constant}) \qquad (4.12)$$

Combining 4.10 and 4.12:

$$\Delta H = q_p \qquad (4.13)$$

Thus for chemical changes at amospheric pressure where the only mechanical work done is expansion against the atmosphere the change in enthalpy is equal to the heat given up or absorbed as a result of the reaction.

The determination of calorific value (see Section 4.2.1) affords a useful illustration of the difference between ΔH and ΔE. The calorific value of a solid or liquid fuel is usually determined using a bomb calorimeter at constant volume, therefore $q_V = \Delta E$ is determined (although of course it is normally expressed on a mass basis rather than on a molar basis as ΔE is).

Constant-pressure calorimeters, on the other hand, are normally used for gaseous fuels and additional energy of expansion (negative or positive) against the atmosphere is involved. Thus $q_p = \Delta H$ is measured.

Heat capacity and specific heat capacity

The addition of heat to a substance normally results in a temperature rise the magnitude of which depends on the mass m and the *specific heat capacity*, c, a constant, characteristic of the substance and of the manner in which the heating takes place, i.e.:

$$\mathrm{d}q = mc\mathrm{d}T \qquad (4.14)$$

The product mc is termed the *heat capacity*. If m is the relative molecular mass then mc is the *molar heat capacity* usually denoted by C. For gases, c may also be expressed in terms of volume and is most commonly encountered in this form.

The specific heat capacity defined in 4.14 is the instantaneous specific heat capacity and cannot be measured directly by thermal means. Instead the mean specific heat capacity $\bar{c}$ is measured over the range $T_2 - T_1$:

$$\bar{c} = q/m(T_2 - T_1)$$

thus

$$\bar{c} = \int_{T_1}^{T_2} c \, \mathrm{d}t/(T_2 - T_1)$$

At constant pressure or volume c varies with temperature and the function is frequently represented by a simple power series

$$c = \alpha + \beta T + \gamma T^2 + \delta T^3 + \ldots$$

From equation 4.9 the heat absorbed at constant volume is equal to the increase ΔE in internal energy E. Thus:

$$C_V = (\partial E/\partial T)_V \tag{4.15}$$

Similarly, at constant pressure:

$$C_p = (\partial H/\partial T)_p \tag{4.15a}$$

In some cases the addition of heat results in no temperature change. This is true, for example, when a phase change occurs. This is *latent heat*, common examples of which are heats of vaporization and fusion. Heats of chemical reaction may also be regarded as latent heats.

Heats of reaction and formation

ΔH is frequently termed the *heat of reaction* at constant pressure and often figures in thermochemical equations, thus:

$$C(s) + O_2(g) = CO_2(g); \qquad \Delta H = -\ 393.5 \text{ kJ} \tag{4.16}$$

i.e. when 12 g of solid carbon react with 32 g of gaseous oxygen to form 44 g of gaseous carbon dioxide, there is a decrease in heat content of 393.5 kJ (i.e. this amount of heat is evolved). A reaction in which ΔH is negative is termed *exothermic*. If heat is absorbed in the reaction, ΔH has a positive value and the reaction is described as *endothermic*.

For reactions in which compounds are formed from their elements the heat of reaction is termed the *heat of formation*, ΔH_f. Equation 4.16 represents such a case and ΔH would formally be written ΔH_f. Consider the combustion of hydrogen:

$$H_2(g) + \tfrac{1}{2}O_2(g) = H_2O(g)$$
$$(\Delta H_{H_2O})_f = H_{H_2O}(g) - H_{H_2}(g) - \tfrac{1}{2}H_{O_2}(g) \tag{4.17}$$

The physical state has to be specified since a different heat of formation would be involved for liquid water. It follows that it is generally necessary to specify a *standard state* for thermodynamic quantities. This is indicated by the superscript $^\ominus$ and represents the most stable form of a substance at room temperature.

Since only heat changes can be determined experimentally it is necessary also to specify some arbitrary datum so that individual enthalpies may be added and subtracted. This is done by assigning the value of zero enthalpy to elements in their standard states at a reference temperature of 25 °C (or 298.15 K which is usually abbreviated as 298).

Thus from equation 4.15 the 'absolute' enthalpy of gaseous water at 1 standard atmosphere and 298.15 K is equal to its heat of formation under these circumstances, i.e.

$$(H_{H_2O}^{\ominus})_{298} = (\Delta H_{H_2O}^{\ominus})_{f,298} \quad (4.18)$$

which would be negative in this case since the reaction is exothermic.

Hess's law states that the heat change in a chemical reaction is the same whether it takes place in one or several steps. Thus the absolute enthalpy of any compound can be obtained by the addition and subtraction of appropriate heats of reaction (or by measurement in a calorimeter).

For enthalpy at temperature T above or below 298.15 K:

$$H_T^{\ominus} = H_{298}^{\ominus} + \int_{298}^{T} C_p \, dT \quad (4.19)$$

For convenience, tabulated data are usually in the form $H_T^{\ominus} - H_{298}^{\ominus}$ and heats of formation, the absolute enthalpy at temperature T being given by:

$$H_T^{\ominus} = \Delta H_{f,298}^{\ominus} + (H_T^{\ominus} - H_{298}^{\ominus}) \quad (4.20)$$

Some simplifications of the generalized statement of the first law have already been made. A further simplification is possible which allows, for example, the final temperature of a flame to be calculated (Section 4.2.6). If an element of flow is regarded as passing through the flame without net gain or loss of heat from the unreacted to the final reacted state then the system is adiabatic, i.e. $\Delta H = q_p = 0$, and all the heat liberated by the reaction simply raises the temperature of the products of combustion. The gases are of course not adiabatic while in the reacting region since there are transport processes operating but if consideration is restricted to the initial and final states the assumption is valid. If radiation losses from the flame are large this relationship does not hold but in small-scale nonluminous flames such losses are small and may be ignored in the calculation of theoretical final flame temperatures (Section 4.2.6).

The second and third laws of thermodynamics, entropy and free energy

The first law, although allowing energy transfer to be considered, provides no answer to the question of whether a particular process will tend to occur or not. For a process to be thermodynamically reversible it must be carried out infinitesimally slowly so that it represents a succession of equilibrium states. However all *natural* or *spontaneous* processes (i.e. those occurring without external interference) are *irreversible.* This is the basis of the second law of thermodynamics.

The principle of irreversibility may be illustrated by considering the expansion of a gas to fill a vacuum or the conduction of heat from a hot to a cold region or the conversion of internal energy into heat in a chemical reaction. In all these cases the reverse process does not occur.

The second law has been stated in many alternative ways. One useful statement of the law introduces the term *entropy*, S, as follows: the entropy of a system and surroundings *together* increases during all natural or irreversible changes whereas for a reversible process the total energy is unchanged.

Entropy may be regarded qualitatively as a measure of the disorganization or randomness of a system. It is best defined in terms of the increase in entropy of a system which is equal to the heat taken up *isothermally and reversibly* divided by the temperature at which it is absorbed, i.e.:

$$\mathrm{d}S = \mathrm{d}q_{\mathrm{rev}}/T \tag{4.21}$$

Equation 4.21 relates to reversible changes and for a process which is not reversible $\mathrm{d}S$ is not equal to $\mathrm{d}q_{\mathrm{rev}}/T$. This appears a serious limitation on the usefulness of $\mathrm{d}S$ since irreversible processes are more interesting. However, changes in entropy are entirely independent of the actual way in which the process is carried out. They depend only on the initial and final states. Thus a reversible and an irreversible process between the same initial and final states have the same entropy change. It is necessary to devise an imaginary reversible process between the same states which can only restore the system to its initial state by the use of work from some outside system.

So far only isothermal processes have been considered. The entropy change for a gas which is heated may be derived as follows:

$$\mathrm{d}S = \mathrm{d}q_{\mathrm{rev}}/T$$

and

$$C_p = \mathrm{d}q/\mathrm{d}T$$

therefore, since $\mathrm{d}q = \mathrm{d}q_{\mathrm{rev}}$:

$$\mathrm{d}S = (C_p/T)\mathrm{d}T$$

and

$$\Delta S = S_{T_2} - S_{T_1} = \int (C_p/T)\mathrm{d}T \tag{4.22}$$

For a limited temperature range over which C_p is constant 4.22 becomes:

$$\Delta S = C_p \ln(T_2/T_1) \tag{4.23}$$

Entropy changes only have been considered so far. It is possible however by assigning a value of zero entropy to a particular state to obtain *absolute entropies* for individual substances. The best organized state of a compound is a perfect crystal, i.e. it has minimum entropy. Thus the third law of thermodynamics which states that all perfect crystals possess zero entropy at absolute zero provides the basis for obtaining absolute entropies.

Substituting $S_{T_1} = 0$ in equation 4.22 gives:

$$S_{T_2} - 0 + S_{T_2} = \int_0^{T_2} (C_p/T)\mathrm{d}T$$

or, in general terms:

$$S_T = \int_0^T (C_p/T)\,\mathrm{d}T \tag{4.24}$$

Using the third law, absolute entropies of substances in their standard state at 25 °C may be determined and are tabulated as $S_{298}^{\ominus}$.

A final thermodynamic function of state which will be considered is the *Gibbs free energy* G. This is probably the most important property in chemical thermodynamics since it provides the criterion for whether a change of any kind will occur or not. Again changes in free energy ΔG are considered rather than absolute values and these provide a simple numerical characterization of chemical and physical changes such that negative values indicate spontaneity, positive values indicate that the reverse change will tend to occur and zero indicates equilibrium.

The Gibbs free energy is defined by:

$$G = H - TS \tag{4.25}$$

Differentiating:

$$\mathrm{d}G = \mathrm{d}H - T\mathrm{d}S - S\mathrm{d}T$$

Considering constant-temperature conditions:

$$\mathrm{d}G = \mathrm{d}H - T\mathrm{d}S \tag{4.26}$$

and for constant pressure, $\mathrm{d}H = \mathrm{d}q$, therefore:

$$\mathrm{d}G = \mathrm{d}q - T\mathrm{d}S$$

From 4.21, $T\mathrm{d}S = \mathrm{d}q_{\mathrm{rev}}$, thus for a reversible process:

$$\mathrm{d}G = \mathrm{d}q_{\mathrm{rev}} - \mathrm{d}q_{\mathrm{rev}} = 0$$

But for spontaneous or irreversible processes, since $\mathrm{d}S > \mathrm{d}q_{\mathrm{irrev}}/T$ and $\mathrm{d}q_{\mathrm{irrev}} < \mathrm{d}q_{\mathrm{rev}}$:

$$\mathrm{d}G = \mathrm{d}q_{\mathrm{irrev}} - \mathrm{d}q_{\mathrm{rev}} < 0$$

Thus G decreases with spontaneous changes until at $G = 0$ no further reaction will take place. By using the change in the Gibbs free energy for a process it is unnecessary to follow changes in the surroundings. Values of ΔH and ΔS alone indicate whether the process is spontaneous or not.

Equation 4.25 is usually written in terms of free energy changes, thus:

$$\Delta G^{\ominus} = \Delta H^{\ominus} - T\Delta S^{\ominus} \quad (T \text{ constant}) \tag{4.27}$$

Values of $\Delta G_{\mathrm{f}}^{\ominus}$, the standard-state free energies of formation are tabulated in common with the other quantities mentioned earlier.

Consider a chemical reaction between gases represented by:

$$a\mathrm{A} + b\mathrm{B} = e\mathrm{E} + f\mathrm{F}$$

then, by definition, when equilibrium has been reached:

$$K = p_{\mathrm{E}}^{e} p_{\mathrm{F}}^{f} / p_{\mathrm{A}}^{a} p_{\mathrm{B}}^{b} \tag{4.28}$$

where K is termed the *equilibrium constant*. K may be written in terms of other concentration units but is most convenient for gases in terms of partial pressures.

Clearly K must be linked with free energy and it may be shown that:

$$\Delta G^{\ominus} = -RT\ln K \tag{4.29}$$

Combining 4.29 and 4.27 yields:

$$\ln K = -(\Delta H^{\ominus}/RT) + (\Delta S^{\ominus}/R) \tag{4.30}$$

which is termed the van't Hoff isochore equation. By regarding $\Delta H^{\ominus}$ and $\Delta S^{\ominus}$ as constant over small temperature ranges this equation may be modified to:

$$\mathrm{d}(\ln K)/\mathrm{d}T = \Delta H^{\ominus}/RT^2 \tag{4.31}$$

to indicate the variation of K with temperature.

The concept of Gibbs free energy is useful in a field of application of considerable interest to gas utilization engineers, that of metal-heating applications in which oxidation and reduction processes occur. For these high-temperature processes a general expression for ΔG at any temperature is required. This is provided by the Gibbs–Helmholtz equation:

$$\Delta G - \Delta H = T(\partial(\Delta G)/\partial T)_p \tag{4.32}$$

Calculations of free energy at high temperatures are lengthy and tedious but graphical data are available for most of the systems of interest. *Ellingham diagrams* consist of plots of $\Delta G^{\ominus}$ versus temperature for the oxidation/reduction reactions of metals and carbon. Application of Ellingham diagrams to furnace atmosphere control is considered in Section 13.2.2.

4.1.3 Chemical kinetics

Chemical reactions take place at a definite rate depending on the conditions, in particular the concentration of reactants and the temperature. The study of the rate at which such reactions occur constitutes the subject of chemical kinetics.

The *rate of a reaction* is simply the time derivative of the concentration of some species involved in the reaction.

Two terms are widely used to classify types of reaction, the *molecularity*, i.e. the number of atoms or molecules taking part in each act leading to chemical reaction, and the *order of reaction*, i.e. the number of atoms or molecules whose concentrations determine the velocity or kinetics of the process.

Thus, for the reaction $A \rightleftharpoons B + C$, the rate of forward reaction at any moment depends on the concentration of A. The reaction is *monomolecular* or *of the first order*. The rate of back reaction depends on the concentrations of B and C and is therefore *bimolecular* or *of the second order*.

The two terms are not necessarily synonymous, unless the reaction actually occurs in detail in the way it is written down. Most reactions involved in flames are bimolecular or termolecular.

Effect of reactant concentration

Reaction rates depend on the concentration of the reactants at any instant. Thus, for the simplest case of a first-order reaction:

$$-\,dc/dt = kc \qquad (4.33)$$

Where c is the concentration of the reacting substance, i.e. the rate of disappearance of the reactant at any instant is proportional to its concentration at that instant. The proportionality constant k is called the *rate constant* (or rate coefficient or specific reaction rate) for the reaction. The value of k is constant only at constant temperature.

The equilibrium constant K referred to in Section 4.1.2 is given by the quotient of the forward and reverse reaction constants k_f/k_r since at equilibrium the forward and reverse velocities are equal.

Collision theory

A bimolecular reaction must involve the coming together of two molecules and it would be expected that the frequency of collision would be important in determining the reaction rate. In fact not all the collisions which occur result in reaction and this is accounted for in the concept of *activation energy* E_a which is the minimum relative energy that the pair of molecules must have in order to react. This accounts for the marked influence of temperature on reaction velocity since a rise in temperature will favour the formation of 'active' molecules.

The probability that a molecule will possess energy in excess of an amount E_a at temperature T is related to the Boltzmann factor of statistical mechanics $\exp(-E_a/RT)$ where R is the molar gas constant.

If A, the *frequency factor*, represents the total frequency of encounters between reactant molecules, irrespective of whether they possess enough energy or not, then the reaction rate will depend on $A\exp(-E_a/RT)$.

By using appropriate units for A, this product may be identified with the rate constant k, i.e.:

$$k = A\exp(-E_a/RT) \qquad (4.34)$$

This is known as the *Arrhenius equation*. Taking logarithms yields:

$$\ln k = -(E_a/RT) + \text{constant} \qquad (4.35)$$

and the value of E_a for a reaction is frequently obtained from the slope of the graph of $\ln k$ versus T^{-1}.

Clearly the activation energy is of paramount importance in determining the rate

Temperature (K)	(°C)	$E_a = 21$ kJ mol^{-1} (5 kcal mol^{-1})	$E_a = 419$ kJ mol^{-1} (100 kcal mol^{-1})
300	27	2.4×10^{-4}	3.0×10^{-73}
1000	727	8.2×10^{-2}	1.9×10^{-22}
1500	1227	0.19	3.4×10^{-15}
2000	1727	0.29	1.4×10^{-11}
2500	2227	0.37	2.1×10^{-9}

Figure 4:1 Values of $\exp(-E_a/RT)$ for various temperatures for high and low activation energies
(From *Flame Structure* by R. M. Fristrom and A. A. Westenberg. Copyright © 1965 McGraw-Hill Inc. Used with permission of McGraw-Hill Book Company)

constant for a particular reaction at a given temperature. A high activation energy (400 kJ mol^{-1} for example) would lead to a negligible value of k at room temperature. This is illustrated in Figure 4:1 which also indicates that the effect of a high activation energy is reduced at high temperatures.

Chain reactions

Reactions between molecules generally have high activation energies and consequently tend to be slow until quite high temperatures are reached. Reactions involving an atom or a radical, however, normally have low activation energies (< 60 kJ mol^{-1}) leading to fast reactions at low temperatures.

Most of the important flame reactions involve radicals and free atoms which participate in *chain reactions*. The chain-reaction theory states that an exothermic reaction, such as that between hydrogen and chlorine, is the result of a chain of reactions involving reactive species.

The reaction between hydrogen and chlorine would normally be written:

$$H_2 + Cl_2 = 2HCl$$

but in terms of the chain-reaction theory might be written:

$$H_2 + Cl = HCl + H$$
$$H + Cl_2 = HCl + Cl$$

The start of this chain requires the presence of Cl atoms which are produced by $Cl_2 \rightarrow 2Cl$ i.e. a *chain-initiating* reaction. The Cl atoms then become *chain carriers*. The reaction of a chain carrier may lead to the formation of two new chain carriers, i.e. *chain branching*, and this will accelerate the reaction rate. Finally *chain-breaking* reactions may occur, e.g.:

$$Cl + Cl = Cl_2$$

Most of the chains involved in flame processes are, of course, considerably more complex than indicated above. Dixon-Lewis *et al.* [2], for example, postulated 17 steps involving the radicals H, OH, O and HO_2 in the apparently simple flame reaction between hydrogen and oxygen.

Recombinations of atoms and radicals occur on solid surfaces and aside from purely thermal effects these may have a quenching effect caused by loss of chain carriers. Similarly traces of *inhibitors* may reduce reaction rates by combining with and destroying chain carriers.

The flames with which this book is concerned are complex reacting systems involving chain reactions as discussed above. They differ, however, in one major respect: the chain-initiating steps are far less important since, as will be discussed later, the radicals required by the chain reaction are continuously supplied by diffusion from downstream areas of high concentration.

4.2 COMBUSTION PROPERTIES AND CALCULATIONS

4.2.1 Calorific value

The most important property a fuel gas possesses is the energy liberated as heat when it is burned. As has been shown earlier this may be expressed as the heats of formation of its combustion products on a molar basis.

In the fuel industries, however, this property is much more commonly expressed as the *calorific value* which is the quantity of heat released by combustion in a *calorimeter* of unit quantity of fuel under given conditions.

The most useful expression of calorific value would be the heat released by complete combustion, under isothermal conditions at a constant pressure of one atmosphere and at a specified reference temperature, of unit quantity of the fuel; the water formed during the combustion being in the liquid state, any sulphur in the fuel being converted to sulphur dioxide and any nitrogen remaining as such.

Complete combustion is possible with gaseous fuels at atmospheric pressure and the calorific value is measured at constant pressure. Solid and liquid fuels on the other hand require higher pressure and determinations of calorific value are made under constant-volume conditions in a *bomb calorimeter*. If the calorific value at constant pressure for these fuels is required then the heat equivalent of the work which would be done by the atmosphere on the products of combustion if the fuel were burned at constant pressure would need to be added to the constant-volume figure. This is not normally done, the calorific value being determined and quoted at constant volume.

Fuel gases which contain hydrogen or hydrocarbons possess two calorific values, the *gross calorific value* C_g and the *net calorific value* C_n depending upon whether the water formed in combustion is in the liquid or vapour phase.

The gross calorific value provides the basis on which charges are made by the gas industry on their consumers and the British gas industry is required by law to declare and maintain (within specific tolerances) this value.

The calorific value of fuel gases is normally measured using the Boys calorimeter or its recording version, the Fairweather calorimeter (see Section 8.2.1).

Calorific values are referred in the British system of measurement to 'standard temperature and pressure conditions' (stp), namely, a temperature of 60 °F and a pressure of 30 inHg (saturated). Under the International System of Units, stp is superseded by 'standard reference conditions' (src) or metric standard conditions (MSC), i.e. a temperature of 15 °C and pressure of 101 325 Pa (dry).

At 15 °C and 101 325 Pa, 1 m^3 of dry gas will occupy 1.017 m^3 when it is saturated with water vapour.

The conversion from standard cubic feet at stp (sft^3) to standard cubic metres ($m^3(st)$) is given by:

$$1\ sft^3 \text{ is equivalent to } 0.02778\ m^3\ (st)$$

(note that this is not a straight conversion from ft^3 to m^3 since the temperatures and hygrometric states are different).

Thus for calorific values:

$$\begin{aligned} 1\ Btu\ sft^{-3} &= 0.03796\ MJ\ m^{-3}\ (st) \\ 1\ MJ\ m^{-3}\ (st) &= 26.34\ Btu\ sft^{-3} \end{aligned}$$

The simplest method of calculating net calorific value from the gross value is to deduct the latent heat (assumed to be 2.455 kJ kg^{-1} or 1055 Btu lb^{-1}) of the water condensed from the products of combustion by cooling to 15 °C or to 60 °F.

The calorific value of a multicomponent fuel gas may be calculated by a simple additive method from the calorific values of its constituents. (Note that this calculation assumes ideal-gas behaviour; for strict accuracy the compressibility factors of the gases, i.e. the ratio of real to ideal-gas volumes should be taken into account—see, for example, reference [3]. However, at the low partial pressures of the constituents, the compressibility factors are very close to unity and a sufficiently accurate result can be obtained from the ideal-gas values.)

Gas	Formula	Ideal-gas calorific values: Btu ft^{-3} at 60 °F and 30 inHg, Saturated	Ideal-gas calorific values: Btu ft^{-3} at 60 °F and 30 inHg, Dry	Ideal-gas calorific values: MJ m^{-3} (st), Dry	Ideal-gas relative density (air = 1) (dry)
Carbon dioxide	CO_2	—	—	—	1.519
Oxygen	O_2	—	—	—	1.104
Nitrogen	N_2	—	—	—	0.967
Hydrogen	H_2	318.7	324.4	12.10	0.0696
Carbon monoxide	CO	315.3	320.8	11.97	0.967
Methane	CH_4	992.9	1010.5	37.69	0.554
Ethane	C_2H_6	1739.6	1770.2	66.03	1.038
Propane	C_3H_8	2475.3	2519.2	93.97	1.552
Butane	C_4H_{10}	3206.7	3263.6	121.74	2.006
Ethylene	C_2H_4	1573.1	1601.0	59.72	0.968
Propylene	C_3H_6	2294.3	2334.7	87.09	1.452

Figure 4:2 Calorific values and relative densities of common gases

Figure 4:2 lists calorific values and relative densities of common gases. The relative density of a gas is given by $d_{gas} = \rho_{gas}/\rho_{air}$ where ρ_{gas} and ρ_{air} are the respective densities at the same conditions of temperature and pressure.

Example
Calculate the gross calorific value of a substitute natural gas produced by the gas recycle hydrogenator process and of the following composition:

	% by volume
CO	3.0
CH_4	34.1
C_2H_6	12.9
H_2	38.4
C_3H_8	11.6

CO	$0.03 \times 11.97 =$	0.36
CH_4	$0.341 \times 37.69 =$	12.85
C_2H_6	$0.129 \times 66.03 =$	8.52
H_2	$0.384 \times 12.10 =$	4.65
C_3H_8	$0.116 \times 93.97 =$	10.90
		37.28

The gross calorific value is 37.28 MJ m^{-3} (st)

4.2.2 Wobbe number

Consider the flow of gas through an orifice (which may be the injector orifice of an aerated burner or the burner port of a post-aerated burner). For a constant pressure drop, $V \propto d^{-\frac{1}{2}}$, where V is the volume rate of flow and d is the relative density (specific gravity) of the gas. The potential heat contained in the gas is proportional to the calorific value C_g. Thus $C_g d^{-\frac{1}{2}}$, which is termed the *Wobbe index* or *Wobbe number*, gives a measure of the relative heat input to a burner at a fixed gas pressure of any fuel gas.

Manufactured gases
Manufactured town gases in Britain were divided into a series of gas groups based on standard ranges of Wobbe numbers as listed in Figure 4:3.

Gas group	Mean Wobbe number (UK units)	Mean Wobbe number (SI units)
G1 } Not now		
G2 } designated		
G3	800 (± 40)	30.3
G4	730 (± 30)	27.6
G5	670 (± 30)	25.4
G6	615 (± 25)	23.3
G7	560 (± 30)	21.2

Figure 4:3 UK gas groups

Natural gas
All of the British natural gases discovered so far fall into a very narrow range of Wobbe number and it has been possible to specify a single gas group of mean Wobbe number equal to 50.7 SI units (1335 UK units) ± 5 per cent variation.

Wobbe number is an important criterion in the interchangeability of gases and in burner design and is considered further in Section 4.11 and in Chapter 7.

The gas groups fit into an international classification of gases based on *gas families*. The International Gas Union (IGU) classification is shown in Figure 4:4.

4.2.3 Calculation of air requirement for complete combustion

To release the potential heat contained in a fuel it is necessary to burn it with a sufficient quantity of air. Insufficient air will lead to loss of potential heat by incomplete combustion whilst an excess may give rise to an unduly large loss of sensible heat. It is important therefore to know the *theoretical* or *stoichiometric* air requirement or air–gas ratio which will give complete combustion of a fuel.

In practice it is usually necessary to use rather more air than the stoichiometric requirement since to obtain complete and rapid combustion, the fuel and air have to be intimately mixed. This cannot be achieved with solid and liquid fuels which means that to ensure complete combustion *excess air* must be used. Gaseous fuels, on the other hand, particularly when the gas and air are *premixed* before admission to the burner, may allow complete combustion to occur using essentially the theoretical air requirement.

Gas family 1 — Wobbe number 17.8–35.8 SI units (470–943 UK units)

3 subdivisions:
- (*a*) Coke-oven gas
- (*b*) Manufactured gas
- (*c*) Hydrocarbon–air mixtures

Gas family 2 — Wobbe number 35.8–53.7 SI units (943–1415 UK units)

Natural gas
2 subdivisions:

(*a*) Wobbe number	35.8–51.6 SI units	(943–1361 UK units)
(*b*) Wobbe number	51.6–53.7 SI units	(1361–1415 UK units)

Gas family 3 — Wobbe number 71.5–87.2 SI units (1885–2300 UK units)

Liquid petroleum gases

Figure 4:4 IGU gas families

Consider the combustion equation for methane:

$$CH_4 + 2O_2 = CO_2 + 2H_2O$$

From Avogadro's law (Section 4.1.2), which states that equal volumes of gases under the same conditions of temperature and pressure contain the same number of molecules, it follows that since:

$$1 \textit{ molecule } CH_4 + 2 \text{ molecules } O_2 = 1 \text{ molecule } CO_2 + 2 \text{ molecules } H_2O$$

then:

$$1 \textit{ volume } CH_4 + 2 \text{ volumes } O_2 = 1 \text{ volume } CO_2 + 2 \text{ volumes } H_2O$$

(assuming ideal-gas behaviour).

Since the compositions of fuel gases are expressed as percentages by volume this leads to considerable simplification of the combustion calculations. In the case of solid or liquid fuels where the composition is normally quoted on a percentage-by-weight basis it is necessary to convert to a molar basis to carry out the calculation.

For gaseous fuels the important combustion equations are shown in Figure 4:5. Since air contains 20.95 per cent oxygen (volume basis) the air requirement for each fuel gas is given by oxygen requirement $\times$ 100/20.95, i.e. each volume of oxygen is accompanied by 3.78 volumes of nitrogen to make up 4.78 volumes of air. For example:

$$CH_4 + 2O_2 + 7.56\ N_2 = CO_2 + 2H_2O + 7.56\ N_2$$

or

$$CH_4 + 9.56 \text{ air} = CO_2 + 2H_2O + 7.56\ N_2$$

Example

Columns (1) and (2) of Figure 4:6 give the constituents of a town gas manufactured by the ICI 500 Process. Columns (3) and (4) show that to burn 1 volume of town gas

(1) Gas	(2) Equation	(3) Stoichiometric O_2 per unit volume of gas	(4) Stoichiometric air per unit volume of gas
CO	$CO + \frac{1}{2}O_2 = CO_2$	0.5	2.39
H_2	$H_2 + \frac{1}{2}O_2 = H_2O$	0.5	2.39
CH_4	$CH_4 + 2O_2 = CO_2 + 2H_2O$	2	9.55
C_2H_6	$C_2H_6 + 3\frac{1}{2}O_2 = 2CO_2 + 3H_2O$	3.5	16.71
C_3H_8	$C_3H_8 + 5O_2 = 3CO_2 + 4H_2O$	5	23.87
C_4H_{10}	$C_4H_8 + 6\frac{1}{2}O_2 = 4CO_2 + 5H_2O$	6.5	31.03
C_2H_4	$C_2H_4 + 3O_2 = 2CO_2 + 2H_2O$	3	14.32
C_3H_6	$C_3H_6 + 4\frac{1}{2}O_2 = 3CO_2 + 3H_2O$	4.5	21.48
CO_2	—	—	—
N_2	—	—	—
O_2		−1.0*	− 4.78*
Air		—	− 1.0

* If the gas contains oxygen, this amount will not need to be taken from the atmosphere and should be subtracted from the total oxygen requirement

Figure 4:5 Stoichiometric oxygen and air requirements for common gases

requires 0.930 volume of O_2. Therefore the stoichiometric air–gas ratio is $0.930 \times 4.78 = 4.45$.

(1) Constituent	(2) Percentage by volume	(3) Stoichiometric O_2/gas ratio	(4) (2) × (3)/100
CO_2	14.8	—	
CO	2.7	0.5	0.0135
H_2	48.9	0.5	0.2445
CH_4	33.6	2.0	0.6720
		Total	0.9300

Figure 4:6 Calculation of oxygen requirement

Excess air

Excess air is normally expressed as a percentage of the theoretical air requirement. Thus, for the example above, if the actual air–gas ratio were 5 : 1 then excess air = $(5.0 - 4.45) \times 100/4.45 = 12.4$ per cent.

4.2.4 Calculation of the analysis and quantity of combustion products

The volume and composition of the products of complete combustion may be calculated from the analysis of the fuel gas as in Figure 4:7, using the same gas as in Figure 4:6.

In addition to CO_2 and H_2O there is also present any N_2 in the original fuel gas (none in the example in Figure 4:7) and the nitrogen in the air used for combustion.

Constituent	Percentage	Equation	Products of combustion (per 100 volumes of gas)	
			CO_2	H_2O (vapour)
CO_2	14.8	—	14.8	—
CO	2.7	$CO \rightarrow CO_2$	2.7	—
H_2	48.9	$H_2 \rightarrow H_2O$	—	48.9
CH_4	33.6	$CH_4 \rightarrow CO_2 + 2H_2O$	33.6	67.2
			51.1	116.1

i.e. 0.511 volumes CO_2 + 1.161 volumes H_2O per unit volume of fuel gas

Figure 4:7 Calculation of composition of combustion products

This amounts to 3.78 × O_2 required which from the previous example was 0.930 volume per unit volume of gas. Thus the N_2 in the combustion products is 3.52 vol/vol.

The volume of gases resulting from the stoichiometric combustion of 1 volume of the town gas is thus:

CO_2	0.51
H_2O	1.16
N_2	3.52
	5.19 volumes

This then is the theoretical volume of combustion products measured under the same conditions of temperature and pressure as the original fuel gas, normally m^3 (st) or sft^3. Their actual volume will, of course, be much greater than this due to their high temperature and this volume may be calculated from the gas laws.

Combustion-product analysis is normally done on a 'dry' basis, that is any water vapour in the gas (in excess of that involved in saturating the gas at 15 °C) is assumed to have been condensed before analysis. Thus on this dry basis the volume of the combustion products would be:

CO_2	0.51
N_2	3.52
	4.03 volumes

and the percentage analysis would be:

$$[CO_2]_0 = 0.51/4.03 = 12.7\%$$
$$[N_2]_0 = 3.52/4.03 = 87.3\%$$

where the subscript 0 indicates stoichiometric conditions.

4.2.5 Interpretation of combustion-product analysis

Analysis of the flue gases from a fuel-fired appliance yields valuable information regarding the efficient operation of the appliance. If incomplete combustion or combustion with an undue excess of air is occurring, this is shown up in flue-gas analysis. Furthermore, flue-gas analysis in conjunction with temperature measurement allows flue losses to be calculated since the analysis allows the excess air and therefore the quantity of flue gases to be calculated. A complete interpretation of flue-gas analysis requires a detailed knowledge of the composition of the fuel used. A number of empirical formulae have been developed which, although not precise, are sufficiently accurate for routine purposes.

Formulae based on CO_2 *content*

Let:

Actual volume of air per unit volume or weight of fuel $= A$

Actual volume of dry combustion products per unit volume or weight of fuel $= V$

Theoretical volume of air per unit volume or weight of fuel $= A_0$

Theoretical volume of dry combustion products per unit volume or weight of fuel $= V_0$

Theoretical percentage of CO_2 in dry combustion products $= [CO_2]_0$

Percentage of CO_2 *actually* present in dry combustion products $= [CO_2]$

Percentage of CO *actually* present in dry combustion products $= [CO]$

Excess air factor i.e.

$$\frac{\text{air actually used}}{\text{theoretical air requirement}} = \frac{A}{A_0} = n$$

Then volume of excess air $= (n - 1) A_0$.
So percentage of excess air $= 100(n - 1) A_0/A_0 = 100(n - 1)$.
Volume of dry flue gases, $V = V_0 + (n - 1)A_0$.

Clearly the quantity of carbon in the flue gases from the combustion of unit quantity of a given fuel is the same whether the theoretical air is used or an excess. Also if any CO is converted to CO_2 the volume would not change. Hence:

$$(V_0 + (n - 1)A_0)([CO_2] + [CO]) = [CO_2]_0\, V_0 \tag{4.36}$$

Rearranging:

$$100\,(n - 1) = \frac{V_0}{A_0}\left(\frac{[CO_2]_0}{[CO_2] + [CO]} - 1\right) 100 \tag{4.37}$$

The ratio V_0/A_0 varies only slightly for a wide range of fuels ranging from 0.90 to 0.96 for coals, cokes, fuel oils and fuel gases and for routine purposes may be taken as unity. Equation 4.37 therefore becomes:

$$\text{Percentage excess air} = 100\,(n - 1) \approx \left(\frac{[CO_2]_0}{[CO_2] + [CO]} - 1\right) 100 \tag{4.38}$$

(Siegert's formula).

Formulae based on O_2 *content*
From the previous section:

$$\text{volume of dry flue gases per unit quantity of fuel} = V_0 + (n - 1)A_0$$
$$\text{volume of air in the flue gases} = (n - 1)A_0$$

Since the oxygen content of air $\approx$ 21 per cent, $[O_2]$, i.e. the percentage of O_2 in the flue gas

$$\approx \frac{21}{100}(n - 1)A_0 \frac{100}{V_0 + (n - 1)A_0}$$

i.e.

$$[O_2] = \frac{21\,(n - 1)\,A_0}{V_0 + (n - 1)A_0}$$

Multiplying across:

$$V_0\,[O_2] + [O_2]\,(n - 1)A_0 = 21\,(n - 1)A_0$$

Therefore:

$$n - 1 = V_0\,[O_2]/A_0\,(21 - [O_2])$$

As before, $V_0/A_0 \approx 1$. Therefore:

$$n - 1 \approx [O_2]/(21 - [O_2])$$

Therefore:

$$\text{percentage excess air} \approx 100\,(n - 1) = 100[O_2]/(21 - [O_2]) \qquad (4.39)$$

Correction for presence of CO. When part of the carbon content of the fuel is burned to CO only, part of the theoretical oxygen requirement is unused and appears as free oxygen in the flue gases. Since $CO_2 = CO + \frac{1}{2}O_2$, the amount of this oxygen is equal to half the amount of CO present. Thus if CO is present, the $[O_2]$ figure in equation 4.39 is given by:

$$[O_2] = \text{observed } [O_2] - \tfrac{1}{2} \text{ observed } [CO]$$

The percentage excess air can thus be calculated directly from the oxygen content of the flue gases whereas the formula based on the carbon-dioxide content requires a knowledge of the stoichiometric carbon-dioxide content.

4.2.6 Calculation of flame temperature

The heat balance
The theoretical burned-gas temperature may be calculated from thermochemical data. For flames with fairly low temperatures, dissociation of the product gases is not appreciable and reasonable estimates of the final burned-gas temperature may be made disregarding dissociation. For this simple case the combustion may be considered in two stages:

1 The reaction is considered to take place at room temperature and the heat liberated in the formation of water and carbon dioxide is determined.

2 The product gases are then assumed to be heated to such a temperature that their enthalpy is equal to the heat of reaction, this temperature being the theoretical flame temperature

A similar calculation can be carried out using calorific value and specific heats to obtain a heat balance in terms of gas volumes. For stoichiometric combustion with air:

$$C_n = (T_b - T_0)(\varphi_{H_2O}\,\bar{c}_{p,\,H_2O} + \varphi_{CO_2}\,\bar{c}_{p,\,CO_2} + \varphi_{N_2}\,\bar{c}_{p,\,N_2}) \tag{4.40}$$

where

C_n = net calorific value (volumetric)
$\bar{c}_{p,\,H_2O}$, etc. = mean heat capacity per unit volume between T_b and T_0
φ_{H_2O}, etc. = volume fraction of H_2O, etc.
T_b = theoretical flame temperature
T_0 = reference temperature

Such a calculation involves an iterative procedure; the flame temperature must be estimated in advance in order to determine the mean specific heat of the product gases then adjusted until the terms balance.

The simple method outlined above is satisfactory for low temperatures but leads to serious inaccuracies for hot flames. For a stoichiometric hydrogen–oxygen flame, the heat of reaction is 57.8 kcal mol^{-1} but the enthalpy of 1 mol of water vapour only attains this value at a temperature around 5000 °C [4] which, in practice, does not actually occur because of dissociation of the water vapour.

Similarly, in high-temperature hydrocarbon–air flames, for example, appreciable quantities of H_2, O_2, OH, O, H and NO occur and the dissociation absorbs considerable energy to limit the flame temperature. This point is illustrated in Figure 4:8 (see also Figure 1:23).

Constituent	At adiabatic flame temperature (2220 K) (1947 °C)	At 1400 K (1127 °C)
N_2	0.70961	0.71616
H_2O	0.18066	0.18751
H_2	0.00365	0.00001
O_2	0.00477	0.00004
O	0.00026	0.00000
OH	0.00308	0.00000
H	0.00045	0.00000
NO	0.00200	0.00001
N	0.00000	0.00000
CO	0.00942	0.00002
CO_2	0.08611	0.09625

Figure 4:8 Equilibrium composition of stoichiometric natural-gas–air combustion gases at 2220 K and 1400 K [5]

Taking dissociation into account complicates the calculation since the temperature must be known to permit calculation of the dissociation of the product gases and therefore their composition but, of course, the temperature itself depends on the composition. Again iteration has to be used and this leads to laborious calculations. Graphical methods and digital computers have been widely used to minimize the effort involved.

Calculation of the equilibrium composition for a given temperature

For a typical combustion process involving carbon, hydrogen, oxygen and nitrogen, the important dissociation equilibria are:

$$CO_2 \rightleftharpoons CO + \tfrac{1}{2}O_2$$
$$H_2O \rightleftharpoons H_2 + \tfrac{1}{2}O_2$$
$$H_2O \rightleftharpoons \tfrac{1}{2}H_2 + OH$$
$$\tfrac{1}{2}H_2 \rightleftharpoons H$$
$$\tfrac{1}{2}O_2 \rightleftharpoons O$$
$$\tfrac{1}{2}N_2 + \tfrac{1}{2}O_2 \rightleftharpoons NO$$

At equilibrium:

(1) $K_1 = p_{CO} \sqrt{p_{O_2}}/p_{CO_2}$ (2) $K_2 = p_{H_2} \sqrt{p_{O_2}}/p_{H_2O}$

(3) $K_3 = p_{OH} \sqrt{p_{H_2}}/p_{H_2O}$ (4) $K_4 = p_H / \sqrt{p_{H_2}}$

(5) $K_5 = p_O/\sqrt{p_{O_2}}$ (6) $K_6 = p_{NO}/\sqrt{p_{N_2}}\sqrt{p_{O_2}}$

where p_{CO} etc. = partial pressures
K_1 etc. = equilibrium constants

Additionally equations can be set up based on the conservation of the numbers of gram atoms of the elements involved:

(7) number of C atoms, $n_C = p_{CO} + p_{CO_2}$
(8) number of H atoms, $n_H = 2p_{H_2O} + 2p_{N_2} + p_H + p_{OH}$
(9) number of O atoms, $n_O = 2p_{CO_2} + p_{CO} = p_{H_2O} + 2p_{O_2} = p_{OH} + p_O + p_{NO}$
(10) number of N atoms, $n_N = 2p_{N_2} + p_{NO}$

n_C etc. are not known as absolute values but their ratios can be obtained from the unburned mixture. Finally the total pressure P is given by:

(11) $P = p_{CO_2} + p_{CO} + p_{H_2O} + p_{O_2} + p_{H_2} + p_{OH} + p_O + p_H + p_{N_2} + p_{NO}$

The ratios n_C/n_H, n_O/n_H are known from the composition of the unburned fuel–air mixture.

Thus, there are 11 equations and 11 unknowns and it is usually necessary to adopt an approximate method of solution or a computer solution.

For occasional calculation of flame-gas composition Gaydon and Wolfhard [4] recommend the method of Damköhler and Edse [6] which involves assuming first two partial pressures (usually p_{H_2O} and p_{CO_2}/p_{CO}) and second the flame temperature. The equilibrium constants K_1 etc. are obtained from tables, or may be calculated from the van't Hoff equation 4.31.

The gram-atom ratios are known from the composition of the unburned mixture. Based on these assumptions all the other partial pressures may be derived e.g. K_1 and p_{CO_2}/p_{CO} determine $\sqrt{p_{O_2}}$ etc. It is necessary then to test the initial assumptions by calculating, say, n_O from equation (9) and P from equation (11) to see if it equals 1.

Successive approximations are then made (still assuming the same flame temperature) until the gram-atoms and total pressure approximate closely to the values they should have. A specimen calculation is presented fully in reference [4].

Several other techniques reduce the considerable number of numerical operations involved. The thermodynamic charts of Hottel, Williams and Satterfield [7] incorporate the equilibrium compositions for a wide range of temperatures for the C, H, O, N system. Weinberg [8] has developed a simplified approach applicable to the C, H, O system in which three of the eight partial pressures are assumed, the other five are calculated then the original three are checked by determinants of the errors in the assumed values.

Clearly high-speed computers lend themselves well to repetitive calculations of this kind and computer solutions have been presented by a number of authors in recent years including Miller and McConnell [9], Harker [10], Harker and Allen [11], Steffenson, Agnew and Olsen [12] and Barrett [13].

Flame-temperature calculation

Having obtained the equilibrium composition (at an assumed temperature) the flame

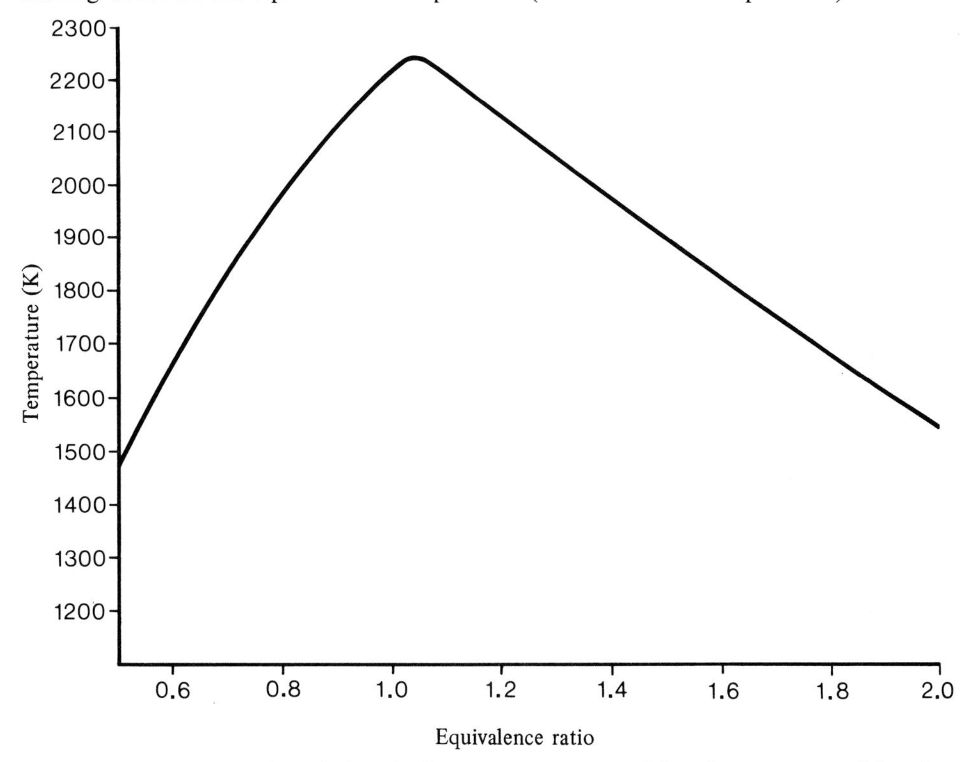

Figure 4:9 Calculated variation in flame temperature with mixture composition for methane–air flames [13]

temperature may now be calculated from a heat balance. First the heat required to raise the burned gases from room temperature to the assumed temperature is calculated from enthalpy data, which are tabulated in, for example, reference [4]. Then the heat produced in the flame is calculated. The main source of heat liberated is in the formation of CO_2, CO and H_2O and is given by the product of their partial pressures and heat of formation. From this heat release must be deducted the heat used in forming free atoms and radicals. The net heat production in the flame is then compared with the heat content of the burned gases and if it is greater, for example, then the true flame temperature must be higher and so the whole procedure must be repeated for a higher temperature.

Again this procedure is best done by computer and of the methods listed above for the calculation of equilibrium compositions most have the additional facility of calculating flame temperature. Figure 4:9 shows the variation in theoretical flame temperature with mixture composition for methane–air flames. The maximum flame temperature lies on the rich side of stoichiometric at an equivalence ratio of about 1.05.

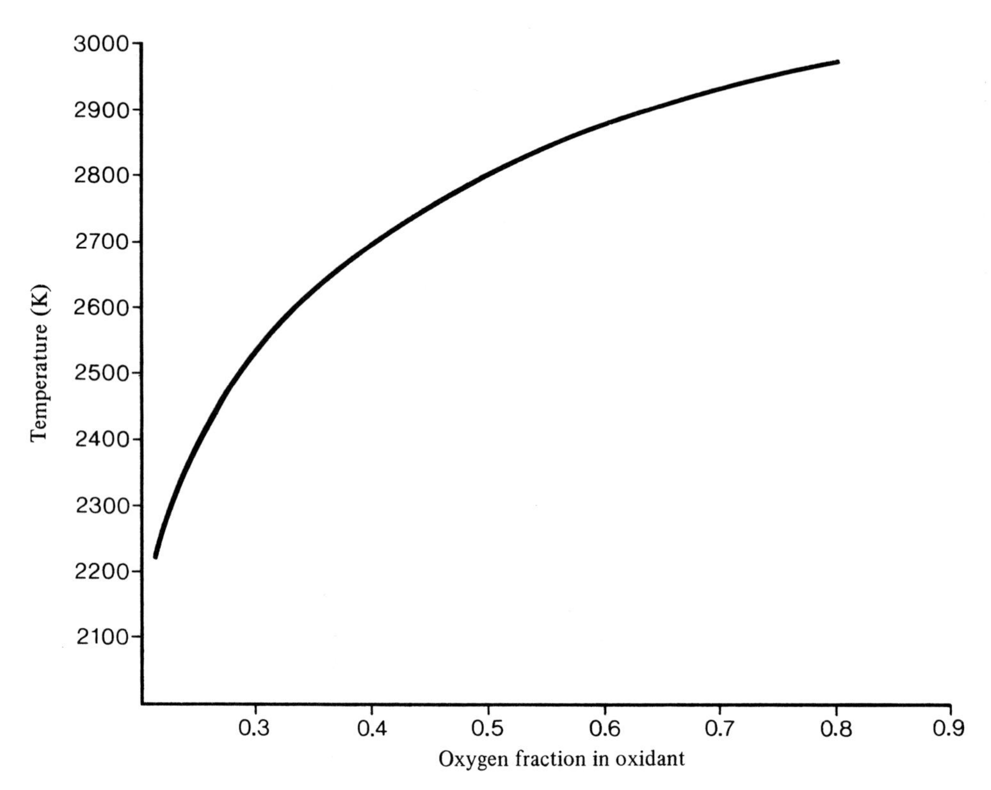

Figure 4:10 Calculated variation in flame temperature with oxygen fraction in the oxidant (CH_4, N_2, O_2 flames) [13]

The effect of oxygen enrichment on flame temperature is indicated by Figure 4:10. The temperature is increased considerably by additions of oxygen but tends to flatten at high oxygen additions due to increasing dissociation at these high temperatures.

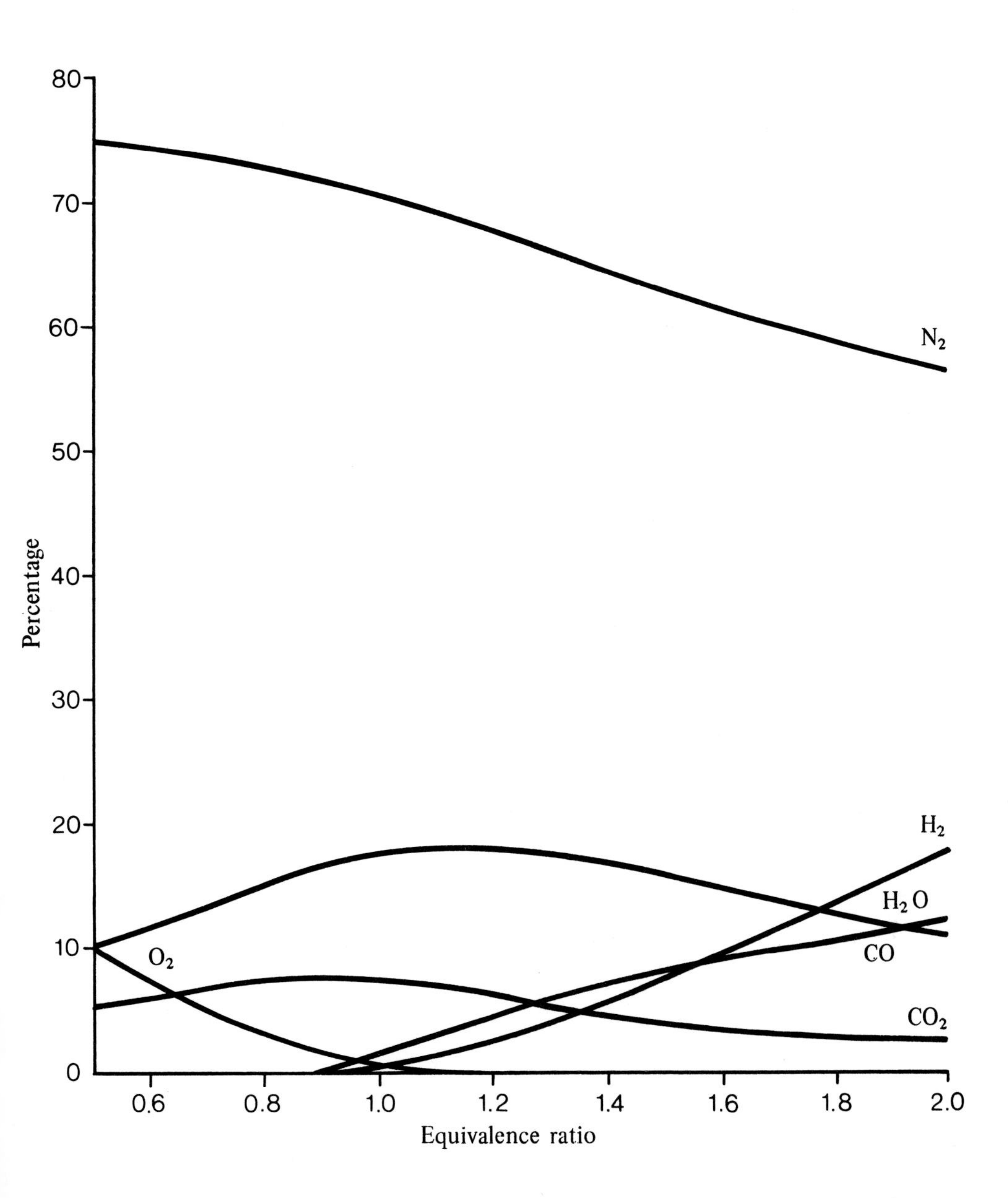

Figure 4:11 Calculated variation in major flame-gas constituents with mixture composition for natural-gas–air flames [13]

The effect of increasing pressure is to increase the flame temperature since increased pressures move the equilibria in favour of the more stable combustion, i.e. towards reduced dissociation. Figures 4:11 and 4:12 show the variation with mixture composition of the major and minor flame-gas constituents respectively for natural-gas–air

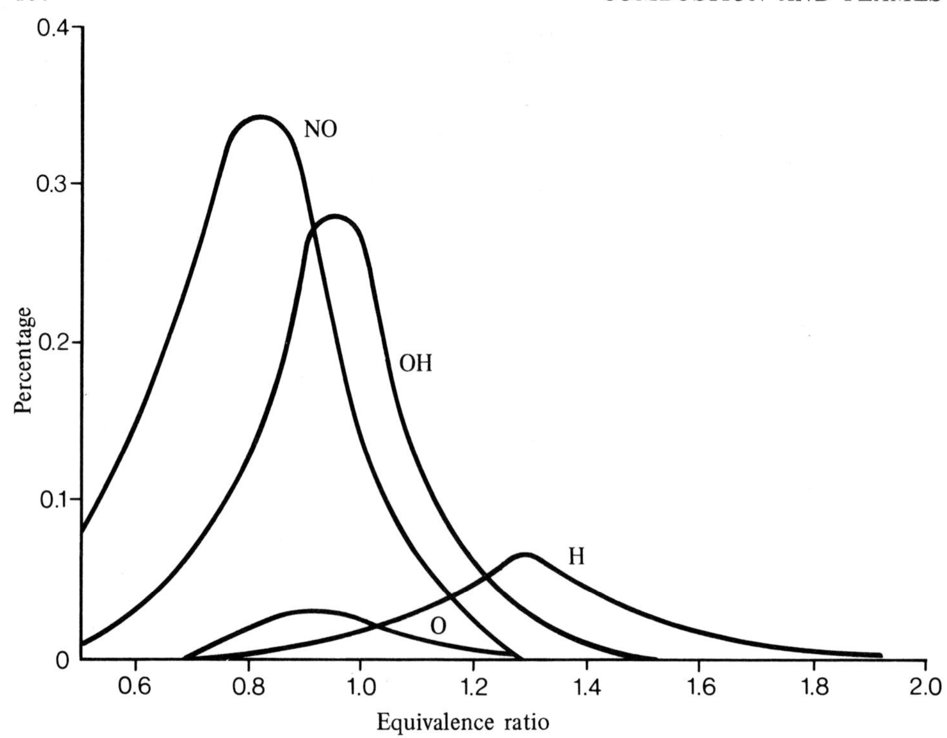

Figure 4:12 Calculated variation in minor flame-gas constituents with mixture composition for natural-gas–air flames [13]

flames. The composition of the natural gas is: 95.03% CH_4, 2.9% C_2H_6, 0.5% C_3H_8, 0.3% C_2H_4, 1.27% N_2.

4.2.7 The heat content of combustion products

Earlier sections of this chapter have presented methods of calculating, for example, stoichiometric air requirements and combustion-product volumes from the fuel-gas analysis. Such calculations are tedious as are calculations of the heat content of combustion products, flame temperatures (Section 4.2.6) and the like. In practice, for calculations of this sort, it is time-saving to adopt one of two alternative approaches:

1 To use generalized, approximate methods applicable to the combustion of virtually any fuel, e.g. Rosin [14, 15].
2 For specific, well-documented gases like British natural gas to use the accurate published data available, e.g. Davies and Toth [5].

Combustion calculations can be simplified by the use of certain statistical relationships between the various factors involved in combustion, for example, between the gross and net calorific values, the stoichiometric air requirement, and the theoretical volume of combustion products. Moreover, for all industrial fuels, the heat content of the theoretical wet combustion gases is approximately a function of temperature alone.

These ideas are embodied in the H–T (originally I–t) diagram of Rosin [15] and it is claimed that the accuracy of calculations based on the H–T diagram suffices for most practical purposes since the deviations are smaller than the errors inherent in sampling and analysis. The only data required are the calorific value of the fuel and the excess air employed in its combustion.

The net calorific value (C_n) and the heat content of the combustion products (H_p) are measures of two heat concentrations which have changed their carriers during combustion.

For complete combustion without losses $H_p = C_n/V$ irrespective of the amount of excess air.

This is particularly useful if it is possible to express V as a function of C_n i.e. $H_p = C_n/f(C_n)$.

Both C_n and V are functions of chemical composition and thereby interrelated and Rosin and Fehling showed statistically that $V_0 \approx C_n/100$ if V_0 is measured in nft^3 * and C_n is measured in Btu nft^{-3}. Therefore: $H_{p0} = C_n/V_0 \approx 100$ Btu nft^{-3} (3.8 MJm^{-3} (st)) (where subscript 0 indicates stoichiometric conditions).

Similarly A_0 can be shown to be $\approx f(C_n)$. So although A_0 and V_0 are linear functions of C_n, the heat content per unit volume of theoretical combustion gas H_{p0} is nearly a constant which is independent of the calorific value of the fuel. The reason for this somewhat surprising conclusion is that all fuels consist basically of carbon and hydrogen which, although they have very different calorific values, have theoretical combustion gas volumes in almost exact inverse proportion. Thus the ratio $C_n/V_0 = H_{p0}$ for carbon and for hydrogen differs only by about $1\frac{1}{2}$ per cent. The H–T diagram consists of two standard heat-content/temperature curves, that for the 'theoretical combustion gas', which is the same for any fuel, and that for air. Thus the heat content of any combustion gas burning with any amount of excess air at any temperature may be obtained simply by reading off the two curves. The H–T diagram is shown in Figure 4:13.

The quantity of heat obtainable by combustion of a given weight or volume of fuel at temperature T of the combustion gas is given by

$$q_T = H_T(\text{theoretical combustion gas})V_0 \tag{4.41}$$

and for fuels with excess air:

$$q_T = H_T(\text{theoretical combustion gas})V_0 + H_T(\text{air})\,[(n-1)A_0] \tag{4.42}$$

Detailed data on combustion-product heat contents are available in compilations produced by the Midland Research Station of the British Gas Corporation together with equilibrium compositions, specific heats, viscosities and heat-transfer parameters for manufactured town gas [16], butane [16], propane [16], and natural gas [5, 17] burning with air. The data cover a wide range of air–gas ratios. In addition, similar data are available for town gas and methane combustion with oxygen and with oxygen-enriched air [18].

The heat contents (MJ m^{-3} (st)) have been obtained from the molar enthalpies

* nft^3 refers to measurement at 0 °C and 760 mmHg.

(Section 4.1.2) of the constituent gases tabulated in the NASA [19] and JANAF [20] compilations by summation and application of the ideal-gas-law conversion factor (1 mol occupies 23.645×10^{-3} m^3 (st)).

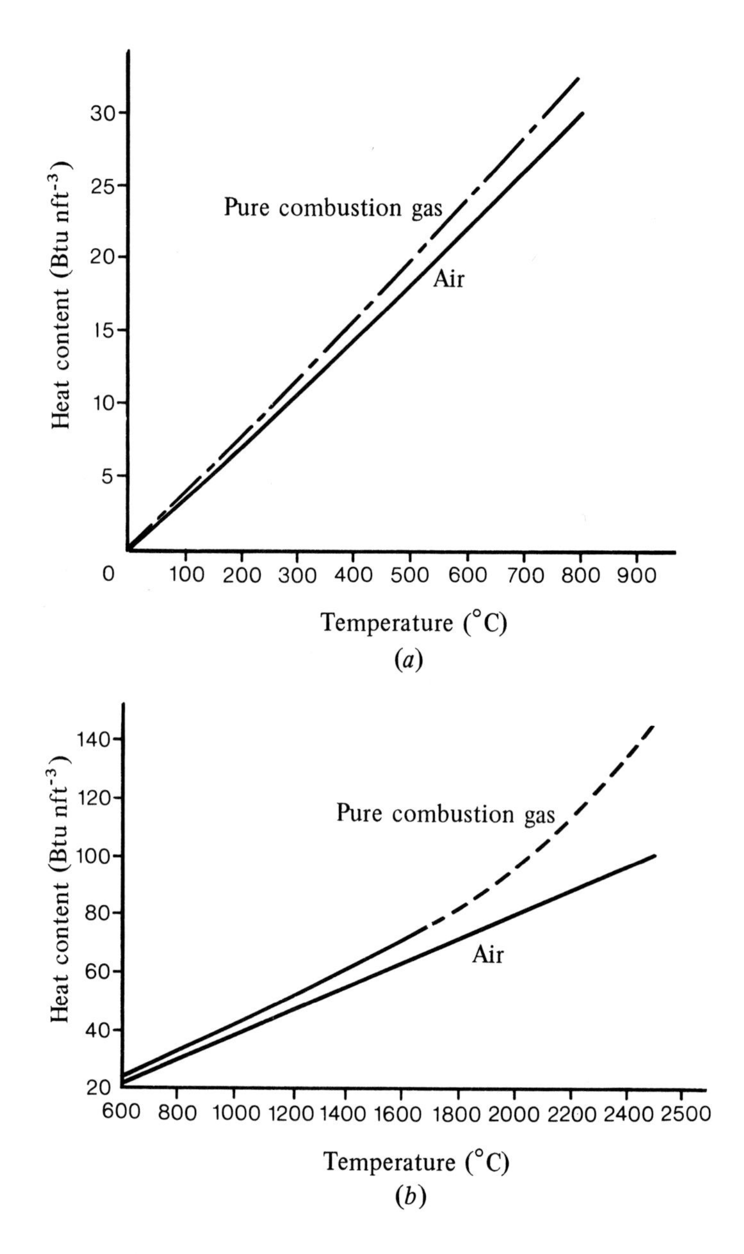

Figure 4:13 *H–T* diagram [14]
(*a*) 0–800 °C (*b*) 600–2500 °C

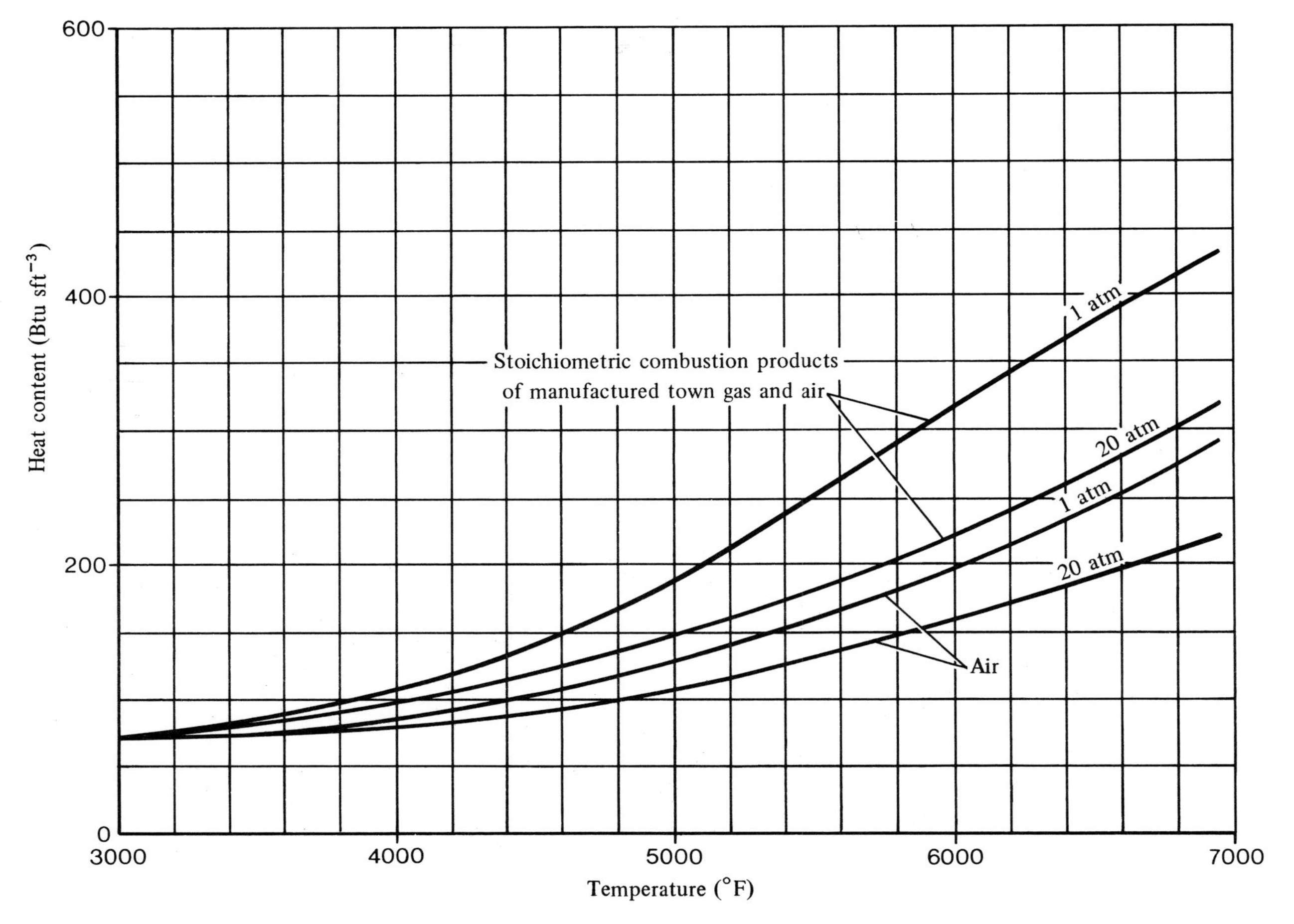

Figure 4:14 Heat content of air and of stoichiometric combustion products of manufactured town gas and air in British units [17]

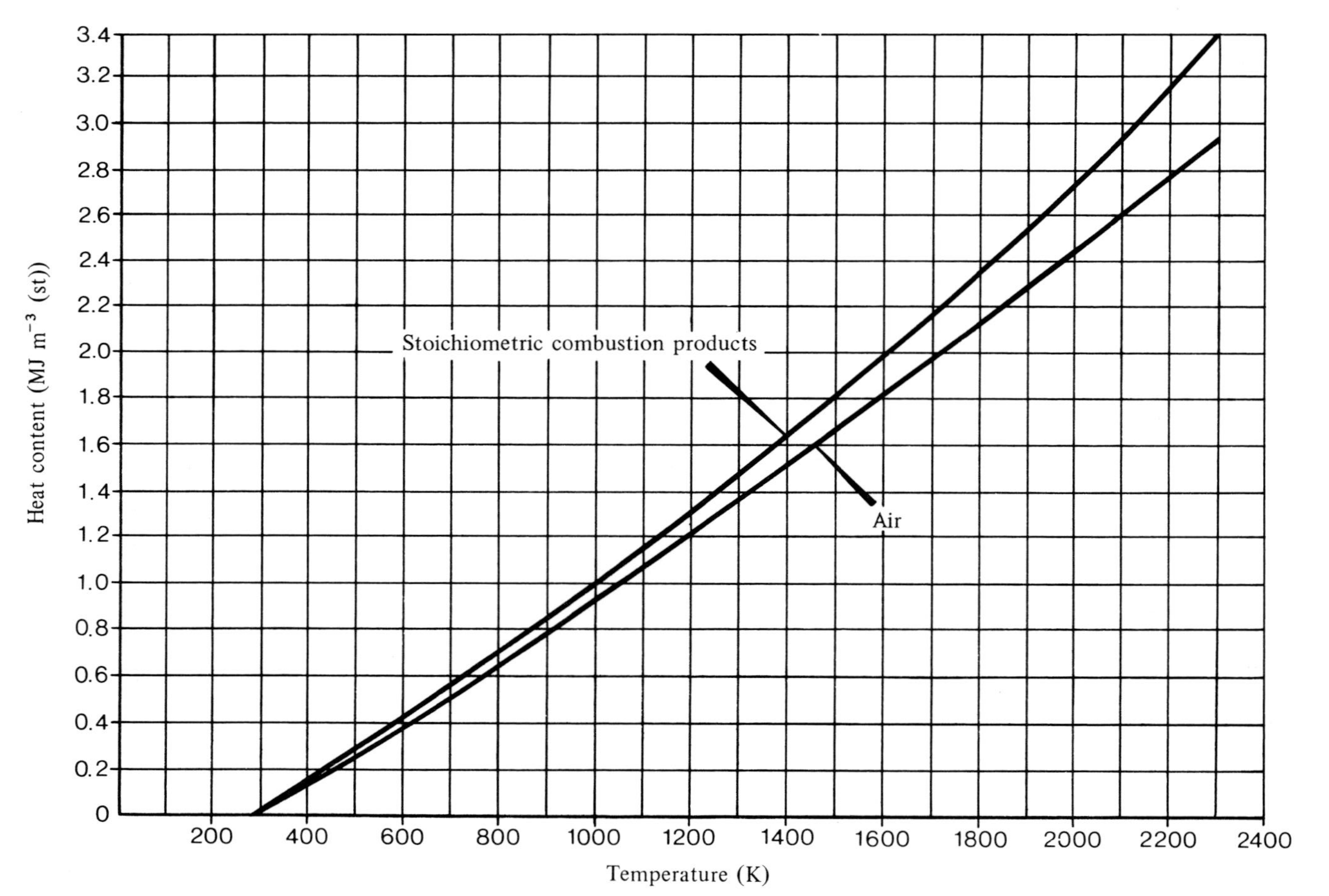

Figure 4:15 Heat content of stoichiometric combustion products of natural gas and of air in SI units [94], data from [5]

Figures 4:14 and 4:15 show the H–T relationships for stoichiometric combustion with air of methane and a G6 manufactured town gas and for air in imperial units and in SI units respectively. The town gas for which data are presented of course differs in composition from other town gases and therefore in calorific value, air requirement and combustion-product volume. This fact does not affect the values of heat content, etc. presented since these will be to all intents and purposes the same for British first-family gases. In fact, as Figure 4:14 indicates, the differences are comparatively small even between town gas and methane.

Simple relationships have been derived for manufactured town gas as follows:

$$C_n = 0.9C_g$$
$$A_0 = C_g/116 \quad \text{where } C_g \text{ is in Btu sft}^{-3} \qquad (4.41)$$

$$A_0 = C_g/4.4 \quad \text{where } C_g \text{ is in MJ m}^{-3} \text{ (st)} \qquad (4.42)$$

$$V_0 = A_0 + 0.7 \qquad (4.43)$$

For natural gas the values of C_g, C_n, A_0 and V_0 used are:

$$C_g = 38.61 \text{ MJ m}^{-3}\text{(st)} \quad (1017 \text{ Btu sft}^{-3})$$
$$C_n = 34.84 \text{ MJ m}^{-3} \text{ (st)} \quad (918 \text{ Btu sft}^{-3})$$
$$A_0 = 9.76 \text{ m}^3\text{(st)m}^{-3}\text{(st)}$$
$$V_0 = 10.79 \text{ m}^3\text{(st)m}^{-3}\text{(st)}$$

Note that V_0 changes with temperature. Below about 1100 °C the products of stoichiometric and lean mixtures are undissociated and V_0 is constant but above 1100 °C V_0 increases as dissociation increases. For purposes of heat-balance calculation the effect can be ignored without serious loss of accuracy.

The usefulness and versatility of these data are best demonstrated by some simple examples.

Example 1

Estimate the stoichiometric air requirement, wet combustion gas volume and adiabatic flame temperature of a manufactured town gas of gross calorific value 480 Btu sft^{-3}.

$$A_0 = 480/116 = 4.14$$

Therefore:

$$V_0 = A_0 + 0.7 = 4.84$$

also $C_n = 0.9C_g = 432$ Btu sft^{-3}.

Since the heat content of the theoretical combustion products $H_0 = C_n/V_0$:

$$H_0 = 432/4.84 = 89.3 \text{ Btu sft}^{-3}$$

which, from Figure 4:14, corresponds to a temperature of 3680 °F.

Example 2

What is the heat availability in the combustion products of natural gas with air for a process operating at 1200 K (927 °C)?

Heat available above 1200 K = heat content of combustion products at flame temperature ($H_{0,f}$) minus heat content at 1200 K ($H_{0,1200}$)

$$\begin{aligned} H_{0,f} &= C_n/V_0 \\ &= 34.84/10.79 = 3.23 \text{ MJ m}^{-3} \text{ (st)} \\ H_{0,1200} &= 1.32 \text{ MJ m}^{-3} \text{ (st)} \quad \text{(from Figure 4:15)} \end{aligned}$$

Therefore available heat above 1200 K = 1.9 MJ m^{-3} (st), i.e. 1.9 × 10.79 = 20.6 MJ are available for each m^3 (st) of natural gas burned.

The heat content of the combustion products leaving the system, i.e. the 'flue loss' based on the gross calorific value is given by:

$$\text{flue loss } \% = (H_T V/C_g)100 + [(C_g - C_n)/C_g]100$$

The first term accounts for the total sensible heat content of the combustion products including that due to water vapour whilst the second term constitutes the latent heat loss.

Thus, in example 2:

$$\begin{aligned} \text{flue loss } \% &= \left(\frac{1.32 \times 10.79}{38.61}\right)100 + \left(\frac{38.61 - 34.84}{38.61}\right)100 \\ &= 36.9\% + 9.8\% \\ &= 46.7\% \end{aligned}$$

Clearly, only by practising some form of heat recovery, which may consist of air-preheating or, in the case of continuous processes, load preheating is it possible to obtain a thermal efficiency greater than 53.3 per cent for such a process (see Chapter 11). In practice, of course, the efficiency would be less than this because of heat losses from the furnace structure.

Example 3

Calculate the amount of natural gas required to heat 5000 kg of steel to 1350 K (1077 °C) in a furnace operating with no excess air.

Assume that the gases leave the furnace at 1400 K (1127 °C) and that there is a 10 per cent loss of available heat through the furnace structure. Take the specific heat of steel to be 630 J kg^{-1} $°C^{-1}$.

As in Example 2, $H_{0,f}$ = 3.23 MJ m^{-3} (st)

Heat content at 1400 K ($H_{0,1400}$) = 1.65 MJ m^{-3}(st)
available heat = 3.23 − 1.65 = 1.58 MJ m^{-3} (st)
less 10% = 1.42 MJ m^{-3} (st)
1 m^3 of natural gas produces 10.79 m^3 (st) of combustion products, therefore
heat available per m^3 of gas = 1.42 × 10.79
= 15.3 MJ m^{-3} (st)
Heat to steel = 5000 × 630 × (1350 − 288) × 10^{-6} = 3345 MJ. Therefore natural gas required = 3345/15.3 = 219 m^3

$$\begin{aligned} \text{flue loss } \% &= \left(\frac{1.65 \times 10.79}{38.61}\right)100 + \left(\frac{38.61 - 34.84}{38.61}\right)100 \\ &= 46.1\% + 9.8\% \\ &= 55.9\% \end{aligned}$$

$$\text{percentage thermal efficiency} = \frac{\text{useful heat}}{\text{heat input}} = \left(\frac{3345}{219 \times 38.61}\right) \times 100$$
$$= 39.6\%$$

or

$$\frac{\text{available heat}}{C_g} = \frac{15.3}{38.61}$$
$$= 39.6\%$$

A number of alternative methods are available for calculating flue losses from natural-gas-fired plant [97–8]. These provide results which are sufficiently accurate for a rapid assessment of losses based on flue-gas temperature measurement, dry CO_2 and/or O_2 analyses and a knowledge of the gross calorific value.

In the case of incomplete combustion, however, these methods in common with the *H–T* approach adopted in this book are inadequate and reference should be made to the tabulated data in reference [5].

4.2.8 Heat balances

The performance of a heating unit is frequently expressed in terms of fuel usage per unit of output, e.g. m^3 (st) of natural gas per kg of steel heated or kg of steam raised. Such a criterion is satisfactory so long as it is desired to compare, say, the performance of furnaces using the same fuel to heat the same material to the same temperature but provides no absolute comparison.

Absolute comparison can be obtained by determining the amount of material which could be heated by unit quantity of fuel in a perfect appliance and comparing this with the amount actually heated. On this basis appliances are compared by the ratio of fuel ideally necessary to that used in practice and this is expressed as efficiency. This provides a useful overall statement of plant efficiency but does not show where inefficiencies occur. For this it is necessary to construct a heat balance which attempts to trace what happens to all the energy at any given stage, i.e. to balance all the components of energy entering and leaving the stage. The heat balance is constructed on the basis of some convenient unit of the process, such as one process cycle, one hour or one unit of processed material. The following information, where appropriate, will be used in the heat balance.

Input to stage or process:

1. The quantity of fuel used.
2. The calorific value of the fuel used. Either gross or net calorific value may be used. Since the latent heat of condensation of water can very rarely be recovered in industrial processes it is perhaps more realistic to use the net value. However, since fuel costs normally relate to the gross value, efficiencies are more commonly based on this. Clearly, efficiencies based on the gross value will be lower than if the net value is used. For the fuel gases of interest in this book the difference is about 10 per cent.
3. The sensible heat of the fuel above a convenient datum, e.g. for SI calculations 25 °C and 60 °F for calculations in British units.
4. The sensible heat content of the combustion air above datum.
5. Electrical or mechanical energy supplied to pumps, fans, etc.
6. Any exothermic heat of reaction (in addition to (2)).
7. The heat content of the material at the start of the process or stage above datum.

Exit stage or process:

1 The sensible and latent heat above datum of the material leaving the process or stage.
2 The total heat content of the combustion products. This consists of their sensible heat content, any unused chemical energy due to incomplete combustion and normally the latent heat in the water content.
3 Heat stored in the furnace structure and furniture.
4 Heat losses from the furnace structure by conduction.
5 Radiation losses through openings.
6 Losses to cooling water where used.
7 Losses to conveyance mechanisms, etc.
8 Heat absorbed in endothermic reactions.

A heat balance may be drawn up on the basis of measurements conducted on an existing furnace or may be predicted from assumptions and data for a proposed furnace. A convenient method of presentation of a heat balance is the *Sankey diagram*. An outline diagram is made of the system upon which is superimposed a stream representing the flow of energy, the width of the stream at any point being drawn in proportion to the amount of energy flowing at that point.

Figure 4:16 shows a typical Sankey diagram.

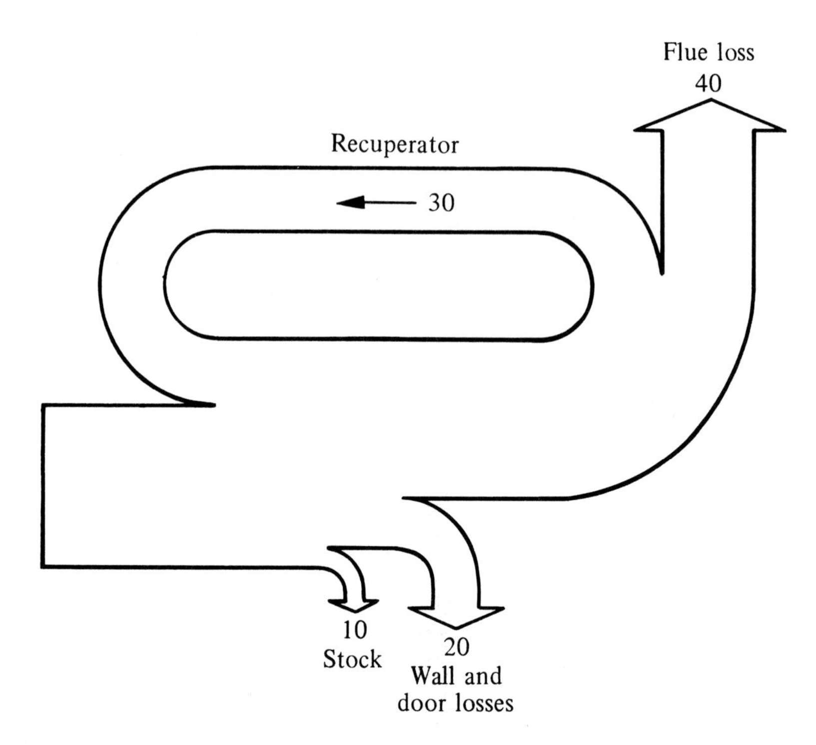

Figure 4:16 Sankey diagram for a furnace with recuperation

4.3 FLAME PROPAGATION

4.3.1 Introduction

A *flame* may be defined as a relatively fast exothermic gas-phase reaction distinguished from other combustion reactions by its ability to propagate subsonically through space. It is usually but not exclusively accompanied by the emission of visible light. This distinguishing property of *spatial propagation* is the result of strong coupling between chemical reaction, fluid flow and the transport processes of mass diffusion and heat conduction. Flame microstructure does not depend on the movement of the flame relative to the observer but on its movement relative to the unburned reactants. The combustion reaction is confined to a zone which is relatively thin and this has been termed the *flame front*, the *combustion wave* or the *combustion zone* by various authors. The first term will be used throughout this book.

Flame propagation may be regarded qualitatively as a positive feedback system between the flame and the reactants supplying it. Heat and active species accelerate chemical reaction, both being produced in the flame front and transported by conduction and diffusion to the fresh reactants. If the feedback exceeds some critical amount, the flame will be self-sustaining, whilst if there is such a small feedback that the acceleration of reaction is not adequate to ensure a volume of gas identical to that from which the transport occurred, then the flame slows up until a steady state is attained. If such a velocity cannot be achieved the flame is extinguished.

The characteristic steady-state velocity of the flame is termed the *burning velocity* S_u which may be defined as the velocity of a plane flame front normal to itself and relative to the unburned reactants. Burning velocity, whilst not providing the detailed characterization offered by studies of flame microstructure, is an extremely convenient and widely used single-parameter description of a complex phenomenon.

Since a flame travels at a characteristic burning velocity a stationary flame may be obtained by passing the reactants into the flame at the same velocity in the reverse direction. In practice flames achieved in this way are unstable and it is necessary to provide a stabilizing device or *burner* which locally interacts with the flow and combustion processes.

Flames are generally classified as *premixed* or *diffusion* flames depending upon whether or not the reactants are mixed before entering the reaction zone, for example, the laboratory Bunsen burner in normal operation produces a premixed flame since entrained air mixes with the gas in its passage up the burner tube. However, if the air shutter is closed, only gas passes out at the burner head and the air for combustion is provided by diffusion from the surrounding air yielding a flame of quite different character, a diffusion flame. If the air supplied premixed with the gas, i.e. the *primary* air, is less than the stoichiometric air requirement, two combustion zones are formed. The outer zone which obtains its combustion air by diffusion is termed the *secondary combustion zone*.

The other method of broad classification relates to the character of the gas flow through the reaction zone. In *laminar* flames all mixing and transport is done by molecular processes whilst in *turbulent* flames these processes are aided but complicated by macroscopic eddying.

The simplest flames and the ones most amenable to theoretical and experimental treatment are those which combine premixed reactants and laminar flow; Sections 4.3, 4.4, 4.5, 4.6 and 4.7 are primarily concerned with flames of this kind. Turbulent flames and diffusion flames are dealt with separately in Sections 4.8 and 4.9 respectively.

4.3.2 Laminar-flame propagation

A convenient theoretical flame model is one in which the surfaces of constant properties (temperature, composition, density, pressure, etc.) lie parallel to each other and flow and diffusion processes occur normal to the flame front and parallel to each other.

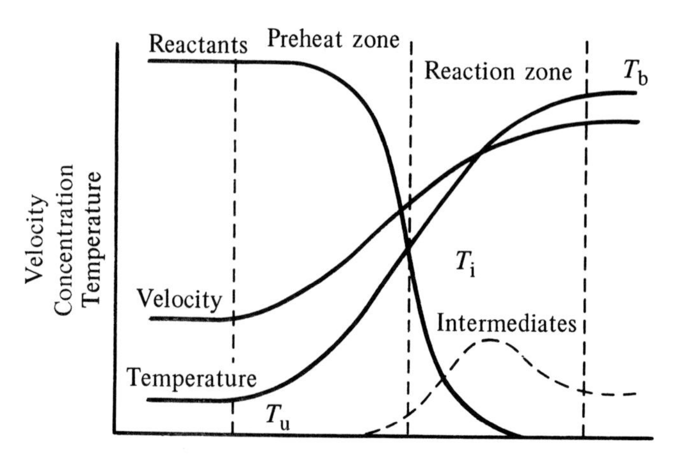

Figure 4:17 Temperature, concentration and velocity profiles through an idealized one-dimensional flame

Figure 4:17 shows velocity, temperature and concentration profiles through such a flame.

An element of the flowing gases can receive heat either from chemical reaction occurring in it or by conduction from the hotter gases downstream.

Two distinct regions of the temperature profile can be distinguished, separated by the point of inflexion at T_i. At low temperatures the heat flow into the element due to conduction is greater than the corresponding heat loss because the gradient is greater on the high temperature side. Beyond T_i where the profile is concave the opposite is true and more heat is lost than is gained but since heat is liberated by chemical reaction in this region the temperature continues to rise but at a progressively slower rate, eventually reaching a constant value T_b, the burned-gas temperature, when the reaction has ceased. Two regions of the temperature profile may be distinguished; T_u to T_i which is termed the *preheat zone* and T_i to T_b which is termed the *reaction zone*.

Analogous considerations apply to changes in molecular concentrations, the loss of reactants by diffusion being symmetrically opposite to the gain in temperature of the element by conduction as shown in Figure 4:17.

Since thermal expansion of the reactants is restricted to the direction normal to the flame front *conservation of mass* requires that:

$$\rho_u u_u = \rho_i u_i = \rho_b u_b \tag{4.44}$$

where subscripts u, i, and b relate to conditions in the unburned gas, at the 'ignition point' and in the burned gas respectively.

Assuming incompressible flow and considering *conservation of momentum* in the gas stream, the momentum change per unit length of the x-coordinate is given by ρu $(\mathrm{d}u/\mathrm{d}x)$ which equals the pressure gradient $-\mathrm{d}P/\mathrm{d}x$. Thus:

$$-\,\mathrm{d}P = \rho u\,\mathrm{d}u$$

but

$$\rho u = \rho_u u_u$$

Integrating over the flame thickness:

$$P_u - P_b = \rho_u u_u (u_b - u_u) \tag{4.45}$$

The velocity u_u which is the velocity of the unburned gas with respect to the flame front is equal to the burning velocity S_u and since:

$$\rho_u S_u = \rho_u u_u = \rho_b u_b$$

equation 4.45 may be written:

$$P_u - P_b = \rho_u S_u^2 [(\rho_u/\rho_b) - 1] \tag{4.46}$$

The pressure difference $P_u - P_b$ across the flame is very small (for a methane–air flame it is about 1 Pa or 0.01 mmHg) and has no effect on the processes occurring in the flame. It may, however, in real flames give rise to distortions in the gas flow pattern which have practical significance (Section 4.3.3).

Some authors [21, 22] have used equation 4.46 to calculate burning velocity from measurements of flame pressure (see Section 4.3.3).

At any point in the flame *conservation of energy* requires that the sum of the heat flow in due to conduction $\lambda\mathrm{d}T/\mathrm{d}x$, the heat flow in due to gas flow $-\,c_pT\rho u$ (negative because the gas is moving in the direction of increasing temperature) and the heat produced by chemical reaction H_rJ must be zero, i.e.:

$$\frac{\mathrm{d}}{\mathrm{d}x}\left(\lambda\frac{\mathrm{d}T}{\mathrm{d}x}\right) - \frac{\mathrm{d}}{\mathrm{d}x}(c_pT\rho u) + H_rJ = 0 \tag{4.47}$$

Combining equations 4.44 and 4.47:

$$\frac{\mathrm{d}}{\mathrm{d}x}\left(\lambda\frac{\mathrm{d}T}{\mathrm{d}x}\right) - \frac{\mathrm{d}}{\mathrm{d}x}(c_pT\rho_uS_u) + H_rJ = 0 \tag{4.48}$$

where c_p is the specific heat at constant pressure
H_r is the heat of reaction
J is the rate of reaction
and λ is the thermal conductivity.

Similarly *conservation of the number of atoms* yields:

$$\frac{d}{dx}\left(D_i \frac{dn_i}{dx}\right) - \frac{d}{dx}(n_i u) - J_i = 0 \tag{4.49}$$

where D_i is the diffusion coefficient of i
n_i is its molar concentration (number density of molecules)
J_i is its rate of reaction.

It would clearly be desirable to be able to calculate the burning velocity of a particular mixture of reactants from a knowledge of their physical and chemical properties. As shown above the problem, in principle, may be stated in terms of a series of differential equations. Unfortunately, the various quantities are closely interdependent, the kinetic and transport data are usually incomplete and insufficient is generally known about the reaction course so that solution is impossible.

Many attempts have been made over the years to solve the equations originally by considerable simplification to reduce the problem to the solution of a single differential equation according to which transport process was considered to be important.

The ignition of the reactants may be caused either by heat conduction or by the diffusion of active species (atoms or radicals). Thermal theories, for example those of Damköhler, Bartholomé, Lewis and von Elbe are those which assume that conduction is responsible and essentially involve a solution of equation 4.48 while diffusion theories (e.g. Tanford and Pease, Manson, etc.) require a solution of equation 4.49. Of course both heat conduction and diffusion occur, and a number of composite theories involving limited solutions of the general equations have been developed (e.g. Lewis and von Elbe, Frank-Kamenetskii, Zeldovich).

Further treatment of flame-propagation theories is outside the scope of this book. However, it is now possible to calculate accurate burning velocities for hydrogen–oxygen and hydrogen–air flames [2] from fundamental reaction kinetic data using modern computational techniques (see Figure 4:22) and similar calculations for hydrocarbon–air flames are a not-too-distant prospect. In the meantime it is necessary to measure burning velocity as an experimental quantity.

The simple one-dimensional flame described earlier provides a convenient theoretical model; however, it is not achieved in practice. Real flames are bounded and finite and because of lateral gas expansion the velocity variation through the flame front is not described by ρu = constant (see Figure 4:18). Fristrom and Westenberg [1] have invoked the concept of quasi-one-dimensionality to describe flames exhibiting lateral expansion. A quantity, for example, the gas velocity through the flame, which varies to some extent across a streamtube is replaced by a quantity averaged across the streamtube.

Thus the modified mass-conservation equation becomes:

$$\rho u a = \text{constant} \tag{4.50}$$

or

$$\rho u A = \rho_u u_u = \rho_u S_u$$

where $A = a/a_u$ the streamtube area ratio at any point relative to the unburned gas boundary. Similarly the other flame equations may be written in this form making them applicable to real flames.

These streamtube area changes have an important bearing on the measurement of

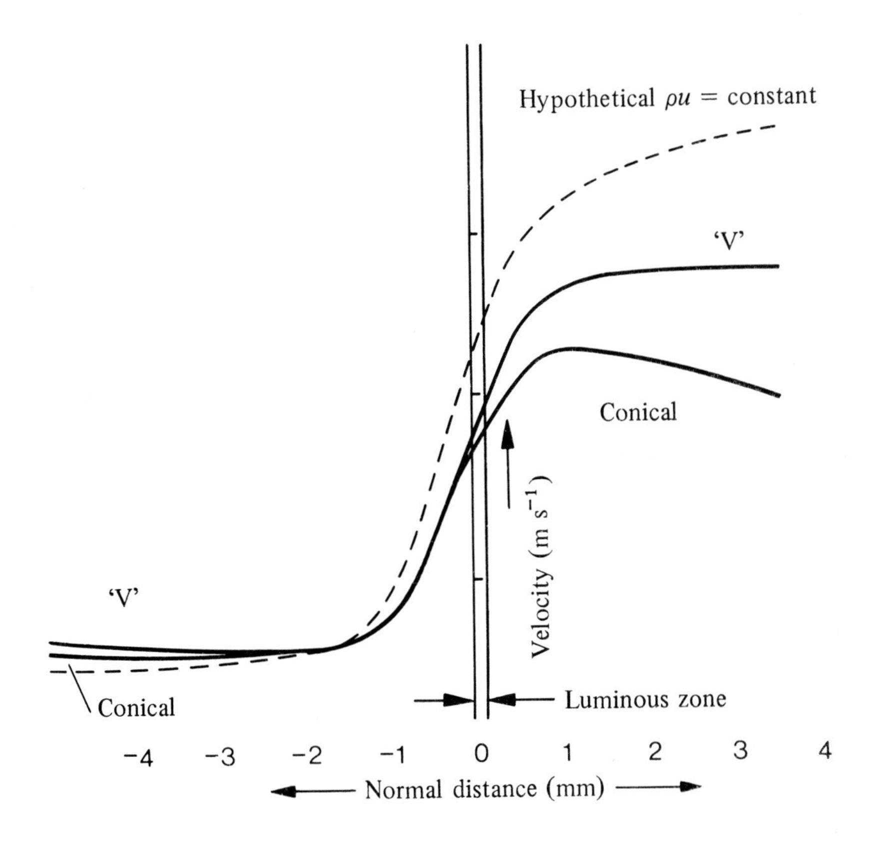

Figure 4:18 Velocity profiles in flames of various geometries [1]
(From *Flame Structure* by R. M. Fristrom and A. A. Westenberg. Copyright © 1965 McGraw-Hill Inc. Used with permission of McGraw-Hill Book Company)

burning velocity; because of the varying approach velocity due to streamtube area changes the question arises of defining the 'initial point' at which the gas velocity becomes *the* burning velocity. Fristrom and Westenberg have indicated that this should be equal to the gas velocity normal to the flame front at the point of first perceptible temperature rise.

4.3.3 The measurement of burning velocity

Introduction
As indicated earlier burning velocity as defined is an invariant characteristic property of a given combustible mixture. It is independent of flame geometry, burner size and flow rate. It depends only on the initial gas composition, temperature and pressure.

However, although its theoretical definition is straightforward, its practical measurement undoubtedly is not. There is as yet no generally accepted standardized method of measurement and the literature reveals considerable discrepancies between the results obtained by the various methods.

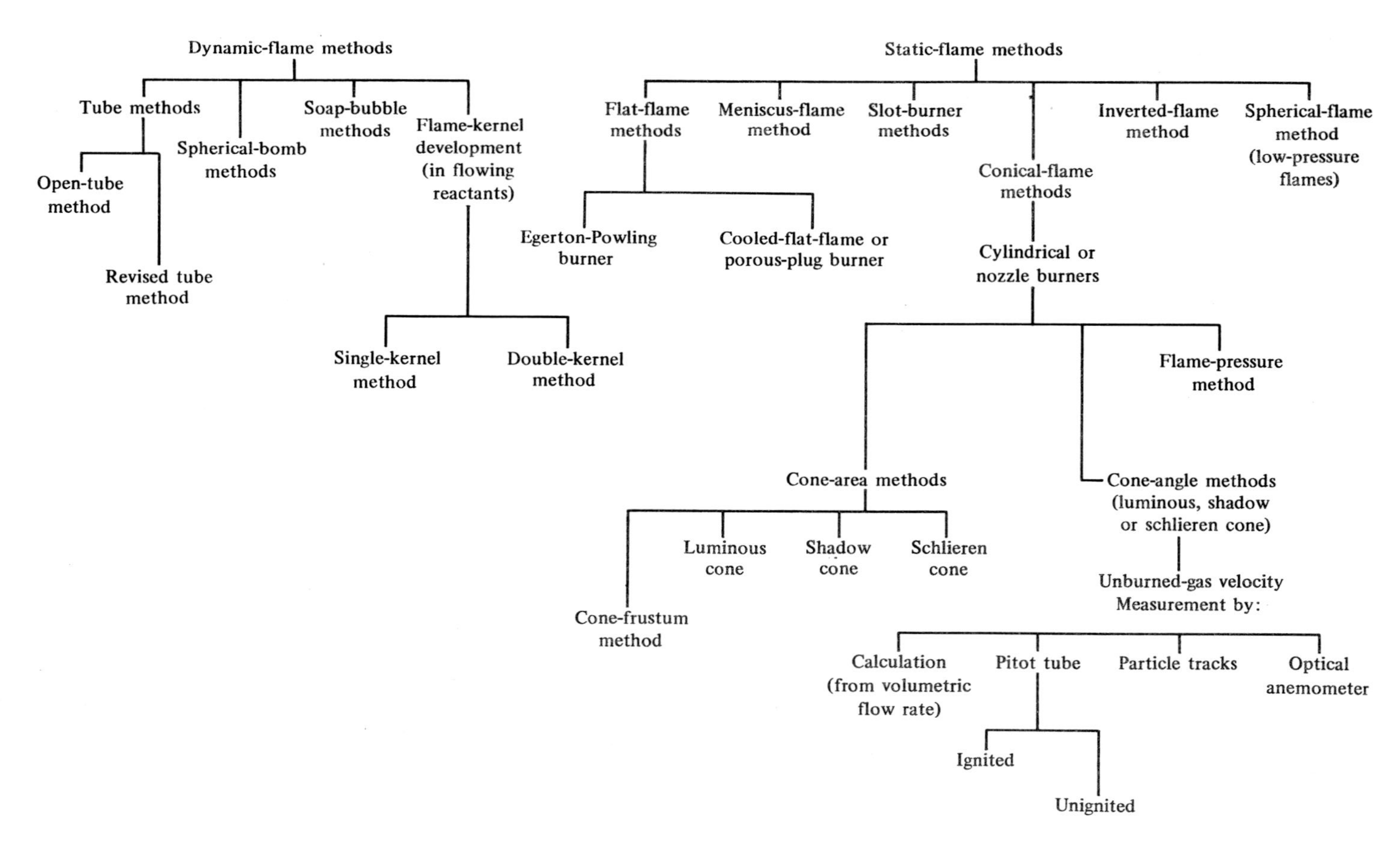

Figure 4:19 Summary of methods of measuring burning velocity

Critical reviews of measurement of burning velocity have recently been published by Andrews and Bradley [23] and Pritchard [24] and these are partially successful in identifying the inaccuracies in and suggesting corrections to some of the published data.

Techniques for measuring burning velocity may be broadly classified under two headings:

1 Static-flame or burner methods in which combustible mixture flows into a flame fixed in space relative to the observer at a velocity equal to the burning velocity.
2 Dynamic-flame or free-flame propagation methods in which the velocity of a flame propagating through an initially quiescent mixture is measured.

Within these two broad divisions, a wide variety of techniques have been employed and an attempt to summarize the most important methods is presented in Figure 4:19.

Static-flame methods
Ideally the flame should consist of a stationary flat flame in free space but such a flame cannot be achieved in practice and instead some stabilizing technique must be applied which interacts with the flame generally as indicated in Section 4.4.

Conical flames. By far the most common flame shape employed has been conical. Early workers used cylindrical tubes which by virtue of their parabolic gas-velocity distribution produce poor approximations to conical flames. More recent measurements have generally involved the use of contraction (Mache-Hebra) nozzles which apart from the boundary regions produce a uniform gas-velocity profile.

Conical flames result which are straight-sided except for the tip, which is rounded due to convergent flame propagation, and the base which is disturbed by quenching near the burner rim.

Most of the early attempts to deduce burning velocity from measurements on conical flames were based on measurements of the flame-cone area, A_f, using the equation:

$$S_u = V_u/A_f \tag{4.51}$$

where V_u is the volumetric flow rate of the unburned mixture. The flame-cone area may be obtained either by assuming it to be a true geometric cone and measuring its height or, more accurately, by graphical integration from photographs.

Cone-area methods, however, suffer from two major disadvantages. The first is that the method produces an average value over the whole cone, including the disturbed base and tip. This may of course be reduced by using large burners but the Reynolds number usually limits this solution.

The second disadvantage is more serious and hinges on the question: at what plane in the flame should A_f be measured? If the flame conformed to the unidimensional model of Section 4.3.2 this question would raise no problem, A_f being constant regardless of the position, and therefore the temperature, in the flame front at which it is measured.

This is, of course, not so and if equation 4.51 is to be used the choice of reference plane becomes a matter of considerable importance. The definition of burning velocity relates to the unburned gas and clearly the reference surface should be at the point of first perceptible temperature rise and of area A_u. Unfortunately this surface, 'the cold-gas cone' cannot be optically visualized.

Three optical methods are available to locate surfaces at different temperatures in the flame front and all three have been used with equation 4.51 to deduce burning velocity.

Early measurements were based on the zone of most intense illumination, the inner and outer boundaries of this *luminous zone* being thought to represent the commencement and completion of combustion. Measurements of A_f were thus based on the inner edge of the luminous zone, A_l. It is now well established of course that the whole of the visible cone lies at a temperature quite close to the maximum flame temperature. Since $A_l > A_u$ this will lead to an underestimate of burning velocity.

The other optically visualized zones, the *shadow image* and the *schlieren image* both make use of an external light source, the light from which is refracted in its passage through the flame to form an image on a screen. The shadow image is formed by light passing through the preheat zone being deflected towards the cooler gas so that the shadow on the screen shows a dark zone adjacent to a bright region. Weinberg [25] has shown that the shadow image corresponds to a flame region at a temperature of $2.19T_u$. Burning velocities based on the shadow image will also be underestimates but less so than those based on the luminous zone.

Schlieren methods employ a knife-edge such that deviations of light due to changes of refractive index in the flame pass the knife-edge and the screen or camera records those regions of the flame which deflect light in the 'right' direction. Weinberg [8] has shown that the schlieren image corresponds to a flame region at a temperature around $1.5T_u$. Thus the schlieren image yields burning velocities closer to the true value than the other optically visualized flame surfaces but there still remains a discrepancy.

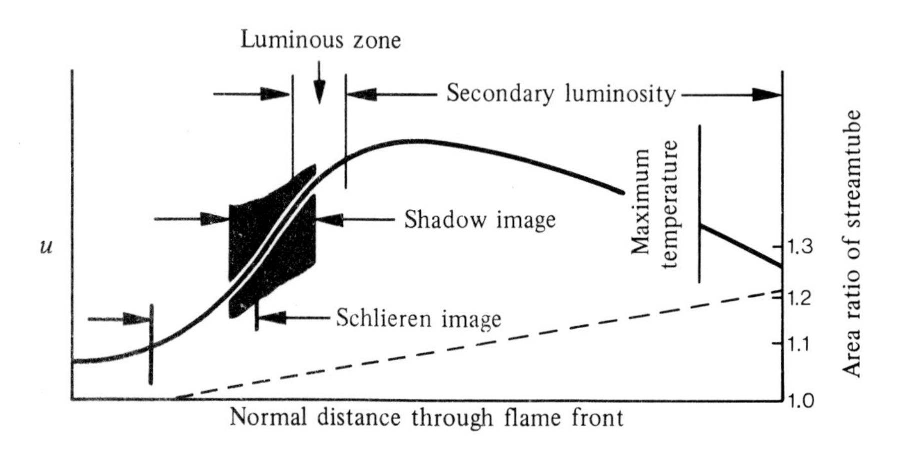

Figure 4:20 Positions of the optically visualized flame surfaces relative to the velocity profile [1]
(From *Flame Structure* by R. M. Fristrom and A. A. Westenberg. Copyright © 1965 McGraw-Hill Inc. Used with permission of McGraw-Hill Book Company)

Figure 4:20 shows the positions of the optically visualized flame surfaces relative to the velocity profile through the flame.

Andrews and Bradley [23] have attempted to quantify the errors in flame-area methods for stoichiometric methane–air flames based on experimentally determined temperature profiles.

Their analysis shows that for these flames, using the luminous surface underestimates burning velocity by about 20 per cent, the shadow surface by about 15 per cent and the schlieren surface by about 10 per cent.

A modification of the flame-area method has been employed [26] in which part of the cone approximating to a frustum and excluding the disturbed areas near the base and tip is selected. The method is still inaccurate, however, because of the problem, referred to above, of identifying a reference surface.

These difficulties disappear (although others appear) if instead of measuring the flame area, the cone angle is measured. Burning velocity is then given by:

$$S_u = u_g \sin \theta \tag{4.52}$$

where u_g is the axial unburned-gas velocity and θ is the angle between the burner axis and the cone flank.

The major difficulty which arises in the use of equation 4.52 is the accurate determination of u_g. Simply obtaining u_g by dividing the flow rate by the burner cross-sectional area leads to considerable inaccuracies due to the effect of the boundary layer and the divergence of the mixture streamlines caused by the flame back pressure.

Pitot-tube traverses of the unignited stream eliminate the first objection but not the second. It is now generally recognized that u_g must be measured directly in the flame. A variety of techniques have been employed. Edmondson and Heap [27] used separate total and static head Pitot tubes of 2 mm diameter inserted through the flank of the flame. Burning velocities obtained were lower than by other recent methods and this may be attributable to the inevitable flame disturbance. Pritchard [24] obtained u_g by measuring the deflection of previously calibrated quartz fibres of diameter 0.005 to 0.02 mm. The same criticisms regarding flame disturbance probably apply.

Most direct measurements of u_g involve the use of the particle-track technique. In this method, which was first applied to flames by Anderson and Fein [28], small solid particles (2–50 μm) are introduced into the gas stream, illuminated intermittently from a bright light source and their tracks photographed at right angles to the plane of illumination. The method provides simultaneous flow visualization and velocity measurement; the particle velocity being obtained by dividing the track length by the time interval between interruptions.

If the flow lines are axial, the approach velocity will be constant and u_g may be measured anywhere in the unburned gas, but if flow divergence occurs a velocity profile must be plotted and the minimum used in the determination of burning velocity since upstream of this point the velocity will vary due to purely geometric influences. The particles must be very small to follow the gas flow accurately and therefore require the brightest possible light sources for their illumination. Repetitive photoflash [1] and laser [24] illumination have been used as well as less bright mercury-vapour lamps. Optical anemometers based on measurements of Doppler frequency shift in scattered laser light offer a recent attractive alternative to the particle-track technique since the method can use extremely small scattering particles (c. 0.5 μm)

[29] and eliminates the tedium and uncertainty which are major drawbacks of the particle-track method.

Measurements of burning velocity carried out on conical flames raise the question of the effect, or otherwise, of flame curvature. Undoubtedly a flame which is concave towards the unburned gas possesses a higher burning velocity than the equivalent plane flame and a number of workers have suggested that conical-flame determinations of burning velocity must be too high because of this effect. Singer [30] eliminated flame curvature by using a rectangular *slot burner* producing tent-shaped flames and claimed that this gave lower burning velocities than cylindrical tube burners.

More recent studies using particle-track measurements [1, 24] indicate that if the radius of curvature is large compared with the flame thickness, as it is in the flank of a typical conical flame used for burning-velocity determination, the effect is very small, usually less than the limits of experimental error. Curvature effects on burning velocity still remain a matter for controversy.

Flat-flame methods. Flat flames approximate most closely to the ideal definition of burning velocity and eliminate any doubts associated with the interpretation of curved flame surfaces.

Two types of flat-flame burner may be distinguished, the *rectified flat-flame burner* or Egerton–Powling burner and the *cooled-flat-flame* or *porous-plug burner*.

In the former case a uniform gas-velocity profile is obtained using a crimped ribbon screen and it is possible at low velocities to obtain an essentially flat stationary flame some distance above the screen [31]. The stabilization appears to take place due to divergence of the streamlines above the burner matrix.

Such burners are incapable of stabilizing flat flames much in excess of 0.08 m s^{-1} and their use in burning-velocity determination is restricted to mixtures near the limits of flammability where they have particular value since other methods are very difficult to apply. Burning velocity is normally obtained simply by dividing the flow rate by the burner area but particle-track studies [32] have shown considerable streamline divergence in the unburned gas suggesting that this simple procedure would lead to error.

In the second method a water-cooled sintered metal disc is used [33] and in contrast to the rectified flat-flame method no attempt is made to balance the stream velocity and burning velocity. Instead the stream velocity is set to some value less than the burning velocity and on ignition the flame moves towards the disc. Heat is progressively transferred to the disc as the flame approaches it lowering the flame temperature and burning velocity.

Consequently the flame adopts an equilibrium position close to the burner disc where it loses just sufficient heat to reduce the burning velocity to the stream velocity. The heat extracted by the cooling water is accurately measured and it is possible to plot curves of mixture velocity versus heat extracted for a given mixture composition. A large amount of heat must be removed when the flow velocity is low, whereas when the flow velocity is increased towards the adiabatic burning velocity of the uncooled flame the heat transferred to the disc decreases.

The adiabatic burning velocity is obtained by extrapolation to zero heat extraction. Botha and Spalding [33], who originated the method, found that the points lay on a straight line and the graph could be extrapolated linearly to obtain the burning velocity. As a result of more recent investigations using a variety of burner sizes and secondary atmospheres, Pritchard, Edmondson and Heap [34, 35] concluded that edge effects lead to erroneous heat extraction rates and that the characteristic curve of heat extraction rate versus mixture velocity is not linear. More accurate results are obtained if reciprocal flame temperatures (calculated from heat extraction rates) are plotted versus the logarithm of the mixture velocity. This relationship appears to be linear for lean and near-stoichiometric flames. The advantages of the cooled-flat-flame method are its simplicity, precision and suitability for an extremely wide range of mixture strengths.

Other static-flame methods. *Inverted flames*, i.e. those stabilized in the wake of obstructions placed in the gas stream have had limited application in burning-velocity measurement. Fristrom [1] has used particle-track measurements in such flames to demonstrate the invariance of burning velocity in flat and moderately curved flame surfaces. Pritchard [24] has used the protected stabilization zone of such flames to measure burning velocity in very lean flames which would blow off normal burners.

Günther and Janisch [36] have recently used *meniscus or button-shaped flames* coupled with particle-track measurements to obtain accurate wide-range burning-velocity data, eliminating doubts about flame-curvature effects.

Static *spherical flames* have been used in the study of low-pressure flames [1] but are of doubtful applicability to atmospheric-pressure flames in which gravitational effects disturb symmetry.

The *flame-pressure method* has been used to deduce burning velocity from the conservation of momentum equation (equation 4.46) and flame-pressure measurements.

However, Edmondson and Heap [37] have pointed out that determinations based on this approach have been in error due to neglect of streamtube expansion in the flame.

For approximate repetitive measurements of burning velocity use has been made in the gas industry of the Union Flame Speed Meter originally developed by Dommer. This is essentially a mechanized version of the cone-area method based on measuring flame height and although convenient to use suffers from the inaccuracies associated with this method.

Dynamic-flame methods

Tube method. In this, the earliest dynamic method, mixture is ignited at the open end of a tube and the course of flame propagation towards the closed end is photographed. The measured *flame speed* (S_f) is not equal to the burning velocity S_u which is given by:

$$S_u = (A_t/A_f)\, S_f \tag{4.53}$$

where A_t and A_f are the tube and flame areas.

In practice the flame aerodynamics vary with distance along the tube and the assumptions of constant S_f and A_f become poor approximations.

Also pressure oscillations which periodically accelerate and retard the flame occur and these may be minimized by providing orifices at both ends of the tube (the revised tube method). Burned gas is discharged and the unburned mixture acquires a velocity (S_g). Burning velocity is given by:

$$S_u = (A_t/A_f)\,(S_f - S_g) \tag{4.54}$$

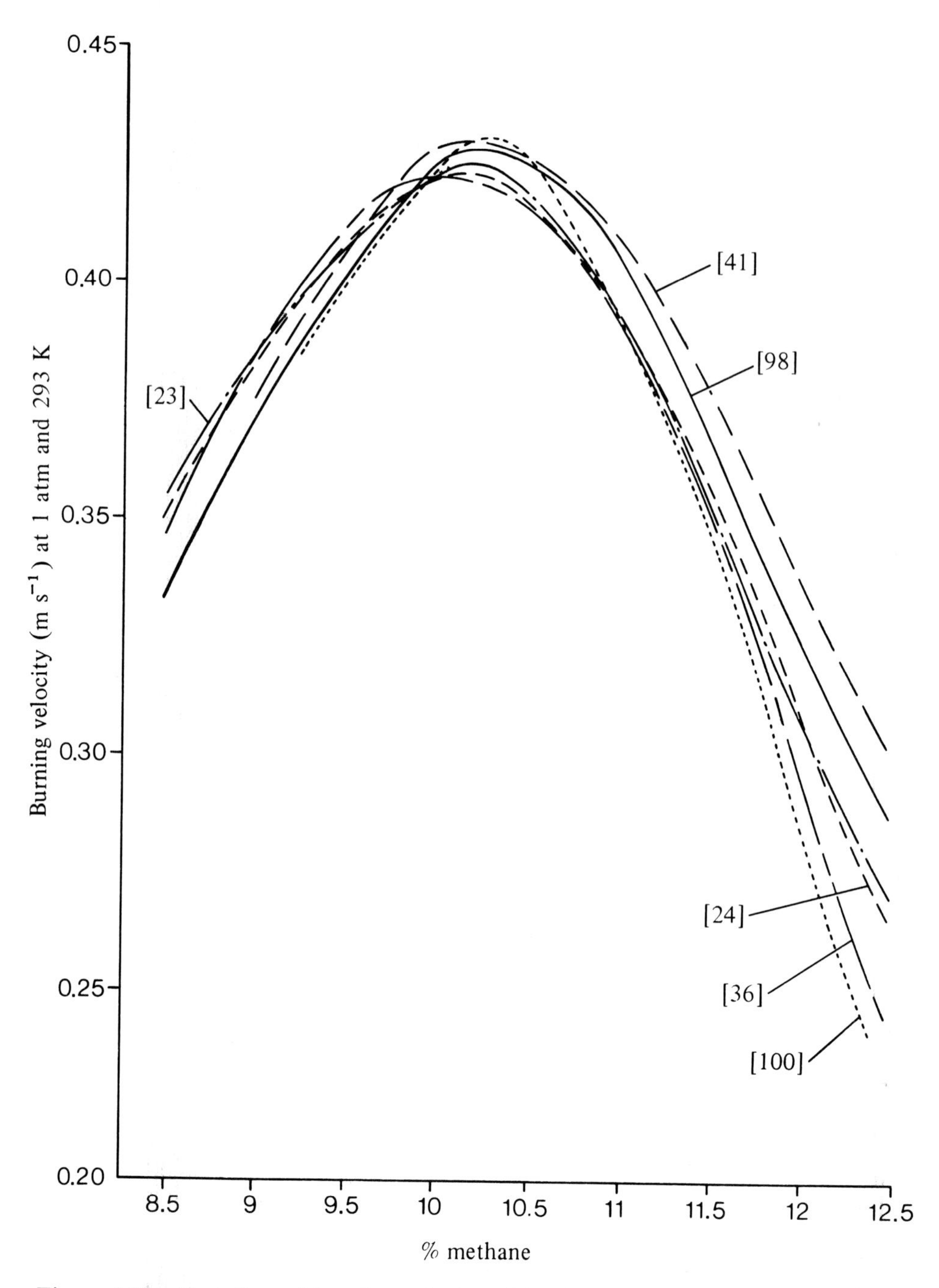

Figure 4:21 Variation of burning velocity with mixture composition for methane–air flames according to various workers

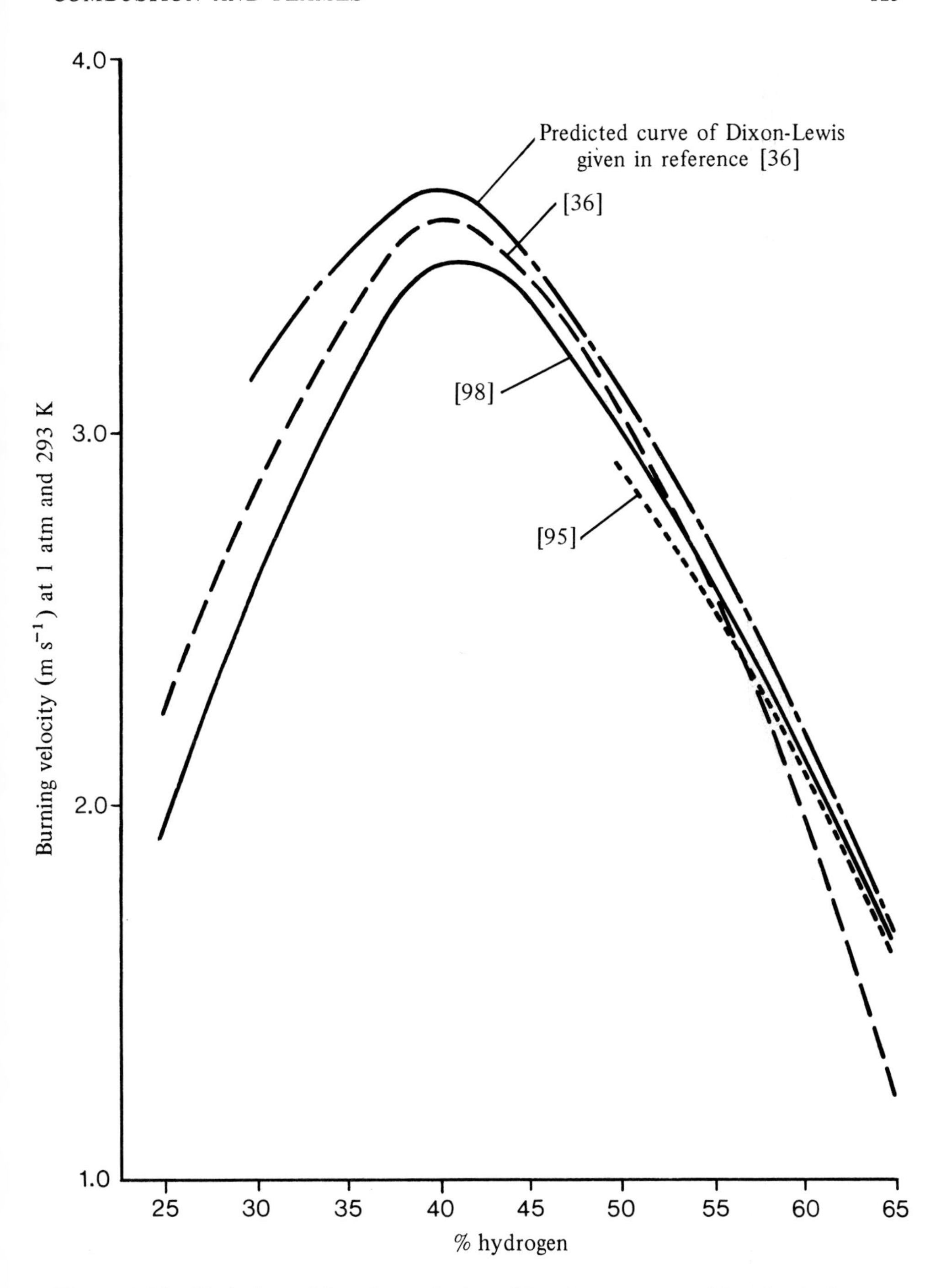

Figure 4:22 Variation of burning velocity with mixture composition for hydrogen–air flames according to various workers

The measurement of A_f is difficult and quenching near the walls leads to inaccuracies. Andrews and Bradley [23] have proposed a modification to equation 4.54 which allows for wall quenching.

Spherical-bomb methods. The rate of propagation of a centrally ignited near-spherical flame in a rigid or (in the case of the *soap-bubble method*) an elastic containing vessel is measured. Flame-curvature effects are avoided by restricting measurements to the well-developed flame occurring in the later stages of explosion. For an explosion in a rigid vessel both the pressure and temperature of the gas increase and the burning velocities obtained relate to the instantaneous values of these parameters. Thus information on the pressure and temperature dependence of burning velocity may be obtained from a single explosion. The major difficulty arises in estimating the mean gas density and this has led to inaccuracies. Direct measurement of unburned gas velocity is possible and Bradley and Hundy [38] have used a hot-wire anemometer for this purpose.

Flame-kernel development. This involves measurements of the rate of flame-kernel development in a flow of combustible mixture. A modification of the method due to Raezer and Olsen [39] avoids the problem of evaluating S_g by using two simultaneously ignited kernels approaching each other. Since the gas velocity on the axis of centres must be zero at the midpoint the flame speed tends to the burning velocity as the kernels approach.

Dynamic-flame methods possess a number of advantages over static-flame methods. They normally require mixture samples of modest size, comparatively large flames may be used and collection of data on temperature and pressure effects is facilitated by some methods. On the other hand, burning velocity is not generally a directly measured quantity and in common with most studies of transient events the measurement techniques are difficult and may call for elaborate experimental apparatus.

Gas	$S_{u,max}$ ($m\ s^{-1}$)	Method	Reference
Methane	0.43	Schlieren cone angle, particle track	[41]
	0.43	Conical flame, particle track	[24]
	0.45	Bomb, hot-wire anemometer	[23]
Ethane	0.487	Schlieren cone angle, particle track	[41]
Propane	0.472	Schlieren cone angle, particle track	[41]
Butane	0.453	Schlieren cone angle, particle track	[41]
Ethylene	0.79	Twin kernel	[39]
Hydrogen	3.57	Particle track, meniscus flame	[36]
	3.45	Particle track, conical flame	[36]
Carbon monoxide	0.175	Particle track, meniscus flame	[36]
	0.19	Particle track, conical flame	[36]
First-family manufactured gases	c. 1.0	—	

Figure 4:23 Maximum burning velocities of common gases with air

Effects of temperature, pressure and vitiation on burning velocity

Mixture temperature. Burning velocity shows a marked positive dependence on initial temperature and a number of investigations of this effect have been carried out.

For a rise in temperature from 160 K to 560 K Dugger and Heimel [40] found that the burning velocity of methane–air flames was given empirically by:

$$S_u = 10 + 0.000342(T_u)^2 \qquad (4.55)$$

A similar quadratic relationship has been proposed by Lindow [41]:

$$S_{u1} = S_{uo} + 0.45 \times 10^{-3}(T_1{}^2 - T_0{}^2) \text{ cm s}^{-1} \qquad (4.56)$$

where S_{u1} and S_{uo} relate to temperatures T_1 and T_0.

Mixture pressure. For hydrogen–air flames there is now fairly definite evidence of a general tendency for a negative dependence of burning velocity on pressure, values of S_u rising with reducing pressure. The effect is small, however, and over the ambient pressure variations experienced in industrial gas utilization will be negligible.

Vitiation. The reduced oxygen content of vitiated combustion air leads to lowered flame temperatures since the amount of inert gas which must be heated to flame temperature per mole of combustible gas burned is greater for vitiated than non-vitiated air. Reduced flame temperatures in turn lead to narrower flammability limits [42] and lowered burning velocities.

Figure 4:24 shows the effect of adding various diluents on the maximum burning velocity of methane–air flames.

Burning velocities of multicomponent mixtures

Unlike many other combustion properties, the burning velocity of a mixture of fuel gases cannot be derived in a simple way from the burning velocities of the constituent combustible gases. Clearly it would be desirable to be able to predict the burning velocity of a gas mixture, for example a substitute natural gas produced by a new process and a number of workers [43–47] have proposed predictive mixing rules. Ganju [48] has recently reviewed these and compared the predictions with measured values for hydrogen–methane–air and ethylene–methane–air mixtures. None of the methods gives accurate predictions over a wide range of mixture composition and there remains the need for a reliable method.

Burning-velocity prediction is an essential part of the study of fuel-gas interchangeability and one method which has been widely used in the British and American gas industries (see Section 4.11) is that due to Weaver [43].

In Weaver's mixing rule a 'flame-speed factor' is derived from tabulated flame-speed coefficients and expresses the burning velocity of the mixture approximately as a percentage of the maximum burning velocity of hydrogen in air which is assigned the value of 100.

Weaver's flame-speed factor S is given by:

$$S = (\varphi_A F_A + \varphi_B F_B + \varphi_C F_C + \ldots)/(A_0 + 5\varphi_i - 18.8\varphi_O + 1) \qquad (4.57)$$

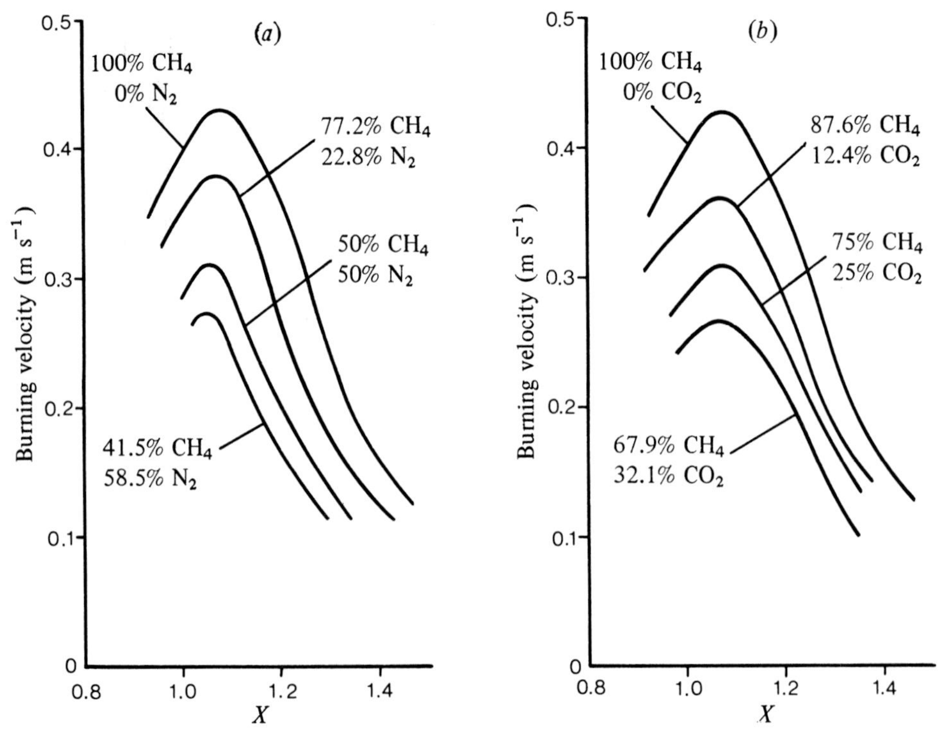

Figure 4:24 The effect of diluents on the maximum burning velocity of methane–air flames [42]

(*a*) N_2

(*b*) CO_2

X = ratio of actual methane content of primary mixture to methane content of a stoichiometric mixture

where φ_A, φ_B, φ_C, etc. are the volumetric fractions of the combustible constituents
F_A, F_B, F_C are the corresponding flame-speed coefficients
A_0 is the stoichiometric air requirement
φ_i is the fraction by volume of inerts
φ_O is the fraction by volume of oxygen in the gas

Values of F are tabulated in Figure 4:25.

Gas	F
Carbon monoxide	61
Hydrogen	339
Methane	148
Ethane	301
Propane	398
Butane	513
Ethylene	454

Figure 4:25 Flame-speed coefficient for common gases [43]

Burning velocities of the pure gases used in the derivation of the coefficients are not in good agreement with modern measurements and a number of approximations are contained in the derivation of S. Weaver's flame-speed factor can only be regarded as an approximate indication of the burning velocity of a mixture but will continue to be used until a better method is found.

Experimental measurements are available in the literature for a limited number of binary and ternary mixtures of hydrogen, methane, carbon monoxide, ethylene and inert gases [48, 49].

4.4 LAMINAR-FLAME STABILIZATION

As indicated earlier, attempting to establish a flame simply by flowing the gases into it at an equal and opposite velocity to the burning velocity does not produce a stable flame. Instead it is necessary to attach the flame to a *flame-holder* or *burner* which locally interacts with the flow and combustion processes to provide a stabilizing zone. The burner may also, of course, fulfil other functions; it may be involved in mixing the fuel and oxidant and it will establish a particular flow pattern.

The effect of the burner rim on the gas-mixture velocity is to reduce it to zero at the walls due to viscous drag whilst its effect on the burning velocity is to reduce this due to *quenching*, i.e. the removal of heat and possibly active species in its vicinity. Thus, close to the rim, the flame's position is determined by the relative magnitudes of the local burning velocity and flow velocity. If the burning velocity is greater than the flow velocity, the flame will move closer to the rim until the burning velocity reduces to equal the local flow velocity. Conversely if the flow velocity exceeds the burning velocity the flame will move downstream until the two become equal. Thus within certain flow velocity limits the flame will be *stabilized* or held in place above the rim. If $u_g > S_u$ and the flame is stabilized at the rim, the flame conforms to the flow pattern such that the burning velocity is balanced by the normal component of gas flow (Figure 4:26(*b*)), i.e. $S_u = u_g \sin\theta$.

Outside the flow velocity limits referred to above various forms of instability are possible: *lightback* or *flashback*, *tilted flames*, *blowoff* and *lift*.

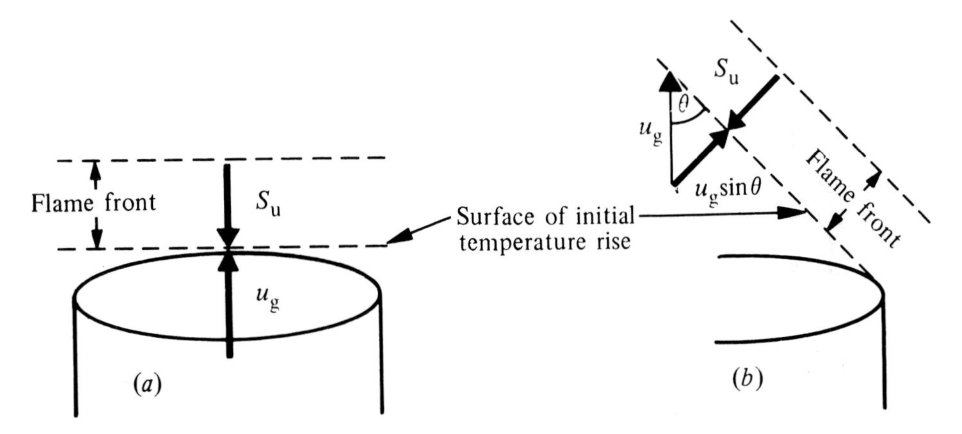

Figure 4:26 Burning velocity and flow velocity for (*a*) flat flame and (*b*) conical flame

Clearly it is desirable to be able to characterize and predict the stability of fuel-gas–oxidant mixtures when burning in a particular situation. A number of methods are available:

1 The use of combustion diagrams
2 The use of critical boundary velocity gradients for lightback (g_F) and blowoff (g_B)
3 The use of the critical flame stretch factor K_B for blowoff.

4.4.1 Combustion diagrams

A simple and very useful method of comparing the stability of different gases is to measure the fuel–oxidant ratio and mixture flow rate at which blowoff and lightback occur for a wide variety of conditions using a standard burner (normally a water-cooled cylindrical tube sufficiently long to ensure laminar flow). The stability limits are then plotted on a graph whose axes are air–gas ratio and mixture flow rate or a function of this (commonly thermal input or flame-port loading). Fuidge, Murch and Pleasance [50] pioneered the use of such diagrams in Britain and included a further limit, the yellow-tipping limit which indicates under-aerated flames which may give rise to incomplete combustion and possibly soot formation. Figure 4:27 indicates the general form of a combustion diagram for a fuel gas.

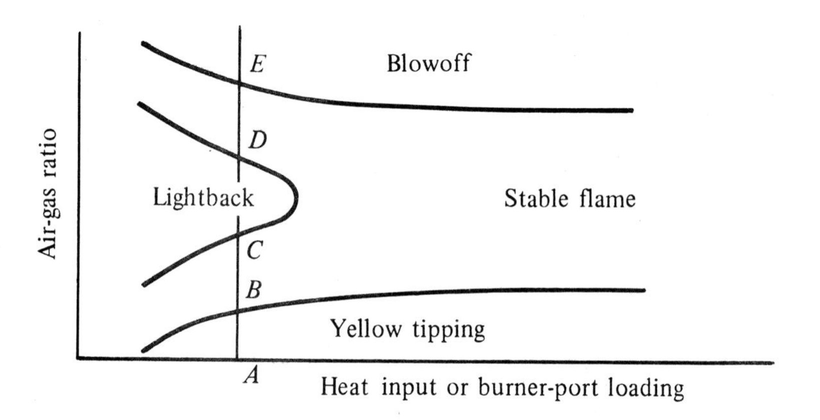

Figure 4:27 Typical combustion diagram

Consider the effect of changing the aeration at a constant value of heat input. At *A* there is no primary aeration and the flame is luminous. Increasing the aeration to *B* produces progressively less luminosity until at *B* the yellow tip just disappears. Between *B* and *C* the primary cone shortens as the burning velocity of the mixture increases until at *C* the burning velocity is sufficiently high to cause lightback. Between *C* and *D* it remains impossible to establish a flame at the burner port at this particular value of heat input until at *D* the burning velocity is sufficiently reduced to allow a flame to be re-established at the burner port. Further increase of aeration produces a progressively leaner flame until at *E* the burning velocity has become so low relative to the unburned-gas velocity that blowoff occurs. The size of the zone within the limits of unsatisfactory performance may be regarded as a measure of the flexibility of the gas.

Figure 4:28 shows combustion diagrams for a typical manufactured town gas, for methane, for hydrogen and for substitute natural gases and natural-gas test gases.

Combustion diagrams of this kind may also be used to indicate the stability of particular *burners* as opposed to gases (see Chapter 7) but such diagrams give little information on the mechanism of flame instability nor do they allow more generalized correlation of instability data. The methods discussed in Sections 4.4.2 and 4.4.3 are based on theories which attempt to describe the mechanism of instability and quantify the stability limits.

A further type of combustion diagram, the interchangeability-prediction diagram is considered in Section 4.11.

4.4.2 Theory of critical boundary velocity gradients

Figure 4:29 represents the profile of a laminar flame close to the burner rim. The unburned-gas velocity is zero at the burner wall and increases to a maximum at the centre. Only the conditions close to the wall are of interest in flame stabilization. This region of interest is normally very small (c. 1 mm) and, since the burner diameter is large in comparison, the gas-velocity profile may be assumed linear.

Imagine a flame established in the flow stream. At some distance from the edge of the stream the burning velocity will have its normal undisturbed value S_{u0} but at points closer to the edge the burning velocity will reduce. If the flame edge is very close to the rim (position 1) the burning velocity in any flow line is smaller than the gas velocity and the flame is driven downstream by the gas flow. As the distance from the rim increases, the loss of heat and active species is reduced and the burning velocity increases. Eventually an equilibrium position (2) is reached in which at some point the burning velocity equals the gas velocity. Should the flame be displaced further downstream (say position 3) the burning velocity becomes larger than the gas velocity at that point and the flame moves back to position 2.

Thus a stable flame exists such that at one point the gas velocity and burning velocity are equal and everywhere else the gas velocity is greater than the burning velocity.

If the profile of unburned-gas velocity is assumed linear then it is possible to write:

$$u_g = gy \tag{4.58}$$

where y is the distance from the stream boundary and g is a constant, the *boundary velocity gradient*, du_g/dy.

If the gas flow is reduced, the velocity profile flattens, i.e. g reduces and the equilibrium position (2) moves closer to the rim. If the gas flow is reduced sufficiently a condition is reached at which the gas velocity becomes less than the burning velocity and the flame propagates into the burner tube. This is *lightback* or *flashback* and the critical value of g at which this occurs is denoted by g_F.

If, on the other hand, the gas flow is progressively increased then the equilibrium position (2) moves downstream. This has two effects on the burning velocity: first increased separation reduces heat losses, increasing the burning velocity, and second, increased separation increases dilution of the mixture with the surrounding atmosphere which reduces the burning velocity. Thus a final equilibrium flame position

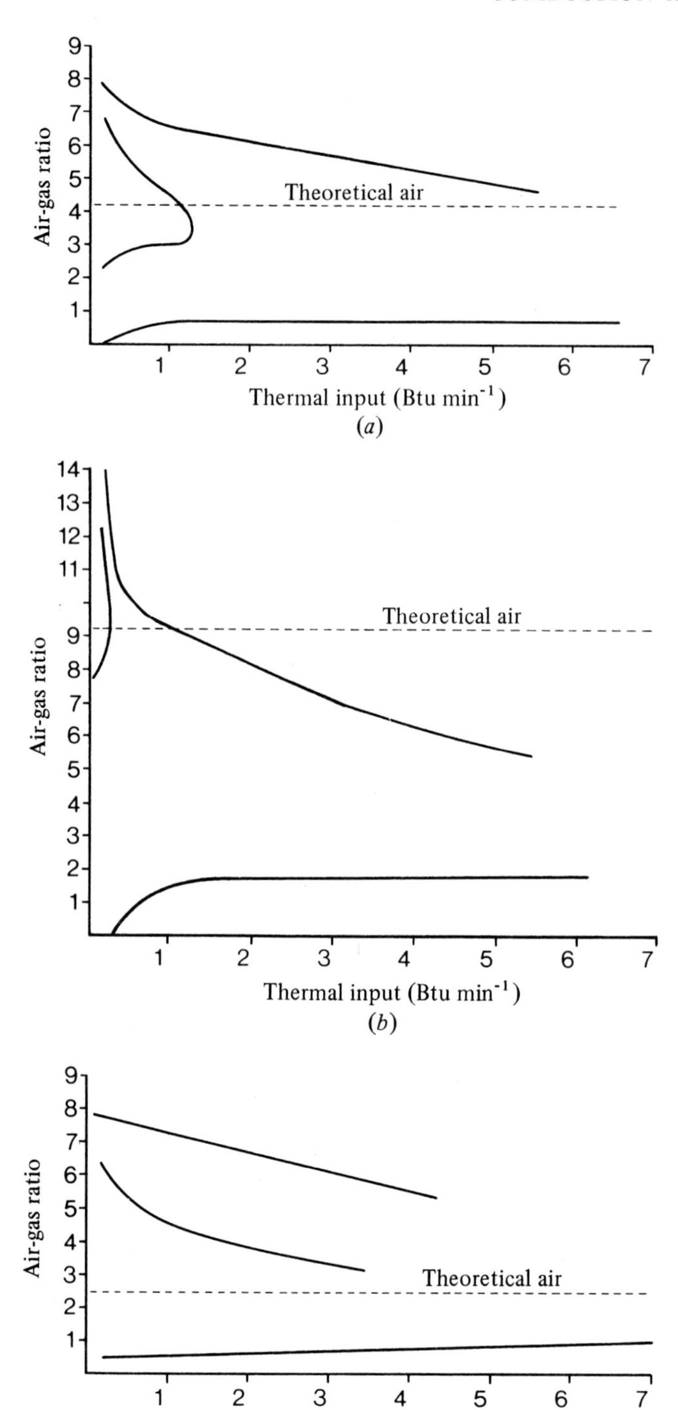
Air-gas ratio
Theoretical air
Thermal input (Btu min⁻¹)
(a)
Air-gas ratio
Theoretical air
Thermal input (Btu min⁻¹)
(b)
Air-gas ratio
Theoretical air
Thermal input (Btu min⁻¹)
(c)

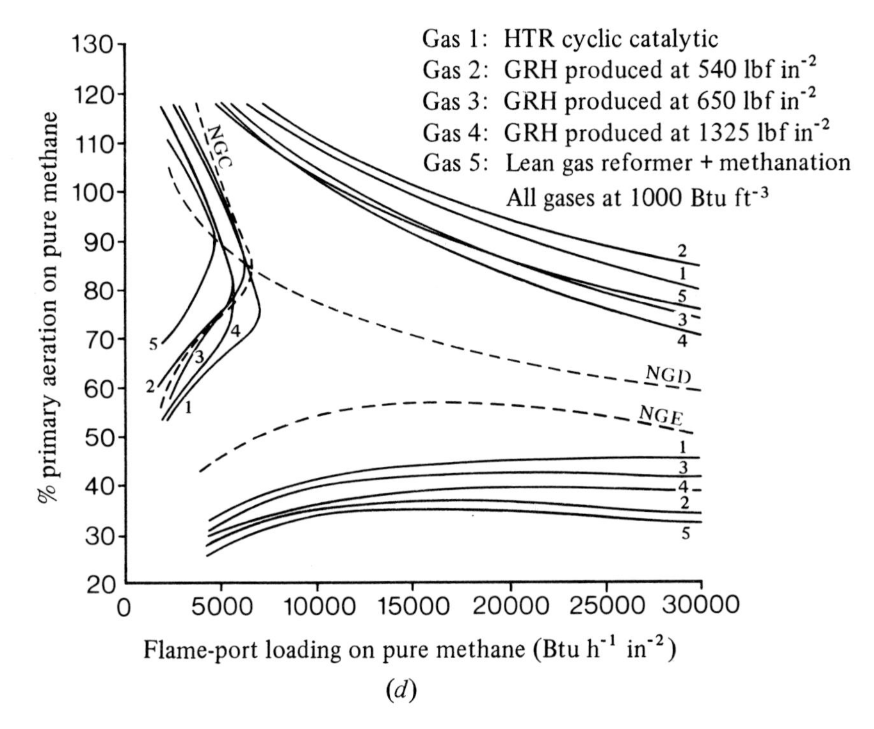

Figure 4:28 Combustion diagrams for various fuel gases [47, 50]
(*a*) First-family town gas
(*b*) Methane
(*c*) Hydrogen
(*d*) Substitute natural gases and natural-gas test gases

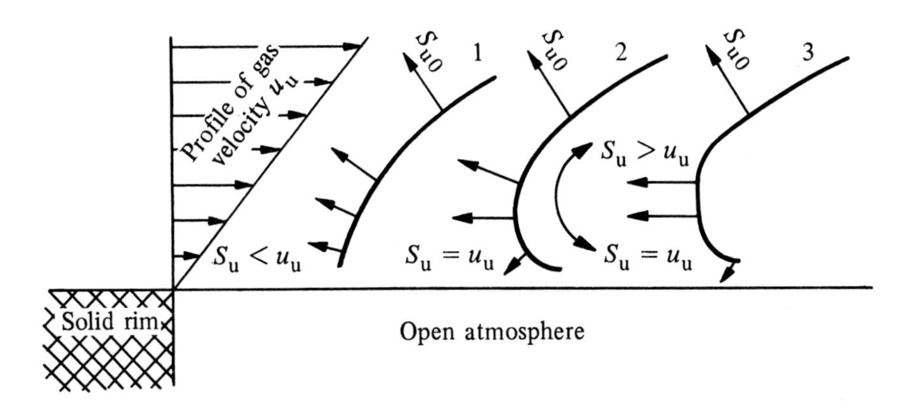

Figure 4:29 Scheme of stabilization of a laminar flame [51]

exists beyond which the second effect predominates and the flame is blown off. The critical boundary velocity gradient at which *blowoff* occurs is denoted by g_B.

Boundary velocity gradients, which have the units s^{-1}, may be calculated from:

$$g = 4\ V_u/\pi\ r^3 \text{ for long cylindrical tubes} \tag{4.59}$$

Where V_u = volumetric flow rate and r = tube radius.

For ports of other configurations use may be made of the more general equation [52]:

$$g = f\,\bar{u}_g(Re)/4r \tag{4.60}$$

where f = Fanning friction factor. Reference [52] lists expressions for f for a wide variety of burner-port shapes.

Critical boundary velocity gradient at lightback g_F

g_F has been shown to be a fundamental parameter independent of burner characteristics such as port diameter, shape and inclination. It is the quotient of two other fundamental parameters, the burning velocity (Section 4.3.3) and the quenching distance (Section 4.6).

Figure 4:30 shows the variation of g_F with mixture composition for natural gas.

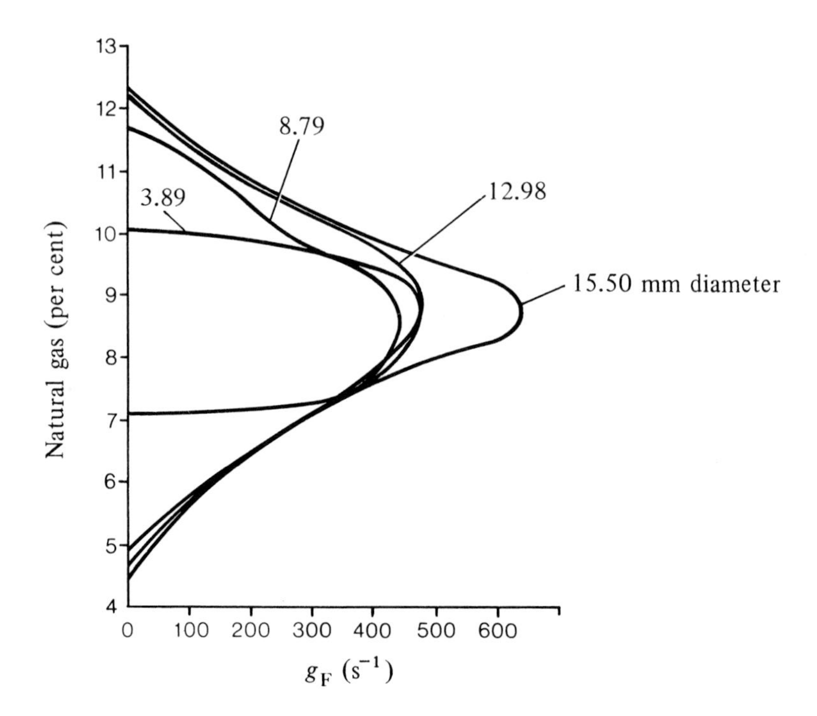

Figure 4:30 Variation of g_F with mixture composition for natural gas and air in cylindrical tubes of various diameters [51]

Clearly it would be an advantage to be able to predict g_F for multicomponent gas mixtures from the mixture composition and a number of predictive methods have been developed including those of Grumer, Harris and Rowe [54], Van Krevelen and Chermin [55], and Caffo and Padovani [56]. France [53] has recently tested these methods against measurements of g_F for the lightback test gas NGC and possible multicomponent alternatives and has found that the Van Krevelen and Chermin method yields the most accurate predictions (within about 6 per cent).

Critical boundary velocity gradient at blowoff g_B
Agreement between values of g_B obtained using a wide range of burners is good as shown by Figure 4:31.

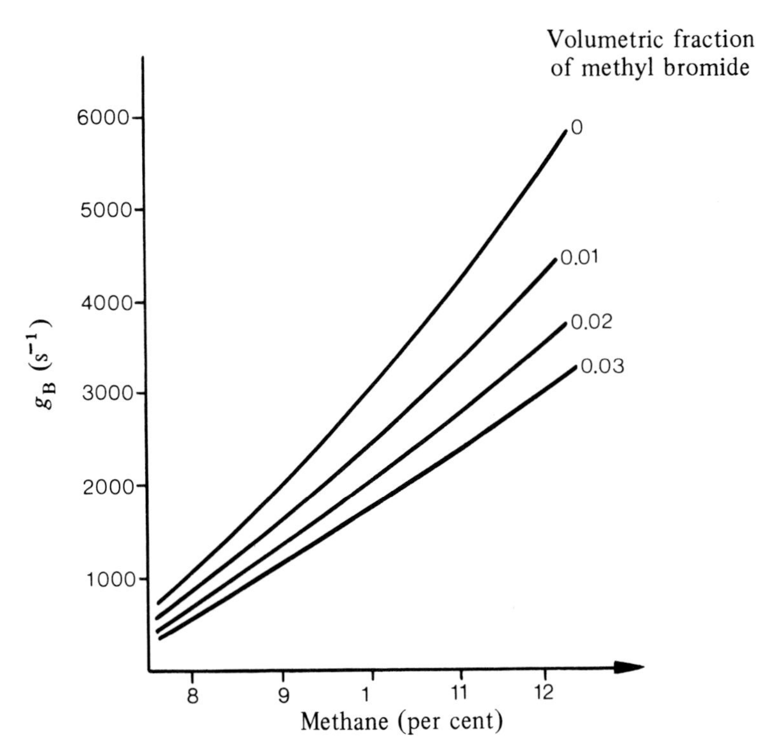

Figure 4:31 Variation of g_B with mixture composition for methane and air with various concentrations of methyl bromide as inhibitor [27]

Flame stability with respect to blowoff is seriously reduced by vitiation as can be seen from Figure 4:32 where β is the ratio of blowoff critical boundary velocity gradients under vitiated and non-vitiated conditions.

The good correlation provided by g_B has discouraged further enquiry into the mechanism of blowoff. However, as Reed [57, 58] has pointed out, there are discrepancies between Lewis and von Elbe's physical picture and some experimental observations concerning the *dead space*, i.e. the dark region between the flame base and the burner rim. In some cases there is no measurable increase in the dead space as the flow rate

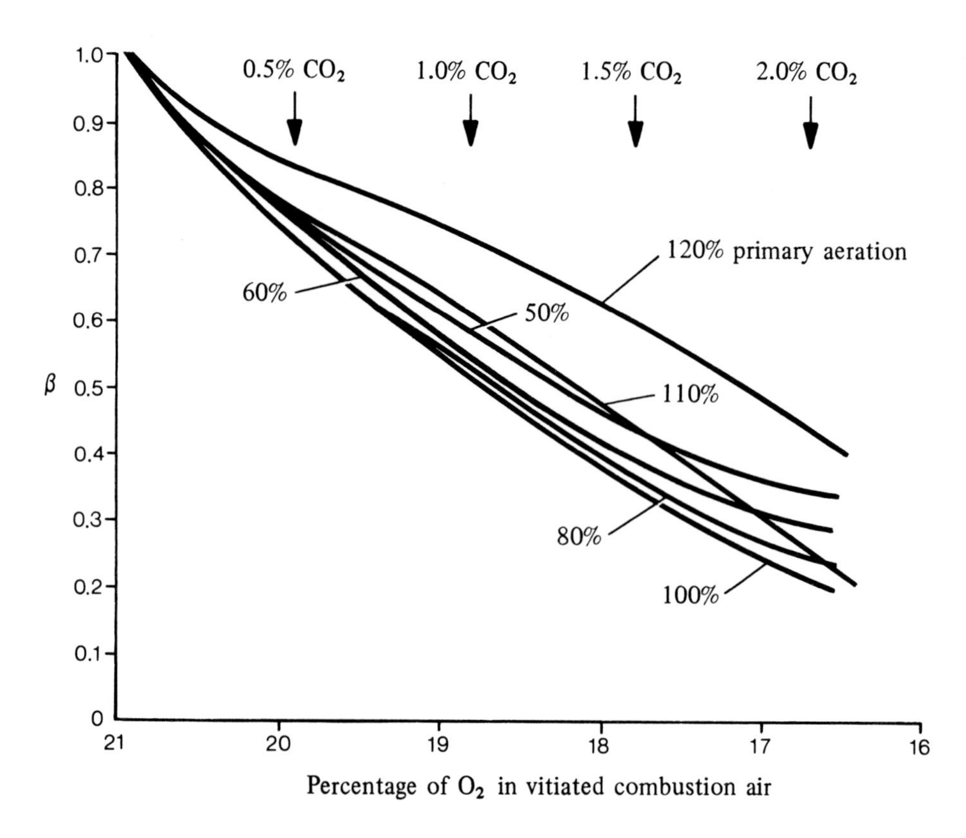

Figure 4:32 The effect on the blowoff stability limit of methane–air flames of vitiating the primary and secondary air supplies with methane combustion products at various primary aerations [42]

is increased towards blowoff, also low-pressure flames have very large dead spaces and would be expected to give a drastic reduction in g_B while in practice the reduction is very small. A further important criticism made by Edmondson and Heap [27] is the fact that the theory fails to predict a quantitative relationship between burning velocity and blowoff flow rate; a relationship which might reasonably be expected to be inherent in a satisfactory theory of blowoff.

There is, however, an alternative mechanism of blowoff, that of aerodynamic quenching or *flame stretch* which leads to an improved correlation of blowoff data.

4.4.3 Flame-stretch theory of blowoff

In addition to rim stabilization, it is possible to stabilize a flame by inserting a solid object (usually referred to as a *flameholder*) into the gas stream. Due to frictional drag the velocity above the object is very low and the flame stabilizes in its wake. If the object takes the form of a wire running along the axis of a tubular burner, the flame becomes an inverted cone. Clearly the blowoff of such flames cannot be described by the theory of boundary velocity gradients since the stabilization zone is at the centre of the gas stream and unaffected by dilution with the surrounding atmosphere.

Lewis and von Elbe considered that the plane-flame model is not applicable to these flames and that blowoff is due to severe curvature, i.e. divergent propagation reducing the burning velocity in the stabilization zone.

Reed [57, 58] proposed that this quenching in divergent propagation or flame stretch applies also to normal burner flames stabilized at the burner rim where the flame curvature and velocity gradients are severe. Figure 4:33 shows velocity vectors in the plane-flame case and the curved-flame case.

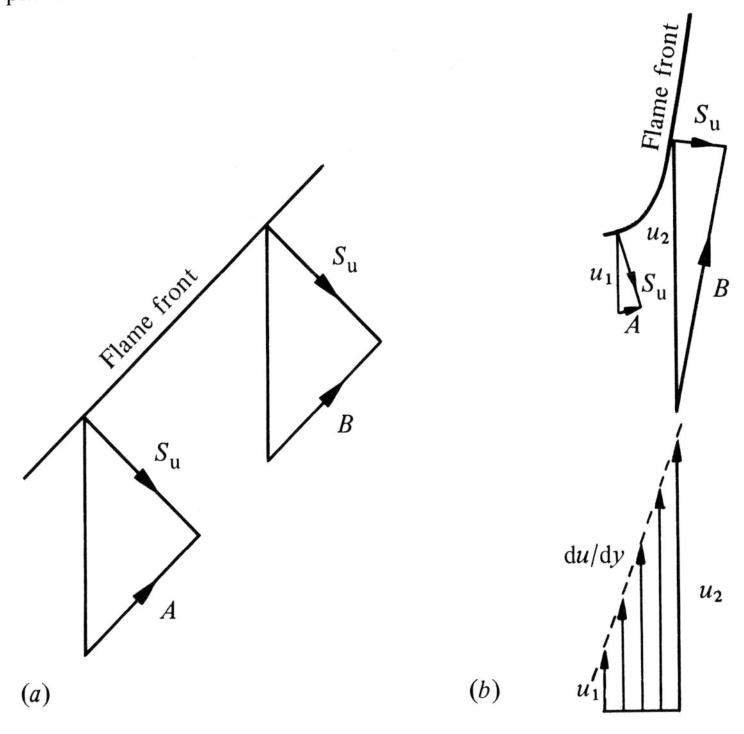

Figure 4:33 Scheme of flame stretch
(*a*) for plane flame
(*b*) for curved flame

In both cases the flame assumes an orientation to the flow such that the component of gas velocity normal to the flame front is equal to the burning velocity. In case (*a*) the vectors A and B parallel to the flame front are equal while in case (*b*), if S_u remains reasonably constant and if u_2 is very much greater than u_1, then vector A is very much less than B.

This increase in flow component parallel to the flame in the case of the curved flame represents an energy flux away from the low-velocity region, i.e. part of the energy conducted into the preheat zone in the low-velocity region is convected back into the reaction zone in the higher-velocity region and the flame is said to be 'stretched'.

This leads to a decrease in temperature and burning velocity in the low-velocity region which may be sufficient to quench the flame. The degree of quenching or stretch may be indicated by the dimensionless *Karlovitz flame-stretch factor* K, the critical value of which at blowoff is denoted by K_B. The simple derivation of K given below is that due to Reed [58]. Alternatively, more complex derivations have been given by Karlovitz [59] and Lewis and von Elbe [51].

Clearly the greater the velocity gradient, the greater the quenching. Therefore:

$$K \propto \mathrm{d}u/\mathrm{d}y$$

Also the effect is upon the preheat zone and so the relevant indication of the scale is the preheat zone thickness η; for a given value of $\mathrm{d}u/\mathrm{d}y$, the quenching effect will be greater for greater values of η. Therefore:

$$K \propto \eta$$

For a given value of $\mathrm{d}u/\mathrm{d}y$ and η the change in component will be smaller the greater the value of u, i.e. the larger the local velocity the less the quenching effect of a given gradient on a fixed preheat zone. Therefore:

$$K = (\mathrm{d}u/\mathrm{d}y)\,(\eta/u) \tag{4.61}$$

A convenient means of expressing η must be found and an approximation to it (denoted by η_0) may be derived from the energy-conservation equation for the preheat zone (in which there is no reaction-energy term) [51]. From equation 4.48:

$$\lambda(\mathrm{d}T/\mathrm{d}x)_i = c_p \rho_u S_u (T_i - T_u)$$

and if η_0 is the subtangent of the temperature profile at the T_i surface, it will approximate to the preheat-zone thickness η and is given by:

$$(\mathrm{d}T/\mathrm{d}x)_i = (T_i - T_u)/\eta_0$$

Therefore

$$\eta_0 = \lambda/c_p \rho_u S_u \tag{4.62}$$

Thus

$$K = (\mathrm{d}u/\mathrm{d}y)\,(\eta_0/u)$$

Therefore

$$K_B = g_B \eta_0/S_u \tag{4.63}$$

i.e.

$$K \propto S_u^{-2}$$

Tests of the flame-stretch theory and its application to practical flames. If blowoff of burner flames results from excessive flame stretch then K_B may be expected to have a constant value regardless of the fuel gas or air–gas ratio.

Figure 4:34 shows Reed's correlation based on burning velocities and blowoff data obtained by a large number of different investigators.

This graph indicates that when only primary combustion occurs, blowoff occurs at a constant value of about 0.23. For gas-rich flames burning in air, the secondary combustion apparently enables the flame to withstand a higher degree of stretch, attributed by Reed to the transfer of heat from the outer diffusion flame cone.

Edmondson and Heap [27] considered the correlation questionable because of the high degree of scatter and carried out a more limited but more precise test using vary-

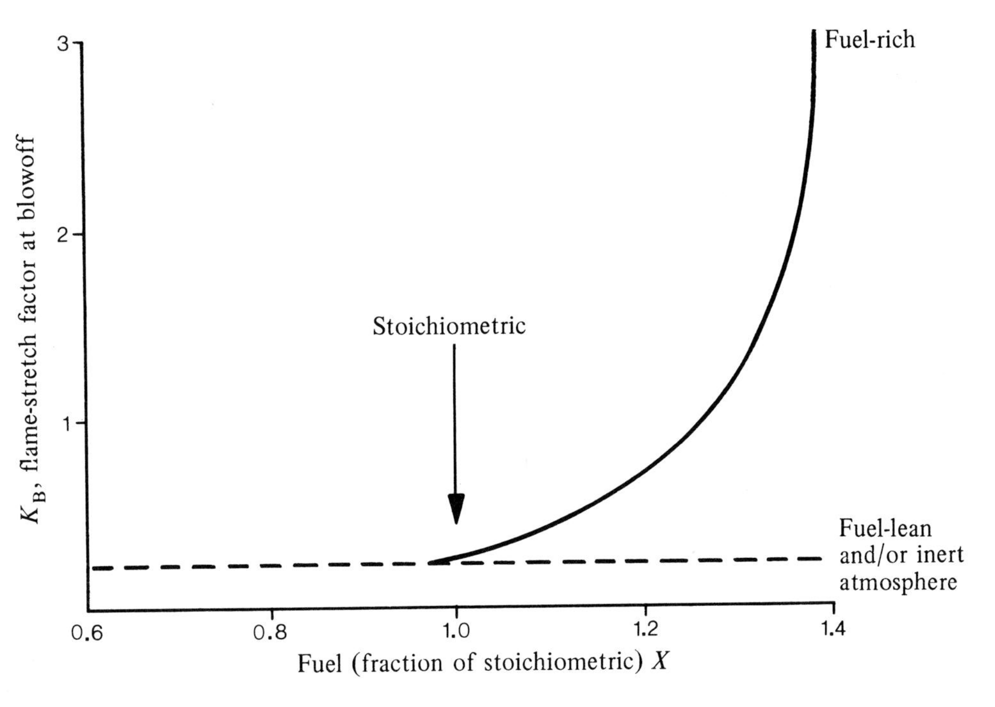

Figure 4:34 Correlation of blowoff data by flame stretch [57]

ing additions of the inhibitor methyl bromide to methane–air flames, thus causing changes in burning velocity without corresponding changes in physical properties.

This led to almost perfect correlation on the flame-stretch basis and the conclusion that the prime cause of blowoff is excessive stretch near the rim. They disagreed with Reed, however, over the interpretation of the steep rise of the curve for rich flames and considered that the effect of secondary combustion is to increase the burning velocity of the primary flame considerably above the normal value. Thus the rise is due to the use of incorrect burning-velocity values in equation 4.63. Furthermore, they regarded interdiffusion of air into the primary zone as having an important complementary effect.

In later work, using wide-range burning-velocity and blowoff data, Edmondson and Heap [60] obtained good correlation for individual gases but found systematic differences in the correlating lines for different gases. K_B was found not to be independent of mixture composition for lean flames and differences were found in values of K_B for gases burning in different secondary atmospheres.

The complications of ambient-atmosphere effects are avoided by using inverted flames and K_B was found to have a constant value close to unity for these flames. Figure 4:35 shows Edmondson and Heap's results for (*a*) burner-stabilized flames and (*b*) inverted flames.

Edmondson and Heap's conclusions are that, unlike inverted flames, the interpretation of blowoff data for normal, burner-stabilized flames in terms of flame stretch is complicated by the difficulty of determining appropriate values of burning velocity and that this fact seriously limits the practical application of the theory.

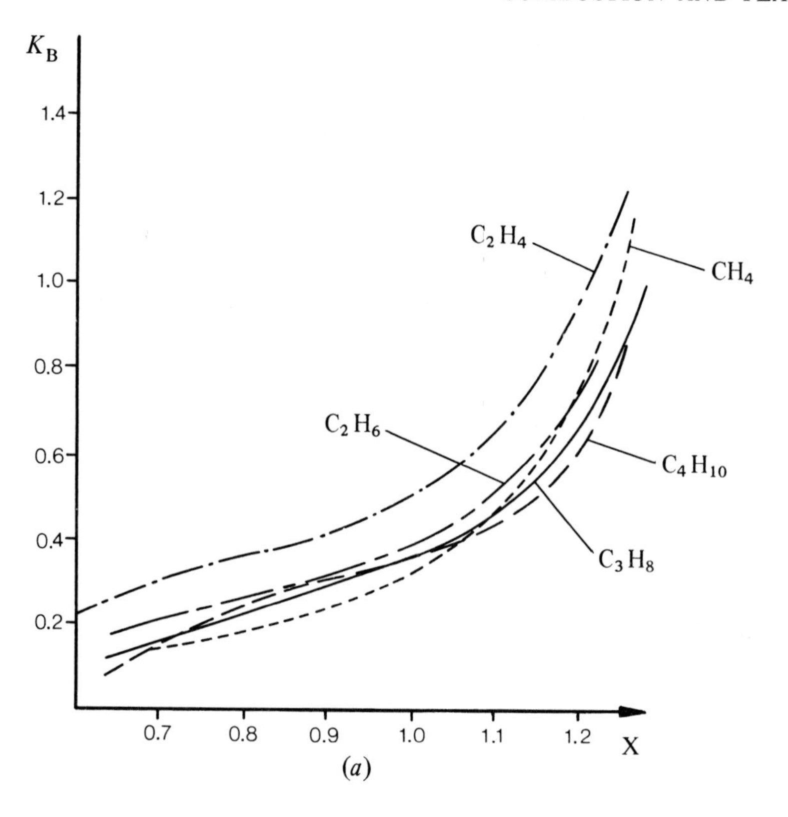

(a)

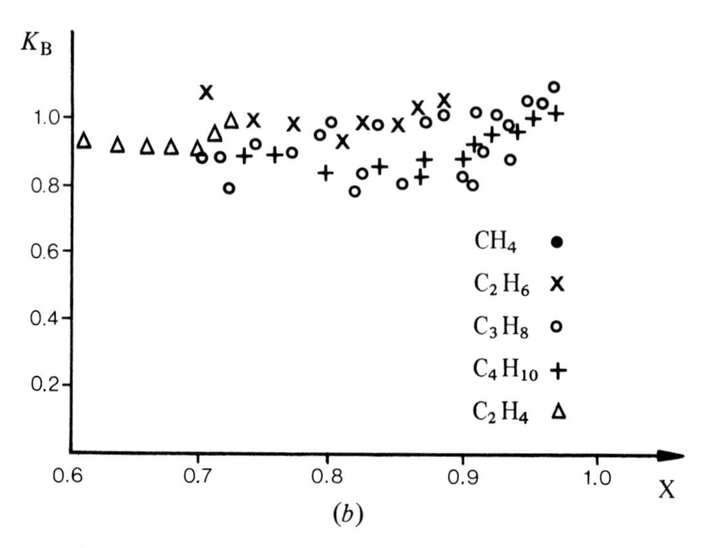

(b)

Figure 4:35 K_B versus mixture composition [60]
(a) for burner-stabilized flames
(b) for inverted flames

4.5 LIMITS OF FLAMMABILITY

For fuel–air or fuel–oxygen mixtures there exist limits of composition between which flame propagation may occur but outside which self-sustaining flames cannot exist.

Standardized experimental methods are in use to determine these limits and although there exists a large body of data and the experimentally derived limits have been proved to be reliable for safety purposes, for example, there is, as yet, no reliable analysis of the phenomenon in existence.

The experimentally determined limits depend somewhat on the apparatus used, values obtained on flat-flame burners differ from those obtained for combustion in tubes, which in turn give different results for upward, downward and horizontal propagation.

One might expect the burning velocity of a mixture to fall smoothly to zero at the flammability limit but, in fact, the limiting burning velocity is found to be finite at around 0.02 to 0.05 $m\,s^{-1}$.

Spalding [61] has proposed that radiation heat loss from the burned gas is the basic cause of flammability limits and if the reaction rate decreases with decreasing temperature more rapidly than does the heat loss, a composition limit of flammability is obtained at which there is a small but finite burning velocity.

Linnett and Simpson [62] have pointed out that convection plays an important part in limit determinations and that there is no evidence that any observed limits are fundamental properties of the fuel–oxidant mixture. If such fundamental limits exist they are probably quite different from the observed convection-dependent limits.

Lewis and von Elbe [51] consider that the convection-initiated quenching which occurs at the flammability limits can be described by the flame-stretch theory, the low burning velocity and thick preheat zones of near-limit flames making them particularly susceptible to flame stretch. Their description accounts for the differences observed in the limits for upward and downward propagation.

The 'conventional' limits for methane–air (i.e. determined using a long, large-diameter tube) are:

upward: 5.35% to 14.85% methane
horizontal: 5.40% to 13.95% methane
downward: 5.95% to 13.35% methane

Such conventional limits appear to obey some practical rules, for example, the lower limit for most gases occurs at a more or less identical stoichiometric composition, corresponding to an adiabatic flame temperature of 1300–1350 °C. However, limits determined by other methods give quite different results. Figure 4:36 gives data on some common gases.

The flammability limits for multicomponent fuel mixtures can be predicted approximately from Le Châtelier's rule:

$$L = (\varphi_A + \varphi_B + \varphi_C + \ldots)/(\varphi_A L_A^{-1} + \varphi_B L_B^{-1} + \varphi_C L_C^{-1} + \ldots) \tag{4.64}$$

where φ_A, φ_B, φ_C, etc. are the volume fractions of the gases and L_A, L_B, L_C are their flammability limits.

This rule predicts most accurately the lower limits.

There are some unexplained pressure effects on the limits; methane for example has an upper limit of about 15 per cent at atmospheric pressure but 46 per cent at 400 atm (40.5 MPa) which, as Gaydon and Wolfhard [4] point out, corresponds to a flame temperature of only about 400 °C.

Gas	Lower limit (%)	Upper limit (%)
Hydrogen	4.0	75.0
Carbon monoxide (moist)	12.5	74.0
Methane	5.3	14.9
Ethane	3.0	12.5
Propane	2.2	9.5
Butane	1.9	8.5
Pentane	1.5	7.8
Ethylene	3.1	32.0
Acetylene	2.5	80.0
Manufactured town gas	4 approx.	40 approx.

Figure 4:36 Limits of flammability for common gases at 1 atmosphere with air

The effect of increased temperature is, as might be expected, to widen the limits, for example, the limits for methane–air at 400 °C are 4.8 per cent to 16.6 per cent.

Combustion with oxygen widens the limits considerably (see Figure 4:37).

Gas	Lower limit (%)	Upper limit (%)
Hydrogen	4.0	94.0
Carbon monoxide (moist)	15.5	94.0
Methane	5.1	61.0
Ethane	3.0	66.0
Ethylene	3.0	80.0

Figure 4:37 Limits of flammability for common gases at 1 atmosphere with oxygen

Vitiation has a considerable narrowing effect on the limits of flammability as evidenced by Figure 4:38.

4.6 QUENCHING

As indicated in Section 4.4 the quenching effect of the burner wall is very important in determining flame stability. In the vicinity of such a heat sink the burning velocity decreases from its normal value S_{u0} to zero near the sink. The dead or dark space represents the closest possible approach of the reaction zone to the sink. Clearly if two heat sinks, for example, two solid plates, approach each other the dead spaces

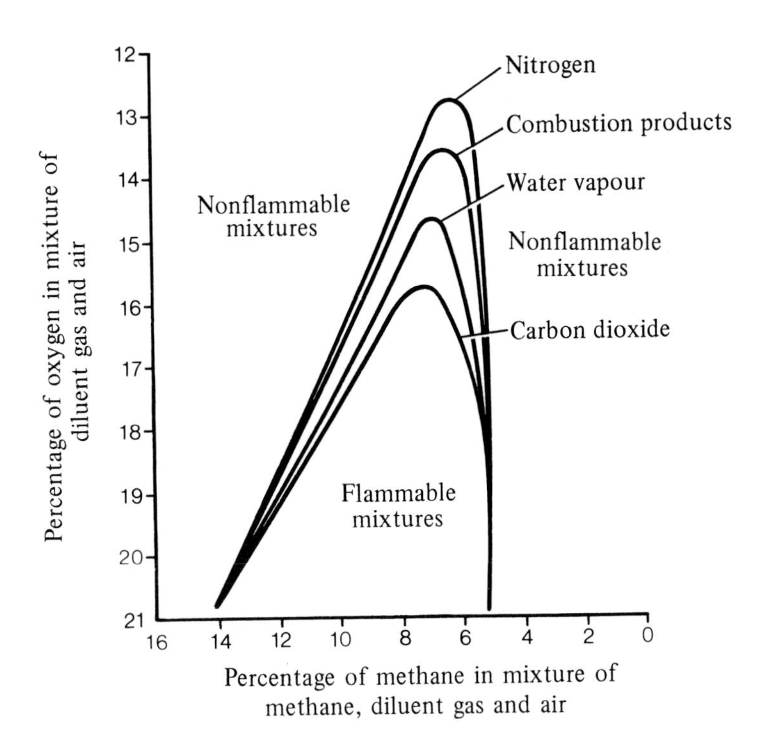

Figure 4:38 The effect on the flammability limits of methane of reducing the oxygen content of combustion air by the addition of four different diluents [42].

will ultimately merge and the flame would be unable to propagate past the quenching zone thus created. This critical separation is the *quenching distance* ($d_{\parallel}$) and the concept forms the basis of the Davy safety-lamp, lightback-resistant flame ports and flame-traps.

The quenching distance for a particular fuel–oxidant mixture may be determined either by the use of parallel plates with adjustable separation and central spark ignition (Blanc *et al.* [63]) or by using a variable-width rectangular burner (Friedman [64]); a flame is established above the burner, the flow is suddenly stopped and the flame will either light back or quench depending on the width of the burner. Similarly quenching may be studied in cylindrical tubes in which case the critical value is the *quenching diameter* D_0 which is somewhat larger than $d_{\parallel}$ (Figure 4:39).

A third quantity, the *depth of penetration of quenching* d_p [51] related to D_0 and $d_{\parallel}$ may be obtained from g_F and S_{u0} thus:

$$d_p = S_{u0}/g_F \tag{4.65}$$

Since the gas velocity near the wall $= gy$ it follows that d_p is the distance from the wall at which the gas velocity at lightback becomes equal to S_{u0}. For $y < d_p$ the burning velocity $< S_{u0}$, consequently d_p may be regarded as a measure of the depth

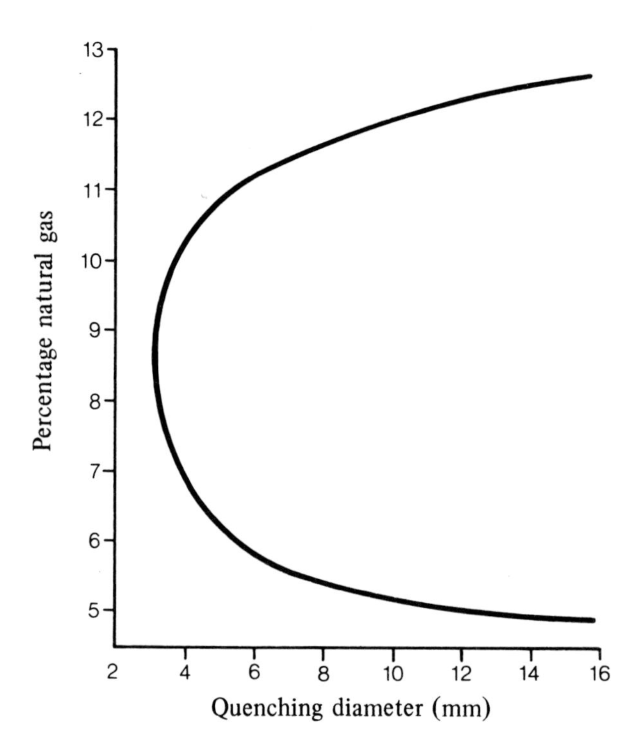

Figure 4:39 Variation of quenching diameter D_0 with mixture composition for natural-gas–air flames [51]

of the quenching effect of a single surface. Thus $d_{\|}$ might be expected to equal $2d_p$ although Lewis and von Elbe found $d_{\|}$ to be somewhat greater than $2d_p$.

Considerable data have been published for quenching or limiting diameters for a wide range of conditions. Potter and Berlad [65] correlated these data with g_F and obtained the relationship:

$$g_F = 2.23(S_{u0}/D_0)^{1.08} \tag{4.66}$$

Quenching dimensions for burner ports other than circular can be obtained from relationships proposed by these workers.

Reed [66] has suggested that an explanation of the limiting dimensions should not rely only on the critical rate of heat and chain-carrier loss but should take into account the flame shape as the quenching limit is approached. The flame, on attempting to light back inside the burner port, tends to invert at its tip and if the diameter of the inverted flame tip approaches the limiting flame diameter D_C, this part of the flame will be extinguished by flame stretch and the quenching diameter will then be:

$$D_0 \approx 2d_p + D_C$$

This could account for Lewis and von Elbe's observation that $2d_p < d_{\|}$.

Quenching effects do not seem to be affected by the nature of the quenching walls.

The thermal conductivity, for example, appears to have no measurable effect on D_0.

Heating the walls reduces the quenching distances and Friedman and Johnson [67] found that D_0 is approximately proportional to $T^{-\frac{1}{2}}$.

Flames established in pipes and vessels may accelerate to detonation and flame traps must be designed, not on the basis of the quenching diameter for normal flame propagation, but taking into account the much more severe conditions existing in such cases (see Section 4.10).

4.7 IGNITION

This topic covers the initiation of combustion by a wide variety of means in reactants in a wide variety of physical conditions. It is usual to consider ignition phenomena in terms of two extreme cases: *homogeneous ignition*, in which ignition occurs simultaneously throughout the reactant volume, and *instantaneous point ignition*, in which a flame develops close to the ignition source (e.g. a spark) and then spreads through the reactant volume. The latter form of ignition is much the more important from a practical point of view and is dealt with here more fully.

4.7.1 Homogeneous ignition

If the temperature of a vessel containing a homogeneous mixture of reactants is raised, a point is reached at which ignition occurs. This is termed the *self-ignition* or *auto-ignition temperature*. The criterion for ignition of this kind is related to the net rate of heat loss or gain in a given volume of the reactants. If the heat loss is less than the rate of heat production due to the reaction then ignition will occur.

Semenov [68] proposed a simple model of self-ignition, in which the temperature T is assumed constant throughout the reactant volume but higher than the wall temperature T_w. Thus the rate of heat loss is given by:

$$Q_{loss} = \lambda A_s(T - T_w)/V \qquad (4.67)$$

and the rate of heat production (see equation 4.34) by:

$$Q_{prod} = H_r f(c) A \exp(-E_a/RT) \qquad (4.68)$$

where λ = thermal conductivity
A_s = surface area of vessel
V = volume of vessel
H_r = heat of reaction
$f(c)$ = an appropriate function of concentration

Ignition occurs above the temperature (the self-ignition temperature) at which the two terms are equal (Figure 4:40).

The variation of self-ignition temperature with pressure exhibits peculiarities which may be identified with abnormal reactions leading to *cool flames* having low velocities (c. 0.05 m s^{-1}) and low temperatures (c. 120 °C).

Homogeneous and two-stage ignition are important in studies of chemical aspects of combustion and in relation to knock in internal-combustion engines but since

Gas	Approximate ignition temperature for stoichiometric mixture (°C)
Methane	700
Ethane	550
Propane	540
Butane	530
Ethylene	540
Hydrogen	550
Carbon monoxide	570

Figure 4:40 Measured ignition temperatures of common gases in air at one atmosphere pressure

they do not approximate to any practical ignition means used in gas engineering, will not be considered further.

4.7.2 Instantaneous point ignition

Normally burner flames are ignited by the use of an external energy source which may be a spark, a hot wire or a pilot flame.

Spark ignition provides a close practical approximation to ideal instantaneous point ignition. The local energy concentration produced is very much in excess of that required for homogeneous ignition and it seems likely that local ignition will always take place in the vicinity of even the weakest sparks. However, it is an experimental fact that, unless the spark energy exceeds a certain critical value, self-propagating combustion does not follow.

This critical value is the *minimum ignition energy* E_{min}.

The small, spherical flame (flame kernel) produced in the vicinity of the spark propagates outwards into the unburned mixture and is convex to it, consequently it is an example of divergent flame propagation and its burning velocity and flame temperature are reduced below the normal values as a result of the severe flame curvature. It follows that there is a minimum diameter (i.e. maximum degree of curvature) below which the flame cannot propagate unaided.

Looked at another way, the volume of hot burned gas inside the flame envelope and available to ignite the reactants is proportional to r^3 whilst the flame area is proportional to r^2, i.e. the contact area between the burned and unburned gases. The former thus reduces more rapidly than the latter with decreasing radius and a radius is reached below which the flame kernel is quenched. The parallel with flame-stretch extinction (Section 4.4.3) is clear.

Thus E_{min} is the minimum amount of energy which the developing flame kernel needs from the ignition source to attain a critical diameter beyond which the flame is able to propagate without further support. Lewis and von Elbe [51] have identified this critical diameter with D_0, the quenching diameter.

Several theories of spark ignition have been proposed with a view to relating E_{min} and D_0 with the fundamental properties of the system and these may be grouped together as activation theories and as thermal theories. The former group regard the spark as a source of activated particles and consider the high local temperatures to be of secondary importance. Thermal theories on the other hand consider ignition purely as a consequence of the temperature distribution in the gas.

Lewis and von Elbe's thermal treatment [51] appears to have had the greatest success in correlating practical results and is based on two relationships.

First, Lewis and von Elbe equate E_{min} with the sensible heat content of a sphere of diameter D_0 the contents of which are burned gases at flame temperature T_b. The second relationship is based on Lewis and von Elbe's *excess enthalpy* concept which is briefly as follows.

In the preheat zone the gases receive heat by conduction from the burned-gas zone while still possessing their as-yet unused chemical energy. Since by conservation of energy the specific enthalpy of the unburned and burned gases must be equal, an excess conductive enthalpy exists in the preheat zone. All flames propagate because excess enthalpy continuously drains into the preheat zone providing the energy to ignite the reactants and, analogously, a spark ignites a mixture if it supplies sufficient energy to provide the excess enthalpy content of the preheat zone in a flame whose critical size is equal to the quenching diameter for that mixture.
Thus

$$E_{min} = D_0{}^2\, h \tag{4.69}$$

Where h is the excess enthalpy per unit area.

This excess enthalpy will also equal the conductive heat flux, i.e.

$$h = \lambda(T_b - T_u)/S_u \tag{4.70}$$

Combining:

$$E_{min} = D_0{}^2\, \lambda(T_b - T_u)/S_u \tag{4.71}$$

Equation 4.71 has been shown to give approximate agreement with experimentally measured values for E_{min}.

Spalding [69] has criticized the excess-enthalpy approach on the basis that diffusion of combustion products into the preheat zone also occurs, and whether an enthalpy excess or deficiency occurs depends on the relative magnitude of heat conduction and product diffusion. This may be expressed by the *Lewis number* (Le) [4]:

$$(Le) = \rho c_p\, D_{eff}/\lambda \tag{4.72}$$

where D_{eff} is the effective mean diffusion coefficient for the combustion products diffusing into the reactants. For many flames $(Le) \approx 1$, i.e. no net enthalpy excess.

Spalding has deduced a similar expression to equation 4.71 from more general ideas which do not require the use of the excess-enthalpy concept.

Before dealing with the spark ignition of natural-gas flames it is worth making a few brief comments about the nature of the spark itself. When a spark passes through a gas the gas changes very quickly from a very good insulator to a good conductor. Applying a high voltage across two electrodes produces an electric field and any charge carriers in the gas (ions or electrons) are accelerated by the field. The moving charged particles may build up sufficient kinetic energy to produce inelastic collisions with molecules resulting in ionization.

An exponential increase in the concentration of charge carriers occurs and the initial very small current is amplified several millions of times to produce a large current dependent on the circuit parameters. This process takes a variable time to build up, depending particularly on the overvoltage applied, i.e. the ratio of applied voltage to breakdown voltage where breakdown voltage is defined as the voltage which if applied for an indefinite time will just cause the gas to break down.

The breakdown voltage depends to some extent on electrode shape, gas composition, temperature and pressure.

Sayers *et al.* [70] found that the breakdown voltage V_b (kilovolts) is given approximately by:

$$V_b = 2.7 + 1.4d \tag{4.73}$$

Where d is the electrode separation in millimetres.

4.7.3 Spark ignition of natural gas

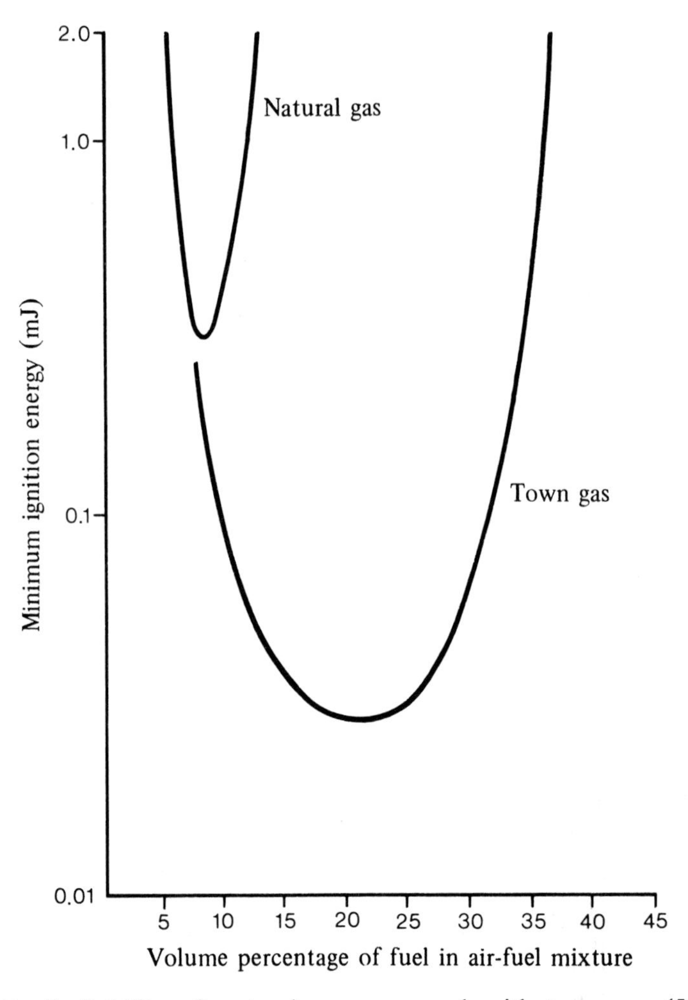

Figure 4.:41 Ignitability of natural gas compared with town gas (50 per cent hydrogen) (logarithmic scale on vertical axis) [70]

Natural gas, in addition to possessing a lower burning velocity and narrower limits of flammability than manufactured town gas is considerably more difficult to ignite.

Comparison of E_{min} for natural gas and manufactured town gas (*Sayers* et al. [70])
Figure 4:41 highlights the much more difficult problem of ignition with natural gas. Not only is E_{min} increased by an order of magnitude, but the composition range over which ignition is possible is considerably reduced.

Effect of electrode separation
According to theory E_{min} is a characteristic property of a combustible gas mixture and ought therefore to be independent of spark-gap size provided this is larger than the critical size (D_0) to which the flame kernel must grow before being able to propagate unaided. If the spacing is less than D_0, heat is lost to the electrodes and more energy is required to ignite the gas.

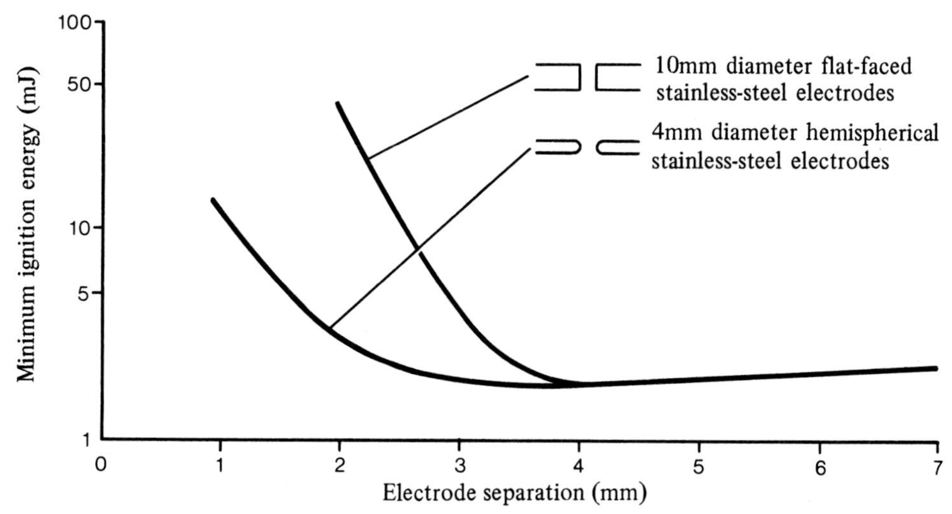

Figure 4:42 Effect of electrode separation on E_{min} for methane–air mixture (11 per cent methane) (logarithmic scale on vertical axis) [70]

This is illustrated in Figure 4:42 in which E_{min} is found to increase considerably below about 4 mm electrode spacing. As might be expected the effect is greater for larger electrodes. Support for Lewis and von Elbe's theory is provided by the good agreement between this figure and the quenching diameter of 5 mm obtained from flashback experiments on cylindrical tubes [51].

Effect of vitiation
The steep dependence of E_{min} on composition is particularly acute if the mixture is vitiated. Sayers *et al.* [70] found that vitiation equivalent to 2.3 per cent CO_2 increased E_{min} by a factor of about 7 and narrowed the ignition limits down to the range 5–9 per cent methane.

Effect of flowing gases and spark location
In the ignition of practical burner flames the situation is considerably more complex

than indicated so far since the flame needs to be ignited and then established at the burner port in the presence of considerable velocity and composition gradients. This aspect has received little attention to date. Sayers *et al.* [70] have plotted 'ignition loops', i.e. lines linking constant values of E_{min} relative to distances from the burner port. If the composition and velocity distribution above a burner port can be calculated it is possible to deduce the region of spark location for a particular spark energy. On this basis Sayers *et al.* have obtained theoretical ignition loops. Comparison with experimentally determined loops show that although agreement is good for a few port diameters downstream, there are considerable discrepancies near the burner port where the spark source is most likely to be placed in practice. Regions of correct spark location for aerated burners have been investigated by Ekins and Wilson [71]. Satisfactory ignition appears to be confined to a small region immediately downstream of the burner port ($<$ 8 mm for a 10 per cent methane–air flame on a 5 mm diameter port). Beyond this the flame is either swept away or stabilizes on the spark electrodes.

4.7.4 Other ignition sources

In the case of spark ignition the ignition source temperature greatly exceeds the flame temperature. In contrast spontaneous ignition temperatures are much below this temperature.

In practical systems where the gas is heated to ignition by a heated source such as a wire filament the minimum wire temperature to give ignition is greater than the spontaneous-ignition temperature. The rate-determining step is the rate of energy supply to the mixture. The minimum source temperature for ignition decreases sharply with heating time from just below the flame temperature to an almost constant value while the energy requirement for ignition increases sharply with time from the minimum ignition energy, then continues to increase slowly due to heat losses from the system.

Filament igniters may be non-catalytic or catalytic. For the former case, in which all the heat required to raise the filament temperature is derived from the electrical supply, the minimum filament temperature to give ignition increases with the gas velocity past the filament, with decreasing wire diameter and with decreasing coil diameter [71]. The variation of minimum filament temperature for ignition with mixture ratio [71] surprisingly shows no maxima or minima, suggesting that the shape of the curves is due to purely thermal effects.

Catalytic filaments do not rely entirely on the electrical supply to heat the wire. Platinum can be used as a cold catalyst to provide ignition for gases containing hydrogen [72] but practical catalytic filaments are normally electrically preheated to promote more rapid catalytic action. For gases not containing hydrogen, for example, natural gas, it seems unlikely that a catalytic reaction would occur. However, if the degree of preheating is sufficient, ignition involving some catalytic action does occur, probably because the mixture becomes air-reformed [71] by an exothermic reaction which in turn assists in promoting ignition in addition to the heat liberated by the catalytic oxidation of the hydrogen formed [71].

Two further ignition devices—pilot flames and flash tubes—are extensively used in gas-ignition practice. Flash tubes are restricted in application largely to domestic gas appliances and as such are outside the scope of this book. Pilot burners, on the other hand, are widely used in industrial practice and are considered in Chapter 7. Practical aspects of ignition including safety considerations are dealt with in Chapter 9.

4.8 TURBULENT PREMIXED FLAMES

When the unburned-gas flow supplying a flame front changes from laminar to turbulent conditions the visible flame changes considerably; the shape changes to a diffuse flame brush, the flame-front thickness appears to increase greatly, the apparent burning velocity increases causing the flame to shorten and the flame starts to emit noise.

As discussed in Chapter 2, particles of gas in turbulent flow have an overall velocity on to which is superimposed a random fluctuating velocity. Due to this randomness turbulence has certain similarities to molecular motion and useful analogies may be drawn but is considerably more complex since the major properties are not characteristics of the medium.

Thus the instantaneous velocity may be expressed as:

$$\begin{aligned} u &= \bar{u} + u' \\ v &= \bar{v} + v' \\ w &= \bar{w} + w' \end{aligned} \tag{4.74}$$

where $\bar{u}$, $\bar{v}$, $\bar{w}$ are the time-average values of the velocity in the x, y and z directions and u', v' and w' are the instantaneous deviations from the mean.

Thus $\bar{u}$, $\bar{v}$ and $\bar{w}$ each equal zero and the root-mean-square values

$$\begin{aligned} u'_{rms} &= \sqrt{\bar{u}^2} \\ v'_{rms} &= \sqrt{\bar{v}^2} \\ w'_{rms} &= \sqrt{\bar{w}^2} \end{aligned}$$

characterize the fluctuating velocity components and are called components of the *intensity of turbulence*.

The pockets of gas produced in turbulent flow are referred to as *turbulence balls* or turbules and their behaviour is somewhat like large molecules. In early theories the distribution of the random movement of these turbulence balls was characterized by the 'mixing length' (in analogy with the mean free path of kinetic theory) i.e. the average distance which a fluid element moves before it loses its identity by mixing with the surrounding fluid.

In more modern theories, statistical correlation coefficients are used to characterize the flow.

In molecular diffusion, the diffusivity is defined by $D = \frac{1}{3} l \bar{u}_m$ where l is the mean free path and $\frac{1}{3}\bar{u}_m$ is the mean molecular (rms) velocity. By analogy, turbulent diffusion is given by:

$$\varepsilon = L u'_{rms} \tag{4.75}$$

where ε is the *eddy diffusivity* or *turbulent exchange quantity* and L is the *scale of turbulence*.

Closer examination of the analogy reveals major differences. Molecular mean free paths are so small and molecular velocities are so high that wall effects and flow velocities have no effect. The opposite is true of turbulent flow in which the principal properties are mainly determined by wall and flow velocity effects.

A major aim of turbulent-flame theories has been to characterize flames using a turbulent burning velocity. This is, however, much more difficult to define and measure than laminar burning velocity, for instance; difficulties arise in the variation of turbulence quantities across the flame, because of the indefinite flame shape and due to the fact that the flame cannot be propagating into stationary unburned gas.

Most methods have used the simple flame-area principle, i.e. burning velocity = volume flow rate divided by flame area, and the values obtained depend considerably on the interpretation of what constitutes the flame area. Damköhler [73] found that the outer edge of the flame brush gave a value close to the laminar burning velocity whilst the inner edge gave the turbulent burning velocity by his definition. Bollinger and Williams [74] used the centre line of the flame brush and showed that turbulent burning velocity as defined increased with Reynolds number.

The effects of turbulence on flame propagation may be subdivided depending on the scale of turbulence.

Small-scale turbulence. When turbulence is small in comparison with the flame thickness, the only effects on the flame are the increased transport properties. In laminar-flow conditions S_u depends upon thermal conductivity, density and heat capacity, thus $S_u \propto \sqrt{(\lambda/c_p\rho)}$. $\lambda/c_p\rho$ is almost identical [4] with the kinematic viscosity ν, i.e. $S_u \propto \sqrt{\nu}$.

Turbulence introduces another transport process which is governed by ε thus:

$$\frac{\text{turbulent burning velocity}}{\text{laminar burning velocity}} = \frac{S_{uT}}{S_{uL}} \approx \sqrt{\frac{\varepsilon + \nu}{\nu}} \approx \sqrt{\frac{\varepsilon}{\nu}} \tag{4.76}$$

Damköhler showed that $\sqrt{(\varepsilon/\nu)} \propto \sqrt{(Re)}$. As Weinberg [8] has pointed out, equation 4.76 is more satisfying theoretically than useful since the thickness of laminar flames is very small. It is unlikely that small-scale turbulence of this order will exist alone, except perhaps close to the burner walls and for very small flames.

Medium-scale turbulence. Turbulence is medium scale when its intensity is sufficient to wrinkle but not break the flame front. Such wrinkled flame fronts are not wrinkled smoothly but show sharp ridges pointing towards the burned gas and smooth curves in the other direction since the former regions by virtue of their concavity have an enhanced burning velocity due to curvature effects (cf. flame stretch, Section 4.4.3). For such a wrinkled flame front

$$S_{uT}/S_{uL} = A_r/A_a \tag{4.77}$$

where A_r = real area taken round all the indentations and A_a = apparent mean surface area, the implication being that had the turbulent burning velocity been arrived at by dividing the flow rate by the correct area, the laminar value would have been obtained. Measurements of the actual wrinkled flame area by Fox and Weinberg [75] have shown that the effective burning velocity over this surface is about 2.5 times greater than the normal laminar burning velocity, the difference being attributed to the effects of small-scale turbulence and of the increased reaction rate in the sharp ridges of the wrinkled flame front.

Large-scale turbulence. Here the flame becomes broken and burning takes place in separate pockets of gas. Theoretical treatment of this situation is clearly even more difficult than for small- and medium-scale turbulence since there no longer exists a coherent flame front.

If the turbulence is extreme it may be sufficient to extinguish the flame. Lewis and von Elbe [51] regard this as due primarily to excessive flame stretch in the curves of the flame surface where the propagation is divergent.

Extinction occurs in the crests of the curves but reignition occurs from adjacent parts of the flame, where the sharp ridges act as centres of reignition (this reignition effect accounts for the widening of the flammability limits in turbulent flames). At higher and higher levels of turbulence disruption of the surfaces becomes more and more frequent and the flame is eventually extinguished.

4.8.1 Flame stabilization in turbulent flow

Practical means of stabilizing turbulent flames involve either the promotion of recirculation allowing hot burned gases to flow back to the flame base or the use of low-velocity pilot flames at the periphery of the main flame producing a zone of continuous reignition. Recirculation may be provided by a sudden enlargement at the burner head, by the use of a bluff-body obstacle placed in the mixture stream producing wake eddies or by the use of swirl to produce reverse flow along the burner axis. The application of these methods is dealt with in Chapter 7.

Pilot-flame stabilization

Using a sufficiently strong pilot allows the main flame to be anchored in a boundary region where otherwise the velocity gradient greatly exceeds the critical boundary velocity gradient for blowoff. A single flame sheet is normally established extending over the pilot and main streams. Lewis and von Elbe [51] and Karlovitz *et al.* [76] have applied the flame-stretch theory to flames of this kind.

Close to the main flame boundary, propagation is divergent to such a degree that flame-stretch quenching will occur unless sufficient heat is supplied by the pilot flame. Lewis and von Elbe regard the growth of the flame surface as the pilot flame propagates into the main stream as analogous to the development of a flame from a spark (see Section 4.7.2) since a spark must supply enough energy to carry the flame kernel through the region of stretch between zero diameter and the minimum free-propagating flame diameter whilst a pilot flame must supply enough energy to carry the flame through the region of stretch at the main stream boundary. There should be order-of-magnitude agreement therefore between the minimum energy requirements for the two cases. Lewis and von Elbe's analysis show that this appears to be the case, although the fact that a spark delivers essentially all its energy to the combustible gas while a pilot flame delivers only a small part of its energy to the unburned gas, wasting most of it to the ambient atmosphere and the downstream gas, makes precise comparisons difficult.

Wake stabilization

If a non-streamlined obstacle is placed in a gas stream and if the flow velocity is sufficiently high then the adverse pressure gradient downstream of the obstacle causes the boundary layer to separate and a recirculating vortex system establishes in the wake of the obstacle (see Section 2.5.1). If the Reynolds number is sufficiently high the vortices increase in size and are shed at regular intervals. When a flame establishes itself in the wake of an obstacle a decrease in downstream stagnation pressure occurs and this acts like a reduction in Reynolds number [77] in that eddy-shedding may cease and a stable vortex-pair be produced. The eddy region is a zone of recirculated burned gas and the mixture stream is ignited and stabilized by this recirculation zone. A steep velocity gradient exists between the recirculation zone and the main stream and according to the flame-stretch theory the flame blows off when the critical value of K is reached. The difficulty is that the flow field of interest is not known in sufficient detail to allow K to be calculated. Zukoski and Marble [78] measured the length of the recirculation zone behind obstacles of various shapes and sizes and found that the blowoff limit is given solely by the time taken for a flow element to sweep past the recirculation zone. This characteristic time τ is given by the length of the wake divided by the flow velocity and was found to be independent of flow conditions and flame-holder geometry but strongly dependent on mixture composition in the manner of a fundamental flame parameter. Lewis and von Elbe [51] applied the flame-stretch concept to the results of Zukoski and Marble and were able to derive, by assuming the critical value of K to be unity, values of τ in good agreement with the experimental results of Zukoski and Marble for lean and stoichiometric mixtures. For rich mixtures the calculated values increasingly exceed the experimental values and this has been attributed to preferential diffusion of oxygen into the stretched zone [51].

Beér and Chigier [77] have considered flame stabilization in terms of the heat balance in the vortex region assuming that concentrations and temperatures are uniform throughout the recirculation zone. Their treatment leads to the conclusion that stability requires:

$$u_{ext}/P^{n-1}\,D = \text{constant} \tag{4.78}$$

which shows that the blowoff velocity (u_{ext}) is proportional to the bluff-body diameter (D) and to the (n − 1)th power of pressure (P) where n is the order of the reaction.

A number of measurements have been made of conditions in the recirculation zone for a variety of bluff-body configurations. These are summarized in the book by Beér and Chigier [77] to which the reader is referred for a detailed treatment of this topic.

In general terms improved stability is obtained in bluff-body stabilized flames by [79]:

1 Increasing the size of the stabilizer
2 Increasing the drag coefficient of the stabilizer
3 Increasing the mixture temperature
4 Increasing the flow pressure
5 Decreasing the main flow turbulence

Stabilization by swirl

Swirling flow produces radial and axial pressure gradients which, if the swirl is sufficiently strong, result in a reverse axial pressure gradient causing reverse flow along the axis and producing an internal recirculation zone. The subject of swirling flow has received a good deal of attention in recent years and a detailed review of this work is to be found in the book by Beér and Chigier [77].

When air is fed tangentially into a burner a spiral flow is set up such that a balance is created between the centrifugal force acting on the air and the pressure forces acting on the burner walls. The low pressure in the centre of the rotating flow recovers downstream of the burner mouth resulting in a reverse axial pressure gradient which, if the degree of swirl is sufficient, causes the flow to reverse and set up a toroidal vortex around the burner axis, the length of which increases with increasing swirl. The effect also causes the outer boundaries of the stream to expand rapidly close to the burner mouth.

Swirl is normally characterized by the dimensionless *swirl number* (S) which is given by [77]:

$$S = G_{\varphi}/G_x r \tag{4.79}$$

where r is the burner nozzle radius, G_{φ} is the axial flux of the angular momentum and G_x is the axial flux of linear momentum.

G_{φ} is given by:

$$\int_0^R wr\,\rho u 2\pi r\,\mathrm{d}r$$

and G_x (excluding the static pressure term) by:

$$2\pi \int_0^R \rho u^2\, r\,\mathrm{d}r$$

where u and w are the axial and tangential components of the velocity in any cross-section of the stream.

Swirl may be generated in burners by tangential entry, by using guide-vanes or by using rotating vanes or tubes (see Chapter 7). For weak swirls (characterized by $S < 0.6$) the reverse pressure gradient is not sufficiently large to cause internal recirculation and flames established in weak swirl flow are of limited practical interest because of their tendency to instability [77]. Strong swirl ($S > 0.6$), on the other hand, causes internal recirculation which increases flame stability considerably. With increasing swirl the angle of spread of the stream increases, entrainment increases and the flame length decreases.

4.9 DIFFUSION FLAMES

Diffusion flames, alternatively known as jet flames, nonaerated flames, postaerated flames or neat-gas flames, differ from premixed flames in that combustion occurs at the interface between the fuel gas and oxidant and the rate of combustion depends more on the rate of mixing than on the much faster combustion reactions. Diffusion flames have received far less attention in fundamental research than premixed flames even though they are extremely important in practice: the Bray jet which dominated the design of town-gas era gas appliances is an example of a laminar diffusion flame while at the other end of the scale, the open-hearth furnace burner is an example of a turbulent diffusion flame.

The major reason for the theoretical neglect of diffusion flames is the absence of some fundamental measurable parameter like burning velocity which can be used to characterize the burning process.

Because mixing exerts the dominating role it is possible to draw a clear distinction between slow-burning laminar diffusion flames in which the mixing occurs solely by molecular diffusion and fast-burning turbulent flames in which the aerodynamics of the system dominate.

Generally speaking, diffusion flames are longer and are considerably more luminous than premixed flames.

4.9.1 Laminar diffusion flames

Probably the most useful and easily measurable parameter is the flame height and a number of treatments have been proposed to relate the flame height to the fundamental properties of the gas.

Mathematical analysis is difficult and rather sweeping approximations are necessary, for example the reaction zone is generally regarded as infinitesimally thin and located at the point where the fuel–oxidant mixture becomes stoichiometric. The assumption is also made that diffusion is rate determining so that the reaction rate is directly proportional to the diffusion rates of fuel and oxygen. This is difficult to apply since the diffusion coefficient varies with composition and temperature.

Jost [80] has derived a simple expression which allows the flame height to be roughly calculated. For a diffusion flame burning in an excess of air, the flame height will be given by the point at which air just diffuses to the jet axis. The distance covered by diffusion is given by the Einstein equation of diffusion:

$$\bar{x}^2 = 2Dt \tag{4.80}$$

where x is the distance, D is the diffusion coefficient and t is the time. For this simple analysis we may write $r^2 = 2Dt$, where r is the tube radius.

Time t is also given by the flame height L_f divided by the gas velocity, u which is assumed constant. Thus:

$$L_f = \tfrac{1}{2}ur^2/D \tag{4.81}$$

The gas flow rate $V = u\pi r^2$ thus:

$$L_f = V/2\pi D$$

In the case of laminar flames where mixing is controlled by molecular diffusion, the diffusion coefficient is clearly independent of the burner dimensions and velocity. Thus it follows from equation 4.81 that flame length (or the length to achieve a specific degree of mixing) is proportional to the volumetric gas flow V.

In the case of turbulent flames D is a turbulent mixing coefficient or eddy diffusivity. This eddy diffusivity (see Section 4.8) is given by the scale of turbulence multiplied by the intensity of turbulence. The scale of turbulence is proportional to the tube radius and the intensity to the average velocity, thus $L_f \propto ur^2/ur$, i.e.:

$$L_f \propto r \tag{4.82}$$

Relations 4.81 and 4.82 illustrate the considerable differences between laminar and turbulent flames, i.e. that the flame length is proportional to the gas velocity and port area for laminar flames but is proportional to the port diameter only for turbulent flames.

An improved derivation by Guyomard [81] assumes that $D \propto T^2$ and yields:

$$L_f = 2V_0T_0/5.76\pi D_0\bar{T}_f \qquad (4.83)$$

where 0 refers to conditions at the inlet and $\bar{T}_f$ is the mean flame temperature.

It should be pointed out that for flames in which carbon formation occurs, which includes most hydrocarbons, the flame height has a less than definite meaning since it is difficult to distinguish between the completion of the combustion of the fuel gas and the carbon particles. With regard to the stability of laminar diffusion flames, the burner does not usually exert as important an influence as in the premixed case since the reaction occurs at the gas–air interface (but see Chapter 7). Lightback is clearly impossible and extremely small flames may be maintained.

Flame flicker is a characteristic feature of many diffusion flames. This is caused by momentum transfer at the jet boundary forming eddies. An eddy of combustion products separates the fuel and oxidant, slows down the diffusion and lengthens the flame. If the eddies are shed the fuel and oxidant come into closer contact and the flame shortens. The flame may not be truly laminar therefore even at low velocities.

Transition to turbulent diffusion flames

Figure 4:43 illustrates the effect of increasing the gas flow rate on the flame characteristics.

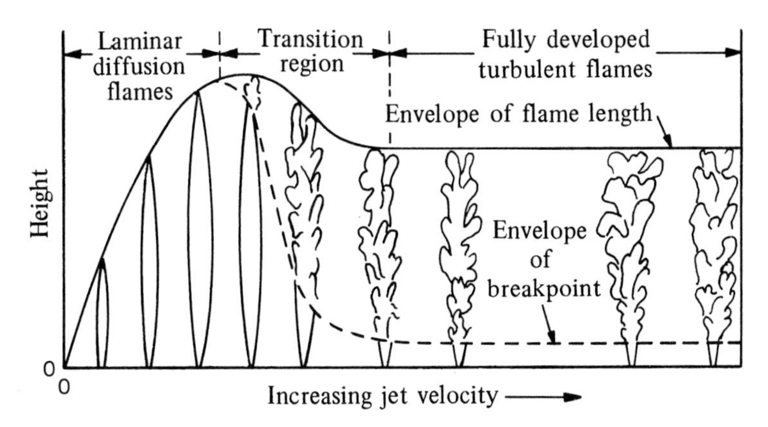

Figure 4:43 Changes in flame type with increasing nozzle velocity [82]

The laminar flame height increases nearly linearly (in confirmation of 4.81) to a maximum at which point breakup starts to appear at the tip. On further increase this breakpoint moves rapidly towards the burner then remains constant near to the burner rim. For fully developed turbulent flow the flame height remains constant (as predicted by 4.82). The noise level increases sharply and the luminosity decreases in the turbulent regime.

It is worth pointing out that laminar and turbulent diffusion flames must be distinguished on the character of flow in the combustion zone rather than in the burner tube. Reynolds number is inversely proportional to kinematic viscosity which increases by a very large factor in the combustion zone. Thus, a Reynolds number of say 7000 to 8000 in the burner tube may fall below the critical value of approximately 2300 and produce a laminar flame [83].

4.9.2 Turbulent diffusion flames

Relation 4.82 indicates that the length of turbulent jet flames is proportional to the burner diameter and independent of the gas velocity. Generally speaking this is borne out by the measurements of Hawthorne, Weddel and Hottel [84] and Wohl, Gazley and Kapp [85] who have attempted to relate flame length to burner diameter on a theoretical basis.

The former approach [84] is based on momentum conservation and takes into account the effects of temperature, density and chemical reaction.

For turbulent free flames without buoyancy effects:

$$L_f/D_n = (5.3/x_r)\sqrt{\{(T_f/\alpha_r T_n)[x_r + (1 - x_r)(M_s/M_n)]\}} \tag{4.84}$$

where L_f = visible flame length
D_n = nozzle diameter
T_f = adiabatic flame temperature
T_n = absolute nozzle fluid temperature
M_s, M_n = molecular weights of surrounding and nozzle fluids
x_r = mole fraction of nozzle fluid in stoichiometric mixture
α_r = moles reactants/mole products

The latter approach is based on eddy diffusivity ε which is defined as the product of mixing length L_1 and turbulence intensity u'_{rms} (Section 4.8). The maximum value of L_1 occurs at the axis and is given approximately by 0.085 D_n, also the value of u'_{rms} at the axis is given approximately by $0.03\bar{u}$, thus ε at the axis is given by

$$\varepsilon = L_1 u'_{rms} = 0.0025\, \bar{u}\, D_n \tag{4.85}$$

Wohl, Gazley and Kapp introduced the arbitrary constant factor f (which is close to unity) to account for the difference between their simple model and practice, thus

$$\varepsilon = 0.0025 f u_0 D_n \tag{4.86}$$

where u_0 is the average gas velocity at the burner nozzle. This approach leads to the following expression for flame length:

$$L_f/D_n = 0.00255 f x_r/16 \tag{4.87}$$

Günther [86] has related flame length with stoichiometric air requirement and gas density in the following semi-empirical relationship:

$$L_f/D_n = 6(A_w + 1)\sqrt{(\rho_u/\rho_f)} \tag{4.88}$$

where A_w = stoichiometric air–fuel ratio (weight basis)
ρ_u = fuel-gas density
ρ_f = flame density

ρ_f is about the same for all gaseous fuels. Therefore, the flame length depends only on burner diameter, gas density and air requirement.

For natural gas L_f/D_n works out as 200, a value which agrees within 10 per cent with practical measurements. Values for other fuels quoted by Beér and Chigier [77] are:

Fuel	L_f/D_n
Carbon monoxide	76
Town gas	136
Hydrogen	147
Propane	296

The expressions given above for the lengths of turbulent diffusion flames relate to free-jet flames, i.e. those burning in the open atmosphere. If the flame is enclosed in a combustion chamber the flame increases in length due to recirculation vitiation and also widens.

Eickhoff and Lenze [87] have shown a linear relationship between the ratio of the lengths of enclosed and free flames and the reciprocal of a recirculation parameter θ (the Thring–Newby number). For further treatment of this aspect the reader is referred to reference [77].

4.9.3 Soot in diffusion flames

Diffusion flames and rich premixed flames of hydrocarbons produce solid carbon which causes the yellow luminosity of such flames and which may, under certain circumstances, escape unburned from the flame and be deposited as soot. Sooting is probably the least well understood of all the combustion phenomena involved in gas utilization. Carbon formation greatly increases radiative heat transfer from a flame (see Chapter 3) and this may be usefully employed in industrial burners especially those burning solid or liquid fuels. Soot deposition, however, is clearly undesirable. If a premixed laminar hydrocarbon flame is supplied with a progressively reducing amount of primary air, a stage is reached at which a region of bright yellow luminosity first appears at the tip of the flame originating from very small soot particles maintained at a high temperature by the adjacent burned gases. The air-gas ratio at which this first occurs is generally termed the *yellow-tip limit* (Section 4.4.1).

With a further reduction in primary air the luminous area increases and dulls in colour changing eventually to orange indicating a progressive lowering of the temperature of the particles.

Figure 4:44 due to Reed and Roper [88] indicates the processes involved in soot formation in a laminar methane diffusion flame.

When the gas temperature at the axis approaches its maximum value and just before the first appearance of soot (say at point *B*) a number of complex reactions occur. These include pyrolysis to acetylene followed by polymerization to polyacetylenes and polyacetylene radicals which in turn through cyclization form polycyclic aromatics and a large number of small soot particles. Once these appear, a variety of species condense on their surfaces and react, increasing the particle size. In addition, the particles agglomerate to much larger particles having a diameter of 10 to 50 nm. As the soot particles grow they become less reactive and their rate of growth slows down. Chain-aggregates may ultimately be formed as large as 500 nm. It is possible

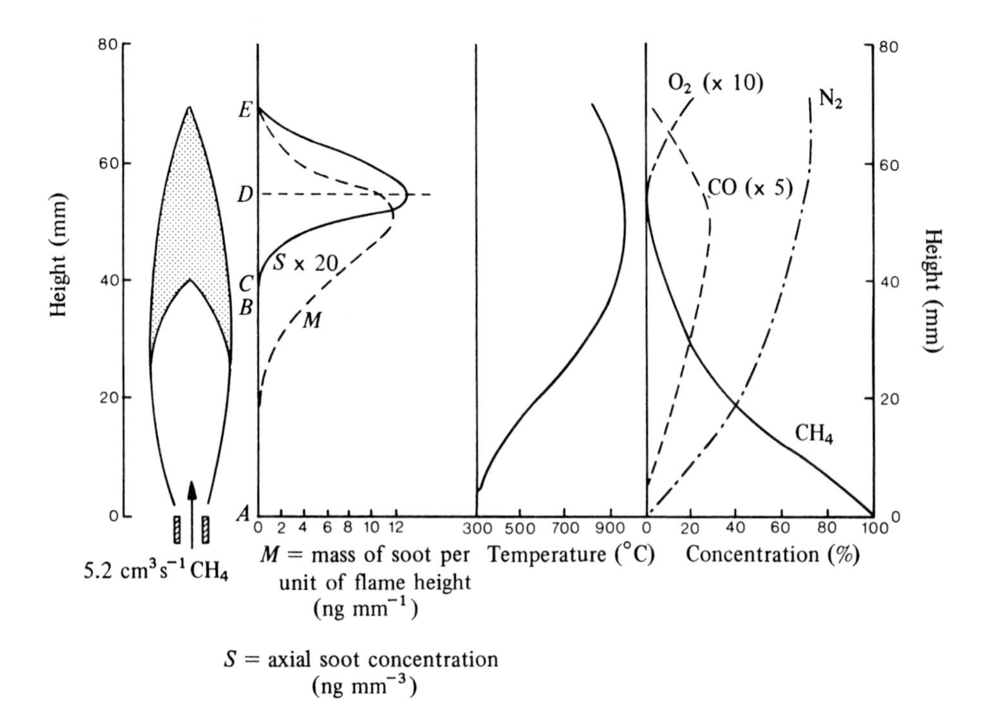

Figure 4:44 Axial temperature, composition and soot-concentration profiles for a laminar methane diffusion flame [88]

that a purely chemical mechanism is not solely responsible for soot formation and that electrostatic forces may play some part [4].

The soot is formed in a reducing region of the flame where the reactions are controlled by the rate of diffusion of oxygen and fuel gas. On the other hand the soot is then burned in a higher region in the flame in the presence of oxygen where the rate of combustion is governed by the reaction rate of the oxidation of the soot particles and on the local oxygen concentration. Figure 4:44 shows that the axial oxygen concentration starts to become significant where the soot concentration starts to decrease (point D). Thus distance AD may be regarded as the height of the diffusion-controlled region, i.e. the height of the diffusion flame, and AE as the height to where the soot concentration falls to zero, i.e. the overall flame height.

In practice, whether or not soot is formed in the flame is less important than whether soot is deposited or not. If the soot concentration is such that the radiation heat loss from the particles is sufficient, combined with the chilling effect of induced secondary air, to reduce the temperature in the soot burn-away region such that soot oxidation ceases, then soot will escape from the flame. Similarly direct flame impingement and secondary air starvation will favour soot deposition. Soot deposited in this way consists of graphite crystallites together with 1 to 3 per cent hydrogen and a wide range of organic compounds such as pyrene, phenanthrene and anthracene; these latter frequently giving it a sticky nature. Figure 4:44 also illustrates that it is quite possible to obtain soot deposition even though the carbon monoxide concentration is within acceptable limits since the carbon monoxide concentration at the end of the soot region is extremely low.

Catalytic effects may sometimes be involved in soot deposition. Iron or nickel carbonyls in very low concentrations, formed by reactions of carbon monoxide with the metals, have led to heavy sooting with manufactured town gas [88].

Carbonyls are unlikely with carbon-monoxide-free natural gases but could cause difficulties with SNG containing a few per cent of carbon monoxide.

Test gases for sooting propensity are discussed in Section 4.11.

Sooting test burners have been used for several years in the British gas industry. The benzene fringe test [89] which related to soot formation from the aromatic content of coal-based town gases was superseded by the sooting test burner developed by Dewhurst and Holbrook [90]. This, however, was intended for use with first-family gases burning in nonaerated flames with air starvation and it is not by any means certain that the results would apply to second-family aerated burners. There is as yet no satisfactory sooting test burner for second-family gases.

The yellow-tip limit gives a good indication of whether or not an aerated burner will give sooting. Olefines require less primary air to prevent yellow tipping than the equivalent paraffins [88]. The primary aeration at the yellow-tipping limit may be calculated for mixed gases with reasonable accuracy [54, 88].

Reed and Roper [88] have proposed a number of simple expressions for soot concentration and flame height to maximum soot concentration for axisymmetric and fishtail diffusion-flame geometries.

Soot formation is significantly reduced when diffusion flames become turbulent and mixing becomes more effective.

4.10 DETONATION IN GASES

In normal flame propagation, molecular diffusion in laminar flames or molecular plus eddy diffusion in turbulent flames provides the feedback mechanism by which the hot reaction products ignite the reactants. The flame front propagates at a comparatively low velocity normally in the range 0.05 to 10 $m\ s^{-1}$.

However, in certain circumstances it is possible for flame fronts to propagate at a much greater velocity than this, e.g. 1 to 3 $km\ s^{-1}$. Such supersonic flames are termed *detonations* and their velocities cannot be explained in terms of the normal transport properties. Ignition in these cases is by rapid compression of the reactants caused by a *shock wave*, the chemical energy released by combustion providing the driving force for the shock wave.

Detonations may be extremely destructive and their occurrence must be avoided. In pipework containing flammable air–gas mixtures from, for example, a premix machine (Chapter 7) there may well be a sufficient distance or *run-up length* for a detonation to develop should lightback into the pipework occur. The use of *flametraps* to guard against hazards of this kind is discussed in Chapter 9.

Detailed discussion of shock waves would be out of place in a book of this kind but since it is impossible to describe detonations without reference to the formation of shock waves a brief consideration of the major points is given below.

Considering first the formation of a shock wave in an incombustible gas, a simple model is provided by an accelerating piston in a long tube containing the gas. The acceleration of the piston may be regarded as consisting of a series of smaller accelerations each of which generates a pressure pulse travelling at the speed of sound which in turn heats the gas adiabatically and imparts a movement to it. Thus each pulse travels faster than its predecessor. These small impulses eventually merge to form a pressure wave moving forward at a velocity greater than that of sound.

In a shock wave the relationships of temperature and pressure are quite different to those exhibited in normal adiabatic, isentropic compression. It is convenient to refer the transition to a coordinate system moving with the shock wave. Since the process is 'stationary' the masses, momenta, and energies before and after the wave must be equal. Thus the conservation equations are:

$$\rho_1 u_1 = \rho_2 u_2 \tag{4.89}$$

$$\rho_1 u_1^2 + P_1 = \rho_2 u_2^2 + P_2 \tag{4.90}$$

$$E_1 + \tfrac{1}{2}u_1^2 + (P_1/\rho_1) = E_2 + \tfrac{1}{2}u_2^2 + (P_2/\rho_2) \tag{4.91}$$

where E_1 and E_2 are the energies per unit mass before and after the shock.

Note that in these equations diffusive processes are regarded as insignificant compared with the flows of mechanical energy and momentum, the exact opposite of the treatment of normal flames.

The pressure rise in the shock wave offers resistance to flow and part of the mechanical energy is irreversibly converted to heat, thus equation 4.91 would not be applicable to flow in which the pressure and volume changes are isentropic (i.e. reversible). Combination of these equations gives the *Rankine–Hugoniot* equation:

$$E_2 - E_1 = \tfrac{1}{2}(P_1 + P_2)(v_1 - v_2) \tag{4.92}$$

where v is the specific volume.

Equation 4.92 allows a *Hugoniot curve*, a graph of pressure versus specific volume, to be constructed which connects states obtainable by a shock transition in a similar way to the P–v curves which relate states which may be obtained by isothermal or adiabatic changes.

The velocity of the wave is given by:

$$u_s = v_1 \left(\frac{P_2 - P_1}{v_1 - v_2}\right)^{\frac{1}{2}} \tag{4.93}$$

Turning now to the case of the detonation wave in which combustion is occurring, the state $(P_2\ v_2)$ must now include the energy liberated by combustion. Thus the Hugoniot curve is displaced to higher values.

It may be shown (see for example [51] and [83]) that one point on the Hugoniot curve, the *Chapman–Jouguet state* represents normal detonation. Points corresponding to values of (Pv) greater than this point, i.e. overdriven detonations tend to be unstable and decay to C–J detonations whilst points corresponding to lower values are either thermodynamically unlikely or correspond to normal flames depending on their position.

For stable C–J detonations, the detonation velocity u_d is given by the simple expression:

$$u_d = (v_1/v_2)a_b \tag{4.94}$$

where a_b is the speed of sound in the burned gas.

Thus the detonation velocity may be calculated simply from the thermodynamic properties of the gases. Experimental measurements of detonation velocities show good agreement with calculated values.

Detonation limits exist, analogous to and within the limits of flammability. Again these limits cannot be explained on the basis of simple detonation theory but may be due to the development of pulsating or spinning propagation [51].

Measured detonation velocities also decrease below the ideal value for comparatively small-diameter tubes, an effect which Fay [91] considers to be due to the increased penetration of the boundary layer causing divergence of the detonation front.

Figure 4:45 lists detonation limits and velocities (for stoichiometric mixtures) initially at atmospheric pressure.

Gases	Detonation limits (%)	Detonation velocity (m s^{-1})
Hydrogen–oxygen	15–90	2840
Carbon monoxide–oxygen	38–90	1790
Acetylene–oxygen	3.5–92	2400
Propane–oxygen	3.1–37	2500
Hydrogen–air	18.2–59	1900
Acetylene–air	4.2–50	1900
Methane–air	—	1800
Propane–air	—	1800
Town gas–air (12.7 mm ($\frac{1}{2}$ in) pipe)	19–22.5	1040
Town gas–air (100 mm (4 in) pipe)	16–28	1980
Town gas–air (extrapolated maximum value)		2040

Figure 4:45 Detonation properties of common gases [83] (data on town gas from [92])

Transition to detonation

If a flame is initiated in the open end of a long tube of sufficiently large diameter the propagation rate which starts initially at the normal burning velocity increases progressively. This acceleration is caused by an increase in the flame area as the flame elongates then becomes turbulent and by an increase in burning velocity as the mixture ahead of the flame is heated and compressed by compression waves produced by the increased pressure buildup in the flame due to increased production of burned gases. These compression waves may coalesce into a shock front which may develop into a detonation. This implies the requirement of a predetonation or run-up length of pipe which might be expected to increase with increasing pipe diameter. Lewis and von Elbe [51] quote measurements of run-up lengths for acetylene detonations of approximately 60 tube diameters for pipe sizes between 6 mm ($\frac{1}{4}$ in) and 100 mm (4 in) (a similar length-diameter ratio to that required for fully developed turbulence to develop in pipe flow). Cubbage [92] found that town-gas–air detonations required longer run-up lengths ranging from 3.4 m for 12.7 mm ($\frac{1}{2}$ in) diameter pipes to 11 m for 100 mm (4 in) pipes. The run-up distance was found to be proportional to the root of the pipe diameter. Cubbage investigated the effects of bends, elbows, etc. on detonations and found that whilst the detonation velocity before and after the fitting was the same there was a considerable reduction of velocity in the fitting. Detonation was generally regained in about 0.9 m. The effect is greater for sudden enlargements. Cubbage's results show that a detonation of 1530 m s^{-1} in a 25 mm (1 in) pipe drops to about 250 m s^{-1} downstream of an enlargement to 100 mm diameter and requires about 3 m to regain detonation velocity. Thus placing a flame arrester element in a pipe enlargement serves two purposes: first it reduces the pressure loss in normal flow and second, should a detonation occur the arrester will need only to stop the much slower flame existing in the enlargement.

4.11 INTERCHANGEABILITY OF GASES

Although gas-burning equipment is capable of tolerating some variation in the properties of the gas supplied, mixtures of fuel gases made regardless of considerations other than the maintenance of the declared calorific value are unlikely to be satisfactory in use.

Clearly there are limits within which satisfactory *interchangeability* of distributed gases may take place and it is important that these limits should be delineated in as accurate and comprehensive a way as possible. Furthermore, it should be possible to predict with reasonable certainty from a knowledge of the chemical composition of a proposed gas whether it would be satisfactorily interchangeable with a currently distributed gas without the necessity for experimental determinations of its combustion characteristics.

4.11.1 Manufactured town gases (first family)

In the 1950s new gas-making techniques led to considerable changes in the combustion characteristics of gas supplied to consumers' appliances in the UK.

Investigations by Gilbert and Prigg [89] enabled predictions to be made of the behaviour of gas appliances designed and adjusted for traditional coal-based gases when burning newer gases to be made. They established the pattern of interchangeability prediction in the UK which has carried forward into the natural-gas era.

Gilbert and Prigg's work was based on a consideration of the following ways in which a new gas may give rise to unsatisfactory performance:

1 The heat input may be too high or too low
2 Aerated burners may light back
3 Flames may lift from aerated or nonaerated burners
4 Incomplete combustion may occur
5 Sooting may occur

Weaver [43] in the USA derived a series of six indexes which relate to the various forms of unsatisfactory behaviour and suggested limits about these to indicate satisfactory interchangeability. Delbourg [93] in France used a diagrammatic method in which Wobbe number is plotted versus 'combustion potential' and an area of satisfactory performance is drawn on the diagram. Gilbert and Prigg's solution was a combination and adaptation of these methods to suit UK conditions.

The main properties of a gas which affect its combustion characteristics are calorific value, specific gravity and burning velocity. Calorific value and specific gravity are combined in the Wobbe number (Section 4.2.2) which determines the heat input to an appliance but which, since air requirement is closely linked with calorific value (see Section 4.2.7) also indirectly indicates other properties. Burning velocity is difficult to express as a single factor since it varies with air–gas ratio and in any case is difficult to measure as indicated in Section 4.3.3. The gas industry in the past has relied on empirical results obtained from test burners which measure a combination of factors. The standard test burner in the UK is the *Aeration Test Burner* which measures the air required to produce a flame of fixed inner cone height on a standard burner, the degree of opening of the air shutter to achieve this being expressed as the aeration number. The prediction of the burning velocity of multicomponent mixtures presents particular difficulties (see Section 4.3.3). Weaver's flame-speed factor provides a useful but inaccurate method of prediction and was adopted by Gilbert and Prigg as one axis of their diagram, Wobbe number being the other. Figure 4:46 shows the

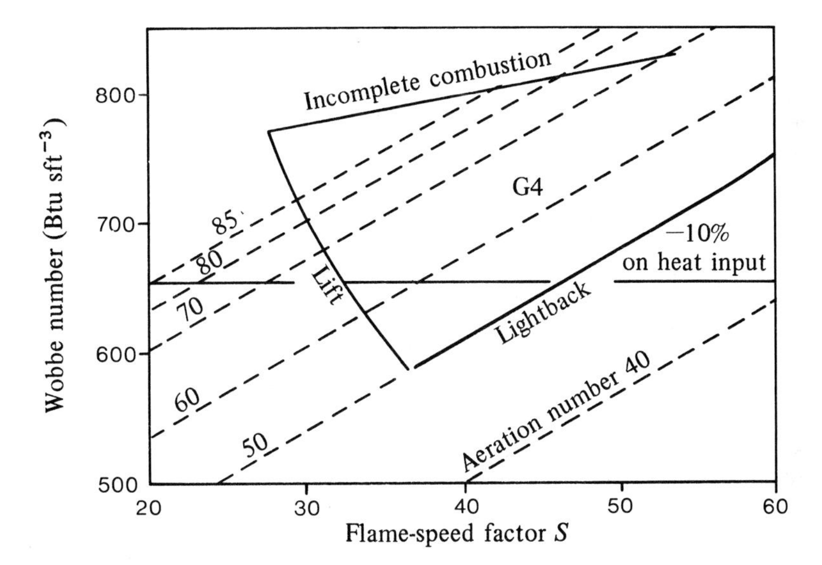

Figure 4:46 Limits of satisfactory performance for G4-adjusted appliances [89]

interchangeability prediction diagram for appliances adjusted to burn gas of group G4 (Wobbe number 700–760 UK units (26.6–28.8 SI units)). Similar diagrams may be plotted for each of the other gas groups.

The Wobbe number and flame-speed factor of any gas may be calculated from its chemical composition and if, when plotted on the diagram, the point lies within the boundaries of satisfactory service the gas may be regarded as interchangeable with the adjustment gas.

The limits were obtained as follows:

1 Heat input: an appliance will normally be satisfactory if the heat input, i.e. Wobbe number, is maintained within $\pm$ 10 per cent of the adjustment gas (i.e. the mean

of the gas group). Only the line of mean Wobbe number -10 per cent is plotted since the upper limit of Wobbe number is dependent on the liability of an appliance to give incomplete combustion and a separate limit is plotted for this.

2 Lightback: the tendency for a gas to light back is conveniently indicated by its aeration number. For G4-adjusted appliances the limiting value is 50 hence the lightback limit may be inserted on the diagram.
3 Lift: the lift limit was based on measurements of the *reversion pressure* of post-aerated jets, i.e. the gas pressure at which a lifted flame reverts to the burner head. Practical experience indicated that distributed gases of reversion pressure greater than 450 Pa (4.5 mbar, 1.8 inH_2O) would be satisfactorily stable. Thus the lift line represents gases of this reversion pressure.
4 Incomplete combustion: this limit corresponds to a CO/CO_2 ratio of 0.02 obtained from measurements on a limited number of appliances corresponding to the British Standard criterion for satisfactory combustion in domestic appliances.
5 Sooting: a tendency to soot is not apparently related to the flame-speed factor and cannot therefore be described by the diagram.

4.11.2 Natural gas

Compared with first family gases, natural gas, consisting largely of methane, has a higher Wobbe number and is relatively slow burning. This latter property affects the design of aerated burners and introduces the problem of flame lift on postaerated burners; on the other hand lightback becomes less of a problem. The narrower flammability range of natural gas makes it more susceptible to vitiation and it is considerably more difficult to ignite.

Clearly, first-family and second-family (natural) gases are by no means interchangeable with each other and first-family gas burners and equipment must be converted to make them suitable for burning natural gas.

The interchangeability of UK natural gases has been examined by Harris and Lovelace [47] who have adapted Gilbert and Prigg's approach to produce similar diagrams.

If only the variations likely to occur in indigenous and imported natural gases needed to be considered, because their composition varies comparatively little, the problem of delineating acceptability limits would not be severe. However the possible use of manufactured substitutes must be considered and these may have very different combustion characteristics. Harris and Lovelace [47] have tabulated possible substitute gases of calorific value 38 MJ m^{-3} (st) (1000 Btu sft^{-3}) which show, for example, variations in hydrogen content from 2.5 to 58 per cent and in flame-speed factor from 14 to 30. Many of these gases contain large amounts of higher hydrocarbons which may lead to an increased tendency to sooting or possibly to the formation of polyhedral or cellular flames in which the inner cone is divided into segments and which display anomalous stability behaviour.

The natural-gas interchangeability-prediction diagram is based largely on the properties of the *test gases* used in approval-testing of domestic appliances. In 1964 the International Gas Union proposed a series of test gases for natural gas appliances, and these have been adopted, with some modification, for British use.

Figure 4:47 lists the composition and properties of the test gases currently in use in the UK.

The suitability of the test gases is under continuous review. The lightback test gas

	NGA (Reference)	NGB (Incomplete combustion)	NGC (Lightback)	NGD(2) (Lift)	NGE (Sooting)
Composition (per cent by volume)					
CH_4	100	87	65	92.5	—
H_2	—	—	35	—	—
N_2	—	—	—	7.5	—
C_3H_8	—	13	—	—	60 + 40 air
Calorific value					
Btu sft^{-3}	995	1187	758	920	1485
MJ m^{-3}(st)	37.8	45.1	28.8	34.9	56.4
Wobbe number					
Btu sft^{-3}	1335	1440	1222	1203	1296
MJ m^{-3}(st)	50.7	54.7	46.4	45.7	49.0
Flame-speed factor	14.0	14.5	26.7	13.4	16.0

Figure 4:47 Natural-gas test gases

NGC, for example, has recently been criticized on the grounds that it does not correspond to a distributable gas since its calorific value is very low [53] and alternatives to it are being considered. A new test gas (NGS) to test for burn-back sooting (see Chapter 7) is also being examined [53].

The prediction diagram (Figure 4:48) is based on the reference gas being pure methane saturated with water vapour and the boundaries of the usable area are fixed by the characteristics of the limit test gas as follows:

1 Heat input: normal limits, mean Wobbe number (1335 UK, 50.7 SI) $\pm$ 5 per cent; emergency limits + 8 per cent–10 per cent.
2 Lightback: although lightback presents no problem with natural gas, possible substitute gases may have high hydrogen contents. The limit must pass through the point representing NGC which has an aeration number of 250; the lightback limit therefore is a line representing gases of aeration number 250.
3 Lift: NGD(2) has a reversion pressure (measured on an *aerated* test burner (see [47])) of 500 Pa (5 mbar, 2 inH_2O) and a line corresponding to this reversion pressure represents the lift limit.
4 Incomplete combustion: this limit line must pass through NGB.
5 Sooting: not possible to include.

The natural-gas prediction diagram, like its town-gas counterpart, is useful but not complete nor sufficiently precise for gases falling close to the boundaries. Major criticisms are the diagram's inability to predict sooting and its reliance on Weaver's flame speed factor.

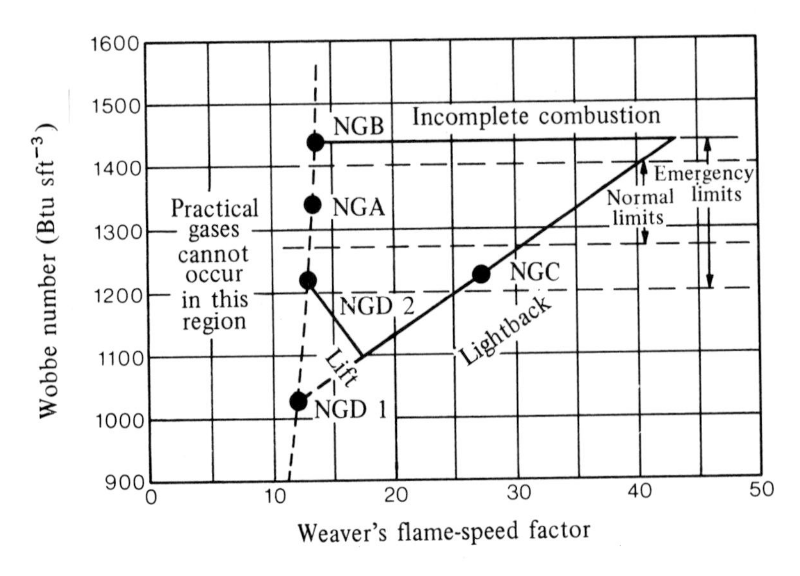

Figure 4:48 Limits of satisfactory performance for UK natural-gas appliances [47]

4.12 FURTHER READING

Chemical thermodynamics and kinetics

Glasstone, S., *Textbook of Physical Chemistry*, 2nd edition, London, Macmillan (1946)

Warn, J. R., *Concise Chemical Thermodynamics in SI Units*, London, Van Nostrand Reinhold (1969)

Combustion properties and calculations

Davies, C., *Calculations in Furnace Technology*, London, Pergamon (1970)

Spiers, H. M. (ed.), *Technical Data on Fuel*, 6th edition, London, British National Committee of the World Power Conference (1962)

Thring, M. W., *The Science of Flames and Furnaces*, 2nd edition, London, Chapman & Hall (1962)

Flames

Bradley, J. N., *Flame and Combustion Phenomena*, London, Methuen/New York, Barnes & Noble (1969)

Fristrom, R. M. and Westenberg, A. A., *Flame Structure*, New York, McGraw-Hill (1965)

Gaydon, A. G. and Wolfhard, H. G., *Flames: Their Structure, Radiation and Temperature*, 3rd edition, London, Chapman & Hall/New York, Barnes & Noble (1970)

Lewis, B. and von Elbe, G., *Combustion, Flames and Explosions of Gases*, 2nd edition, New York, Academic Press (1961)

Weinberg, F. J., *Optics of Flames*, London, Butterworth (1963)

REFERENCES

1 Fristrom, R. M. and Westenberg, A. A., *Flame Structure*, New York, McGraw-Hill (1965)
2 Dixon-Lewis, G., Garside, J. E., Kilham, J. K., Roberts, A. L. and Williams, A., 'Fundamentals of methane combustion', *Inst. Gas Eng. Commun.*, 861 (1971)
3 *Gas Making and Natural Gas*, London, BP Trading (1972)
4 Gaydon, A. G. and Wolfhard, H. G., *Flames: Their Structure, Radiation and Temperature*, 3rd edition, London, Chapman & Hall (1970)
5 Davies, R. M. and Toth, H. E., 'Equilibrium compositions, heat contents and physical properties of natural-gas–air combustion products in SI units', British Gas Corporation Midlands Research Station Internal Report N 187 (1972)
6 Damköhler, G. and Edse, R., *Z. Elektrochem.*, **49** (1943), 178
7 Hottel, H. C., Williams, G. C., and Satterfield, C. N., *Thermodynamic Charts for Combustion Processes*, New York, Wiley (1949)
8 Weinberg, F. J., *Optics of Flames*, London, Butterworth (1963)
9 Miller, P. A. and McConnell, S. G., *J. Inst. Fuel*, **45** (1972), 43
10 Harker, J. H., *J. Inst. Fuel*, **40** (1967), 206
11 Harker, J. H. and Allen, D. A., *J. Inst. Fuel*, **42** (1969), 183
12 Steffensen, R. J., Agnew, J. T. and Olsen, R. A., *Engineering Bulletin of Purdue University*, No. 122 (1966)
13 Barrett, J. H., 'Flame temperatures, calculated and measured', Project Report, BSc Course in Gas Engineering, University of Salford (1972)
14 Rosin, P. O., *J. Inst. Fuel*, **19** (1945), 53
15 Spiers, H. M. (ed.), *Technical Data on Fuel*, London, British National Committee of the World Power Conference, 1962, p. 98
16 Francis, W. E. and Toth, H. E., 'Equilibrium compositions and heat contents for combustion products of town gas and its constituents', Gas Council Industrial Gas Development Committee Special Report No. 2 (1963)
17 Toth, H. E., 'Equilibrium compositions and physical properties for products of combustion of methane and natural gases with air', Gas Council Midlands Research Station Report ER 97 (1966)
18 Davies, R. M. and Toth, H. E., 'Equilibrium compositions, heat contents and physical properties of town-gas–oxygen and methane–oxygen combustion products', Gas Council Industrial Gas Development Committee Special Report No 5 (1966)
19 McBride, B. J. *et al.*, 'Thermodynamic properties to 6000 K for 210 substances involving the first 18 elements', NASA SP-3011 (1963)
20 *JANAF Chemical Tables*, Dow Chemical Co., Michigan (1965 *et seq*)
21 Manson, N., *C.R. Acad. Sci. Paris*, **220** (1945), 734
22 von Elbe, G. and Mentser, M., *J. Chem. Phys.* (1945), 89
23 Andrews, G. E. and Bradley, D., *Combust. Flame*, **18** (1972), 133
24 Pritchard, R., 'Laminar burning velocities of methane–air flames', PhD Thesis, University of Salford (1972)
25 Weinberg, F. J., *Proc. R. Soc.*, **235** (1956), 510
26 Dery, R. J., quoted by Lewis and von Elbe [51]
27 Edmondson, H. and Heap, M. P., Twelfth Symposium (International) on Combustion, 807. Pittsburgh, The Combustion Institute, 1969
28 Anderson, J. W. and Fein, R. S., *J. Chem. Phys.*, **17** (1949), 1268

29 Durst, F., Melling, A. and Whitelaw, J. H., *Combust. Flame*, **18** (1972), 197
30 Singer, J. M., Fourth Symposium (International) on Combustion, 352. Baltimore, Md., Williams & Wilkins, (1953)
31 Powling, J., *Fuel*, **28** (1949), 25
32 Levy, A. and Weinberg, F. J., Seventh Symposium (International) on Combustion, 296, London, Butterworth (1959)
33 Botha, J. P. and Spalding, D. B., *Proc. R. Soc.*, **A 225** (1954), 71
34 Edmondson, H., Heap, M. P. and Pritchard, R., *Combust. Flame*, **14** (1970), 195
35 Pritchard, R., Edmondson, H. and Heap, M. P., *Combust. Flame*, **18** (1972), 13
36 Günther, R. and Janisch, G., *Combust. Flame*, **19** (1972), 49
37 Edmondson, H. and Heap, M. P., *Combust. Flame*, **14** (1970), 405
38 Bradley, D. and Hundy, G., Thirteenth Symposium (International) on Combustion, 575, Pittsburgh, Pa., The Combustion Institute (1971)
39 Raezer, S. D. and Olsen, H. L., *Combust. Flame*, **6** (1962), 227
40 Dugger, G. L. and Heimel, S., NACA Tech. Note 2624 (1952)
41 Lindow, R., 'Zur Bestimmung der laminaren Flammengeschwindigkeit', Dissertation, Technischen Hochschule, Karlsruhe (1966)
42 Reed, S. B. and Wakefield, R. P., *Gas Council Res. Commun.* GC162 (1969); *J. Inst. Gas Eng.*, **10** (1970), 77
43 Weaver, E. R., 'Formulas and graphs for representing the interchangeability of fuel gases', *J. Res. Natl. Bur. Stand.*, **46,** No. 3 (1951), paper 2193
44 Payman, W. and Wheeler, R. V., *Fuel*, **1** (1922), 185
45 Spalding, D. B., *Fuel*, **35** (1956), 347
46 Yumlu, V. S., *Combust. Flame*, **11** (1967), 190
47 Harris, J. A. and Lovelace, D. E., *Gas Council Research Commun.* GC142 (1967), *J. Inst. Gas Eng.*, **8** (1968), 169
48 Ganju, P., 'The prediction and measurement of laminar burning velocities of multicomponent mixtures of fuel gases with air', MSc. Thesis, University of Salford (1973)
49 Scholte, T. G. and Vaags, P. B., *Combust. Flame*, **3** (1959), 511
50 Fuidge, G. H., Murch, W. O. and Pleasance, B., *Trans. Inst. Gas Eng.*, **89** (1940), 283
51 Lewis, B. and von Elbe, G., *Combustion, Flames and Explosions of Gases*, 2nd edition, New York, Academic Press (1961)
52 Grumer, J., Harris, M. E. and Schultz, H., Fourth Symposium on Combustion, 695, Baltimore, Md., Williams & Wilkins (1953)
53 France, D. H., 'The effect of substitute natural gas on appliance design', MSc. Dissertation, University of Salford (1973)
54 Grumer, J., Harris, M. E. and Rowe, V. R., *U.S. Bur. Mines Rpt. Invest.*, 5225 (1956)
55 Van Krevelen, D. W. and Chermin, H. A. G., Seventh Symposium on Combustion, 358, London, Butterworths (1959)
56 Caffo, E. and Padovani, C., *Combust. Flame*, **7** (1963), 331
57 Reed, S. B., *Combust. Flame*, **11** (1967), 177
58 Reed, S. B., *Gas Council Res. Commun.* GC 141 (1968); *J. Inst. Gas Eng.*, **8** (1968), 157
59 Karlovitz, B., Denniston, D. W., Knapschaefer, D. H. and Wells, F. E., Fourth Symposium on Combustion, 613. Baltimore, Md., Williams & Wilkins (1953)
60 Edmondson, H. and Heap, M. P., *Inst. Gas Eng. Res. Commun.*, 838 (1970); *J. Inst. Gas Eng.*, **11** (1971), 305

61 Spalding. D. B., *Proc. R. Soc.*, **A 240** (1957), 83
62 Linnett, J. W. and Simpson, C. J., Sixth Symposium on Combustion, 20, New York, Reinhold, (1956)
63 Blanc, M. V., Guest, P. G., von Elbe, G. and Lewis, B., *J. Chem. Phys.*, **15** (1947), 798
64 Friedman, R., Third Symposium on Combustion, 110, Baltimore, Md., Williams & Wilkins (1949)
65 Potter, A. E. and Berlad, A. L., *J. Phys. Chem.*, **60** (1956), 97
66 Reed, S. B., *Combust. Flame*, **13** (1969), 583
67 Friedman, R. and Johnson, W. C., *J. Appl. Phys.*, **21** (1950), 791
68 Semenov, N. N., *Z. Phys.*, **48** (1928), 571
69 Spalding, D. B., *Some Fundamentals of Combustion*, London, Butterworth (1955)
70 Sayers, J. F., Tewari, G. P., Wilson, J. R. and Jessen, P. F., *Gas Council Res. Commun.* GC171 (1970); *J. Inst. Gas Eng.*, **11** (1971), 322
71 Ekins, B. and Wilson, J. R., *Gas Council Res. Commun.* GC185 (1972)
72 Ekins, B. and Brown, A. M., *Gas Council Res. Commun.* GC146 (1967); *J. Inst. Gas Eng.*, **8** (1968), 223
73 Damköhler, G., *Z. Elektrochem.*, **46** (1940), 601
74 Bollinger, L. M. and Williams, D. T., Third Symposium on Combustion, Flames and Explosion Phenomena, 178, Baltimore, Md., Williams & Wilkins (1949)
75 Fox, M. D. and Weinberg, F. J., *Proc. R. Soc.*, **268** (1962), 222
76 Karlovitz, B., Denniston, D. W., Knapschaefer, D. H. and Wells, F. E., Fourth Symposium on Combustion, 613, Baltimore, Md., Williams & Wilkins (1953)
77 Beér, J. M. and Chigier, N. A., *Combustion Aerodynamics*, London, Applied Science Publishers (1972)
78 Zukoski, E. E. and Marble, F. E., *Combustion Researches and Reviews, 1955*, 167, London, Butterworth (1955)
79 Bespalov, I. V., quoted by Beér and Chigier [77]
80 Jost, W., *Explosion and Combustion Processes in Gases*, translated by H. O. Croft, New York, McGraw-Hill (1946)
81 Guyomard, F., *C.R. Acad. Sci. Paris*, **234** (1952), 67, quoted by Bradley [83]
82 Hottel, H. C. and Hawthorne, W. R., Third Symposium on Combustion, Flames and Explosion Phenomena, 254, Baltimore, Md., Williams & Wilkins (1949)
83 Bradley, J. N., *Flame and Combustion Phenomena*, London, Methuen/New York, Barnes & Noble (1970)
84 Hawthorne, W. R., Weddel, D. S. and Hottel, H. C., Third Symposium on Combustion, Flames and Explosion Phenomena, 102, Baltimore, Md., Williams & Wilkins (1949)
85 Wohl, K., Gazley, C. and Kapp, N., Third Symposium on Combustion, Flames and Explosion Phenomena, 288, Baltimore, Md., Williams & Wilkins (1949)
86 Günther, R., *Int. Z. Gaswärme*, **15** (1966), 376
87 Eickhoff, H. and Lenze, B., *Chem. Ing. Techn.*, **41** (1969), 1095, quoted by Beér and Chigier [77]
88 Reed, S. B. and Roper, F. G., *Gas Council Res. Commun.* GC186 (1971)
89 Gilbert, M. G. and Prigg, J. A., *Gas Council Res. Commun.* GC35 (1956); *Trans. Inst. Gas Eng.*, **106** (1956–7), 530
90 Dewhurst, J. R. and Holbrook, C. G., *Inst. Gas Eng. Commun.*, 695 (1966); *J. Inst. Gas Eng.*, **6** (1966), 387
91 Fay, J. A., *Phys. Fluids*, **2** (1959), 283

92 Cubbage, P. A., *Gas Council Res. Commun.* GC63 (1959); *Trans. Inst. Gas Eng.*, **109** (1959–60), 543
93 Delbourg, P., 'Indices d'interchangeabilité et caractéristiques d'utilisation des combustibles gazeux', *Ass. Tech. Ind. Gaz France* (1953)
94 Pritchard, R., 'Natural gas properties', Symposium on the Utilization of Natural Gas in the Chemical Process Industries. Joint Symposium, University of Salford and Northwest Branch of the Institution of Chemical Engineers, April 1974
95 Edmondson, H. and Heap, M. P., *Combust. Flame*, **16** (1971), 161
96 North Thames Gas Board, 'A nomogram for the estimation of percentage flue loss for natural gas' (n.d.)
97 North Western Gas Board Industrial Development Centre, 'Flue loss chart for North Sea gas', reproduced in Proffitt, R., 'Some aspects of natural gas conservation', Manchester and District JGA (November 1975)
98 France, D. H. and Pritchard, R., 'Laminar burning velocities using a laser-Doppler anemometer', *J. Inst. Fuel*, **49** (1976), 79
99 Heap, M. P., 'The blowoff of burner flames', PhD thesis, University of Salford (1970)
100 Reed, S. B., Mineur, J. and McNaughton, J., *J. Inst. Fuel*, **44** (1971), 149

5

Acoustics

5.1 NATURE AND MEASUREMENT OF SOUND

The sensation of sound is caused by small fluctuations in air pressure at the human ear. The magnitude of the pressure variations determines the *loudness* of the sound, and their frequency determines the *pitch*, which is higher with increasing frequency. If the pressure fluctuations are regular, that is, if the pattern of fluctuations repeats itself regularly, the sound has a musical quality, but if the fluctuations are random the effect is a noise. In particular, if the pressure variation is sinusoidal a pure musical tone is obtained.

The standard A used in modern orchestras is a sinusoidal pressure variation of 440 Hz (cycles per second) and is used as a tuning signal by the BBC. The physiological effect of doubling the frequency of a pure tone is to produce a similar sensation; thus sinusoidal pressure variations of 110, 220, 440, 880, 1760 Hz, etc., all sound as the note A, but of different pitch, Similarly, frequencies which bear a simple fractional ratio to each other produce sounds which blend together pleasantly to produce a harmony; for example the notes C, E, G which form the common chord of C major have frequencies 264, 330, 396 Hz giving ratios:

$$\frac{G}{C} = \frac{3}{2}, \quad \frac{E}{C} = \frac{5}{4}$$

Musical instruments do not produce pure tones but mixtures of these simply related tones or harmonies, and the characteristic sound quality of an instrument is determined by the harmonics present. For example, Figure 5:1 shows some of the ways in which a stretched string can vibrate and the resulting mixture of harmonics.

A rotating fan will clearly produce a regular pressure variation due to the passage of each blade tip past the outlet port, and if there are m blades rotating at n revolutions per second then the basic frequency is $n \times m$ Hz. However, there is nothing in the geometry of the system to produce a pure sinusoidal variation. The form of the variation could be analysed into its sine components (Fourier components) and it would be found to contain a large number of components, some not harmonically related to the basic frequency and, therefore, producing a degree of non-musical quality.

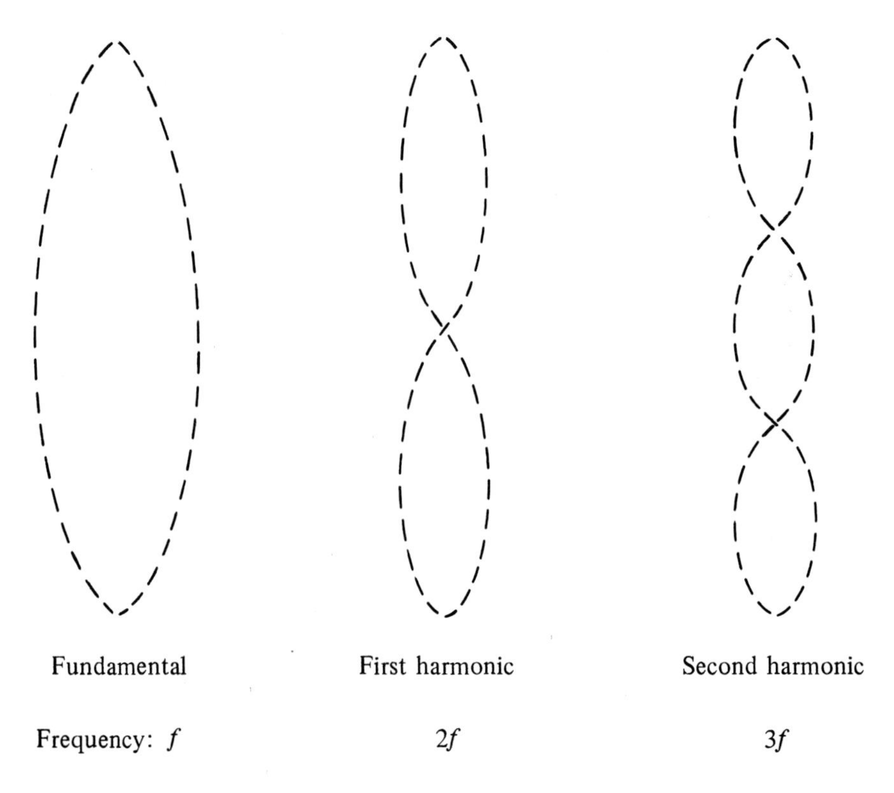

Figure 5:1 Some vibration modes of stretched string

5.1.1 Sound measurement

Pressure fluctuations may be detected by means of a microphone in which the fluctuations cause corresponding movements in a diaphragm. These are converted into electrical signals which are amplified in a sound-level meter. The meter may also contain electrical filters which permit the passage of signals of given frequency ranges only, thus enabling the separate frequencies to be measured.

A microphone measures the root-mean-square (rms) pressure variation, given by:

$$\bar{p}^2 = T^{-1} \int_0^T \{p(t)\}^2 \, \mathrm{d}t \tag{5.1}$$

where $\bar{p}$ is the rms pressure, $p(t)$ is the fluctuating pressure, and T is the period of the fluctuations. For example, for a simple harmonic pressure variation:

$$p(t) = p_{\max} \cos\omega t$$

Therefore

$$\bar{p}^2 = \frac{\omega}{2\pi} \int_0^T p_{\max}{}^2 \cos^2\omega t \mathrm{d}t$$

since $T = \dfrac{2\pi}{\omega}$, and this gives:

$$\bar{p}^2 = \tfrac{1}{2}\, p_{max}{}^2 \tag{5.2}$$

The sensitivity of the human ear to pressure variations of the order of 1000 Hz ranges between 2×10^{-5} Pa (threshold of hearing) to about 20 Pa (onset of pain), and the sensation of loudness is approximately proportional to the logarithm of $\bar{p}$ so that it is convenient to use a logarithmic scale for describing the intensity of a sound. The energy per second per unit cross-section carried by a sound wave is:

$$I = \bar{p}^2/\rho_0\, a \tag{5.3}$$

where ρ_0 is the density of the medium and a is the velocity of sound in the medium. Equation 5.3 is true for sound waves which are sufficiently far from their source to be considered as plane waves. For waves in air at 20 °C and 1 atm the value of the characteristic impedance $\rho_0\, a = 415$ Pa s m^{-1} and the threshold value of $\bar{p} = 2 \times 10^{-5}$ Pa corresponds to $I = 10^{-12}$ W m^{-2} which is the threshold intensity. Hence we may define:

$$\text{intensity level (IL)} = 10\log_{10}(I/10^{-12}) \text{ decibels (dB)} \tag{5.4}$$

and

$$\text{sound pressure level (SPL)} = 10\log_{10}\{\bar{p}^2/(2 \times 10^{-5})^2\} \text{ dB} \tag{5.5}$$

When $\rho_0 a$ has the characteristic value for air given above, intensity level and sound pressure level are almost exactly the same.

The total acoustic power radiated by a sound source is clearly given by:

$$W = \int_A I \,\mathrm{d}A \tag{5.6}$$

where A is the total area over which the source is radiating. If, for example, the source radiated uniformly over a hemisphere, then $I = W/2\pi r^2$, so that at a radius r:

$$\begin{aligned} \text{IL} &= 10\log_{10}\{(W/2\pi r^2)/10^{-12}\} \text{ dB} \\ &= 10\log_{10}\{(W/2\pi)/10^{-12}\} - 20\log_{10} r \text{ dB} \end{aligned}$$

and it therefore follows that in this case a doubling of r reduces IL by 6 dB.

If the sound source were a long straight pipe for which the acoustic power was W per unit length as in Figure 5:2 then $I = WL/2\pi rL$ and it follows that in this case a doubling of r causes a reduction of 3 dB in IL.

It is clear that the SPL at a given point due to several sources is *not* the sum of the individual SPLs at that point, but must be obtained by adding together the intensities (equation 5.3) obtained from the antilogarithms of the ILs (equation 5.4). It will be found that an additional source producing an IL that is 10 dB less than the existing IL at a point does not significantly alter the IL.

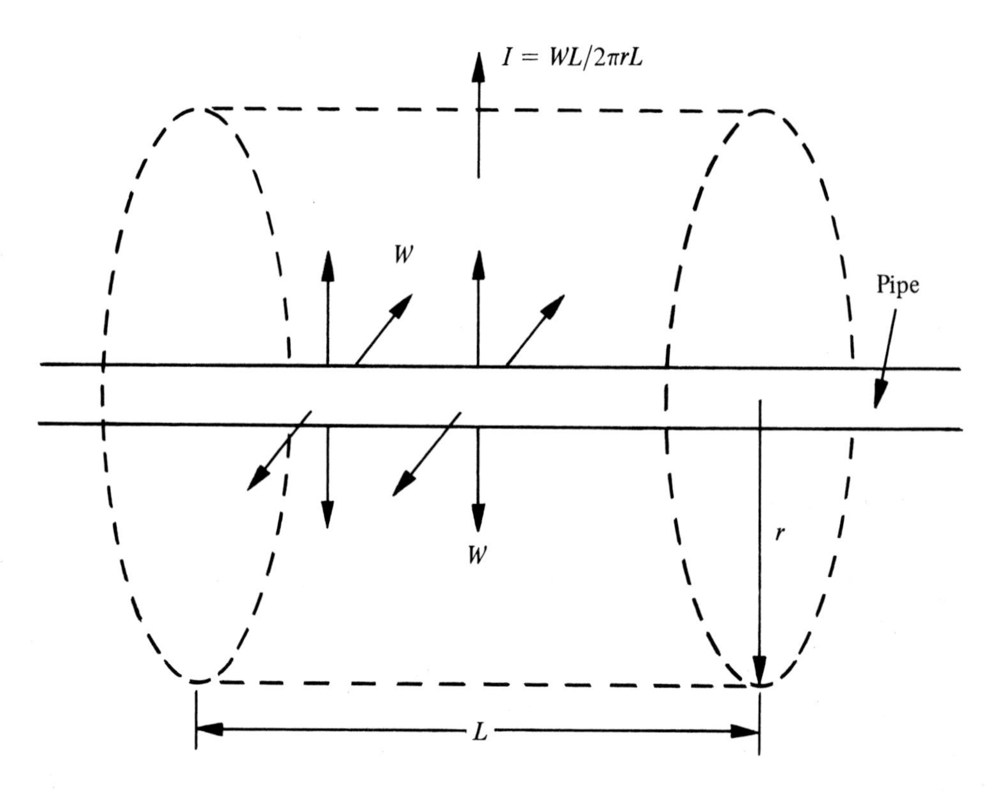

Figure 5:2 Uniform sound radiation from long source

5.2 PROPAGATION OF SOUND

The equation describing the motion of a pressure fluctuation in a gas may be derived by writing down the equations of continuity, motion and state using spherical coordinates, and remembering that the pressure fluctuations are small compared with the absolute pressure in the gas.

Continuity. Referring to Figure 5:3 where u is the velocity of the gas particles in the wave, and ρ is the small change in the density ρ_0

$$-\frac{\partial}{\partial r}[(\rho_0 + \rho)\, ur^2]\,\mathrm{d}r = r^2\,\mathrm{d}r\,\frac{\partial}{\partial t}(\rho_0 + \rho)$$

or

$$(\partial u/\partial r) + (2u/r) + \rho_0^{-1}\,(\partial \rho/\partial t) = 0 \tag{5.7}$$

State. Conditions in the wave are assumed to be adiabatic, so from equation 2.23:

$$(P_0 + p)/(\rho_0 + \rho)^\gamma = P_0/\rho_0^\gamma$$

i.e.

$$(P_0 + p)/P_0 = (\rho_0 + \rho)^\gamma/\rho_0^\gamma = \rho_0^\gamma\,[1 + (\rho/\rho_0)]^\gamma/\rho_0^\gamma$$

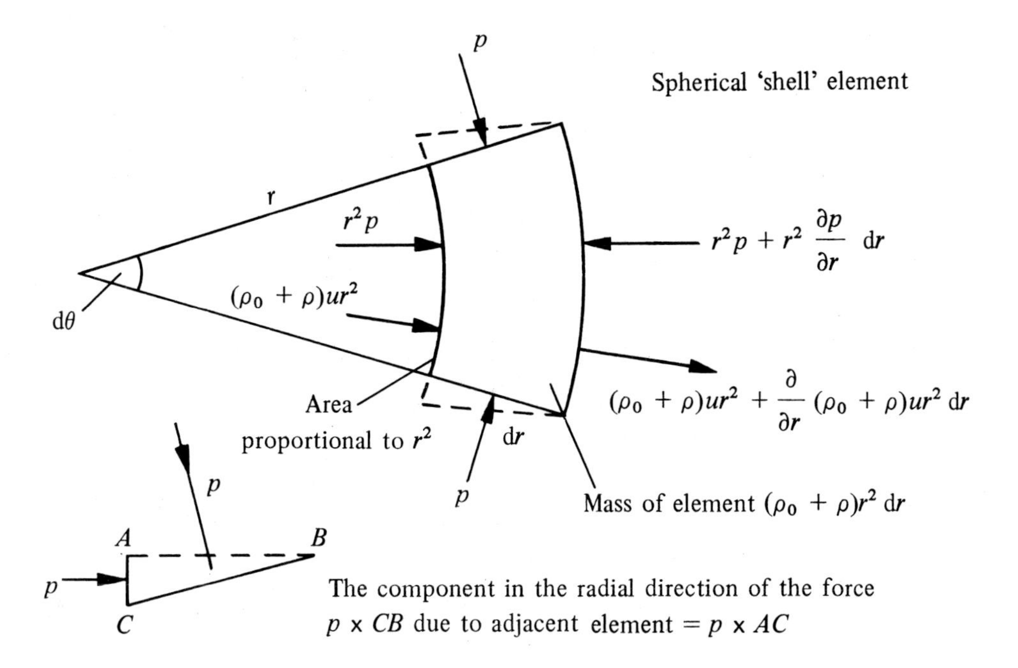

Figure 5:3 Elemental shell in spherical coordinates

and using two terms of the binomial expansion, since ρ/ρ_0 is small:

$$1 + (p/P_0) = 1 + \gamma(\rho/\rho_0)$$

or

$$p = \gamma(P_0/\rho_0)\rho \tag{5.8}$$

Motion. This was dealt with in Section 2.2.1. In the present case the motion is not steady so that the term $\partial u/\partial t$ in equation 2.14 cannot be ignored. Furthermore, it may be shown that for the small fluctuations occurring in sound waves $u(\partial u/\partial r) \ll \partial u/\partial t$ so the troublesome nonlinear first term will be neglected, whereupon the equation of motion becomes:

$$(\partial p/\partial r) + \rho_0 (\partial u/\partial t) = 0 \tag{5.9}$$

Differentiating 5.7 with respect to t, and 5.9 with respect to r, and eliminating $\partial^2 u/\partial r\, \partial t$ between them gives:

$$\frac{1}{\rho_0}\left(\frac{\partial^2 p}{\partial r^2}\right) - \frac{2}{r}\left(\frac{\partial u}{\partial t}\right) - \frac{1}{\rho_0}\left(\frac{\partial^2 \rho}{\partial t^2}\right) = 0$$

Then 5.8 can be used to eliminate ρ and 5.9 to eliminate u, giving finally:

$$\frac{\partial^2 p}{\partial r^2} + \frac{2}{r}\left(\frac{\partial p}{\partial r}\right) - \frac{\rho_0}{\gamma P_0}\left(\frac{\partial^2 p}{\partial t^2}\right) = 0 \tag{5.10}$$

The general solution of 5.10, which may be verified by direct differentiation of 5.11, is:

$$p = Kr^{-1}\,[f_1(t - ra^{-1}) + f_2(t + ra^{-1})] \tag{5.11}$$

where $a = (\gamma P_0/\rho_0)^{\frac{1}{2}}$ and this clearly implies that the pressure is the same at the same values of $(t \pm ra^{-1})$, i.e. that the pressure fluctuation moves through the gas with a velocity a which is therefore called the velocity of sound for the particular gas.

In equation 5.11 the f_1 term represents a disturbance travelling away from the source in the positive r direction while the f_2 term is one travelling towards the origin in the negative r direction. In open space where reflected sound is not being considered, only the f_1 term is needed.

Consider the case where the fluctuation is sinusoidal, say:

$$\begin{aligned} p &= Kr^{-1}\cos\omega(t - ra^{-1}) \\ &= Kr^{-1}\cos(\omega t - kr) \end{aligned} \tag{5.12}$$

where $k = \omega/a$ is called the circular wavenumber. Then, from 5.9, $\partial u/\partial t = \rho_0^{-1}$ $(\partial p/\partial r)$ whence 5.12 gives:

$$\partial u/\partial t = -\rho_0^{-1}Kr^{-1}\omega a^{-1}\sin\omega(t - ra^{-1}) + \rho_0^{-1}Kr^{-2}\cos\omega(t - ra^{-1})$$

and

$$u = (K/\rho_0 ar)\cos\omega(t - ra^{-1}) + (K/\omega\rho_0 r^2)\sin\omega(t - ra^{-1}) \tag{5.13}$$

At a sufficiently large value of r the second term of 5.13 becomes negligible. This is called the *far-field* region, and it will be seen from 5.12 and 5.13 that p and u vary inversely with r in the far field.

The constant K in 5.12 and 5.13 would, of course, be evaluated from given boundary conditions. For example, the sound source might be a small sphere of gas of mean radius r_0 oscillating with a surface velocity given by $u_0\cos\omega t$. This is called a monopole source and is of importance in the theoretical treatment of combustion noise.

5.2.1 Energy intensity and density

The energy intensity is the rate at which the pressure does work, per unit cross-section, and is the time-averaged product of pressure and velocity. For the case discussed above it is the product of equations 5.12 and 5.13. Only the cos term of equation 5.13 needs to be taken, since the time-averaged value of $\cos\theta\sin\theta$ is 0. Noting from 5.12 and 5.13 that $u = p/a\rho_0$ then:

$$I = T^{-1}\int_0^T pu\,\mathrm{d}t = T^{-1}\int_0^T (p^2/\rho_0 a)\cos^2\omega(t - ra^{-1})\,\mathrm{d}t = \bar{p}^2/\rho_0 a$$

as previously given in equation 5.3.

The energy density is the energy contained within unit volume of fluid and is clearly given by

$$D = I/a = \bar{p}^2/\rho_0 a^2 \tag{5.14}$$

5.3 SOUND WAVES IN ONE DIMENSION

Sound waves frequently occur in enclosures which are long relative to their cross-section, such as pipes, ducts and combustion chambers. In these, the sound propagation is effectively one-dimensional. Also, reflection of sound waves can occur so that waves often travel in both directions along the pipe.

Applying the same method as in Section 5.2 to one-dimensional waves:

$$(\partial^2 p/\partial x^2) - a^{-2}(\partial^2 p/\partial t^2) = 0 \tag{5.15}$$

and

$$\partial u/\partial t = -\rho_0^{-1}(\partial p/\partial t) \tag{5.16}$$

The general solution of 5.15 is:

$$p = K[f_1(t - xa^{-1}) + f_2(t + xa^{-1})] \tag{5.17}$$

which is similar to equation 5.11 but shows that the pressure fluctuations do not diminish with distance from the source in the one-dimensional case.

In considering the propagation of simple harmonic acoustic waves in ducts it is very convenient to represent the pressure as a complex number. For example, for a wave travelling in the positive x-direction, a perfectly general expression is

$$p_+ = A_+\cos(\omega t - kx + \varphi_+) \tag{5.18}$$

where the phase angle φ_+ allows initial boundary conditions to be met. Equation 5.18 may be written as:

$$p_+ = A_+\, e^{j\varphi}\, e^{j(\omega t - kx)} = A_+\, e^{j(\omega t - kx)} \tag{5.19}$$

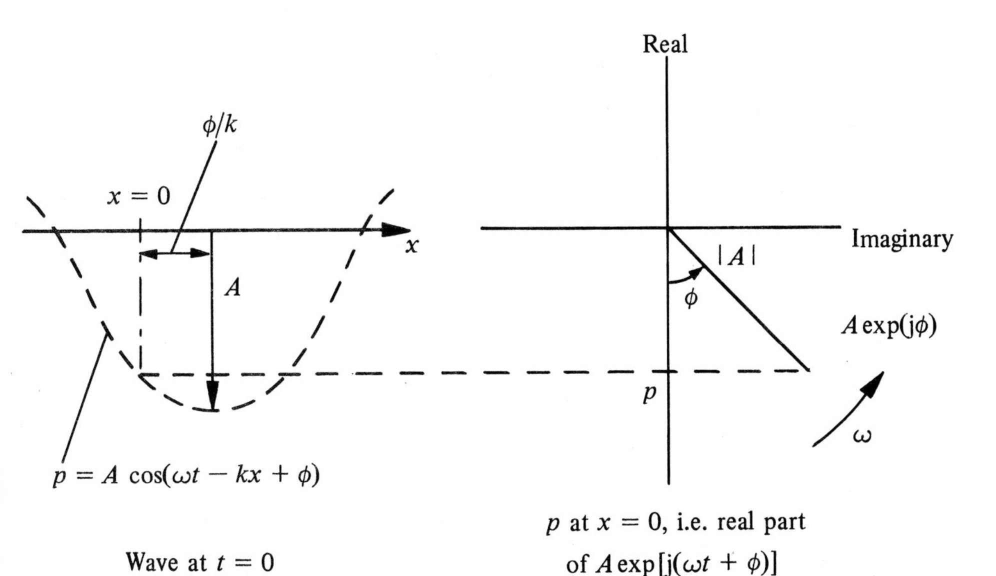

Figure 5:4 Representation of wave as a complex number

where p_+ is the complex pressure, and the actual pressure is given by the real part of 5.19. The meaning of equation 5.19 is shown in Figure 5:4. By using equation 5.16 in equation 5.19:

$$u_+ = (A_+/\rho_0 a)e^{j(\omega t - kx)} = p_+/\rho_0 a \tag{5.20}$$

Similarly, for the wave in the negative direction:

$$p_- = A_- \; e^{j(\omega t + kx)} \tag{5.19a}$$

and

$$u_- = -p_-/\rho_0 a \tag{5.20a}$$

Clearly, at any point in the duct:

$$p = p_+ + p_-$$

and

$$u = u_+ + u_-$$

It is also useful to define an acoustic impedance for the duct as:

$$Z = p/Su \tag{5.21}$$

where S is the cross-sectional area of the duct.

The mean square pressure is as found in equation 5.2:

$$\bar{p}^2 = \tfrac{1}{2}A^2 \tag{5.22}$$

and the power associated with the wave is, from equation 5.6, equal to the intensity × cross-sectional area, i.e.:

$$W = A^2S/2\rho_0 a = A^2/2Z \tag{5.23}$$

where the quantities refer to the wave under consideration at the particular place in the duct.

The above equations 5.18 to 5.23 are used in considering a wide variety of prob-

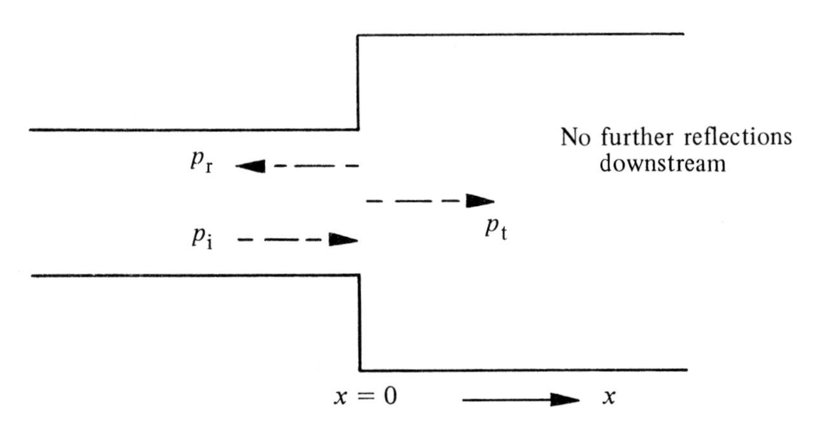

Figure 5:5 Reflection of wave at change of cross-section in a duct

lems of sound behaviour in ducts, such as the effect of junction and other discontinuities, and some reference will be made to these in later sections in the chapter. A specific example is considered next.

5.3.1 Effect of a change in duct cross-sectional area

This is shown in Figure 5:5 in which a wave p_i in the positive direction is partly reflected (p_r) and partly transmitted (p_t) at the junction. Then:

$$p_t = p_i + p_r \tag{5.24}$$

and by continuity:

$$u_t S_2 = u_i S_1 + u_r S_1$$

or from 5.20 and 5.20a:

$$\frac{p_t S_2}{\rho_0 a} = \frac{p_i S_1}{\rho_0 a} - \frac{p_r S_1}{\rho_0 a} \tag{5.25}$$

Eliminating p_r between 5.24 and 5.25 gives:

$$p_t/p_i = 2S_1/(S_1 + S_2) \tag{5.26}$$

and using 5.23 a power transmission coefficient can be defined as:

$$\begin{aligned} \alpha_t = W_t/W_i &= p_t^2 S_2/p_i^2 S_1 = 4S_1^2 S_2/S_1 (S_1 + S_2)^2 \\ &= 4/(1 + S_2 S_1^{-1})(1 + S_1 S_2^{-1}) \end{aligned} \tag{5.27}$$

Hence only a fraction of the incident power is transmitted past a change in the cross-sectional area of a duct, and this may be used to reduce noise transmitted along a duct (see Section 5.6).

5.3.2 Standing waves

Waves travelling in opposite directions along a duct, such as the p_i and p_r of Section 5.3.2, form a standing wave in which the pressure at any section varies sinusoidally with an amplitude which is a function of x. The pressure is given by:

$$p = A_i \, e^{j(\omega t - kx)} + A_r \, e^{j(\omega t + kx)} \tag{5.28}$$

and the form of the standing wave is as shown in Figure 5:6.

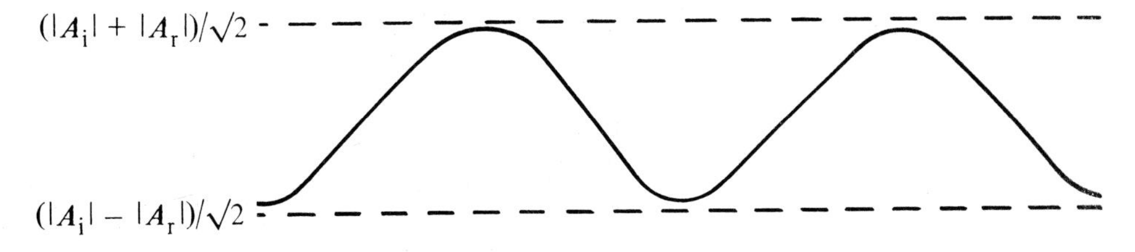

Figure 5:6 Standing wave in duct

A particularly important case is where the duct has a free open end, such as a burner or a flue. The results of Section 5.3.1 show that p_t will be small and that $p_i \approx -p_r$. In this case:

$$p = p_i + p_r = A\cos(\omega t - kx) - A\cos(\omega t + kx) = 2A\sin kx\sin\omega t \tag{5.29}$$

and

$$u = (p_i - p_r)/\rho_0 a = 2A\cos kx\cos\omega t \tag{5.30}$$

Clearly, the amplitude of the pressure variation is zero wherever $\sin kx = 0$, that is at $x = 0$, $\pm\pi/k$, $\pm 2\pi/k$, etc., and these points are called pressure nodes. At values of $x = \pm\pi/2k$, $\pm 3\pi/2k$, etc., the pressure variation has a maximum value of $2A$, and these are the pressure antinodes. Similarly, velocity nodes and antinodes coincide with pressure antinodes and nodes, respectively. Figure 5:7 shows the form of the

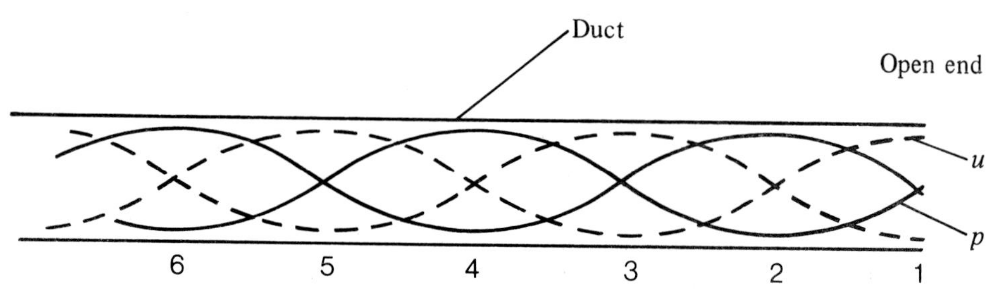

Figure 5:7 Standing wave when $|A_i| = |A_r|$ (equation 5.30)

standing wave in this case, and it is clear that the same pattern of standing waves could exist if the tube were open at any pressure node, or closed at any velocity node.

The foregoing theory becomes inaccurate when the fluid is moving since the term $u(\partial u/\partial x)$ cannot be ignored, and further complications occur when there are temperature gradients in the fluid.

5.4 NOISE AND THE SENSATION OF LOUDNESS

Noise, i.e. unwanted sound, is a nuisance, and since nuisance is a rather subjective matter it is desirable, although difficult, to correlate 'nuisance value' with the precise acoustical measurements of sound pressure level and frequency distribution.

Many experiments have been carried out in which people with normal hearing have been asked to compare the loudness of pure (i.e. single-frequency harmonic) tones of different frequencies and intensities. The results are shown in Figure 5:8 which reveals that frequencies in the region 2–5 KHz produce the maximum sensation of loudness, but the irregular shape of the 'equal loudness' contours in this region is due to the distortion of the sound field by the head of the listener, and will therefore differ for sources of different shapes, e.g. point sources or large diffuse sources. The given

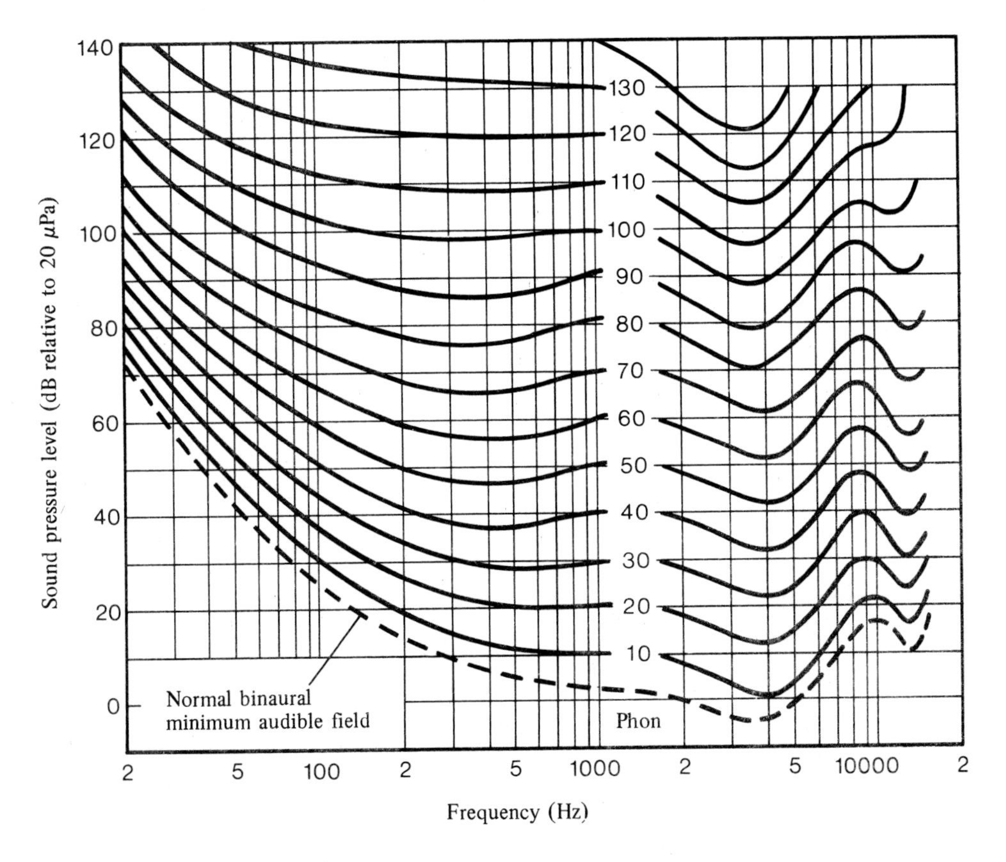

Figure 5:8 Equal-loudness contours [1]

curves are for free-field conditions, and will be somewhat different if the sound is presented to the listener through earphones. The unit of loudness is the *phon*, which is defined as the SPL of a 1 KHz tone having the same loudness as the given tone.

The disadvantage of the phon is that like the decibel (Section 5.1.1) it is not linearly related to loudness, and therefore another unit called the *sone* has been defined. This is a loudness index which is linearly related to the sensation of loudness so that two individual noises of, say, 3 sones and 5 sones respectively produce a total noise of 8 sones. The units are related by:

$$\left.\begin{aligned} \text{(phons)} &= 40 + 10 \log_2 \text{(sones)} \\ \text{or} \qquad \log_{10} \text{(sones)} &\approx 0.03 \times \text{(phons)} - 1.2 \end{aligned}\right\} \tag{5.31}$$

(see reference [2]).

A further complication arises because of the phenomenon of *masking* whereby the presence of a loud sound renders inaudible a less loud sound. For example, it is a common experience that speech or music may not be heard in the presence of a loud sound such as that of a lorry or aircraft. For pure tones of around 1 KHz a sound is masked by another which is about 22 dB higher.

5.4.1 Loudness of broad-band noise

Usually we are subjected to noises covering a wide range of frequencies coming from several directions. Lines of equal loudness similar to those of Figure 5:8 can be drawn for diffuse sounds covering complete octave bands instead of pure tones. The SPL of the noise to be evaluated is measured in each octave band, and this information, together with the equal-loudness contours, may be used to calculate a noise index.

In Stevens' method [3, 4] the measured level in each octave band is converted to a loudness level in sones using the equal-loudness contours, and the total noise index is computed using the equation:

$$s_t = s_m + 0.3\,(\sum s - s_m) \tag{5.32}$$

where s_t is the total loudness, s_m is the loudness of the loudest octave band, and Σs is the sum of the loudness of all the octave bands, all given in sones. This formula takes into account the masking effect due to the loudest octave band.

Another method used in the gas industry is to plot the measured noise distribution on the same graph as the equal-loudness contours, and the noise rating number N of the whole noise is taken as the highest individual octave-band noise rating, i.e. N is the phon value of the loudness contour which just touches the given distribution curve. This is shown in Figure 5:9. Details of this method are given in reference [5].

The above methods involve fairly detailed measurement of the noise in octave bands which is essential to the engineer who is concerned with designing methods of reducing the noise level. A much simpler method for evaluating the total noise level is to incorporate into the sound-level meter electrical weighting networks which reduce the effect of high and low frequencies relative to the mid-frequencies in a similar manner to the ear. Three such networks are in common use corresponding roughly to the shapes of the 40, 70 and 100 phon contours of Figure 5:8 and called the A, B and C networks respectively. Thus, a single measurement with the A network in the meter gives a noise level given in db(A), and this is probably the most frequently used method of noise assessment.

5.5 SOURCES OF NOISE IN COMBUSTION SYSTEMS

The three main sources of noise are:

1 Machinery, e.g. pumps and fans
2 Gas flow, e.g. in valves, pipes, and injectors
3 Flames

Theoretical understanding of noise production by these sources is quite limited, and only a brief account can be given here.

5.5.1 Fan noise

Fans and their characteristics are dealt with in Chapter 10. The noise produced by fans comprises:

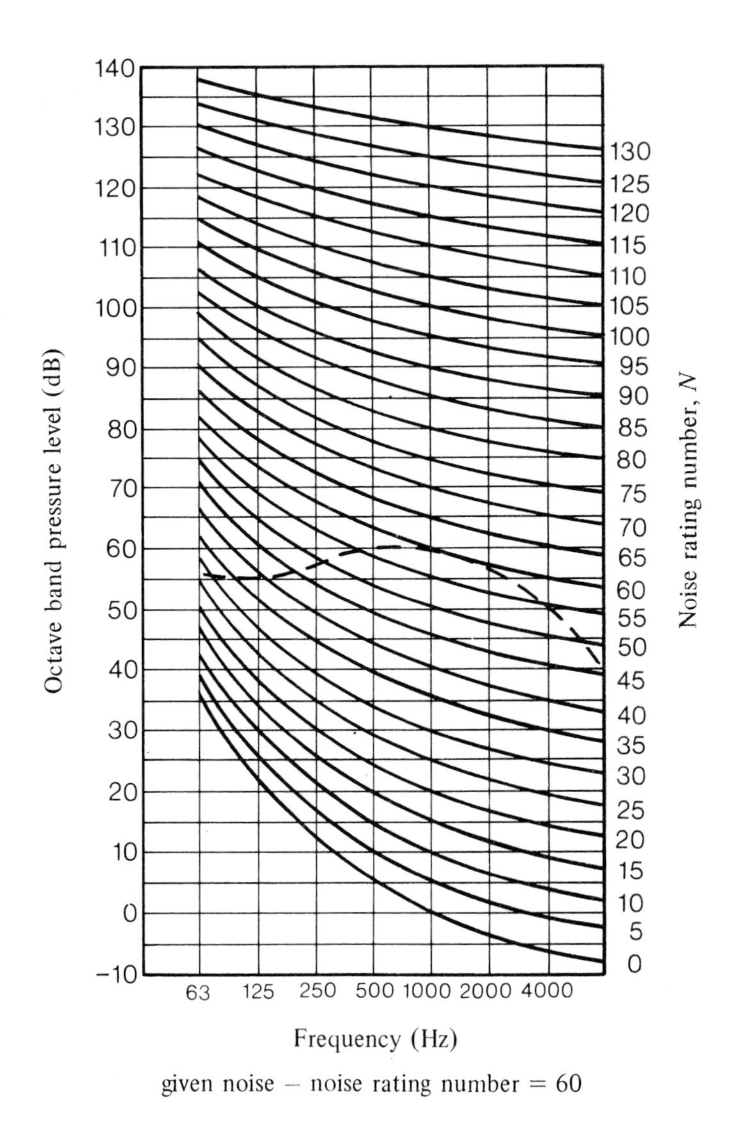

Figure 5:9 Noise rating number of given noise [1]

1 Mechanical noise of bearings, unbalanced rotating parts and associated resonances of casing and ducts
2 Noise due to the random shedding of vortices by the blades, and the generation of turbulence
3 Noise produced by the regular passage of the blades past the fixed vanes or outlet ports

The mechanical type of noise should not be a problem in a well maintained and properly installed system. This may involve the use of vibration isolators, the use of flexible ducting, and the damping of vibrations by means of bitumen-type coatings on casings and ductwork. As a result of (2) and (3) the noise spectrum from fans is essentially broad band due to (2) with certain discrete frequencies due to (3) superimposed on it. The broad-band noise is at a minimum when the fan is operating at its point of maximum efficiency and when care is taken to avoid the generation of unnecessary turbulence at inlet or outlet. For example, a bell-mouthed fan entry is better than a sharp-edged entry.

The discrete frequencies depend on the number of moving blade/fixed vane coincidences per second, and this part of the noise output is reduced if the ratio, number of blades/number of vanes is not an integer, and if the clearance between blades and vanes is increased.

A larger-diameter low-speed fan will generally produce less noise than a smaller high-speed fan working at the same output and efficiency.

5.5.2 Flow noise

Flow in ducts and valves produces noise only when turbulence is present since it is only in turbulent flow that fluctuations in velocity and pressure occur. This is mainly caused by sudden changes in direction and cross-section and, in fact, wherever any dissipation of energy occurs. Thus more noise is produced in a partially closed than in a fully opened valve. Since turbulence is essentially a random fluctuation, the noise associated with flow is broad band. Very substantial noise will be produced in shock waves but these will not occur in the low-velocity flow conditions associated with combustion systems. A great deal of energy is dissipated in a jet of gas expanding into a relatively stationary gas, and the injector is a common source of noise. Noise is not radiated uniformly in all directions from a jet but is mainly in the region 20°–60° from the jet axis. The acoustic power radiated is proportional to u^8A where u is the jet velocity and A its cross-sectional area, so the noise increases rapidly as the injector pressure increases. The noise is broad band with a peak frequency in the region of u/D where D is the nozzle diameter.

5.5.3 Combustion noise

The noise generated by a flame is due to local random fluctuations in the rate of reaction which cause fluctuations in the local velocity and pressure changes in the combustion zone. Perfectly laminar flames in a homogeneous gas are practically silent, and the noise from a flame is dependent upon turbulence in the gas stream and local variations in mixture composition. The noise is broad band covering a rather wider range of frequencies and having a peak at a rather lower frequency than the aerodynamic noises. The term 'combustion roar' is often used to describe this noise. The noise output from premixed flames varies linearly with the Reynolds number of the cold gas leaving the burner port, and with the burning velocity, S_u, of the mixture if lean, but approximately with $\sqrt{(S_u)}$ if rich. As would be expected, considerable noise reduction can be achieved by inserting flow straighteners in the gas stream entering the burner.

Turbulent diffusion flames present a more complex problem, since both mixing and combustion take place in the combustion zone, and wide variations occur with different designs, probably associated with difference in the rapidity and homogeneity of mixing. The most important factor affecting noise from a given burner shape appears to be the total velocity of the gas plus air.

This result has been obtained from laboratory tests on packaged burners which show also that for a fixed total flow rate the maximum noise output occurs at stoichiometric conditions. The effect of total flow rate is, however, more important for this type of burner so that at a given gas input the noise increases with excess air.

The design of the burner head affects the noise output, and a large burner with low air-supply pressure produces the least noise.

When placed in the boiler the burner noise is strongly attenuated and may be no more than that from the fan. The use of silencing materials in the air passages and burner fan inlet can considerably reduce the noise from the installation.

Combustion oscillations

It often occurs that a loud noise consisting mainly of a single frequency is emitted by a combustion system. In such cases it is found that standing waves are set up in the burner, combustion chamber, furnace and flue and it is the interaction of the wave with the flame which causes pulsations in the fuel supply and combustion process in such a manner as to feed energy continuously to the wave. There are obvious similarities to the production of a sound in an organ pipe, hence the term 'organ-pipe resonance' used to describe the phenomenon. The phenomenon is influenced by the impedances in the system and may be controlled by altering, for example, the terminal impedance or by modifying the system using quarter-wave tubes. There is a certain similarity between this type of oscillation and the instability which may occur in a feedback system, but the characteristics of the 'amplifier', i.e. the flame, are likely to be nonlinear and are certainly not understood at present.

In some cases the phenomenon is a true resonance in which a particular frequency produced in one part of the system, for example from a fan, interacts with a column of gas in the burner or combustion system. A similar phenomenon is the interactive coupling which can occur between two similar noise sources such as twin burners in adjacent fire tubes. These may be satisfactory when used singly but noisy when used together.

The periodic shedding of vortices at obstructions in the flow system can also be the cause of this type of noise.

There are, therefore, many ways in which the single-frequency noise can arise, and many possible remedies, including:

1. Use of a less intense, i.e. longer, flame
2. Improvement of flame stability
3. Ensuring that no acoustic frequencies in the combustion part of the system coincide with each other or with natural frequencies in the mechanical part of the system. This may require dimensional changes in the system
4. Avoidance of vortex-shedding obstructions upstream of the flame
5. Preventing standing-wave formation by drilling relief holes near pressure maxima (antinodes), or by using quarter-wave tubes or Helmholtz resonators.

5.6 METHODS OF NOISE REDUCTION

Figure 5:10 illustrates the various basic methods of noise reduction which may be

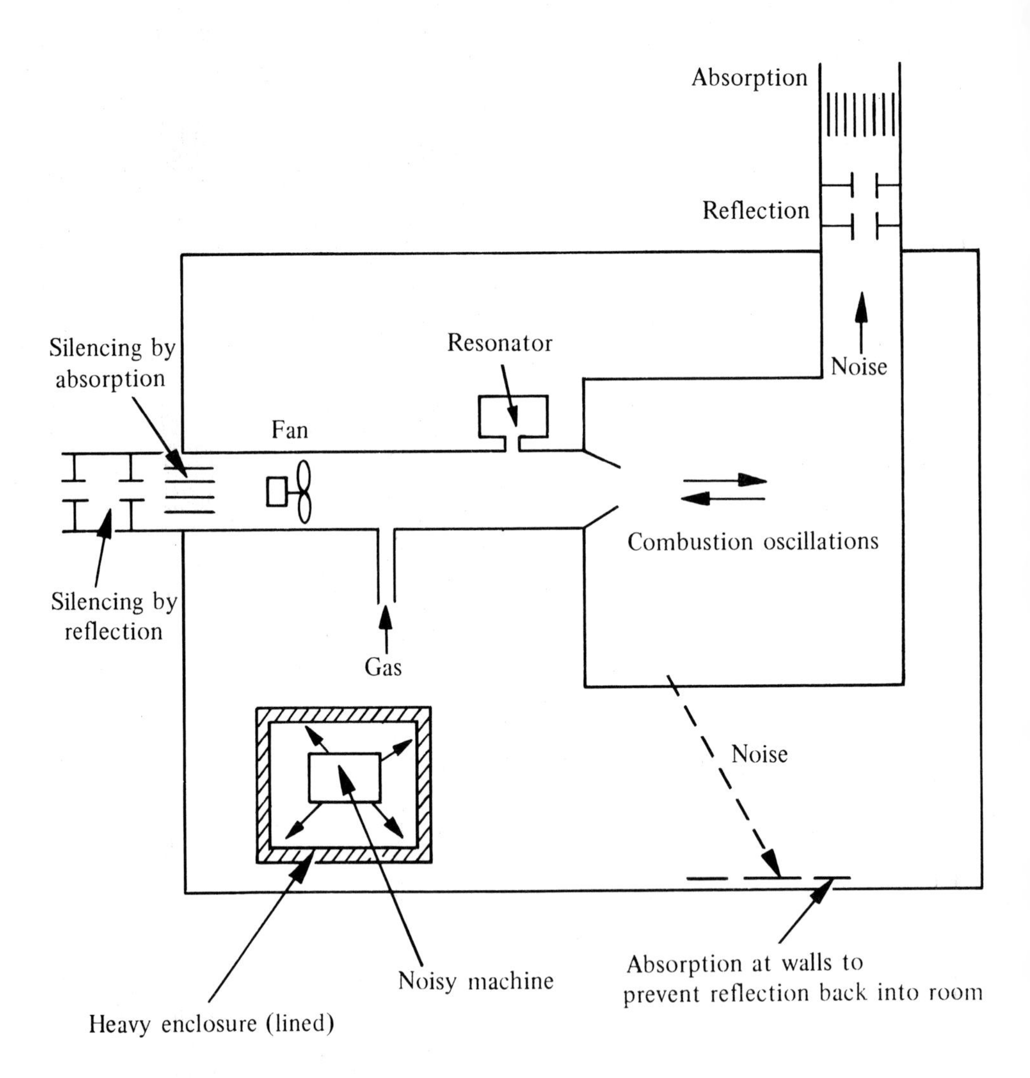

Figure 5:10 Methods of noise reduction

used. It is clear that these are likely to be fairly costly, and therefore the more the noise can be reduced at its source, the better.

5.6.1 Enclosures

Noisy machines such as compressors and gas turbines may be completely enclosed in a chamber. It is clear that since sound energy is continuously leaving the machine it must ultimately be converted into heat energy so that it is essential to line the inside of the enclosure with a porous material which will dissipate energy by friction, or incorporate into the wall a material exhibiting viscous damping in order to prevent the sound intensity in the chamber from building up to a large value.

The sound insulation value of a wall is given approximately by:

$$R = 20\log_{10}(\omega M/\rho_0 a) - 5\,\text{dB} \tag{5.33}$$

where R is the reduction in sound passing through the wall, ω is angular frequency and M is the mass per unit area of the wall. Hence a wall needs to be of heavy material, and is less effective in reducing the transmission of low-frequency noise. Somewhat better results can be obtained by using a double-wall construction, since for completely separated walls the total R is the sum of the individual R values, whereas doubling M for a single wall increases R by only 6 dB, but it is difficult to achieve values of R of much above 50 dB.

Problems are also encountered in providing means of access to the machine. Noise may also be brought out of the enclosure through pipes connected to the machine, and flexible connections are essential. Even partial enclosures or large barriers may be of some use in reducing nuisance due to high-frequency noise.

5.6.2 Sound absorption

This is the process whereby the energy of sound waves impinging on a material is degraded to random heat energy, and since for almost all sound-reduction methods the sound energy must eventually be so converted it follows that sound-absorbing materials must play a very important part in sound-reduction techniques.

The essential mechanism is that the sound wave travels into the material and the particle velocity is reduced therein by fluid friction. There is, therefore, an optimum porosity of the material such that the wave can readily enter it (i.e. not be totally reflected from the surface) and yet have sufficient flow resistance to reduce the particle velocity substantially.

Absorption will clearly be greater if the region of maximum particle velocity of the wave is within the absorbing material which means that for a given thickness of material, high frequencies (i.e. shorter wavelengths) are more readily absorbed. Absorption of lower frequencies is improved by having thicker layers of absorbing material and by having a gap between the material and the wall on which it is mounted.

The effectiveness of a sound-absorbing material is described by an absorption coefficient α where:

$$\alpha = \frac{\text{incident energy}}{\text{emergent energy}} \text{ of wave} \tag{5.34}$$

The value of α depends on the angle of incidence of the wave, the thickness of the material, and the frequency, and is often quoted for random incidence, i.e. for waves impinging in all directions as in a room, although for duct linings the value for grazing incidence is more appropriate. Some typical values are given in Figure 1:30.

Absorbent materials are used in two main ways:

1 To line walls and ceilings of rooms in which noise is generated in order to reduce the sound intensity in the room

2 As an internal lining to ducts to absorb noise generated in the duct and thus to prevent sound being radiated from the ends of the duct

Many semi-rigid and flexible materials are used as wall linings. When applied direct to the wall or ceiling they have maximum absorption at high frequencies. Greater absorption at low frequencies may be achieved by mounting the lining away from the wall and by covering the surface of the porous lining with a perforated sheet. The absorption characteristics for these types are shown in Figure 5:11.

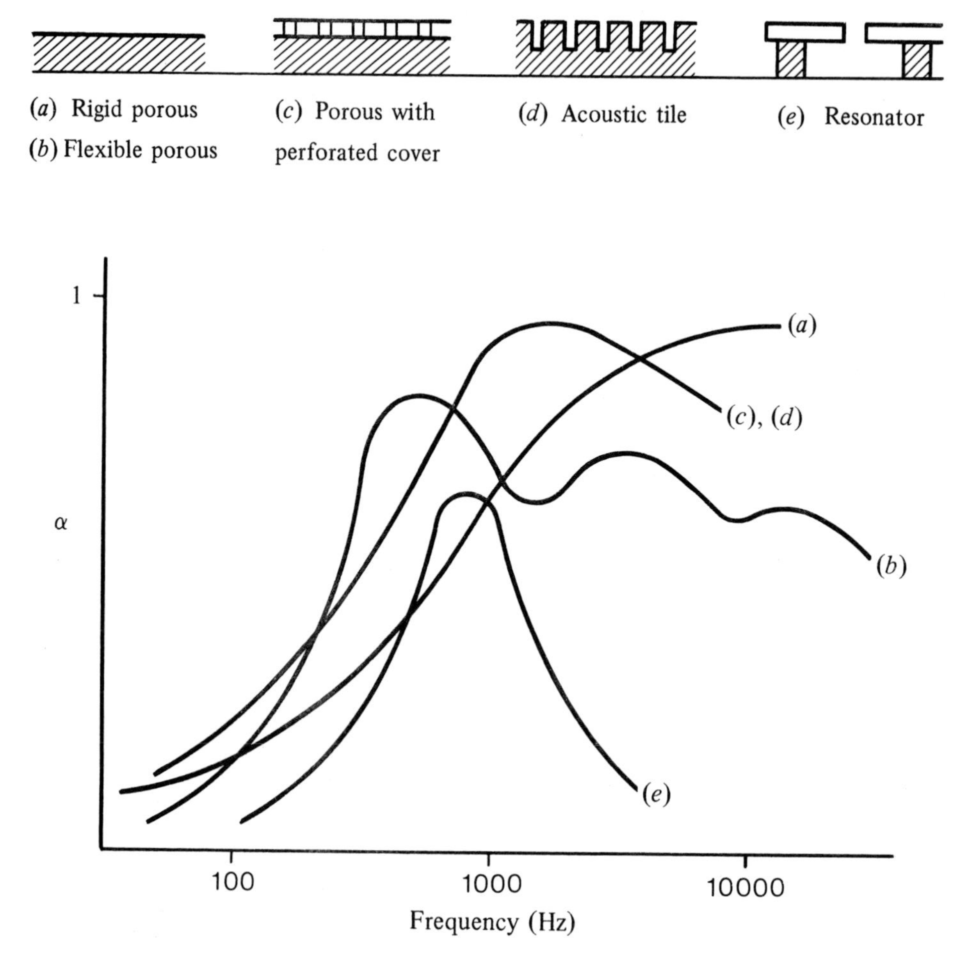

Figure 5:11 Absorption coefficients for acoustic materials

There are also panels designed as Helmholtz resonators (see Section 5.6.3) which can be made to absorb sound in any desired frequency range.

When used in ducts, as great a surface area of absorbing material as possible should be placed in contact with the sound, and a common method of achieving this is to provide additional surface in the form of 'splitters' as shown in Figure 5:12. This method is used in many simple silencers placed on fan and burner air intakes or on

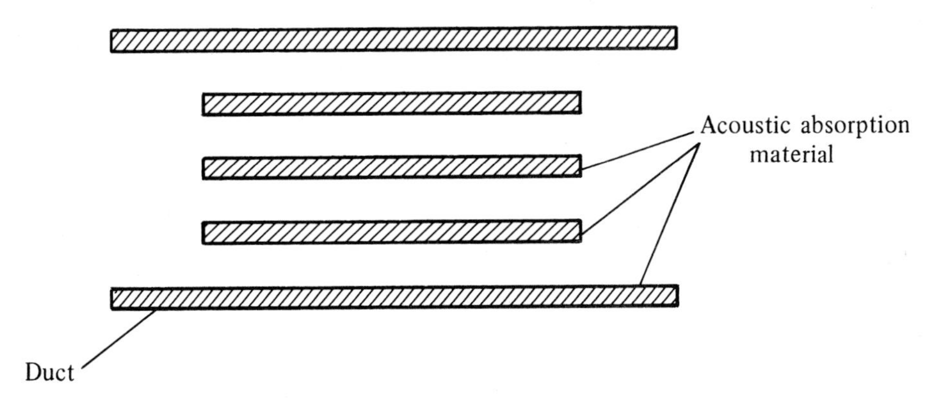

Figure 5:12 'Splitters' for increasing absorption in duct

exhaust systems. Sometimes it may be necessary to use a heat-resisting material and if this is a loose material, e.g. rock wool, it can be secured behind a mesh.

The attenuation (i.e. sound reduction) obtained by this means is given by

$$\text{attenuation} = 4.2\,\alpha^{1.4}\,L/D\ \text{dB} \tag{5.35}$$

where L is the length of lined duct and D is its diameter or hydraulic mean depth, i.e. 4 × cross-sectional area/perimeter.

Some sound reduction can also be achieved by introducing a flow resistance over the whole cross-section of the duct by making the gas flow through a porous material, and this has sometimes been effective in eliminating combustion oscillations.

5.6.3 Acoustical filters or mufflers

In this type of silencer used in ducts, changes in impedance in the duct are used to prevent the sound wave from freely passing and to partially reflect it back to its source. The two basic methods used are changes in cross-sectional area in the duct, and the provision of side branches. The theory of such devices is reasonably well understood but is complex and specialized, and only a brief outline can be given here.

Figure 5:13 Silencer element

The simplest silencer unit employing these principles is shown in Figure 5:13 and the theory of its operation is essentially that given in Section 5.3.1. The maximum attenuation occurs for wavelengths of 4 × length of element, and a range of frequencies can be filtered out by having sections of different lengths in series. The attenuation is also increased by increasing the ratio of outside pipe diameter to inside pipe diameter. Such a device also produces a resistance to the steady flow of the fluid which may be calculated by the conventional theory of fluids.

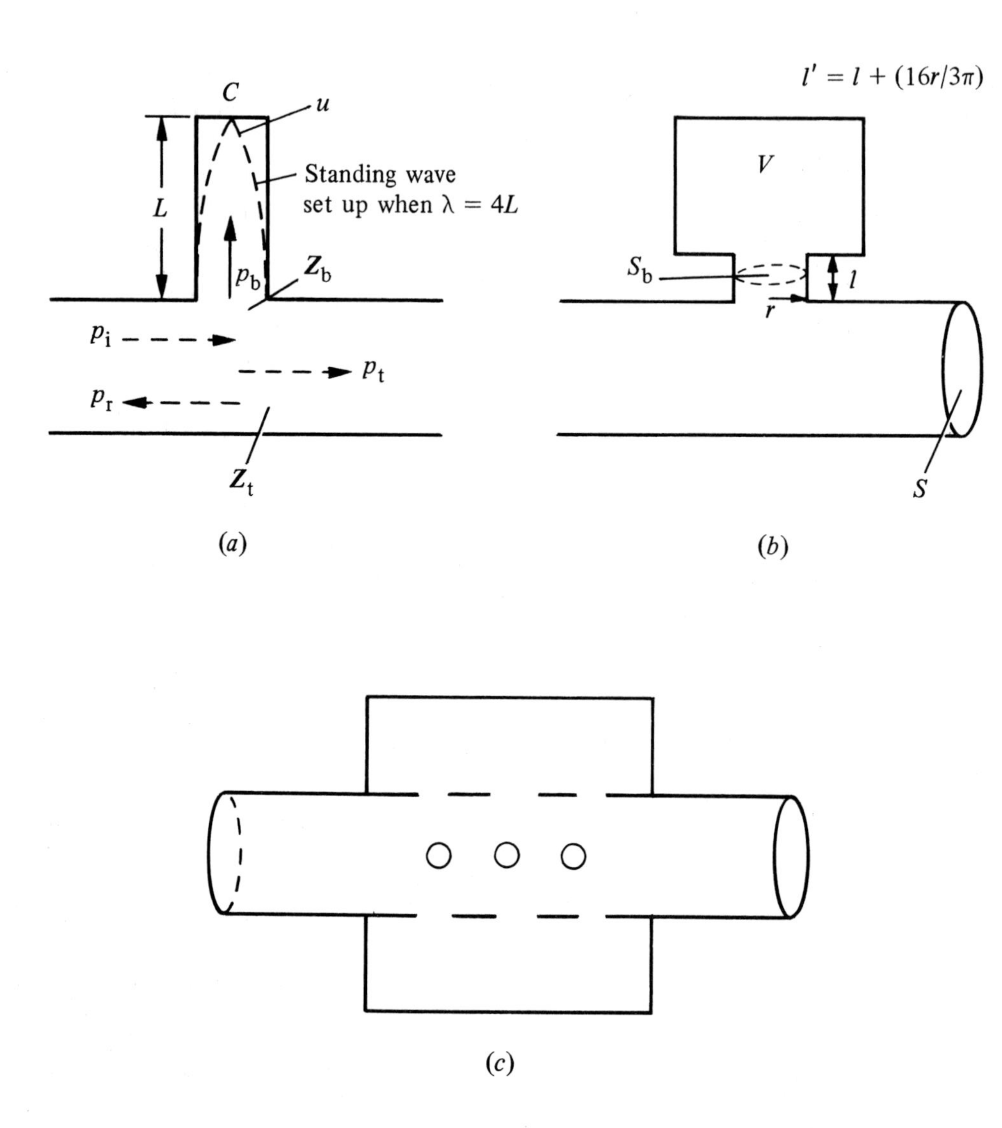

Figure 5:14 (*a*) Quarter-wave tube (*b*) and (*c*) Helmholtz resonators

The two most important side-branch filters are:

1 Quarter-wave tubes
2 Helmholtz resonators

The general form of these is shown in Figures 5:14 (*a*) and (*b*) respectively. A rather neater form of the Helmholtz resonator is shown in Figure 5:14 (*c*) and this kind of design is used in many motor-car exhaust silencers.

Quarter-wave tube
This will completely eliminate any transmitted wave (p_t in Figure 5:14 (*a*)) that has a wavelength of $4L$, but has very little effect on other frequencies. For this reason it is only of use in eliminating a single-frequency sound such as may occur in combustion oscillations, and it can be remarkably effective in this kind of application. In practice it is a good idea to replace the closed end C by a movable piston so that the device can be 'tuned' exactly.

By writing down continuity of pressure and flow at the side branch, as in Section 5.3.1, it is easy to show that:

$$\alpha_t = |\mathbf{Z}_b|^2/|\tfrac{1}{2}\mathbf{Z}_t + \mathbf{Z}_b|^2 \tag{5.36}$$

where $\mathbf{Z}_t$ and $\mathbf{Z}_b$ are the acoustic impedances of tube and side branches respectively. The vertical bars denote the moduli of what are generally complex numbers. The attenuation, i.e. reduction is sound transmitted, is $10\log\alpha_t^{-1}$ dB, i.e.:

$$\text{attenuation due to side branch} = 10\log_{10}(|\tfrac{1}{2}\mathbf{Z}_t + \mathbf{Z}_b|^2/\mathbf{Z}_b^2)\ \text{dB} \tag{5.37}$$

When the frequency is such that a standing wave is set up in the side tube as shown in Figure 5:14 (*a*) then $\mathbf{Z}_b = 0$.

Helmholtz resonator
This attenuates sound over a range of frequencies and is therefore of more general application than the quarter-wave tube. It has been successfully incorporated into a burner mixing tube to reduce mixing noise. It can be shown that:

$$\mathbf{Z}_b = \mathrm{j}(\omega\rho_0 l' S_b^{-1} - \rho_0 a^2 \omega^{-1} V^{-1})$$

and using this in 5.37 with a little rearrangement gives the convenient form:

$$\text{attenuation due to Helmholtz resonator} = 10\log_{10}\left\{1 + \left[\frac{\sqrt{(S_b V/l')}/2S}{ff_r^{-2} - f_r f^{-1}}\right]^2\right\}\ \text{dB} \tag{5.38}$$

where $f_r = (a/2\pi)\sqrt{(S_b/l'V)}$ is the resonant frequency at which the attenuation is a maximum. Figure 5:15 shows the form of the attenuation for different values of the geometrical quantity $\sqrt{(S_b V/l')}/2S$. Thus by choosing the desired value of f_r and the desired shape of attenuation curve the dimensions of the resonator can be determined. The above equations become inaccurate if the length of the resonator exceeds about $\frac{1}{8}$ wavelength. Improvements in performance are achieved by lining the chamber with an acoustic absorbing material, and the whole device acts analogously to the vibration absorber often used in mechanical engineering.

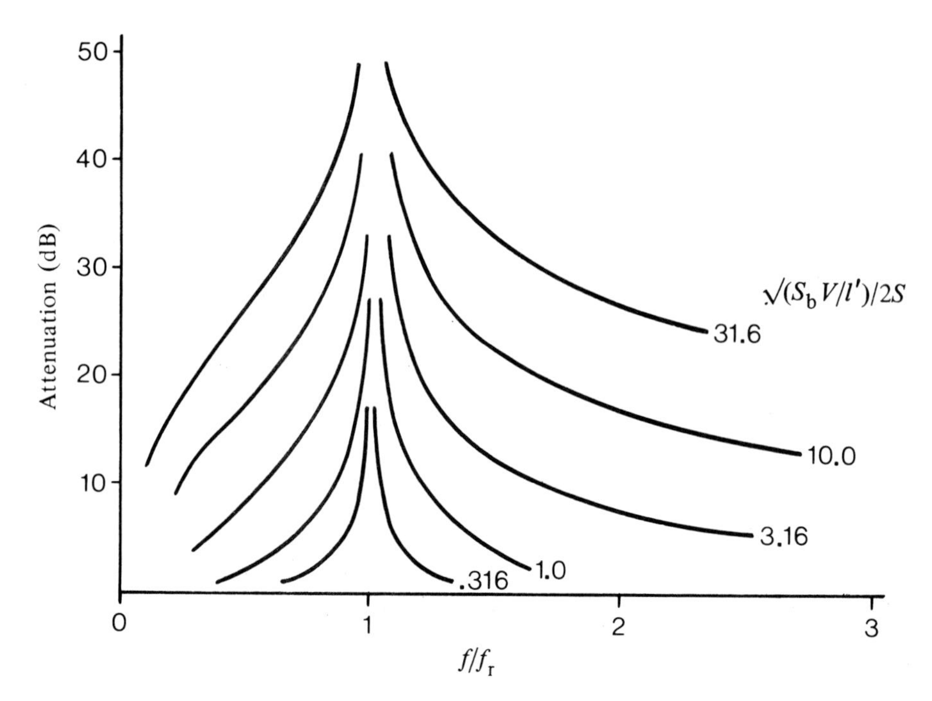

Figure 5:15 Attenuation curves for Helmholtz resonator (equation 5.38)

5.7 FURTHER READING

British Standards Institution, *British Standard 3383 : 1961. Normal Equal-Loudness Contours for Pure Tones and Normal Threshold of Hearing under Free-Field Listening Conditions* (1961)

Harris, C. M. (ed.), *Handbook of Noise Control*, New York, McGraw-Hill (1957)

Morse, P. M. and Ingard, K. U., *Theoretical Acoustics*, New York, McGraw-Hill (1968)

Noise and the Industrial Use of Fuel. Papers presented at a symposium, Southampton, January 1973, Institute of Fuel and Institute of Sound and Vibration Research

REFERENCES

1 Ford, R. D., *Introduction to Acoustics*, Amsterdam, Elsevier (1970)

2 British Standards Institution, *British Standard 3045 : 1958. The Relation between the Sone Scale of Loudness and the Phon Scale of Loudness Level* (1958)

3 British Standards Institution, *British Standard 4198 : 1967. Method for Calculating Loudness* (1967)

4 International Organization for Standardization, *ISO/R 532. Method for Calculating Loudness Level* (1966)

5 Proceedings of Conference on Control of Noise, London, HMSO (1962)

6

Refractory materials

6.1 INTRODUCTION

Refractory materials are indispensable to the use of gas in most of the processes and industrial applications with which the industrial gas engineer is concerned. In its working environment a refractory material has to withstand high temperatures without undergoing major structural change or chemical decomposition. The properties required from a refractory material are:

1. As high a fusion point as possible
2. Resistance at working temperature to deformation under load
3. Minimum volume change with temperature
4. Adequate mechanical strength and insensitivity to temperature change
5. High resistance to wear
6. Adequate chemical resistance to slag attack
7. Suitable thermal properties (e.g. for insulation)
8. Electrical non-conductance, if necessary
9. Uniform and constant composition

A primary requirement is obviously a high melting point. Seven main types of chemical substances can be considered as potential refractories: oxides, carbides, borides, silicides, nitrides, sulphides and certain elements (e.g. carbon and some metals). However, for large-scale applications most of these are ruled out for reasons such as cost and scarcity, chemical instability in oxidizing atmospheres, reaction with liquid metals and dissociation at high temperatures.

Although there are important exceptions (e.g. silicon carbide and carbon) we are left, in general, with only a limited number of oxides available in the quantities required for commercial application. It seems inevitable, therefore, that most refractories will continue to be based on the six oxides Al_2O_3, SiO_2, MgO, CaO, Cr_2O_3 and ZrO_2 and their compounds of appropriate refractoriness (see Figure 6:1).

Oxide	Chemical formula	Fusion point (°C)
Alumina	Al_2O_3	2050
Chrome	Cr_2O_3	1990
Chromite	$FeOCr_2O_3$	2180
Lime	CaO	2570
Magnesia	MgO	2800
Silica	SiO_2	1712
Zirconium dioxide	ZrO_2	2687
Titania	TiO_2	1825

Figure 6:1 Fusion points of metal oxides

For small-scale and laboratory work the range of usable materials widens to include rare oxides, e.g. thorium oxide, ThO_2 melting point 3300 °C; oxides of yttrium 2410 °C, hafnium 2810 °C, cerium 2600 °C and beryllium 2530 °C [1].

6.2 TRENDS IN REFRACTORY PRODUCTION

The temperatures and severity of conditions of use employed by industry are, in general, increasing. This trend has increased the demand on manufacturers for improved refractories. Substantial advances have been made during the past 20 years, based largely on achieving improvements in the quality of raw materials used and in methods of manufacture.

The marked improvements in the properties of existing materials referred to above have been made by producing synthetic high-purity raw materials, by selective quarrying and beneficiation of naturally occurring materials. Also, by controlling the impurities present, the high-temperature properties can be optimized.

In refractory production, improved density of the grog (a refractory clay that has been specially calcined and crushed for use as a non-plastic constituent of a refractory batch; strictly grog refers to used material treated as described) has been obtained by increasing the dead-burning temperature, some grogs being fused to almost theoretical density [2]. Improved plant for processing and pressing the bricks and the use of tunnel firing kilns with maximum temperatures in the range 1700 to 1800 °C, have all contributed to the production of materials of superior quality, particularly in relation to density, strength and volume stability.

Pure oxides of high melting point represent the ideal, and chemical purity is the aim of modern developments in commercial materials, although concessions are needed to produce strong bonding in the material or are dictated by impurities that cannot be economically removed. In single-oxide systems chemical purity ensures maximum refractoriness, i.e. the melting point of the material, and the structure obtained when such oxides are fired at appropriately high temperatures, bring corresponding maxima in physical properties such as mechanical strength. In multiple-oxide systems the melting points of compounds in the system are never higher than those of the components. Also, 'impurities', i.e. substances other than the principal refractory base, may form relatively low-melting complexes when the raw material is fired and these will impair the refractory properties. These complexes usually develop as a matrix which may persist as a glass on cooling and surround the more refractory grains. The effect is to produce a melting range rather than a sharp melting point and a sensitivity to deformation under load usually far greater than would occur with the pure material in which the crystals are directly bonded to each other [1].

6.3 PROPERTIES OF REFRACTORIES [3]

Refractoriness is the power to resist, without fusing, the effect of high temperatures. In practice it is measured by comparing the behaviour of a small cone of material with that of other standard cones which have known softening points when they are heated under carefully controlled conditions. The result is reported as the pyrometric cone equivalent (PCE).

Refractoriness under load is of more practical significance than refractoriness. A test block is heated gradually under a load in a suitable furnace. The temperatures at which the specimen starts to deform or sag and eventually fails are reported usually in the form of a graph. A load of 2 kg cm^{-2} is used for dense refractories and 1 kg cm^{-2} for porous materials.

Creep describes the heat-activated process of the plastic deformation of crystals under stress. One method of measurement determines the creep under axial loading of a specimen by making continuous measurements of the dimensions at a constant temperature and varying load.

Volume stability. (*a*) Permanent changes—'after contraction' and 'after expansion'—occurring during heat soaking are determined by heating the materials in a test furnace at various temperatures with a specified heating cycle, and measuring the dimensions before and after.

(*b*) Reversible thermal expansion, γ determines the level of stresses formed in a refractory structure. The volume coefficient of expansion is taken to be three times the linear coefficient α. A number of methods have been used to measure α such as the interferometer, fused quartz tube, dilatometers and comparator methods.

Thermal-shock (spalling) resistance is the capacity of refractories to retain their original form without cracking, splitting or flaking when subjected to sudden temperature changes. Again, a number of tests have been devised to check spalling resistance. The simplest method is to heat a specimen in a furnace to about 1300 °C and then drop it into cold water. The number of times the specimen stands this treatment without spalling, is a rough measure of its thermal-shock resistance. Various heating and cooling rates are recommended, depending on the type of material. A better method is to use a panel test in which bricks are assembled to form one side of a test furnace, which is then heated to reproduce the conditions in a normal furnace. The method is satisfactory, but is laborious, costly and takes a long time.

Slag and metal resistance describes the behaviour of a refractory when subjected to attack by molten slags or metals. Methods of determining slag resistance in the laboratory do not accurately reproduce service conditions. They usually involve drilling holes in specimens, filling these with powdered slag and melting under various conditions in furnaces. Sections of the specimens are then examined, possibly with some quantitative measurement of the area of slag attack. More recent methods have tried to reproduce service conditions more closely.

Porosity and permeability. Two kinds of pores can be defined, open and closed. The open, or apparent porosity, is determined by measuring the volume of water which a test piece of known bulk volume will absorb. The closed-pore volume is the difference between the true porosity and the apparent porosity. Measurement of true porosity involves density measurements on powdered material.

The permeability is measured by finding the volume of air which will pass through a specimen of known dimensions in a given time with a known pressure drop across the specimen.

Mechanical strength. The measurable strength of refractories is much lower than the theoretical strength calculated from interatomic bond strengths of the ideal crystal lattice. Strength is important in both the furnace structure and in carriage and handling. Tests are carried out to determine compressive strength, abrasion and impact resistance.

Thermal conductivity, λ, is the capacity of a refractory material to conduct heat. The property is measured by the coefficient of thermal conductivity, which is the quantity of heat passing through a wall of standard thickness over a standard area with a difference in temperature at the hot face and cold face specified by the test.

Thermal diffusivity, or temperature conductivity, or thermometric conductivity, is equal to the thermal conductivity λ divided by the specific heat capacity per unit volume.

Thermal capacity is the quantity of heat stored in a body.

Electrical properties. At low temperatures most refractories are dielectrics, but during heating their electrical conductivity rises in most cases and above 950 °C they are quite good conductors.

Volatilization. This effect is important in certain types of refractory structure, because of the resulting changes in structure and the possible effects of contamination of the materials being processed. The relevant factors are temperature, vapour pressure and surface area.

6.4 PHASE-EQUILIBRIUM STUDIES ON REFRACTORY SYSTEMS

In studies of heterogeneous equilibria, phase diagrams can be used to answer questions arising, and present information [3] on:

1 The mineral components of the materials at any temperature
2 The temperature at which liquid just forms
3 The alterations in the concentrations of liquid and the liquid compositions with temperature
4 The order of crystallization of the phases during cooling, their compositions and amounts
5 The chemical solubility of components or phases at different temperatures

The raw materials used in refractory technology are, as previously stated, multi-component mixtures because of the inevitable impurities. The phase equilibrium diagrams, therefore, become very complex and are difficult to carry beyond the three-component stage unless the axes are made to represent the concentrations of known compounds rather than the component oxides. In practice the problem can be simplified by selecting a large part of the total composition. Thus, the study is restricted to a binary or ternary system.

Information used in constructing phase diagrams can be obtained by chilling the system, followed by determination of the mineralogical composition, using petrographic or X-ray techniques. The sudden cooling fixes the crystalline phases and the liquid phase is changed into the particular glassy phase existing at the stated temperature.

It must be noted that the information obtained from such diagrams applies only to equilibrium conditions. They do not provide information about the viscosity of the melt, and the chemical solubility considerations do not take account of porosity or capillary action producing changes in the composition as the fluxing elements move across temperature gradients. Since most of the refractories used in practical applications are not in an equilibrium state, allowances must be made.

The diagrams do, however, help to provide explanation and prediction of the effects taking place in refractories under conditions of service and thus give guidelines for improvement in properties, e.g. slag resistance.

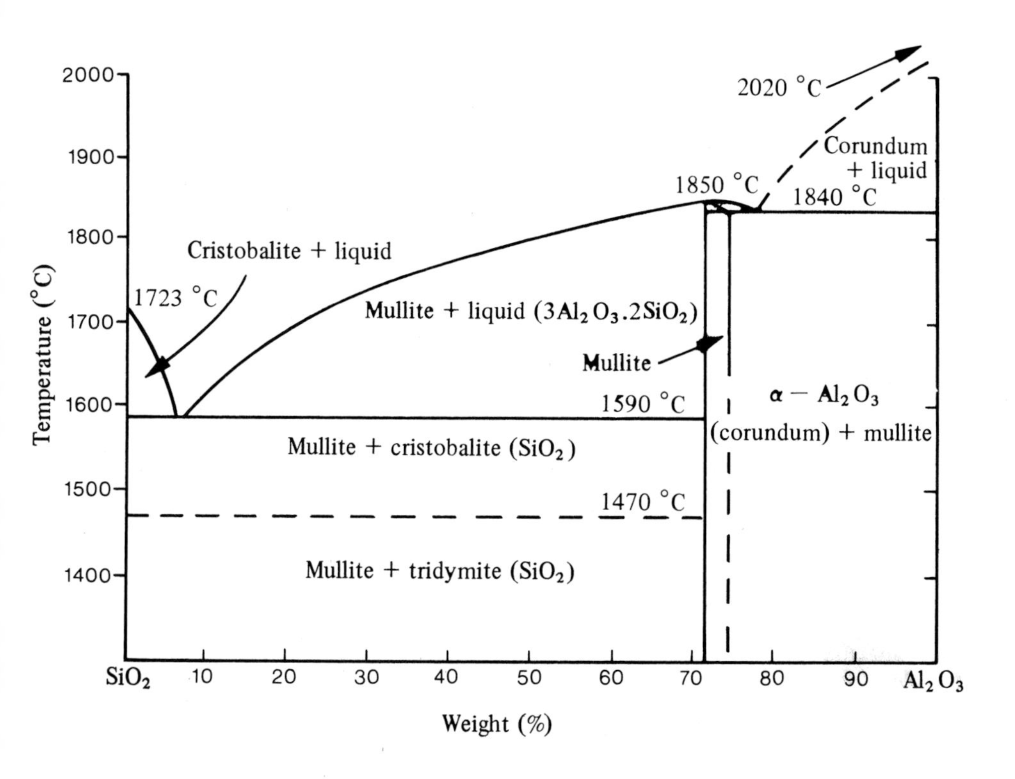

Figure 6:2 Phase diagram for the system Al_2O_3–SiO_2 [1]

The simplest system to illustrate the application of phase diagrams to refractories is that of silica–alumina (SiO_2–Al_2O_3), Figure 6:2 [1]. This diagram contains the essence of the behaviour and properties of a wide range of important refractories, those comprised essentially of silica or alumina as well as fireclays that contain both in various amounts. It can be seen that pure alumina (2020 °C) and pure silica (1723 °C) both have high melting points, but as soon as either silica is admixed with alumina or alumina with silica, the temperatures of total melting are lowered. Within the limits shown in Figure 6:2, e.g. 1840 °C for alumina-rich and 1590 °C for silica-rich compositions, a liquid phase coexists with the solid. Below these temperatures the compositions are wholly solid consisting of two crystalline phases, except for the narrow region in which mullite alone appears. The diagram illustrates how reaction between two refractory oxides produces lower-melting liquid phases. Also, in broad terms it shows that the refractory properties of Al_2O_3–SiO_2 mixtures containing more than 5 per cent Al_2O_3 increase directly with the content of Al_2O_3. In practice, however, as has been stated, the presence of impurities makes the system more complex.

Further information can be obtained by reference to refractories textbooks or to the American Ceramic Society's publication, *Phase Diagrams for Ceramicists*.

6.5 REFRACTORIES USED IN INDUSTRY

6.5.1 Refractory materials

The natural and synthetic materials used in the manufacture of the refractories discussed in this section are [3]:

1 Fireclay, less than 78% SiO_2 and less than 38% Al_2O_3
2 High alumina, 45–100% alumina
 (*a*) Mullite-siliceous, 45–70% alumina
 (*b*) Mullite-alumina, 70–95% alumina
 (*c*) Corundum, 95–100% alumina
3 Silica, more than 93% SiO_2
4 Magnesite, more than 85% MgO
5 Magnesite-chromite, more than 50% magnesite
6 Chrome-magnesite, 70% chrome, 30% magnesite
7 Chrome, Cr_2O_3 and at least 10% magnesia
8 Dolomite, CaO 52.5%, MgO 42%, SiO_2 1.2–1.5%, Fe_2O_3 1.8%, Al_2O_3 2–2.5%
9 Carbon, sulphur should be less than 2.5%, ash less than 4–5%
10 Non oxide materials

6.5.2 Refractory bricks

Figure 6:3 shows standard shapes for refractory bricks. These can be produced from any of the refractory materials referred to in 6.5.1. Figure 6:4 gives the typical compositions and properties of a number of common refractory materials.

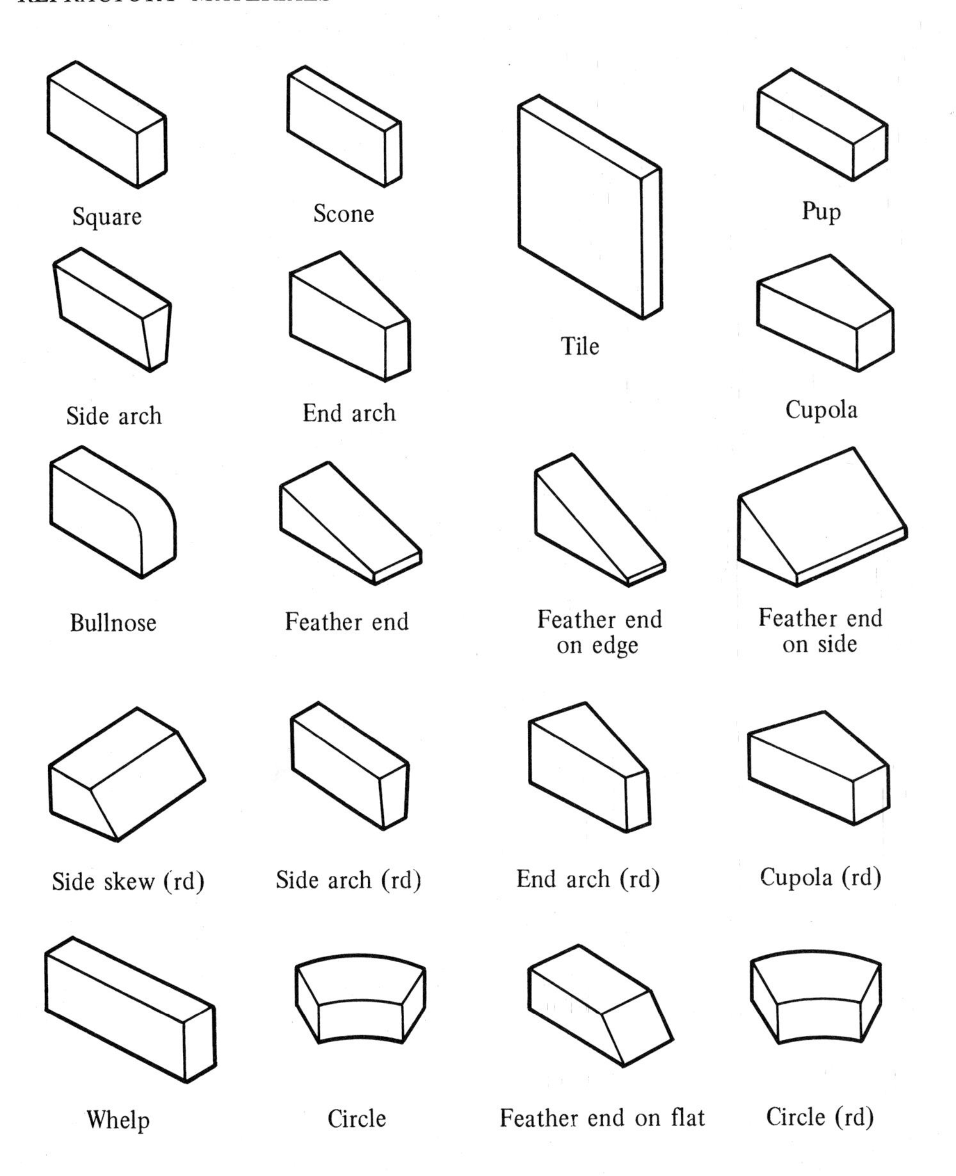

Figure 6:3 Standard shapes for refractory bricks

6.5.3 Monolithic refractories

This description is given to a very wide range of materials which are suitable for the production of shapes, especially *in situ* in new or existing structures and in burner manufacture. Three basic forms are usually distinguished: castables, mouldables and ramming mixtures.

		Firebrick	Aluminous firebrick	Siliceous or semi-silica
Refractoriness	Cone	28–31	32–34	29–31
	°C	1630–1700	1710–1760	1650–1690
Refractoriness under load of 193 kN m^{-2} (28 lbf in^{-2})	°C	1440–1520	1550–1620	1550–1600
Bulk density	kg m^{-3}	1920–2080	1920–2080	1840–2000
	lb ft^{-3}	120–130	120–130	115–125
Cold crushing strength	MN m^{-2}	17–38	17–38	17–38
	lbf in^{-2}	2500–5500	2500–5500	2500–5500
Approximate thermal conductivity at 1000 °C	W m^{-1} °C^{-1}	1.1	1.2	1.4
	Btu in ft^{-2} h^{-1} °F^{-1}	8	8.5	10
Typical composition	SiO_2	57.8	51.3	85.0
	Al_2O_3	36.3	43.1	12.5
	Fe_2O_3	2.4	2.1	0.8
	Cr_2O_3	—	—	—
	TiO_2	1.5	1.9	0.5
	CaO	0.9	0.5	0.2
	MgO	0.2	0.3	0.2
	Alkalis	1.1	1.0	0.6
	SiC	—	—	—
	B_2O_3	—	—	—

Figure 6:4 Typical composition and properties of some common refractory materials

Silica	Sillimanite	Magnesite	Chrome magnesite	Stabilized dolomite	Silicon carbide	Ceramic fibre
31–33	35–36	38	38	38	38	Max. temp. 1260 °C
1690–1730	1770–1790	1850	1850	1850	1850	
1660–1700	1700	1500–1630	1500–1700	1550–1650	1700	—
1680–1840	2000–2160	2640–2880	2800–3360	2480–2640	2400–2640	60–130
115–125	105–115	125–135	165–180	175–210	155–165	4–8
14–38	34–48	34–55	21–34	41–55	>55	—
2000–5500	5000–7000	5000–8000	3000–5000	6000–8000	>8000	—
1.7	1.6	3.6	1.6	2.2	11.7	0.07–0.22
12	11	25	11	15	81	0.5–1.5
95.7	34.0	3.1	4.5	14.7	4.5	50–54
1.1	61.4	2.2	10.3	2.2	1.4	43–47
0.7	0.7	4.5	12.7	4.0	1.0	0.8–1.8
—	—	—	30.2	—	—	—
0.1	1.9	—	—	—	—	1.2–3.5
2.1	0.9	2.5	1.0	39.0	—	0.1–1.0
0.1	0.4	87.5	41.0	40.6	—	Trace
0.2	0.7	—	—	—	—	0.2–2.0
—	—	—	—	—	93.2	—
—	—	—	—	—	—	0.6–1.0

Castable materials
These consist of a refractory aggregate mixed with a suitable cement. They are used as indicated above in furnace construction, either after pre-moulding and setting, or are cast in position. The hydraulic bond serves to provide the necessary strength until, on firing, the hydraulic bond is destroyed by heat and is replaced by a ceramic bond. The selection of materials to ensure the development in service of such a bond is very important. Aluminous cements such as cement fondu, and Secar grades which are basically calcium aluminates and are serviceable at higher temperatures are usually used for this purpose. Figure 6:5 shows some typical compositions and service temperatures.

Composition	Service temperature range	Properties
Alag + cement fondu	1200 °C	High strength, high abrasion resistance
Calcined 42–45% alumina clay + cement fondu	1250 °C	High strength
As above but with 'leaner' cement mix	1300–1400 °C	General purpose
Calcined 42–45% alumina clay + Secar 250	1500 °C	
Bauxite + cement fondu; impure fused alumina + cement fondu	1550–1600 °C	
Molochite + Secar	1550–1600 °C	Low ferric oxide for reducing conditions
Sillimanite + Secar	1550–1600 °C	
Brown fused alumina + Secar	1650–1700 °C	
Pure white fused alumina + Secar	1750–1800 °C	

Figure 6:5 Castable compositions and service temperatures

For insulation purposes, castable mixtures contain lightweight aggregates based on materials such as diatomite, perlite, bubble alumina, calcined porous clay, etc.

Mouldable materials
These consist of refractory aggregates bonded with clays and water. They are available in workable condition and sometimes include a cold setting bond. As with castable compositions, a ceramic bond develops as the temperature is increased. Typical compositions contain 40, 60 or 80 per cent alumina.

Ramming materials
These are similar to the mouldable compositions, but have a finer grain size which produces a closer texture in the finished product. The materials may be based on silica, alumina-silicates, basic materials and on silicon carbide.

Precast shapes
The materials described as monolithic refractories are also used to produce components larger than the conventional fired bricks and shapes for use in furnace construction and repair as an alternative to *in situ* work.

The selection of materials for precast shapes involves consideration of the properties of the precast mixes when fired, e.g. refractoriness, dimensional behaviour under load at working temperature and resistance to chemical effects in operation.

6.5.4 Fibre materials

Fibrous materials such as asbestos, glass wool, mineral and slag wools have been available for many years.

Ceramic fibres are a more recent development in the refractories field. They have extended the temperature range of the earlier fibre materials. Ceramic fibres can now be prepared from alumina, silicon carbide, boron nitride and kaolin. The first material developed was alumino-silicate fibre prepared commercially from kaolin. Manufacture involves melting the raw materials in a furnace from which a thin molten stream at controlled temperature falls on to a spinning disc or into a high-velocity blast of air or steam, the molten material being spun or blown into a mass of fibre. This mass is then compacted by rollers to produce the blanket product.

These fibres are available in many other forms such as paper, rope, blocks and as castable or ramming mixes (see Figure 6:4). Work on the practical assessment of ceramic fibre materials has been carried out by the Joint Refractories Committee [4] latterly contributing towards the compilation of a British Standard for testing methods for refractory fibres.

The use of these materials is also discussed in Sections 6.6 and 6.7. References [5–8] give more details of the properties and uses of ceramic-fibre materials particularly with regard to fuel and power savings. Descriptions are given of installations where big reductions in fuel demand have been achieved.

6.5.5 Non-oxide materials

These materials have very high melting points, are very hard and may possess useful electrical, chemical and magnetic properties. They can be divided into three types, metallic, nonmetallic and mixed. The most important material in this class is silicon carbide, which is used in large quantities for making kiln furniture, crucibles, ladles and for furnace construction.

In Figure 6:6 the materials are categorized with their general properties [3].

Type	Compounds of	General properties
Metalloid	Zirconium, molybdenum, cobalt, nickel with carbon, boron, nitrogen, silicon, phosphorus and sulphur	Metal-like electrical conductivity, high thermal conductivity and high thermal expansivity, mechanically weaker than metals, much harder, lower thermal-shock resistance, much higher melting points
Nonmetallic	Boron, nitrogen, carbon and silicon	High electrical resistance, low thermal conductivity, high hardness and spalling resistance, low thermal expansivity, good slag resistance
Mixed or intermediate	Aluminium or beryllium with nitrogen, carbon and boron	Intermediate between the metalloid and nonmetallic groups

Figure 6:6 Non-oxide refractories

6.6 INSULATING MATERIALS

The retention and transmission of heat by furnace materials is a complicated process which is affected by a number of variables and factors which greatly influence design, choice of material and service. The factors referred to include shape, size, proposed duty and type of service, e.g. batch or continuous. The variables can be changing temperatures and atmospheric conditions, physical and chemical changes that occur in the refractory linings or charges of furnaces during processing [3].

Temperature range	Materials
Low < 900 °C	Asbestos, diatomite and vermiculite products
Medium 900–1200 °C	Fireclays, kaolins, silica, basic materials, high-alumina materials, alumina, zirconia and other oxides
High > 1200 °C	Non-oxides

Figure 6:7 Insulating materials and temperature ranges

A wide range of insulating materials is available (Figure 6:7) and choice of a particular one will depend on the maximum temperature and conditions of service and also cost. The cost depends on purity and nature of the raw material and also the method of fabrication.

The structure of a refractory material, especially its porosity, is possibly more important than chemical composition in determining its ability to behave as an insulant. In designing heat insulation the aim is to reduce the thermal conductivity of the material as much as possible. This is usually done by enclosing air in the material to give maximum porosity consistent with other properties such as mechanical strength remaining adequate. The two main methods for obtaining porosity are:

1 The combustible additive method, in which organic materials may be used such as sawdust or coke. The material must burn out readily during the firing process. The grain-size distribution of the combustible additives is important in determining the pore size and the general porous structure of the finished refractory.
2 The foaming method, which employs a chemical foaming agent in addition to a combustible additive and can produce materials with densities in the range 250–900 kg m^{-3} (15–55 lb ft^{-3}).

A third method involving the production of gases by chemical reaction to induce the pores is sometimes used.

The pore-size distribution, the manner in which it was obtained, the ratio of crystalline and vitreous phases present and the anisotropic nature of the material, all exert an effect on the amount of heat transmitted.

In general, for any given chemical-mineral composition the higher the porosity (the lower the bulk density) then the greater the insulating ability (the lower the thermal conductivity) of the material. Figure 6:8 shows the type of material, form, density and thermal conductivities at various temperatures.

Type of brick	Density (kg m^{-3})	Thermal conductivity (W m^{-1} °C^{-1}) at temperature of:					
		0 °C	200 °C	400 °C	600 °C	800 °C	1000 °C
HT Insulating	775	0.17	0.21	0.24	0.29	0.34	0.38
HT Insulating	925	0.24	0.28	0.31	0.36	0.41	0.49
HT Insulating	1100	0.36	0.40	0.44	0.48	0.52	0.60
Porous Silli-manite	1100	0.23	0.26	0.28	0.30	0.35	0.42
Insulating	350	0.054	0.090	0.126	0.158	0.187	—
Insulating	450	0.083	0.113	0.142	0.170	0.192	—
Insulating	550	0.100	0.128	0.156	0.181	0.202	—
Insulating	750	0.140	0.163	0.187	0.208	0.226	—
Insulating	900	0.177	0.188	0.200	0.214	0.228	—

Figure 6:8 Thermal conductivities of refractory insulating bricks

Refractory insulating materials used in furnace construction can be referred to as hot-face insulation when forming the internal lining or as backup insulation when used as an outer layer to a dense refractory lining. They are available in all the forms described in Section 6.5 and reference will be made to their use in furnace construction in Sections 6.7 and 6.8.

6.7 FURNACE CONSTRUCTION

The refractory construction surrounding a flame used for heating stock constitutes a furnace. The structure prevents excess air being drawn into the combustion chamber which would otherwise dilute the fuel/air mixture producing a lower flame temperature and lower combustion efficiency. It also prevents radiation losses from the stock to the surroundings which would result in low heat-transfer efficiency.

The furnace enclosure becomes hot due to heat interchange between itself, the stock and the flame. The refractory materials used in construction must be able to withstand the temperatures attained, they must also have the required mechanical properties so that the furnace is rigidly self-supporting and resistant to the process and operating conditions [9].

6.7.1 Furnace walls

Lining construction follows installation of the furnace casing, which usually has a steel sheet and joist construction. This structure must withstand the forces arising from thermal expansion of the wall material and also support the lateral thrusts exerted by the furnace crown. The joists are designed to achieve this and are interconnected across the top of the crown and often also, through the furnace substructure, by tie bars. Spring washers may be used in the structure to assist in uniform expansion and contraction.

Bricklaying must be carried out very carefully. The joints should be as thin as possible. The strength of these joints depends on the dimensional accuracy of the bricks or shapes and on the quality of the mortar. Two types of mortar may be used. In the first type, known as heat-setting mortar, the bond develops only at temperatures of the order of 1100–1200 °C. The second type is called air-setting mortar, which develops a strong bond on air-drying, the strength being maintained throughout the working temperature range. Expansion of the wall in service is allowed for by incorporating regularly spaced expansion joints, sized in accordance with the expansion coefficients of the materials used and the anticipated temperatures of utilization, e.g. with firebrick the joints should be about 0.40 per cent of the total length and with magnesite about 2 per cent. During building the joints are filled with a combustible material such as cardboard which burns out during the first heating.

The load on the bricks at any point is calculated directly from the weight of the bricks above it. The general principle, where the refractories are insulated and used to the limit of their temperature range, is to hang as much of the weight as possible onto the outer steel structure with brackets. These brackets are kept from overheating by leaving gaps in the insulation or by water cooling.

Internal walls, however, cannot incorporate any metal reinforcing [10]. Here the basic principles of construction are to obtain gas-tightness by using small shapes with labyrinth seals or monolithic construction. Strength is obtained by careful design eliminating any long spans of unsupported refractory. High conductivity is desirable and is obtained by using dense bricks of low porosity or other materials with high thermal conductivities.

The use of precast shapes, fibre materials or monolithic construction in building furnace walls can have a number of advantages over the conventional brick wall:

1 Time and money may be saved on installation of complex structures
2 The number of joints is reduced, but due regard must be paid to providing in the shapes or monoliths used, adequate resistance to thermal stresses
3 Furnace designs can be changed to incorporate shapes and profiles which would be difficult to achieve in conventional bricks or shapes.

6.7.2 Furnace roofs

The two main ways of covering a furnace chamber are to use a masonry arch or some type of suspended structure.

A masonry arch is constructed of wedge-shaped bricks. It spans the chamber and is held up by an inclined pinching force from the sides, supplied by the skewback bricks. These are supported in a steel channel along the top of the side walls and held together by vertical steel girders using horizontal tie rods. In particular cases by using specially shaped bricks a flat roof (jack arch) can be constructed, but generally the curved arch is employed. This is more stable and can span greater widths, but it does dictate the internal profile of the furnace, since the centre is appreciably higher than the edges. The arches can be circular or of an inverted catenary shape [10]. An inverted catenary can be stable with any ratio of rise to span and is also a more suitable shape for carrying insulation. In the design of arched roofs, allowance must be made for the fact that refractory material should always be installed under compression and not subjected to shear or tension. Analysis of the structure is complicated by the nonuniform stressing due to differential expansion of the two faces which is caused by heat being supplied to only one side. This stress can be reduced by insulation which will reduce the temperature gradient. The expansion of the arch as a whole causes it to rise and squeeze the hot face. The ideal solution to this type of stress is to allow the skewbacks to move apart in proportion to the expansion, so that the arch does not rise, this can be achieved by putting spring members in the support structure to keep the compression more or less constant.

A number of alternative designs are available for a suspended roof structure. The bricks can be hung individually from steel bars, either by means of tie rods hooked into the bricks, by clamps fitted over projections in the bricks or by water-cooled tubes through the bricks. In another type a number of bricks or a shaped panel can be supported on one hanger. Suspended roofs can span a furnace of any desired width and they can accommodate any special contours which allows the designer greater control of the internal shape. The main disadvantage is the difficulty of insulation because of the possibility of overheating the roof supports. This can be alleviated by using refractory or stainless-steel hangers of relatively small area which are independent of the main bricks [5]. Alternatively the roof can be constructed entirely of high-temperature insulating material. Here, as indicated for furnace walls, a monolithic type of structure can have advantages over conventional methods of construction.

6.7.3 Furnace hearths

The hearth is very often the most important and most ill-treated component of a furnace. It must accept the weight of the charge, withstand the wear caused by placing the charge, be resistant to the chemical effects of contact with the charge and anything that is produced during the process. The selection of materials and the method of construction for a hearth are probably the most difficult aspects of design.

The chemical effects referred to above result from the formation of various scales produced during the melting or heating of metallic charges. In time the refractory lining will become thickly covered by a layer of such a scale. In furnaces operating at higher temperatures, these scales can fuse and set up eutectic systems with the refractory material. Much of the success of a given refractory brick and jointing material in hearth construction will depend on their ability to resist the flow of these liquids, which have low fusion points, into the main structure, since once this happens the life of the material is significantly, sometimes disastrously, shortened.

The hearth then can be said to be the furnace component to which the application of monolithic construction will be most beneficial. Monolithic hearths are usually rammed *in situ* over a firebrick bed lining. The materials used are chosen to suit the particular conditions, and usually produce surfaces which are resistant to abrasion, scale attack and thermal spalling, after being carefully burned-in at a temperature higher than the proposed operating temperature.

6.7.4 Furnace doors

Most doors with an area greater than 10 m^2 (108 ft^2) are mechanically operated, so there is the obvious necessity of keeping the deadweight of the lined door down to a practical minimum. Door casings can be of cast construction, which is usually limited to small and medium-sized furnaces, or fabricated from steel sections. Stability of the door casing is very important as it must form a good seal by seating onto the furnace casing plates forming the door frame. Any distortion will allow leakage of hot gases which will cause more distortion. At the higher end of the temperature range, door casings and frames will usually be water cooled to minimize thermal distortion.

The refractory lining chosen will depend on door size, since this will affect the stability of the lining which may require anchoring to the door casing, operating temperature and conditions in the furnace. The material selected may be used in the form of bricks or precast panels or be used in monolithic construction. Ceramic fibres in panel or blanket form are being used increasingly for this purpose.

6.7.5 Burners

Work has been carried out by the Joint Refractories Research Committee on the choice of refractory and method of construction of gas burners, particularly for tunnel-burner applications [4].

Aluminium titanate refractories were tried, but had a low strength, a tendency for permanent expansion on repeated heating and instability on prolonged heating in the range 800–1200 °C. The addition of 10 per cent china clay increased the modulus of rupture but instability in the range 850–1250 °C was still present. The most satisfactory results were obtained using fused alumina and a calcium aluminate cement. Small additions of magnesia can be made to the mix.

All the burners tested suffered from cracking caused by thermal shock, shrinkage of refractory at operating temperatures and thermal stress resulting from burner geometry and temperature distribution.

The conclusions drawn were that the formation of cracks was inevitable in both single- and double-case burners, but that crack-formation effects could be minimized in practical burners in three ways:

1. By using reinforcement in the form of a heat-resistant metal mesh. If the metal is kept away from the hottest part of the burner, it serves to encase the refractory and minimize the spread of any cracks that form
2. By using reinforcement, consisting of metal fibres which will help to prevent localized disruptions spreading to form major cracks
3. By using ceramic fibres, which have a twofold function in that they can act both as crack arresters and also provide good insulation. This means that they are eminently suitable to serve as a backing material for a thin-walled tube which acts as a tunnel. However, the relatively low strength of the fibres may create handling difficulties

6.8 FURNACE TYPES

6.8.1 The batch furnace

Here the refractories should have the ability to withstand the required operating temperature and frequent rapid heating and cooling cycles without spalling [9]. Traditionally, multilayer walls were used having a working face of dense, strong refractory backed by one or more layers of insulating refractories chosen to withstand the interface temperatures predicted by solution of the heat-conduction equations. A typical combination would be 42 Alumina brick backed by porous fireclay insulating brick and diatomite insulating brick. In areas of severe mechanical wear or where contact with scale or slag might occur, high alumina or basic refractory can be used for the front half of the firebrick layer. The more recent monolithic type of construction has advantages in installation time, mechanical stability and thickness. Such a lining rammed *in situ* would typically consist of a layer of high-temperature mouldable backed by a layer of slab insulation or insulating castable. The smaller thickness is a consequence of the lower thermal conductivity of rammed products compared with brickwork. This also results in lower mean temperatures through the walls with consequent reduction of heat storage although heat losses may be higher.

These linings have relatively low steady-state heat losses but their thermal capacity is high; occasionally this may be desirable, but often the furnace cycle times are such that a low thermal capacity is required. Heat-treatment furnaces are typical examples where controlled heating and cooling rates have to be adhered to and furnaces with low thermal capacities are attractive. High-temperature insulation bricks backed by diatomite bricks are widely used.

Refractory fibre blankets (Section 6.5.4) offer great scope for resistance to thermal fluctuation and for fuel economy. A suitable lining would consist of a ceramic-fibre blanket backed by a mineral-fibre blanket, mounted directly on the inside of a metal furnace casing.

6.8.2 The continuous furnace

In furnaces with a large space above the stock, it is easy for heat interchange to occur and the situation approximates to that in batch furnaces. Also, because of the total mass of brickwork in the furnace structure, there is the possibility that the heat capacity in the hot region is much higher than that of the stock. The first statement implies that such a furnace will be difficult to control under conditions of variable load (see Section 6.8.3) and the second that if a production holdup occurs, there is a possibility of overheating the stock [9].

This latter possibility could be overcome if the furnace structure had a low thermal capacity. In reheating-furnace construction a traditional lining would consist of a layer of firebrick, backed by a layer of high-temperature insulating brick and a layer of low-temperature insulating brick. Monolithic construction has also been applied on these furnaces, a layer of mouldable or castable refractory backed by a layer of slab insulation being a typical combination. The possibility of significant reductions in lining thickness is attractive as it can reduce the thermal capacity of the structure and also in some cases permit increased throughput.

6.8.3 New designs of furnace

In the last section it was stated that conventionally designed continuous furnaces may be difficult to control under conditions of variable load. This problem can be considerably reduced by reducing the chamber height so that the possibility of radiant interchange between the roof and other parts of the furnace is significantly reduced [9]. This greatly increases the amount of radiation from the roof to the stock.

Both the problem of control and that due to the thermal capacity of the furnace can be further reduced by employing convection as a principal mode of heat transfer in the furnace, rather than radiation which is the dominant mechanism in conventional types.

In recent years, much of the work on furnaces carried out by the Midlands Research Station of the British Gas Corporation [11] has been on the development of convection as the main mode of heat transfer. This work has been aimed at producing heating units that can be incorporated by industry into production lines, whereas in the past the use of conventional furnaces has meant that the rate and pattern of working has been governed by the operating characteristics of the furnace.

Four main types of heating unit have been developed: the single-cell furnace, the continuous counter-flow furnace, the part-bar heater and the whole-bar heater. The common features of these furnaces are that the heating of the stock is rapid and the thermal inertia is low. These are achieved as indicated above by promoting convection as the main mode of heat transfer. This has been achieved by using high-velocity tunnel burners in which combustion is completed within a refractory tunnel. This has enabled the chamber to be made a close fit round the stock making the best use of the momentum of the combustion products. To summarize, the advantages of these heating techniques are:

1 Rapid start-up from cold

2 Short residence times at high temperature (this greatly reduces detrimental metallurgical effects)

3 High outputs from relatively small efficient units that can be easily automated

High-alumina (90 per cent) refractories in the form of prefired shapes or as ramming compositions are used in the burners. The refractory lining for the furnace must be chosen to suit the particular duty envisaged.

Radiation can be retained as the dominant mode of heat transfer in furnaces with little free space by using refractory surfaces heated by flames. This can be achieved by using radiant roof burners positioned so that the hot gases flow over shaped refractory panels which then radiate to the stock. Again, high-alumina ramming materials are suitable for large radiant burners. Smaller burner components are made from fired refractories of a range of aluminosilicate compositions and zircon. Porous-plate burners can also be used, arranged in the roof of a furnace enclosure which is effectively constructed from them such that the roof height is low. This also results in an effectively low thermal inertia, due to the steep temperature gradient through the porous plate. The effect can be enhanced on gas cutoff by leaving the combustion air on which would cool the refractory and therefore reduce radiation. The refractories used must be porous, able to withstand the hot-face temperature and the large temperature gradients, and be spall resistant. Insulating brick material of 26 grade is suitable from temperature considerations but care must be taken to ensure uniform texture and controlled permeability.

More information on refractory materials for use in furnaces is given in Figure 6:9. Reference to the selection of refractory materials for specific furnace applications is made in Chapter 13.

Material	Maximum temperature of use (°C)	Liability to spall	Resistance to basic slag	Resistance to acid slag	Resistance to oxidizing atmosphere	Resistance to reducing atmosphere	Resistance to molten metal
Fireclay	1300–1450	small or nil	low	moderate	very high	high up to 1350 °C	high up to 1300 °C
Semi-silica	1300–1400	small	low	moderate	very high	high up to 1400 °C	high up to 1300 °C
Mullite or sillimanite	1600–1800	very small	high especially to alkaline vapours	high	very high	high	high
Al_2O_3 as corundum	1500–1900	small	moderate	high	very high	high up to 1800 °C	high
SiO_2	1550–1720	high < 600 °C low > 600 °C	alkaline vapours low	high	very high	high up to 1700 °C	high to Zn, Cu, Sn
MgO magnesite	1700–1800	high	very high	low	very high	high	high in absence of oxides and carbides

Figure 6:9 Properties of refractory materials for use in furnaces

Material	Maximum tempera- ture of use (°C)	Liability to spall	Resistance to basic slag	Resistance to acid slag	Resistance to oxidizing atmosphere	Resistance to reducing atmosphere	Resistance to molten metal
$FeO. Cr_2O_3$ chromite	1400–1500	high	high	moderate	very high	low at 1500 °C	high
Chrome magnesite	1700–1750	small	very high	moderate	very high	moderate	high to Fe, Cu, Sn, Pb, Bi, Sb
SiC	1500–1600	very low	very low	low	decomposes at 1100 °C but protec- tive film forms, can be used up to 1500 °C	high	low to vapours; high to Zn, Cu
Carbon	above 2000	nil	very high	high	destroyed	high	high but forms carbides with some metals
Zirconia	2500	very low	very high	very high	very high	very high	very high

Figure 6:9 *continued*

REFERENCES

1 Roberts, A. L., *J. Inst. Fuel*, **45** (1972), 507
2 Spencer, D. R. F., *J. Inst. Fuel*, **45** (1972), 597
3 Shaw, K., *Refractories and Their Uses*, London, Applied Science Publishers (1972)
4 Joint Refractories Committee of the Gas Council and the British Ceramic Research Association, Reports, *Gas Council Res. Commun.* GC111; *J. Inst. Gas Eng.*, **5** (1965), 618; *Gas Council Res. Commun.* GC118; *J. Inst. Gas Eng.*, **6** (1966), 677; *Gas Council Res. Commun.* GC127; *J. Inst. Gas Eng.*, **7** (1967), 534; *Gas Council Res. Commun.* GC144; *J. Inst. Gas Eng.*, **8** (1968), 500; *Gas Council Res. Commun.* GC148; *J. Inst. Gas Eng.*, **9** (1969), 352; *Gas Council Res. Commun.* GC159; *J. Inst. Gas Eng.*, **10** (1970), 392; *Gas Council Res. Commun.* GC174; *J. Inst. Gas Eng.*, **11** (1971), 407; *Gas Council Res. Commun.* GC206; summary in *J. Inst. Gas Eng.*, **12** (1972), 279
5 *Ind. Process Heat.*, **13,** No. 10 (1973), 20
6 Morganite Ceramic Fibres Ltd., *Properties of Triton Kaowool Ceramic Fibre*
7 Beatson, C., 'How to save power in furnaces with lightweight refractory', *The Engineer*, **241,** No. 6229 (April 1975), 48
8 Fryatt, J., 'Major fuel savings with ceramic fibre', *Maint. Eng.* (London), **18,** No. 3 (1974), 58
9 Fitzgerald, F. and Lakin, J. R., *J. Inst. Fuel*, **45** (1972), 551
10 Thring, M. W., *The Science of Flames and Furnaces*, 2nd edition, London, Chapman & Hall (1962)
11 Masters, J., Oeppen, B., Towler, C. J. and Wong, P. F. Y., *Inst. Gas Eng. Res. Commun.*, 849 (1971); *J. Inst. Gas Eng.*, **11** (1971), 695

7

Gas burners and burner systems

7.1 INTRODUCTION

In Chapter 4 the role of the burner in providing flame stabilization was discussed. In addition to this, practical burners may fulfil a variety of other functions such as air or gas entrainment, mixing or establishment of a particular flow pattern.

Provided combustion is complete, the heat liberated by combustion of unit quantity of gas is the same however the gas is burned. However, industrial, commercial and domestic applications for gas cover an extremely wide range of requirements with regard to flame temperature, shape, size and aeration, for example, which need to be satisfied for optimum performance. These requirements are met by an increasingly large range of burners produced by commercial manufacturers or developed by the industry's research and development organizations. It is not the purpose of this chapter to provide a comprehensive list of manufacturers' products but to deal with the principles and practice of burner design, illustrated where appropriate by reference to commercially available products.

Burners for industrial and commercial applications may be crudely but conveniently classified under five main headings governed principally by the means adopted of mixing gas and air.

1 *Postaerated or diffusion-flame burners.* In these, also known as *neat gas, non-aerated* or *luminous* burners no premixing of air and gas takes place. All air for combustion arrives at the reaction zone from the surrounding atmosphere by diffusion or entrainment after the burner nozzle. Laminar-flow and turbulent-flow types of postaerated burners may be distinguished.
2 *Atmospheric burners.* Here a stream of gas emerging from an orifice into a mixture tube entrains primary air from the surrounding atmosphere. If the gas is supplied at the normal supply pressure of a few millibars the primary aeration is normally less than stoichiometric and an adequate secondary air supply is required. Such burners are normally termed *low-pressure aerated burners.* The Bunsen burner is the commonest example. If the gas-supply pressure is higher, stoichiometric proportions are easily achieved and the burner becomes a *high-pressure gas burner* or *high-pressure atmospheric burner.*

3 *Air-blast burners*. In these burners all or part of the air required for combustion is supplied at pressures above atmospheric by means of a fan, compressor, etc., the gas usually being supplied at normal supply pressure or governed down to atmospheric pressure. Two distinct types of air-blast burner may be distinguished:

(*a*) *air-blast premix burners* in which the air supply entrains gas in an injector and a gas–air mixture is delivered to the burner head.

(*b*) *nozzle-mixing air-blast burners* in which turbulent mixing takes place at the burner head.

4 *Machine-premix burner systems*. Here gas and air are drawn through a proportioning device or carburettor into a compressor or a fan where mixing takes place, the mixture being delivered at high pressure to the burners.

5 *Other burner systems*. The preceding four groups comprise most of the burner systems in industrial and commercial gas usage. There are, however, a number of combustion techniques which do not fit conveniently into this classification. Such systems include surface-combustion burners, pulsating combustors and catalytic combustors. Such systems are considered individually towards the end of this chapter together with some special applications of types (1) to (4), for example radiant tubes, packaged burners and dual-fuel burners.

7.2 POSTAERATED BURNERS

As indicated above, all the air for combustion is obtained from the surrounding atmosphere, i.e. there is no primary air. In laminar postaerated flames the mixing process is essentially molecular diffusion whilst in turbulent postaerated flames eddy diffusion is the dominant process. The fundamentals of diffusion flames have been considered in Section 4.9. It will be recalled that the major features of laminar diffusion flames are their luminosity, freedom from lightback and consequently high turndown ratio, and extremely low noise levels. Additional advantages offered over aerated burners are their freedom from linting and port blockage, their compactness,

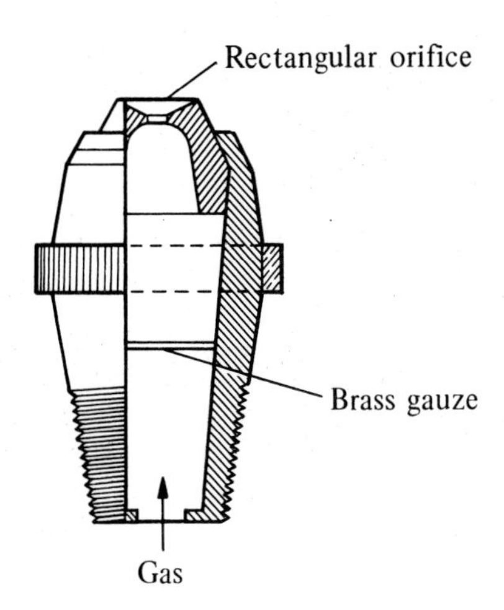

Figure 7:1 Bray industrial burner (Geo. Bray & Co. Ltd)

the well-defined flame they produce and their relative insensitivity to changes in combustion characteristics.

These advantages are such that postaerated burners and in particular the 'Bray jet' dominated the design of most domestic, commercial and low-temperature industrial appliances during the era of manufactured town gas. However, such burners give unsatisfactory performance on natural gas and on conversion have been replaced by aerated burners. Considerable research effort has gone into the design of postaerated burners for natural gas but these have not yet become re-established.

Because of the considerable differences in design, first-family and second-family laminar postaerated burners are dealt with separately in the following sections.

7.2.1 Laminar postaerated burners for first-family gases

The Bray burner (Figure 7:1) was the most widely used example of this burner type. It consists of a threaded brass body into which is set a ceramic tip. This is provided with an accurately profiled orifice which produces a thin fan-shaped flame having a high ratio of surface area to volume so as to facilitate diffusion. The ceramic (steatite) tip offers an inexpensive way of providing reproducible and complex orifice geometries in addition to being resistant to heat, corrosion and erosion. Brass jets of this type are recommended for use up to 400 °C but above this temperature heat-resistant steel one-piece jets operating on a similar principle may be used up to 800 °C.

Bray jets operate normally at pressures of 250 to 625 Pa (2.5 to 6.25 mbar, 1 to 2.5 inH_2O), the industrial burners ranging in cold gas rates from 0.018 $m^3\,h^{-1}$ (0.62 $ft^3\,h^{-1}$) to 0.39 $m^3\,h^{-1}$ (13.9 $ft^3\,h^{-1}$) for gas of relative density 0.5 at 380 Pa (3.8 mbar, 1.5 inH_2O). Note that the quoted cold gas rate becomes reduced in practice depending on the temperature reached by the jet tip. This burndown must be allowed for in burner sizing. The manufacturers consider an allowance of 15 per cent should cover the majority of cases. Priestley [1] quotes Lloyd [2] as suggesting the reduction amounts to 1 or 2 per cent for a single jet, 3 or 4 per cent for a bar burner and can be 20 per cent or more in situations where air circulation is restricted and in high-temperature appliances. High burndown rates present particular problems in applications with restricted combustion space since for the first few minutes after lighting, the burner may be seriously overloaded and incomplete combustion and sooting may result. In addition to burners producing fan-shaped flames, ceramic-tipped jets are available producing cylindrical flames, normally for use as pilots either single-jetted or double-jetted for cross-lighting purposes.

7.2.2 Laminar postaerated burners for natural gas

Postaerated burners designed for first-family gases do not perform satisfactorily on natural gas. Soft flames of fluctuating shape are formed with much decreased stability due primarily to the lower burning velocity and narrower range of flammability.

These difficulties prompted a re-examination of the mechanism of stabilization of first-family gas diffusion flames. Garside [3] considered that one essential to the production of a stable flame is the formation of a zone at the base of the flame in which gas/air mixing takes place and that the stabilization mechanism is essentially similar to that of a premixed flame involving a balance between flow and burning velocities. Edmondson [4, 5] studied the behaviour of the flame base region using first-family and natural gas on Bray jets and showed that when first-family gas is burned air is entrained downwards countercurrent to the emergent gas stream resulting in rapid mixing in a region of low flow velocities and the base of the flame develops below the tip of the jet. Conversely on natural gas the flame base is always located above the top of the Bray jet, countercurrent air flow does not appear to occur and the flame is prone to lift (reversion pressure 75 Pa (0.75 mbar, 0.3 inH_2O). The absence of reverse flow with natural gas was attributed to the fact that for the same thermal input the volume of natural gas is halved so that the tendency to create reverse flow is much reduced and may be overcome by convective effects. Modification of the jet superstructure resulted in improved stability but not a commercially viable burner. Other attempts to modify first-family burners involved the use of retention flames [6] and stabilizing rods [7], none of which were satisfactory.

Later developments broke away from the wedge-shaped orifice which worked so well on first-family gases and made use of slots, 'pinholes' and symmetrical tube arrays. An example of the slot type is the Uniplane burner [8] in which the slot is located at the base of a metal shield up which the gas flows entraining air before combustion. A major difficulty with this burner which has led to its abandonment is the tendency for high-burning-velocity substitute natural gases to burn on the slot with inadequate air entrainment leading to burn-back sooting. The other two approaches have led to acceptable commercial designs which are described below.

Pinhole burners

Early designs of pinhole burners were developed by the VEG Gas Instituut in the Netherlands and have been described by van der Linden [6] and Dutton *et al.* [8]. These consist essentially of a centre-fed pressed-steel flat-topped box with a row of small holes, typically 0.3 to 0.4 mm diameter which form the main flame ports. The vertical cylindrical main flames are stabilized by adjacent small auxiliary flames.

Pinhole burners of a similar type are currently produced by several UK manufacturers. Geo. Bray produce a range of neat natural-gas modules incorporating either single or double rows of flame ports of overall length 86.3 to 124.8 mm and of heat output (hot) 2.1 to 2.9 kW (7100 to 10 000 Btu h^{-1}) at 1250 Pa (12.5 mbar, 5 inH_2O) pressure. Satisfactory burner pressures are in the range 1–2 kPa (10–20 mbar, 4–8 inH_2O). As with all pinhole burners they suffer from severe burndown estimated at 33 per cent by the manufacturers. However, because of their low mass most of the burndown occurs in the first minute of operation. Optimum performance is obtained when combustion air flows upwards between adjacent modules; cross draughts may disturb the retention flames allowing the main flames to lift and become noisy.

The Adaptojet (Figure 7:2) is of similar design. The Aeromatic Pinhole Burner uses a single row of main flame ports and, it is claimed, operates at very low temperatures minimizing burndown. The burner rating is variable within the range 1.2–17 kW m^{-1} (100–1500 Btu h^{-1} in^{-1}).

Other developments of the pinhole principle have been towards individual post-aerated 'jets', for example the Bray neat-gas heating jet in which a number of pinhole orifices in a domed head produce a fanned-out flame array, the flames being stabilized by a shroud at the base which promotes reverse flow.

Matrix burners

Matrix burners represent a fresh approach to the problem of burning neat gas. The principle was first developed by Desty and co-workers at the British Petroleum Research Centre [9, 10] and the first matrix burner consisted of a symmetrical array

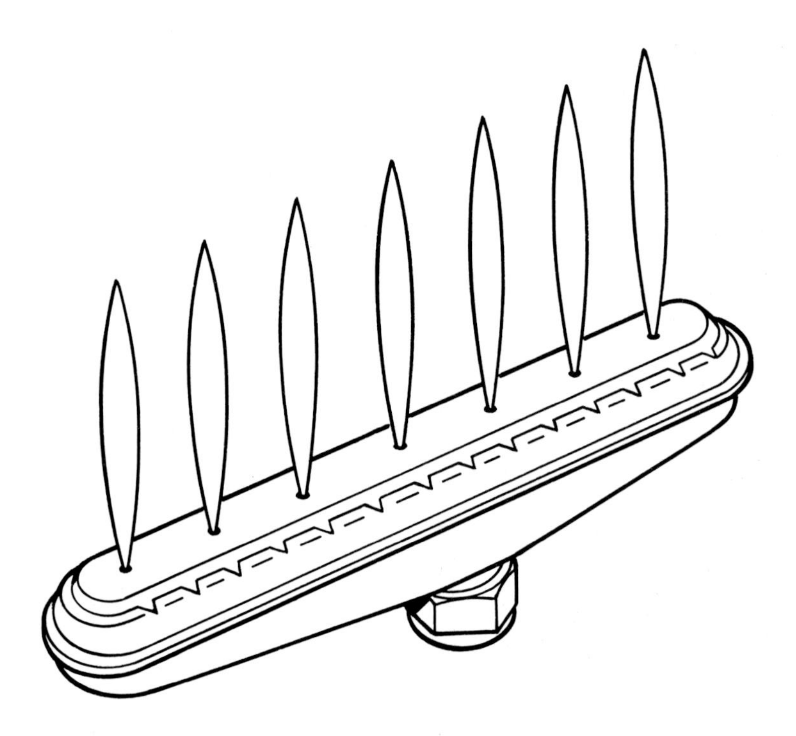

Figure 7:2 Adaptojet pinhole burner (Adaptogas Ltd)

of short hexagonal-ended tubes. Gas fed into the free space between the tubes emerges through the very small interconnected slots formed between hexagonal tube ends on the burner top, the bottom ends being sealed. The exit flow resistance is higher than in the transverse direction so that a uniform gas distribution at the burner face is achieved. Air flow takes place through the tubes under the influence of flue draught or from a low-pressure axial-flow fan (Figure 7:3). The burner is in principle capable of being extended indefinitely in two dimensions. Burners of this type were claimed to burn satisfactorily virtually any fuel gas ranging from propane to hydrogen with a flame height of only a few millimetres.

Clearly the principle of the matrix burner is an attractive one, in particular for use in compact domestic appliances and is, at the time of writing, undergoing development work and field trials. Development work has largely been directed towards reliable designs capable of low-cost mass production, the design of special appliances to accept the matrix burner and solutions to combustion problems peculiar to the matrix burner.

Figure 7:4 (*a*) shows the effect of excess air on effective flame height; this being defined as the burner/heat-exchanger separation at which the flue products achieve the British Standard acceptable CO/CO_2 ratio of 0.02. With a typical operating excess air rate of 100 per cent the separation may be as little as 5 mm.

Figure 7:4 (*b*) shows the variation of CO/CO_2 ratio with burner loading and air

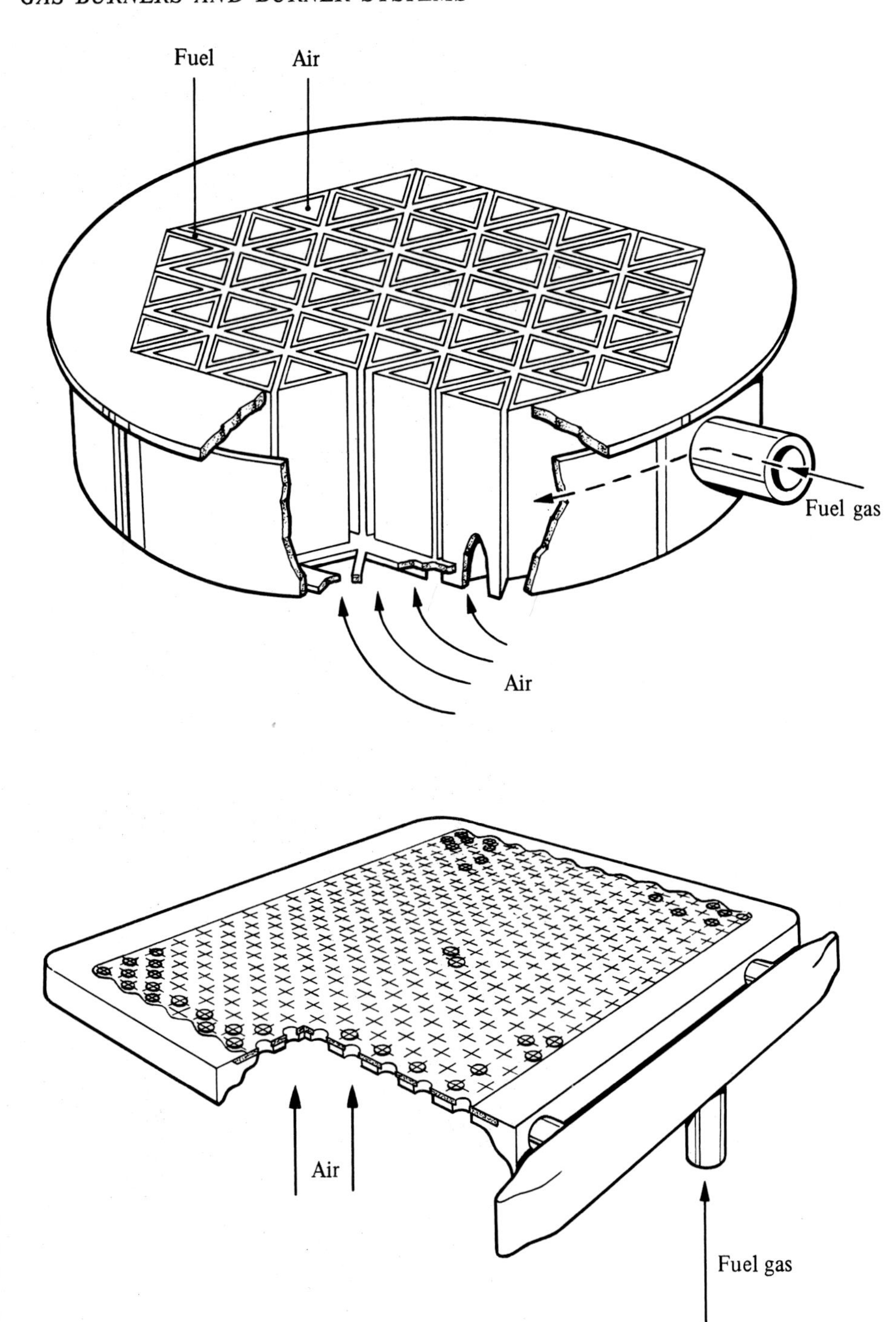

Figure 7:3 Typical matrix burner designs

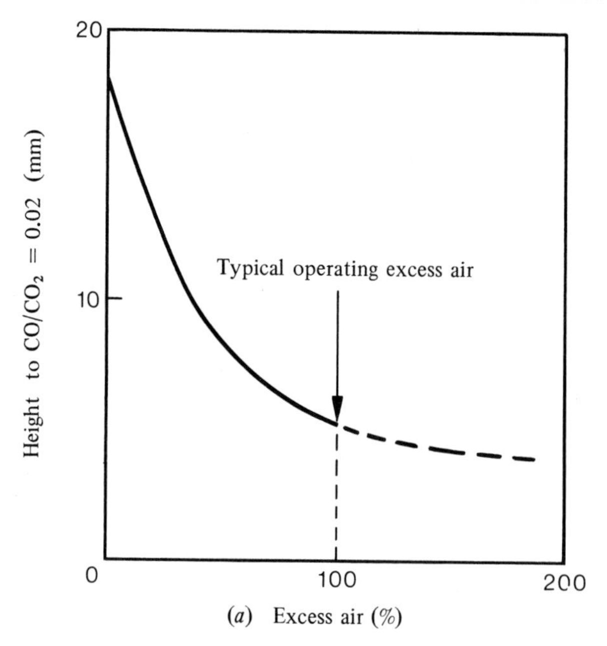

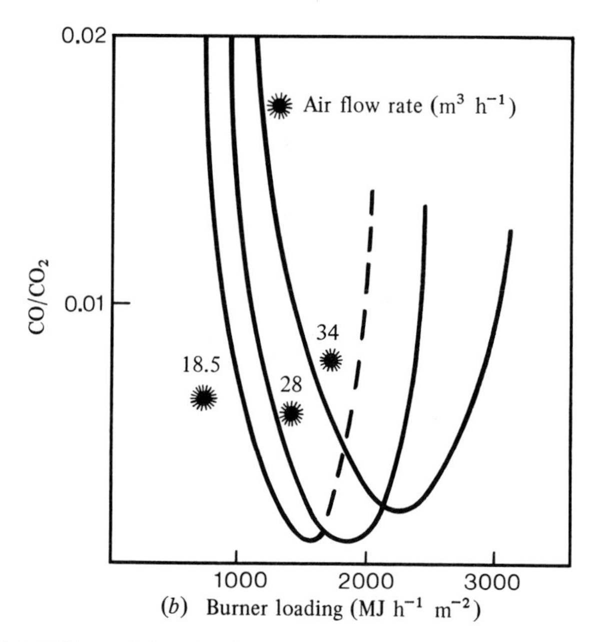

Figure 7:4 (*a*) Effect of level of excess air on flame height for a typical matrix burner [11]
(*b*) Effect of air flow rate and burner loading on completeness of combustion for a typical matrix burner/heat exchanger combination

flow rate and indicates that there is an optimum burner loading corresponding to each air flow rate.

In addition to affecting the completeness of combustion the combination of air and gas flow rates affects the burner temperature and consequently its durability. When air and gas velocities are high the flame is lifted a millimetre or so above the burner giving an insulating layer of cool gas above the burner.

Also related to the position of the flame relative to the burner top is the phenomenon of burn-back sooting which has presented an important difficulty in the development of matrix burners. On gases with low burning velocity, like natural gas, the separation between flame base and burner top allows some time for gas/air intermixing and the resultant flame is blue without yellow-tipping. If, however, hydrogen-rich substitute natural gases are used the flame base 'burns back' closer to the burner top causing the burner temperature to rise and more importantly allowing reduced time for intermixing leading to yellow-tipping and in some instances soot deposition. The general solution to this problem [11] is to arrange for gas and air flows to be sufficiently high for the flames to be lifted even when using gases with high burning velocity. The production difficulties of achieving a uniform array of very small slots at low cost on a mass-production basis have been considerable. A metal-foam version of the matrix burner was developed in which the gas diffused through a metal-foam plate into which cylindrical air tubes were set at intervals [12]. Nonuniform foam permeability proved the major problem and further developments by the same company (Dunlop Ltd.) have been based on perforated sheet metal. The basic elements of the burner are two formed metal sheets forming the matrix burner together with additional top and bottom plates which may be included to promote air turbulence and to increase stability for high heat outputs.

The gas orifices are formed by broaching the top plate after the forming and perforation operations have been completed.

As with other postaerated burners, matrix burners suffer from burndown, a typical figure being 40 per cent. Because of the low gas orifice pressure drop this may be reduced significantly by placing a restrictor upstream of the burner. The Dunlop matrix burner may be provided with an automatic burndown control which reduces burndown to less than 10 per cent. The Aeromatic Linear Slot Burner uses stainless-steel blades attached to a mild-steel distribution chamber forming a grid array of slotted burner ports. Heat outputs of up to 1.8 MW m^{-2} (4000 Btu h^{-1} in^{-2}) are obtained on natural draught although higher ratings on fanned air are obtainable. Other manufacturers have developed further the hexagonal configuration of the early prototypes, in general by increasing the size of the air ducts and subdividing them using perforated plates, 'spiders', etc.

7.2.3 Turbulent postaerated burners

The most widely used first-family large-scale postaerated burner was the Hypact burner. This is essentially a scaled-up Bray industrial jet but operates in the turbulent-flow regime. It consists of an internally screwed chromized-steel cylindrical socket with a profiled orifice milled in the closed end. It is particularly stable on turndown in forced or induced draughts. The flame produced is roughly fan-shaped and is relatively noisy. Turndown ratios of several hundreds to one are obtainable and burner pressures of more than 140 kPa (1.4 bar, 20 lbf in^{-2}) [13] have been used without blowoff. In common with turbulent-diffusion flames produced on cylindrical nozzles (Section 4.9.2) the flame length for a given burner is virtually unaffected by gas rate. Hypact burners are able to withstand temperatures up to 850 °C and range

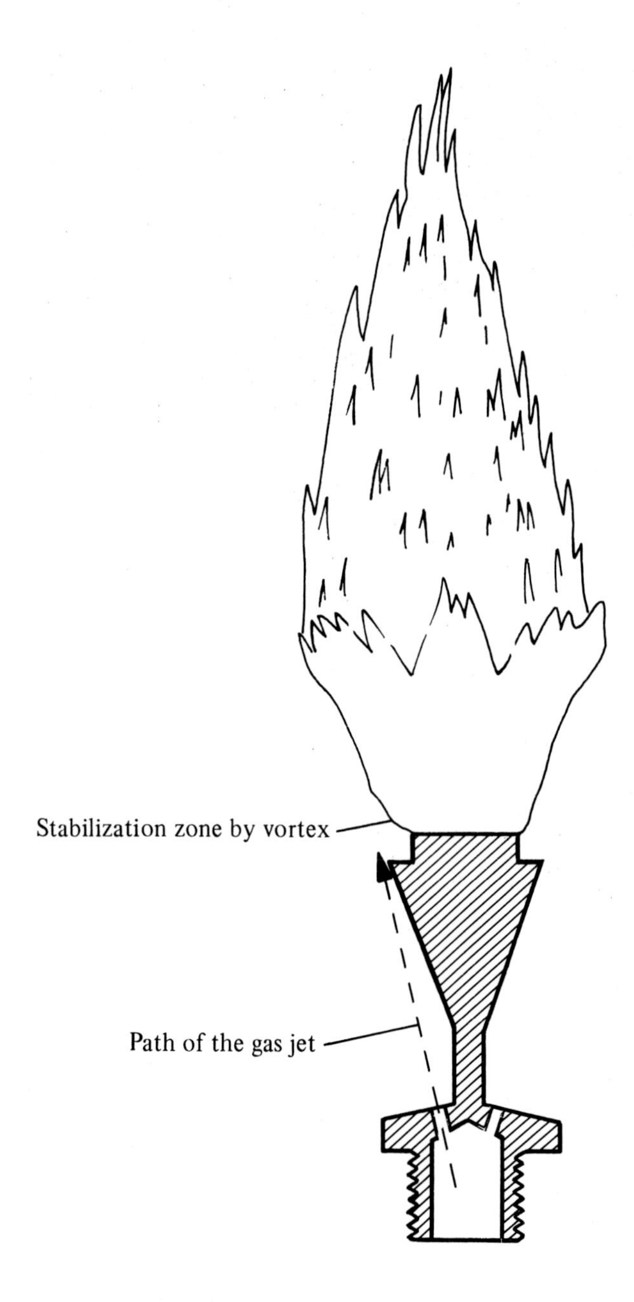

Figure 7:5 Free-jet burner [16]

in size from 8.8 kW (30 000 Btu h^{-1}) to 440 kW (1 500 000 Btu h^{-1}) at 250 Pa (2.5 mbar, 1 inH_2O). They have found wide application in processes such as air and liquid heating and low-temperature metal melting.

Hypact burners are unstable on natural gas and need to be modified or replaced on conversion. North Thames Gas devised a modification which consists of a wedge-frustum shroud made of wire gauze shaped to the profile of the flame base. The Aeromatic Se-Neat burner was developed originally by South Eastern Gas as a conversion replacement for the Hypact burner. It consists of a square-section header with a central wedge-shaped deflector. Two opposed gas jets are directed towards and deflected by the deflector so that a fan-shaped flame is formed along the main axis of the header. High aeration is achieved due to entrainment between the jet and the deflector and turbulent mixing in the coalesced streams. Conversion is simply achieved by replacing the two screw-in jets.

Ratings range from 5.9 kW (20 000 Btu h^{-1}) to 117 kW (400 000 Btu h^{-1}) for natural gas.

The 'target impact burner' has been investigated by a number of workers in Britain and in the USA but does not appear to have been developed commercially. Simmonds and Willett [14] have described the operation of this simple system in which neat gas at high pressure is directed through a circular orifice to impinge on a target which

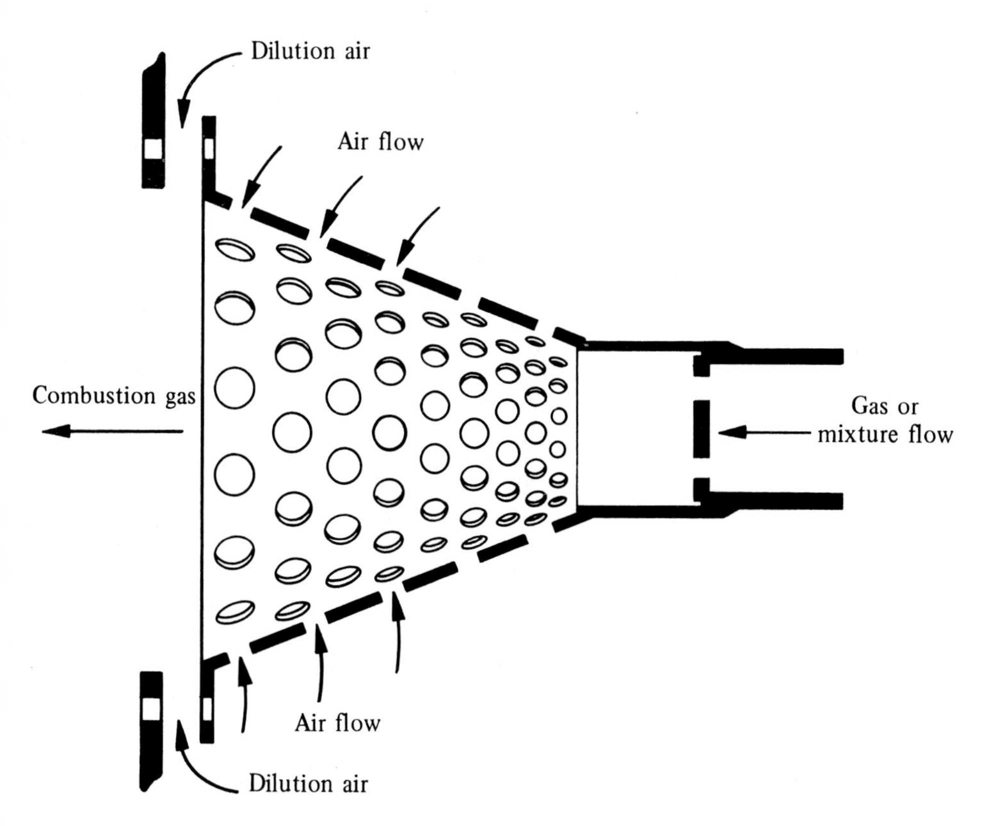

Figure 7:6 West Midlands Gas Fire Cone Burner (Landon Kingsway Ltd)

converts the jet into a thin sheet of gas in the form of a disc producing a well-aerated disc-shaped flame. Stable flames can be achieved with first-family gases at gas pressures in excess of 400 kPa (4 bar, 60 lbf in^{-2}) with appropriate orifice/target separation. Simmonds and Willett restricted their investigation to first-family gases but the American Gas Association have studied similar burners using natural gas and liquefied petroleum gases [15]. Target impact burners have advantages of simple design, small flame volume, variable flame shape depending upon target design and in some cases the target may be the surface which it is required to heat.

A 'free-jet burner' developed in France has been described by Laligand [16] in which neat-gas jets emitted from small holes at the burner base develop and entrain air in the free space around a conical or prism-shaped solid producing a flame which stabilizes at an eddy-producing step as shown in Figure 7:5.

As the burner operates without a mixture tube or flame ports, a large proportion of the initial momentum is retained by the flame gases, stability is possible for burner pressures up to several bars and a turndown ratio of 30 : 1 is claimed. As might be expected the burners tend to be noisy. They have found their widest application in France in immersion-tube water heating.

Suction burners, which are used mainly for low-temperature oven heating, are usually postaerated. The air for combustion is induced either by natural draught or more commonly by an extraction fan in the exhaust ducting. Examples of proprietary designs are the Keith Blackman Suction Burner and the West Midlands Gas Fire Cone Burner (Landon Kingsway Ltd.) shown in Figure 7:6. The latter may operate on neat gas or a partially aerated mixture, primary air being drawn through the holes in the cone which increase in size as the open end is approached. Diluent air is introduced via a shutter on the burner mounting.

7.3 ATMOSPHERIC AERATED BURNERS

7.3.1 Introduction

In an atmospheric burner, primary air is entrained by momentum sharing between the gas jet and the surrounding air. The amount of air induced in this way is generally about 50 to 70 per cent of the stoichiometric air requirement, although it is feasible to entrain the whole of the air requirement in this way.

Two types of atmospheric burners may be distinguished, those in which the gas is used at normal supply pressures of 625 to 2000 Pa (6.25 to 20 mbar, 2.5 to 8 inH_2O) and those in which the gas is supplied from a compressor or high-pressure supply. The importance of the former type hardly needs stressing. The forerunner is, of course, the Bunsen burner which heralded a new phase in gas-utilization development when it was first introduced. Low-pressure atmospheric burners are now used on most domestic appliances and have very many low-temperature industrial applications. High-pressure atmospheric burners are restricted to industrial applications and have only been adopted on a limited scale in Britain. They are considered briefly towards the end of this section. Figure 7:7 shows the essential features of an atmospheric burner and indicates the terminology used in this section.

The theory of entrainment by the use of injectors is developed in Chapter 10 and applied to practical burners in this section.

Analytical and experimental investigation of the mechanism of atmospheric

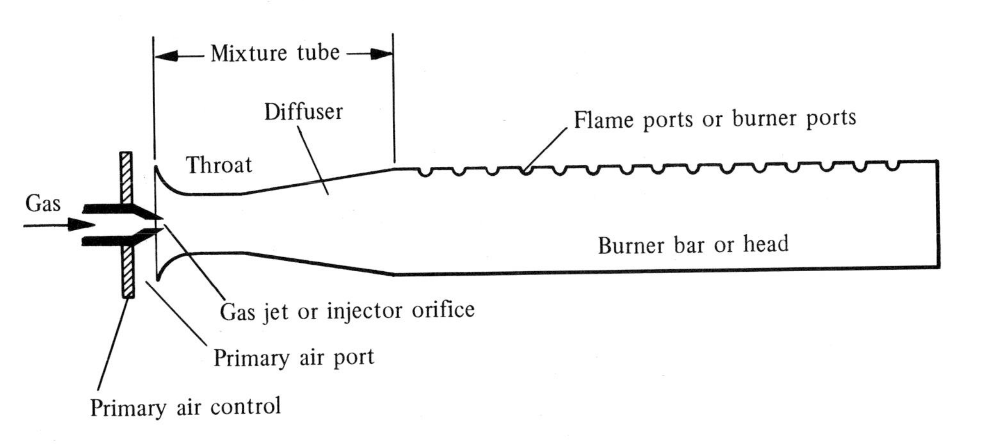

Figure 7:7 Atmospheric-burner terminology

injectors has a long history dating back to Bunsen. The first significant advances were made by Silver [17] and by von Elbe and Grumer [18] in 1948. The latter authors used a force–momentum balance to calculate entrainment. They considered entrainment to be produced by the gas mixing with the surrounding air at constant momentum, this process continuing until the mixture had just sufficient kinetic energy to overcome the resistance of the burner and flame ports. The work was empirically biased but laid down guidelines for ratios connecting the injector orifice, throat and port dimensions. Prigg [19] adopted this approach in developing a design procedure for British town-gas burners. Simmonds [20] presented a comprehensive theory of atmospheric burners without diffusers, including the effects of buoyancy at low flows without the use of empirical coefficients. Francis [21] produced a theoretical treatment of jet pumps and injectors working under conditions of incompressible flow which can readily be applied to atmospheric gas burners. Francis was able to derive the conditions for optimum injector performance and to show that considerable generalization and simplification are possible for injectors provided that consideration is restricted to optimum performance conditions. A similar approach has been adopted in Chapter 10 of this book.

7.3.2 The gas jet or injector orifice

The integrated form of the Bernoulli equation (equation 2.16) governs the discharge of gas through the injector orifice. This may be written:

$$V_g = 1655\, A_o \sqrt{(p_o/d)} \tag{7.1}$$

where
V_g = theoretical gas flow rate ($\mathrm{sft^3\ h^{-1}}$)
A_o = injector orifice area ($\mathrm{in^2}$)
p_o = static gas pressure (gauge) at orifice ($\mathrm{inH_2O}$)
d = relative density of gas (air = 1)

In practice the discharge rate is less than this depending on the orifice geometry and the Reynolds number. It is usual to express the ratio of actual to theoretical discharge rate as the discharge coefficient C_{do}. This is the product of two terms, the stream contraction ratio (C_c) and the friction coefficient. For a well designed nozzle, stream-contraction effects are negligible and C_{do} becomes equivalent to the friction coefficient and $C_c = 1$. Thus $A_j \approx A_o$. For this reason A_o is used in this chapter instead of the A_j of Chapter 10. In practice, for an orifice length/diameter ratio close to unity and an approach angle between 30° and 60°, C_{do} lies between 0.85 and 0.95 [22] for turbulent flow. Discharge coefficients for a wide range of orifice geometries are available in the literature [19, 23]. For values of (Re) less than 2000, C_{do} reduces markedly and may be as low as 0.6 at low turndown flow rates [19].

Thus actual gas rate V_g (sft^3 h^{-1}) is given by:

$$V_g = 1655\, C_{do}\, A_o \sqrt{(p_o/d)} \tag{7.2}$$

The potential heat flow $Q = V_g C_g$ (where C_g is the gross calorific value) is given by:

$$(Q/\text{Btu h}^{-1}) = 1655\, C_{do} A_o C_g \sqrt{(p_o/d)}$$

and since Wobbe number $W = C_g/\sqrt{d}$

$$(Q/\text{Btu h}^{-1}) = 1655\, A_o\, C_{do}\, W \sqrt{p_o} \tag{7.3}$$

Similarly, in SI units:

$$(Q/\text{W}) = 12.55\, A_o\, C_{do}\, W \sqrt{p_o} \tag{7.4}$$

where p_o is in mbar, A_o in mm^2, and W in MJ m^{-3} (st).

Figure 7:8 shows the relationship between jet diameter and heat input for British natural gas ($W = 50.7$ SI units, 1335 UK units) at a supply pressure of 1.5 kPa (15 mbar, 6 inH_2O) assuming $C_{do} = 0.9$.

Injector noise

Noise in atmospheric burners stems mainly from the injector and is generated aerodynamically in the shear layer between the high-velocity gas jet and the surrounding air. The noise produced normally presents no problem in industrial applications but may be unacceptable in some domestic appliances. Division of the gas-jet stream into a number of smaller-diameter jets using a multihole injector is a common solution since this reduces turbulence without sacrificing momentum [24, 25]. Multiple injectors are normally formed in ceramic materials which allow small orifices of the correct shape and a smooth internal finish to be produced.

The gas modulus

For the same heat input and the same degree of primary aeration virtually the same volume of primary air needs to be entrained in an atmospheric burner whatever the fuel gas (Section 4.2.7). The volume of entraining gas, however, is inversely proportional to the calorific value. Thus, on conversion from first-family to second-family gas about one-half of the volume of gas must entrain the same volume of air. For burners of similar resistance this smaller flow must be given the same momentum by

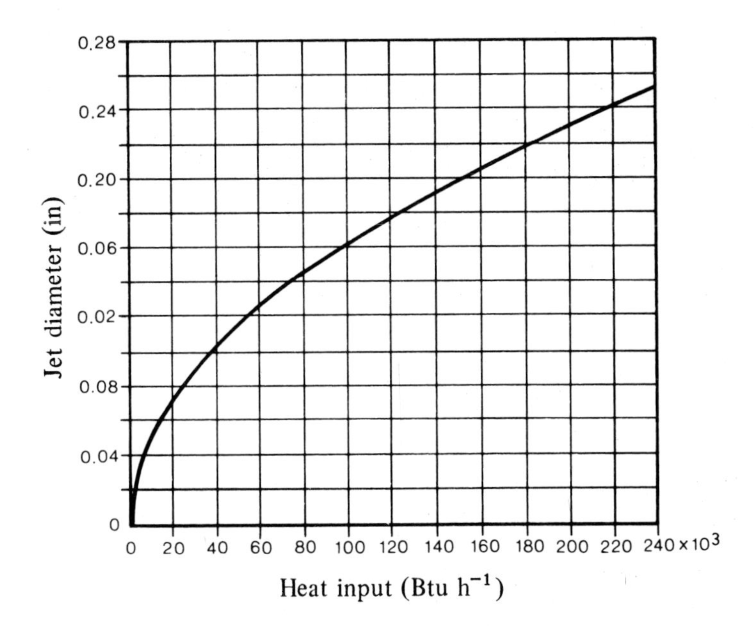

Figure 7:8 Relationship between jet diameter and heat input for natural gas [22]

using a higher pressure. Van der Linden [26] has shown that this pressure may be calculated from the *gas modulus* derived as follows [27].

Consider two fuel gases denoted by subscripts 1 and 2. For constant momentum:

$$V_1^2 \rho_1/A_{o1} = V_2^2 \rho_2/A_{o2}$$

i.e.

$$A_{o1}/A_{o2} = (\rho_1/\rho_2)(V_1/V_2)^2$$

For constant heat input, $V_1/V_2 = C_{g2}/C_{g1}$. Therefore:

$$(\rho_1/\rho_2)(C_{g2}/C_{g1})^2 = (W_2/W_1)^2 = A_{o1}/A_{o2} \quad (7.5)$$

Potential heat input through an injector is given by equation 7.3:

$$Q = A_o W\sqrt{p_o} \times \text{constant}$$

Therefore:

$$A_{o1}W_1\sqrt{p_{o1}} = A_{o2}W_2\sqrt{p_{o2}}$$

Therefore:

$$A_{o1}/A_{o2} = (W_2/W_1)\sqrt{(p_{o2}/p_{o1})} \quad (7.6)$$

Combining 7.6 and 7.5 yields:

$$W_2/W_1 = \sqrt{(p_{o2}/p_{o1})} \quad (7.7)$$

i.e. the ratio $\sqrt{p}/W$ is a constant, the gas modulus.

Thus, for conversion from first-family gas of Wobbe number 27.8 SI units (730 UK units) at a pressure of 625 Pa (6.25 mbar, 2.5 inH_2O) to natural gas at 50.7 SI units (1335 UK units), the gas pressure required is 2.1 kPa (21 mbar, 8.4 inH_2O). This is the basis on which conversion pressures have been calculated in the UK. It should be noted, however, that due to stability considerations, many natural-gas burners will operate at a lower flame-port loading (Section 7.3.4) and consequently have a lower flow resistance. Thus acceptable aeration may be achieved at much lower pressures than that yielded by the gas modulus. For example, for a flame-port loading of 9 MW m^{-2} (20 000 Btu in^{-2} h^{-1}) Harris and Prigg [27] indicate that 1 kPa (10 mbar, 4 inH_2O) would be adequate.

The axial separation of the injector orifice from the burner throat is not critical but in general is 2 to 3 throat diameters. The alignment of the jet relative to the mixture tube is important, however, and significant reductions in aeration occur for quite small departures from the axial position.

7.3.3 Primary aeration

The percentage primary air required and the means by which this aeration is achieved exert an overriding influence on the design of atmospheric burners.

In Chapter 10 it is shown that the relationship between primary-air entrainment, gas rate and burner dimensions is given by equation 10.4:

$$[(p_{go} - p_{gp}) A_t^2/\rho_a V_g^2] - \tfrac{1}{2}(r + 1)(r + d)[(C_{dp}^{-2} + C_F)(A_t/A_p)^2 + 1] + d(A_t/A_o) = 0$$

where:
r = ratio of entrained to jet fluid, i.e. primary air/gas ratio
V_g = volume rate of flow of gas
C_{dp} = discharge coefficient of burner ports
A_o = injector orifice area (see Section 7.4.2)
A_t = throat area
A_p = burner port area
C_F = friction-loss coefficient, where $\Delta p_{fric} = \tfrac{1}{2}C_F u_p^2$
d = relative density of gas (air = 1)
ρ_a = density of air

If the two pressures in the first term are identical, representing the common case where a burner fires into a combustion chamber at essentially atmospheric pressure, the term involving V_g disappears indicating that there is an inherent self-proportioning effect yielding (in principle) a constant air–gas ratio for all throughputs and equation 10.4 reduces to:

$$(r + 1)(r + d) = 2d(A_t/A_o)[1 + K(A_t/A_p)^2]^{-1} \quad (10.5a)$$

where $K = C_{dp}^{-2} + C_F$.

Thus a graph of primary air–gas ratio versus gas pressure would be expected to be like curve (*a*) in Figure 7:9.

In practice the form of the graph is as shown in (*b*), the decrease in aeration at low flow rates being caused by changes in C_{dp} and C_F at low Reynolds numbers.

As shown in Chapter 10, the maximum value of r is obtained when:

$$A_t/A_p = K^{-\frac{1}{2}} \tag{10.8}$$

which, for typical values of C_{dp} and C_F yields an optimum throat area of 0.45 to 0.65 of the burner port area. This maximizes the kinetic head at the exit of the throat, i.e. the gauge pressure at this point is zero.

For a burner with the optimal throat area, equation 10.5 reduces [19] to:

$$r \approx \sqrt{d}\,[\sqrt{(A_t/A_o)} - 1] \tag{10.10a}$$

an expression widely used in burner design.

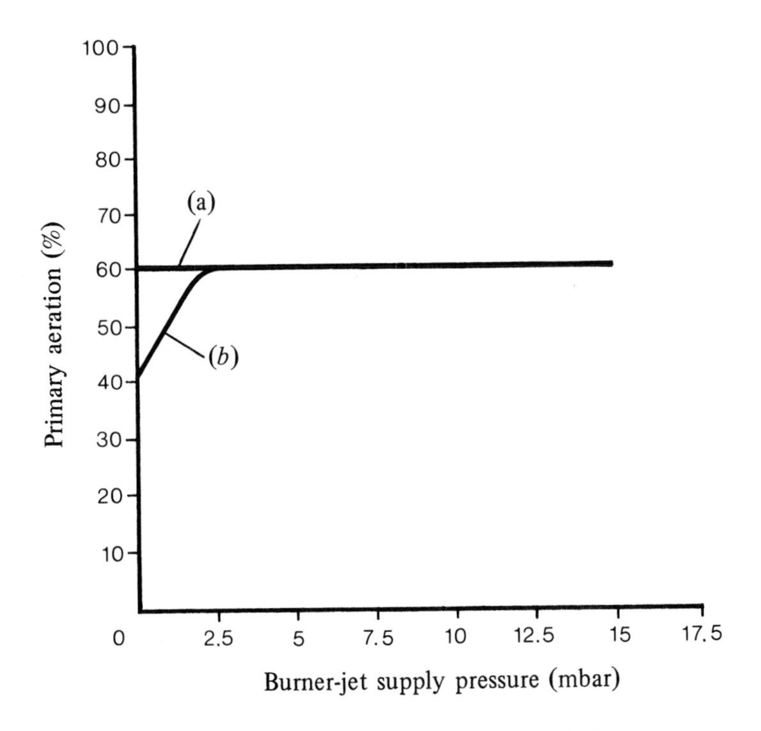

Figure 7:9 Variation of air–gas ratio with pressure for a typical aerated burner [22]

The above treatment is valid for burners in which the mixture pressure is identical for each burner port. In a bar burner, however, in which the burner ports are spaced along the length of the mixture supply tube the situation is more complex since the mixture flow and pressure vary along the tube. The mixture pressure from the injector must be equated to the pressure at the first burner port which can only be calculated from a knowledge of the pressure distribution along the burner bar. As indicated in Section 10.1.6, the progressive removal of mixture through the burner ports is equivalent to a reduction in C_F. Goodwin, Hoggarth and Reay [22] have taken this effect into account by introducing the correction factor F into the expression:

$$A_p/A_o = (r + 1)(r + d)\sqrt{[F(1 + C_F)]}/dC_{dp} \tag{10.16a}$$

For a simple injector without a diffuser the term $(1 + C_F)$ is replaced by $(2 + C_F)$ [22]. (Note that in reference [22], C_F is replaced by a differently defined coefficient S.)

Figure 7:10 shows the variation of F with burner bar area A_b and port area A_p for a range of ratios of length to diameter.

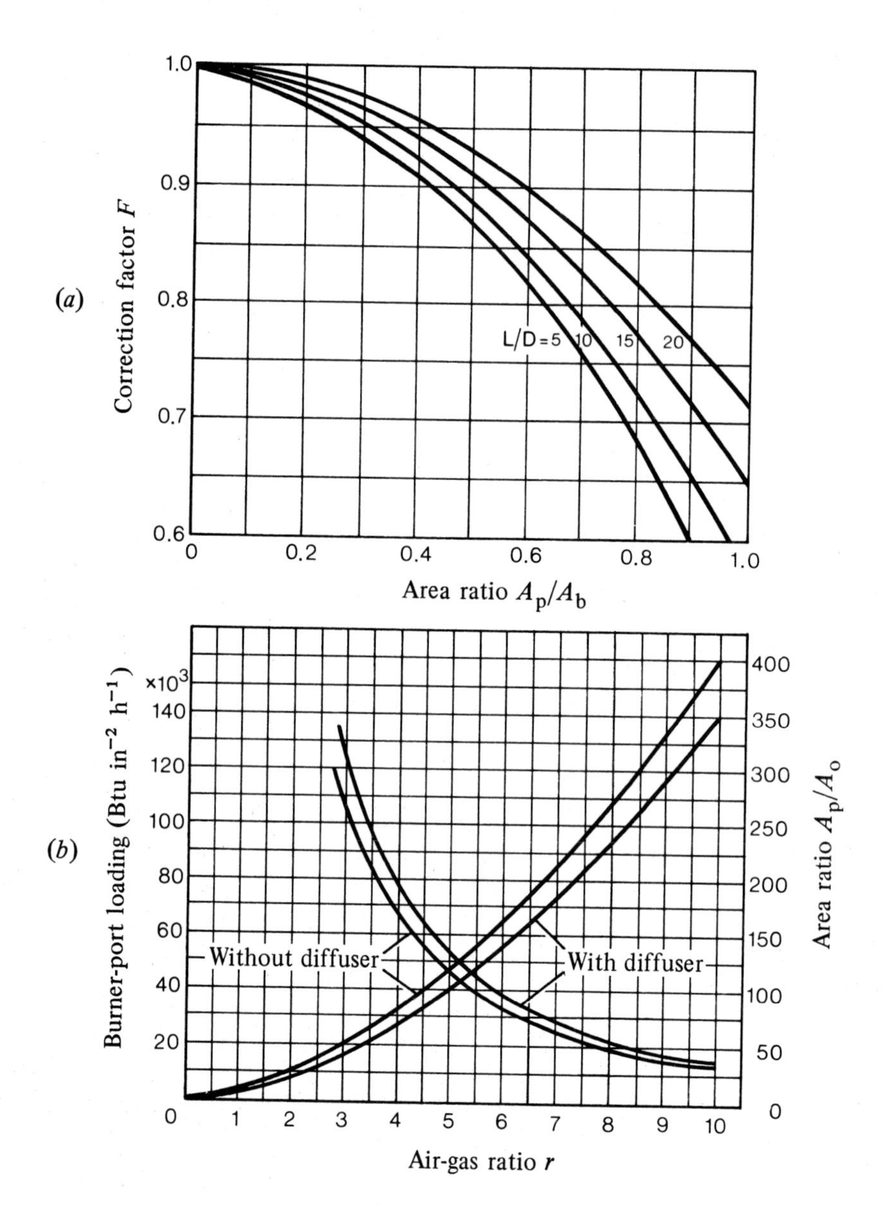

Figure 7:10 (*a*) Relationship between F and A_p/A_b [22]
(*b*) Relationships between A_p/A_o and r and between burner-port loading and r for natural gas of relative density 0.56 [22]

Clearly the major factor determining the degree of primary aeration is the total area of the flame ports for a particular duty, r increases with values of A_p/A_o.

The burner-port area for a particular gas rate is conveniently expressed as the *burner-port loading* or *flame-port loading*, i.e. the heat input to the burner divided by the total flame-port area $V_g C_g/A_p$ (MW m^{-2} or Btu in^{-2} h^{-1}).

In addition to the effects on air entrainment, changing the burner-port loading will also have an important effect on flame stability. Thus reducing the burner-port loading (i.e. reducing the mixture velocity at the flame ports) will, in principle, increase the resistance of a burner to blowoff and decrease its resistance to lightback particularly if the reduction in burner-port loading is achieved by providing larger burner ports.

Atmospheric burners designed for town gas normally operate at 40 to 60 per cent primary aeration ($r = 2.0$ to 2.5) and this is achieved with burner-port loadings of 18–32 MW m^{-2} (40 000 to 70 000 Btu h^{-1} in^{-2}) i.e. $A_p/A_o = 25$–30. Here the limit to stable flames is set by the tendency for the flames to light back with high aerations and large flame ports especially on turndown.

In the case of natural-gas burners it is clear from Figure 4:18 that blowoff would occur at burner-port loadings considerably lower than in the town-gas case and it is necessary to reduce the loading to 9–13.5 MW m^{-2} (20 000–30 000 Btu h^{-1} in^{-2}) to achieve stability. Burners with low flame loadings are widely used in the USA, but in Britain the more common solution to the problem of lift has been to use high flame-port loadings in conjunction with some external means of flame stabilization, e.g. the use of auxiliary flames. Burner-port sizing and flame retention are considered in Section 7.3.4.

The use of low flame-port loadings does, however, increase the amount of primary air entrainable and it is not difficult to achieve stoichiometric primary aeration at normal gas pressures by this means. This approach has not been generally favoured (although the advantages of fully aerated burners are considerable, i.e. freedom from secondary air vitiation, short flames, etc.) on the grounds that the physical size of the burner is inevitably increased and that the burner is less stable with regard to lightback. On the other hand if ribbon burners, in which an array of small flame ports is formed by sandwiching corrugated metal strips between straight ones [28] or punched metal burner heads [29] are employed, the resistance to lightback is high and in some applications the increase in injector size may be offset by the much smaller spacing necessary between the burner head and the heated object. Clearly burners of this kind require stoichiometric primary aeration anyway since the inner flames are shielded from secondary air by the outer flames.

7.3.4 Flame-port design

First-family gases

Small flame ports are normally employed as they minimize the tendency to light back. On the other hand large numbers of small drilled ports are expensive to produce and may be prone to blockage. Prigg [19] quotes a suitable compromise size of 2.5–3.1 mm (0.10–0.125 in). Easy cross-lighting from port to port is necessary to prevent delayed ignition and Prigg [19] has shown that the cross-lighting distance is approximately proportional to the flame-port diameter.

Figure 7:11 indicates maximum port spacing for satisfactory cross-lighting for various burner-port sizes using town gas of 18.7 MJ m^{-3} (500 Btu ft^{-3}) in free air.

Port diameter (in)	Port diameter (mm)	Maximum distance between centres Full rate (in)
1/16	1.6	0.35
3/32	2.4	0.45
1/8	3.2	0.65
5/32	4.0	0.90
3/16	4.8	1.10
7/32	5.6	1.25

Figure 7:11 Maximum port spacing for satisfactory cross-lighting [19]

Natural gas
Either low efflux velocities resulting from large flame-port areas are used or some external means of flame retention must be provided. These two approaches are considered below.

Burners with large flame-port areas. In general, to avoid lift with primary aerations of about 50 per cent, burners should be designed to have a flame-port loading of approximately 9 MW m^{-2} (20 000 Btu h^{-1} in^{-2}). Close spacing of the flame ports, large flame ports, and ledges above or below the flames will increase flame stability.

Goodwin, Hoggarth and Reay [22] have applied Lewis and von Elbe's critical boundary velocity gradient theory (Section 4.4.2) to the blowoff of drilled-port bar burners without auxiliary flame retention with limited success. The boundary velocity gradient is not easy to calculate with precision for short burner ports where the flow profile is still developing. However, using the Campbell and Slattery [30] solution which expresses boundary velocity gradient in terms of Reynolds number and friction loss coefficient, reasonable predictions for a range of single-port geometries were obtained. In the case of drilled multiport burners, blowoff is more difficult to define since it does not occur simultaneously from all ports; however, the results of Goodwin *et al.* show some improvement in lift resistance when multiple burner ports are used due primarily to the mutual stabilizing influence of adjacent flames.

The prediction of lightback using the boundary velocity gradient theory of Lewis and von Elbe has been well established in the case of burners giving fully established laminar flow as discussed in Chapter 4. However, practical burner ports have sharp-edged entrances and small length-to-diameter ratios and present difficulties in the application of the simple theory. Townsend [31] has obtained lightback curves for a range of sharp-edged ports having length-to-diameter ratios from 2.0 to 0.15 and has found that for small ratios the presence of a flame appreciably altered the flow conditions so that the lightback curves could not be satisfactorily predicted from the velocity profile measured in the absence of a flame.

Flame retention. To prevent lift from burners with high flame-port loading, two widely used techniques are either to promote the recirculation of hot flame-gas at the base of the flame or alternatively to provide low-velocity auxiliary flames close to the base of the main flame. The former technique is widely used in air blast burners and is considered under that heading; however, it has been applied to atmospheric bar burners by fixing a continuous metal strip at the base of the main flame (Figure 7:12 (*b*)) [22]. Because of practical drawbacks, such as dirt accumulation and buckling of the strip [22], auxiliary flame retention is the commonest solution to the problem of lift. In the case of open-ended burners, resistance to blowoff may be improved by sinking the burner tip in a cup of significantly larger bore than the burner tip (commonly twice the bore and in depth twice the burner bore) thus promoting recirculation eddies at the flame base. Alternatively proprietary non-blowoff tips incorporating auxiliary flames may be employed (see Section 7.4.2). Another technique applied by Vignes [32] to French domestic-cooker burners is termed velocity gradient retention in which the mixture flows through narrow slots shaped so that the effective port depth is greater at the bottom of each slot than the top. Thus the efflux velocity at the lower end is reduced and provides stabilization. The more normal method, however, is to bypass a proportion of the mixture flowing to the burner ports through small metering orifices and thence to separate auxiliary ports of greater cross-section so that the efflux velocity is low. Typical auxiliary flame retention arrangements applied to industrial bar burners are shown in Figure 7:12.

The mechanism of auxiliary flame stabilization has been investigated by a number of workers. It appears that the main flame is protected from the flow of secondary air which is replaced by products of incomplete combustion which protect the mixture, increasing its burning velocity and providing a source of free radicals.

The proportion of total gas flow necessary for the auxiliary flames is dependent on the air–gas ratio, burner rating, turndown and port spacing [33]. Goodwin *et al.* [22] have adopted a figure of 20 per cent of total flow with the auxiliary ports not more than 6.5 mm (0.25 in) from the main ports in their industrial bar-burner designs. Harris and Prigg [27] quote Continental practice as using up to about 15 per cent of total flow with 2 to 3 mm (0.079 to 0.118 in) separation. Their own studies on domestic-cooker burners indicated that 8 per cent auxiliary gas gave adequate retention with a spacing of 2.5 mm (0.1 in).

Burner-port sizing and spacing. For industrial bar burners with high flame-port loading incorporating auxiliary flame retention, Goodwin *et al.* [22] recommend main flame port diameters of 2.5 to 4.8 mm ($\frac{1}{10}$ to $\frac{3}{16}$ in.) and auxiliary metering orifices of around 1.6 mm ($\frac{1}{16}$ in.).

If the auxiliary flame takes the form of a single sheet the main-flame spacing is not critical and may be considerably greater than one port diameter apart (which is recommended for burners without auxiliary flame retention) [34]. Burners designed for low flame-port loading at high primary aeration, particularly those requiring a high turndown ratio, require considerably smaller port diameters to provide protection against the possibility of lightback on high-burning-velocity SNG.

Flame height. In applications in which atmospheric burner flames come into close proximity with solid surfaces flame height becomes important. Clearly, impingement and breaking of the primary cone will cause carbon monoxide formation and should be avoided. Prediction of the primary-cone height is relatively straightforward as indicated below. Normally, however, the primary mixture in an atmospheric burner is gas rich and this gives rise to a secondary combustion zone.

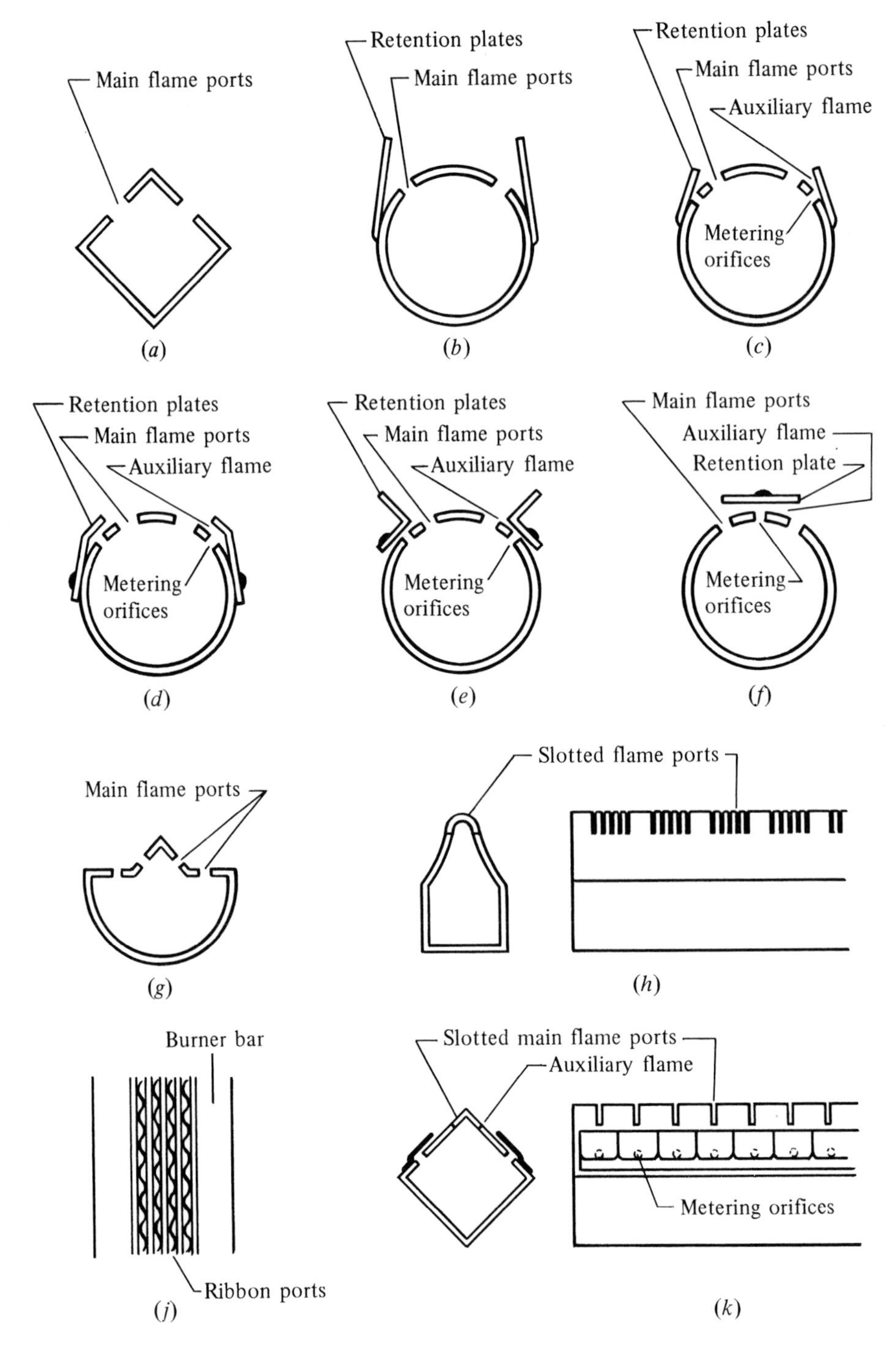

Figure 7:12 Burner-port configurations [22]

Consequently the degree to which this secondary zone may be allowed to impinge on a cold surface is also of importance. Since the secondary zone is less well defined this presents a more difficult problem which has usually been approached by defining, and attempting to predict, the flame height as the separation between the burner head and the solid surface for a CO/CO_2 ratio of 0.02 in the combustion products. Shnidman [23] quotes empirical expressions for both inner and outer flame heights obtained by the American Gas Association for burners of low flame-port loading.

Reed [35] and Hess [36] have approached the prediction of the height of the inner cone more fundamentally and the former has shown that for very small flame-port length-diameter ratios the height of the primary flame cone (L_p) is approximated closely by:

$$L_p = \tfrac{1}{2} D_p \, (U/S_u) \tag{7.8}$$

where D_p = flame-port diameter, and for very large ratios:

$$L_p = \tfrac{2}{3} D_p \, (U/S_u) \tag{7.8a}$$

Predictions of the outer cone height L_{co} defined as indicated earlier have been made by Roper [37, 38, 39]. Measured flame heights for a range of industrial bar burners are quoted by Reay [40].

7.3.5 Design of mixture tube and throat

In designing the mixture tube and throat it is generally desirable to use the minimum diameters consistent with obtaining the desired primary aeration and port loading and, in the case of bar burners, with avoiding excessive variations in flame height along the burner.

A simple relationship between the throat cross-sectional area and the flame-port area has been quoted earlier as:

$$A_t \approx 0.45 \text{ to } 0.65 \, A_p$$

A more detailed treatment [21] includes the effects of the presence or absence of a diffuser and the type of burner head on the optimum throat area. In the case of a bar burner the variation of flow and pressure along the tube require the introduction of a further factor. Thus five cases may be distinguished:

1 Parallel throat with diffuser; drilled port burner head [21]:

$$A_t/A_p = C_{dp} \, (1 + C_F)^{\frac{1}{2}} \tag{7.9}$$

2 Parallel throat, no diffuser; drilled port burner head [21]:

$$A_t/A_p = C_{dp} \, (2 + C_F)^{\frac{1}{2}} \tag{7.10}$$

3 Parallel throat, no diffuser; mixture-tube acting as the flame port, i.e. simple open-ended burner [21]:

$$A_t/A_p = 1 \tag{7.11}$$

4 Bar burner with diffuser [22]:

$$A_t/A_p = C_{dp}\,[(1 + C_F) + \tfrac{1}{4}K_{av}]^{\frac{1}{2}} \tag{7.12}$$

5 Bar burner without diffuser [22]:

$$A_t/A_p = C_{dp}\,[(2 + C_F) + K_{av}]^{\frac{1}{2}} \tag{7.13}$$

where K_{av} is the average pressure rise along the bar in velocity heads. Goodwin *et al.* [22] quote values of K_{av} as 0.90 and 0.66 for bars with ratios of flame-run length to bar diameter of 10 and 20 respectively. Calculated values of A_t/A_p for bars of various length-to-diameter ratios are given in Figure 7:13.

L_f/D_b	Without diffuser	With diffuser
10	1.15	0.81
15	1.13	0.80
20	1.10	0.79
25	1.07	0.78
30	1.035	0.76
35	1.00	0.75

Figure 7:13 Area ratios A_t/A_p for various ratios of flame run (L_f) to bar diameter (D_b)

If no diffuser is used, either a parallel throat inside the burner bar may be used, or the bar itself may be sized to be equal to the calculated throat diameter in which case a simple burner results where the burner bar itself acts as the mixing throat. As Goodwin *et al.* [22] point out, in practice the nearest nominal pipe size would be chosen producing only a small effect on the aeration obtained. In a parallel injector of this kind the maximum static pressure rise occurs at about six mixture-tube diameters and the first burner port should be no closer than ten tube diameters from the injector orifice [38] to ensure adequate mixing and to smooth flow irregularities.

In the case of a throat with diffuser, again about six throat diameters length appears optimal plus a convergent approach section of some 2 to 2½ throat diameters as shown in Figure 7:14. The burner-tube diameter should lie between 1.25 and 1.55 throat diameters [22] to ensure maximum conversion to static pressure. Prigg [19] recommended that the burner-tube area should not be less than 1.5 A_p which yields a similar result. Francis [21] has estimated a total value of $C_F \approx 0.4$ for the combined pressure loss in the throat and diffuser.

The design of the air port or primary air inlet is not particularly critical. It should be as large as possible and offer as little flow resistance as possible. An air-port area of $> 2A_p$ meets these requirements [19].

It is generally desirable to provide aeration control on industrial burners and this may be provided by an air shutter, by screwing up the injector orifice close to the mixture tube entry or by means of a screw inserted in the burner throat. The last mentioned is not recommended since it may lead to gas spillage at low inputs.

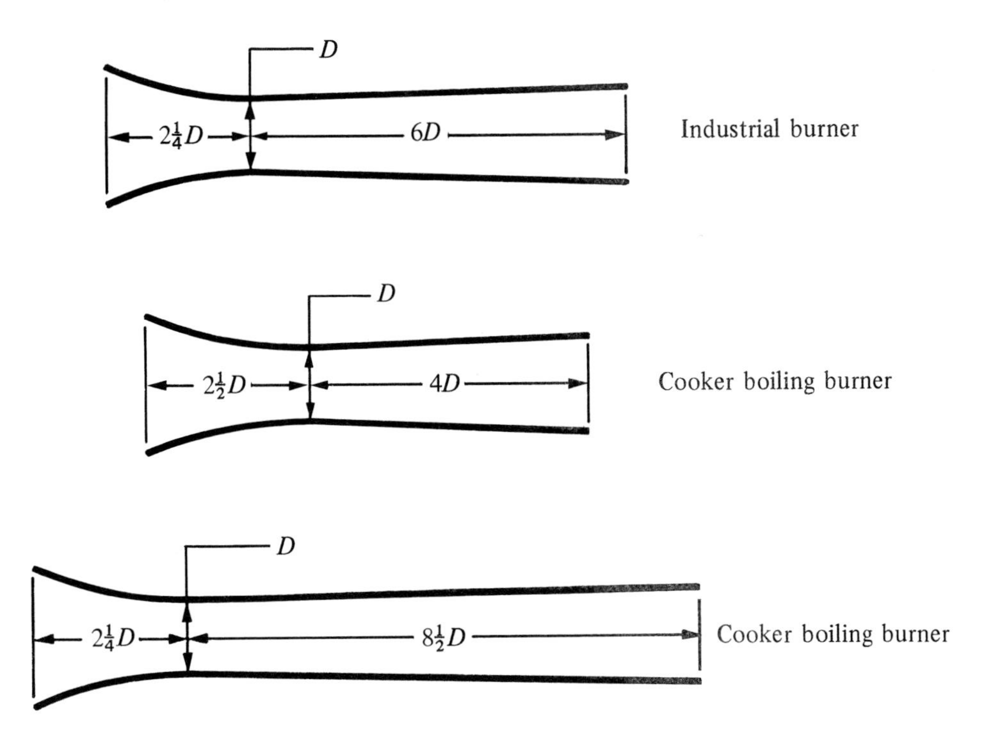

Figure 7:14 Throat shapes [19]

7.3.6 Bar-burner design procedure

An example of industrial bar-burner design is given below to illustrate the application of the foregoing design information.

Determine the principal dimensions of a drilled tubular natural-gas bar burner of high port loading incorporating a diffuser and using auxiliary flame retention for a heat release rate of 15 kW (51 200 Btu h^{-1}), a flame run length of 280 mm (11 in) and a primary aeration of 60 per cent stoichiometric.

(i) Orifice sizing.

From equation 3.4 heat input Q is given by:

$$Q = 12.55\ A_o C_{do} W \sqrt{p_o}$$

and $W = 50.7$ MJ m^{-3} (st), $p_o = 15$ mbar, $C_{do} = 0.9$.

$$\text{So:}\quad 15\,000 = 12.55 \times 0.9 \times 50.7 \times \sqrt{15} \times A_o$$

Therefore: $A_o = 6.763\ mm^2$

Therefore: $D_o = 2.9$ mm (0.11 in)

(Alternatively this value may be read off Figure 7:8):

(ii) Burner-port area

60 per cent stoichiometric primary aeration is equivalent to an air–gas ratio of 5.8. Thus from equation 10.16a:

$$A_p/A_o = (r + 1)(r + d)\sqrt{[F(1 + C_F)]}/dC_{dp}$$

Given that $d = 0.56$ and assuming $C_{dp} = 0.65$, $F = 0.9$ and $C_F = 0.3$. Thus:

$$A_p/A_o = 129$$

The same result may be read off Figure 7:10 (*b*). It follows that $A_p = 872$ mm^2 (1.35 in^2).

(iii) Assuming that 20 per cent of the total flow is supplied to the auxiliary ports, the area of the main ports totals $0.8 \times 872 = 698$ mm^2 and the area of the metering orifice for the auxiliary ports totals $0.2 \times 872 = 174$ mm^2. Using 3.2 mm diameter main ports, of area 8.04 mm^2, $698/8.04 = 87$ are required.

If 1.6 mm diameter metering orifices are used, the same number is required.

(iv) The main burner ports may be spaced about 1 diameter apart in two rows giving a flame run length of about 282 mm, as required. The auxiliary ports would be placed at the same spacing in two rows adjacent to the main ports as in Figure 7:12 (*c*), (*d*) or (*e*).

(v) Throat sizing.

From equation 7.12:

$$A_t/A_p = C_{dp}\,[(1 + C_F) + \tfrac{1}{4}K_{av}]^{\frac{1}{2}}$$

Given that $K_{av} = 0.90$ for an assumed flame-run/bar-diameter ratio of 10:

$$\begin{aligned} A_t/A_p &= 0.65\,[(1 + 0.3) + \tfrac{1}{4} \times 0.9]^{\frac{1}{2}} \\ &= 0.80 \end{aligned}$$

Alternatively A_t/A_p may be read from Figure 7:13. It follows that $A_t = 698$ mm^2 giving a throat diameter of 29.8 mm.

(vi) The burner bar diameter is normally 1.25 to 1.55 times the throat diameter thus a nominal pipe size of $1\frac{1}{2}$ in (38 mm) would be chosen.

(vii) The burner-port loading obtained with this burner is $15\,000/872 \times 10^{-6} = 17.2$ MW m^{-2} (38 000 Btu h^{-1} in^{-2}).

7.3.7 Commercially available low-pressure aerated burners

The foregoing sections have dealt with the design of low-pressure aerated burners and application of this information will allow satisfactory burners to be designed for a wide variety of applications. This is not to imply, however, that in practice burners are always designed in this way and constructed for specific applications. There is a wide range of proprietary injectors, burner heads, bars and jets and complete burner assemblies available for the gas engineer to draw on in plant design. The purpose of this section is not to catalogue these products (and it is by no means complete in this respect) but to indicate features of particular interest in a number of commercial products.

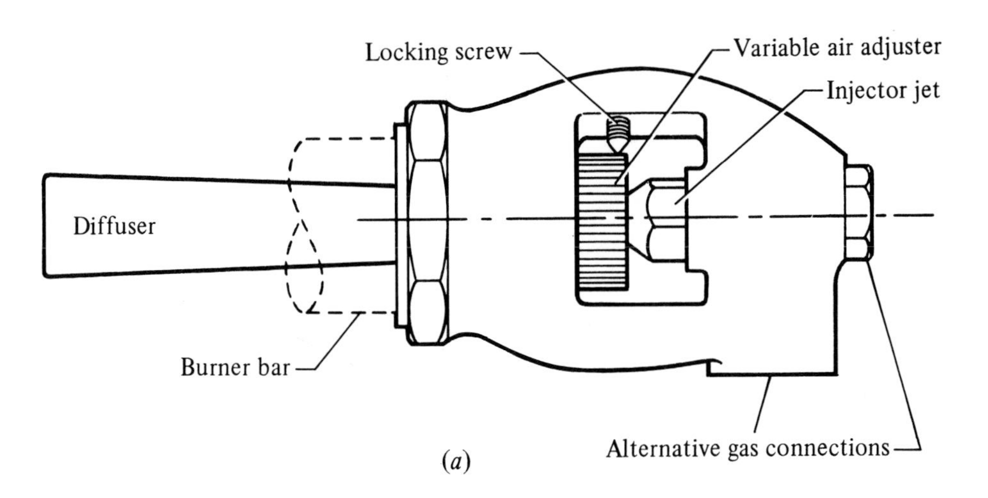
Locking screw
Variable air adjuster
Injector jet
Diffuser
Burner bar
Alternative gas connections

(a)

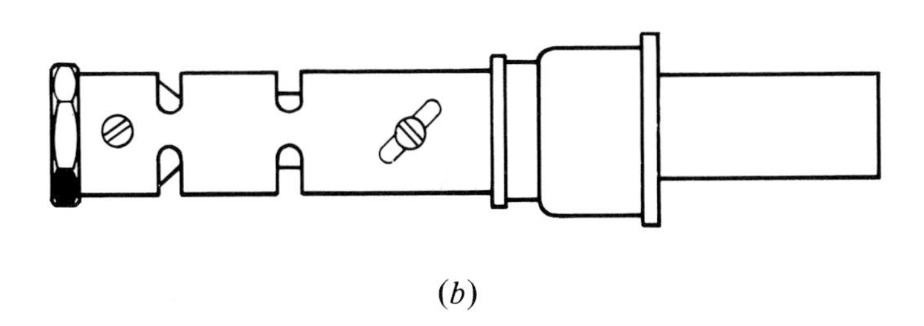

(b)

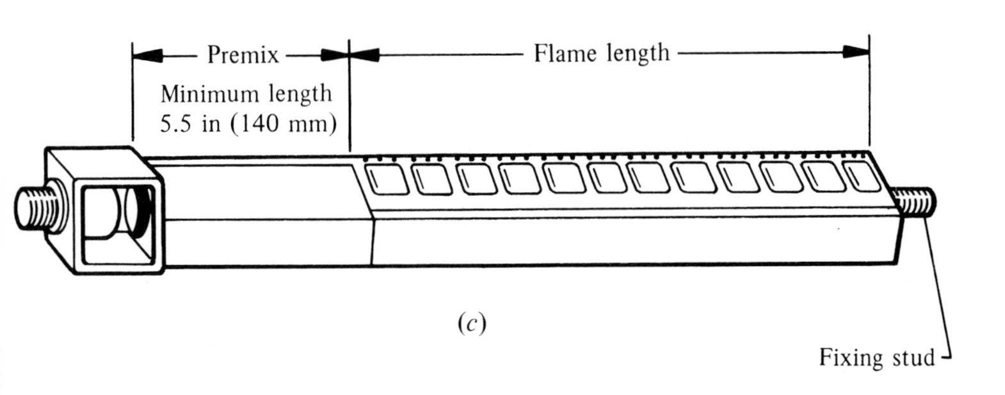
Premix
Flame length
Minimum length
5.5 in (140 mm)
Fixing stud

(c)

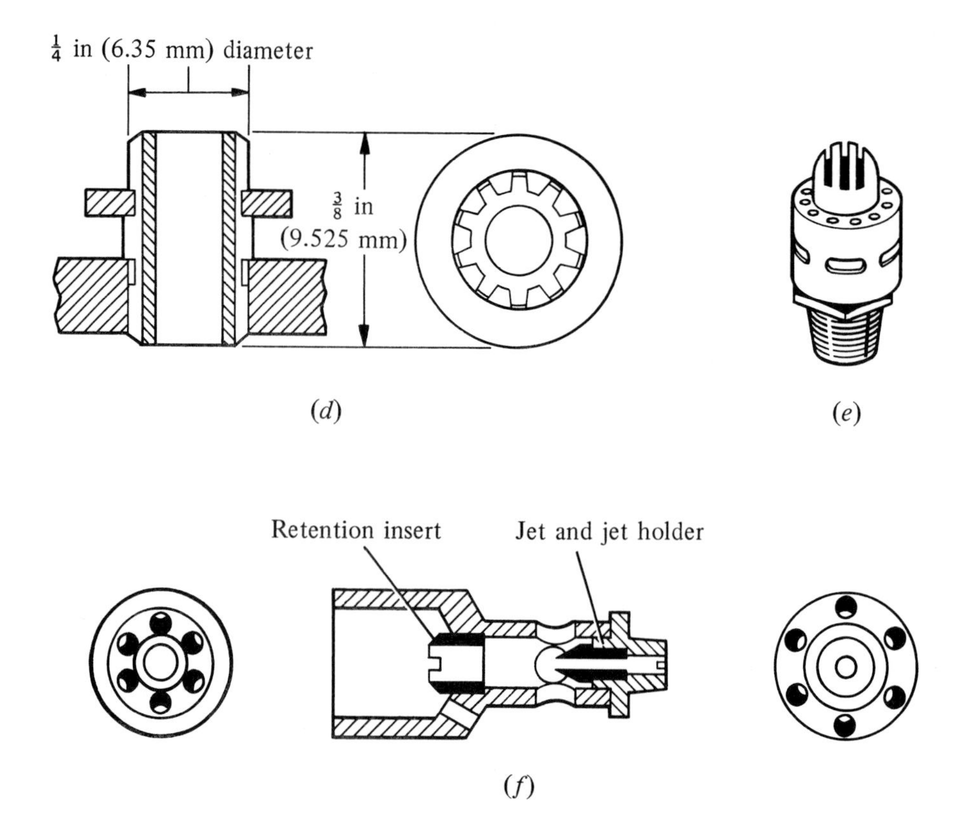

Figure 7:15 Proprietary low-pressure aerated burners and injectors
(*a*) Amal injector type 345 (Amal Ltd)
(*b*) Aeromatic two-stage injector (The Aeromatic Co. Ltd)
(*c*) Tekni MHR type multigas burner (Teknigas Ltd)
(*d*) Aeromatic Insertajet (The Aeromatic Co. Ltd)
(*e*) Bray flat-flame aerated jet (Geo. Bray & Co. Ltd)
(*f*) Aeromatic Segas burner (The Aeromatic Co. Ltd)

Injector assemblies incorporating injector orifices and throat diffusers of similar design to the recommendations given earlier are produced by a number of manufacturers including Amal and Teknigas (Figure 7:15 (*a*)) in mixture tube sizes from ¾ in BSP to 2 in BSP. An injector assembly incorporating two-stage air entrainment is produced by Aeromatic and advantages claimed for this system include high aeration and turndown ratio (Figure 7:15 (*b*)).

Many manufacturers produce complete burner assemblies primarily for domestic equipment and normally of the universal or multigas type. These are burners which by a change of injector orifice, a change of burner pressure and, in some cases, a change of aeration, can be used to burn virtually any fuel gas. The burner ports are normally provided with auxiliary flame retention and are commonly narrow slots or punched circular holes. Figure 7:15 (*c*) indicates a typical arrangement. A common feature of this type of burner is the incorporation of the mixing tube diffuser inside the burner body so as to reduce the overall burner length.

A convenient means of converting aerated town-gas bar burners to natural gas at high burner-port loadings is to insert tap-in replacement flame-port tips into the drilled out existing flame ports. Each Insertajet (Figure 7-15 (*d*)) incorporates auxiliary flame retention and is rated at about 0.15 kW (500 Btu h^{-1}).

The conversion of town-gas bar burners using postaerated jets (Section 7.3.1) to natural gas may be done as an alternative to replacing the whole burner with an aerated one using individually aerated screw-in jets as replacements. Typical of these is the Bray flat-flame aerated jet shown in Figure 7:15 (*e*) which incorporates flame retention ports. The heat input per jet ranges from about 0.44 to 1.3 kW (1500 to 4500 Btu h^{-1}).

Similar in principle but much larger (2.9 to 29 kW) (10 000 to 100 000 Btu h^{-1}) is the Aeromatic Segas burner (Figure 7:15 (*f*)) which has found particular application in the conversion of vertical fire-tube boilers.

7.3.8 High-pressure atmospheric burners

If gas is available at pressure it may be used in an atmospheric burner with a number of potential advantages over gas at normal supply pressures:

1 Stoichiometric primary aeration is easily obtained using smaller burner cross-sections producing high-temperature flames free from the need for secondary air.
2 High burner-port loadings are possible.
3 Control valves and pipework sizes may be reduced.
4 High turndown ratios at constant primary aeration are attainable.

Burners of this type have not been widely used in Britain due to the lack of availability of high supply pressures and their disadvantages listed below compared with air-blast systems. Town-gas systems in use have normally used a positive-displacement compressor raising the pressure to 20.6 to 69 kPa (206 to 690 mbar, 3 to 10 lbf in^{-2}) and this introduces complications with regard to operational safety (see Chapter 9). Higher pressures are recommended for natural gas, normally 69 kPa (690 mbar, 10 lbf in^{-2}) to 100 kPa (1 bar, 15 lbf in^{-2}).

Many of the same benefits are provided by air-blast systems which, by using simple single-stage air fans avoid these complications. As Francis [21] has pointed out, raising the gas pressure to 20.6 kPA (206 mbar, 3 lbf in^{-2}) raises the maximum flame-port heat release to about 45 MW m^{-2} (100 000 Btu h^{-1} in^{-2}) but the maximum mixture pressure is about 250 Pa (2.5 mbar, 1 inH_2O) which is about a tenth of that available from an air-blast system using fan air at 6.9 kPa (69 mbar, 1 lbf in^{-2}).

The increasing use of high-pressure distribution systems suggests the possibility of still higher gas pressures being available. Francis [21], using the compressible-flow equation, has derived the relationship between gas pressure and heat release per unit area of flame port for a high-pressure atmosphere injector operating on town gas at stoichiometric air–gas ratio shown in Figure 7:16.

The figure indicates the large increase in pressure required to produce only a modest increase in flame-port heat release; a hundredfold pressure increase from 20.6 kPa (206 mbar, 3 lbf in^{-2}) to 2.06 MPa (20.6 bar, 300 lbf in^{-2}) yields only a fivefold

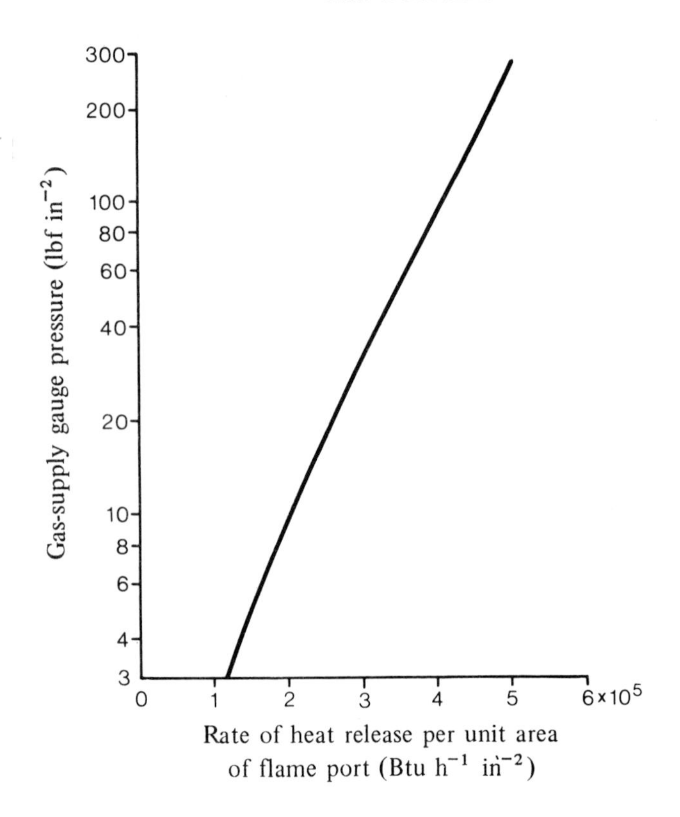

Figure 7:16 High-pressure atmospheric burners: variation of heat-release rate per unit area of flame port with gas supply pressure for town gas at stoichiometric conditions [21]

increase in heat release [21]. It is significant that this flame-port heat release is still only about 50 per cent greater than that obtained from a normal air-blast system.

The design principles of high-pressure atmospheric burners are essentially the same as for low-pressure burners. Because of the relatively high burner-port loadings employed, flame-retention provision is necessary, frequently of the types employed in conjunction with air-blast systems and discussed in the section devoted to these burners.

7.4 PREMIX AIR-BLAST BURNERS

Forced-draught premix burners have been adopted widely in industrial gas practice for the following main reasons:

1 Stoichiometric proportions are easily attained
2 Control of air–gas ratio is relatively simple and precise
3 They possess an inherent self-proportioning facility if the gas is entrained from a supply at atmospheric (or combustion-chamber) pressure

4 Throughput may be varied by a single control valve on the air supply
5 High mixture pressures are obtained using modest air-supply pressures resulting in high turndown ratios
6 High burner-port loadings may be obtained
7 The combustion products may be given sufficient momentum to produce more positive hot-gas distribution in the combustion chamber or furnace and less reliance on natural draught than low-pressure systems
8 Air–gas mixing is good, minimizing flame-length and combustion-space requirements
9 Easy interchangeability of fuel gases

Air-blast pressure burners are conveniently considered as two separate components, the injector assembly and the burner head.

7.4.1 Air-blast injectors

Air, normally at pressures in the range 2.5 to 7.5 kPa (25 to 75 mbar, 10 to 30 inH_2O) is fed to the orifice of an injector inducing gas at normal supply pressure or, if a 'zero governor' is employed, at atmospheric pressure via a sidelimb. Both simple cylindrical injectors and those incorporating diffusers are employed.

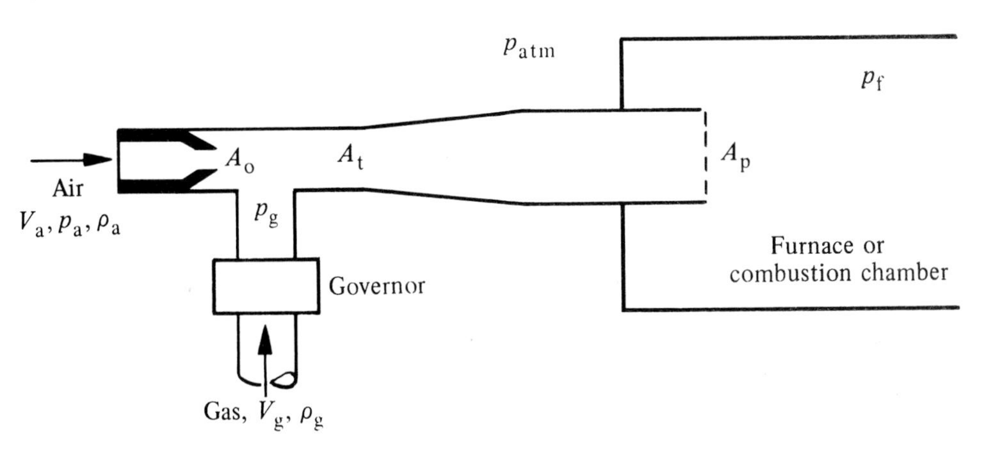

Figure 7:17 Air-blast injector: nomenclature

The size of the air orifice is determined from equation 7.2 by substituting $d = 1$ and $V_a = rV_g$ where V_a is the desired air flow rate for an air–gas ratio of r:

$$V_a = rV_g = 1655\ C_{dao}A_{ao}\sqrt{p_{ao}} \tag{7.14}$$

and in SI units

$$12.55 \times 10^{-6}\ C_{dao}\ A_{ao}\sqrt{p_{ao}} \tag{7.14a}$$

C_{dao} normally lies between 0.85 and 0.95. Francis and Jackson [41] have quoted a value of 0.95 for a nozzle of less than 30° included angle followed by a parallel section less than one diameter in length.

Using the nomenclature of Figure 7:17, equation 10.4 may be rewritten:

$$[(p_g - p_f)A_t^2/\rho_g V_a^2] - \tfrac{1}{2}(r+1)(r+d)[(C_{dp}^{-2} + C_F)(A_t/A_p)^2 + 1] + d(A_t/A_0) = 0 \qquad (7.15)$$

If $p_g = p_f$ then r is independent of V_a, i.e. there is an inherent self-proportioning action, a constant air–gas ratio is given for all throughputs and throughput is controlled very simply by a single control valve on the air supply. Furnace pressure p_f is commonly atmospheric and, if air–gas ratio control is required, p_g is reduced to atmospheric via a zero-pressure governor. Pressure governors are considered in Chapter 9 but since zero-pressure governors are restricted to this application it is appropriate to consider them here.

Zero-pressure governors are essentially low-pressure governors in which the weight of the diaphragm and valve parts is supported by a tension spring or in which the diaphragm is vertical and the valve mechanism horizontal to eliminate the action of gravity on the valve parts. Any increase in outlet pressure raises the diaphragm and partially closes the valve tending to restore the outlet pressure to zero gauge pressure. Slight adjustments may be made to the spring tension to vary the outlet pressure from atmospheric if desired.

Figure 7:18 (*b*) shows the vertical diaphragm type which consists of two main parts, the zero governor proper which is a standard unit for all governor sizes and a diaphragm valve which is in effect a relay-valve (Section 9.4.3) connected to the zero-governor valve via an integral weep pipe. There is no provision for altering the outlet pressure in this design.

It should be noted that since the performance of zero governors is adversely affected by too high or widely fluctuating inlet pressures they should be preceded by a conventional low-pressure (appliance) governor. Further information on zero governors is given in references [42, 43, 44].

The basic control scheme for a self-proportioning air-blast burner is shown in Figure 7:19. Although an injector may be designed to produce virtually any desired air–gas ratio it is normal to slightly over-design the injector (or to select an appropriate proprietary injector which will normally be over-sized) and to restrict the flow of gas into the injector by means of a ratio-setting valve. This may be a separate valve as shown or may be integrated into the injector body (in which case it is termed an obturator). The proportioning accuracy of the system obviously depends on the performance of the zero governor and, less obviously, on the flow characteristics of the ratio-setting valve which are affected by changes in friction factor in the valve with varying flow [44]. The air–gas ratio is set at the low flow rate by adjusting the zero governor spring (if present) and at high flow rate by the ratio-setting valve. In the case of the non-adjustable type of zero governor there is a marked tendency for the mixture to become progressively gas-weak as the throughput is reduced [44]. This is not due to any defect in the zero governor, which maintains atmospheric pressure with considerable accuracy, but to changes in friction factor in the ratio-setting valve with changes in Reynolds number. In practice this tendency may offer advantages, particularly with fast-burning gases, since, by reducing the tendency to lightback it may allow the turndown range to be increased. In contrast, with the adjustable type, the ratio may be set independently at low and high flow rates by adjusting the spring and ratio-setting valve respectively. Changes in friction factor are compensated in this way and a reasonably constant air–gas ratio obtained over the working range. Regarding the choice of ratio-setting valve, reference [43] suggests quadrant or diaphragm valves which should not be of such a size that they are nearly

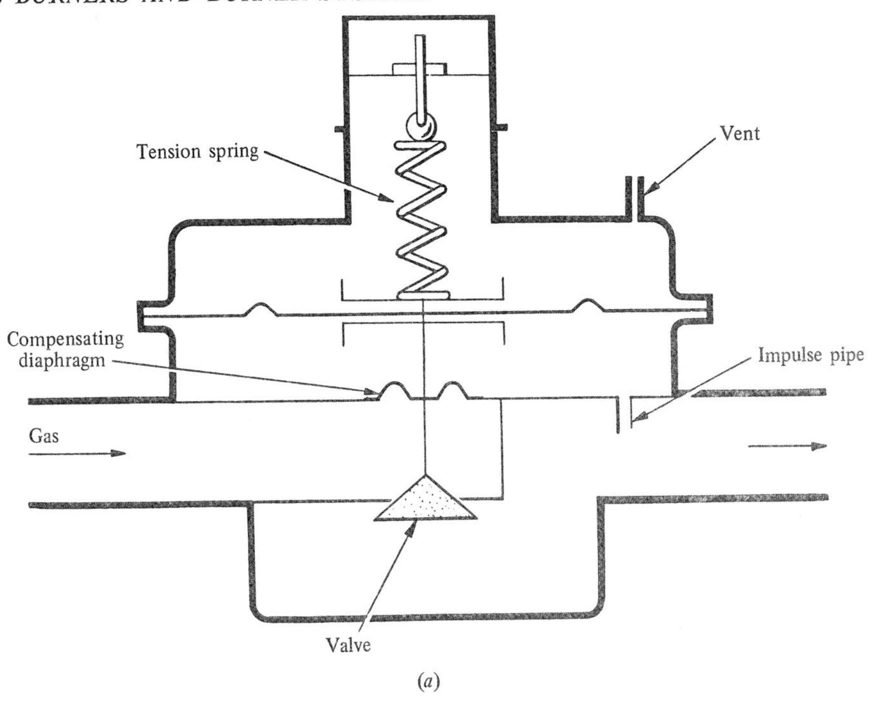

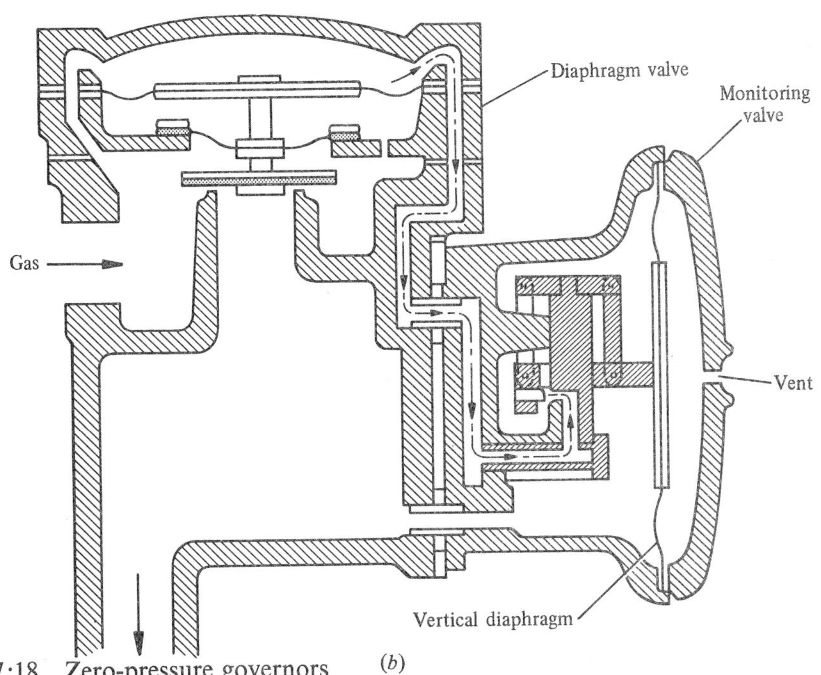

Figure 7:18 Zero-pressure governors
(*a*) Tension-spring type
(*b*) Vertical-diaphragm type

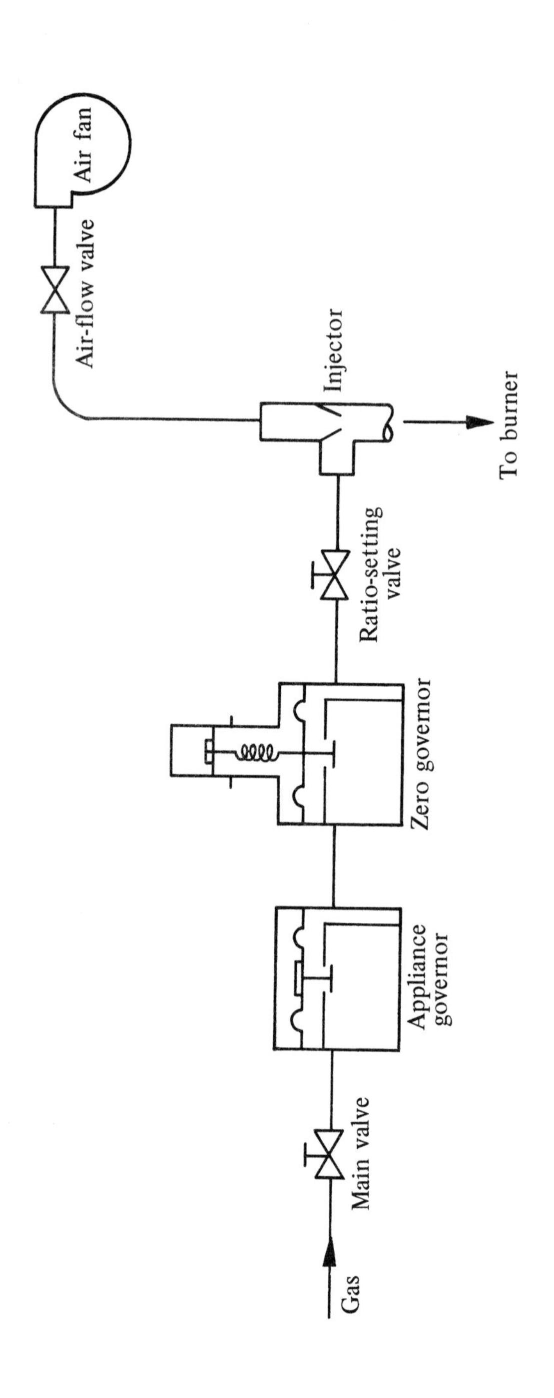

Figure 7:19 Self-proportioning air-blast burner: basic control scheme. A non-return valve must be fitted in the gas supply line

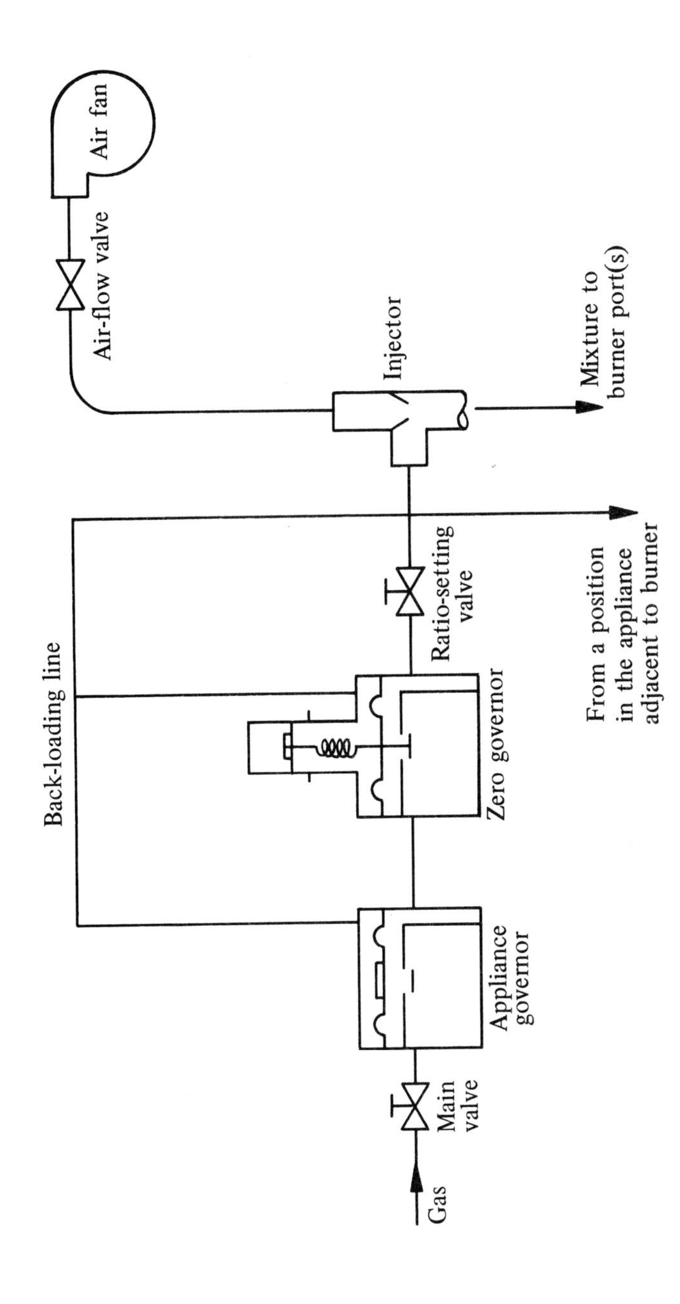

Figure 7:20 Self-proportioning air-blast burner with appliance-pressure back-loading. A non-return valve must be fitted in the gas supply line

closed when in the set position. Edmondson and Clark [44] found that all types of external ratio-setting valves gave inferior performance to that obtained with an integral obturator although diaphragm valves approached this performance most closely.

Unlike postaerated and low-pressure atmospheric burners, air-blast burners are frequently required to fire into combustion chambers in which the pressure differs significantly from atmospheric. Thus to maintain self-proportioning, the reference pressure for the zero governor (and the appliance governor) is not atmospheric but the combustion-chamber pressure, i.e. $p_g = p_f > p_{atm}$. Changes in p_f are easily compensated for by *back-loading* the diaphragms with the combustion-chamber pressure. This technique is indicated in Figure 7:20. The zero-governor inlet pressure should be at least 75 Pa (0.75 mbar, 0.3 inH_2O) higher than the appliance chamber pressure [43].

Alternatively, mixture-pressure back-loading may be used which provides complete compensation for changes in back pressure downstream of the mixture tube. However, since the zero-governor pressure must be slightly higher than the back-loading pressure, gas boosting is normally required [43].

If an injector incorporating a diffuser is used this will normally be selected from a range of commercially available designs which give a mixture pressure of approximately 40 per cent of the air pressure [45]. If it is required to construct this type of injector the throat cross-section should be twice the air-nozzle area, the diffuser taper 8–15° included angle and the throat length four to six throat diameters [45]. Recommended connection sizes [45] are as follows: the air-pipe size should be not less than 2½ times the air-nozzle diameter (this allows for a maximum pressure loss not greater than 5 per cent of air pressure and is equivalent to two 90° bends and 30 diameters of straight pipe); the mixture outlet connection one size larger than the air connection and the gas connection one size smaller. These recommendations relate to town-gas injectors and the last-mentioned size may be reduced in the case of natural gas. Parallel tube injector dimensions have been optimized for air-blast tunnel burners and further information is given under that heading (Section 7.5).

7.4.2 Air-blast burner heads

Since air-blast burners produce high mixture velocities, means need to be provided to prevent blowoff. This may be done by the use of (*a*) a refractory tunnel; (*b*) a flame-retention cup incorporating either auxiliary flames or a sudden enlargement; or (*c*) flame impingement (impact burner).

Tunnel burners

In tunnel burners a refractory tunnel is employed, in which combustion is substantially completed, stabilization resulting partly from the shielding and heating effect of the tunnel which quickly becomes incandescent and partly as a result of the sudden enlargement which exists between the tip of the mixture tube and the base of the tunnel.

In addition to providing flame stability, tunnel burners offer major advantages over conventional systems which have led to their widespread use and which have had a profound influence on the development of gas-fired plant in recent years. The tunnel dimensions may be chosen such that the burner emits a high-velocity stream of burned gas at near flame temperature with little or no flame at the tunnel exit. Optimization of the relative areas of air nozzle, mixture throat and tunnel cross-sections enables the most efficient use to be made of the available air pressure in terms of tunnel exit velocity. Burners of this type are generally termed high-velocity tunnel burners although the terms jet burner and high-intensity combustor are also used.

Their major uses are in providing very high convective heat-transfer rates for local heating processes, in promoting positive hot-gas circulation in furnace chambers and in producing high levels of jet-driven recirculation thereby increasing convective heat transfer and minimizing temperature differentials. Many examples of the use of these techniques will be found elsewhere in this book.

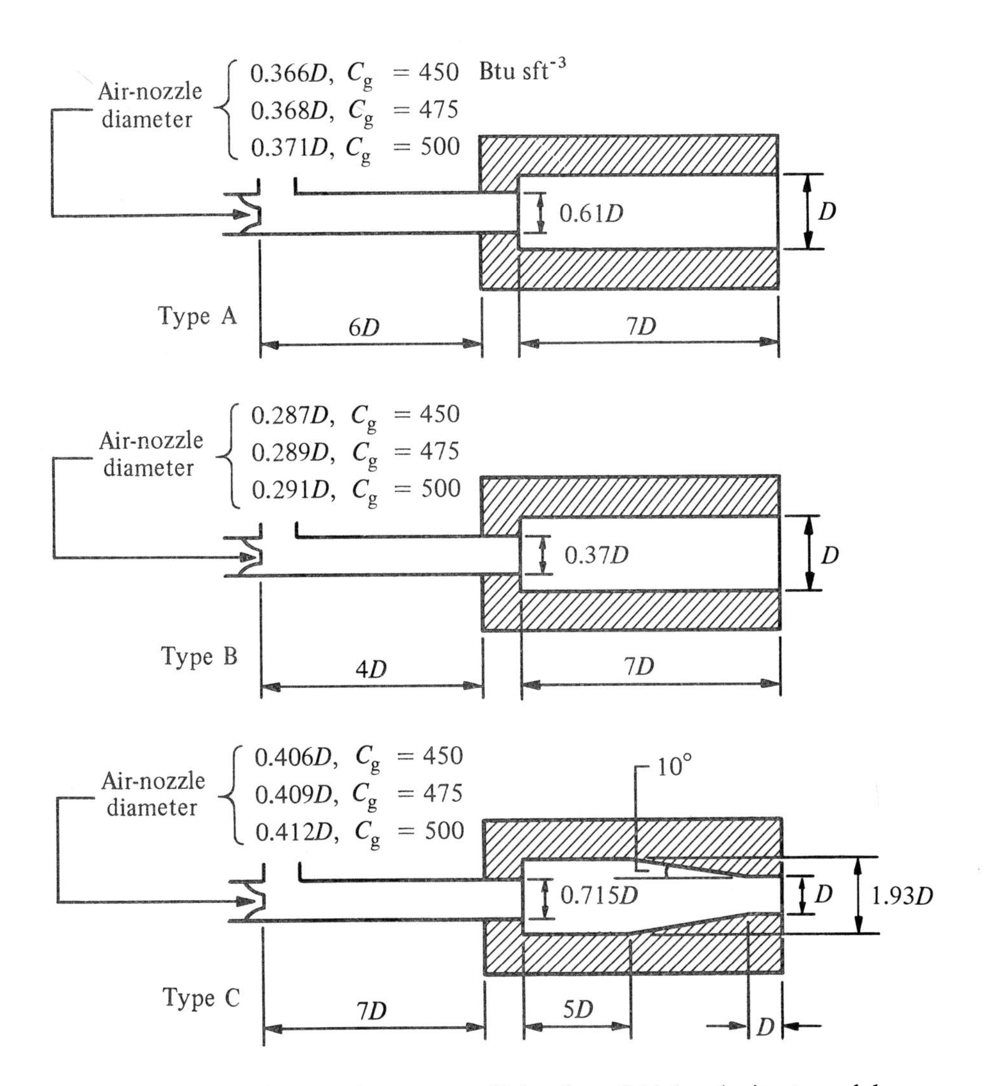

Figure 7:21 Dimensions and pressure efficiencies of high-velocity tunnel burners for first-family gases

Francis [41, 46] has developed an optimized-design procedure for high-velocity tunnel burners operating on town gas based on maximizing the pressure efficiency, defined as the ratio of the dynamic pressure of combustion products at the tunnel exit to the dynamic pressure of air in the air nozzle.

Referring to Figure 7:21, the dynamic pressure of air in the air nozzle is:

$$\rho_a V_a^2/2g A_{ao}^2 = \rho_a r^2 V_g^2/2gA_{ao}^2$$

and of exit combustion products:

$$\rho_t V_t^2/2gA_t^2 = \rho_a (r + d)(r + 0.7)V_g^2 T_t/2gA_t^2 T_o$$

where subscripts a, g and t refer to air and gas, and to combustion products at tunnel-exit conditions, respectively.

Thus pressure efficiency:

$$\eta_p = (r + d)(r + 0.7)r^{-2}(T_t/T_o)(A_{ao}^2/A_t^2) \qquad (7.16)$$

Figure 7:21 shows the optimized dimensions and pressure efficiencies quoted by Francis [46] for various injector/tunnel combinations.

The choice of an appropriate tunnel burner for a particular application is based on the requirements of exit velocity, turndown and combustion intensity. Hoggarth [47] has summarized the performance of the various types with respect to these properties.

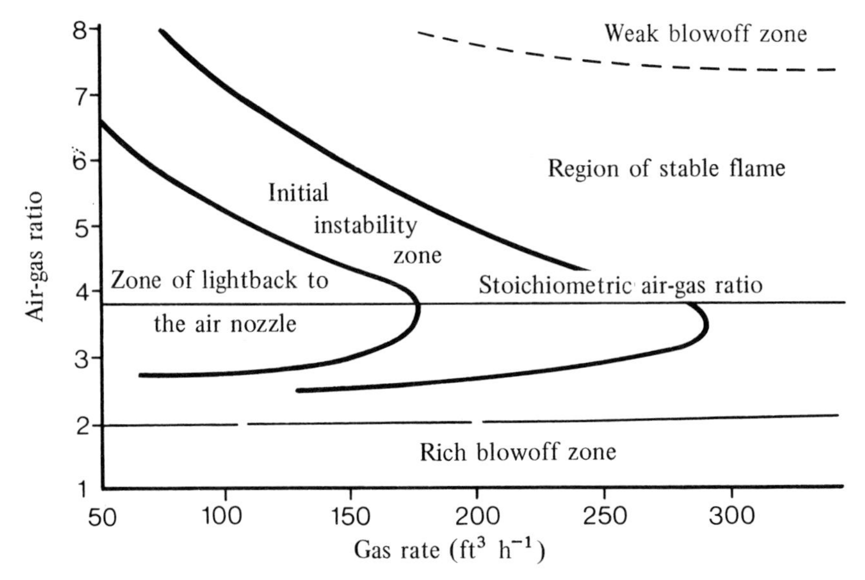

Figure 7:22 Typical stability diagram for a tunnel burner (first-family gas) [48]

Stability on first-family gases. Regarding stability, town-gas tunnel burners present no blowoff problems as Figure 7:22 indicates. Lightback may occur on turndown if the air pressure is very low but adequate turndown ratios are obtained with small burner sizes (tunnel exit $<$ 75 mm diameter). As the size of single mixture-tube burners is increased, however, the turndown ratio is progressively reduced since the lightback flow rate increases with size more rapidly than the maximum flow rate. Francis and Hoggarth [48] considered that this effect was due to increased burning velocities promoted by high turbulence levels in the mixture tube. They found it possible to reduce this turbulence in high-throughput tunnel burners by the use of multiple mixture tubes, multiple drilled nozzle plates or annular mixture nozzles, thus in effect grouping together several burners firing into a single tunnel to obtain the throughput of a large burner whilst retaining the turndown facility of the smaller burners. Francis and Hoggarth [48] have developed a comprehensive design procedure for burners of this type and the reader is referred to their paper for further details.

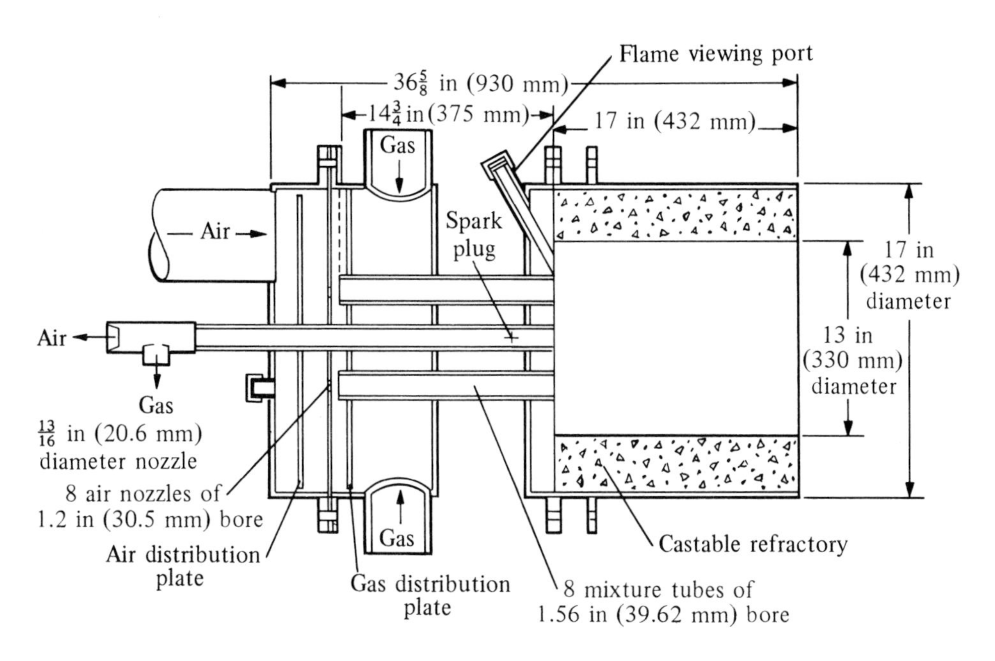

Figure 7:23 Multiple-mixture-tube tunnel burner [47]

Figure 7:23 shows a typical multiple-mixture-tube burner incorporating a separately fed pilot burner.

Stability on natural gas. In the design of premix tunnel burners for town gas, as indicated above, the dimensions are chosen to provide either the maximum mixture-tube velocity to prevent lightback or to maximize the tunnel-exit velocity. Little consideration has been given to the effect of the burner dimensions on flame blow-off since this is not a problem with town gas. With natural gas however, the burner dimensions to achieve acceptable flame stability become much more critical. Cox [49, 50] has investigated the stability of tunnel burners with mixture-tube sizes in the range 12.7 mm (0.5 in) to 50.8 mm (2.0 in) with varying parallel tunnel dimensions. The results show a wide variation in performance with tunnel size. Since the tests were carried out firing into the open atmosphere, the weak blowoff results were used as the stability criterion. Figure 7:24 shows that maximum stability is obtained, for all mixture-tube sizes, when the ratio of tunnel diameter to mixture-tube diameter is between 2.5 and 3.0 : 1.

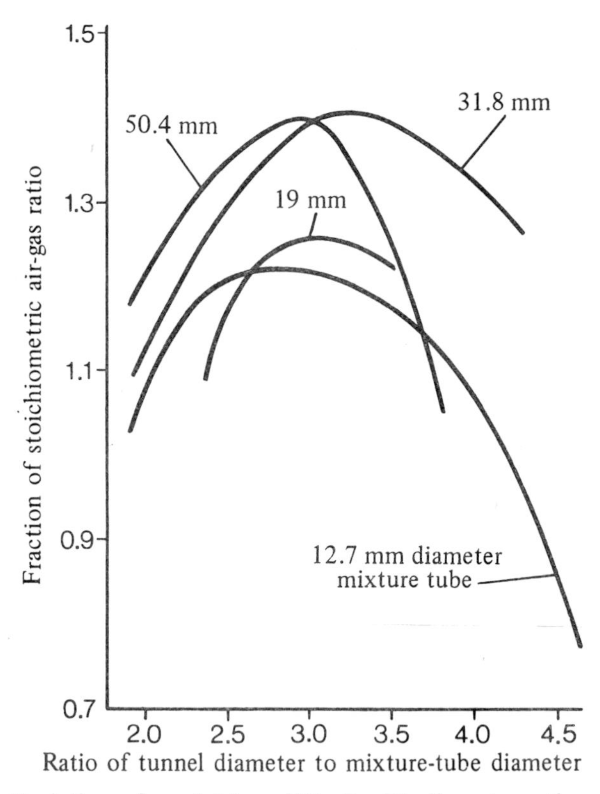

Figure 7:24 Variation of weak-blowoff limit with diameter ratio

These results were obtained using tunnel lengths of 7 to 10 mixture-tube diameters. However, the optimum tunnel diameter shows a dependence on tunnel length. There is an overall improvement with increasing tunnel length and longer tunnels are more tolerant to changes in diameter. Cox [50] has obtained the following empirical relationship between the optimum ratio of tunnel diameter to mixture-tube diameter D_t/D_m and tunnel length l_t:

$$(D_t/D_m)^2 = 0.95 l_t/D_m + 1 \tag{7.17}$$

Multiple-mixture-tube tunnel burners designed for town gas give good flame stability on natural gas, probably due to the high level of recirculation produced by a number of closely grouped mixture tubes and possibly because of mixture composition and velocity variations between individual tubes.

Tunnel materials. Refractory tunnels may either be in the form of commercially available prefired burner blocks or cast directly in the burner casing using a castable mix of suitable refractoriness and thermal shock resistance. Normally these are based on calcium aluminate cement with aggregates of sillimanite, mullite or fused alumina. Reductions in outside surface temperature and noise emission are obtained by the use of porous refractory materials, and castables using bubble alumina aggregates have been widely used (see Chapter 6). The Joint Refractories Committee of the Gas Council and the British Ceramic Research Association undertook investigations of refractories for high-temperature tunnel burners over a number of years and the results are referred to in Chapter 6.

Tunnel-burner noise. A disadvantage of tunnel burners is their high noise level. It is inevitable that a burner producing a turbulent jet of high-velocity combustion products will give rise to a certain amount of broad-band noise (see Section 5.5). In addition tunnel burners may emit noise of high intensity mainly at single frequency, i.e. 'organ-pipe resonance'. Tunnel length is the most critical factor in determining whether resonance of this kind will occur. Kilham, Jackson and Smith [51], from studies of acoustically simplified single- and multiple-mixture-port burners, concluded that the shorter the tunnel, the less likely is the burner to be objectionally resonant. Francis [46] obtained reductions of 15 dB by reducing tunnel lengths from 12 to 7 tunnel diameters. In the case of tunnels formed from porous refractories the overall noise level was significantly reduced ($\approx$ 10–23 dB less) and virtually independent of tunnel length up to 16 tunnel diameters [46].

Other flame-retention methods

The principle of using a sudden enlargement to create a low-pressure area thereby causing recirculation of hot gases to the base of the flame is the principal means of providing stabilization in tunnel burners. The same approach is adopted in some air-blast burners, particularly in the smaller size range, using a short metal cup to retain the flame. This may rely on the sudden enlargement to provide stabilization or, as in most proprietary non-blowoff tips, may incorporate auxiliary flames in the expansion. Results reported by Cox, Hoggarth and Reay [52] for natural gas indicate that for a $1\frac{1}{4}$ in (32 mm) mixture-tube diameter and cup length of 4 in (100 mm) the optimum stability is obtained with a 2 : 1 diameter ratio of sudden enlargement. It seems probable that the best stability is obtained using as great an enlargement as possible provided that the flame fills the cup before the end. Priestley [1] has recommended an expansion ratio of 3 for mixture-tube diameters up to 1 in (25 mm) and 1.5 over 4 in (100 mm).

Stability is further improved in designs of non-blowoff tips that incorporate auxiliary flames.

A final stabilization method is incorporated in the 'impact burner' in which the flame from a short refractory nozzle is made to impinge obliquely onto a bed of broken refractory material which becomes incandescent and acts as a reignition source.

7.5 NOZZLE-MIXING AND TUNNEL-MIXING AIR-BLAST BURNERS

In nozzle-mixing burners air–gas mixing and combustion take place simultaneously at the burner nozzle. If, as is commonly the case, a tunnel or quarl is employed and simultaneous mixing and combustion occur in the tunnel, the burner is termed a tunnel-mixing burner. Although such burners have been available for many years,

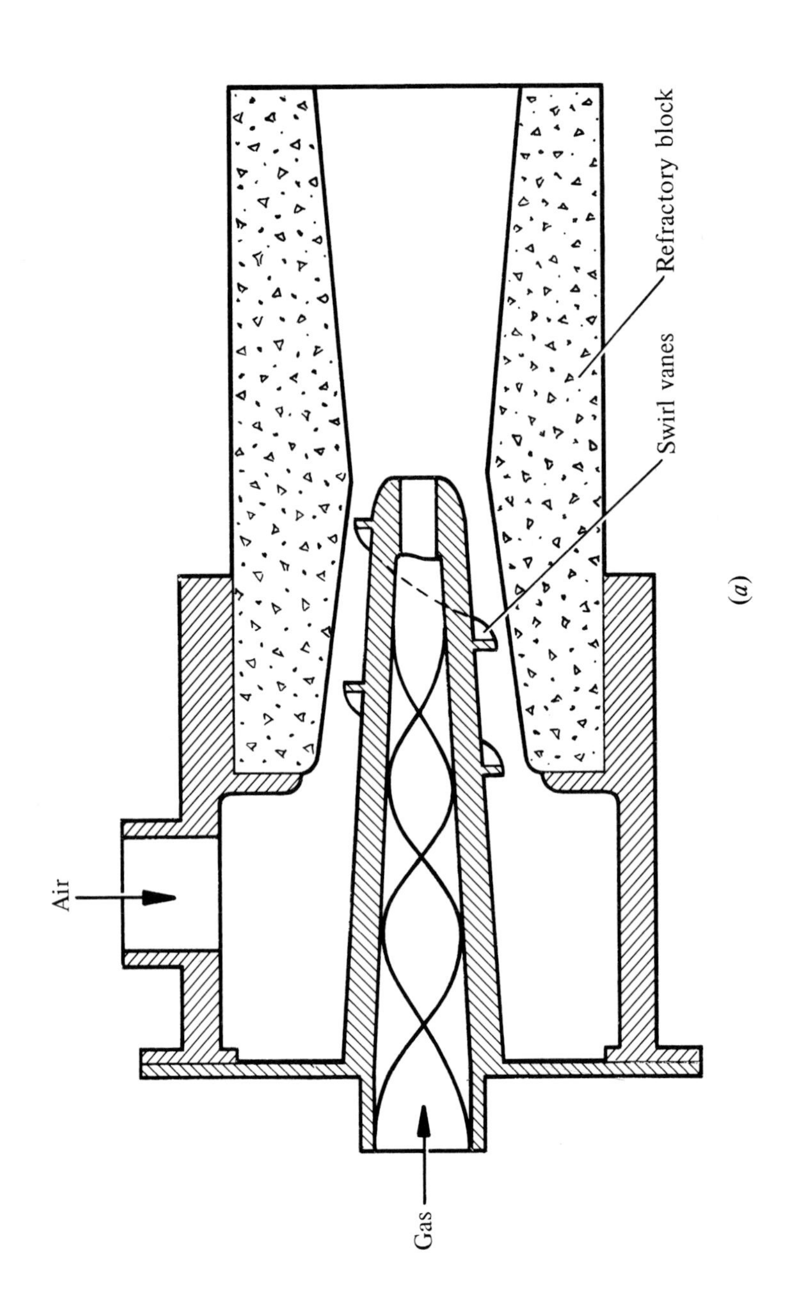

(a)

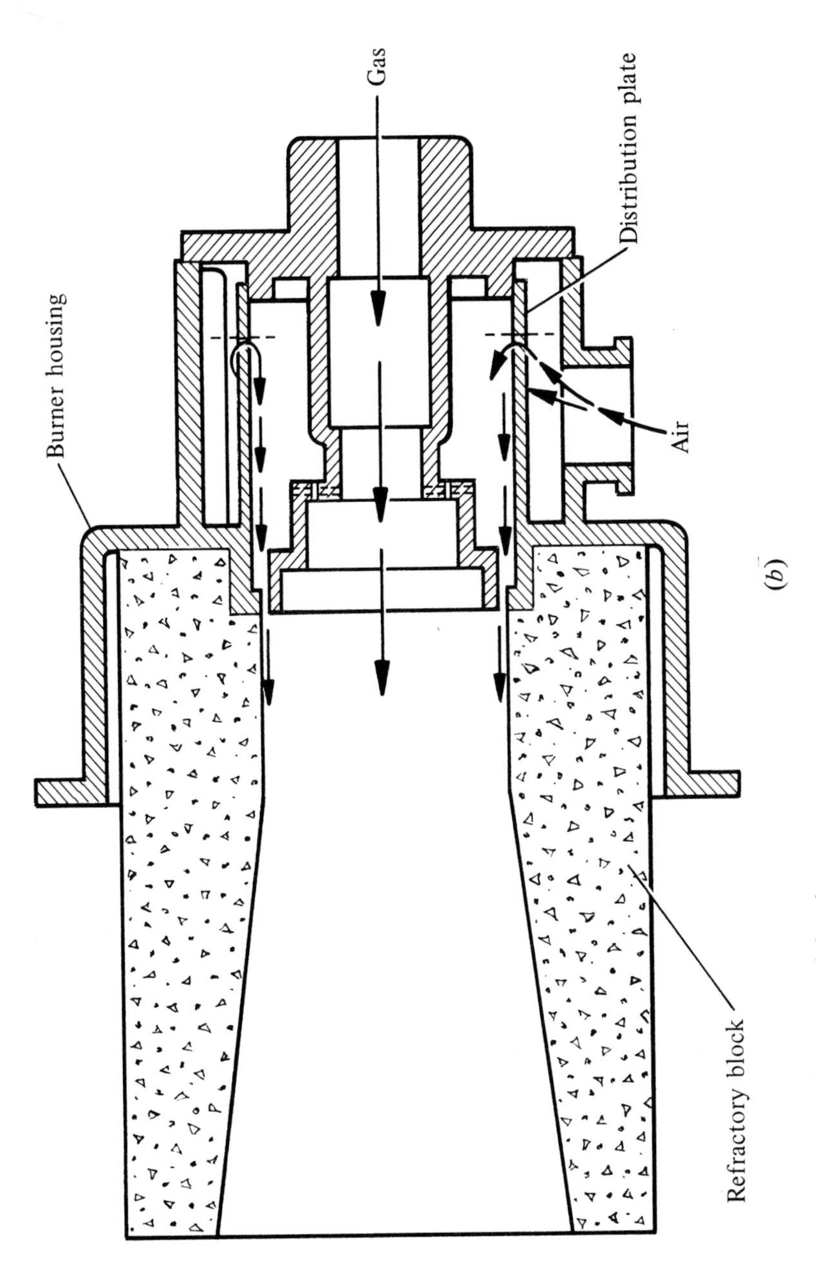

Figure 7:25 Examples of tunnel-mixing burners
(*a*) Eddy Ray
(*b*) Stordy-Hauck NMG 110

their usage has increased markedly in recent years due to a number of advantages possessed over premix air-blast systems:

1 Lightback is eliminated
2 Lower air pressure may be used
3 Preheated air may be employed
4 Flame stability with respect to lift is greatly improved and burners are capable of operating at high excess air levels
5 Conversion from town gas to natural gas is facilitated, the only necessary alteration being a change of air–gas ratio

With regard to (4) (and consequently (5)) tunnel-mixing burners derive their great stability from the fact that mixing is non-uniform and over a wide range of overall air–gas ratios there exist local areas of mixture close to stoichiometric proportions which provide ignition sources for the remainder of the mixture. In addition, low-velocity and reverse-flow regions occur which contribute to improved stability.

An extremely wide variety of proprietary nozzle-mixing burners are now available arising from the attempts of numerous designers to produce burners appropriate to particular heating requirements. An increasingly popular variant integrates the burner with throughput, ratio, ignition and flame-detection controls into a packaged system. Packaged burners are treated separately in Section 7.10. Nozzle-mixing burners also constitute the gas-burner part of dual-fuel burners which again have a section (7.11) devoted to their design.

An advantage of nozzle-mixing burners is that they can operate satisfactorily with preheated air. Nozzle-mixing burners incorporating integral recuperators are finding increasing industrial usage. Self-recuperative burners are considered in Section 7.5.3. To obtain satisfactory combustion in nozzle-mixing burners, mixing must be rapid and this may be achieved by the use of a high relative velocity between the air and gas streams and by the provision of a large surface area for mixing, usually by discharging the air through an annulus or a series of jets.

Figures 7:25 and 7:26 indicate the principles of operation of a selection of tunnel-mixing burners.

In Figure 7:25 (*a*) opposing gas and air swirls are employed to assist mixing. In practice the swirl achieved in the air annulus is of a low order and mainly axial flow results [53]. In Figure 7:25 (*b*) the air enters in three different positions; tangentially through holes in the gas inlet, axially through jets in the collar surrounding the gas inlet, and axially through the annulus surrounding the gas inlet. The arrangement is simpler in Figure 7:26 (*a*) in which air is directed towards the axis through a series of jets. The use of air jets provides a larger surface area for mixing than an annulus and also promotes combustion-product recirculation to improve flame stability. Figure 7:26 (*b*) uses air jets to provide a large contact area and incorporates a further development of the tunnel-mixing idea in that by a choice of suitable air-nozzle and tunnel areas the burner possesses an inherent self-proportioning action thus simplifying the control of air–gas ratio over the turndown range. Parallel, converging and slot-

Figure 7:26 Examples of tunnel-mixing burners
(*a*) Urquhart-Bloom NM
(*b*) BGC Midlands Research Station self-proportioning design
(*c*) Eclipse flat-flame burner

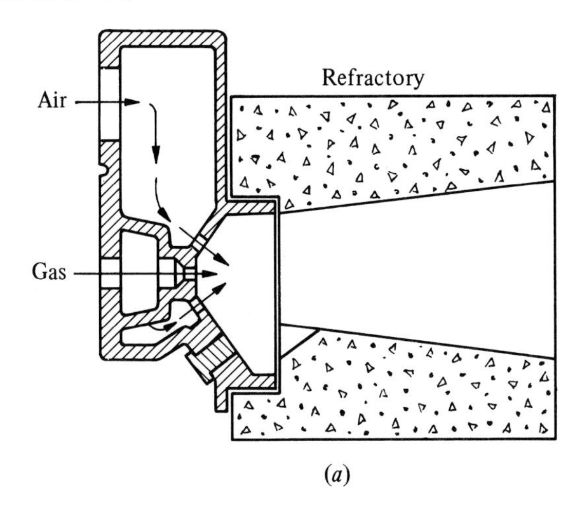
Refractory
Air
Gas

(a)

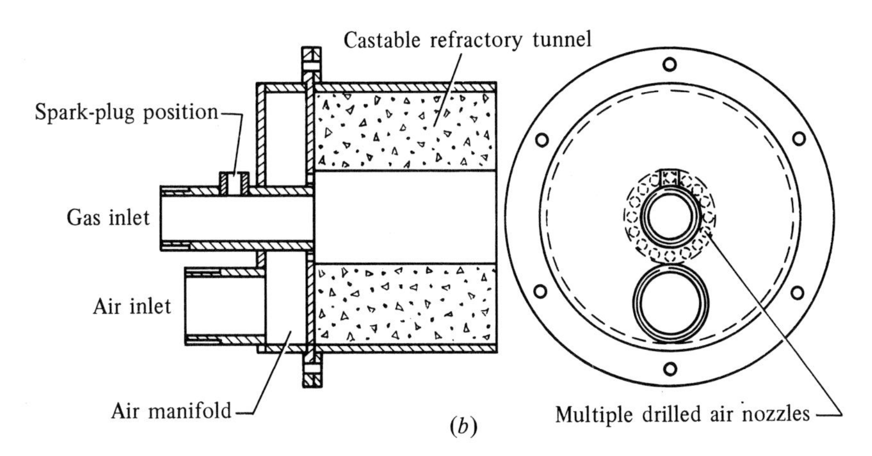
Castable refractory tunnel
Spark-plug position
Gas inlet
Air inlet
Air manifold
Multiple drilled air nozzles

(b)

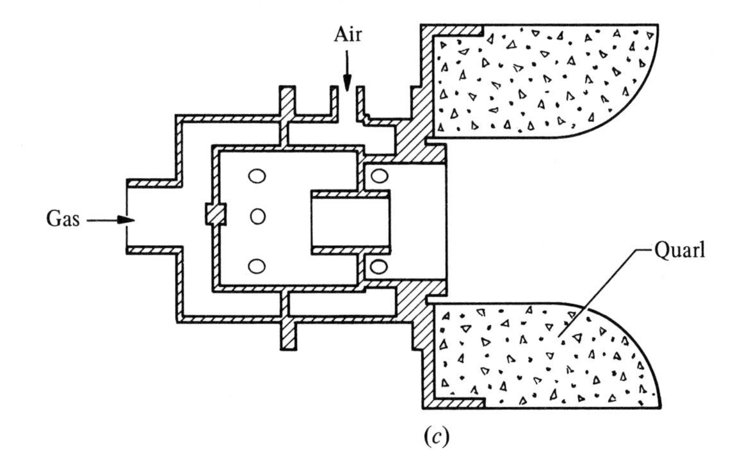
Air
Gas
Quarl

(c)

shaped tunnels have all been developed from the original design and ranges from 14.7 kW (50 000 Btu h^{-1}) to in excess of 1.47 MW (5×10^6 Btu h^{-1}) have been employed.

Figure 7:26 (*c*) is representative of designs produced by a number of burner manufacturers as 'radiant-cup', 'radiant-cone' or 'flat-flame' burners. Air enters tangentially via a side orifice plate which produces a vortex directing it across the gas stream giving rapid mixing. Flame stabilization is assisted by recirculation of hot gases into the low-pressure region at the burner axis (see Section 4.8.1).

The flame gases follow the contour of the flared burner quarl resulting in a large flat flame disc with little forward gas velocity. Burners of this type may be placed close to furnace stock without direct flame impingement.

In the 'Coanda' nozzle-mixing burner (Hotwork Development Ltd) the tunnel is designed such that the combustion products turn through 90° to fire parallel with a side wall or to follow the profile of a curved furnace lining. This offers a number of advantages in, for example, the replacement of multi-burner installations in box furnaces and in tangential firing of circular furnaces.

7.5.1 Design of self-proportioning tunnel-mixing burners

Hoggarth and Aris [53] have developed a theoretical treatment and design procedure for burners of this type. By summing the pressure losses through the burner system they obtained:

$$(p_g - p_2)/(p_a - p_1) = (r + d)(r + 0.7)r^{-2}(T_2/T_o)(A_1/A_2)^2(C_{F12} + 2) - 2(A_1/A_2) + dr^{-2}(A_1/A_gC_{dg})^2 \qquad (7.18)$$

using the nomenclature of Figure 7:27.

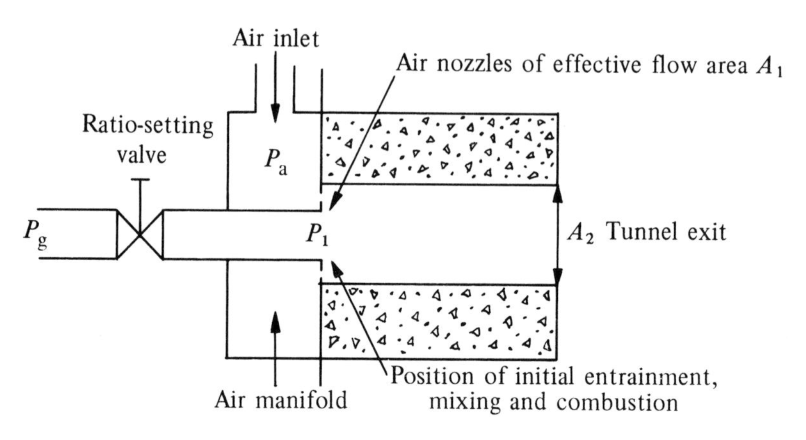

Figure 7:27 Diagram of tunnel-mixing burner showing dimensions and pressures used in the derivation of the design equations [53]

If $(p_g - p_2)$ is maintained at zero by means of a zero governor then a range of A_1/A_2 values will satisfy the equation for stoichiometric conditions. Two values of A_1/A_2 are of particular significance since these correspond to (*a*) the minimum gas restriction and therefore maximum exit velocity and (*b*) maximum gas restriction and therefore maximum suction p_1 at the gas inlet.

In the *maximum exit velocity case*, equation 7.18 reduces to:

$$A_1/A_2 = 2r^2T_o/(r + d)(r + 0.7)(C_{F12} + 2) \qquad (7.19)$$

and in the *maximum suction case*:

$$A_1/A_2 = r^2\,T_o/(r + d)(r + 0.7)(C_{F12} + 2) \qquad (7.19a)$$

For maximum velocity and maximum suction, the area ratios differ by a factor of 2, i.e. the exit velocity is halved and the pressure efficiency is reduced by four. Equation 7.18 indicates that perfect self-proportioning occurs when $p_g - p_2$ is zero and all the terms on the right-hand side of the equation remain constant at all flow rates. In practice, this is not so; C_g, T_2 and C_{F12} may all vary with flow so that to obtain reasonably constant proportioning over the turndown range the ratio-setting valve opening (A_g) is set at high flow rates and the zero-governor outlet pressure (p_g) at low flow rate (cf. premix air-blast system, Section 7.4.1).

For use with preheated air, the air-nozzle size must be increased for a given s.t.p.

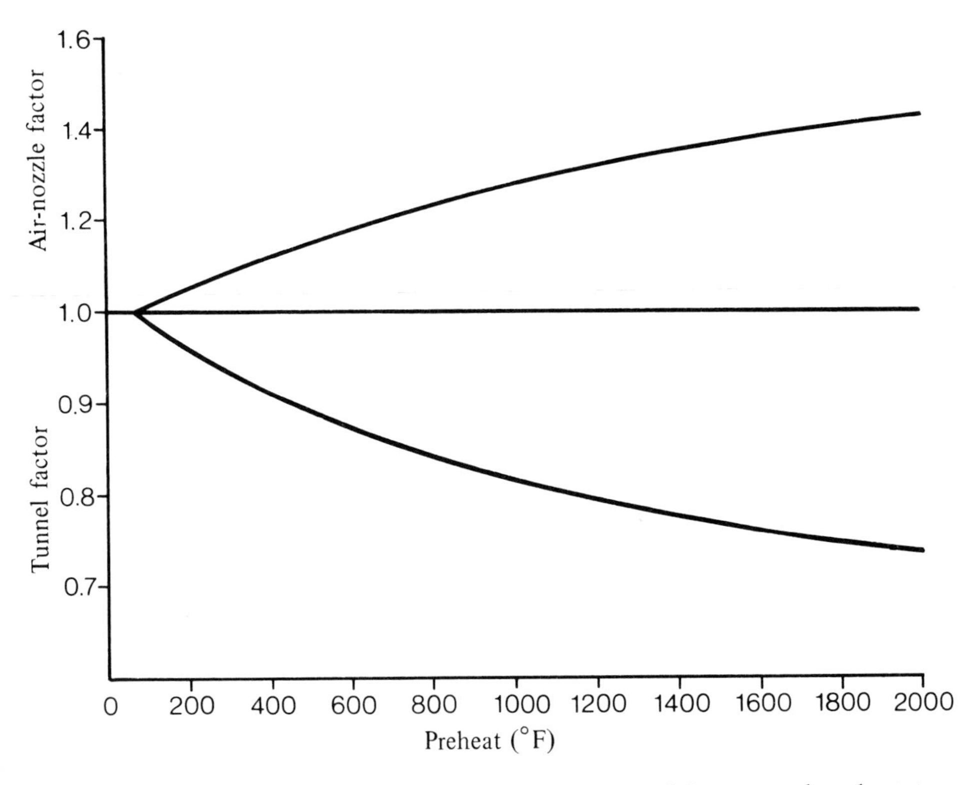

Figure 7:28 Relationship between air-nozzle and tunnel factors and preheat temperature [53]

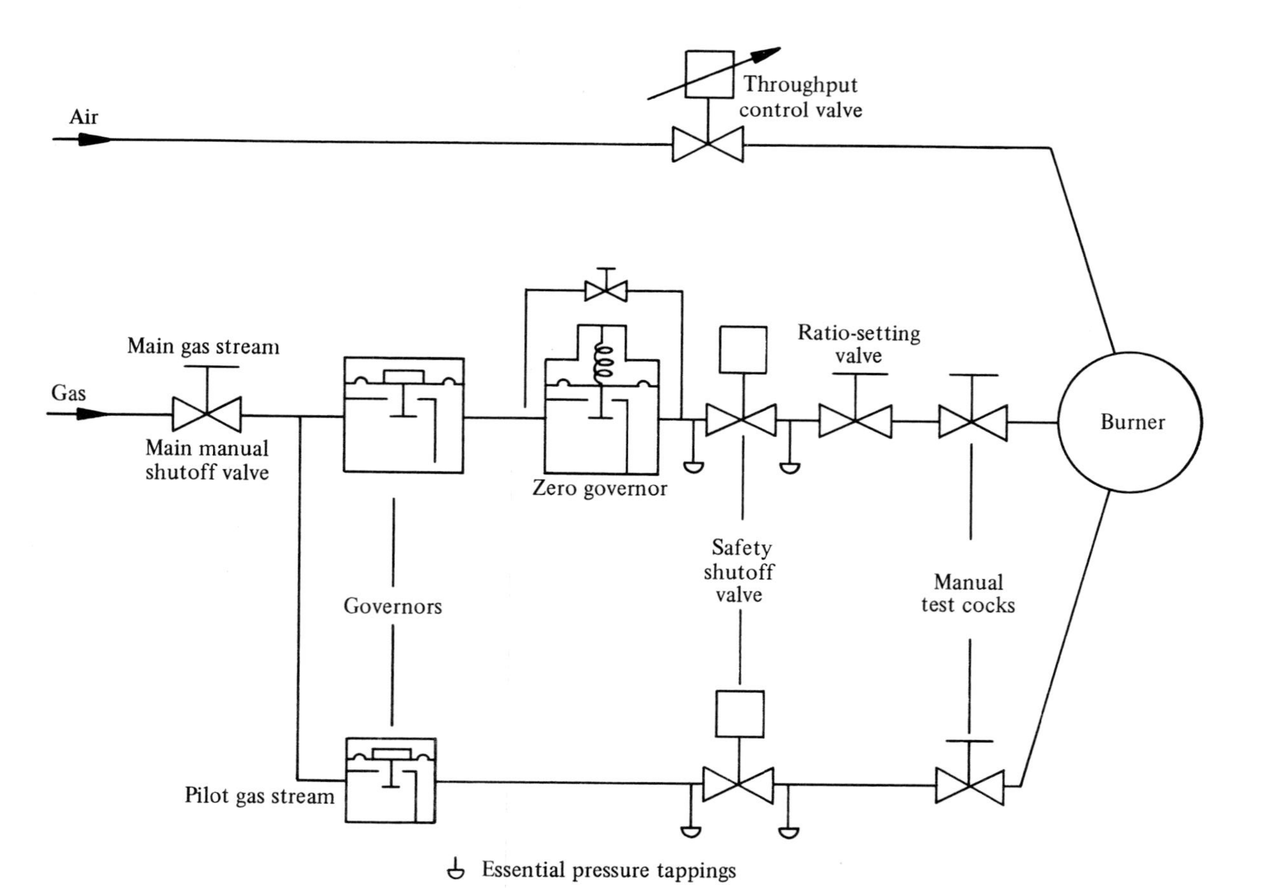

Figure 7:29 Valve train for single-valve throughput control for self-proportioning tunnel-mixing burners

flow rate. Conversely the tunnel size may be reduced since the airstream momentum is increased. This allows higher exit velocities to be attained but will require higher gas pressures when starting from cold. Figure 7:28 shows the correction factors for the air-nozzle and tunnel sizes for varying preheat temperatures.

7.5.2 Air–gas ratio control of nozzle-mixing burners

In the case of nozzle- or tunnel-mixing burners which are capable of entraining gas from a zero-governed gas supply the pressure-air technique allowing single-valve throughput control is applicable. This is identical in principle with that used for premix air-blast systems and the recommended valve train is as shown in Figure 7:29. Compensation for furnace-chamber pressure fluctuations may be incorporated by back-loading both the zero and appliance governors from the furnace chamber.

For burners which possess no inherent self-proportioning action one of the methods normally classified as *pressure-air/pressure-gas techniques* [43, 54] would be employed. In these the relationship between flow rate, the area of an orifice and the differential pressure across it leads to two basic methods of throughput and air–gas ratio control; the *linked-valve method* in which constant differential pressures are maintained across linked valves with suitably matched flow characteristics and the *fixed metering orifice method* in which a constant relation is maintained between the differential pressures across fixed metering orifices in the air and gas lines.

Linked-valve method

The air and gas flows may be controlled using two valves linked mechanically or electrically so that the air and gas flows may be adjusted simultaneously on throughput control. Changes in back pressure are, however, often appreciable compared with the gas supply pressure and proportioning suffers accordingly. Similarly if preheated air is used the air–gas ratio will be affected due to the air-density change at the burner nozzle. If, however, the differential pressure across each valve is held constant then, by choosing suitable valves the air–gas ratio may be maintained within narrow limits. This may be achieved by installing a back-loaded governor upstream of each valve. If the air valve is installed upstream of a recuperator (if used) the air is controlled at essentially constant temperature and therefore density. Figure 7:30 (*a*) illustrates this technique applied to mechanically linked valves and Figure 7:30 (*b*) to weep-relay valves.

As an alternative to back-pressure compensation where this is known to be relatively small, adequate control is possible using a high constant upstream pressure. In practice this is usually achieved by using air and gas fans operating on the constant-pressure portion of their characteristics [54] and by dropping a large proportion of the supply pressure across the valves. When proportional or modulating control is required it is necessary to match the air and gas flows over the whole throughput range by using adjustable-port or adjustable-flow valves (see Section 9.4.2).

Fixed metering orifice method

Balanced differential pressures. Figure 7:30 (*c*) shows a flow ratio controller which is based on maintaining the differential pressures across orifice or venturi flow-metering elements in the gas and air lines. Measurement of differential pressure may be carried out by means of a ring-balance, diaphragm or bell-type device and any deviation from the balanced condition results in a hydraulic, pneumatic or electrical signal to a valve controlling the flow in one of the flow lines. Throughput control is by means of a valve in the other line. The system is sophisticated and costly and tends to be most suitable for proportional control of a single burner system subject to air preheat or back-pressure variations [54].

Pressure-divider methods. If there is little or no injection or pressure interaction between the air and gas streams, for example in the case of a simple concentric-tube burner firing into a large combustion chamber, single-valve throughput control is possible using the pressure-divider technique (Figure 7:30 (*d*)). A zero governor on the gas line is loaded with a fraction of the air-supply pressure via pressure-dividing orifices, one of which is used as a ratio-setting valve. The system is suitable for any form of throughput control by appropriate selection of an air-control valve, proportional control being particularly easy by means of a single linear valve in the air supply. Alternatively if boosted gas is available the arrangement shown in Figure 7:30 (*e*) may be used in which the pressure division is effectively carried out in the gas line by dropping a proportion of the gas pressure across an orifice which can be the ratio-setting valve. Compensation for variation in combustion-chamber pressure is automatically provided in this arrangement but can be catered for in the arrangement shown in Figure 7:30 (*d*) by terminating the pressure-divider bleed line in the combustion chamber close to the burner.

Further information on air–gas proportioning techniques may be found in Section 2 (Flow Controls) of the *Industrial Gas Controls Handbook* [43] and in reference [54]. The valves and other components employed in these systems are described in Chapters 8 and 9.

7.5.3 Recuperative nozzle-mixing burners

In most industrial heating processes a large proportion of the heat supplied is wasted in the hot flue gases. Waste-heat recovery methods are employed (see Chapter 11) but their use has been restricted mainly to large-scale process plant because of the cost and size of the equipment required. However, smaller furnaces and kilns are often the least efficient. This indicates the need for a simple compact system in which the burner, flue and recuperator are integrated into a single unit. Such *recuperative burners* have been developed as a result of work carried out by the Midlands Research Station of the British Gas Corporation and elsewhere.

In the MRS design, air preheats of up to 600 °C are attainable in the recuperator section which occupies about a tenth of the volume of a conventional stack recuperator [55] and installation of the unit involves little more than installation of a conventional burner.

Air preheating may provide other benefits in addition to improving thermal efficiency: flame temperatures are increased (e.g. a 600 °C air preheat raises the

---

Figure 7:30 Valve trains for air–gas ratio and throughput controls for nozzle-mixing burners

(*a*) Using mechanically linked valves
(*b*) Using weep relay valves
(*c*) Using balanced differential pressures
(*d*) Pressure-divider technique
(*e*) Pressure-divider technique using boosted gas

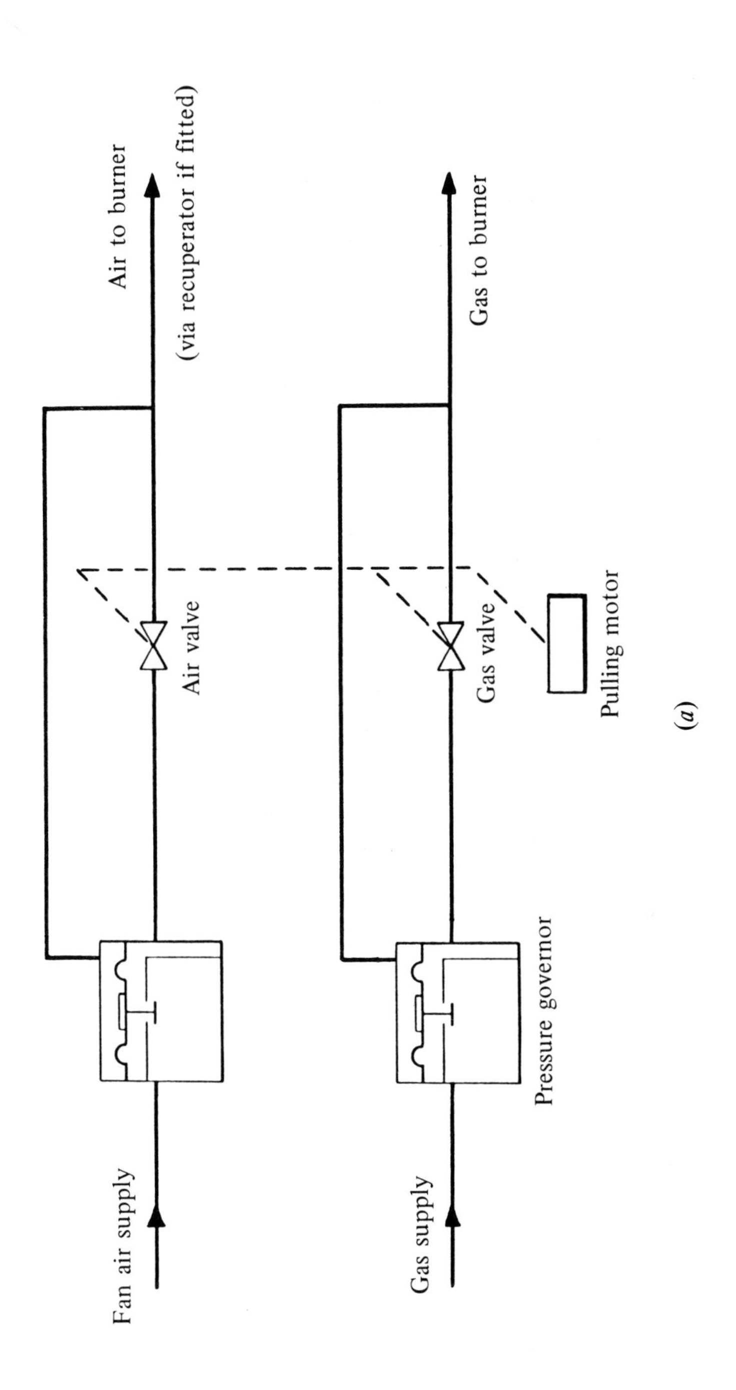
Air to burner
(via recuperator if fitted)
Gas to burner
Air valve
Gas valve
Pulling motor
(a)
Pressure governor
Fan air supply
Gas supply

Figure 7:30

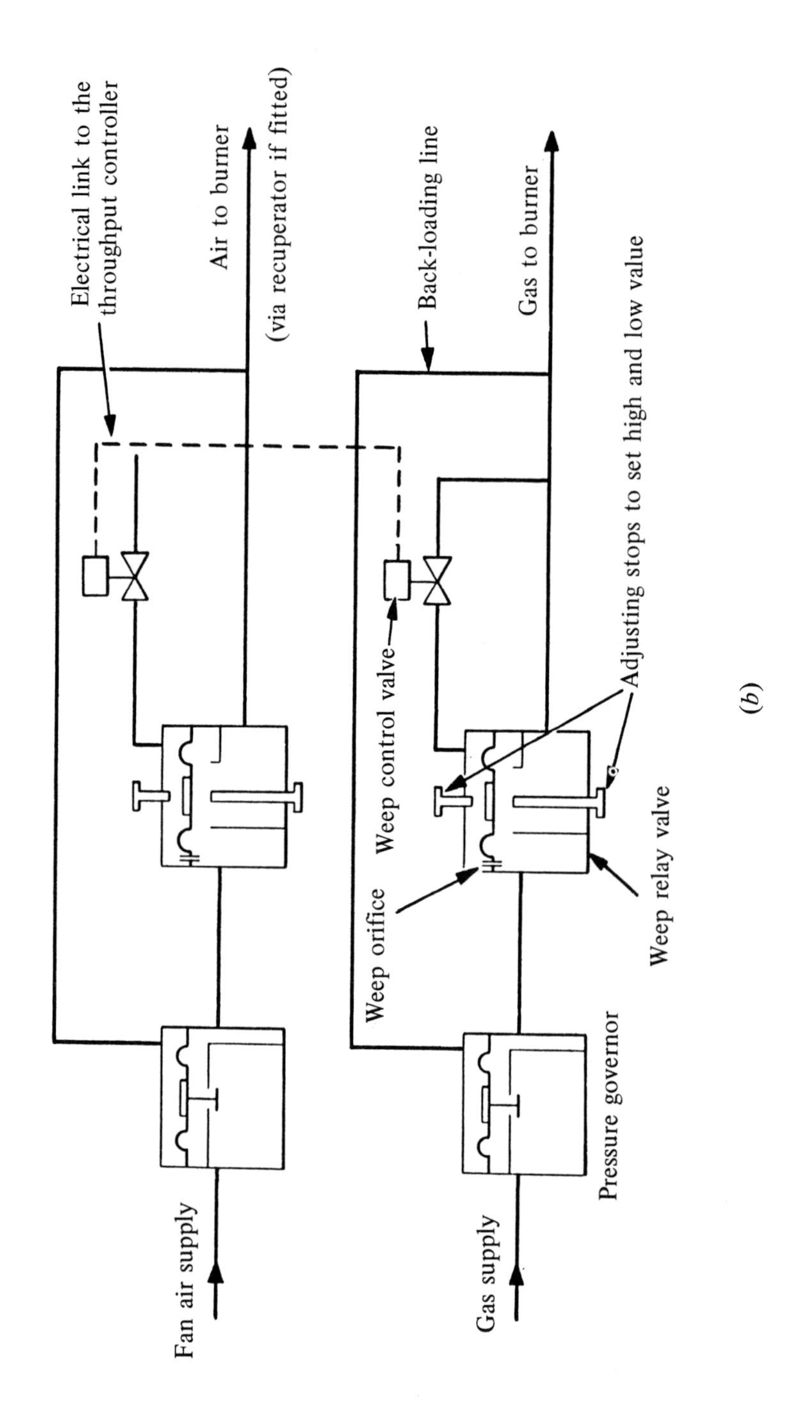
Electrical link to the throughput controller
Air to burner
(via recuperator if fitted)
Back-loading line
Gas to burner
Adjusting stops to set high and low value
Weep control valve
Weep orifice
Weep relay valve
Pressure governor
Fan air supply
Gas supply
(b)

Figure 7:30

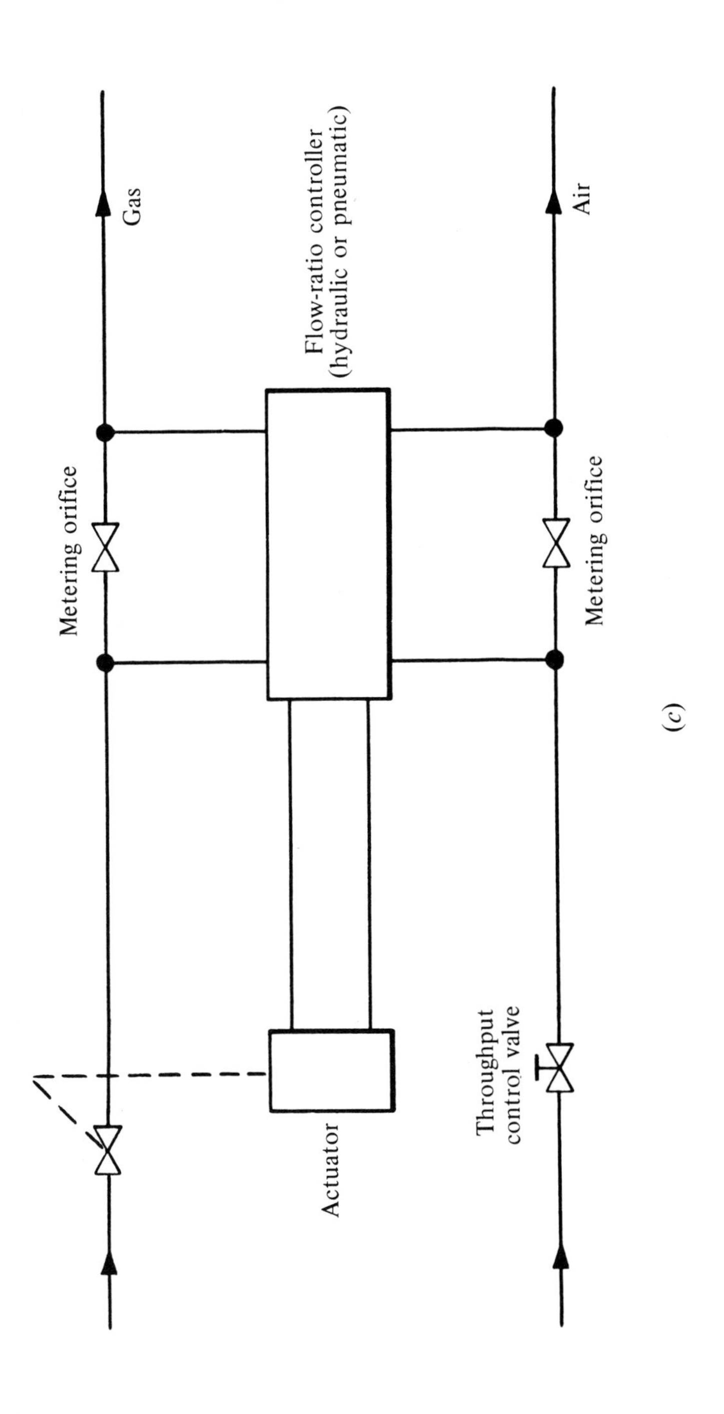
Gas
Flow-ratio controller
(hydraulic or pneumatic)
Air
Metering orifice
Metering orifice
(c)
Throughput
control valve
Actuator

Figure 7:30

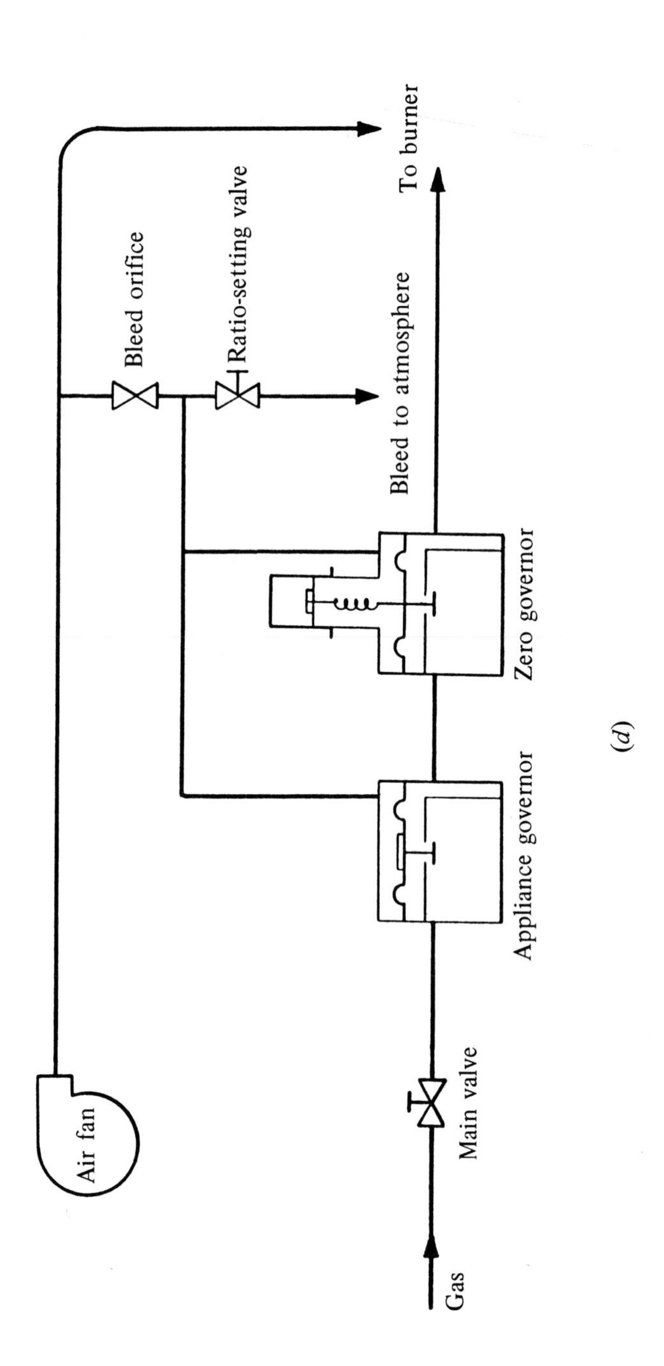
To burner
Bleed orifice
Ratio-setting valve
Bleed to atmosphere
To burner
Zero governor
Appliance governor
Air fan
Main valve
Gas
(d)

Figure 7:30

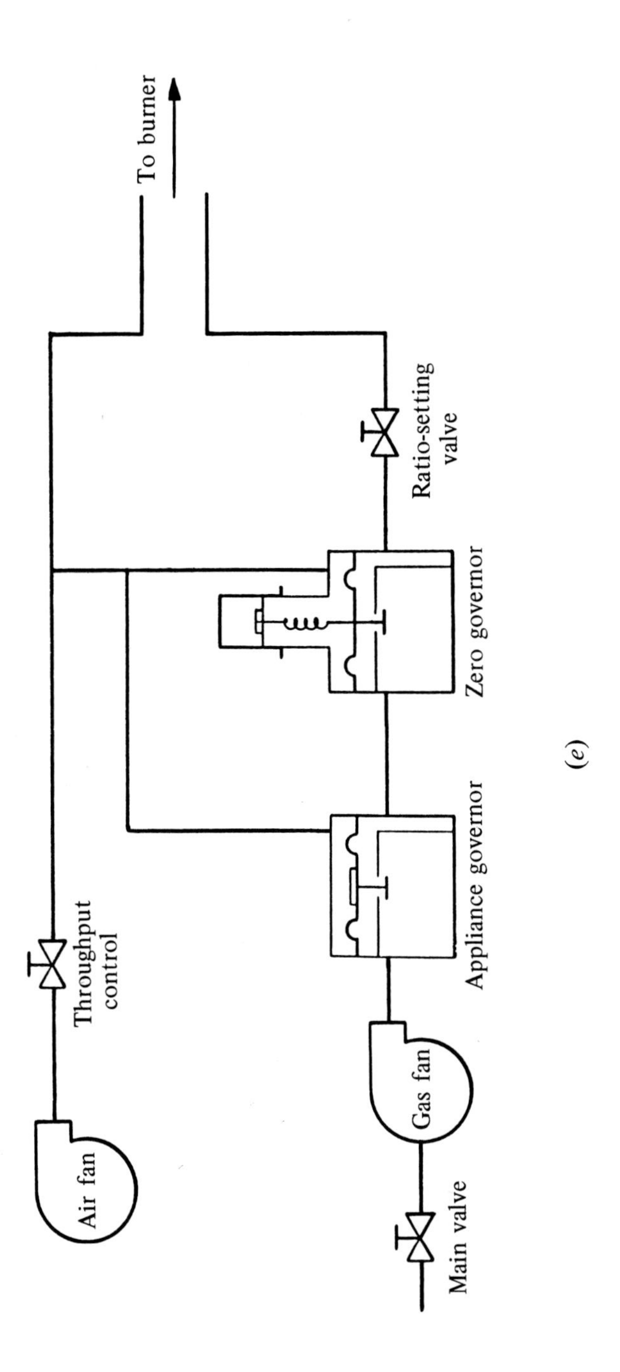
To burner
Ratio-setting valve
Zero governor
Appliance governor
Throughput control
Gas fan
Air fan
Main valve
(e)

Figure 7:30

temperature of a natural-gas flame by approximately 220 °C) leading to increased heat transfer rates and if large fuel savings are made a lower installation cost may result from the use of smaller controls and fans [56]. Additional savings may be made

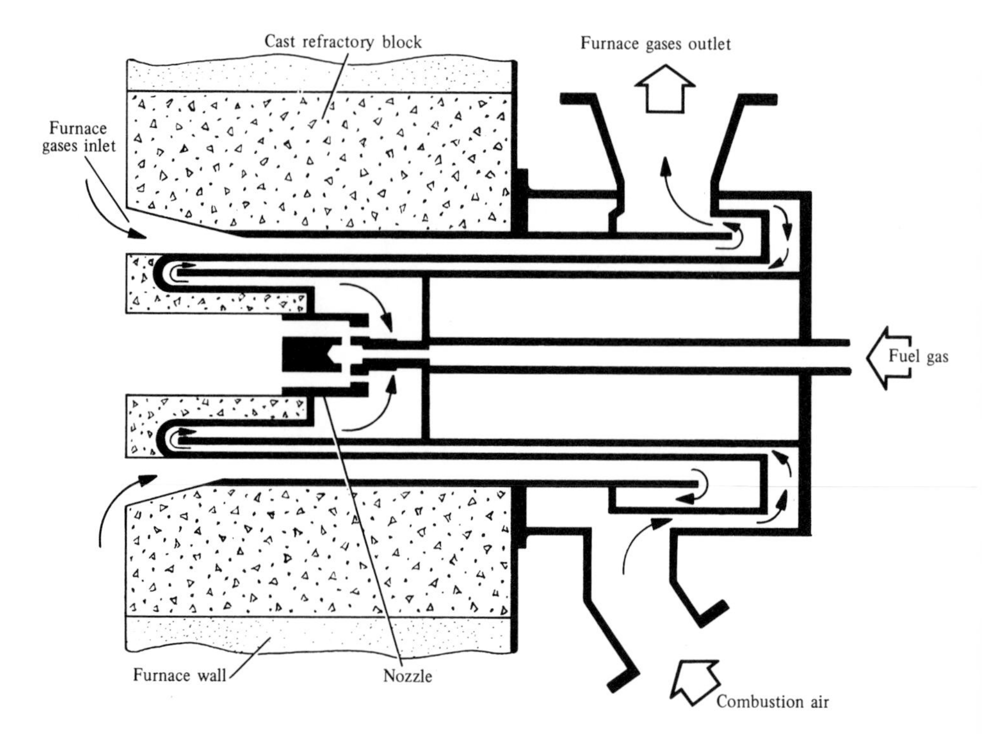

Figure 7:31 MRS recuperative burner

by dispensing with furnace flue ducts. Figure 7:31 shows the MRS recuperative burner which consists essentially of a high-velocity tunnel-mixing burner inside a counter-flow heat exchanger which supplies preheated combustion air to the burner. The design shown differs from that described in [56] in a number of respects with a view to increasing its working temperature, e.g. all the metal components in contact with the flue products are now air-cooled. The air for combustion enters at a manifold surrounding the flue annulus then passes forward along the air annulus keeping the external surface cool. Combustion products are drawn counter-flow around the outside of the air annulus. At the furnace end of the recuperator the airstream turns through 180° and flows to the burner nozzle. In general all the combustion products are extracted through the recuperator using an eductor which allows control of the furnace pressure.

Tunnel exit velocities are in excess of 50 m s^{-1} (160 ft s^{-1}). The burner can operate with more than 100 per cent excess air if required and possesses a turndown ratio of about 10 : 1.

The use of recuperative burners becomes feasible if within an acceptable timescale the total savings can be balanced against the additional costs of the burner and ancillaries. In the present context of increasing fuel prices and the need for energy conservation the application of these burners becomes more and more desirable.

In order to calculate the potential fuel savings a method of predicting the performance of recuperative burners is necessary and enables the optimum burner(s) to be designed for a specific application. Harrison *et al.* [56] have developed such a technique.

In many applications of recuperative burners installation may simply consist of replacing existing burners but in applications where considerable recirculation may be promoted their use may make possible a significant reduction in the number of burners required. Here physical and mathematical modelling provide useful information on optimizing the burner/furnace-chamber configuration and in predicting the furnace performance [57]. Chapter 12 describes the use of physical and mathematical modelling in applications of this kind.

At the time of writing MRS recuperative burners have been applied to rapid billet-heating furnaces, steel-reheating furnaces, aluminium-melting crucible furnaces and intermittent ceramic kilns showing fuel savings of 30 to 50 per cent and reference to these applications is made in the appropriate sections of Chapter 13 of this book. In addition recuperative burners are applied to radiant tubes and these are considered in Section 7.9 of this chapter.

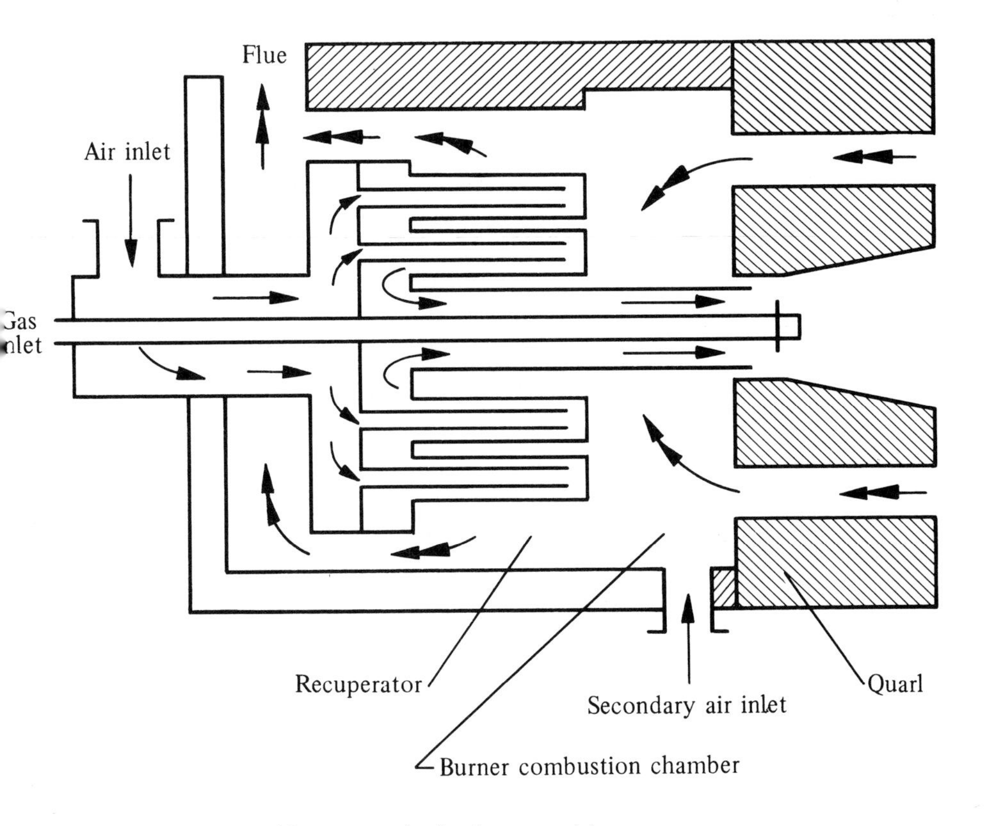

Figure 7.32 Emgas self-recuperative/scale-control burner

An alternative design has been developed by East Midlands Gas which may be used as a conventional self-recuperative burner or as a gas-rich recuperative burner which may be applied for controlling the rate of oxidation when, for example, heating steels to rolling temperatures (see Sections 13.2 and 13.5).

For a furnace temperature of 1200 °C an air preheat of up to 600 °C can be achieved for stoichiometric operation and up to 800 °C for gas-rich operation. Figure 7:32 shows the general arrangement of the burner. For gas-rich operation secondary air is introduced into the burner combustion chamber to complete the combustion of the waste gases before they are vented to the atmosphere. Alternatively air may be introduced to dilute and cool the combustion products for self-recuperative operation in very-high-temperature applications, possibly as high as 1700 °C.

7.6 MACHINE-PREMIX SYSTEMS

In mechanical premixing machines gas and air are drawn through a metering and mixing device into a positive-displacement rotary compressor or fan where the mixture is compressed before being delivered to the burner head(s). Types using rotary compressors are more commonly used in British industrial gas practice providing a mixture of substantially constant air–gas ratio at pressures generally up to 35 kPa (350 mbar, 5 lbf in^{-2}). Machine-premix systems find their widest application in supplying controlled-atmosphere generation plant (Chapter 13) and in providing a supply to large numbers of small burners, for example working-flame burners (Chapter 15).

Two modes of operation may be distinguished:

1. *Partial premix* in which the air–gas mixture is beyond the rich limit of flammability and may be safely distributed to burner heads which may incorporate atmospheric injectors to provide stoichiometric operation.
2. *Full premix* in which stoichiometric mixtures are produced.

With full premix, strict safety measures are essential, in particular the incorporation of flame traps, to eliminate the hazards involved. Safety requirements for premix systems are considered in Sections 9.2 and 9.4.

7.6.1 Rotary compressor premix systems

Commercially available premix machines using rotary compressors (e.g. Selas Mixing Machine, Nu-way Eclipse Consta-mix Machine) use a mixer which is based on the principle of maintaining constant differential pressures across variable-area metering orifices.

Figure 7:33 shows the layout of a typical commercial premix machine using a positive-displacement compressor [43].

The mixer consists of two adjustable-port valves linked mechanically and actuated by a horizontal diaphragm exposed to atmospheric pressure on its underside. Gas is drawn into the valve from a zero-governed supply and air is drawn in from the atmosphere. The air–gas ratio is set by presetting the valve port-area ratio, normally by rotating the valve plug. Throughput is controlled using a valve or valves in the mixture outlet pipe. The accuracy of air–gas ratio control over the throughput range

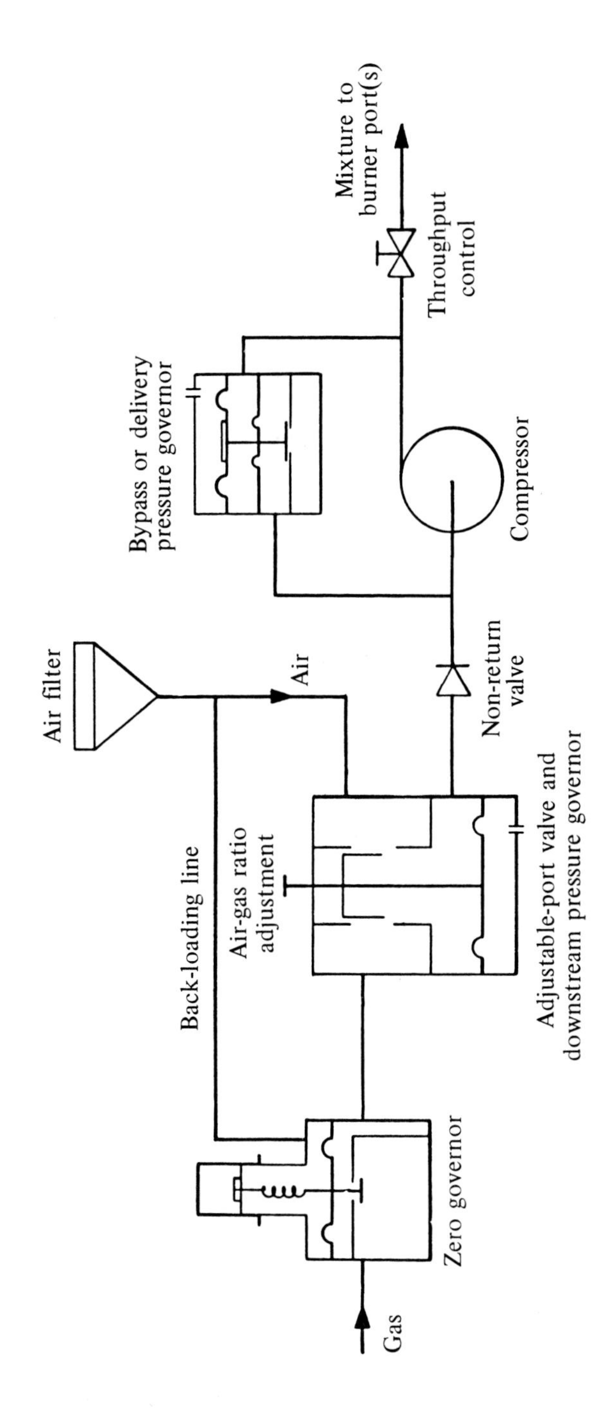

Figure 7:33 Schematic diagram of premix machine. It is essential to have an anti-suction valve and a low-pressure cutoff device on the gas supply line

is limited by changes in the shapes of the valve openings and therefore flow coefficients. Hancock [54] has reported measurements of air–gas ratio using town gas on a commercially available premix machine which indicate a control accuracy of ± 2 per cent over a 7.5 : 1 turndown ratio for stoichiometric conditions reducing to ± 8 per cent for a ratio of 1.6 : 1 at a turndown ratio of 12 : 1. The bypass governor shown in Figure 7:33 is to prevent excessive downstream pressures due to restriction downstream. When the force acting on its diaphragm exceeds the weight of the valve and added weights the valve opens feeding mixture back to the compressor inlet.

Note that to comply with the requirements of the Gas Act 1972, 4th Schedule, it is essential to install a low-pressure cutoff device in the supply line (see Sections 9.2 and 9.4.6).

7.6.2 Fan-premix systems

Fan-premix systems are normally limited to mixture delivery pressures up to 14 kPa (140 mbar, 2 lbf in^{-2}). They possess the advantage over air-blast methods of producing gas–air mixtures that the fan is a highly efficient mixing device, the total fan pressure being available as mixture pressure. As Hancock has pointed out [54] the fan pressure is density-dependent and will be less for gas–air mixtures than for air alone. For example, the fan pressure for a mixture of 1.5 : 1 air–gas ratio is reduced by about 20 per cent. For the same conditions of stoichiometry the effect is clearly much less in the case of natural gas than town gas.

Fan-premix systems may incorporate variable-area, constant-pressure proportioning, as described in Section 7.6.1, or may involve constant-area, variable-pressure techniques [54]. In these the fan draws in air from the atmosphere through a metering device and gas through another metering device from a zero-governed supply. Throughput is controlled by a valve in the mixture pipe and proportioning is achieved by suitable design of the metering elements which may be orifices or linear-flow elements.

Orifice mixers

Since both gas and air are supplied at atmospheric pressure and the metering elements have a common downstream pressure it is clear that the gas and air elements have equal differential pressures. Thus air–gas ratio r

$$= V_a/V_g = (C_{da}\, A_a/C_{dg}\, A_g)\sqrt{(\rho_g/\rho_a)}$$

i.e.

$$r = (C_{da}\, A_a/C_{dg}\, A_g)\sqrt{d} \tag{7.20}$$

where d is the relative density (specific gravity) of the gas.

Thus if the discharge coefficients are equal a simple orifice mixer will produce a constant air–gas ratio independent of throughput. Pressure drops are normally less than 750 Pa (7.5 mbar, 3 inH_2O) and the discharge coefficient is very sensitive to variations in Reynolds number in this range. One solution to this difficulty is provided by the adjustable-slot mixer shown in Figure 7:34 in which area-ratio adjustment is provided by rotating the sleeve thereby varying the length of the air slot. Since Reynolds number is determined solely by slot width the air slot and the gas orifice can be sized such that the Reynolds numbers are equalized.

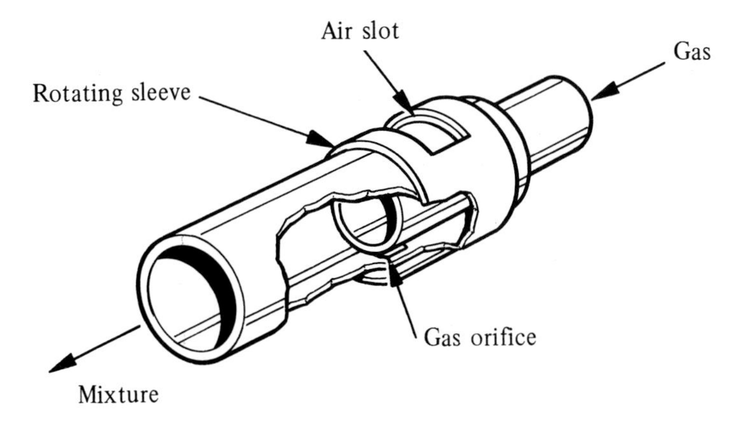

Figure 7:34 Adjustable slot mixer

Linear-flow mixers
In the orifice types described above the flow through the elements is proportional to the square root of the differential pressure across them and the accuracy of proportioning is governed by the accuracy of control of the differential pressure across the air and gas orifices. This in practice is governed by the ability of the zero governor to maintain the gas at atmospheric pressure, which worsens with decreasing differential pressure. If linear-flow elements in which the flow is *directly* proportional to differential pressure are used then the turndown is significantly increased.

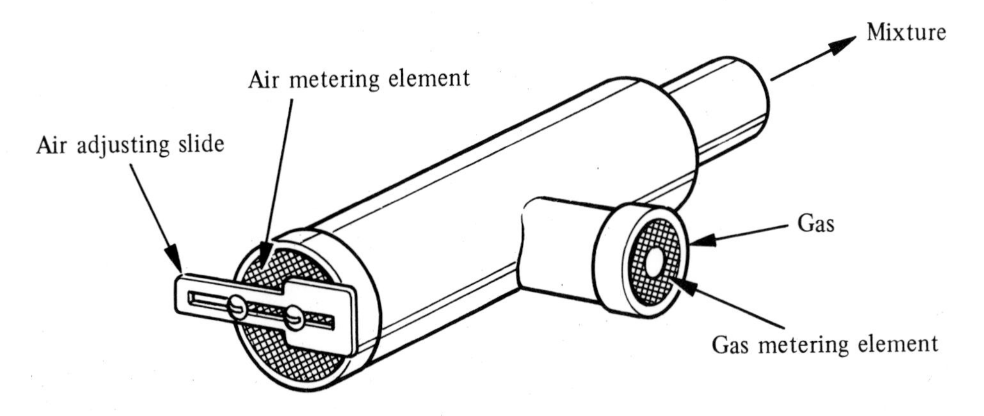

Figure 7:35 Linear-flow mixer

A convenient form of linear-flow element is the crimped-metal flame-arrester element described in Section 9.4.10 in which the flow is streamline and the flow rate is governed by the Poiseuille equation (2.39). Figure 7:35 shows a linear-flow mixer which is capable of maintaining a constant air–gas ratio $\pm$ 4 per cent for a turndown range of 15 : 1.

An alternative proportioning method commonly used in the USA may be classified as a variable-pressure, variable-area method [54] and makes use of a proportioning valve (e.g. the Eclipse Variport Mixer). Control is by two linked valves on the fan inlet, the gas entering from a governed supply at normal pressures. The air valve is a rotating shutter whilst the gas valve incorporates an adjustable strip-cam, the profile of which may be varied by adjusting a number of set screws along its length. The gas valve may be set to give the correct air–gas ratio at a number of points over the throughput range.

7.7 RADIANT BURNERS

7.7.1 Introduction

In radiant burners a high proportion of the energy supplied in the fuel is converted into radiant energy. The temperature and material of construction of the burner's radiant surface govern the heat radiated from it. Clearly a perfect black body (Chapter 3) produces the highest possible emission for a given temperature and also the maximum emission over the whole range of wavelengths in which heat is radiated. At high temperatures the radiation encroaches increasingly on the visible band of the emission spectrum, first being visible at about 500 °C (dull red) passing through bright red and orange to white in the region of 1500 °C [58]. In general, refractory materials form the radiating surfaces of radiant burners and these behave neither as 'black bodies' nor strictly as 'grey bodies', the emission differing appreciably at different wavelengths.

In the early part of this century Bone and others [59] observed that gas–air mixtures flowing through the pores of refractory tiles burned intensely on the surface of the tile maintaining its surface incandescent without the *apparent* presence of flames. Bone considered that the major characteristic of this phenomenon, termed 'surface combustion', was the catalytic effect of the refractory material. Later work, reviewed in [58], has suggested that catalytic effects do not occur but that thermal conductivity is the important property rather than surface activity and that any acceleration of combustion occurs because of preheating of the mixture stream in the pores of the refractory [58]. No evidence of surface activity was apparent in studies of multiple-channel Schwank blocks [58] but Coles and Bagge [60], in referring to porous-medium burners in which the pore diameters are small concluded that the combustion mechanism is a combination of both gaseous and surface catalytic combustion.

7.7.2 Proprietary radiant burners

The wide variety of types of radiant burners now in use have been reviewed by Kilham and Lanigan [58] and by Coles and Bagge [60] and may be conveniently classified as follows.

Porous-medium burners

In this type, which is a direct development of Bone's system, gas–air mixture, either from an atmospheric injector or from an air-blast mixer, flows through a porous tile and burns at the surface producing a uniformly radiating area (Figure 7:36 (*a*)).

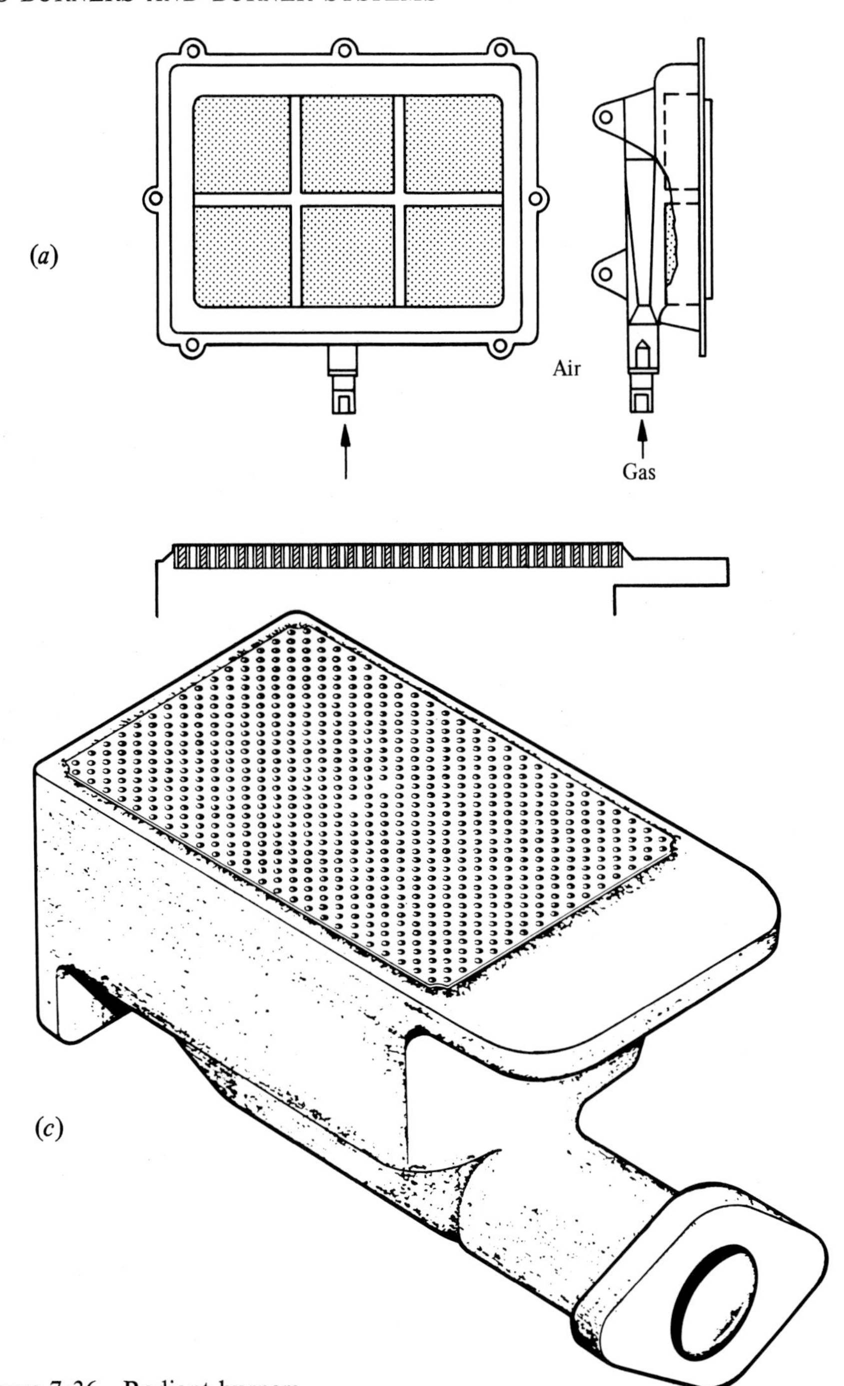

Figure 7:36 Radiant burners
(a) Porous-medium radiant burner
(c) Schwank burner

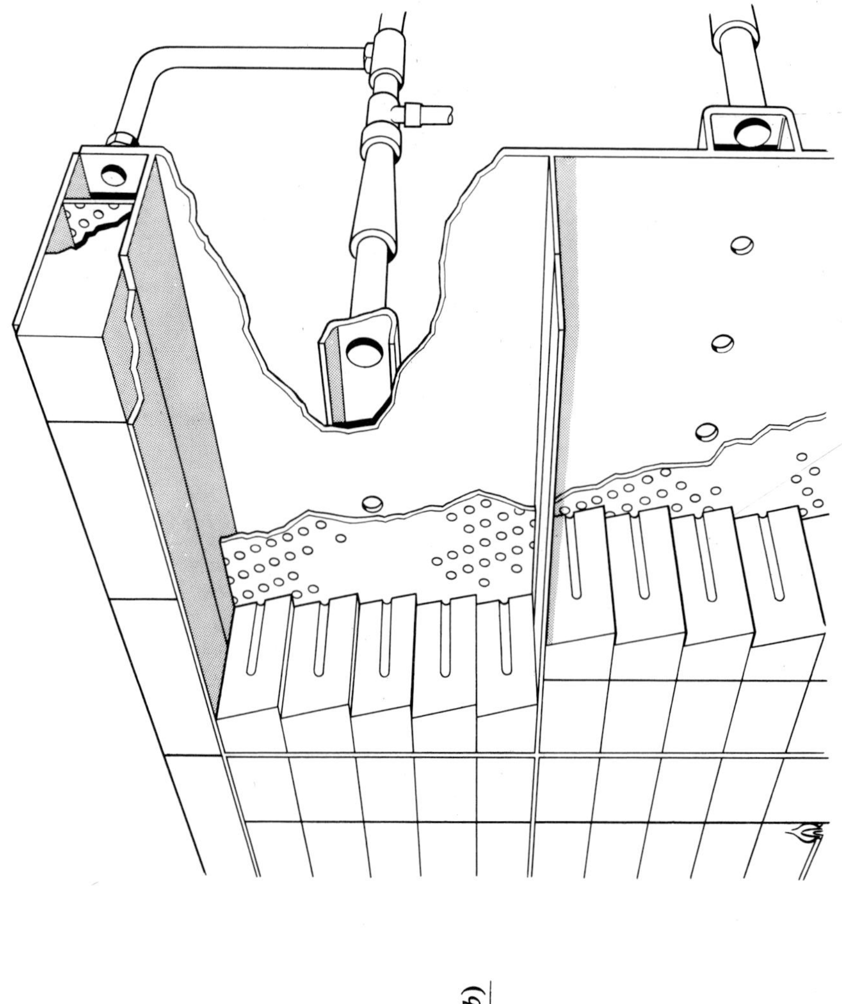

(*b*)

Figure 7:36 (*b*) Luminous-wall system (Maywick)

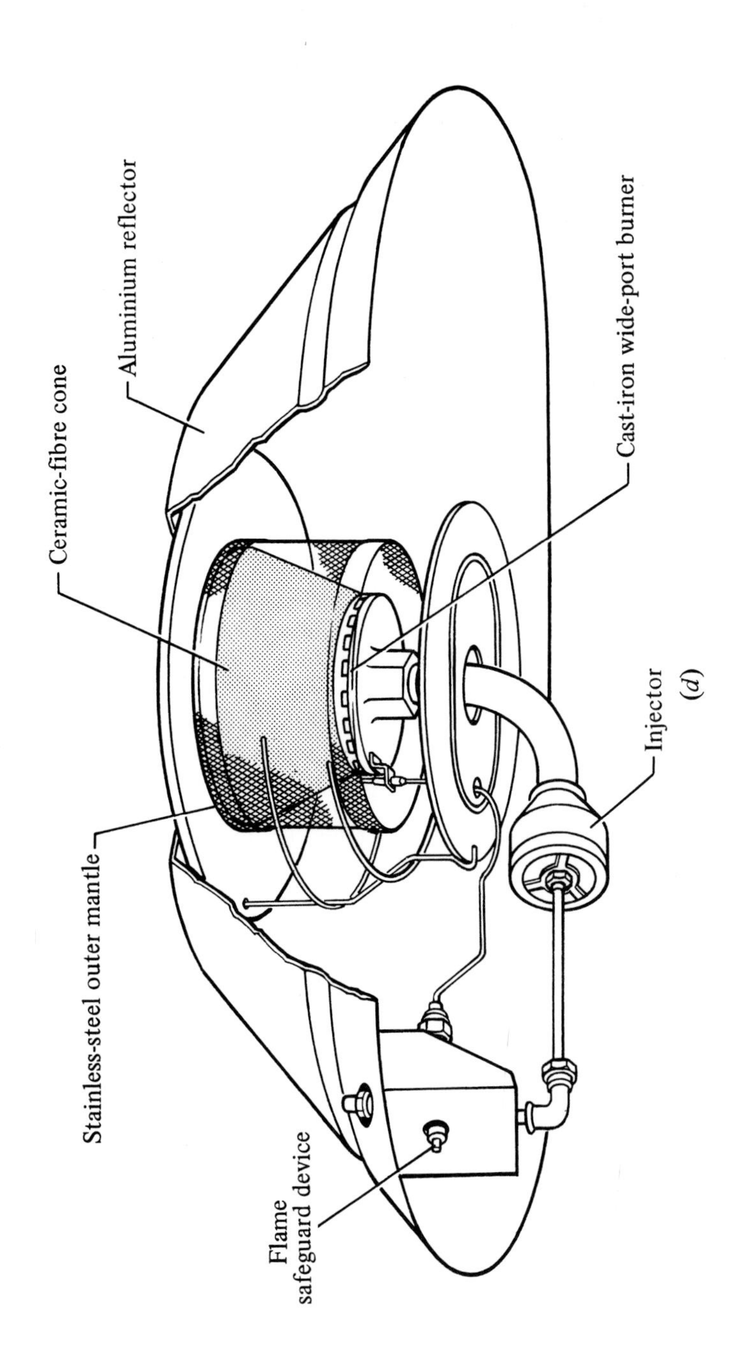

Figure 7:36 (*d*) Wire-gauze burner (Maywick-Titan)

The surface temperature of the type shown is normally about 860 °C although using different materials higher temperatures are possible. An extension of this idea is the luminous-wall design (Figure 7:36 (*b*)) in which a large proportion of the furnace structure is constructed of porous burner-bricks. These comprise standard-size porous bricks containing drilled holes to within 50 mm (2 in) of the surface. Gas–air mixture from air-blast injectors passes into a plenum chamber and thence into the back of the bricks through which it flows to burn uniformly at the surface. In becoming preheated the gas–air mixture cools the refractory behind the hot face and only a thin surface layer reaches a high temperature. Thus wall heat-losses are minimized and rapid heating and cooling of the structure are facilitated.

Heat-input rates are typically 250 kW m^{-2} (80 000 Btu h^{-1} ft^{-2}) although a maximum of 470 kW m^{-2} (150 000 Btu h^{-1} ft^{-2}) is attainable. The surface temperature is normally limited to 1100 °C.

Combustion in channels or perforations in the burner block

In this group combustion occurs in a series of channels or perforations through the burner block. A widely used example is the Schwank burner shown in Figure 7:36 (*c*) in which the gas–air mixture is supplied from an atmospheric injector and passes through perforations of about 1.4 mm (0.055 in) diameter giving a surface temperature of about 850 °C. The radiating area is built up from individual 65 mm × 45 mm (2.56 in × 1.77 in) ceramic plates.

A detailed investigation of the combustion mechanism of the Schwank burner has been conducted by Kilham and Lanigan [58] who have shown that on town gas the flame base is located about 1.5 mm below the surface of the plate. No evidence was found of surface activity, the flame front occurring at an equilibrium position where the local mixture velocity is equal to or greater than the burning velocity, this in turn being affected by the channel diameter and by the amount of mixture preheat which is related to the amount of heat conducted back from the surface. On conversion to natural gas the flame tends to lift off the burner surface, the flame front no longer being below the surface and therefore preheating the mixture less. This may be overcome by placing a metal gauze 5 mm or so above the block surface to reflect heat and thereby stabilize the flame by increased preheat.

Other multiple-tunnel designs incorporate much larger channels up to 20 mm diameter. The flame is much more apparent in these designs but the burner still transfers much of its heat by radiation. Burners of this type may be operated up to 1350 °C with a maximum heat input of 1.6 MW m^{-2} (500 000 Btu ft^{-2} h^{-1}) face area.

Radiant-cup burners, flat-flame burners, etc.

This group differs from the types described above in that gas–air mixture does not flow through pores or tunnels in the radiating block. Instead the radiating surface is a cup or cone-shaped refractory quarl fed centrally with gas–air mixture, or with separate gas and air supplies in nozzle-mixing designs. Designs often incorporate spin which serves the function of stabilizing the flame (Section 4.8.1) and spreading the mixture into a layer over the refractory surface where combustion takes place. 'Flat-flame' burners of this type are described in Section 7.5 of this chapter.

Radiant tubes

In applications where it is required to separate the combustion products from the stock, an alternative to using muffle furnaces is to employ internally-fired metal tubes, i.e. radiant tubes. Since these transfer heat largely by radiation they have been included in this classification. They are, however, sufficiently important and different from the other radiant-burner types described in this section to justify separate consideration in Section 7.9.

Wire-mesh burners

Wire meshes instead of porous blocks have been used in burners for low-temperature applications. In a Swiss design described in [58] two layers of gauze are employed, the inner layer which remains relatively cool serving to prevent lightback. The interior of such burners is normally polished to reflect radiation falling on it. Surface temperatures in excess of 900 °C have been achieved but they are limited in use to relatively low-temperature environments to avoid overheating of the burners.

The burner shown in Figure 7:36 (*d*) is used in overhead radiant space heaters and consists of an atmospheric ring burner firing onto a ceramic-fibre truncated cone operating at temperatures up to 1200 °C. An outer cylindrical shield of stainless-steel mesh which operates at 850 °C forms the radiant surface.

Catalytic burners

It is possible to burn gases without visible flame at the surface of a catalyst at very low temperatures (about 450 °C for natural gas) but with the same heat release as normal high-temperature combustion. Flueless catalytic space heaters using LPG are widely used for domestic and camping purposes on the Continent, and in the USA and USSR catalytic heaters using either LPG or natural gas are used for non-domestic heating including livestock rearing houses and cargo trucks [61].

Catalytic combustion of natural gas presents some difficulties and is not currently used in commercially available equipment in the UK. The idea is an attractive one, however, and studies are currently being undertaken at BGC's Watson House Research Station of its applicability in space heating.

In principle the system consists of a porous pad of catalyst-impregnated fibres through which neat gas passes. Oxygen from the atmosphere diffuses through the pad in the reverse direction and reaction occurs at the catalyst surface. Heat is transferred by radiation from the surface and the products of combustion are forced by the gas flow into the atmosphere.

Catalytic systems have the advantage that since carbon monoxide is not a combustion intermediate as is the case with flames they can work under conditions of air starvation without danger. They possess the further advantage that relatively high radiant efficiencies may be obtained. There are two important disadvantages however. First, at the present stage of development, some gas always passes through the pad unburnt and methane is the most difficult hydrocarbon in this respect. The best pads currently under development consume up to 95 per cent of the methane supplied but the figure may be as low as 76 per cent for methane burning on pads designed for LPG [61]. The second disadvantage is that the catalyst requires heating to a certain threshold temperature (about 250 °C for natural gas) before the reaction becomes self-sustaining. Preheating may be achieved by temporarily burning the emerging gas as an unsteady diffusion flame at the surface or by the use of an electric heating element embedded in the pad. Current development work is directed towards improving the catalyst to reduce the temperature at which the reaction becomes self-sustaining.

7.7.3 Performance of radiant burners

A convenient means of indicating the performance of radiant burners is provided by the radiant efficiency, i.e.:

$$\frac{\text{radiation output}}{\text{heat input}}$$

Coles and Bagge [60] have calculated ideal radiant efficiencies based on the net calorific value for various surface temperatures by equating the heat lost by radiation and in the burned gases with the heat input. Their analysis assumed black-body radiation and the absence of other heat losses, e.g. from the sides of the burner. Figure 7:37 shows the variation of radiant efficiency with surface temperature for an ideal radiant heater.

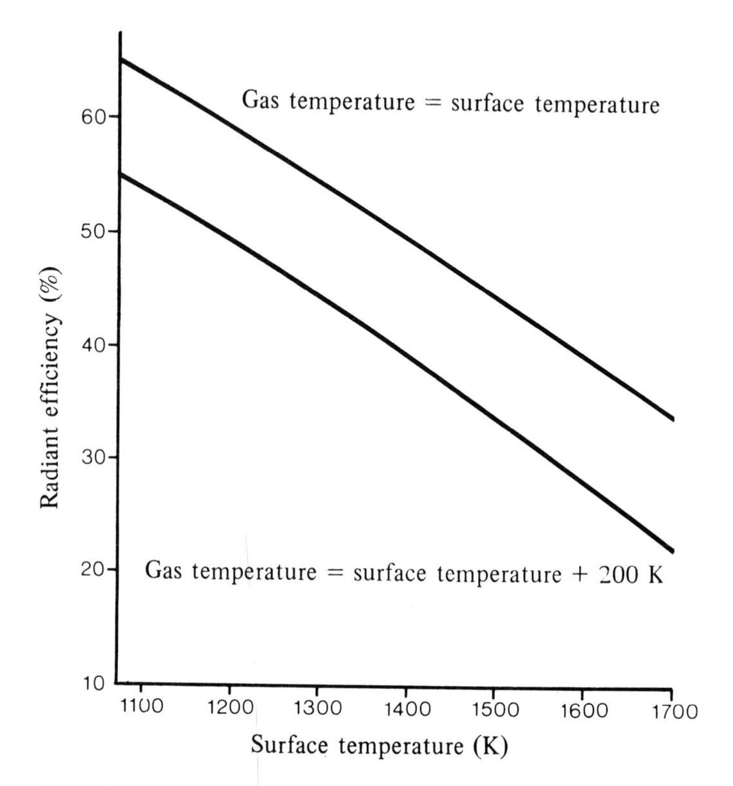

Figure 7:37 Maximum radiant efficiency based on net calorific value of an ideal gas-fired radiant heater at different temperatures [60]

Figure 7:38 indicates radiant efficiencies and surface temperatures of a range of commercially available burners operating in the open air.

Coles and Bagge [60] have compared the heat-transfer rates attainable using radiant systems with high-speed convection techniques for rapid-heating processes and concluded that there was a need for radiative furnaces operating up to 1400 °C at

Type	Surface temperature	Radiant efficiency based on net CV		Maximum temperature burner can withstand
		Typical	Ideal	
	(°C)	(%)	(%)	(°C)
Schwank	850	45	63	900
Large multiple-tunnel burner	900	33	60	1330
Radiant cup	1200	10–20	46	1500–1700
Porous medium	950	54	58	1030
Radiant tube (recuperative)	1000	50	56	1000

Figure 7:38 Radiant efficiencies and surface temperatures for various types of burner operating in the open air

high radiant efficiencies. The porous-medium burner was chosen for further development since its radiant efficiency, as indicated above, is close to the ideal and it possesses low thermal inertia. This work has led to the development of the Shell PR burner described by Coles and Wilbraham [62] which is shown in Figure 7:39.

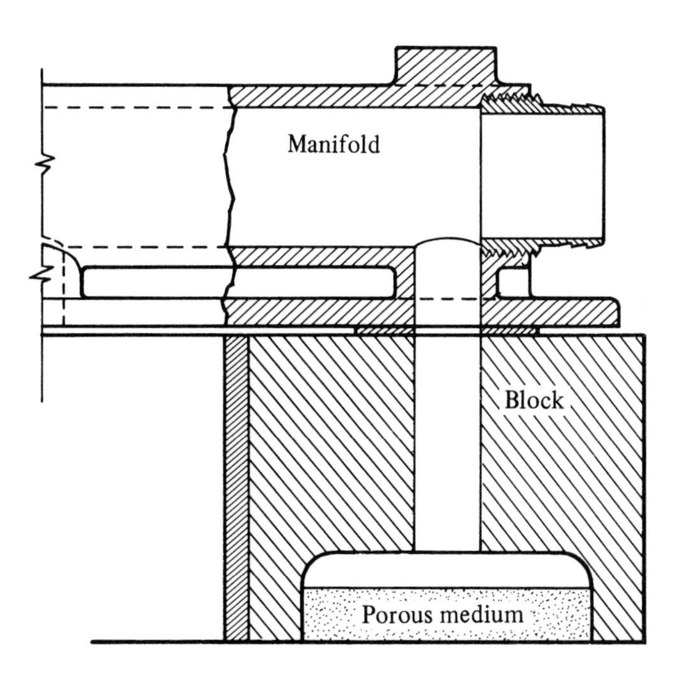

Figure 7:39 General design of Shell PR burner with manifold

The porous-medium element which is mounted in an impermeable block 150 mm square by 115 mm deep is made from either zirconia or sillimanite. In common with other porous-medium types, three modes of combustion may be distinguished: *free-flame combustion* in which small multiple flames burn above the panel surface which remains cool, *surface combustion* which occurs beneath the panel surface and which is the desired mode and *unstable interstitial combustion* in which the flame propagates back through the medium until lightback finally occurs.

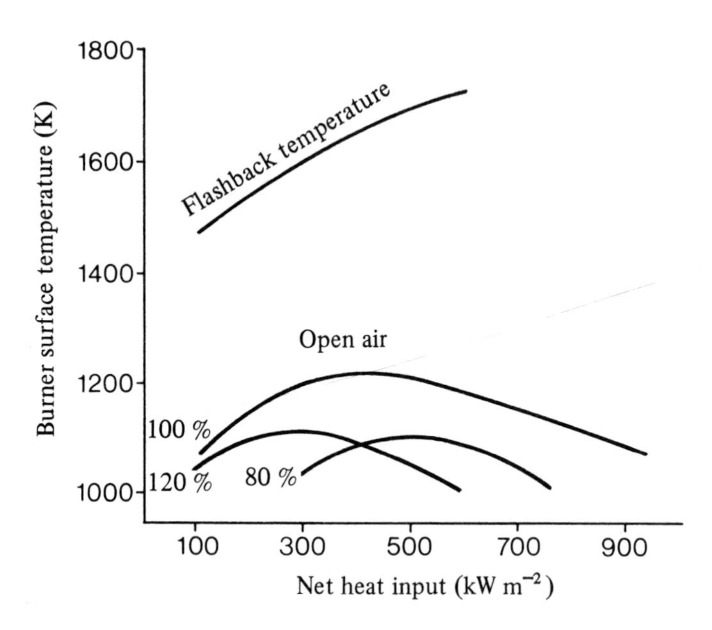

Figure 7:40 Relationship between temperature and heat input for Shell Zircon PR burner in open air and at flashback for 80, 100 and 120 per cent of stoichiometric natural gas/air [62]

Figure 7:40 shows the operating characteristics of the Zircon PR burner using stoichiometric natural-gas–air and propane–air mixtures.

Surface combustion occurs between about 80 per cent and 130 per cent stoichiometric proportions for natural-gas–air. Mixture-pressure requirements clearly depend upon the material permeability, panel thickness, mixture flow rate and panel surface temperature but typically range from 500 Pa (5 mbar, 2 inH_2O) to 2.5 kPa (25 mbar, 10 inH_2O). The maximum surface temperature attainable in furnace operation is about 1430 °C.

7.8 PULSATING COMBUSTORS

In pulsating-combustor systems, combustion and therefore heat release occur periodically and the pressure waves produced give rise to high sound levels at the frequency of the combustion oscillations. Oscillatory combustion sometimes occurs in conventional burner systems, particularly those using high combustion intensities, and normally constitutes a serious nuisance. In pulsating combustors, however, the system is deliberately designed to produce oscillations, the energy in the pressure waves being used to induce the combustion air and to eject the combustion products. The system operates such that a periodic heat release is obtained at a point corresponding to a pressure antinode of a fundamental acoustic mode for the system, the oscillation being sustained if the periodic heat release is in phase with pressure variation.

Interest in practical applications for pulsating combustion date back to the 1930s and the work of Schmidt which led to the development of the V-1 flying bomb. In more recent years development has been directed towards more peaceful applications including immersion-tube water heating [63], portable heaters, ice-melters, vehicle heaters and grain dryers [64]. A wide variety of fuels have been employed including LPG, natural gas, gasoline and pulverized coal.

Briffa *et al.* [64] have listed the advantages and disadvantages of pulsating combustors as follows:

Advantages

1. Relatively simple construction
2. Combustion intensity is high since the process approximates to constant-volume combustion
3. Velocity fluctuations in the exhaust tube(s) increase convective heat-transfer coefficients appreciably reducing heat-transfer surface requirements
4. The pressure developed in the combustion chamber facilitates removal of the combustion products. This allows heat-exchange tubes of high length/diameter ratio to be used increasing efficiencies

Disadvantages

1. Noise levels are high
2. Air inlet valve design presents some difficulties
3. Turndown ratios are limited

The simplest form of pulsating combustor is shown in Figure 7:41 (*a*) and consists basically of a resonance tube several metres long and gas and air inlet arrangements. The inlet end of the resonance tube forms the combustion chamber into which air is drawn from the atmosphere through a valve which may be a flap valve as shown or an aerodynamic valve without moving parts.

The system operates as follows:

1. Gas and air flow into the combustion chamber and are ignited by a spark
2. Combustion occurs, the increase in pressure closes the inlet valve and drives the burnt gases out through the resonance tube
3. Inertia creates a low pressure in the combustion chamber which opens the inlet valve and more air and gas enter
4. Reignition occurs from the previous cycle and the process continues. The ignition spark is not required after this stage

This simple type of combustor is commonly designated the Schmidt burner [63] or quarter-wave tube burner [64]. In operation it resonates as an organ pipe (see Section

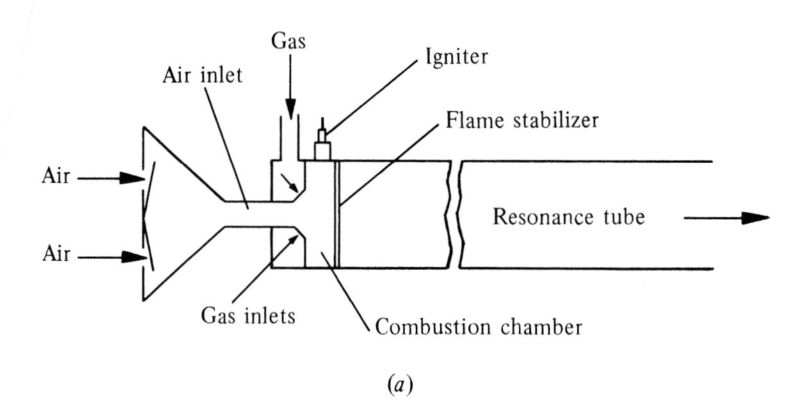

(*a*)

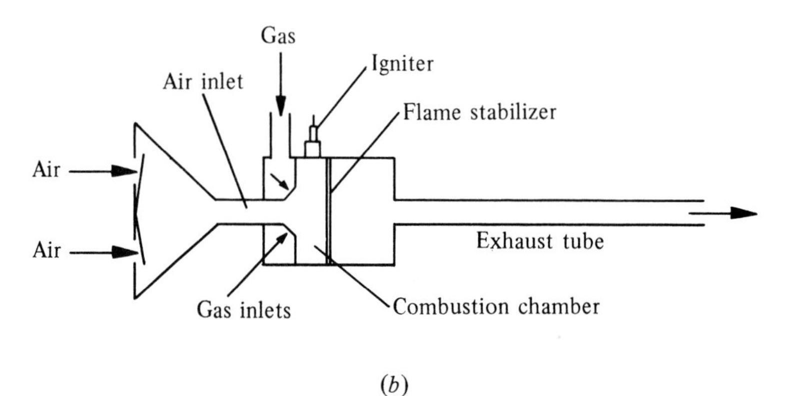

(*b*)

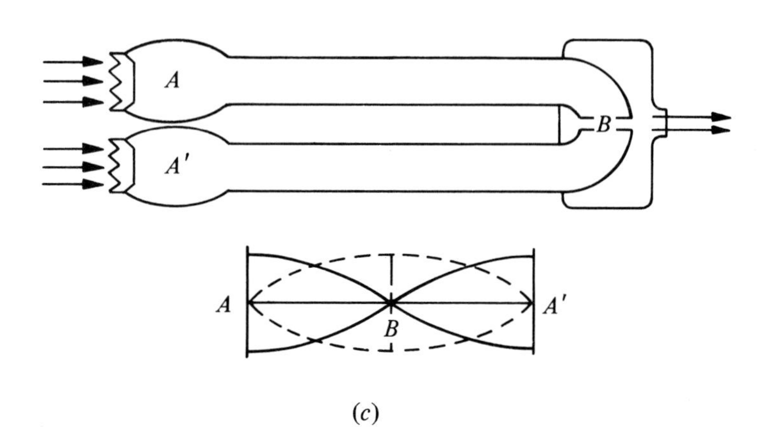

(*c*)

Figure 7:41 Pulsating combustors [64]
(*a*) Simple pulse burner (Schmidt type)
(*b*) Helmholtz pulse burner
(*c*) Schmidt burner, interference operation

5.3.2) with a pressure antinode at its closed end and a velocity antinode at its open end. The frequency of the fundamental mode of a pipe open at one end and closed at the other is given by

$$f = a/4L_r \tag{7.21}$$

where a = speed of sound, L_r = length of resonator.

For a more rigorous treatment the steady component of flow in the tube and the variation in gas temperature down the tube should be considered (see [65]). The pressure and velocity amplitudes are given by equations 5.29 and 5.30 of Chapter 5. In practice exact quarter-wave resonance only occurs with a very small air-inlet pipe and frequency transitions are normal with practical systems [63] in which, as the tube length is progressively increased, the frequency is more or less maintained by frequency jumps which add an extra half wavelength to the standing-wave pattern. An odd number of quarter wavelengths is always present maintaining a pressure node at the exit and a pressure antinode at the combustion chamber.

Figure 7:41 (*b*) shows an alternative arrangement, normally designated the Helmholtz type, in which the combustion chamber is of much larger diameter than the resonance tube. The combustion chamber is almost a closed cavity and forms with the air inlet a resonator of the Helmholtz type which governs the working frequency of the burner, the exhaust tube ideally being tuned to resonate at the same frequency [63] although stable pulsating combustion may still be obtained if the exhaust is untuned. In the practical systems investigated by Francis *et al.* [63] the fundamental frequency decreased with increasing length, the effect being more marked for increases in the combustion-chamber length than the exhaust tube. The sharp frequency jumps obtained with the Schmidt burner did not occur.

In practice some form of stabilizing device appears necessary. This can take the form of a small-diameter rod [63] across the combustion chamber which may act as a bluff body or may consist of extended spark electrodes [64]. The design of the air-inlet valve presents difficulties: satisfactory mechanical reed-valves using rubberized fabrics have been developed [63] but these may have a limited working life. On the other hand aerodynamic valves such as the Borda mouth [63] tend to allow back flow of combustion products.

In pulsating combustors a much larger volume of combustion products may be driven through a given size of exhaust tube than in the natural-draught case. Consequently significantly enhanced convective heat transfer may be expected. In addition the presence of the pulsations themselves leads to further increases in convection, this increase being related to the magnitude of the oscillating velocity of the gas stream and hence to the pressure amplitude in the acoustic wave. Hanby [65] quotes increases of over 100 per cent in heat-transfer coefficient over the corresponding steady-flow case. Clearly the design of a compact pulsating combustion/heat-transfer unit requires the use of high-pressure amplitude in the system which in turn implies that noise reduction becomes more significant the higher the performance of the combustor [65].

Francis *et al.* [63] used conventional 'straight-through' absorption silencers on the exhaust and inlet connections producing an average sound-level attenuation of about 15 dB. Inlet silencing using aerodynamic valves was found to be impracticable since combustion products emitted by the valve were trapped in the silencer and drawn back into the combustion chamber on the next half-cycle.

An alternative approach to sound reduction has been investigated by Briffa *et al.*

[64] in which two pulsating combustors are arranged to run with a phase difference of 180°. Theoretically, complete cancellation can be obtained if the pressure fluctuations produced by both tubes have the same frequency and are symmetrical about the same mean pressure. Out-of-phase operation was obtained both by interference operation in which the exit ends of two tubes are joined together as shown in Figure 7:41 (*c*) or by using fluid-logic phasing by which combustion is arranged to occur alternately in two tubes. Schmidt-type resonators are more suitable for noise cancellation than the Helmholtz type since they produce a more uniform sinusoidal waveform. Briffa *et al.* [64] obtained reductions of about 15 dB and 30 dB for Helmholtz and Schmidt types respectively.

The application of pulsating combustion to immersion-tube tank heating is considered briefly in Chapter 15.

For further information on pulsating combustors the reader is referred to references [63–7].

7.9 RADIANT TUBES

Many industrial heat-treatment processes require the use of controlled atmospheres (Chapter 13) and must therefore be indirectly heated, i.e. the products of combustion from the heating burners must be kept separate from the stock. There are two basic methods by which separation may be achieved, either by the use of muffle furnaces or by the use of radiant tubes. Radiant tubes are basically internally fired tubular heaters and heat transfer to the stock and furnace chamber is predominantly by radiation although in instances where the furnace atmosphere is recirculated convection may be significant.

For many applications radiant tubes offer advantages over muffles in being less prone to leakage and more readily zoned.

The design of radiant tubes is mainly governed by the need to use the tube materials close to the temperatures at which oxidation and creep reduce their working life. Radiant tubes are frequently required to provide furnace temperatures of about 950 °C to 1050 °C and for the alloy steels normally used in their construction it is clearly important that no part of the tube is appreciably above the mean tube temperature. Thus the requirement of tube-temperature uniformity exerts a dominating influence over the design of the combustion system.

Two basic approaches have been adopted to obtain temperature uniformity:

1 By controlling the combustion and heat-release pattern along the tube, usually by using a long largely laminar diffusion flame giving a slow progressive heat release pattern
2 By using jet-driven recirculation to re-entrain flue gases so that the high-temperature flame gases are diluted by large amounts of gases at a temperature near to that of the furnace.

7.9.1 Non-recirculating types

Considering the first category, the simplest arrangement is the straight-through tube Figure 7:42 (*a*), which locates in coaxial holes in the opposite walls of the furnace. Despite its simplicity and low capital cost the straight-through tube has a number of

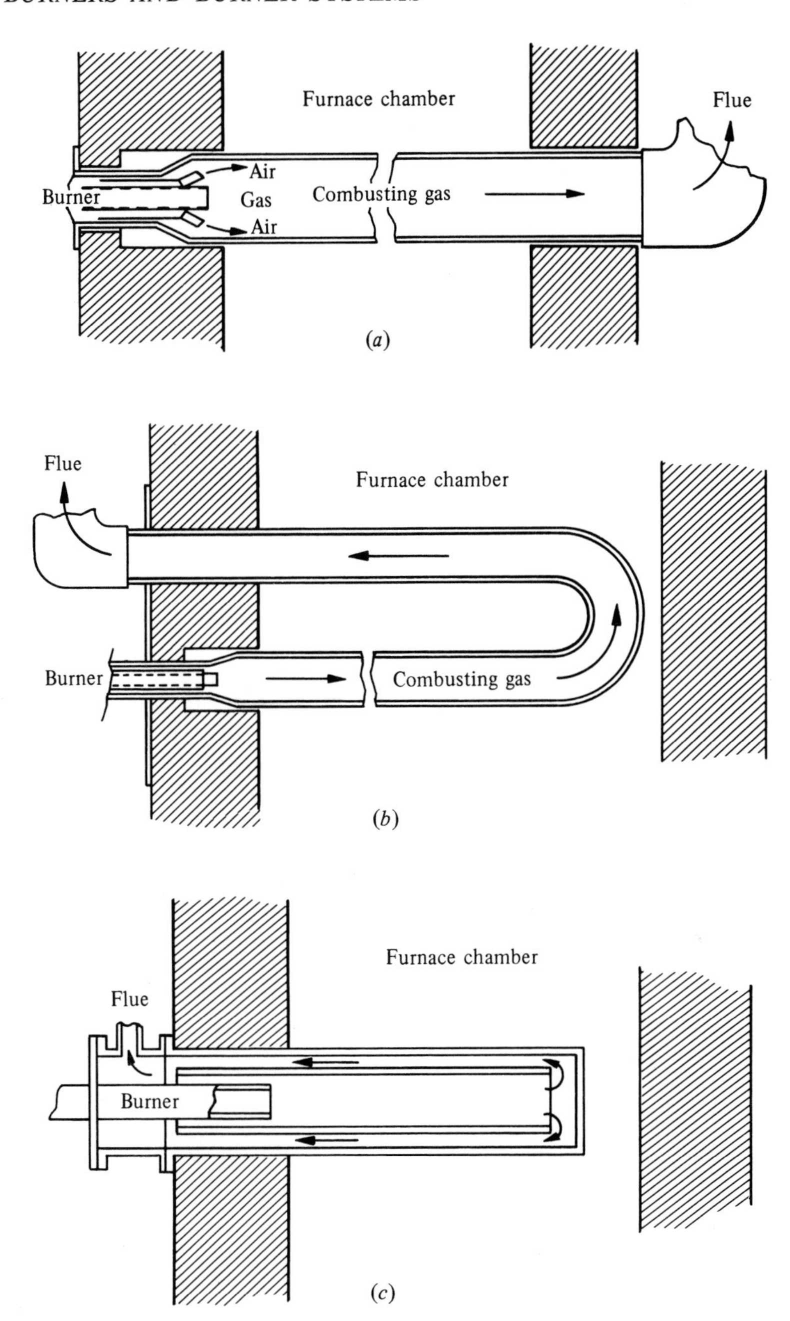

Figure 7:42 Non-recirculating radiant tubes
(*a*) Straight-through tube
(*b*) U-tube
(*c*) Indugas tube (Salem-Brosius Ltd)

important disadvantages: the temperature uniformity is poor (up to 150 °C variation [68], the thermal efficiency is low (30–40 per cent), fixing is required on both sides of the furnace leading to sealing problems or the need for expansion joints and if recuperation is employed the preheated air must be conveyed from one side of the furnace to the other. Ceramic straight-through tubes exist, notably the Ipsen tube in which sealing is achieved by the use of a metallic bellows.

The problems of two-wall fixing are largely eliminated by the use of a U-tube arrangement (Figure 7:42 (*b*)). Doubling the tube length doubles the heat-transfer-surface area and consequently the gas rate leading to higher convective heat-transfer rates to the tube wall. The bend can assist in mixing but is also likely to give rise to 'hot spots'. Thermal efficiency (with special mixing devices) may be as high as 48 per cent [69] at normal operating temperatures. Temperature variation along the tube may be up to 300 °C although the juxtaposition of the hot and cold legs leads to relatively uniform heat transfer from the tube.

Variations of the U-tube design include a parabolic arrangement (the Lee–Wilson tube), 'W-tubes' and other multiple-leg tubes sometimes in different planes. Similar considerations apply to these designs.

A second group of non-recirculating radiant tubes may be described as double-pass single-ended types in which the burner and flue are combined at one end and an internal tube transmits the hot gases to the closed end of the outer tube where they turn through 180° and travel back through the annulus.

The Indugas tube (Figure 7:42 (*c*)) is an example of this type and uses a short-flame high-velocity burner. Due to the counter-flow arrangement the outer-tube temperature uniformity is good (10–70 °C). These tubes require only single-hole fixing in the furnace wall and, being single-ended, eliminate thermal stresses. Additionally the tube interior can be replaced in service without loss of furnace atmosphere. The Hazen tube of American origin consists of three coaxial tubes. Gas and primary air flow to the far end of the tube where initial combustion occurs. The products of partial combustion return down an annulus formed between the perforated air tube and the outer tube. Secondary air to complete combustion flows through the perforations in the air tube allowing further combustion to take place along the length of the tube. This design shares the advantages of single-hole fixing and good temperature uniformity with the Indugas design. It is, however, more elaborate in construction.

7.9.2 Recirculating types

Better temperature uniformity and increased efficiency may be obtained if the combustion products are recirculated more than once through the radiant tube before passing to the flue. The flame gases are 'swamped' by several times their own volume of flue gases at a temperature close to that of the furnace. Additionally the higher mass flow rate in the tube increases convective heat transfer to the tube wall and reduces the differential between the flue-gas and tube temperatures increasing the thermal efficiency. High levels of recirculation are readily achieved using high-exit-velocity tunnel burners of premix or tunnel-mixing design.

Figure 7:43 (*a*) shows a 'P-type' recirculating radiant tube (Wellman-Incandescent Jetube). The burner employed is an all-metal air-cooled jet burner providing sufficient recirculation to restrict gas temperatures in the tube to less than 100 °C above the furnace temperature. Tube heat-release rates can be up to 63 kW m^{-2} (20 000 Btu $ft^{-2} h^{-1}$) and a gas path length of up to 6.7 m (22 ft). Recuperative versions are available.

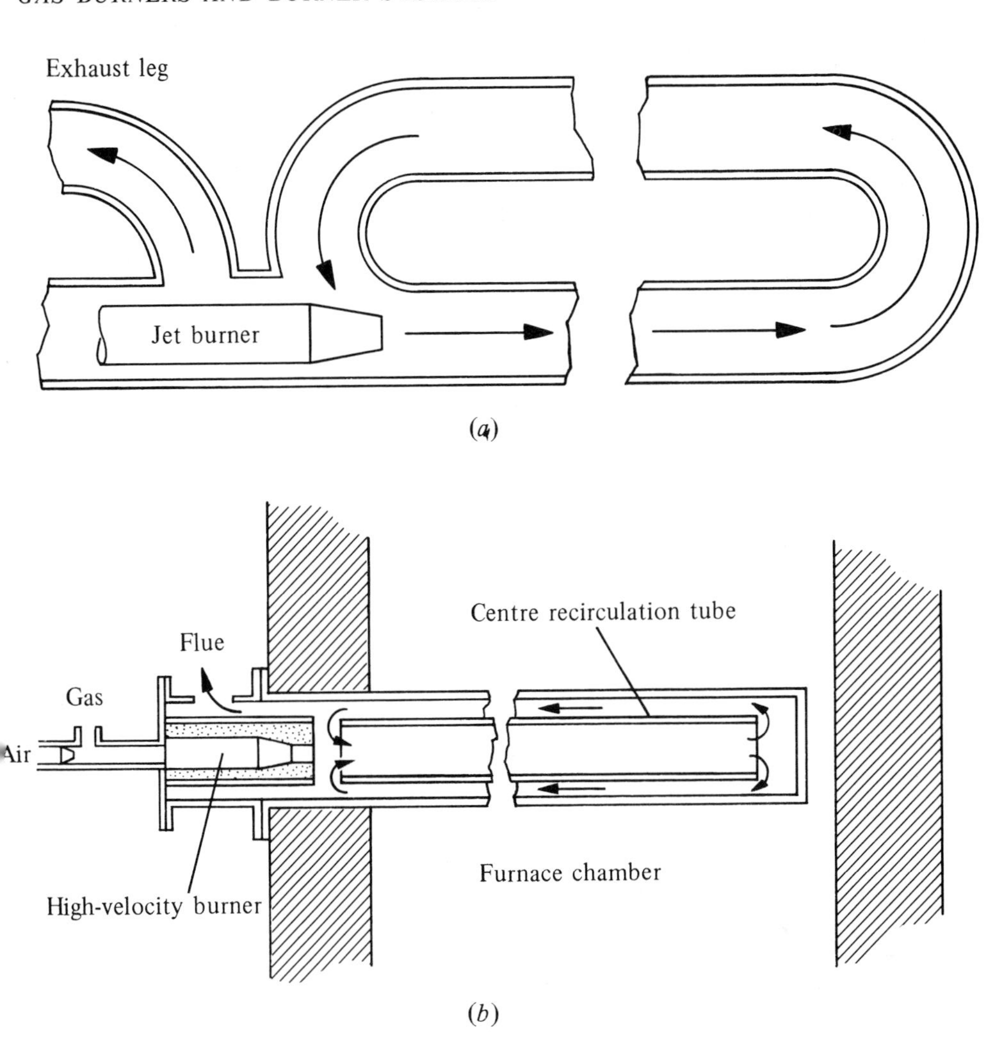

Figure 7:43 Recirculating radiant tubes
(*a*) P-type
(*b*) MRS single-ended type

Figure 7:43 (*b*) shows the single-ended recirculating radiant tube developed by the British Gas Corporation's Midland Research Station. In its original non-recuperative form the tube was heated by a MRS type A high-velocity tunnel burner (Section 7.4.2). The development and design of this type have been described in detail by Bridgens and Francis [70].

Temperature measurements by Bridgens and Francis [70] indicate that the temperature uniformity along the tube (on town-gas firing) was ± 5 °C at surface temperatures up to 1230 °C and that the metal inner tube operates 60 or 70 °C hotter than the outer tube. With furnace temperatures around 1000 °C the tube operates about 50 °C higher than the furnace temperature for normal heat-input ratings which represents a thermal efficiency of around 50 per cent, i.e. approaching the maximum possible without recuperation.

7.9.3 Recuperative types

Clearly the logical development of the above design is to combine its virtues with those of an integral recuperative burner (Section 7.5.3) increasing the efficiency to the order of 65 to 70 per cent. The single-ended design lends itself particularly well to recuperation since the combustion-air inlet and flue-gas outlet are close to each other in an inherently counter-flow arrangement at the same end of the tube.

Increasing the efficiency to around 63 per cent requires a combustion-air preheat of the order of 500 °C which in turn means that the premix tunnel burner of the earlier type must be replaced by a tunnel-mixing design for stability reasons.

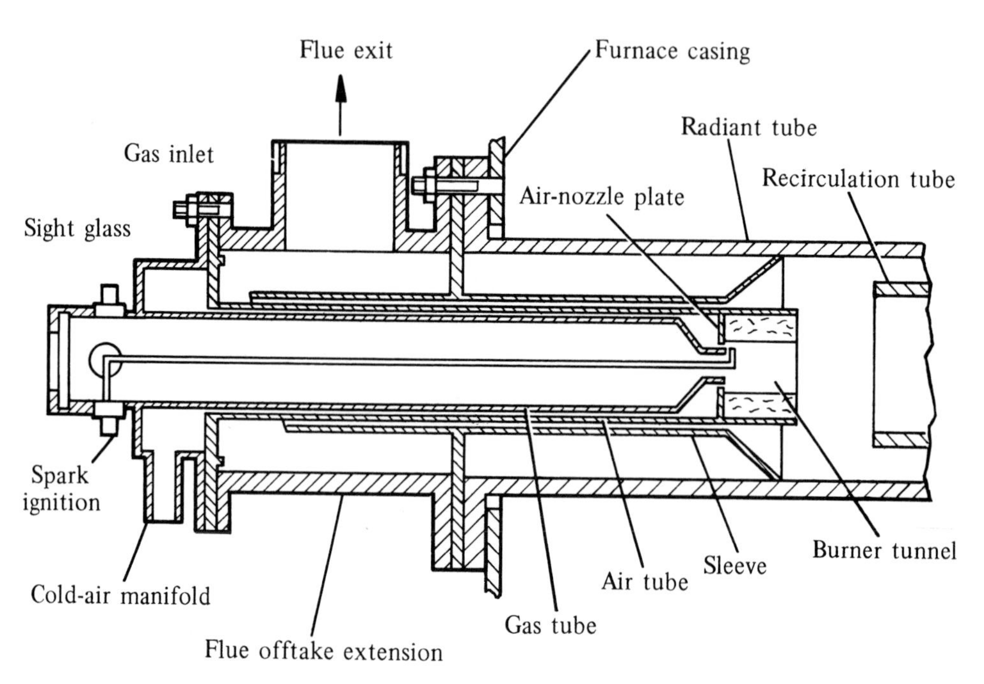

Figure 7:44 Recuperative single-ended recirculating radiant tube

Figure 7:44 shows the basic features of the recuperative single-ended radiant tube developed by the British Gas Corporation's Midland Research Station [71]. Chatwin *et al.* [71] have presented an analysis of heat transfer, thermal efficiency and pressure losses in tubes of this design and the reader is referred to their paper for a detailed treatment. However, the main points of interest are as follows.

At normal values of heat flux the differential between the tube-surface temperature and the temperature of the combustion products entering the recuperator is low ($\approx$ 40 °C at 1000 °C). This indicates the effectiveness of heat transfer in the tube and implies that the recuperator materials need be no better than those of the tube itself.

At a tube temperature of 1000 °C the air-preheat temperature varies from 450 °C at high throughputs to about 650 °C at low throughputs. The gas temperature rise is about half that of the air and no difficulties have been experienced with gas cracking.

Two ways of defining efficiency may be distinguished:

η_1, the efficiency of heat transfer through the radiant tube, i.e.

$$\frac{\text{useful heat}}{\text{heat supplied}} = \frac{rH_{\text{ae}} + H_{\text{ge}} - H_{\text{pi}}V}{C_{\text{g}}}$$

and η_2, the efficiency of heat transferred from the combustion products, i.e.

$$\frac{\text{heat available}}{\text{heat supplied}} = \frac{C_{\text{n}} - H_{\text{pf}}V}{C_{\text{g}}}$$

where subscripts a = air
g = gas
e = exit recuperator
i = inlet recuperator
f = flue exit

and H is heat content per unit volume.

The experimental results of Chatwin *et al.* [71] show that the heat lost by the combustion products in the recuperator is always about 20 per cent greater than the heat gained by the gas and air. The difference is due to heat transfer into the furnace wall and accounts for the few per cent difference between η_1 and η_2. As would be expected, efficiencies decrease with rising temperature and throughput. At 900 °C and a heat flux of 22 kW m^{-2} (7000 Btu ft^{-2} h^{-1}) η_1 is about 66 per cent and η_2 70 per cent.

Pressure losses are relatively low, always being less than 250 Pa (2.5 mbar, 1 inH_2O) between the inside of the tube and the flue offtake. A simple air ejector driven from the combustion air provides extraction and maintains the tube pressure below atmospheric.

Regarding radiant tube materials, the majority in operation at present are manufactured from nickel–chrome alloy steels with maximum service temperatures in the region of 1050–1200 °C. For operation at higher temperatures ceramic materials must be considered but the choice is difficult. High mechanical strength, resistance to thermal shock and freedom from porosity are all essential properties and, whilst silicon carbide, silicon nitride and recrystallized alumina appear to be possibilities, they present difficulties with regard to cost and availability in suitable tubular forms.

7.10 AUTOMATIC PACKAGED BURNERS

A major development in the firing of boilers and other low-temperature plant during the last decade has been the evolution of a wide range of forced-draught automatic packaged burners in which the burner, fan, throughput and air–gas ratio controls and ignition and safety systems are integrated into a single unit.

In 1966 the Gas Council published *Standards for Automatic Gas Burners, Forced- and Induced-Draught* [72] revised in 1970 [73], which lays down safety requirements relating to, for example, permissible energy release during ignition, purging and safety shutoff valves for automatic burners of rating above 44 kW (150 000 Btu h^{-1}). More recently the philosophy of the standards has been extended in the *Interim Codes of Practice for Large Gas and Dual-Fuel Burners* [74] to cover large burner installations having a total thermal rating greater than 3 MW (102 therm h^{-1}).

These requirements are outside the scope of the present chapter and are considered in Chapter 9. Consideration here is restricted to the burner head and to throughput and air–gas ratio control.

Packaged burners are almost exclusively of the nozzle-mixing type and use metallic burner heads incorporating a wide variety of stabilizing and mixing arrangements. The burner design needs to be such that it provides safe and reliable ignition, flame detection, cross-lighting, flame stability, acceptable combustion quality and freedom from excessive noise [75]. In order to satisfy these requirements a number of components need to be integrated into the head assembly, e.g. the ignition source, pilot burner and flame detector and their placing and security of fixing are crucial for reliable operation.

Figure 7:45 shows in outline a variety of burner-head configurations representing typical proprietary burners. The coverage is not by any means comprehensive and relates only to those designs tested by Hoggarth, Jones and Pomfret [75]. Ratings range from 65 to 410 kW (250 000 to 1 400 000 Btu h^{-1}) and the fan pressure lift (for 10 per cent excess air) from 160 to 1930 Pa (1.6 to 19.3 mbar, 0.6 to 7.7 inH_2O).

In type (*a*) gas is discharged from a series of orifices at right-angles to the burner axis over the surface of the air-distribution plate. No separate pilot burner is used, ignition occurring on turned-down main flame. In type (*b*) provision is made for changing the flame shape from short and broad to long and narrow by adjusting the gap between the disc and the gas-tube face using washers. Type (*c*) incorporates two-stage mixing of gas and air. A separate nozzle-mixing pilot burner is fitted coaxially with the main burner. The burner head of type (*d*) consists of an open-ended gas pipe with a sudden enlargement at the exit. Orifices in the wall of the gas pipe provide partial premixing of air and gas. In type (*e*) the gas enters via a series of short converging tubes between which flows air from a central supply. A swirl vane is placed immediately downstream of the gas tubes. Burner (*f*) is a dual-fuel burner but only the gas-firing components are illustrated. The main gas flow is via nozzles from an annular gas pipe surrounding the air duct (which accommodates the oil-firing equipment when required). In burner (*h*) the head consists of a central gas-supply pipe with radial orifices. For first-family gas these are simple drillings but for second-family gas they take the form of short pipes. Air is swirled by vanes attached to the gas pipe upstream of the gas orifices. The gas jets in type (*i*) are long slots over which the swirled air supply flows.

Whether the throughput control mode is on-off, high-low or proportional depends on the process requirements and the cost of the system. In general the smaller burner sizes use on-off control using the main safety shutoff valve as the control valve. In the larger sizes some burners are provided with high-low-off systems but proportional systems are rare. With the possible exception of air heaters, the types of plant fired by packaged burners have high thermal inertia and on-off control is normally adequate [75].

Air–gas ratio setting is obviously important with regard to satisfactory combustion performance and plant efficiency and is easily obtained on on-off systems by adjusting and locking the butterfly valve or shutter which controls the air flow and setting

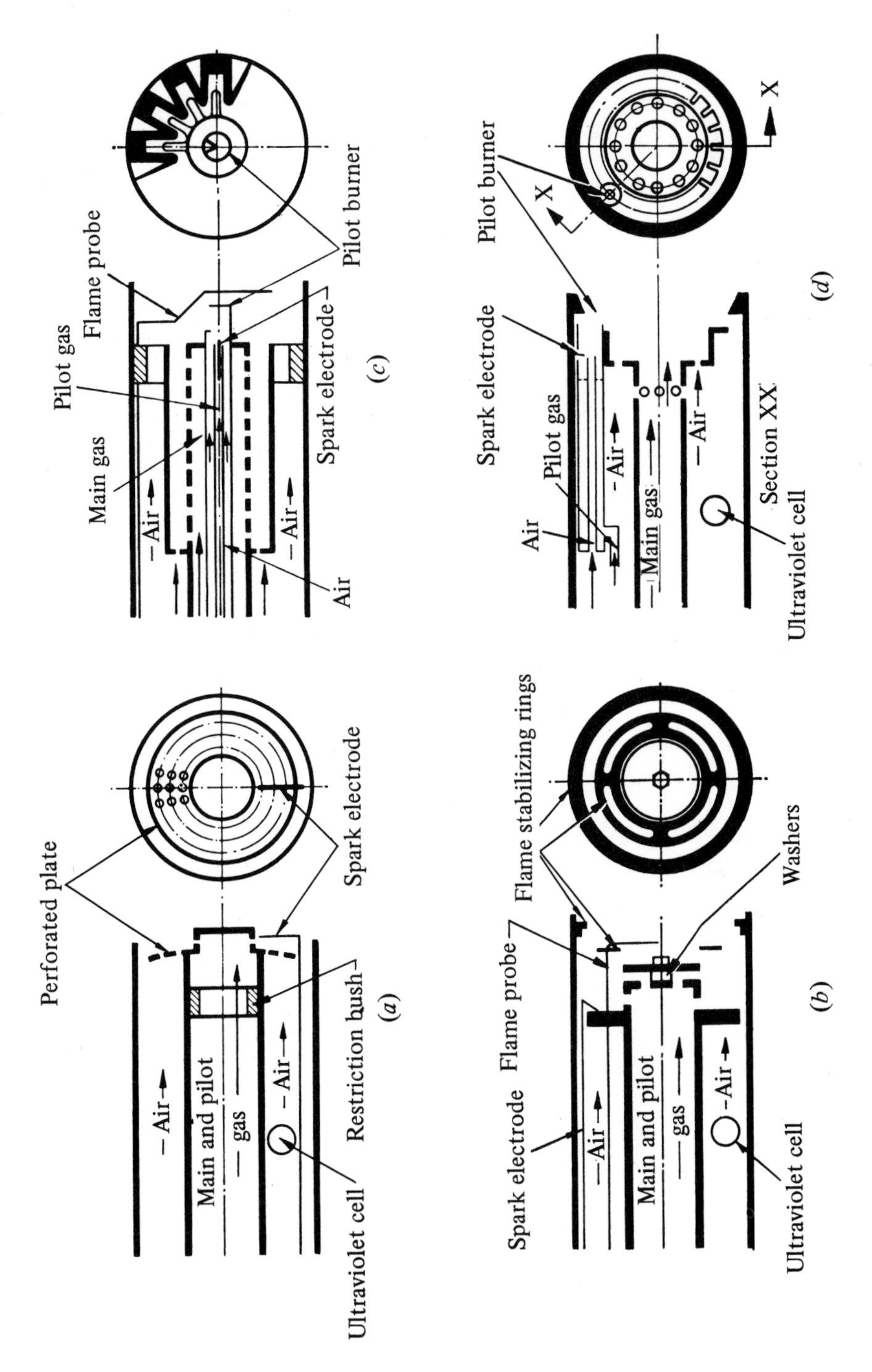

Figure 7:45 Typical proprietary packaged burner head assemblies [75]

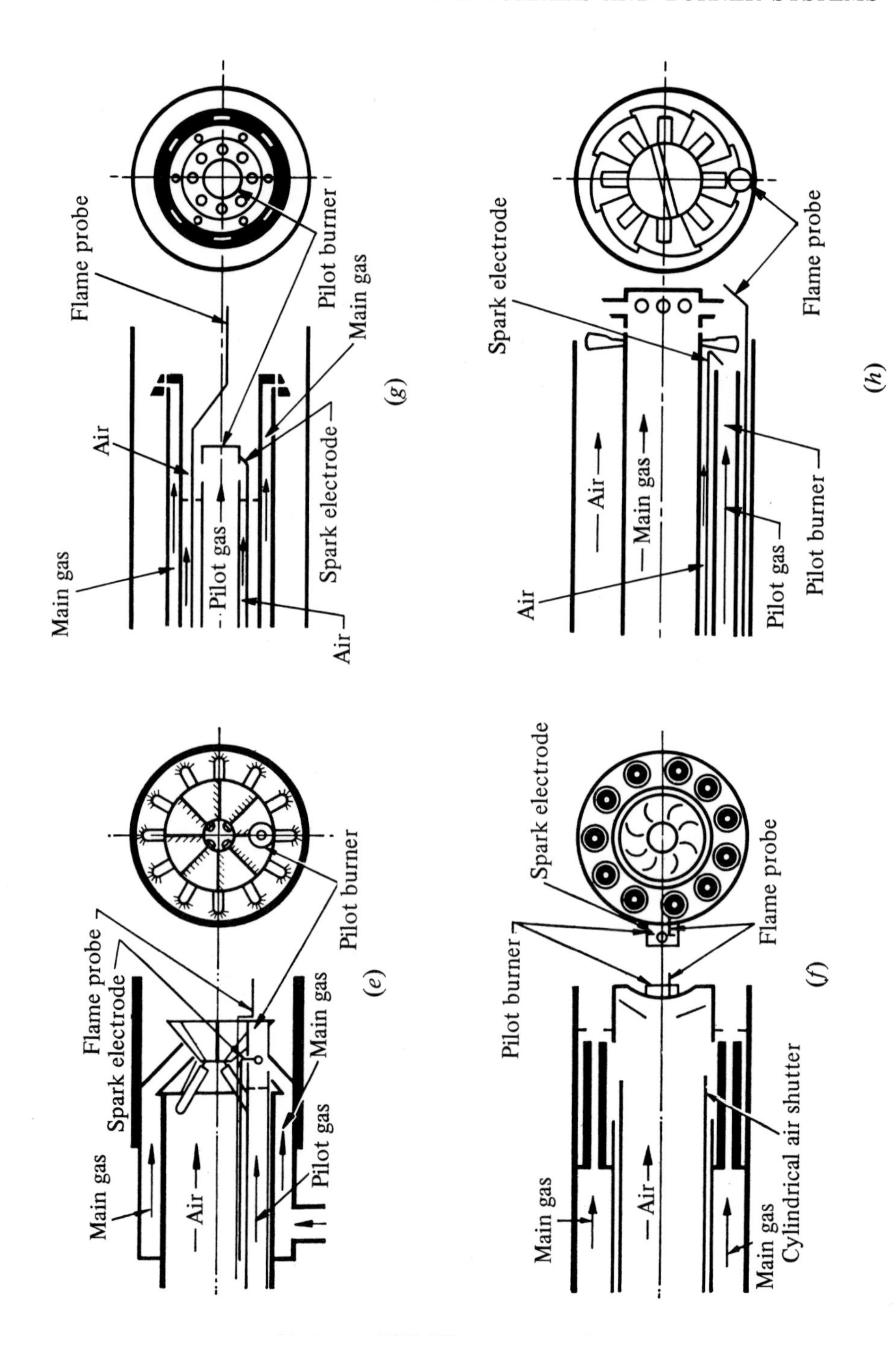

Figure 7:45 *continued*

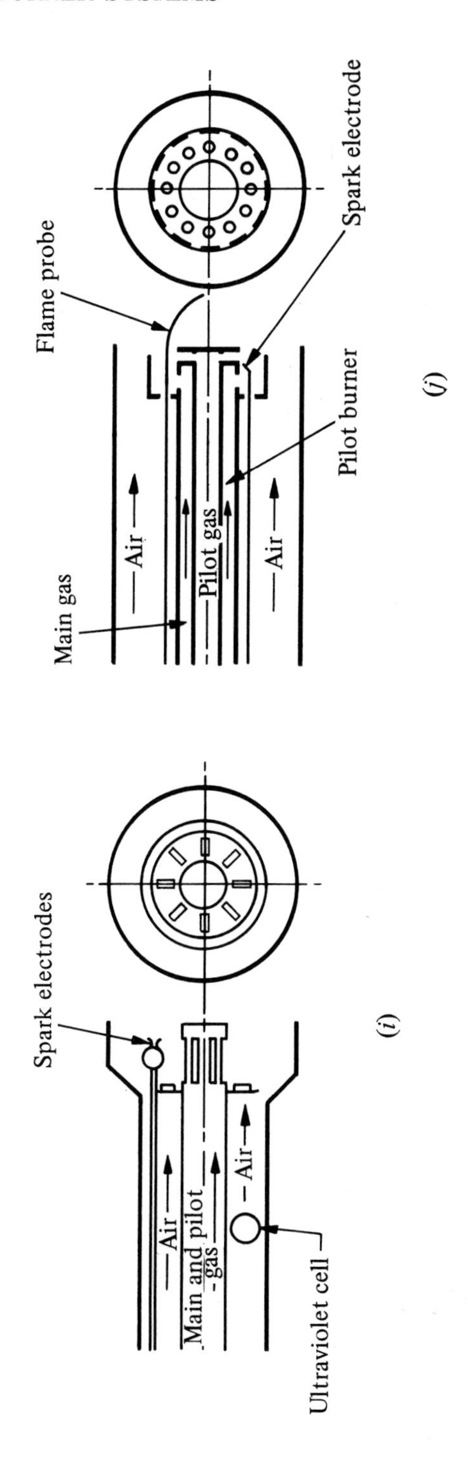

Figure 7:45 *continued*

the gas governor. Butterfly valves are much more sensitive in adjustment than sliding shutters due to their nonlinear opening characteristic and the latter type are preferable for setting the air-flow rate. For high-low-off control, mechanically linked valve arrangements are normally used (Section 7.5.2) and the link mechanism and valve characteristics need to be chosen to prevent serious off-ratio operation occurring on transition from high to low fire and vice versa.

Since packaged burners are of the nozzle-mixing type, slightly more than the stoichiometric air requirement is needed for complete combustion. There is no British Standard requirement for CO/CO_2 ratio in industrial and commercial equipment. The BS 1250 [76] limit for domestic gas appliances is a CO/CO_2 ratio of 0.02, corresponding to about 2400 ppm CO in the flue gases. Between 200 and 500 ppm CO is currently regarded as an acceptable flue-gas concentration.

In [75] the effects of combustion-chamber pressure variations are considered and whilst these have little effect on flame stability and noise the effect on combustion quality may be significant since this alters the air flow through the burner. It is recommended that the pressure conditions under which the burner is initially commissioned are maintained wherever possible by the use of flue breaks or draught stabilizers in the flue.

7.11 DUAL-FUEL BURNERS

7.11.1 Introduction

A substantial proportion of the gas supplied to industry in Britain has been sold on an interruptible basis. This enables the gas industry to match seasonal variations in supply and demand. At times of very high demand such consumers may be requested to cease or reduce gas consumption and operate their plant on an alternative standby fuel. Since the supply profile approximates to the demand profile, interruptions can occur at any time of the year although they are more likely to occur during the winter months. Additionally, interruptible gas allows the gas industry to optimize the operation of its transmission and distribution system. The consumer benefits from the advantageous terms the gas industry is able to offer compared with 'firm' gas. Provision for interruptible operation may be made by the installation of separate burners for the two fuels but is much more commonly made by the use of dual-fuel burners, i.e. burners designed to burn gas or oil (or less commonly gas or coal) separately, the equipment being capable of rapid and efficient switchover to the alternative fuel. Dual-fuel burners have found their widest application in industrial boiler firing and their application to boilers is considered in Chapter 14.

Over the last few years the gas industry has expended considerable effort on studies of the design and performance of dual-fuel systems with particular regard to safety and reliability. *Standards for Automatic Gas Burners, Forced- and Induced-Draught* [73] and the *Interim Codes of Practice for Large Gas and Dual-Fuel Burners* [74] deal primarily with requirements of safety and control and are considered in Chapter 9. Consideration here is restricted to the design of the burner head and the general performance of dual-fuel burners.

Dual-fuel burners may be conveniently divided into two classes: *packaged burners* supplied as factory-assembled units incorporating the fan and controls and *register burners* using separate fans, the burners and controls being site-assembled. The former group find their widest application in shell boiler firing whilst the latter are generally fitted to water tube boilers. In general dual-fuel burners have been developed around existing oil burners. Since the oil atomizer is normally sited on the centreline of the burner assembly, the gas must be introduced in the space between the oil atomizer and the outside of the burner. Similar considerations apply in the much less common case of gas/coal burners in which the pulverized fuel duct lies on the burner axis. A dual-fuel burner may be regarded as consisting of three parts: the combined burner components, the oil burner components (or coal burner components) and the gas burner components.

7.11.2 Combined burner components

The common components for both systems include the air-supply duct which in the case of packaged burners is attached to the burner body, vanes or plates to control the air direction, an air manifold or 'windbox' and a refractory tunnel or quarl. In addition a pilot burner or burners must be provided and carefully located so as to ignite both oil and gas reliably. The pilot burner will be integrated into the control programme of both fuel systems when automatic control is used. Flame detectors also need to be particularly carefully located so that the pilot and main gas and oil (or coal) flames are detected. The fact that two fuel-supply systems are incorporated into one burner housing leads to some degree of mutual interference. The gas system is much more tolerant with regard to mixing and combustion than the alternative system and the effect of introducing a small oil atomizer on the centreline of the gas-supply system presents few problems. Conversely the introduction of a relatively much larger gas supply into an oil burner may interfere with the mixing pattern and the solution has generally been to dispose the gas inlets out towards the periphery of the burner.

7.11.3 Oil-burner components

These may be classified according to the method by which the oil is atomized, the atomizer serving the function of disintegrating the liquid stream and dispersing the resulting droplets in a controlled way into the airstream. A wide variety of atomizing devices exist and a comprehensive treatment is inappropriate in a book of this kind. Coverage will be restricted to a brief description of the types commonly used in dual-fuel burners.

The larger dual-fuel packaged burners have in general developed around existing *rotary-cup* atomizers. Some *medium-pressure air* atomizers have been used and *pressure-jet* atomizers are used in the small sizes. Register burners, which are normally of larger capacity than packaged burners use *steam-blast* atomization, medium-pressure air or pressure-jet systems [77].

Pressure-jet atomizer

Figure 7:45 (*a*) shows a typical design. The oil enters through angled ports into a swirl chamber and leaves the discharge annulus as a conical sheet which breaks up into droplets. Oil pressures are typically 0.35 to 1.4 MPa (3.5 to 14 bar, 50 to 200 lbf in^{-2}) but may be as high as 7.5 MPa (75 bar, 1100 lbf in^{-2}). The oil viscosity for optimum operation is around 15 cSt (84 $\text{in}^2\ \text{h}^{-1}$). Simple pressure-jet systems suffer from a poor turndown range but this is improved in systems incorporating spill-

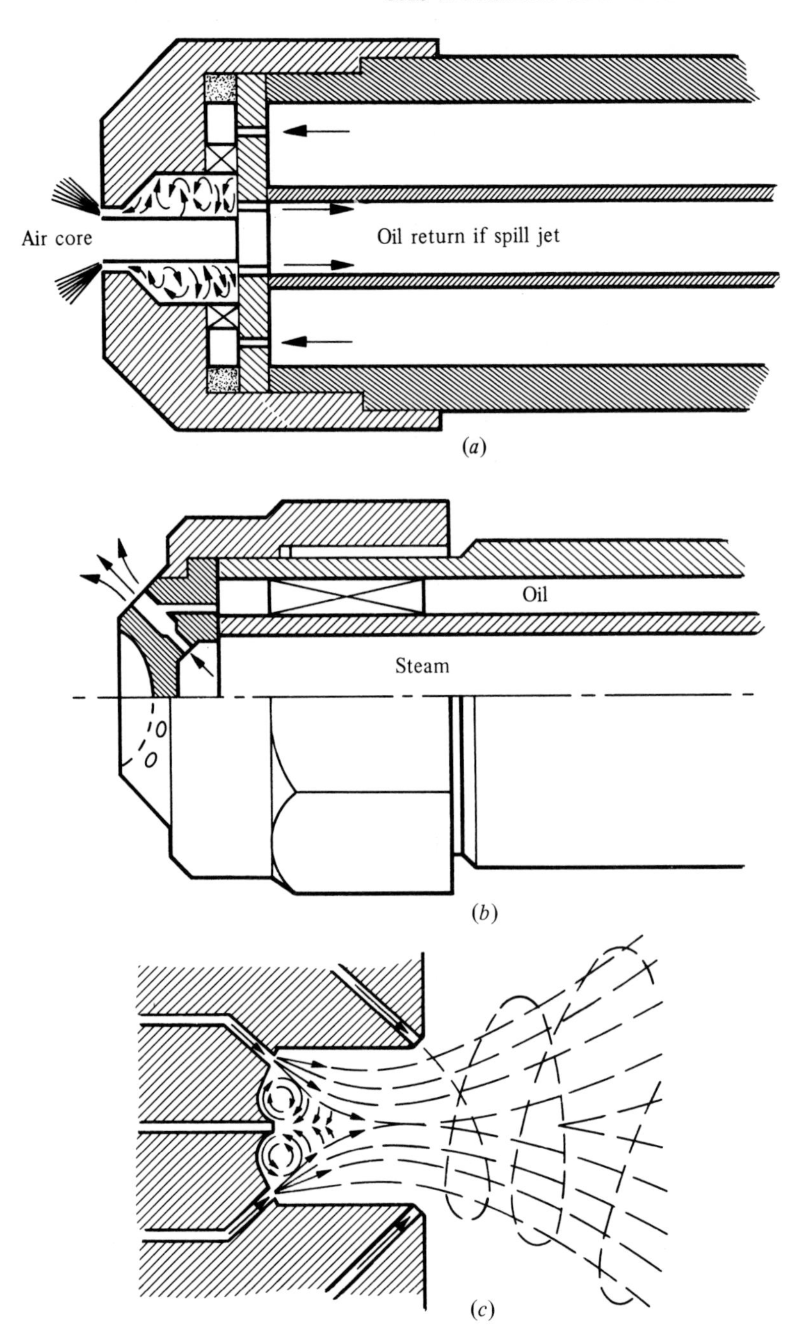

Figure 7:46 Oil-atomizing systems
(*a*) Pressure-jet
(*b*) Steam-blast
(*c*) Medium-pressure air

return in which a constant quantity of oil is fed to the swirl chamber regardless of throughput, maintaining rotation at a high level and thereby holding the film thickness at the final orifice at a minimum level. On turndown the surplus oil is returned to the pump suction. Although the droplet size is small on turndown the spray angle increases and this may cause problems. The major problem however is the return of hot oil (on heavy-oil units) to the pump suction which presents maintenance problems.

Steam-blast atomizers

Some of the earliest atomizers used were steam-blast types in which steam was directed onto the oil jet which it broke up by impact. These rapidly went out of favour with the introduction of the pressure-jet system since the steam consumption was often an appreciable proportion of the boiler output. The design has been greatly improved in recent years and very low steam consumptions (0.2–0.5 per cent output) are now possible. Figure 7:46 (*b*) shows a modern steam-blast atomizer in which the oil is sheared by the steam blowing across an oil film. The turndown ratio of steam atomizers can be very high (6 : 1 to 15 : 1) and the installation is usually cheap since large atomizing air fans or compressors are not required. However the absence of steam on lighting-up complicates matters. This may be overcome by the provision of an air compressor, by the use of a small auxiliary kerosene-fired steam boiler or by the use of a small pressure-jet auxiliary burner.

Medium- and low-pressure air atomizers

This is perhaps the most versatile design of atomizer. A wide range of spray angles is possible with a good turndown ratio (5 : 1 to 10 : 1). The main disadvantage is the cost of the compressed air, most proprietary designs using air at 70–100 kPa (0.7–1.0 bar, 10–15 lbf in^{-2}) comprising 1 to 2 per cent of the stoichiometric air requirement. This is satisfactory for relatively small burners but may become expensive in capital and running costs for large units. Higher pressures are in general uneconomical. For shell boilers medium-pressure air burners are technically attractive, particularly for the smaller sizes, since they are capable of achieving high heat-release rates with minimum stack-solids emission and the turndown ratio is good. Figure 7:46 (*c*) shows a typical design.

Low-pressure air atomizers use air at around 6 kPa (60 mbar, 25 inH_2O) from a single-stage fan for atomization, the atomizing air constituting 15–25 per cent of the stoichiometric air requirement. This type is not widely used on boiler plant.

Rotary-cup atomizers

In these designs the oil is spread onto the surface of a cup rotating at 3500 to 6000 rpm via a hollow shaft. This ensures an even oil distribution and also initiates atomization. Air at low pressure (4–5 kPa, 40–50 mbar, 15–20 inH_2O) from an integral fan atomizes the sheet. Approximately 10 to 20 per cent of the stoichiometric air requirement is provided as atomizing air. The cup may be rotated either by a small air turbine or directly motor driven. Rotary-cup burners have advantages of low cost, relative insensitivity to oil viscosity and good combustion performance. The main disadvantage is that the cup has to be cleaned regularly which makes uninterrupted running difficult. Figure 7.47 (*b*) shows a typical design.

For detailed consideration of the principles of atomization and their application in practical burners the reader is referred to references [78–80].

Stabilization of oil flames
A variety of techniques are employed, all directed towards obtaining a region of low velocity and some degree of reverse flow. The simplest arrangement is a bluff-body baffle (Section 4.8.1) mounted just upstream of the atomizer head. If the baffle is sufficiently large, adequate recirculation is created to provide flame stability. Flames tend to be long and narrow, however, and fuel is drawn back onto the face of the baffle possibly leading to overheating and carbon deposition. This may be overcome by perforating the baffle but if carried to excess the reversal zone may be destroyed.

Axial swirlers are sometimes employed in which the air passes through a baffle incorporating swirl causing a reversal zone (Section 4.8.1) centralized on the axis of the burner. The diverging air flow from the swirler also diverges the air flowing around the swirler and increases the diameter of the recirculation zone to greater than the baffle diameter causing the flame to become wider. Radial swirlers incorporate no baffle as such but direct the air tangentially towards the burner axis some distance upstream of the oil nozzle. The final exit path of the air depends on the angle and number of swirl vanes and the air velocity. This provides a method of controlling the flame shape but has the disadvantage on large sizes that the reverse-flow region tends to be wide, extending upstream towards the swirl vanes.

In the case of rotary-cup burners the oil leaves in the form of a hollow cone and reversal may take place within the cone to provide stabilization.

7.11.4 Gas-burner components

Gas-burner systems used in dual-fuel burners are invariably of the nozzle-mixing type. Unlike normal nozzle-mixing gas burners in which a number of options are open with regard to the introduction of gas and air, in dual-fuel applications there are considerable limitations. As mentioned above, the oil atomizer invariably lies on the centreline and is therefore normally in the centre of the stabilizing zone. The gas can be introduced in the centre using an interchangeable gun or an annular supply but this is not possible, for example, with a rotary-cup burner and in other designs centre gas feed becomes more difficult as the burner capacity increases [81]. Thus, as previously noted, the gas is commonly introduced by means of ring manifolds or via 'spuds' in the space between the atomiser and the outer periphery of the register. In this position the air flow will probably have a poor stabilizing effect and it becomes necessary to introduce local stabilization means or to use low efflux velocities and to ensure good mixing by careful positioning of the gas jets [81].

Figure 7:47 shows a number of proprietary dual-fuel burner designs illustrating a variety of methods of introducing the gas.

The burner shown in Figure 7:47 (*a*) is essentially a tunnel-mixing gas burner with the addition of a low-pressure air atomizer in the centre. It is a 'combination burner' in that it can burn gas or oil separately or simultaneously. It is restricted in use to light distillate fuel oils up to 35 seconds Redwood I at 100 °F (37.8 °C). Burners of this

---

Figure 7:47 Typical dual-fuel burners
(*a*) System using an axial annular air nozzle (Stordy-Hauck NMC series)
(*b*) System using multiple axial air orifices (Saacke)
(*c*) System using multiple gas orifices and radial swirl (Peabody)
(*d*) System using multiple gas nozzles and axial swirl (Babcock and Wilcox)
(*e*) Gas/coal dual-fuel burner

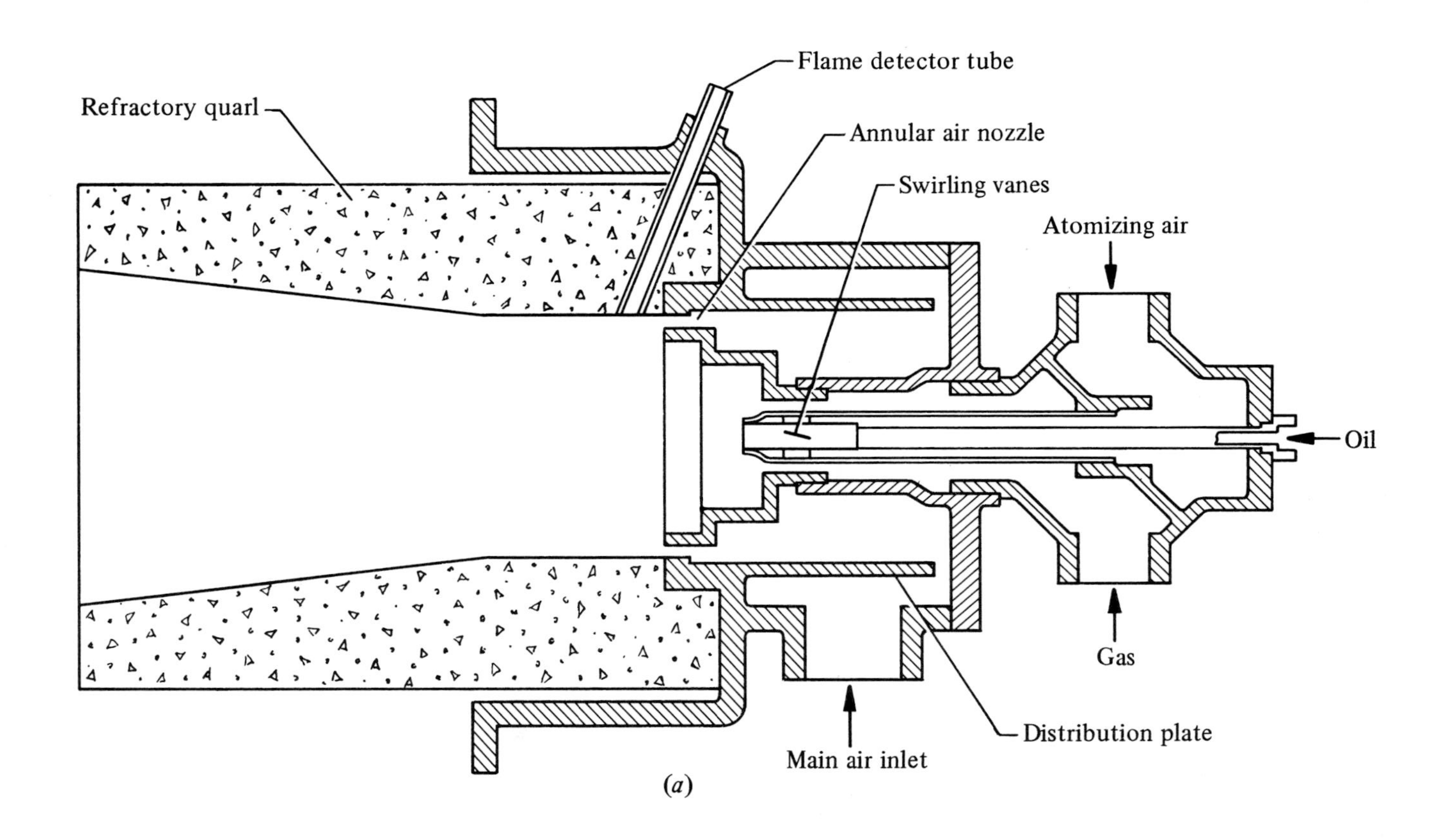
Refractory quarl
Flame detector tube
Annular air nozzle
Swirling vanes
Atomizing air
Oil
Gas
Distribution plate
Main air inlet
(a)

Figure 7:47

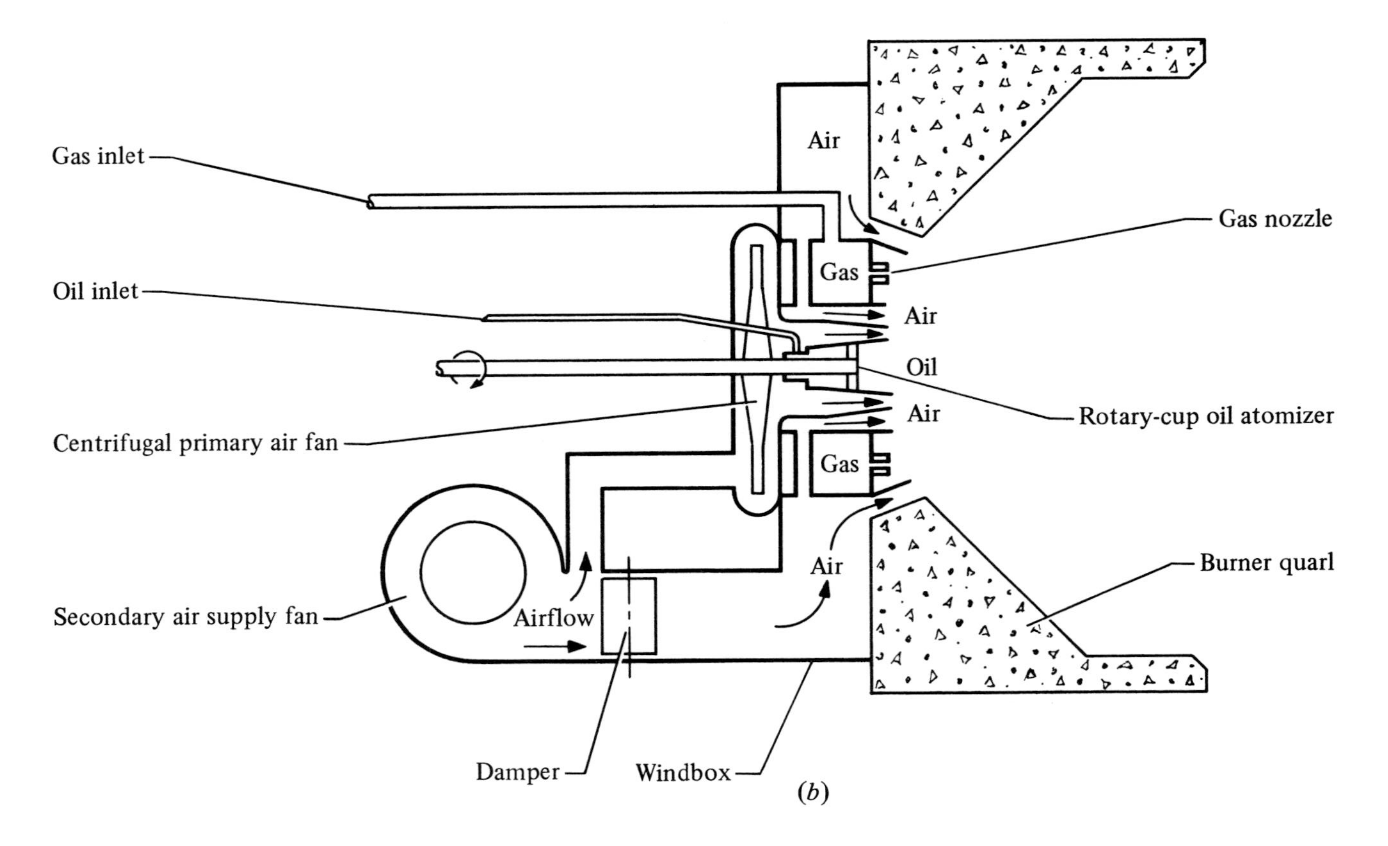
Gas inlet
Oil inlet
Centrifugal primary air fan
Secondary air supply fan
Air
Gas
Air
Oil
Air
Gas
Air
Airflow
Gas nozzle
Rotary-cup oil atomizer
Burner quarl
Damper
Windbox
(b)

Figure 7:47

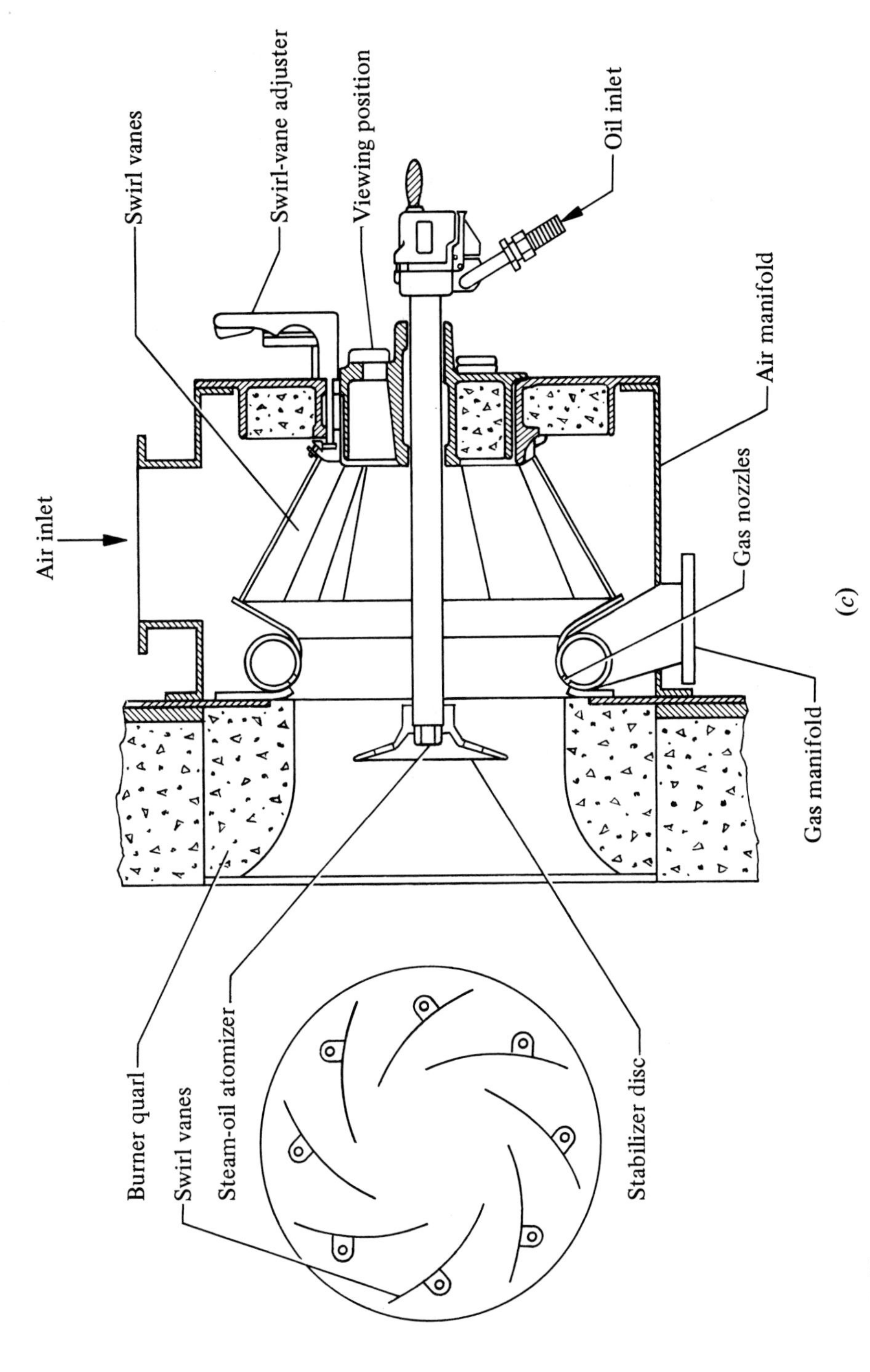
Air inlet
Swirl vanes
Swirl-vane adjuster
Viewing position
Oil inlet
Air manifold
Gas nozzles
Gas manifold
Burner quarl
Swirl vanes
Steam-oil atomizer
Stabilizer disc

(*c*)

Figure 7:47

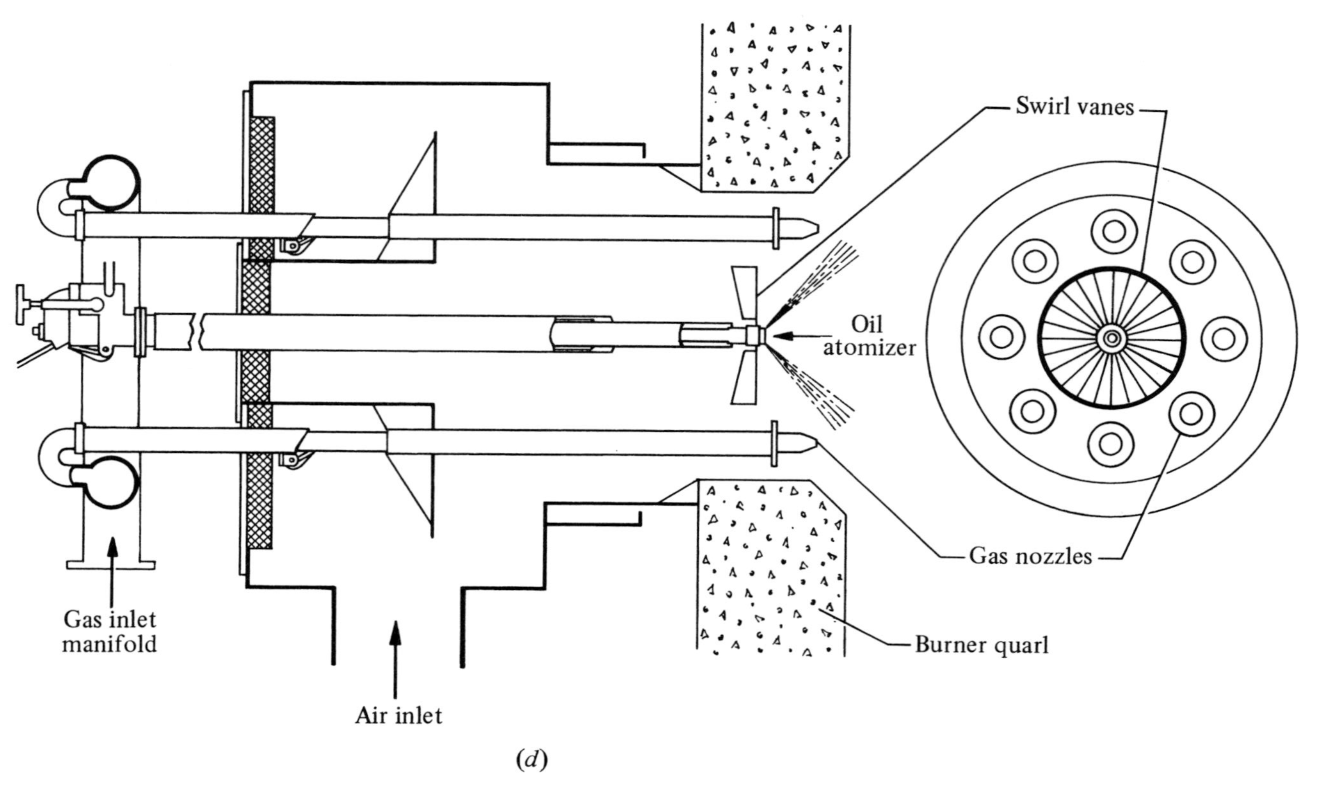
Swirl vanes
Oil
atomizer
Gas nozzles
Burner quarl
Gas inlet
manifold
Air inlet
(d)

Figure 7:47

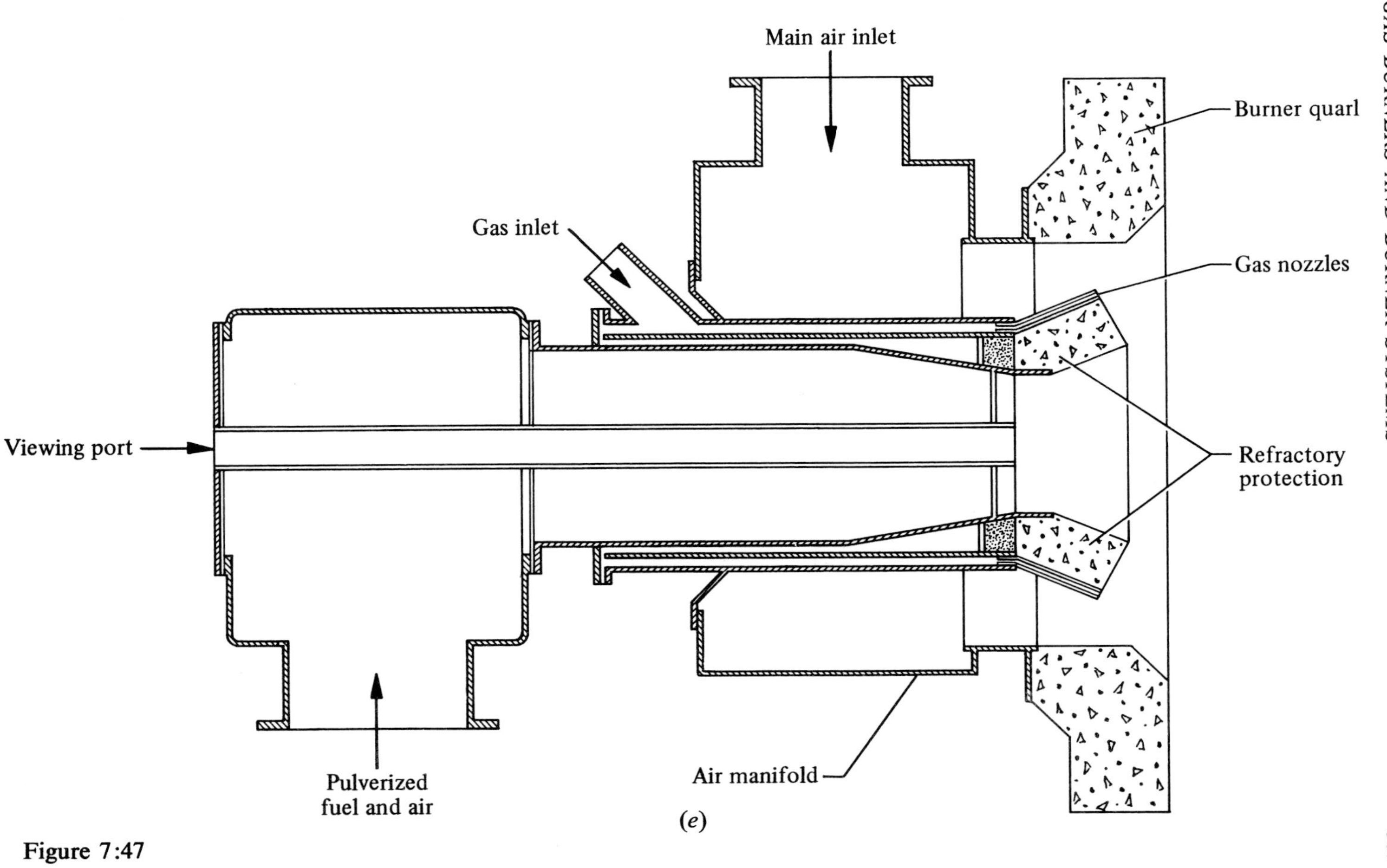
Main air inlet
Burner quarl
Gas inlet
Gas nozzles
Viewing port
Refractory protection
Pulverized fuel and air
Air manifold

(*e*)

Figure 7:47

type find their widest application in furnace and kiln firing. No swirl is employed, stabilization being provided by the tunnel.

In the design shown in Figure 7:47 (*b*), which uses a rotary-cup oil atomizer, the gas is introduced via a ring of gas nozzles whilst the annular airstream is also divided into a number of separate streams to promote rapid mixing.

In designs incorporating airstream swirl the gas nozzles are sited so that gas is drawn into the recirculating airstream. In Figure 7:47 (*c*) a combination of radial swirl and baffle stabilization is employed in conjunction with a steam-blast atomizer. In Figure 7:47 (*d*) a number of separate gas nozzles with individual baffles surround an oil atomizer with axial swirl. In arrangements like this incorporating widely spaced nozzles, particular care needs to be taken to ensure that smooth cross-ignition is achieved at all operational flow rates [82].

Figure 7:47 (*e*) shows a gas/coal dual-fuel burner.

7.11.5 Dual-fuel burner performance

Hoggarth *et al.* [77] and Pomfret [83] have reported the results of tests carried out on six proprietary burners of a variety of designs covering a range of ratings from 3.5 MW (12×10^6 Btu h^{-1}) to 29 MW (100×10^6 Btu h^{-1}) with regard to flame establishment, stability and off-ratio operation, flame shape and size, combustion quality and minimum excess-air requirement. In general, flame establishment was found to be satisfactory although in the case of register burners using widely spaced gas spuds the location of the pilot flame was critical.

Good air and gas distribution were shown to be extremely important. Tests in a shell boiler have shown that at the end of the firetube where combustion should have been complete considerable stratification may occur if the gas and air distribution at

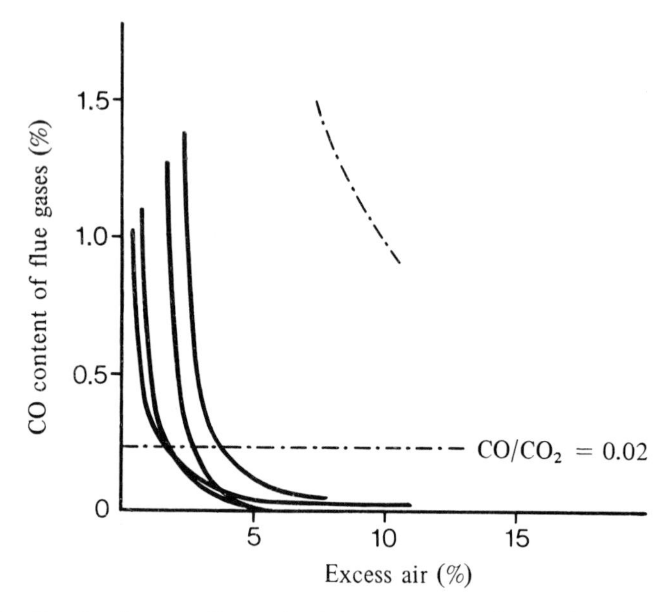

Figure 7:48 Combustion quality and minimum excess air required for gas firing [83]

the burner is poor. For example 4–5 per cent oxygen in one quadrant with 0.5–1.0 per cent carbon monoxide in the diametrically opposite quadrant have been found at the end of a circular fire tube.

Figure 7:48 shows the relationship between the excess air supplied and the carbon monoxide level in the combustion products on gas firing for the six burners tested.

There is a characteristic sharp increase in carbon monoxide as the excess air is reduced below a certain value and this takes place over a small change in excess air (3–5 per cent).

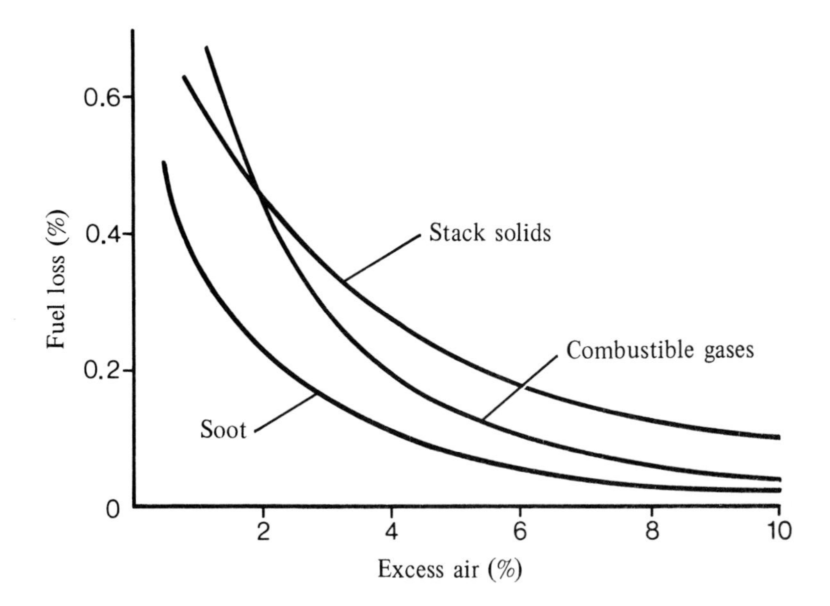

Figure 7:49 Combustion quality and minimum excess-air requirement on oil firing (register burners, water tube boilers) [83]

Similar considerations apply in the case of oil firing in which the performance is normally indicated in terms of stack solids and combustible gases as shown in Figure 7:49. For large water-tube boilers it is possible to obtain less than 0.2 per cent solids emission at 1.5 to 2.5 per cent excess air but for most burners used in industrial water-tube boilers a more realistic range would be 2.5 to 10 per cent excess air.

Clearly, in addition to the burner-head design the accuracy of the air–gas ratio control system exerts an important influence on the selection of suitable excess-air levels.

Operating excess-air levels for a variety of burner-head/air–gas ratio control combinations for gas firing have been proposed by Hoggarth *et al.* [83] ranging from 3.5–8.5 per cent for register burners with closed-loop control to 10–18 per cent for packaged burners with linked valves. On oil firing with good designs of register burners it is possible to operate at the same excess-air levels as gas but in most packaged systems higher excess levels are required (15–30 per cent at maximum load). The difference in excess-air requirement on oil and gas of this type of burner presents difficulties in dual-fuel operation. The excess-air level may be set for gas and the burner downrated on oil firing, the excess air may be set for oil in which case the performance would be degraded for most of the time or, as a more satisfactory long-term solution, the ratio-control linkages may be designed to set the appropriate excess-air level for each fuel automatically.

7.12 GAS–OXYGEN BURNERS

High heat-transfer rates may be obtained using high-velocity jets of combustion products at flame temperature as previously discussed. Using preheated air allows higher flame temperatures and therefore correspondingly greater heat-transfer rates to be attained. If oxygen is used in place of air even higher flame temperatures and heat-transfer rates may be achieved. The high cost of oxygen in the past limited the application of oxy-gas burners to small-scale specialized applications (for example, flame cutting) but improved oxygen production methods and the development of oxygen-based steel-making processes have meant that relatively inexpensive oxygen is available in large quantities in the steel industry. Much of the development work on large-scale oxy-gas burners has been associated with steel-making processes (see Section 13.3.2). However, oxy-gas burners are applicable to other industries in which high temperatures and heat-transfer rates are beneficial in process heating.

7.12.1 Properties of oxy-gas flames

Two main factors which affect heat transfer from flames to solid objects are:

1 The flame temperature, since this determines the temperature difference
2 The amount of dissociation of the flame-gas molecules into atoms and radicals

The maximum theoretical flame temperature (assuming no preheat) is around 2700 °C for town gas or natural-gas–oxygen flames compared with about 2000 °C for gas–air flames (see Section 4.2.6). However, heat-transfer rates are higher than may be accounted for by this difference in temperature 'head'. At temperatures above 1400 °C combustion gases are in an increasingly dissociated state. Thus when heat is transferred to colder surfaces, chemical energy released by the recombination of the dissociated species augments the normal energy transfer of the stable molecules [84]. The main difference between air–gas and oxy-gas flames lies in the different degrees of dissociation of the combustion products, the former containing only a small proportion whilst about 25 per cent of the products of a stoichiometric oxy-gas flame are dissociated [85]. Figure 7:50 illustrates the difference in heat-transfer rates taking into account the effects of flame temperature, dissociation and difference of mass flow rate.

The results of the diffusion and recombination of species may be regarded as an increase in the effective thermal conductivity of the combustion products which may be several times greater than the non-reacting or 'frozen' conductivity. Assuming that $\lambda_e/\lambda_f \approx C_{pe}/C_{pf}$ where λ_e and λ_f are the effective and frozen thermal conductivities and C_{pe} and C_{pf} are the effective and frozen specific heat capacities, effective conductivities may be obtained which can then be used in heat-transfer calculations. Chen and McGrath [84] have summarized the literature on convection in reacting systems and the reader is referred to their paper. Included in reference [87] are calculated values of effective thermal conductivities, specific heats and Prandtl numbers relating to high-temperature natural-gas combustion products.

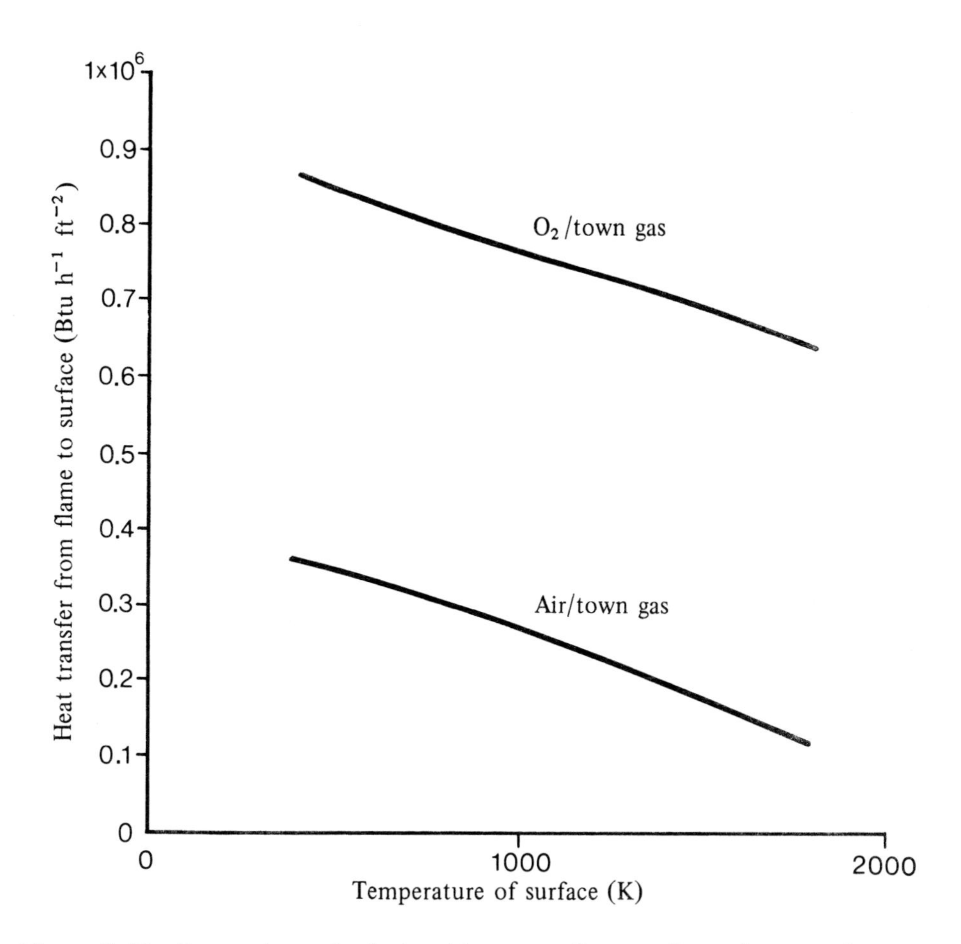

Figure 7:50 Comparison of calculated heat-transfer rates from air–gas and oxygen–gas flames [86]

7.12.2 Types of gas–oxygen burners

Although premixing the fuel gas and oxygen provides the most effective form of mixing this technique is mainly restricted to individual burner ratings below about 30 kW (100 000 Btu h^{-1}) used for welding, cutting or glassworking. Since the burning velocity is very high (typically four or five times as great as gas–air flames) lightback is the limiting consideration.

Small-scale oxy-gas burners are described in Section 15.4.

Larger oxy-gas burners may be designated face-mixing or post-mixing [88]. In the former case simple concentric tubes carrying gas and oxygen may be used to produce

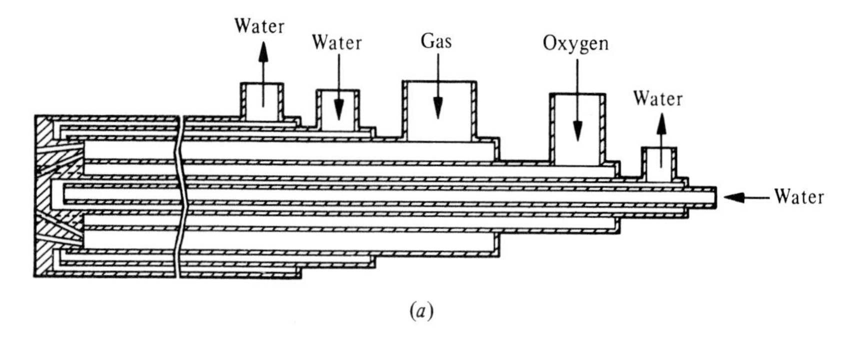

(a)

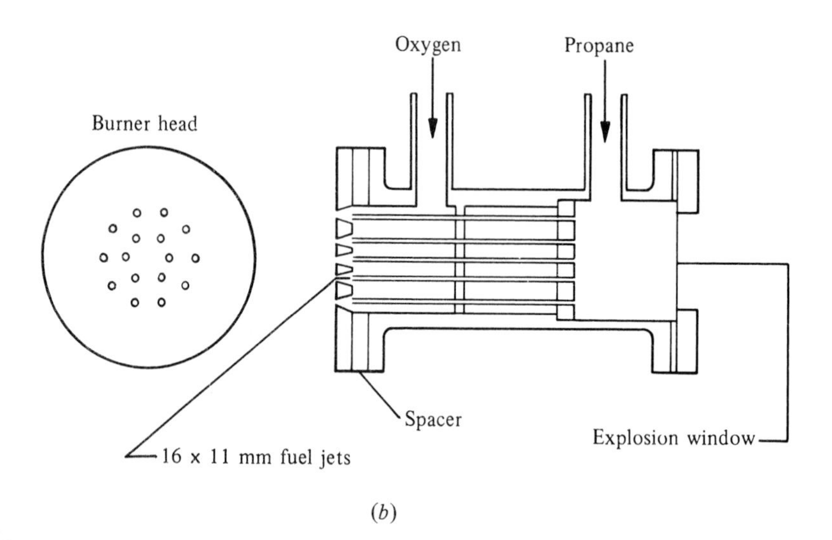

(b)

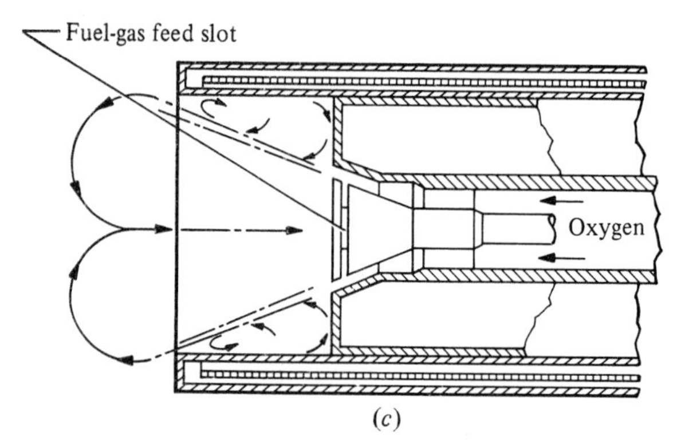

(c)

Figure 7:51 Oxygen–gas burners
(a) Post-mixing oxygen–gas burner, MRS design [85]
(b) Concentric-tube face-mixing burner [88]
(c) Shell toroidal burner [90]

a long small-diameter flame or an array of smaller concentric tubes may be provided to produce, in effect, a shorter wider flame. The inner concentric tubes may be recessed to improve mixing characteristics. In post-mixing burners, angled fuel and oxygen jets are arranged to impinge downstream of the burner head. The jet angles give considerable scope for varying the flame shape.

With all designs the burner must bring the gas and oxygen streams into contact with as much interfacial area as possible for stability and low noise levels.

Figure 7:51 (*a*) shows a post-mixing water-cooled burner designed by Midlands Research Station with a rating of 1.3 MW (4.5×10^6 Btu h^{-1}) for steel melting trials.

Baxendale *et al.* [88] have compared the performance of face-mixing concentric-jet burners with post-mixing designs on propane–oxygen over a wide range of equivalence ratios. In general the concentric-jet burner produced much higher combustion intensities (50–800 MW m^{-3}) than the impinging jet types due probably to the recessed central jets producing partial premixing.

The most recent designs of face-mixing burners incorporate a conical divergence at the end of the oxygen tube to encourage reverse flow of gas into the root of the oxygen tube and provide satisfactory stabilization on natural gas [28].

In the Shell toroidal burner [89], oxygen is supplied at a sufficiently high pressure to reach sonic velocity as it leaves the annulus surrounding a conical flame stabilizer and this provides intense combustion. The flame stabilizer, burner head and outer shroud are all water-cooled (Figure 7:51 (*c*)).

REFERENCES

1 Priestley, J. J., *Industrial Gas Heating (Design and Application)*, London, Ernest Benn (1973)

2 Lloyd, P., *Theory of Industrial Gas Heating*, London, Walter King (1938)

3 Garside, J. E., *Inst. Gas Eng. Commun.*, 702 (1965); *J. Inst. Gas Eng.*, **6** (1966), 357

4 Edmondson, H. and Garside, J. E., Tenth Symposium (International) on Combustion, Pittsburgh, Pa., The Combustion Institute (1965)

5 Edmondson, H., *J. Inst. Fuel*, **37** (1964), 506

6 van der Linden, A., *Inst. Gas Eng. Commun.*, 729; *J. Inst. Gas Eng.*, **7** (1967), 195

7 Culshaw, G. W. and Prigg, J. A., *J. Inst. Gas Eng.*, **10** (1970), 303

8 Dutton, B. C., Harris, J. A. and Kavarana, B. J., *Gas Council Res. Commun.* GC179 (1970); *J. Inst. Gas Eng.*, **11** (1971), 476

9 Desty, D. H. and Whitehead, D. M., *New Scientist* (January 1970)

10 British Patent 1205432

11 Radcliffe, S. W., BGC *Watson House Bulletin*, **37**, No. 271 (July–August 1973), 173

12 'A landmark in gas burners', *Gas World* (14 March 1970)

13 Gas Council, Industrial Gas Development Committee Report 344/48, U 0514

14 Simmonds, W. A. and Willett, A. L., *Gas Research Board Commun.* GRB65 (1951); *Trans. Inst. Gas Eng.*, **101** (1951–2), 517

15 *American Gas Assoc. Res. Bull.*, No. 62

16 Laligand, A., 'Examples of new burners developed in France', Session 1, Paper 6, Proc. 2nd Conf. on Natural Gas Research and Technology, AGA/IGT, Atlanta (1972)

17 Silver, R. S., *Gas Research Board Publication* GRB39/23 (1948)
18 von Elbe, G. and Grumer, J., *Ind. Eng. Chem.*, 40 (1948)
19 Prigg, J. A., *British Junior Gas Assoc. Jt. Proc.*, **38** (1954–5), 165
20 Simmonds, W. A., *Gas Council Res. Commun.* GC20 (1954); *Trans. Inst. Gas Eng.*, **104** (1954–5), 557
21 Francis, W. E., *Gas Council Res. Commun.* GC101 (1964); *J. Inst. Gas Eng.*, **4** (1964), 399
22 Goodwin, C. J., Hoggarth, M. L. and Reay, D., *Gas Council Res. Commun.* GC165 (1969); *J. Inst. Gas Eng.*, **10** (1970), 324
23 Shnidman, L., *Gaseous Fuels*, New York, AGA (1954)
24 Sutherland, A., *Gas Council Res. Commun.* GC161 (1969); *J. Inst. Gas Eng.*, **10** (1970), 445
25 Dance, E. W. G. and Sutherland, A., I.G.U./E6–67, 10th Int. Gas Conference (1967)
26 van der Linden, A., I.G.U./34–61, 8th Int. Gas Conference (1961)
27 Harris, J. and Prigg, J. A., *Gas Council Res. Commun.* GC115 (1964); *J. Inst. Gas Eng.*, **5** (1965), 203
28 Downie, J. M. and Hoggarth, M. L., *J. Inst. Fuel*, **47** (1974), 124
29 Britten-Austin, R. H., 'Stability and combustion performance of a low pressure high aeration natural gas burner', MSc. Dissertation, University of Salford (1973)
30 Campbell, W. D. and Slattery, J. C., *J. Basic Eng.; Trans. American Soc. Mech. Eng.*, Ser. D, **85** (1963)
31 Townsend, C., BGC Watson House Report WH/TN/R&D/72/219 (1972)
32 Vignes, R., Communication to the Conference of the A.T.I.G.F., Paris (1971)
33 Melmoux, J., Association Technique du Gaz en France, Congress (1958)
34 Griffiths, J. C. and Weber, E. J., *American Gas Assoc. Res. Bull.*, **77** (1958)
35 Read, S. B., *Combust. Flame*, **13** (1969), 583
36 Hess, K., 9th Int. Gas Congress, The Hague, IGU/E1–64 (1964)
37 Roper, F. G., BGC Watson House Report R & D/72/59 (1972)
38 Roper, F. G., BGC Watson House Report R & D/72/180 (1972)
39 Roper, F. G., BGC Watson House Report R & D/73/90 (1973)
40 Reay, D., *J. Inst. Fuel*, **45** (1972), 462
41 Francis, W. E., *Gas Council Res. Commun.* GC44 (1956), *Trans. Inst. Gas Eng.*, **107,** (1957), 555
42 Gas Council I.G.D.C. Report 644/57,U 951 (1957)
43 Gas Council, *Industrial Gas Controls Handbook* (Report No. 763/69, U9) (1969)
44 Clark, M. E., 'Air–gas ratio control for air blast tunnel burners', University of Salford, Hons. BSc. in Gas Engineering, Project (1969)
45 Gas Council I.G.D.C. Report 729/60, U 0515 (1960)
46 Francis, W. E., *Gas Council Res. Commun.* GC48 (1958); *Trans. Inst. Gas Eng.*, 108 (1958–9), 414
47 Hoggarth, M. L., Proceedings of First Symposium on the Industrial Uses of Town Gas, Paper 1, Institute of Fuel (1963)
48 Francis, W. E. and Hoggarth, M. L., *Gas Council Res. Commun.* 68; *J. Inst. Gas Eng.*, **1** (1961), 320
49 Cox, R. W., Gas Council MRS Ext. Rept. ER 100 (1967)
50 Cox, R. W., Gas Council MRS Ext. Rept. ER 121 (1968)

51 Kilham, J. K., Jackson, E. G. and Smith, T. J. B., *Gas Council Res. Commun.* GC74 (1960); *J. Inst. Gas Eng.*, **1** (1961), 251
52 Cox, R. W., Hoggarth, M. L. and Reay, D., *J. Inst. Fuel*, **40** (1967), 498
53 Hoggarth, M. L. and Aris, P. F., Paper 2, Industrial Gas Conference, Cranfield (1966)
54 Hancock, R. A., *Gas Council Res. Commun.* GC110; *J. Inst. Gas Eng.*, **5** (1965), 470
55 Bryan, D. J., Masters, J. and Webb, R. J., *Inst. Gas Eng. Commun.*, 952 (1974)
56 Harrison, W. P., Oeppen, B. and Sourbutts, S., *Gas Council Res. Commun.* GC164; *J. Inst. Gas Eng.*, **10** (1970), 538
57 Davies, R. M., *J. Inst. Fuel*, **44** (1971), 453
58 Kilham, J. K. and Lanigan, E. P., *Gas Council Res. Commun.* GC167 (1969); *J. Inst. Gas Eng.*, **10** (1970), 700
59 Bone, W. A. and McCourt, C. D., *J. Franklin Inst.*, **101** (1912)
60 Coles, K. F. and Bagge, L. P., *Inst. Gas Eng. Commun.*, 839 (1970); *J. Inst. Gas Eng.*, **11** (1971), 387
61 Radcliffe, S. W., *BGC, Watson House Bulletin*, **37** (1973), 173
62 Coles, K. F. and Wilbraham, K. J., *Inst. Gas Eng. Commun.*, 953 (1974)
63 Francis, W. E., Hoggarth, M. L. and Reay, D., *Gas Council Res. Commun.* GC91 (1962); *J. Inst. Gas Eng.*, **3** (1963), 301
64 Briffa, F. E. J., Staddon, P. W., Phillips, R. N. and Romaine, D. R., *Inst. Gas Eng. Commun.*, 860 (1971)
65 Hanby, V. I., *J. Inst. Fuel*, **44** (1971), 595
66 Reynst, F. H., *Pulsating Combustion*, Oxford, Pergamon Press (1961)
67 Brown, D. J. (ed.) *Proceedings of the First International Symposium on Pulsating Combustion*, University of Sheffield (1971)
68 Chatwin, G. R., 'Gas fired radiant tubes', Gas Council Cranfield Conference (1964)
69 The Gas Council, *Gas Fired Radiant Tubes*, reprinted from *Flambeau*, 6, No. 2 (September 1968)
70 Bridgens, J. L. and Francis, W. E., 'The development of a single ended recirculating radiant tube', Institute of Fuel and Institution of Gas Engineers Joint Conference on Industrial Uses of Town Gas (1963)
71 Chatwin, G. R., Francis, W. E., Harrison, W. P. and Lawrence, M. N., *Gas Council Res. Commun.* GC124 (1966), *J. Inst. Gas Eng.*, **6** (1966), 207
72 Gas Council *Standards for Automatic Gas Burners, Forced and Induced Draught* (Report No. 762/66, U051, U9) (1966)
73 Gas Council, Report No. 765/70, U0515, U9, U96, second edition (1970)
74 Gas Council, *Interim Codes of Practice for Large Gas and Dual-Fuel Burners* (Report Nos. 764/70, 766/70, 767/71, 768/71) (1970 and 1971)
75 Hoggarth, M. L., Jones, D. A. and Pomfret, K. F., *Gas Council Res. Commun.* GC182 (1971)
76 British Standards Institution, *British Standard 1250: Part 1: 1966, Specification for Domestic Appliances Burning Town Gas: General Requirements* (1966)
77 Hoggarth, M. L., Pomfret, K. F. and Spittle, P., *Gas Council Res. Commun.* GC196 (1972); *J. Inst. Gas Eng.*, **12** (1972), 276
78 Hedley, A. B., Nuruzzaman, A. S. M. and Martin, G. F. *J. Inst. Fuel*, **44** (1971), 38
79 Eisenklam, P., *J. Inst. Fuel*, **34** (1961), 136

80 Pope, J. G., *J. Inst. Fuel*, **34** (1961), 290
81 Fitzsimons, W. A., 'Changeover of industrial boilers to natural-gas firing', Institute of Fuel Conference, University of Aston (March 1972)
82 Hoggarth, M. L., 'Dual-fuel burners for boiler plant', Institution of Heating and Ventilating Engineers Symposium (1969)
83 Pomfret, K. F., B.G.C. MRS Ext. Rept. 195 (1972)
84 Chen, D. C. and McGrath, I. A., *J. Inst. Fuel*, **42** (1969), 12
85 Arnold, G. D., *J. Inst. Fuel*, **40** (1967), 117
86 Arnold, G. D., MRS Pub. No. 86 (1966)
87 Davies, R. M. and Toth, H. E., 'Equilibrium compositions, heat contents and physical properties of natural-gas–air combustion products in SI units', British Gas Corporation Midlands Research Station Internal Report N187 (1972)
88 Baxendale, D. N., Horne, W. and Williams, A., *J. Inst. Fuel*, **47** (1974), 139
89 Bagge, L. P. and Shipmand, R. H., *J. Inst. Fuel*, **45** (1972), 391
90 Lashley, H. and Cowan, H., 'Development of oxy-fuel burners in electric arc steelmaking', Combustion Engineering Association, Scotland (September 1964)

8

Measurement

8.1 INTRODUCTION

This chapter is divided into four main sections which cover the measurement of gas quality, temperature and heat flux, pressure and flow.

An industrial gas engineer must be conversant with the applications of the above measurement parameters to distributed gas, gas and air supplies to equipment on consumers' premises, products of combustion in industrial equipment, such as furnaces, process heaters, etc., and in space-heating applications. He must also be capable of measuring comfort conditions in space heating including psychrometry measurements which are also important in many industries, see Chapter 3. The range of temperature, pressure and flow measurement encountered by the industrial engineer is also very wide. In view of this, the final section of the chapter deals specifically with the metering of large industrial and commercial loads.

8.2 MEASUREMENT OF GAS QUALITY

The criteria of quality to be covered are:

1 Calorific value
2 Relative density
3 Wobbe number
4 Component analysis

8.2.1 Calorific value

This has been a very important criterion since the 1920 Gas Regulation Act [1]. Under the provisions of this Act, Gas Referees were appointed to supervise the official testing procedures. General instructions were published, governing the use of approved calorimeters and associated equipment.

These instructions or directions have been amended on a number of occasions to take account of changing situations and equipment. Apart from the 1972 Gas Act the most important recent one is *The Method of Testing the Calorific Value of Natural Gas (37.26–44.71 MJ m^{-3}) (1000–1200 Btu ft^{-3})*, 1966, with the amendments reported in reference [2].

The approved types of calorimeter are:

1 Standard Boys Non-Recording Calorimeter
2 Fairweather Recording Calorimeter Mk II and Mk IV
3 Cambridge-Thomas Recording Calorimeter Mk I and Mk II

All these types are in current use. Each has faults which have been highlighted by use on natural gas [2]. Nevertheless the instruments remain basically dependable for use in testing. A description of all these calorimeters is given in reference [1], but for completeness here, a brief description will be given of the Thomas Calorimeter (see Figure 8:1), as installed at Bacton.

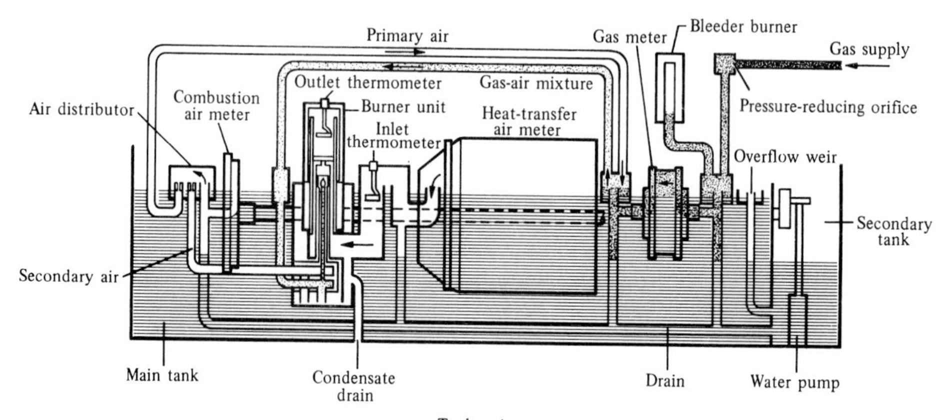

Tank unit

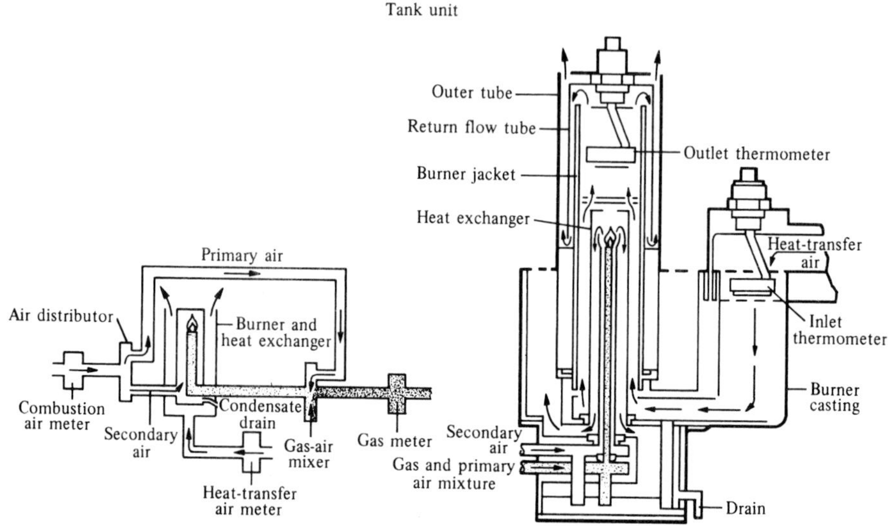

The burner unit and simplified diagram of gas and air flow

Figure 8:1 The Cambridge-Thomas gas calorimeter

The gas under test is burned under controlled conditions and the heat generated is transferred to a stream of heat-transfer air. The temperature of this air is increased in direct proportion to the calorific value of the gas. The streams of gas, heat-transfer air, and combustion air, are maintained in a volumetric ratio by three separate motor-driven wet meters geared together and working in a common water seal. These streams are thus saturated before being metered under the same conditions of temperature, barometric pressure and humidity thus making the recorder readings independent of these variables.

The combustion products do not come into contact with the heat-transfer air and they leave the instrument practically at calorimeter room temperature, so that the true total or gross calorific value is recorded without requiring any correction. The meter drums are driven by an induction motor through suitable gearing, and any change in speed due to normal variations of voltage and frequency does not affect the accuracy of the results. The use of a static water tank with a reserve tank reduces the water requirements to 1 to 2 kg per week and a constant level is ensured by using a weir and pump system. The gas is burned in a closed burner supported in a casing which is immersed to within 6 to 10 mm of the top in water to obviate errors due to rapid changes in room temperature or draughts. This also stabilizes the cold-balance settings of the recorder circuit. To reduce heat losses, a polished double-walled jacket is fitted round the heat exchanger, the burner itself being enveloped by a sheath of heated air. To eliminate errors caused by residual heat losses the calorimeter is tested on hydrogen, under conditions such that the temperature rise in the burner is equal to that obtained using the gas normally tested. The measuring circuit is then adjusted until the recorded rise in temperature is equal to the theoretical rise expected.

This test can be repeated at any time. The measuring circuit in the recorder is an equal-ratio bridge with the two resistance thermometers in adjacent arms. The recorder can be calibrated directly in calorific value or in percentage deviation from the desired value and the instrument is easily adapted for distant reading or control purposes.

8.2.2 Relative density

The relative density (or specific gravity) of a gas has been defined as the ratio of the density of the gas to the density of dry air under the same conditions of temperature and pressure [1]. This means that an instrument working on the principle of comparing the density of a gas to that of air will produce readings as defined, only if the reference tank is filled with air of a specified composition.

The definition also allows the relative density of any gas to vary slightly because of the deviations of real gases from the ideal-gas law. When any two gases undergo equal changes in temperature and/or pressure, their compressibility factors do not change by equal amounts and so their density ratio will not be quite constant. It is better to define relative density as the ratio of the gas density to the density of dry air at the reference conditions of 100 kPa (1 bar) and 0 °C which eliminates the above variations.

A number of methods and instruments are available for the determination of the specific gravity of gas [1]. The two described here, which are in use at Bacton [3], are:

1 Ac-Me Balance
2 Debro recorder

Ac-Me balance
This is a pressure-balance type of instrument. A beam carrying a bulb and counterweight is brought to balance successively in air and in gas by adjusting the balance-case pressure. Absolute pressures are determined by means of a barometer and a mercury manometer. The relative density of the gas is calculated from the ratio of the absolute pressures. The instrument can be used in the laboratory and the field for spot determinations. During a test the temperature should remain constant otherwise a correction must be applied. The accuracy of this instrument is within the region of 0.002 or $\pm$ 0.33 per cent.

Debro recorder
This instrument depends upon the principle that the output pressure of a fan is proportional to the density of the gas being pumped. It comprises two identical bells suspended from a balance beam and sealed in an oil bath. Dry air and dry gas are delivered by identical fans driven by a common motor, taking air and gas at virtually atmospheric pressure and delivering to the bells at pressures ranging from 0.5 to 1 kPa (50 to 100 mmH_2O). The difference in pressure between the air and gas is recorded by the deflection of a balance beam. The instrument must be checked using dry air before carrying out a determination. It gives good accuracy and can indicate variations of 0.001 [2]. Other methods for measuring the relative density of gases are described in references [1] and [4].

8.2.3 Wobbe number

The maintenance of the combustion performance of gases with regard to the heating output and air requirement of burners is influenced essentially by the ratio $W = C_g/\sqrt{d}$ where C_g is gross calorific value and d is relative density (specific gravity). W is known as the Wobbe index or number. Therefore, particularly in the use of gas and gas–air mixtures in industrial processes which require careful heat control, it is necessary to measure and control the Wobbe number. A number of instruments for recording Wobbe number are available and descriptions can be found in reference [1], together with instruments for measurement of aeration number, cone height, reversion pressure and sooting propensity.

The Wobbe recorder in use at Bacton is a Sigma instrument and satisfactory results are obtained provided that the instrument is standardized against known calorific value and relative density at regular intervals.

8.2.4 Component analysis

Introduction
The techniques available for analysing gases are numerous and include chemical absorption, spectrochemistry, mass spectrometry, chromatography, and determination of thermal conductivity, heat of reaction and paramagnetism. Much of the available equipment is complicated and often rather bulky and is used in laboratories and research stations. Full details of the above techniques can be found in a number of textbooks such as [4] and [5]. Instruments for particular applications can be selected from manufacturers' publications.

The industrial gas engineer is more likely to be concerned with combustion measurements such as air–gas ratio, CO/CO_2 ratios, CO_2 content of combustion products, nitrogen oxides (NO_x), excess O_2 and humidity. The equipment chosen, particularly for on-site work, must be portable, reliable and reasonably accurate. Brief descriptions of a number of instruments are given below.

The Draeger gas analyser [6] [7]
This is a simple, rapid-acting device suitable for the determination of a large variety of gases that may be present in air, gas or flue gas. It is used for the determination of carbon monoxide, carbon dioxide, oxides of nitrogen and excess oxygen in the products of combustion from appliances and furnaces, etc.

The detector consists of a hand-held bellows pump into which is fitted a tube containing a solid absorbent suitably impregnated with a reagent which reacts with the constituent being determined. The pump is designed to draw 100 ml of the air

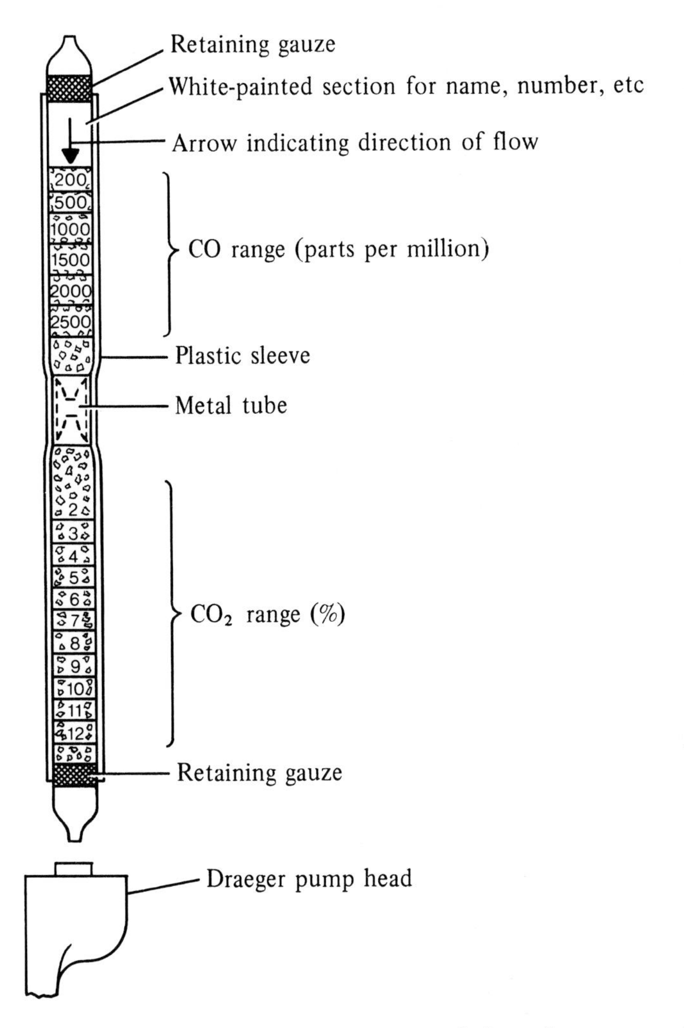

Figure 8:2 Combined CO—CO_2 detector tube consisting of separate tubes linked by a metal tube

sample through the detector tube at a prescribed rate. The normal tubes are of glass and are about 130 mm long and 6.4 mm diameter filled with a granulated material impregnated with the appropriate reagent. To carry out a test, the ends of the sealed tube are broken off, the tube is inserted into the pump, the bellows squeezed shut and then allowed to open under its own tension, this being designed to draw the sample through at the correct rate.

The tube shown in Figure 8:2 is a combined CO–CO_2 detector and is intended for sampling from the primary flue of conventionally flued appliances. It has a range of 0 to 2500 ppm for CO (0–0.25 per cent) and 2 to 12 per cent for CO_2. In cases where sampling from the primary flue is impractical and the sample is obtained from the secondary flue the combined tube ranges may not be adequate, so separate CO and CO_2 tubes may have to be used. Use of the combined tube will eliminate errors due to changes in flue-gas composition which may occur during the interval between taking separate CO and CO_2 samples. Full details of the equipment, operating instructions and tubes available can be found in the manufacturers' literature. If all precautions are taken the accuracy of a combined tube should be within $\pm$ 10 per cent of the true value.

The Fyrite detector [6]

This consists essentially of a graduated tube containing a liquid which absorbs the gas being measured. Determination of the concentration is made by turning the tube upside down and noting the change in liquid level when it is returned to the upright position. Several inversions are necessary until a constant reading is obtained. Separate tubes are required for the determination of carbon dioxide and oxygen, and the solutions have to be changed when exhausted. It is important to ensure that the sampling line is purged with the atmosphere under test. This applies to all sampling systems and analysis equipment.

Paramagnetic oxygen analyser [5]

This instrument makes use of the property, possessed by oxygen, nitric oxide and

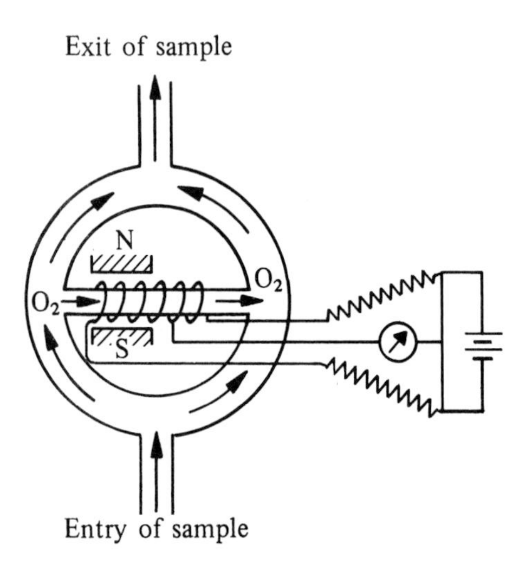

Figure 8:3 Principle of the paramagnetic analyser [5]

nitric peroxide, of being paramagnetic. The magnetic susceptibility χ_m is positive and is a function of temperature in accordance with Curie's law $\chi_m = C/T$, where C is a constant and T is the absolute temperature.

The method of operation shown in Figure 8:3 is that the paramagnetic gas of the mixture to be analysed is drawn into the field of a magnet from which it is expelled by heating. χ_m falling with temperature. Measurements are made of the resistance of hot wires placed in the flow stream of gas formed in this way. The resistance is proportional to the concentration of paramagnetic gas. The temperature of the wires is also affected by a number of other factors including the temperature, pressure and flow rate of the gas mixture being analysed, the composition, viscosity and thermal conductivity of the gas mixture, and the stability of the electrical supply. These must all be effectively controlled and allowed for in an instrument. The gas mixture being analysed should also be dried before it enters the analyser to prevent condensation and because water vapour has different physical characteristics to the majority of other gases. Errors will be kept small if the conditions of measurement and calibration are very similar, this error should not exceed 1 to 2 per cent of maximum scale value. The response time for 90 per cent of the signal is of the order of 10 seconds including the time taken for the sample to reach the measurement chamber.

Infrared analyser [5]
The majority of compound gases possess the property of selectively absorbing certain long-wave (infrared) radiation. The principle of infrared analysis is shown in Figure

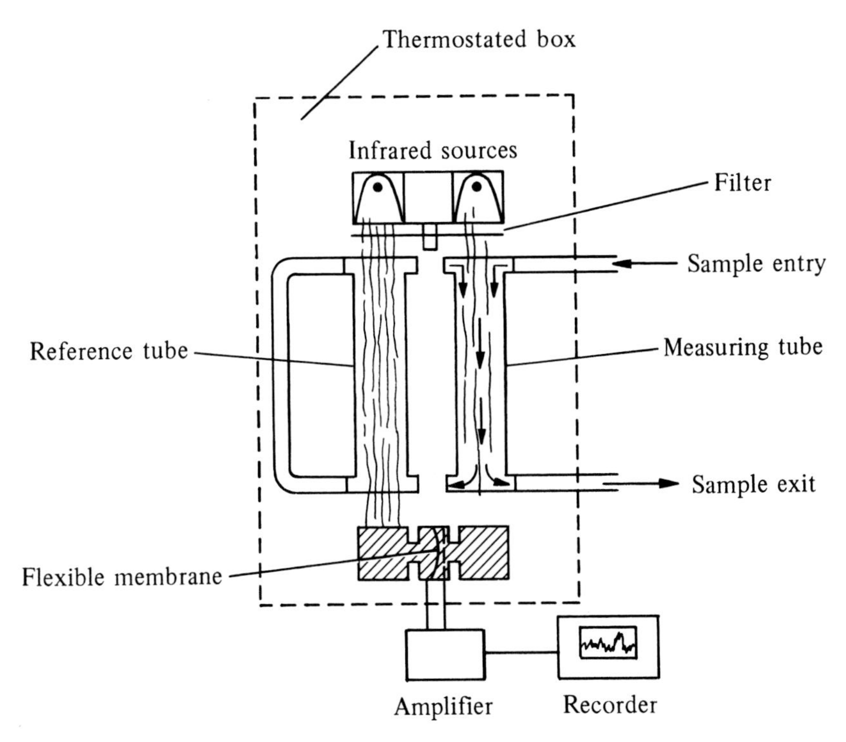

Figure 8:4 Principle of infrared analysis [5]

8:4. The radiation emitted by an infrared source, filtered if necessary, traverses the measurement chamber containing the gas to be analysed. Absorption occurs, at wavelengths characteristic of the gas, in proportion to the product of the thickness of the chamber and the partial pressure of the gas. The residual radiation is then absorbed by a receiving cell filled with the constituent to be analysed, which heats up, increases in pressure and causes the displacement of a pressure diaphragm. This displacement is measured by a variable-capacitance transducer and the signal obtained is proportional to the concentration of the constituent gas being sought. The following four points are very important:

1 Only the wavelengths absorbed by the constituent to be analysed are taken into account by the receiving cell which is filled with the same gas. The apparatus is, therefore, selective, although the source is not monochromatic, except when the mixture introduced contains other constituents whose absorption bands interfere with those of the gas being analysed. In such gases the radiation emitted by the source must be filtered. This is done by placing a tube containing the parasitic gases in front of the measurement tube. This holds back the radiation in the common bands. Compensation of this kind is not possible for particles (grey bodies which absorb all wavelengths) or for water vapour (condensation, complex spectrum), so that the gases must be carefully dried and freed from dust before being introduced into an analyser.
2 To compensate for the effects of variable ambient conditions, particularly temperature, commercial analysers have two identical tubes one of which is filled with a non-absorbing gas, nitrogen, to act as a reference. The two receiving cells are built as one unit, the pressures in them acting in opposition across the diaphragm.
3 The measurement ranges vary from 0–0.01 per cent to 0–100 per cent, according to the thickness of the measurement tubes. The tubes are interchangeable and when using them it is necessary to ensure that pressure multiplied by thickness is constant. To achieve this the gas is introduced into the tube through an annulus and at a low rate which gives the required response time to avoid over-pressure. The analyser is also operated with sample exhaust to atmosphere, so that the pressure inside it is very close to atmospheric pressure. The correction factor to be applied to the result is then the ratio of the pressure at the time of measurement to that at the time of calibration.
4 Commercial analysers are thermostatically controlled and fitted with voltage and frequency stabilizers. The possible errors do not exceed a few per cent at the top of the scale, and the response time, which depends on the flow of gas fed into the apparatus, can be less than one second.

The gases which can be analysed cover nearly all the compound gases. This type is particularly applicable in the gas industry for the analysis of CO, CO_2, CH_4 and oxides of nitrogen.

Psychrometers

The ability to measure the amount of water vapour in gaseous atmospheres is essential in many industries. Examples are storage of materials, metal heat treatment, heating and ventilating. Psychrometry was studied in Chapter 3 and here brief descriptions of a number of instruments available to measure humidity are given.

Wet and dry bulb hygrometer [8]. This type of instrument consists of two thermometers, one of which measures the ambient temperature, the other being covered with a damp wick from which water evaporates, so lowering the bulb temperature. Such instruments depend upon a minimum air flow ($3\ \mathrm{m\,s^{-1}}$) across both wet and dry bulbs in order to obtain accurate results. A number of instruments have been designed to achieve this air movement. Two hygrometers suitable for laboratory or meteorological use are the sling hygrometer and the Assman hygrometer. Industrial types generally consist of more robust equipment, utilizing mercury-in-steel or electrical-resistance thermometers and continuous recording facilities (Figure 8:5).

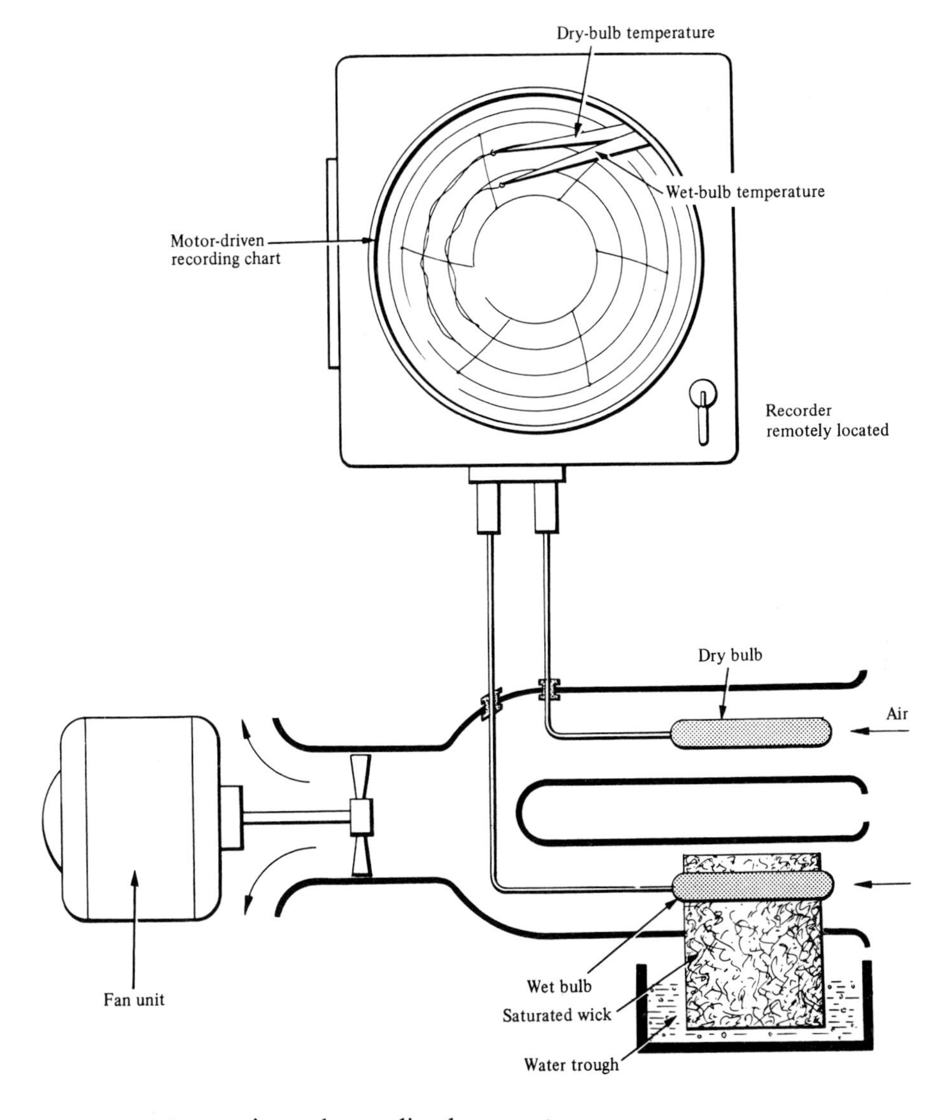

Figure 8:5 Mercury-in-steel recording hygrometer

The accuracy of a resistance type can be within $\pm$ 2 per cent or better. The main disadvantage is that the wet bulb must be kept wet.

Hair hygrometer [8]. Human hair possesses the property of changing length as the moisture content of a gas changes. This property can be used in a simple hygrometer for indicating or recording purposes. They are used in heating and air-conditioning applications. Accuracy is about $\pm$ 4 per cent. The disadvantages are that they cannot monitor rapid changes in humidity, they need correction if used outside the range 4–27 °C and they are sensitive to vibration.

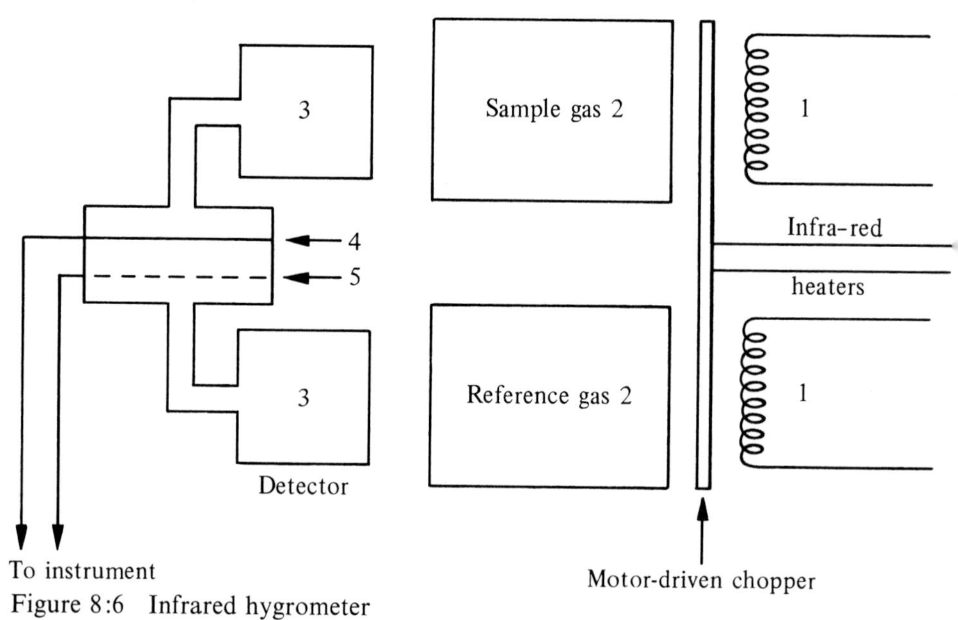

Figure 8:6 Infrared hygrometer

Infrared hygrometer [8]. This instrument is shown in Figure 8:6. The samples do not absorb radiation to the same extent; this results in an unequal heating effect in the detectors, causing a pressure difference which acts on the diaphragm. The resultant variation in electrostatic capacity can be used to indicate or record the absolute humidity.

Expansion-type dew-point meter [8]. If a gas is compressed isothermally to a higher pressure and then instantaneously released, it will expand and cool. If the ratio of the two pressures is high enough and the temperature drop is sufficient to bring the gas below its dew point, moisture will condense out of the gas. The point at which a mist is just seen to appear is the dew point. Its value is obtained from the pressure ratio and the initial temperature from tables:

$$\text{final temperature} = \text{initial temperature} \left(\frac{\text{final pressure}}{\text{initial pressure}}\right)^{(\gamma-1)/\gamma}$$

$$\text{where } \gamma = \frac{\text{specific heat at constant pressure}}{\text{specific heat at constant volume}}$$

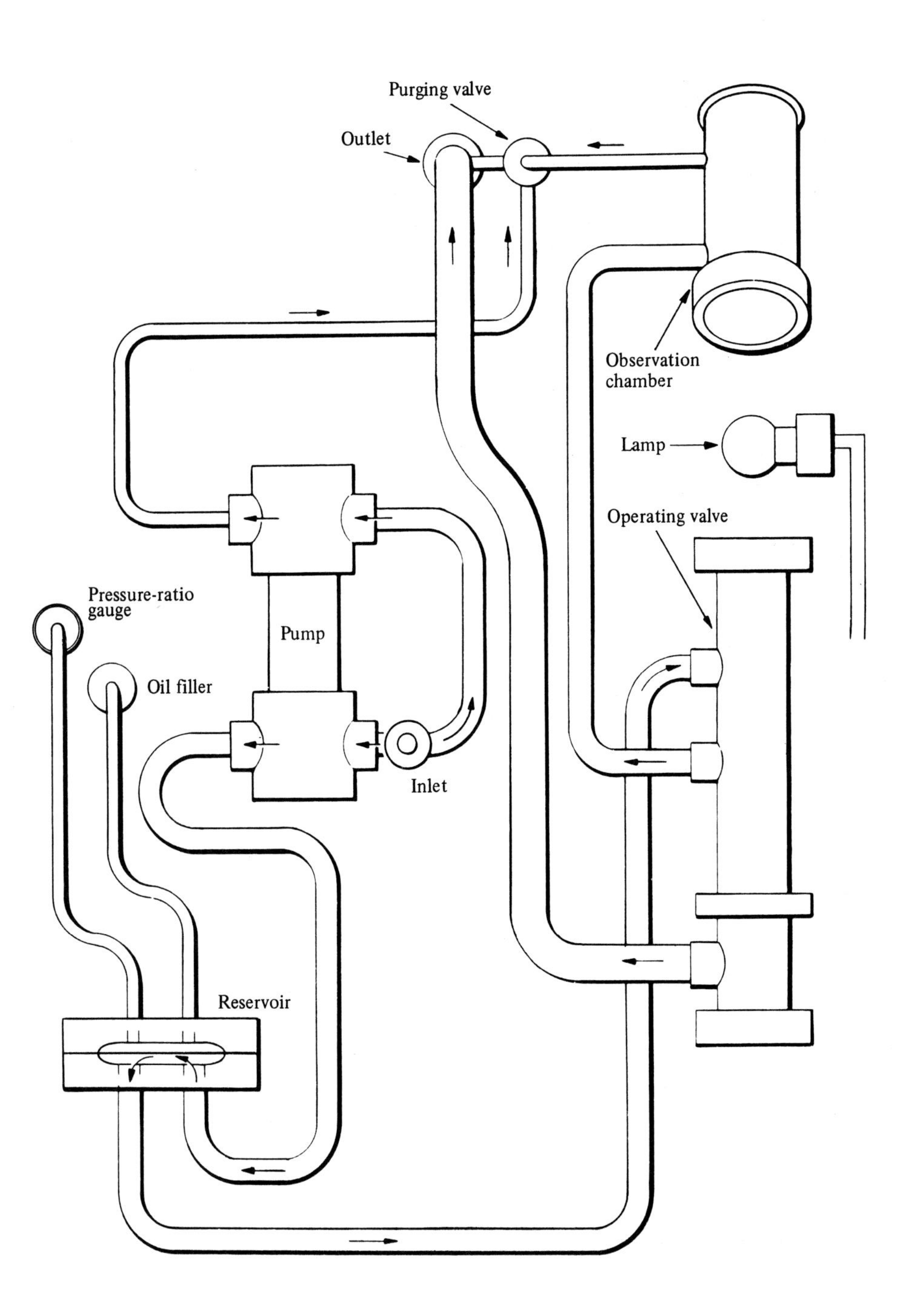

Figure 8:7 Expansion-type dew-point meter

The expression $(\gamma - 1)/\gamma$ is referred to as the Q-value. It depends on the mixture ratio and constituents of the gas or atmosphere.

Figure 8:7 shows the layout of an instrument of this type. They are portable and have a wide range of operation at ambient temperature, being capable of measuring dew points down to − 60 °C. The main disadvantage is that a continuous record cannot be obtained.

Conductivity-type dew-point meter [8]. This type, Figure 8:8, consists of a sample chamber containing an electrically insulating material, one side of which is cooled.

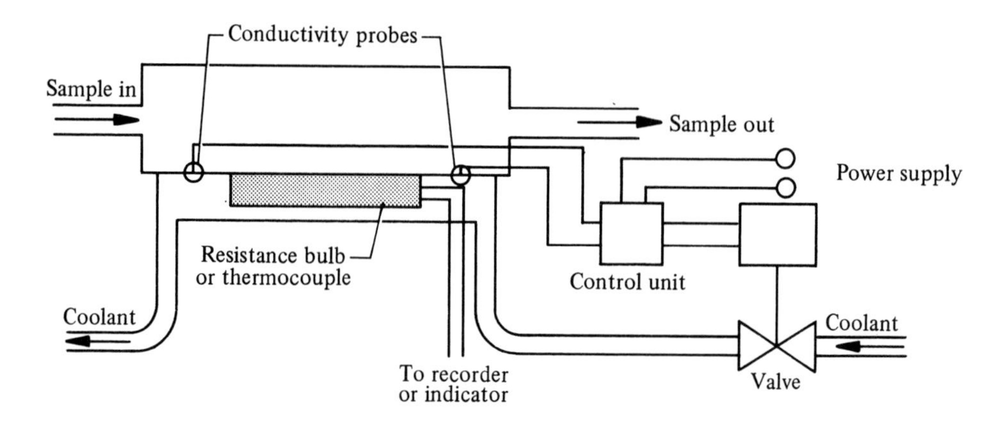

Figure 8:8 Conductivity-type dew-point meter

The probes measure the change in conductivity arising when moisture condenses on the surface. By suitable control the surface is kept at a temperature at which condensation just takes place. This temperature is measured by a resistance bulb or thermocouple and is the dew point of the gas sample.

8.3 MEASUREMENT OF TEMPERATURE AND HEAT FLUX

8.3.1 Introduction

The achievement of maximum efficiency in existing plant or optimum design of new plant, requires an accurate knowledge of several parameters. In many cases this can only be accurately obtained by direct measurement. This is particularly true for temperature and heat flux.

A range of instruments have been developed specifically for measurements in industrial furnaces, combustion chambers or other process heating requirements. These instruments are covered in this section together with a more general survey of temperature-measurement techniques.

8.3.2 The International Practical Temperature Scale

The thermodynamic scale of temperature, which relates the concept of temperature to the laws of thermodynamics is the accepted international standard. The realization of this scale is a precise laboratory technique, so in order to produce a practical scale, an international body meets at regular intervals to assign values to certain easily reproduced physical phenomena. Instruments and methods for interpolating between these points are also specified [9]. Figures 8:9 and 8:10 show the present fixed points and interpolation methods.

Fixed point	Assigned value °C	K
Triple point of hydrogen	– 259.34	13.81
Boiling point of hydrogen (25/76 standard atmosphere)	– 256.103	17.042
Boiling point of hydrogen	– 252.87	20.28
Boiling point of neon	– 246.048	27.102
Triple point of oxygen	– 218.789	54.361
Boiling point of oxygen	– 182.962	90.188
Triple point of water	0.01	273.16
Boiling point of water	100.00	373.15
Freezing point of zinc	419.58	692.73
Freezing point of silver	961.93	1235.08
Freezing point of gold	1064.43	1337.58

All defined points are at one standard atmosphere pressure except the triple points and one hydrogen boiling point

Figure 8:9 Defined fixed points of IPTS-68

Temperature range	Method
– 259.34 °C to 630.74 °C	Platinum resistance thermometer
630.74 °C to 1064.43 °C	Platinum/platinum 10% rhodium thermocouple
Above 1064.43 °C	Disappearing filament optical pyrometer

Figure 8:10 Interpolation methods in IPTS-68

8.3.3 Thermometers in use (practical thermometers)

In most forms of temperature measurement, the variation of some property of a substance with temperature is studied and a mathematical relationship established between the variation of the property and the temperature.

8.3.4 Expansion thermometers

Expansion of solids

When a solid is heated, it increases in volume. The increase in length of any side of a solid will depend on its original length, l_0, the rise in temperature, t and the coefficient of linear expansion, α. The new length, l_t will be given by:

$$l_t = l_0 + l_0 \alpha t = l_0 (1 + \alpha t)$$

Rod thermostats [10]. It is easier to measure a small differential in the expansion of two dissimilar materials than to measure the actual expansion of one material.

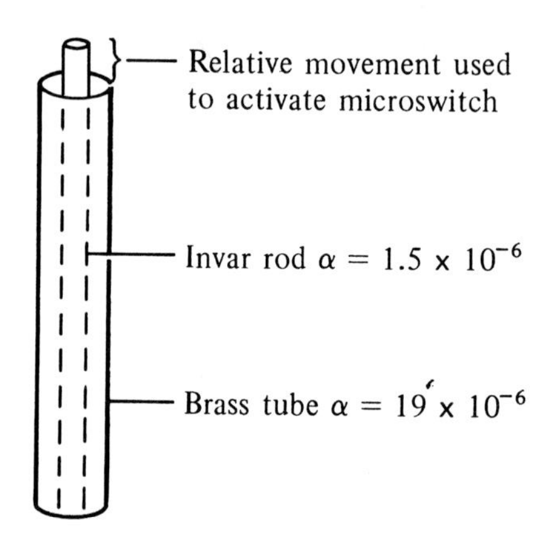

Figure 8:11 Principle of a thermostat sensor

Figure 8:11 shows the principle of a thermostat sensor. The tube will expand with respect to the end of the rod with an apparent coefficient of 17.5×10^{-6}. The length of the stem governs the sensitivity of the device. The movement of the tube around the rod is used to operate a microswitch or valve to control the heat input to the system.

Bimetallic-strip thermometers [10]. These are produced by fixing two strips of dissimilar metals such as brass and Invar rigidly together as shown in Figure 8:12. As

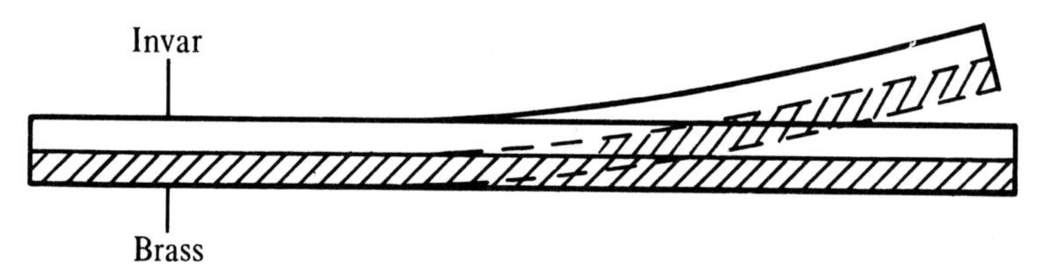

Figure 8:12 Bimetallic strip

temperature rises the strip will curve on the side of the material with the lower expansion coefficient. If one end is fixed the movement of the other end can be used to indicate temperature variation. If the bimetallic strip is made in the form of a long thin sandwich this can be coiled to provide a compact sensor which on expansion or contraction produces a rotary movement. If one end is fixed and a pointer connected to the other then a fairly accurate, responsive thermometer can be produced as seen

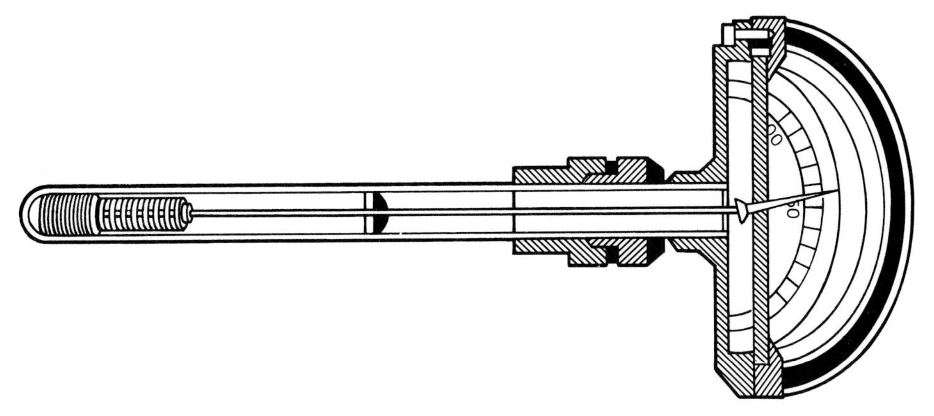

Figure 8:13 Bimetallic-strip thermometer

in Figure 8:13. They can also be used to operate a switch in a heater circuit to provide a simple closed-loop control system or to provide compensation for changes in atmospheric temperature e.g. in Bourdon tubes or aneroid barometers.

The expansion thermometers described are inexpensive, robust, direct reading and need no outside power. They cannot be remotely indicated, are not very accurate and have a narrow operating temperature range. Their operation depends on thickness, elastic properties and difference in expansion coefficients of the materials together with the difference in temperature.

Expansion of liquids

This type of thermometer uses the expansion of liquid in the thermometer, allowance being made for the expansion of the containing vessel. The mean coefficient of expansion between two temperatures t_1 and t_2 is given by the ratio of the increase in volume per degree rise of temperature to the original volume, i.e.:

$$(V_{t_2} - V_{t_1})/V_{t_1}(t_2 - t_1)$$

where V_{t_1} = volume at temperature t_1

V_{t_2} = volume at temperature t_2

Liquid-in-glass thermometers. These are simple to use, relatively accurate and cheap. However, for industrial use they are fragile and cannot always be located at points where they can be conveniently read. Figure 8:14 shows the temperature ranges and filling liquids used. Whenever possible, thermometers should be calibrated, standardized and used totally immersed, with sufficient time allowed for equilibrium to be reached. Selection depends on the application and accuracy required.

Liquid	Temperature range (°C)
Mercury	− 38 to + 600
Thallium	− 55 to + 600
Alcohol	− 80 to + 50
Toluene	− 90 to + 50
Pentane	−200 to + 30

Figure 8:14 Temperature ranges of thermometers [11]

Liquid-in-metal thermometers [10]. To overcome the disadvantages of fragility and location in industrial use, liquid-in-metal thermometers have been developed. The construction is shown in Figure 8:15. The bulb and capillary are completely filled with liquid. The change in volume on thermal expansion is indicated by a Bourdon tube. Liquid-in-metal thermometers can be made using a variety of alloys and filling liquids to suit the range and conditions of use. Sheaths or pockets may be used to protect the bulb from corrosion or pressure and a number of bulb shapes are available. Similarly, many designs of Bourdon tube and transmission linkage are possible. Where capillary tubes are of appreciable length compensation is needed for changes

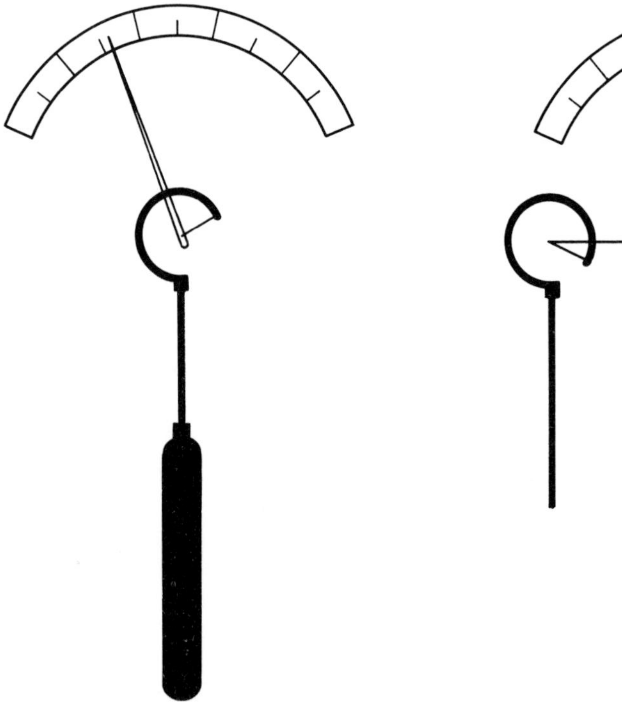

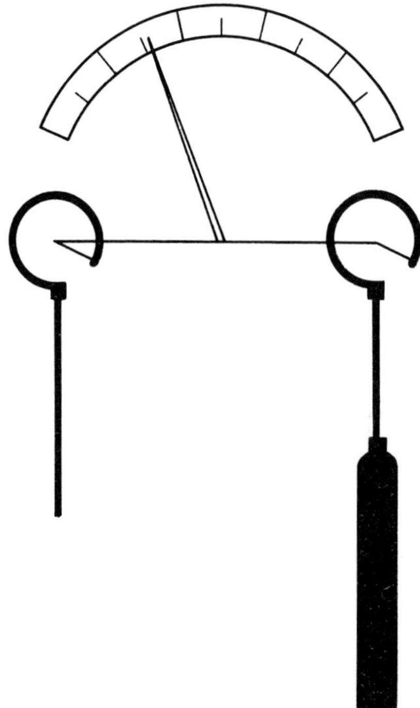

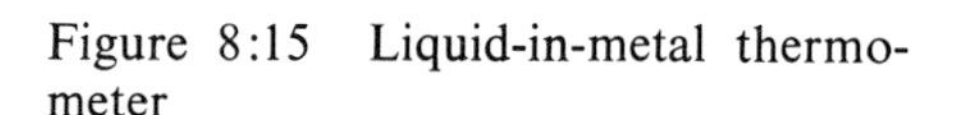

Figure 8:15 Liquid-in-metal thermometer

Figure 8:16 Liquid-in-metal thermometer with compensation for changes in ambient temperature along the capillary

caused by variation in ambient temperature. This can be achieved by a system such as that illustrated in Figure 8:16.

Expansion of gases

Gas thermometer [10]. If a gas is maintained at constant volume its pressure will increase by 1/273 of the pressure at 0 °C for every 1 °C rise in its temperature. This effect is much greater than the coefficient of expansion of solids and liquids and is the basis of the gas thermometer.

The construction is similar to the liquid-in-metal type, consisting of a bulb, capillary and Bourdon tube containing a known volume of an inert gas (nitrogen) which closely obeys the gas laws. The effect of temperature on the gas changes the pressure and this is indicated by the movement of the Bourdon tube. If the change in volume of the tube as it uncurls can be neglected a fairly linear scale is produced. Response is more rapid than liquid types and the range (— 50 °C to 500 °C) is governed by gas condensation at the lower end and metal properties at the high end. The range of a particular instrument depends on the Bourdon tube and pressure of filling (10 MPa (100 bar) at 0 °C) and is usually 100 °C. Compensation for changes in ambient temperature is by bimetallic strips.

8.3.5 Change-of-state thermometers

Vapour-pressure thermometers [10]. The vapour pressure of a volatile liquid is directly proportional to its temperature. By using an expansion thermometer system to enclose a volatile liquid and a space into which it can vaporize a thermometer which uses this effect is produced. Figure 8:17 shows a dual-filled type. The vapour pressure depends only on the temperature of the interface which is normally within the bulb. It is therefore independent of the volume of the bulb, the capillary and the ambient conditions in which the capillary and Bourdon tube are situated.

The relationship between vapour pressure and temperature is not linear. The thermometer scale must have wider divisions at the higher-temperature end. However, it is possible to linearize the scale over a predetermined portion.

The number of liquids which are suitable for use in vapour-pressure thermometers is large, but each is effective only over a fairly limited range. The range and duty must be decided before a choice of liquid and materials can be made.

Condensation in the capillary and Bourdon tube occurs if the temperature being measured is greater than the temperature of the capillary and Bourdon tube. This will transmit the pressure change adequately, but where the temperature being measured fluctuates about the ambient temperature, response becomes very slow because as the temperature being measured decreases the condensed liquid in the Bourdon tube and capillary will have to evaporate before the pressure stabilizes at the vapour pressure corresponding to the temperature being measured.

This deficiency is overcome by using a dual-filled thermometer in which a heavy involatile liquid such as mercury fills the Bourdon tube, capillary and part of the bulb, the rest of the bulb space is taken up by the volatile liquid and its vapour. Variations in bulb temperature are then transferred as fast as the vapour pressure equilibrates at the liquid–vapour interface and no problems of condensation occur.

The vapour pressure thermometer is cheaper than a mercury-in-steel type, it does

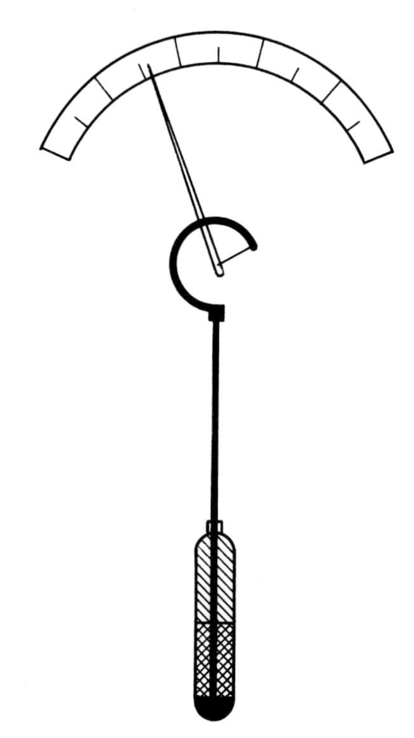

Figure 8:17 Dual-filled vapour-pressure thermometer

not suffer from ambient-temperature effects and can be made with a smaller bulb than most other types of expansion thermometer.

Pyrometric cones, bars and rings [12, 13]. The melting points of mixtures of certain minerals can be used to determine temperature. This property is used particularly in the ceramic industry to determine kiln temperatures. The minerals, being similar in nature to the ceramic ware, behave in a manner which indicates what the behaviour of the pottery under similar conditions is likely to be. The mixtures, which consist of silicate minerals such as china clay, talc and quartz, together with other minerals such as calcium carbonate, are made up in the form of cones, rings or bars. The cones are known as Harrison or Seger cones and by varying their composition, a range of temperatures between 600 °C and 2000 °C may be covered in convenient steps. A series of cones are placed in the kiln. Those of lower melting point will melt, one will just bend over. This cone indicates the temperature of the kiln, which is confirmed by the fact that the cone of next higher melting point is not affected. Such cones really give a comparative indication of the effect of temperature and time, often referred to as 'heat work'.

In order to obtain maximum accuracy, which is of the order of $\pm$ 10 °C the cones must be heated at a controlled rate.

The bars known as Watkin heat recorders are used in the same way as cones. Their deformation is observed and related to temperature.

The rings are called Buller's rings and indicate the work done by the heat by their radial contraction, which is dependent on the temperature. After firing and cooling, the ring is dropped into a gauge which registers the temperature corresponding to the ring size.

8.3.6 Electrical methods of temperature measurement

Resistance thermometers [14]. The variation of the electrical resistance of a metal with temperature provides a very convenient and accurate means of measuring the temperature. The metals used are platinum, nickel and copper.

The relationship between electrical resistance and temperature has the form:

$$R_t = R_0 (1 + at + bt^2)$$

but over narrow ranges of temperature bt^2 is small and a linear relation provides a good approximation. From these relationships the value of the temperature coefficient of resistance can be obtained. For a particular thermometer the values of the constants are determined experimentally from measurements of resistance at fixed temperatures to cover the range required. The difference between the resistance at 0 °C and at 100 °C is called the 'fundamental interval' (FI) for the thermometer.

Resistance thermometers consist of spirals of fine wire wound on insulating formers. The winding, which must be free from strain, is annealed at a temperature greater than the maximum temperature of its designed range. One typical type for industrial use consists of platinum wire on a ceramic spool 75 mm long and 6 mm diameter. The wire is wound non-inductively in a twin spiral groove and is covered with a ceramic glaze for protection. It is then enclosed in a suitable outer sheath. The FI can be made to be say 40, 50 or 100 ohms, so that the change in resistance is large compared with the change in the leads. The leads may be copper or constantan. All possible precautions are taken to avoid contact resistance and thermoelectric effects.

Platinum resistance thermometers provide the most accurate means of measuring temperature up to 600 °C. The accuracy of a laboratory instrument is ± 0.01 °C at 500 °C, that of an industrial instrument better than ± 0.5 per cent. Resistance thermometers can be obtained to read up to 1000 °C and above, although the main application is for medium temperatures where high precision is needed. Thermometers with nickel elements are also available for use up to 250 °C. They are not as accurate as platinum types but are less expensive. Speeds of response are of the same order as thermocouples but usually slower due to bulb size. The main advantage of resistance thermometers over thermocouples is that they require no cold junction measurement.

Thermistors [12]. These consist of an element made from a semiconducting material which has a negative temperature coefficient about ten times greater than that of copper or platinum. They are, therefore, considerably more sensitive than a normal resistance element. The resistivity of the thermistor material is also much higher than that of any metal, so the size can be made very small, giving a very rapid response. They can be used for temperatures up to about 1000 °C, but the usual range is lower (– 100 °C to 300 °C).

Thermistors are made from metal oxides or mixtures of oxides. The oxides used are those of cobalt, copper, iron, magnesium, manganese, nickel, tin, titanium, uranium and zinc. The powdered oxides are compressed into shape and sintered. This gives a dense ceramic body. The contacts can be wires embedded before firing or baked-on metal ceramic coatings. The element is usually a bead of 0.25 to 0.5 mm diameter.

The temperature T and resistance R are related by the expression:

$$R = a\mathrm{e}^{b/T}$$

where a and b are nearly constant over small ranges.

A wide range of thermistor elements are available [12] which makes selection for a particular application simple. In selecting an element, consideration must also be given to making the method of indication compatible with any existing system. Thermistors must not be used at temperatures greater than the specified temperature; they must not be stressed or vibrated excessively. Used in this manner with properly designed measuring circuits, temperature differences as small as 0.001 °C can be measured.

Thermoelectric thermometers [14]. In thermoelectric thermometers, two wires of dissimilar metals are connected together to form a loop with two junctions as in Figure 8:18. When heat is supplied to one junction so that the junctions have different temperatures an e.m.f. is generated and the current which flows in the loop can be detected by a galvanometer. The magnitude of the e.m.f. depends on the temperature difference between the junctions. Such a system is known as a thermocouple. In practice one junction, the reference junction, is maintained at a constant temperature and the e.m.f. generated then measures the temperature of the other junction. The e.m.f. can be measured by a millivoltmeter or by a potentiometer as discussed on pages 340–342. The relationship between the thermo-e.m.f. and the difference in temperature of the junctions is approximately parabolic of the form $E = at + bt^2$.

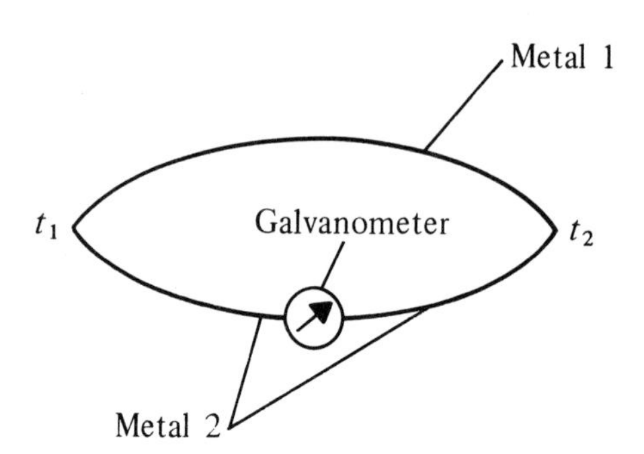

Figure 8:18 Thermoelectric effect

A system of this kind can be used as a thermometer if suitable materials can be found to produce measurable e.m.f.s. The e.m.f. need not be large but the larger the e.m.f. the better the sensitivity. Another consideration is the relationship between the e.m.f. and the temperature difference. A couple with a linear relationship is often preferable to one with a marked parabolic relationship within the useful range.

The actual application of thermocouples to the measurement of temperature depends upon the laws of thermoelectricity. The law of intermediate temperature is illustrated in Figure 8:19 and states that the e.m.f. generated in a thermocouple with

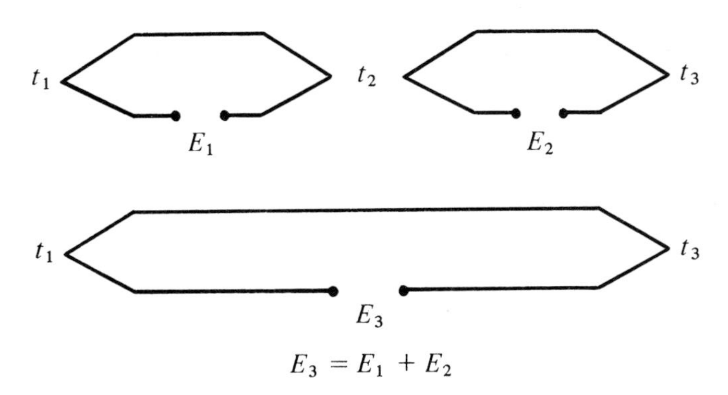

Figure 8:19 Law of intermediate temperatures

junctions at temperatures t_1 and t_3 is equal to the sum of the e.m.f.s generated by similar thermocouples, one acting between temperatures t_1 and t_2 and the other between t_2 and t_3, where t_2 lies between t_1 and t_3. The law of intermediate metals is similarly illustrated in Figure 8:20 which shows that the basic thermocouple consists

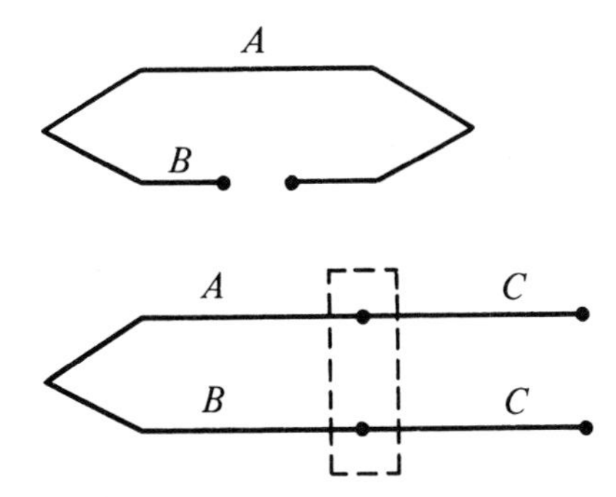

Figure 8:20 Law of intermediate metals

of a loop made up of two metals A and B. If a third metal is introduced then three junctions are formed. The e.m.f. is unaltered if the two new junctions BC and CA are at the same temperature. The value of this law is that it enables an instrument to be introduced into the circuit to measure the e.m.f. produced.

The choice of materials for thermocouples is governed by a number of factors. They must develop a relatively large e.m.f., which does not drift from the calibrated value when in use. The composition must not change and the material must be able to withstand the maximum temperature of use. Also the material must be of consistent manufacture to enable replacement of elements without recalibration. Usually, the thermocouple is rugged and easy to install, and, if properly connected, it indicates the true temperature of its junction. However, care must be taken to ensure that the correct compensating cable is used and that correction for the temperature of the cold junction is made. For industrial applications thermocouples may be divided into (*a*) base-metal thermocouples and (*b*) rare-metal thermocouples.

Base-metal thermocouples [9]. The different types that are readily available are listed in Figure 8:21, which shows the type, maximum temperature, and accuracy.

Element	Maximum temperature (°C)		Accuracy	Calibration standard
	Continuous	Spot reading		
Copper/constantan	400	500	$\pm\frac{3}{4}\%$	BS1828: 1961
Iron/constantan	800	1100	± 3 °C below 400 °C $\pm\frac{3}{4}\%$ above 400 °C	BS1829: 1962
Chromel/Alumel T1/T2	1100	1300	± 3 °C below 400 °C $\pm\frac{3}{4}\%$ above 400 °C	BS1827: 1952

Figure 8:21 Base-metal thermocouples

Copper–constantan elements are normally only used for laboratory measurements. Iron–constantan elements can be used up to 800 °C under reducing conditions, but they deteriorate rapidly under oxidizing conditions, or in the presence of acids. Nickel–chrome–aluminium alloys are suitable for use up to 1000–1200 °C under oxidizing conditions. These couples are, however, susceptible to attack by sulphur compounds, cyanide, carbon and hydrogen. Alternate reducing and oxidizing conditions can cause low readings, due to the formation of a nickel layer which partially shorts the hot junction.

The thermocouples can be obtained in a mineral-packed type of construction which gives strong, small-diameter flexible thermocouples. A variety of sheath materials and mineral insulating materials are available to suit different applications.

Element	Maximum temperature (°C)		Accuracy
	Continuous	Spot reading	
Pt 13% Rh V Pt Pt 10% Rh V Pt	1600	1750	BS 1826: 1952 ± 1 °C at 1000 °C ± 5 °C at 1500 °C 1% above 1700 °C
Pt 20% Rh V Pt 5% Rh Pt 30% Rh V Pt 6% Rh	1700	1800	
Pt 40% Rh V Pt 20% Rh	1800	1880	± 8 °C to 1800 °C
Ir V Ir 40% Rh	Up to 1900/2000 °C in oxidizing atmosphere		$\pm 1\%$
W VW 26% Re	Up to 2800 °C in neutral or reducing atmosphere		$\pm 5\%$

Figure 8:22 Rare-metal thermocouples

Rare-metal thermocouples [9]. The more common types are listed in Figure 8:22. The platinum and platinum–rhodium couples are very widely used. It is always advisable to use them with good-quality impervious sheaths, as they are susceptible to contamination by dirt and grease. Contamination is particularly severe in the presence of silica, and can also be caused by metal vapours, hydrogen and other furnace gases.

Pt 13% Rh V Pt and Pt 10% Rh V Pt are the most common rare-metal thermocouples for use up to 1600 °C. Prolonged use at high temperature will, however, lead to errors caused by migration of rhodium through the junction. The Pt 20% Rh V Pt 5% Rh and Pt 30% Rh V Pt 6% Rh couples have been developed to minimize this migration error. They are also mechanically stronger at high temperatures. The latter couple requires no compensating leads for cold-junction temperatures up to 70–100 °C. The Pt 40% Rh V Pt 20% Rh couple can be used continuously for tem-

Material	Approximate maximum temperature (°C)	Properties and applications
Mild steel	550	Neutral, slightly oxidizing or slightly reducing atmospheres. Oil baths, water or steam
Stainless steel (En 58B or En58J)	800	Instead of mild steel if rusting cannot be tolerated, e.g. food industry
Calorized mild steel	800	Sulphurous, oxidizing and reducing atmospheres. Tin, inert salt baths, heat-treatment furnaces. Not molten aluminium or lead if antimony present
20% chrome/iron	950	Sulphurous atmospheres, heat-treatment furnaces in most atmospheres
Inconel 75% nickel/15–17% chrome/10% iron (max.)	1100	General-purpose use, good hot strength. Not recommended for sulphurous atmospheres although a small percentage can be tolerated in oxidizing conditions
Cronite (50–60% nickel)	1100	Cast sheath, good hot strength. High resistance to oxidation and abrasion
Nitrogen-loaded 27% chrome/iron	1150	High temperatures in sulphurous atmospheres
80% nickel/20% chrome	1200	Similar properties to 75% nickel/15–17% chrome. Very good hot strength. Furnace atmospheres and low-temperature salt baths
24% chrome/iron + about 1% each silicon, manganese, aluminium	1200	Very good in sulphurous atmospheres. Excellent oxidization resistance but hot strength limited above 1000 °C
Kanthal DS	1200	Practically immune to sulphur. Must be heated in air for 8 hours at 1050 °C before using in reducing conditions. Better in neutral or oxidizing atmospheres
Kanthal A1	1350	High-temperature version of Kanthal DS, but hot strength not so good

Figure 8:23 Metal thermocouple sheaths

peratures up to 1800 °C. Rhodium migration is low and compensating cable is not required.

In most industrial applications it is advisable to protect the thermocouple from contamination, oxidation and mechanical damage. This is usually achieved as already mentioned by mounting the element within a protecting tube or sheath. The material used varies with the temperature and application. Figure 8:23 lists a summary of available materials.

Measurement of the electrical variable. In resistance thermometers [14], Wheatstone-bridge circuits are used in which the current flows through the resistances in the bridge and also through the thermometer bulb. This generates heat in the thermometer and causes its temperature indication to change. This effect must be reduced to a minimum by reducing the current through the thermometer to as low a value as possible. The 'null' method which gives the highest accuracy and sensitivity of measurement requires that the bridge is balanced when a reading is taken, the balance being obtained by adjustment of the other resistances in the bridge. This method will give spot readings only and in order to obtain a continuous indication some method of self-balancing must be incorporated. Another method uses the 'out-of-balance' current, the resistances being adjusted to give zero reading on the galvanometer at some definite temperature so that any deviation from this will result in a deflection. By comparison with a standard resistance thermometer the galvanometer deflection can be calibrated to indicate temperature directly. This method requires a constant supply voltage for the bridge.

The bridge measures the resistance of the leads connecting it to the thermometer as well as that of the thermometer bulb, so the thermometer must be calibrated with the leads attached. Also, any change in the temperature of the leads would appear as a change in the temperature of the thermometer winding. If the leads were made from thick copper wire of very low resistance, the effect would be negligible. However, in practice this would be inconvenient and expensive. Instead, compensation for the resistance of the leads may be provided as shown in Figure 8:24. This shows a set of dummy leads introduced into the arm AD of the bridge. These leads are identical with the actual leads of the thermometer element and are laid alongside them; they

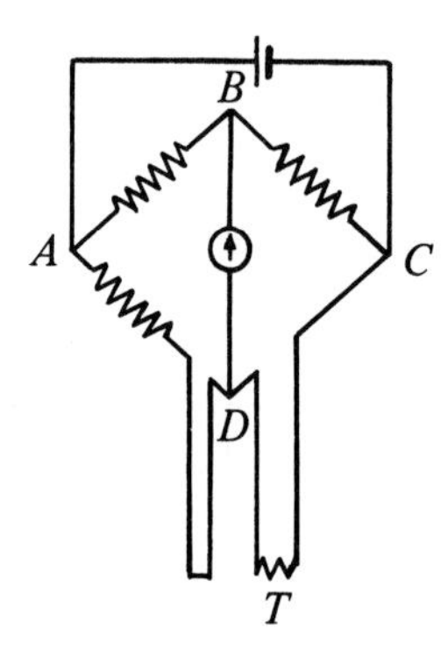

Figure 8:24 Compensation using dummy leads

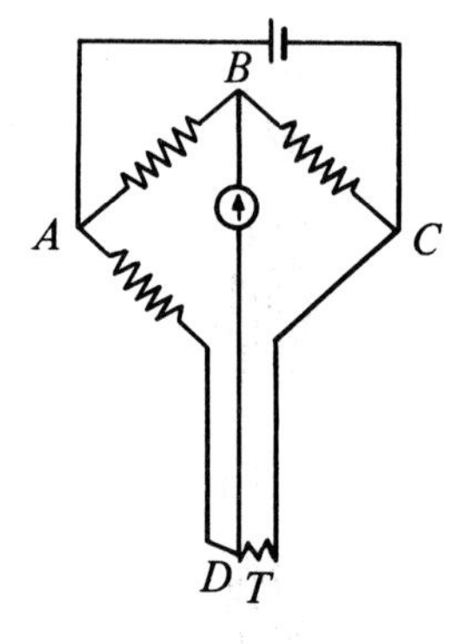

Figure 8:25 Compensation by the three-lead method

have the same resistance and will undergo the same changes in resistance. To obtain complete compensation the bridge must be operated with a 1 : 1 ratio in the upper arms. The three-lead connection shown in Figure 8:25 is more economical. One lead is in the galvanometer arm which at balance carries no current, the other two are in the lower arms of the bridge and cancel each other. As previously stated, for accurate measurement contact resistances and thermo-e.m.f.s in the system must be reduced to a minimum.

Traditional practice in resistance thermometry has favoured the use of null-balance bridges which are usually resistive, but can be capacitive, or inductive. More recently constant-current circuits have become available, enabling resistance-thermometer elements to be used with standard voltage-measuring instruments [15].

Data processing apparatus and computers need much larger voltage inputs than are conventionally available from metallic resistance elements. Future trends may be towards elements of much higher resistance to give the required output directly with negligible self-heating, however, it is more likely that elements with the BS1904 specification will still be used with amplifying stages or pulse-coding techniques.

In *thermocouples* [14, 16] the simplest way of measuring the e.m.f. is to introduce a millivoltmeter (moving-coil galvanometer) into the circuit. A number of methods can be used to attach the leads to the meter, with a copper–constantan couple the wires can simply be extended to the meter. However, with rare-metal couples this would be much too expensive, so other materials are introduced into the circuit. Figure 8:26

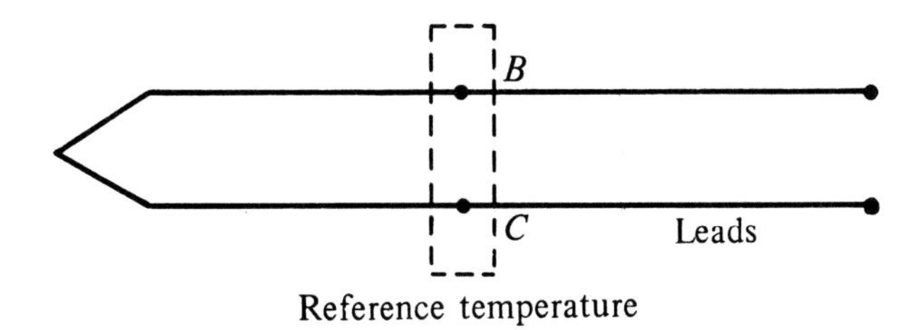

Figure 8:26 Location of reference temperature

shows a couple with additional leads. In effect, the reference junction is moved to the location of the additional junctions. If these are close to the system whose temperature is being measured then the reference temperature will be quite high and liable to fluctuation. It is preferable to move the reference junction to the meter without the use of leads of the couple materials. To achieve this, compensating leads are used. These are leads of materials with thermoelectric properties similar to those of the couple materials, but which are less expensive. Where a very long run is involved, the compensating leads need only be taken to a point at which the temperature can be maintained at a steady known value, with the run completed to the meter using copper leads. This method involves the use of a thermostatically controlled chamber to house the reference junction and to maintain it at a value just above the highest ambient temperature. An alternative is to allow the junction temperature to vary and insert an electrical cold-junction temperature compensator into the circuit at the junction. The measuring system described is a direct-deflection method using the millivoltmeter to measure current. This can be used to indicate e.m.f. or temperature only if the electrical resistance of the complete circuit, junction leads and meter has a constant specified value. Meters for this type of work have their external resistance value marked on them.

The e.m.f. can also be measured by a potentiometer designed to cover the range of generated e.m.f.s. At the point of balance no current flows from the source of the e.m.f. being measured, which means that the resistance of the leads connecting the thermocouple to the potentiometer is not important. The potentiometric method gives better sensitivity, a faster response and is the most accurate. It gives better zero repression and is better in control systems.

8.3.7 Radiation and optical pyrometry

When thermal radiation impinges on a body, some fraction α may be absorbed, some reflected (ρ) and some transmitted (τ):

$$\alpha + \rho + \tau = 1$$

A body which neither reflects nor transmits radiation, has an absorptance of unity and is defined as a 'black body' [14].

Kirchhoff's law states that the radiation emitted by any body at a certain temperature is equal to the radiation emitted by a black body at the same temperature multiplied by the absorptance. A black body also emits the highest possible radiation at any temperature. If the emissivity ε of a body is defined as the ratio of the radiation

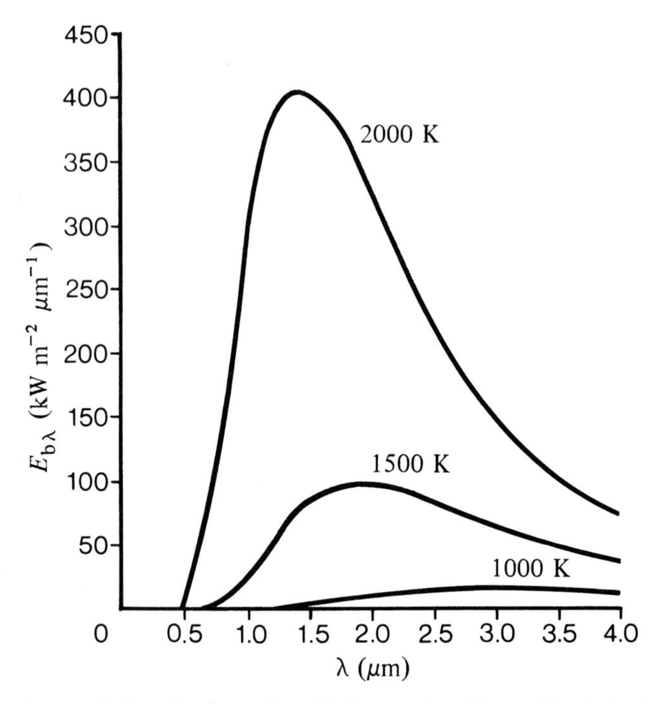

Figure 8:27 Spectral distribution of radiation emitted by a black body [17]

(per unit area per unit time) emitted by the body at a certain temperature, to the radiation of a black body at the same temperature, then the absorptance α will equal the emissivity.

The dependence of the spectral distribution of radiation emitted by a black body on the temperature is described by Planck's law and is illustrated in Figure 8:27. $E_{b\lambda}$ is the amount of energy, per unit of wavelength, radiated per second in a narrow waveband centred on wavelength λ, per unit area of a black body at thermodynamic (absolute) temperature T:

$$E_{b\lambda} = c_1\lambda^{-5}/[\exp(c_2/\lambda T) - 1]$$

where c_1 and c_2 are constants equal to 3.74×10^{-19} kW m^2 and 1.439×10^{-2} mK respectively [17]. This shows that there is a wavelength for each temperature at which the spectral radiation density is a maximum and this wavelength shifts towards the shorter wavelengths at higher temperatures. The product of the absolute temperature and the wavelength at which this maximum occurs is a constant (Wien's displacement law). On the basis of Planck's law the measurement of radiation density makes it possible to determine the temperature accurately in any desired spectral range if the body radiates as a black body. This does not restrict radiation pyrometry as much as might be expected, because if any body is placed in a cavity enclosure then it will radiate as if it were a black body. In practice many industrial furnaces approximate quite closely to black-body radiators. The emissivity deviates more or less from unity in any body which is not located inside such a cavity. If this emissivity is independent of wavelength within a certain spectral range, the body is called a grey-body radiator. Planck's law applies but all radiation densities are reduced according to the emissivity, so if the emission is known the temperature can be determined. In most bodies the emissivity depends on the wavelength to a greater or lesser extent and the temperature can only be determined if the measurements are made in a spectral range for which the emissivity is known.

Radiation pyrometers are used when the temperature is too high or the conditions too severe for a sheathed thermocouple to withstand or when it is difficult or inconvenient to put a thermocouple at the point at which the temperature is required.

Basically all the instruments consist of four components [17]:

1 A detecting element which converts the radiant energy to electrical energy
2 An optical system which ensures that radiation from the target either falls or is focused on the detecting element
3 An optical system which selects only radiation in the required waveband for detection
4 An amplifier and recorder or indicator which measures and displays the output of the detector

Total-radiation pyrometers. A number of designs are available commercially with reproducibility and accuracy at least as good as base-metal thermocouples. The detector is usually a thermopile, which consists of a number of thermocouples connected in series having one set of junctions blackened to improve absorption (the hot junctions) and the other set shielded from the radiation are cold junctions. The instrument is compensated for changes in ambient temperature, often by thermistors. The principle of operation of a typical instrument is shown in Figure 8:28. This uses

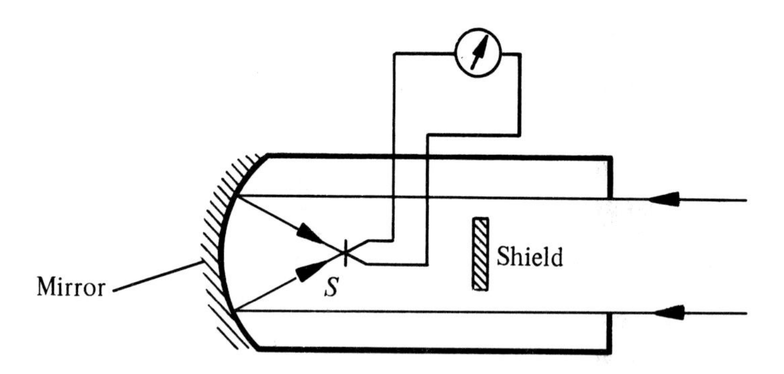

Figure 8:28 Féry total-radiation pyrometer

a concave mirror to concentrate the radiation onto the detector disc. An auxiliary optical system ensures that the whole area of the detector receives radiation. Leads from the detector are led from the pyrometer to the measuring instrument. Modern instruments can be used in control systems. They are suitable for many temperature measurements within furnace enclosures. Other detectors [9, 17] particularly thermistors are now being used in total-radiation pyrometers.

Partial-radiation pyrometers [9]. These are similar in construction to the total-radiation instruments, but are sensitive only to a restricted waveband. In some cases the limited waveband is achieved by using an optical filter with a total-radiation detector. It is more usual, however, for the waveband to be determined by the detector itself. Detectors in use are photoelectric, photoconductive and photoemissive types. These respond mainly to visible radiation, or the near infrared. The response times are very short which makes them suitable for detecting fast temperature changes. Their higher sensitivity at short wavelengths gives them an advantage over the thermopile for the measurement of high temperatures. Much smaller objects can be viewed and the pyrometer can be made smaller. One of the major advantages of the partial-radiation type is the possibility of selecting the sensitive waveband to suit a particular requirement (such as measuring the internal temperature of plate glass).

Reflected radiation can cause serious errors in radiation pyrometry at temperatures below 800 °C. This is particularly serious if the target is a good reflector, such as aluminium. These errors can be minimized by making a pyrometer insensitive to the visible waveband (below 0.8 μm).

Chopped-radiation pyrometers [9]. In chopped-radiation pyrometers, the radiation entering is periodically interrupted by a mechanical device, which is usually a rotating disc with alternate open and opaque segments. It is therefore possible to use AC amplification and the more sensitive but less stable detectors, such as photomultipliers and photovoltaic cells. The detector output is standardized, either manually or automatically, to a built-in standard source. As with the partial-radiation pyrometers, the sensitive waveband can be chosen to suit particular applications. A typical application of this type of instrument is to measure the temperature of aluminium extrusions.

Ratio (two-colour) pyrometers [9]. A ratio pyrometer is basically two partial-radiation pyrometers built into one instrument. The two halves are sensitive to different wavebands and the temperature is obtained from the ratio of the outputs. The advantages of the method and the necessary operating requirements are:

1 The reading is independent of the target emissivity, provided that the emissivity is the same in both wavebands
2 The reading is independent of partial obscuration of the target and of atmospheric absorption, provided that the absorption is not wavelength dependent in the operating wavebands
3 The reading is independent of target size, provided that the target surroundings are at least 300 °C colder than the target

Ratio pyrometers are more dependent on detector stability and are more expensive than single-waveband partial-radiation pyrometers. They are used in cement kilns and for the measurement of the temperature of liquid metals.

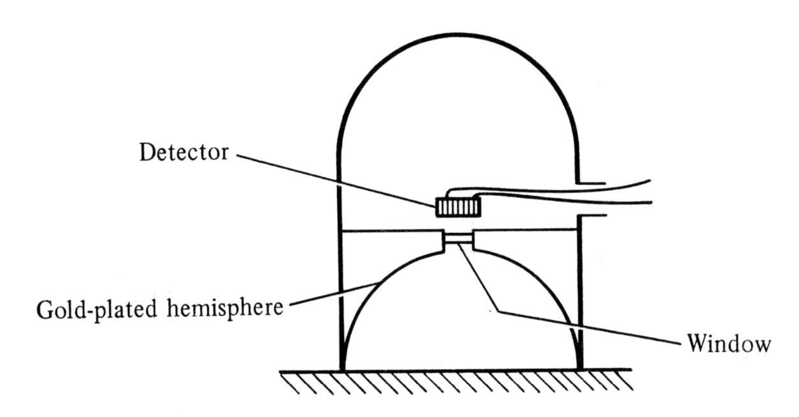

Figure 8:29 Hemispherical pyrometer [17]

Reflecting-hemisphere pyrometers [17]. Figure 8:29 shows the principles of construction of a reflecting-hemisphere instrument. The temperature is determined without knowing the emissivity of the surface, since it is measured under essentially black-body conditions. The radiation detector is mounted at the top of the gold-plated reflecting cup which is placed on the hot surface. These instruments are suitable for spot temperature measurement up to 1300 °C on all surfaces except unoxidized metals. A thermopile or a silicon solar cell can be used as the detector. The solar cell is faster (response time 1 ms) and is better able to withstand repeated operation at high temperatures. It is less sensitive to emissivity errors, but is more expensive.

Optical pyrometers [14]. The radiation from a source at a high temperature extends into the visible region. Within this region a given wavelength has a fixed colour and the energy of radiation is interpreted as intensity or brightness. If therefore the brightness of the light of a given colour emitted by a hot source is measured, it provides an indication of the temperature.

In an optical pyrometer the wavelength of the radiation accepted is restricted by means of a coloured filter to a narrow range (about 0.65 μm) and the brightness is measured by comparison with a standard lamp of controlled brightness (which has been calibrated against a black-body source). The most common type in use is the disappearing-filament pyrometer, which is shown diagrammatically in Figure 8:30. An image of the radiating source is produced by a lens and made to coincide with the filament of an electric lamp. The current is measured by a precision potentiometer.

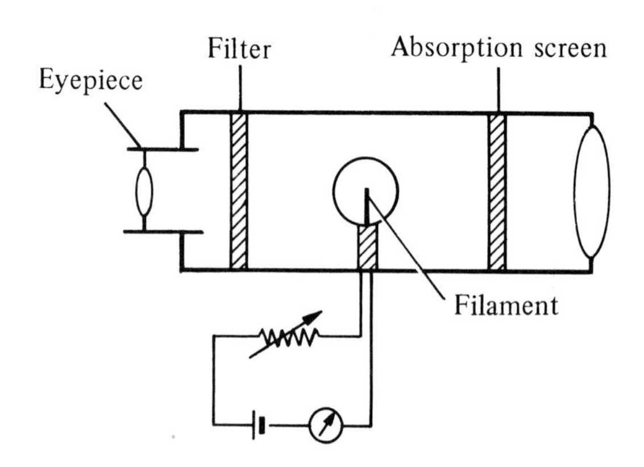

Figure 8:30 Disappearing-filament pyrometer

The range of temperature which can be measured is from about 700 °C up to about 1400 °C, this being the maximum allowable filament temperature. The range can be extended, however, by introducing an absorption screen near the objective lens. A known fraction of the energy from the source then enters the pyrometer for comparison. A good disappearing-filament pyrometer is one of the most accurate methods of making a spot measurement of a high-temperature enclosure ($\pm$0.5 per cent of the maximum measuring range). Self-balancing types are available in which the human eye is replaced by a photocell. These can have accuracies of $\pm$0.05 °C at 1000 °C and are used for high-precision standards work. Further information on radiation pyrometers can be obtained from British Standard 1041 : Part 5 [18].

8.3.8 Measurement of the bulk temperature of solids and liquids

This is one of the easier temperature-measuring problems encountered in industry. Provided that the sensor is completely surrounded by the medium and immersed to an adequate depth, no corrections are required for transient mass- or heat-transfer effects, conduction effects, radiation effects or velocity effects.

The most common methods of measuring industrial bulk temperatures are thermocouples and resistance thermometers (Section 8.3.6).

8.3.9 Measurement of the surface temperature of solids and liquids

Industrially, particularly when questions of heat transfer arise, there are many important cases in which the surface temperature is the desired parameter. In fact, the surface is often the only portion of the system that is accessible for measurement.

Where the surface is accessible, measurement can be made with surface pyrometers such as the radiation and optical methods discussed in Section 8.3.7, or contact pyrometers based on the use of thermocouples or resistance thermometers discussed in Section 8.3.6 These instruments are simple to use and are suitable for discontinuous measurements.

Where the surface is not accessible, which is generally the case in industrial environments particularly for solid surfaces, or when continuous measurements are required, it is necessary to fix the detecting element in place before heating the surface [5]. The elements used are usually thermocouples. The method is simple but precautions are necessary to avoid errors, which can be introduced through changes in local heat-transfer conditions, uncertainty about the position of the hot junction, and heat losses by conduction along the thermocouple wires. The couples should therefore have as small a diameter as possible and be well insulated. For refractory walls the hot junction can be incorporated in an embedded metal plate of high thermal conductivity and in which the temperature will not be appreciably affected by conduction losses. On metal walls the couple can be fixed to the wall or put into a small cavity in the surface.

8.3.10 Measurement of the temperature of gases [9]

This is very difficult as a gas is rarely in thermal equilibrium with its surroundings and in furnaces its spectral characteristics are largely unknown. Only a few of the many instruments available for the measurement of gas temperatures are suitable for industrial furnace applications:

1 The suction pyrometer
2 The venturi-pneumatic pyrometer
3 The Schmidt radiation method

The suction pyrometer
A thermocouple placed in a hot gas will indicate a temperature somewhere between that of the gas and its surroundings. In furnace work this can result in large errors. The suction pyrometer is designed to minimize these errors by:

1 Protecting the thermocouple from radiation errors by shielding
2 Aspirating the gas over the thermocouple at high speed to increase the heat transfer

The efficiency of a suction pyrometer is defined as the difference in indicated temperature with the suction on and off, divided by the error with the suction off. This is usually presented as a percentage. A number of instruments are available to cover temperature ranges up to 1800 °C. The main disadvantages are liability to blockage of the shield system by dirty gases, deterioration of the thermocouple and a slow response speed in some cases.

Two suction pyrometers have been developed specifically for industrial gas engineering use [19, 20]. They are:

1 A suction pyrometer for temperatures up to 1100 °C
2 A suction pyrometer for temperatures above 1100 °C

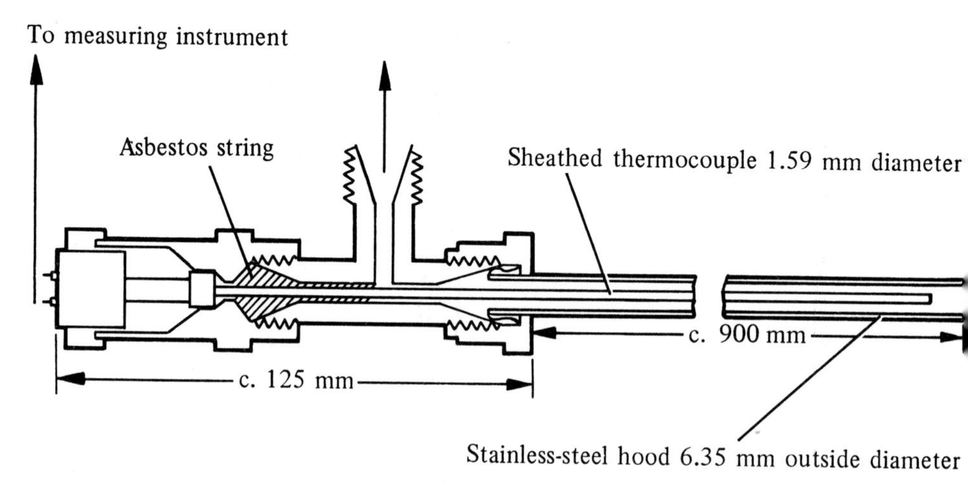

Figure 8:31 Metal suction pyrometer

The *metal suction pyrometer* is shown in Figure 8:31. Gas is drawn through the stainless-steel tube and then passes out of the instrument through the T-piece to the suction line. The e.m.f. generated by the thermocouple is fed to a socket housed in the assembly at the rear of the T-piece. Compensating cable is used to connect the pyrometer to the measuring instrument. The instrument is calibrated for use at a flow rate of about 30 m s^{-1}. This is a relatively low rate and is due to the small size of the instrument, which also means that it can be widely used on equipment of all sizes. The high response speed of the instrument under suction and non-suction conditions means that it can be used during heating-up periods or for investigating gas streams where there are fluctuations of the order of 100 °C or less per minute.

The equipment is portable, robust, is calibrated for use up to 1100 °C and requires only an AC supply for its operation.

The *water-cooled metal suction pyrometer* is shown in Figure 8:32. It was developed as a high-temperature accessory for use with the portable suction pyrometer described above. The instrument is calibrated for use up to 1600 °C, but by extrapolation this can be extended to 1800 °C. To use the instrument it is only necessary to provide a 14.5 mm clearance hole leading to the position at which a temperature reading is required, insert the pyrometer and take a single reading with suction applied. The true gas-stream temperature corresponding to this reading may then be found from the calibration curve. The instrument has a very rapid speed of response which enables any cyclic temperature variations likely to be encountered in normal industrial practice to be measured. The only additional equipment required is that concerned with the supply of water, which may be supplied from the mains, from a tank using a pump, or a pump and reservoir for closed-circuit cooling. The water flow must be sufficient to prevent boiling to avoid instrument damage. It must, however, be a high enough flow to limit any error in indicated temperature caused by the temperature of the water jacket becoming too high.

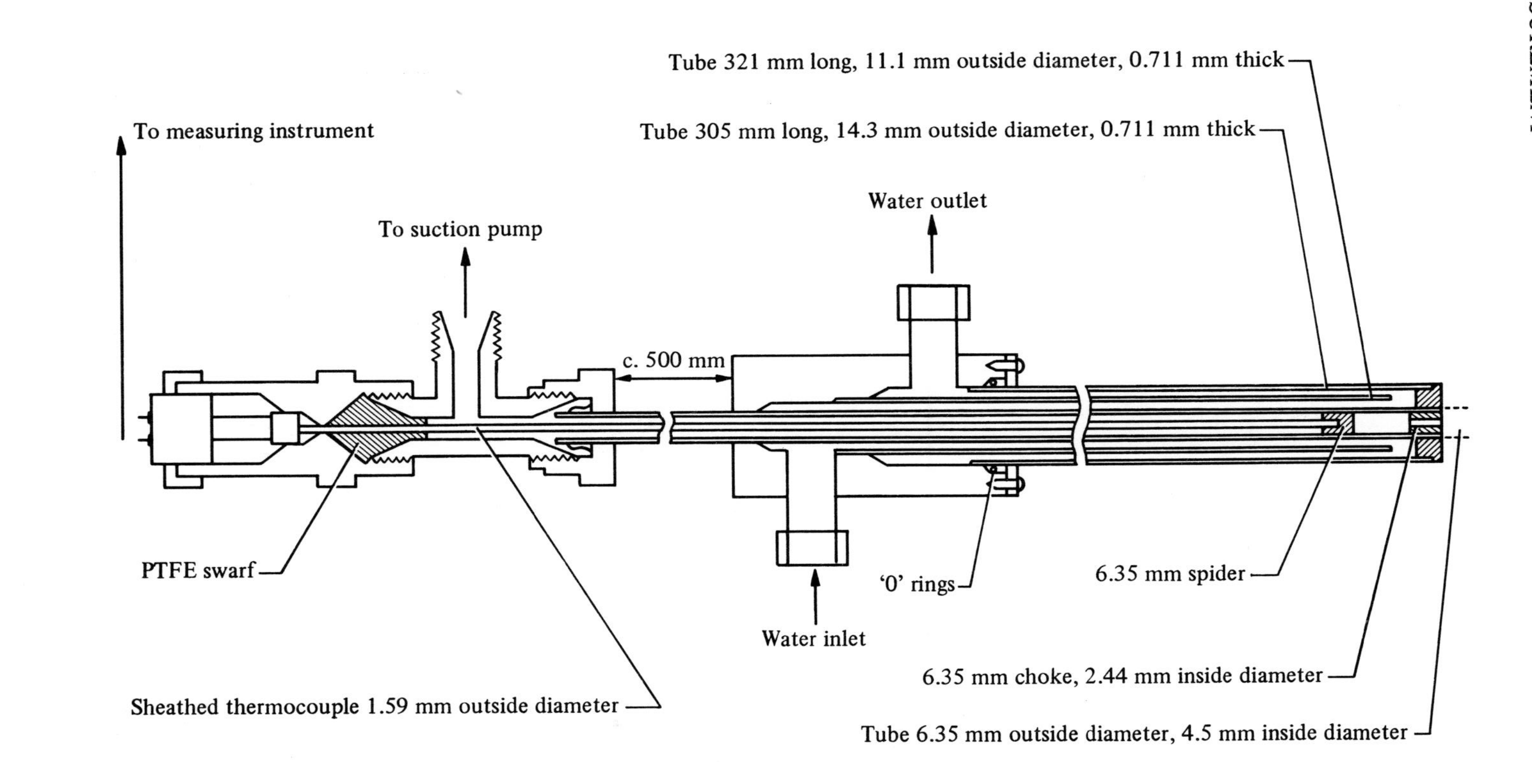

Figure 8:32 Water-cooled metal suction pyrometer

The venturi-pneumatic pyrometer [21]
This instrument is used for applications where conditions are too severe for the suction type or where a rapid response is required. No part of the system attains the hot-gas temperature, the upper temperature limit is fixed by the probe cooling requirements and the level at which the 'perfect-gas' assumptions become invalid. This limit is normally 2500 °C.

The gas temperature is measured by comparing the gas density at the unknown temperature with its density at a known lower temperature. The density measurements are achieved by venturi restrictions. If ΔP_h and ΔP_c are the pressure drops across the 'hot' and 'cold' venturis respectively, and if gas temperatures at these points are T_h K and T_c K, then:

$$T_h \propto (\Delta P_h/\Delta P_c)T_c$$

The Schmidt radiation method [21]
This method overcomes the difficulties caused by the partially transparent nature of the gas and provides a measure of the effective temperature and emissivity of the gas. Two readings are taken through the gas with a total-radiation pyrometer (Section 8.3.7), one with a cold background the other with a hot background at a known radiation temperature. Alternatively a twin-beam pyrometer can be used, one beam sighted on a hole in the furnace and the other on a hot region of the lining which contains a thermocouple. The temperature obtained is a mean value along the optical path through the gas and this can be unacceptable if large temperature gradients are involved. The accuracy also depends on the emissivity of the gas. The method is usually only preferred if probes cannot be used.

8.3.11 Measurement of heat flux [21]

It is frequently desirable to make a direct measurement of the heat flux incident upon a point in a furnace wall. The instruments described are available commercially, but they are usually made specially to suit a particular application.

Plug-type heat-flux meter
Figure 8:33 shows the construction. The heat incident upon the plug face is conducted through the plug to the water cooling at the rear. A guard-ring system ensures the laminar flow of heat through the plug and the whole assembly is mounted in a water-cooled probe. Measurement of heat flux is made by two thermocouples mounted with their hot junctions a distance apart on the axis of the plug, the heat flow through the plug and hence the heat flux incident on the plug face can be determined from:

$$Q = A\lambda_m (T_2 - T_1)/l$$

where Q = total heat flux
A = cross-sectional area of the plug
λ_m = mean thermal conductivity of the plug material between the two thermocouples
T_2, T_1 = thermocouple temperatures
l = distance between the thermocouples

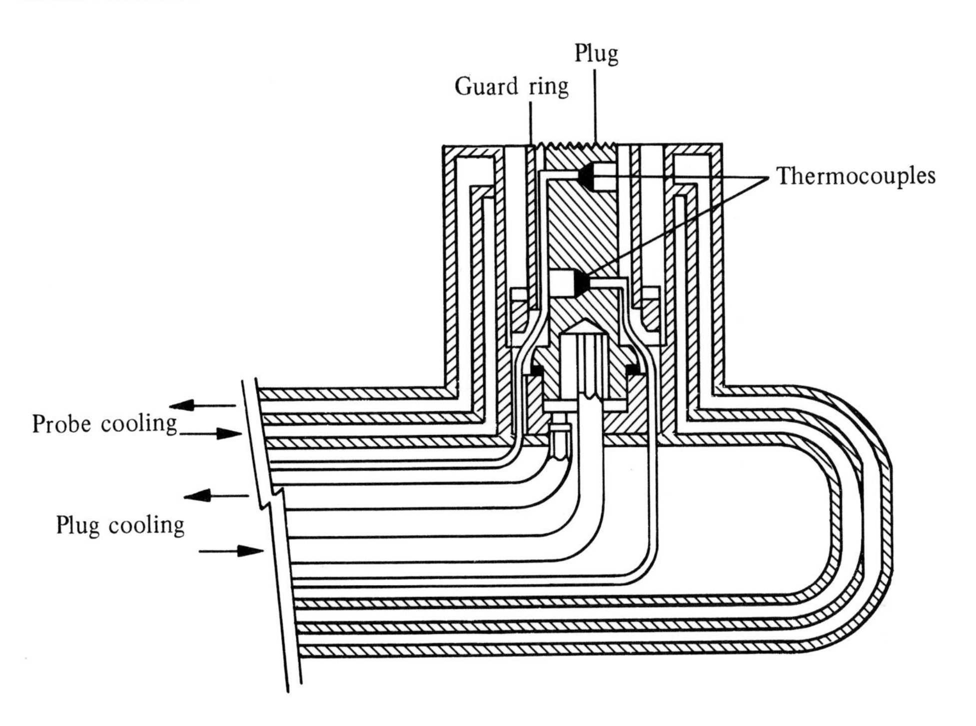

Figure 8:33 Plug-type heat-flux meter [21]

In practice it is usually difficult to define l and λ_m precisely and guard-ring systems are not perfect so the instrument is usually calibrated by means of a black-body furnace. This type of instrument measures the total heat flux incident upon its face from a solid angle of 2π.

Radial-disc heat-flux meter [21]
This type also measures the total heat flux. Figure 8:34 shows a typical construction.

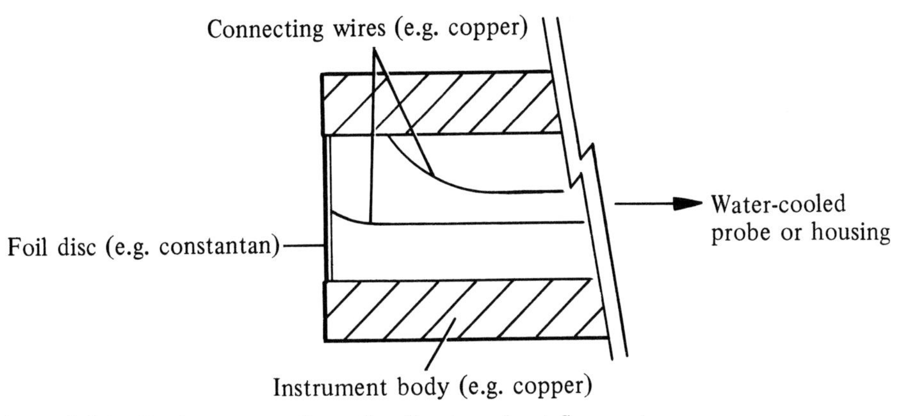

Figure 8:34 Basic construction of a disc-type heat-flux meter

Here, the heat flux incident on the disc is conducted away radially to the instrument body which is water-cooled. The output is obtained from the differential thermocouple formed by the junction of the disc circumference with the instrument body and the wire attached to the disc centre. The instrument is simple in construction and can be used in positions inaccessible to the plug type. It is, however, usually less accurate, less robust and each instrument requires individual calibration.

Ellipsoidal radiometer [21]
Figure 8:35 shows the construction of this instrument which is designed to measure the total radiation from a solid angle of 2π. Radiation that passes through the instrument aperture is focused by the ellipsoidal mirror onto the blackened hemispherical detector which is effectively the front surface of a very simple heat-flow plug. The output is obtained from thermocouple junctions on the plug axis. A nitrogen purge is fed to the ellipsoid through small holes around its mirror diameter to prevent entry of furnace gases and particles. Calibration is carried out by means of a black-body furnace.

The response time is of the order of 1 minute which means that these instruments cannot monitor rapid changes of incident radiation.

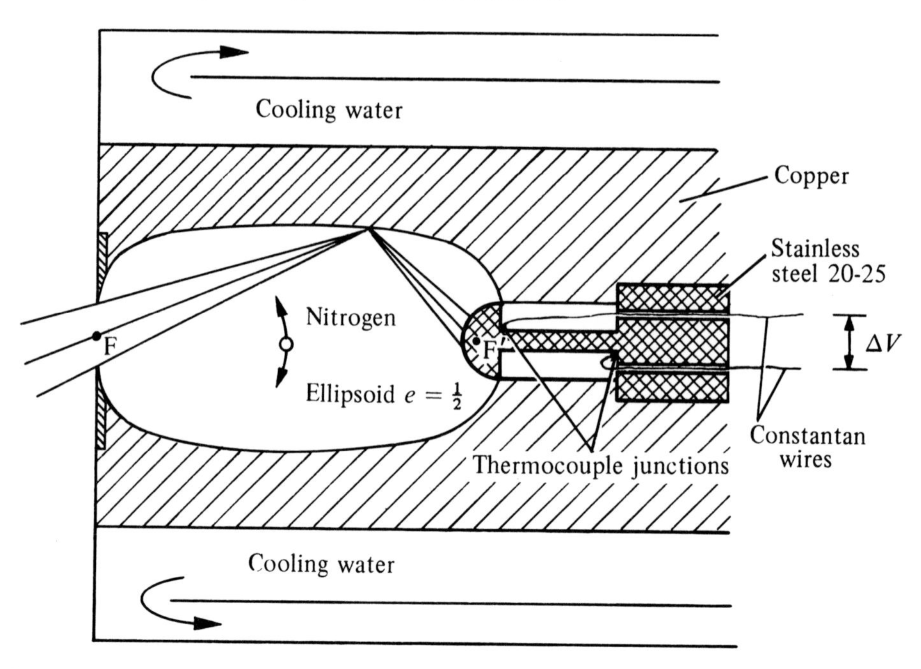

Figure 8:35 Ellipsoidal radiometer [5]

8.4 MEASUREMENT OF PRESSURE

8.4.1 Introduction

Pressure is force per unit area and the units of pressure are the pascal ($1\ \mathrm{N\ m^{-2}}$) and the bar (10^5 Pa). Before considering the actual methods of measuring pressure, it is necessary to clarify the terms absolute pressure, differential pressure and gauge pressure. The absolute pressure of a fluid is the difference between the fluid pressure and the pressure in a complete vacuum. Differential pressure refers to the measurement of the difference between any two pressures and gauge pressure is a differential-pressure measurement where one pressure is the pressure of the atmosphere.

Pressure may be measured directly in two ways:

1 Pressure due to a fluid may be balanced by the pressure produced by a column of liquid of known density
2 Pressure due to a fluid may be allowed to act over a known area, producing a force whose magnitude depends on the pressure

Other methods of measuring pressure depend upon indirect means.

8.4.2 Liquid-column pressure gauges (manometers)

U-tube manometer

A simple manometer is shown in Figure 8:36. It consists of a glass U-tube containing liquid. The pressures P_A and P_B whose difference is required are applied to the twc arms of the U-tube. If P_B is greater than P_A the liquid level will fall in B and rise in A.

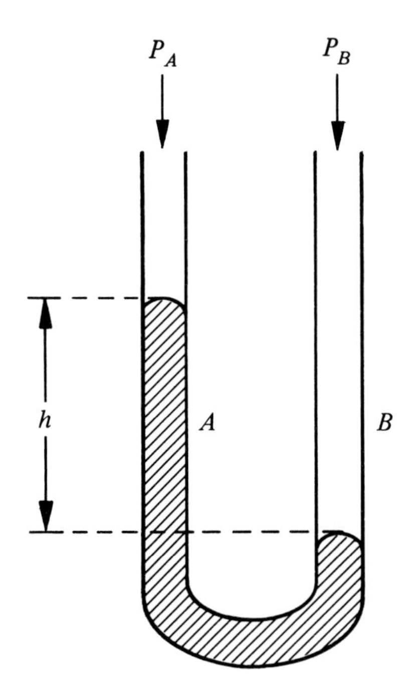

Figure 8:36 Simple manometer

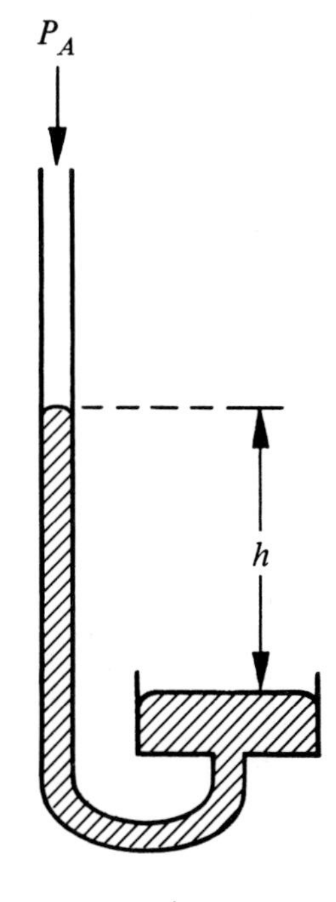

Figure 8:37 Well-type manometer

Equilibrium is attained when the pressure difference is balanced by the column of liquid of height h:

$$P_B - P_A = g\rho h$$

where g is the acceleration due to gravity and ρ is the density of the manometer liquid. The difference in heights of the liquid in the two arms can be used as an indication of the pressure if the liquid is specified, e.g. 20 mm water (2 mbar or 200 Pa). In the simple type of manometer the liquid levels in both limbs have to be observed and the difference taken.

Well-type manometer

This is a more direct reading instrument than the U-tube manometer and is shown in Figure 8:37. The well has a cross-section larger than that of the indicating tube so that when the level in the well falls slightly, the level in the indicating tube rises appreciably. A direct reading of h can be taken from the scale which is designed to correct for the error introduced by the fall of level in the well.

A number of variations of the simple liquid gauges described above are available for industrial and laboratory use. These include inclined-limb, differential-liquid and micro-manometers (see below).

The type of liquid used in the gauges will depend upon the pressure and the nature of the fluid whose pressure is being measured. They can be used in pressure measurement from very low pressures or differential pressures up to about 800 mm of mercury (1 bar or 0.1 MPa). They are also used as a standard in the calibration of other pressure-measuring elements.

Inclined-tube manometer

The construction of the instrument is shown in Figure 8:38. If the limb is inclined at a slope of 1 : 20 to the horizontal, a rise of h mm in the level of the liquid in the tube will produce a movement of $20h$ mm along the tube which makes it suitable for measuring small differential pressures. Gauges of this type usually have a range up to 40 mmH_2O (4 mbar or 400 Pa) and can be read to 2 mm. The inclined tube may be straight or curved, in the latter case the gauge can be calibrated to give a flow measurement directly.

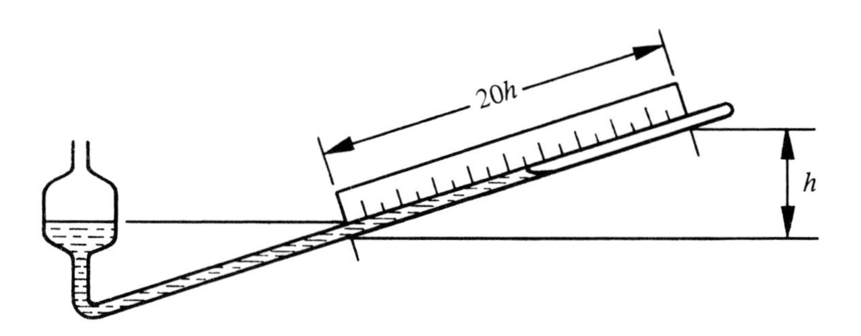

Figure 8:38 Inclined-tube manometer

Differential liquid manometer [12]
This manometer utilizes the movement of an interface between two liquids which have different relative densities (specific gravities), different colours and do not mix. Figure 8:39 shows the construction. The cross-sectional areas of the wells are usually

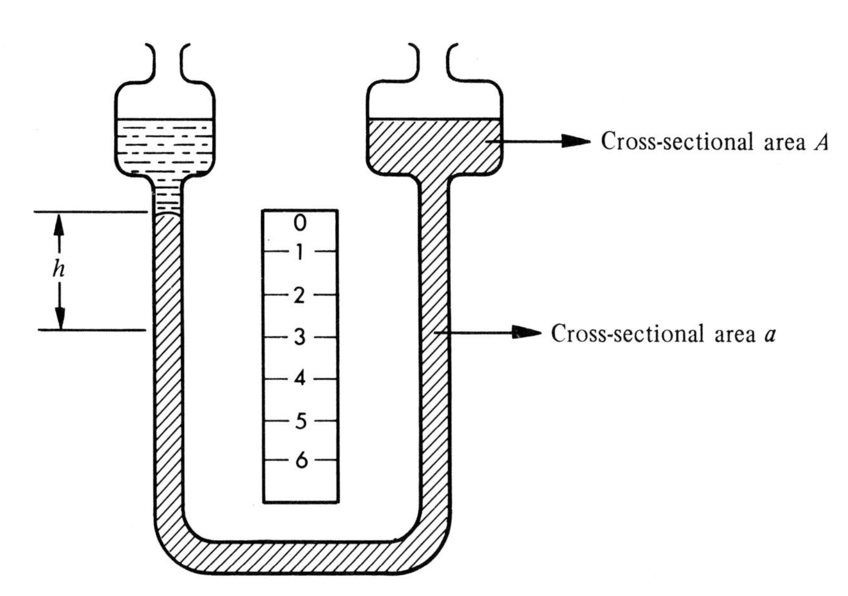

Figure 8:39 Differential liquid manometer

about 10 times larger than that of the U-tube. A change in the level of h mm of the interface represents a change of ha/A in the wells and represents a change in applied pressure of $haA^{-1}(d_1 + d_2) + h(d_2 - d_1)$. Therefore for a differential pressure of 1 mmH_2O:

$$h = [aA^{-1}(d_1 + d_2) + (d_2 - d_1)]^{-1}$$

By suitable choice of liquids h may be made to equal 5 mm for 1 mm pressure differential, enabling differences in pressure of 0.04 mmH_2O (0.4 Pa or 0.004 mbar) to be measured.

Micro-manometer
A very useful micro-manometer has been designed by the National Physical Laboratory [22]. It has a sensitivity of 24.5 mPa and an accuracy of 4.9 mPa. It consists of a shallow U-tube system, the left-hand limb being in the form of an inclined sight tube. The right-hand limb consists of a reservoir which may be raised or lowered so as to tilt the instrument about an axis coinciding with the centre line of the sight tube. The height of the reservoir in relation to the sight tube is adjusted by means of a vertical micrometer screw. The gauge is constructed so that the opposing effects of temperature change on the surface tension and density of the gauge liquid largely cancel out. A microscope is provided for the precision sighting of the null point. Ranges available are 0–3 mbar (0–300 Pa) and 0–50 mbar (0–5 kPa).

8.4.3 Force-balance pressure gauges [12]

The force produced by a pressure may be measured (*a*) by balancing it against a known weight or (*b*) by the strain or deformation it produces in an elastic medium.

Measurement by balance against a known weight
The pressure gauges operating on this principle are:

1. The piston type
2. The ring-balance type
3. The bell type

The piston type. In this type of instrument, the force produced on a piston of known area is measured directly by the weight it will support. In Figure 8:40 if the pressure acting on the piston is p Pa and the area over which it acts is A m^2 the force produced will be pa N. If there is no friction then $pA = G$, the weight supported by the piston. The system shown is the deadweight pressure tester which is used as a standard of pressure measurement for testing other types of instrument. The type shown can be used up to 55 MPa (550 bar).

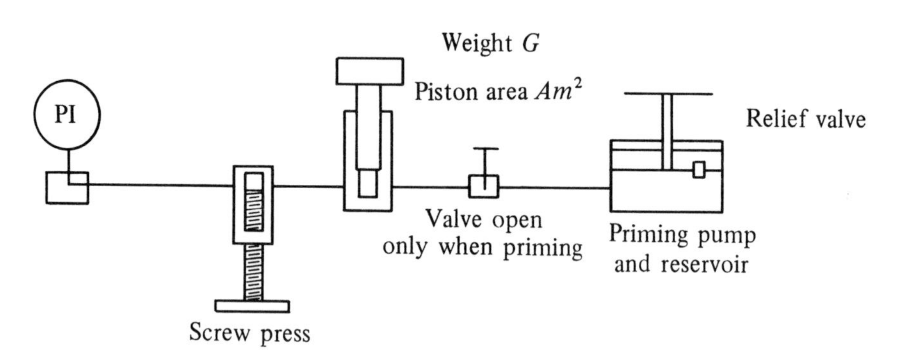

Figure 8:40 Piston-type pressure gauge [12]

The ring-balance type. This type of instrument is very useful for the measurement of low differential pressures of the order of 100 Pa (1 mbar). Figure 8:41 shows the principle of operation. The operating force depends on the difference between the pressures on the two sides of the partition. The calibration of the instrument can be changed by changing the control weight. The range is fixed by the nature and thickness of the material of construction and the pressure which the flexible connections will stand. The range of differential is fixed by the size of the ring and the nature and quantity of the sealing liquid.

The bell type. In the bell type of gauge the force produced by the difference of pressures on the inside and outside of the bell is balanced against a weight or the force produced by the compression of a spring.

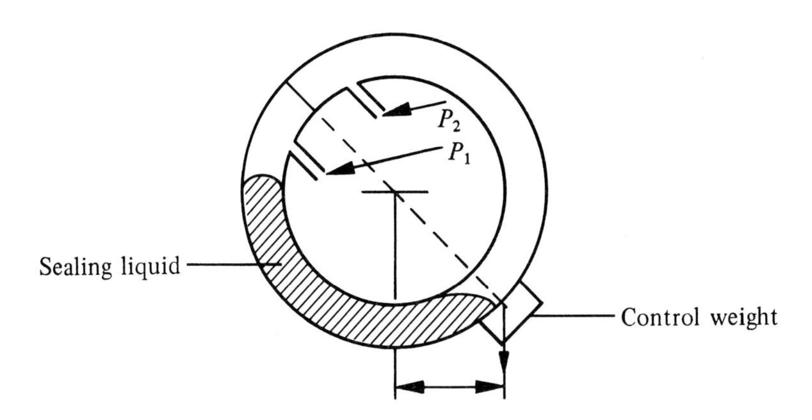

Figure 8:41 Ring-balance type pressure gauge

Figures 8:42 and 8:43 show the principle of operation of a thick-wall type and a thin-wall type respectively. In the thick-wall instrument the resultant force causes the bell to rise until equilibrium is reached. The thickness, density, cross-sectional area and the density of the sealing liquid are determined by the range for which the instrument is to be used. In the thin-wall instrument, the controlling force is obtained by a spring. The effects due to displacement are small and if the sealing liquid is not very dense they can be neglected. The travel of the bell is proportional to the differential pressure to be measured. This type of instrument needs no overload device. The range will be determined by the modulus of elasticity of the spring and by the density of the sealing liquid. They are very useful for differential pressure measurement at low static pressures, changes of 25 mPa (0.00025 mbar) can be detected. Differential ranges of 0–250 Pa (0–2.5 mbar) and 0–3 kPa (0–30 mbar) are typical at static pressures up to about 1 MPa (10 bar).

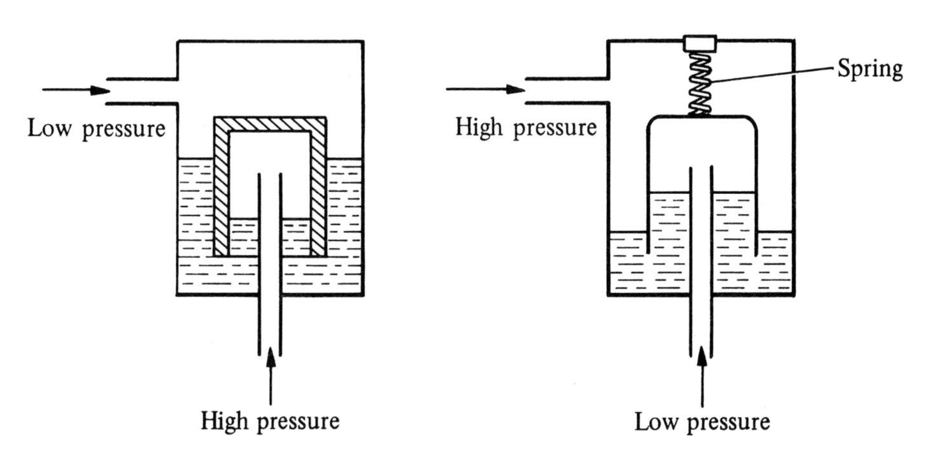

Figure 8:42 Bell-type pressure gauge (thick wall) [12]

Figure 8:43 Bell-type pressure gauge (thin wall) [12]

Measurement of strain or deformation
The pressure gauges employing this principle are:

1 The Bourdon-tube type
2 The diaphragm type

These pressure elements are mechanical devices which are deformed by the applied pressure. They possess elasticity and when deformed the stresses establish equilibrium with the applied pressure. The choice and design of the type of element used depends on the magnitude of the pressure to be measured.

The Bourdon-tube type. This type of instrument is very widely used. It consists of a narrow-bore tube of elliptical cross-section sealed at one end as shown diagrammatically in Figure 8:44. The pressure is applied at the other end which is open and fixed. The tube is formed into an arc of a curve, a flat spiral or a helix. When the pressure is applied the effect of the forces in the tube is to straighten it so that the closed end is displaced. This displacement is magnified and indicated on a circular dial by means of mechanical linkages.

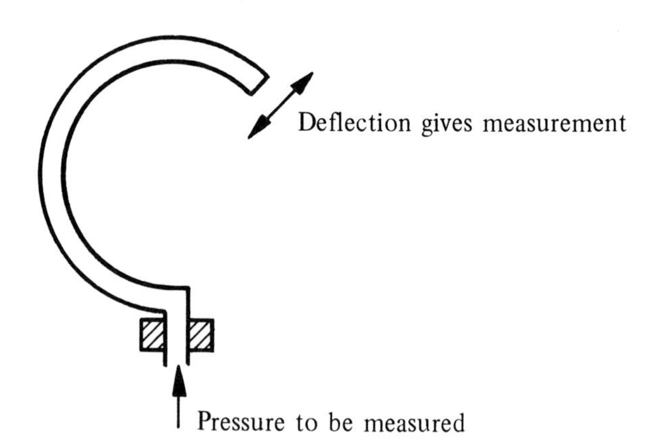

Figure 8:44 Bourdon tube

The tubes are made from a variety of materials in a variety of thicknesses. The material chosen depends upon the nature of the fluid whose pressure is being measured and the thickness on the range of measurement required. The actual dimensions of the tube used will determine the force available to drive the pointer mechanism which should be large enough to make any frictional force negligible. Ranges from 0.1 MPa (1 bar) up to 130 MPa (1300 bar) are available. High-precision types can be produced with a sensitivity of 0.0125 per cent of full scale, an accuracy of ± 0.1 per cent of full scale and a temperature effect of 0.15 per cent of the range per 20 °C. Normal working pressure should not be greater than 60 per cent of the maximum pressure indicated. Figure 8:45 gives the standard range for industrial gauges.

The diaphragm type. The movement of a diaphragm is a convenient way of sensing a pressure differential. The unknown pressure is applied to one side of the diaphragm whose edge is rigidly fixed and the displacement of the centre of the diaphragm is

Service	Range	Maximum working pressure for which the gauge is suitable	
		Steady pressure up to approximately 75% full-scale range	Fluctuating pressure up to approximately 65% full-scale range
	(bar)	(bar)	(bar)
Vacuum	− 1 to 0		
Combined vacuum and pressure	− 1 to + 1.5	1.1	1
	− 1 to + 3	2.2	2
	− 1 to + 5	4	3.2
	− 1 to + 9	7	6
Pressure	1	0.8	0.6
	1.6	1.2	1
	2.5	1.8	1.6
	4	3	2.6
	6	4.5	4
	10	7.5	6.6
	16	12	10
	25	18	16
	40	30	26
	60	45	40
	100	75	65
	160	120	100
	250	180	160
	400	300	260
	600	450	400
	1000	750	650

Figure 8:45 Standard ranges for industrial gauges (BS 1780 : Part 2 : 1971)

transmitted via a suitable joint and a magnification linkage to the pointer of the instrument. The principle is shown in Figure 8:46. Corrugated discs give a larger displacement and can be conveniently combined to form a capsule. A stack of capsules will give even greater deflection.

The materials of construction are chosen to suit the application and the instruments can be used above or below atmospheric pressure. They are better than Bourdon gauges for work below 0.1 MPa (1 bar) and they can measure fluctuating pressure.

Examples of the lower-range instruments are the altimeter and aneroid barometer. Similar instruments built up from diaphragms usually about 100 mm diameter and from 0.005 mm thickness upwards are made to cover ranges from 250 Pa (2.5 mbar) up to 0.4 MPa (4 bar).

An alternative to the diaphragm stack is the bellows element which can be put into instruments to measure differential, absolute or gauge pressure. The 'flexibility' of a bellows is proportional to the number of convolutions and inversely proportional to the wall thickness and modulus of elasticity of the bellows material.

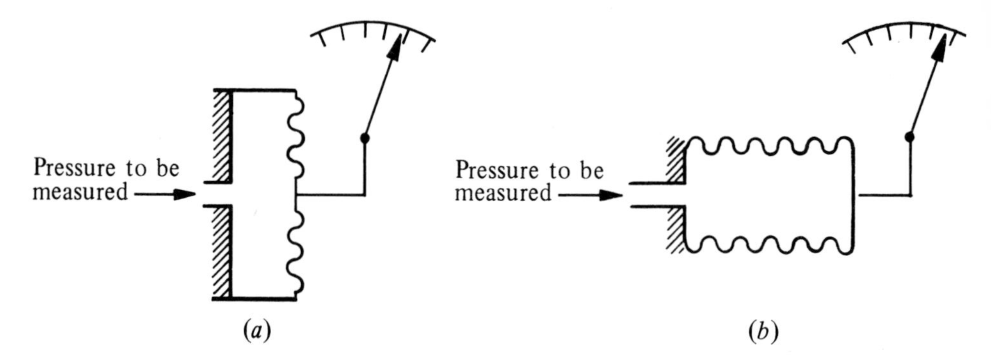

Figure 8:46 Diaphragm-type pressure gauge
(*a*) Diaphragm pressure element
(*b*) Bellows pressure element

The 'compression modulus' of a bellows varies directly as the modulus of elasticity of the bellows material and the cube of the wall thickness. It is also inversely proportional to the number of convolutions and to the square of the outside diameter of the bellows. The spring rate can be increased by fitting an internal spring to oppose the force compressing the bellows. The calibration can be varied by suitable choice of spring. These instruments can be used to measure differential pressure in a line having a high static pressure by enclosing the bellows in a high-pressure casing.

8.4.4 Pressure transducers

Mechanical transducers [8]
These consist of a pressure-sensing element composed of metal diaphragm, capsule and other device capable of translating applied pressure into mechanical displacement (Figure 8:47). This displacement is generally applied to the slider of a potentiometer to produce an electric signal. Typical pressure ranges are −500 Pa to + 500 Pa (− 5 mbar to + 5 mbar) or 0 to 300 kPa (3 bar), but they will operate at pressures up to 35 MPa (350 bar), the range depending on the design of the element used. Accuracy is normally ±1 per cent. Other types are available for high-temperature operation; up to 300 °C without external cooling or up to 1100 °C when fitted with special adaptors.

Electrical transducers
These convert response of a pressure-sensing element into an electrical output and are much better than liquid manometers for remote reading, automatic recording or other signal-processing requirements. Delays in response due to long pressure leads, etc. are avoided. They are essential for measuring rapidly fluctuating pressures.

For steady-pressure measurement various types of pressure transducer are available (Figure 8:48) in which movement of a pressure-actuated diaphragm or bellows effects a corresponding change in voltage, resistance, reluctance, inductance or capacitance.

In one type, movement of a diaphragm is transmitted to a wiper of a potentiometer by means of a magnifying mechanical linkage. In the resistance type a strain

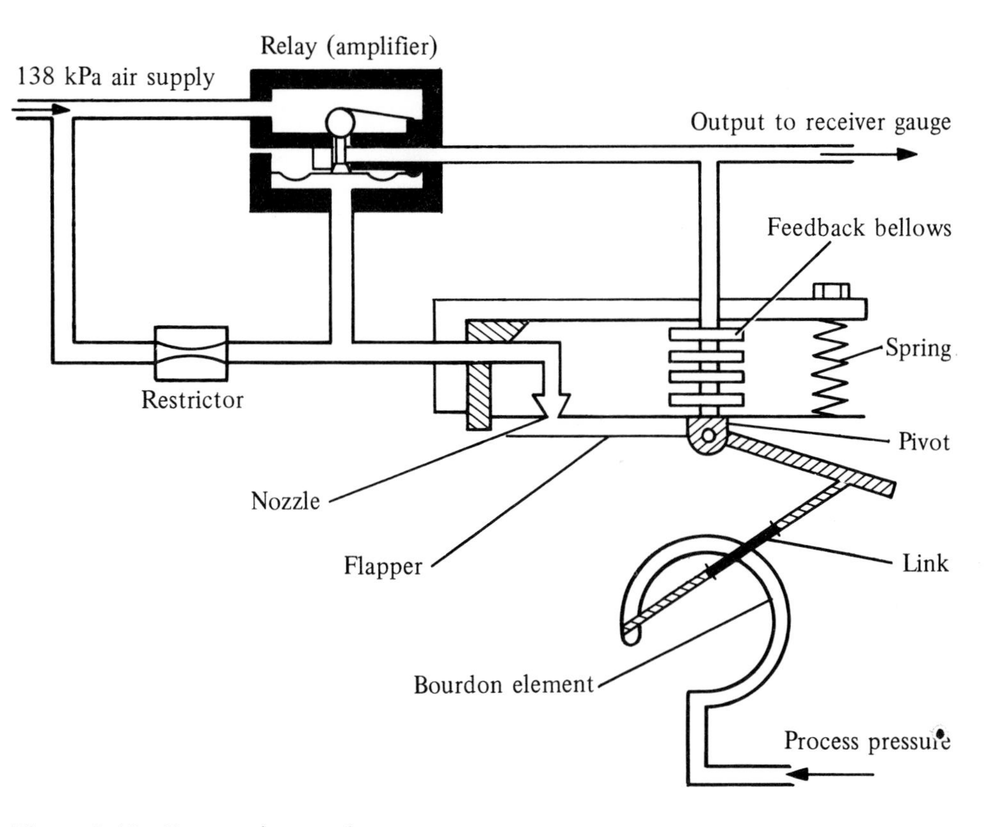

Figure 8:47 Pneumatic transducer

gauge is bonded to the diaphragm or stretched between it and a fixed point. Changes of reluctance or inductance are produced by attaching to the diaphragm a piece of magnetic material or an electrical coil respectively, so that diaphragm movements vary the air gap in a magnetic or electric circuit. In every case the effects produced by the diaphragm movements are measured by standard electrical methods.

For fluctuating pressures, a small transducer is essential to ensure adequate frequency response and also enable siting close to the pressure sensing element, so avoiding long leads. Capacitor microphones are suitable (sizes up to 5 mm diameter). Voltage is supplied from a high-impedance source so that the charge on the capacitor remains effectively constant. The fluctuating capacitance shows as a fluctuating voltage which can be amplified directly.

Pressure transducers have to be calibrated since they are affected by ambient temperature changes (0.02–0.03 per cent maximum output for potentiometer, strain-gauge and reluctance types and 0.1–0.15 per cent for capacitance and capacitor-microphone types). They also give errors if subjected to acceleration.

They are more convenient for measuring large pressure differences than liquid manometers. Their usefulness on low pressure differences is limited by calibration drift and mechanical friction in some types. For steady applied pressure the lower limit is 10 Pa (0.1 mbar).

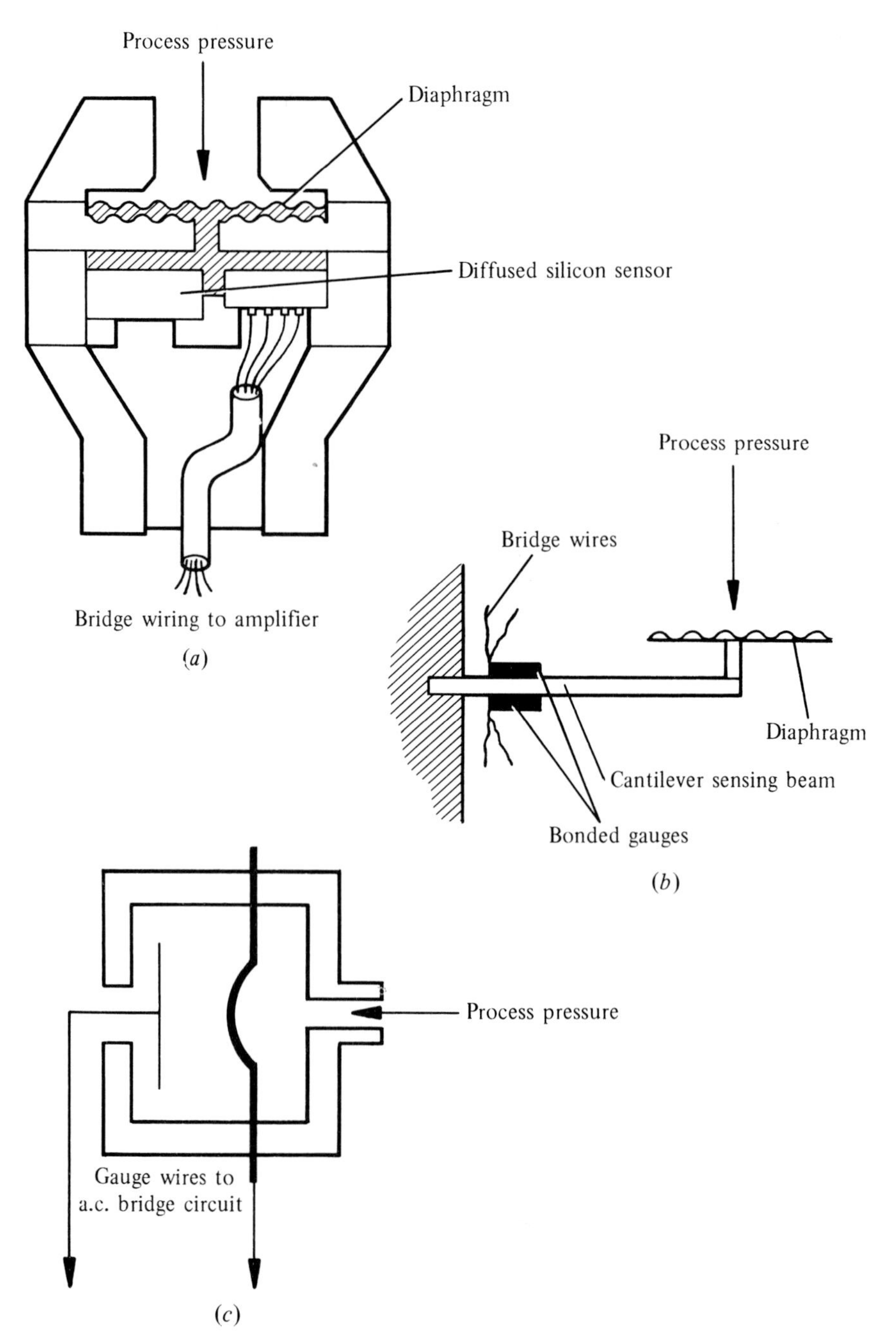

Figure 8:48 Electrical transducers
(*a*) Piezo-resistive silicon chip
(*b*) Bonded-foil gauge
(*c*) Capacitive transducer

8.5 MEASUREMENT OF FLOW

8.5.1 Introduction

Instruments used in the measurement of flow can be divided into two main classes:

1 Quantity meters
2 Rate-of-flow meters

In the first class the total quantity which flows in a given time is measured, to obtain the average flow rate this quantity must be divided by the time taken. In the second class, the actual flow rate is measured; if the total quantity which flows in a given time is required, the quantities flowing in each small interval of time must be integrated.

The instruments in these two classes are then further subdivided according to the nature of their primary elements. Each instrument is considered to have two parts, a primary element and a secondary element. The primary element is defined as that portion of the instrument which converts the quantity being measured into a variable which operates the secondary element. The secondary element then measures this variable which is arranged to depend on the quantity being measured.

Quantity meters for gas measurement, i.e. volumetric meters, are known as positive displacement meters. These are covered in Sections 8.5.2 (diaphragm meters) and 8.5.3 (rotary displacement meters).

Rate-of-flow meters can be divided into those which use direct velocity measurement and those using pressure-differential measurement. The direct velocity types are covered in Section 8.5.4. (turbine meters) and Section 8.5.5 (anemometers). Pressure-differential types (Section 8.5.6) are subdivided into (*a*) constant-area, variable-head methods and (*b*) variable-area, constant-head methods.

8.5.2 Diaphragm meters [23]

The principle of measurement is positive displacement. Figure 8:49 shows the sequence of operation. Four chambers in the meter fill and empty in turn. The exact volume of the compartments is known, so by counting the number of displacements the volume is measured. Sliding valves control the entrance and exit of the gas to the compartments. Compartments 2 and 3 are encased by bellows which expand and contract.

Tinplate meters are usually only suitable for pressures up to 5 kPa (50 mbar) although a special range is available up to a capacity of 170 $m^3 h^{-1}$ for working pressures up to 35 kPA. The new steel case that will be available up to a capacity of 160 $m^3 h^{-1}$ will be able to withstand working pressures up to 20 kPa as standard. Meters with aluminium die-cast cases are available with capacities up to 280 $m^3 h^{-1}$ at pressures up to 700 kPa. The specifications for these meters are contained in BS 4161 parts 1 to 5 [24–28] and their installation is covered in BSCP 331 part 2 [29]. The IGE Recommendations on Gas Measurement Practice [30] are discussed in Section 8.6.

Meters of the types mentioned above are installed in a large number of commercial and industrial consumers' premises. The revenue received by the gas industry from

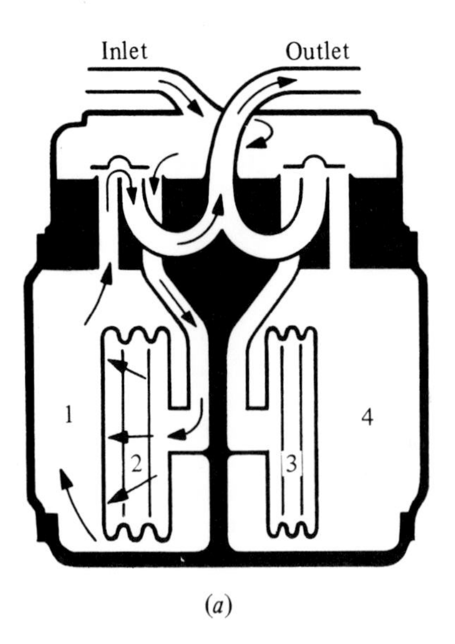

(a)

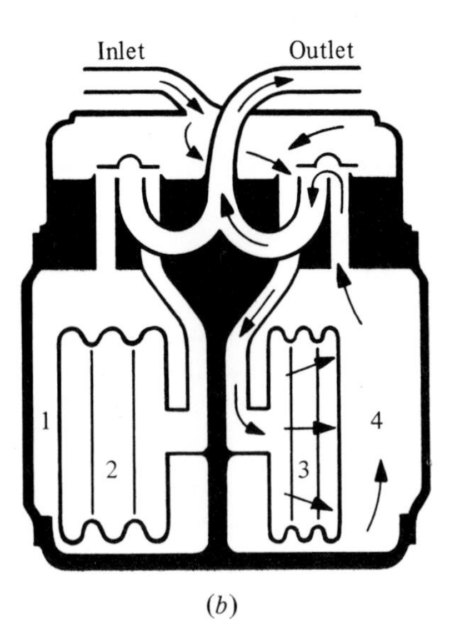

(b)

(c)

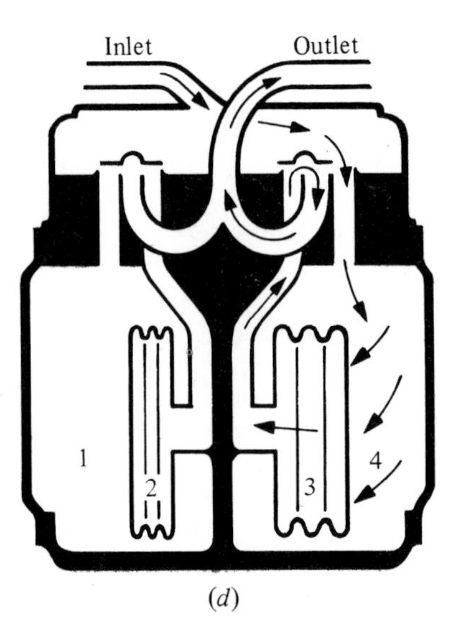

(d)

Figure 8:49 Sequence of operation of a diaphragm meter
(*a*) Chamber 1 is emptying, 2 is filling, 3 is empty and 4 has just filled
(*b*) Chamber 1 is now empty, 2 is full, 3 is filling and 4 is emptying
(*c*) Chamber 1 is filling, 2 is emptying, 3 has filled and 4 has emptied
(*d*) Chamber 1 is now completely filled, 2 is empty, 3 is emptying and 4 is filling

these consumers is measured solely by these meters and therefore their performance in the field is very important.

Investigations have shown that older types of meters suffer from an increase in pressure absorption caused by breakdown of lubrication on the meter linkages and sliding valves [23]. This causes the diaphragms to balloon out at the end of a stroke and the resultant increase in swept volume produces slow registration, particularly on very low flow rates.

On current designs, the use of new materials and design features has been directed towards reducing pressure absorption, meter size for a given rating, maintenance and production costs and increasing strength and durability. References [23] and [31] deal more fully with resistance–power ratio and the materials used for meter construction, diaphragms and dry lubrication.

Equipment (Section 8.6.3) that will automatically register correction for variation in pressure and temperature is available. Provision to take this equipment is built into aluminium-case meters. However, tinplate and steel-case meters are not easily adapted to take correcting equipment.

Remote-reading systems have recently been developed and most meter designs are capable of being modified to take the necessary index unit if at any time in the future a remote-reading system is operated.

8.5.3 Rotary displacement meters

A rotary displacement meter is defined as 'a positive displacement meter in which the measuring compartment is formed between the walls of a stationary chamber and a rotating element or elements making substantially gas-tight contact with the walls' [32].

Two different types of rotary displacement meter are available:

1 The Roots type
2 The rotary vane type

The Roots meter [33]

This is shown diagrammatically in Figure 8:50. The two impellers are geared together by timing gears which ensure that they contrarotate at the same speed. The impellers intermesh closely at all times which ensures that gas can only pass forward from the inlet to the outlet through the measuring chambers which are formed between the sides of the impellers and the walls of the casing. There is no contact at any time between the impellers or between the impellers and the casing. The clearances between the various parts are as little as 0.05 mm for all positions of the impellers on small meters. On very large low-pressure meters the clearances may exceed 0.2 mm.

The impeller shafts run in ball or roller bearings which are amply dimensioned for the comparatively light duty they have to perform. The pressure drop across a meter amounts to 100–500 Pa (1–5 mbar).

The flowing gas drives the impellers round because of the unbalanced load on one or other impeller, when pressure is applied at the inlet. In the diagram the upper impeller is uniformly loaded and has no turning force on it. On the other hand the upper side of the lower impeller has a greater pressure on the half exposed to the inlet pressure, than that exposed to the outlet pressure. If the inlet pressure is high enough the impeller will rotate anticlockwise discharging the contents of the measur-

Figure 8:50 Roots-type meter

ing chamber to the outlet. The timing gears cause the upper impeller to rotate clockwise at the same time. At the halfway point the lower impeller will be pressure balanced, but the upper impeller will now be out of balance and rotation will continue.

Work will be done in causing the impellers to rotate because of friction in the bearings and gears and viscous drag in the clearances. The energy required is taken from the gas and is shown as a pressure loss in the gas. This pressure loss causes gas to pass forward through the clearances round the impellers and is known as 'slip'.

There is an additional pressure loss between the inlet and outlet of the meter due to the peculiar pattern of flow of gas through the meter. This is known as the 'aerodynamic' pressure loss.

In practical meters the percentage slip is substantially constant from less than 10 per cent of the normal rated maximum throughput up to the normal rated throughput. The slip is normally of the order of 1 ± 0.5 per cent of the actual throughput over the load range 5–10 per cent to 100 per cent. It is taken care of in the meter counter which is geared to place the calibration curve within the required tolerance.

Rotary-vane meter [33]

This type is shown in Figure 8:51 and is sometimes referred to as the 'vane-and-gate' or 'rotary-piston' type. Basically it consists of an annular measuring chamber formed between the casing and a static central cylinder, with vanes rotating around the annular space with equal volumes of gas between successive vanes. A second rotor, driven by the vanes through timing gears, acts as a gate to allow the vanes to return from the meter outlet to the meter inlet while preventing gas from bypassing the metering annulus.

There are small clearances round the rotating parts where large-area seals minimize slip at low flow rates.

Depending upon the make of meter the vanes and gate rotate in the same or opposite directions.

Construction

The casings, impellers, rotors and shafts of both types are mainly cast iron or steel

Figure 8:51 Rotary-vane meter

with aluminium impellers on the smaller meters. However, there are some smaller meters that are constructed with aluminium cases and rotors. If necessary special materials or material coatings can be used. The design and construction is such that there is no change in measurement accuracy with time, the calibration curve can be used indefinitely unless the meter case becomes distorted through use at excessive pressures (see under 'Meter performance', below).

Meter performance
A theoretical treatment of the behaviour of rotary displacement meters is given in [33] and is compared with practical tests. A brief summary is given here of the factors which affect meter performance [34].

Pressure loss. How the pressure loss occurs and is built up is described on page 366. The broad picture for performance is that at low speeds, low throughputs and low pressures, the differential pressure is almost constant. It increases only slightly as meter speed increases. At high speeds, high throughputs and high pressure, the relationship between speed and differential pressure approaches the square law. At the maximum meter rating the pressure loss, when operating at low pressure, is approximately 500 Pa (5 mbar) irrespective of make or type.

Calibration
A 'hand slip calibration' may be carried out to obtain a speed/differential-pressure curve, speed/slip curve and percentage-slip/throughput (error) curve.

The slip may be measured directly by passing air through a blocked meter and measuring it. A slip/differential-pressure curve is produced.

A 'direct calibration' would be the ideal way. This has not been possible in the past because of the cost and the need for a suitable reference meter. The British Gas Corporation now has facilities for testing meters at high pressure and high loads on natural gas.

Accuracy. This depends on:

1 The meter itself

2 The physical properties of the gas being measured
3 The loading
4 Outside influences

The meter. The exact volumetric capacity of the measuring chambers must be known. The meter must be constructed as accurately as possible to the design specifications.

Physical properties of the gas. The main characteristics concerned are the density and viscosity. They vary from one gas to another and for one gas with temperature and pressure. They control, together with the mechanical loss, the actual pressure difference between inlet and outlet, depending on the load and the amount of slip due to the pressure difference from inlet to outlet.

Loading. Accuracy is maintained over the range 10 to 100 per cent of the meter rating.

Outside influences. Tests have shown that pipeline configuration as such has little effect on the overall accuracy at low pressures. However, resonance effects associated with specific loadings on the upstream pipeline can cause detectable errors.

Use. Rotary-displacement meters have been accepted for the sale of gas since 1955 [33] and can easily meet the 'badging' requirements (Section 8.6) being accurate to within 1 per cent over the range 10 to 100 per cent of maximum flow.

Most meters can be used beyond their rated capacity if the pressure available is sufficient to absorb the increased pressure loss. The accuracy is maintained. The only problems are noise and wear on gears.

High-pressure meters

Accuracy. Meters are usually calibrated on air at low pressure. When calibrated on gas the slip increases at low flows; at large flows the difference disappears. Testing at pressure shows that slip tends to be reduced and can become very small. It is concluded that meters calibrated at low pressure can be used at high pressure provided that no distortion occurs.

Distortion of the casing. Meters are available for pressures up to 10 MPa (100 bar, 1450 $\mathrm{lbf\,in^{-2}}$); they are heavy because of case thickness but are relatively small even for very large volumes. For the higher pressures the pressure on the measuring chamber must be relieved in some way and for the lower pressures the casings are strengthened by ribs to prevent distortion.

Indication. High-pressure operation brings manufacturing difficulties due to the difficulty of keeping glands tight. The counter may be fitted inside the casing, magnetic couplings can be used, or alternatively, electrical methods may be adopted, particularly where the readings are corrected for operating conditions or are transmitted.

Capacities. Almost all rotary displacement meters are suitable for use up to pressures of 850 kPa (8.5 bar, 123 $\mathrm{lbf\,in^{-2}}$) in a capacity range up to 6500 $\mathrm{m^3\,h^{-1}}$. Certain makes and models of meters are available for higher pressures up to a maximum of 10 MPa (100 bar, 1450 $\mathrm{lbf\,in^{-2}}$) at which the greatest capacity is 30 750 $\mathrm{m^3(st)\,h^{-1}}$ (1 107 000 $\mathrm{sft^3\,h^{-1}}$). At the other end of the scale the smallest meters can register down to 3 $\mathrm{m^3\,h^{-1}}$ when operating on low pressure.

8.5.4 Turbine meters [35]

A gas turbine meter is a velocity device for measuring flow volume, in which the direction of flow is parallel to the rotor axis of the meter and where the speed of rotation is proportional to the rate of flow.

In-line type
For gas measurement an axial-flow meter must be designed to have low non-fluid retarding torques so that rotor slip caused by these torques is always small within a reasonable flow range. Figure 8:52 is a drawing showing the basic measurement

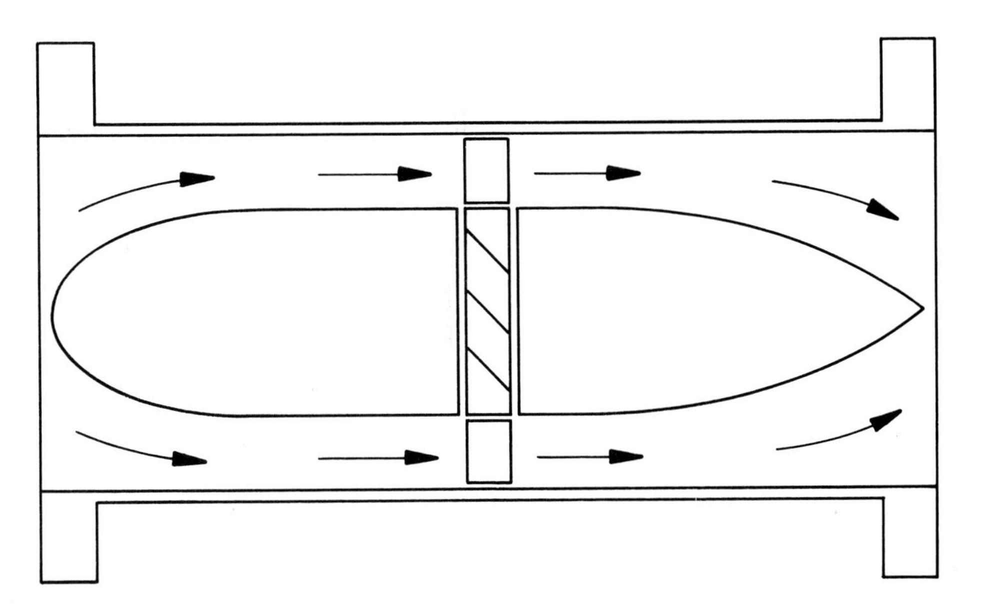

Figure 8:52 Axial-flow turbine meter

elements. Additionally, a drive and index is required to convert the gas velocity directly into volume. The relationship between volume and output is linear. Commercial meters basically follow the theoretical design shown in Figure 8:52. They consist of a body, an internal rotor mechanism with inlet and outlet diffusers, a magnetic drive and an index or instrument to provide totalization. The materials used vary with different types, usually meter bodies are made of ductile iron, aluminium or steel, rotor assemblies of plastic (e.g. Delrin, Rilsan, etc.) and/or aluminium and diffusers of plastic or aluminium.

The meter obtains its driving energy from the kinetic energy of the gas, which is caused to increase as the gas passes through the meter by reducing the cross-sectional area of the flow stream. This is done by diffusers which direct the gas to the outer periphery of the meter body. To keep the total pressure drop as low as possible, the flow passages are contoured similarly to a venturi. To create the highest torque for a given force the rotor must be constructed with the largest possible effective radius.

The theoretical design can be based either on a 'momentum' approach or an 'air-foil' approach [35]. In both cases a simplified expression for the percentage registration is

$$P_r = 1 - K(Q_n + Q_f)/\rho V^2 = 1 - (KQ_r/Q_d)$$

where P_r = percentage registration
K = constant
Q_n = retarding torques due to non-fluid forces (friction)
Q_f = retarding torques due to fluid forces (turbulence)
ρ = density of fluid
V = volumetric flow rate
Q_r = total retarding forces
Q_d = maximum available driving torque at a given flow rate

The above expression for available driving torque, shows that the meter characteristics also depend on the retarding torques. The percentage slip is directly proportional to the retarding torque which is made up of:

1 Retarding torques due to non-fluid forces such as rotor-bearing friction and mechanical loading from the index or instrument used with the meter. These torques can be assumed constant for a given meter.
2 Retarding torques due to fluid forces such as fluid-skin friction drag on the rotor which is a direct function of the Reynolds number and a second fluid drag due to turbulence of the gas or anything which results in loss of energy as the gas moves past the rotor. The torque due to turbulence is proportional to the mass of the fluid times the velocity squared, as is the driving torque, so the retarding torque due to turbulence causes a constant percentage slip throughout the flow range of the meter.

Capacity. Units are available in nominal pipe sizes at low pressure of 50 mm (2 in) having a range of 14–113 $m^3 h^{-1}$, up to 1015 mm (40 in) with a maximum capacity of 45 300 $m^3 h^{-1}$ and various pressure ratings are available from 250 kPa (2.5 bar, 36 lbf in^{-2}) to 10 MPa (100 bar, 1450 lbf in^{-2}). The capacity increases as the pressure increases approximately by the number of bars increase.

The maximum capacity rating is established by three factors:

1 The pressure drop across the meter. On low-pressure installations this is usually in the region of 500–800 Pa (5–8 mbar), whilst on higher pressures it may rise to approximately 5 kPa (50 mbar).
2 The end thrust of the rotor shaft and bearing caused by the differential pressure
3 The rotational frequency of the rotor assembly

The minimum capacity of a turbine meter is the lowest rate that will result in an acceptable accuracy. All types of meter will start to rotate at flows below the recom-

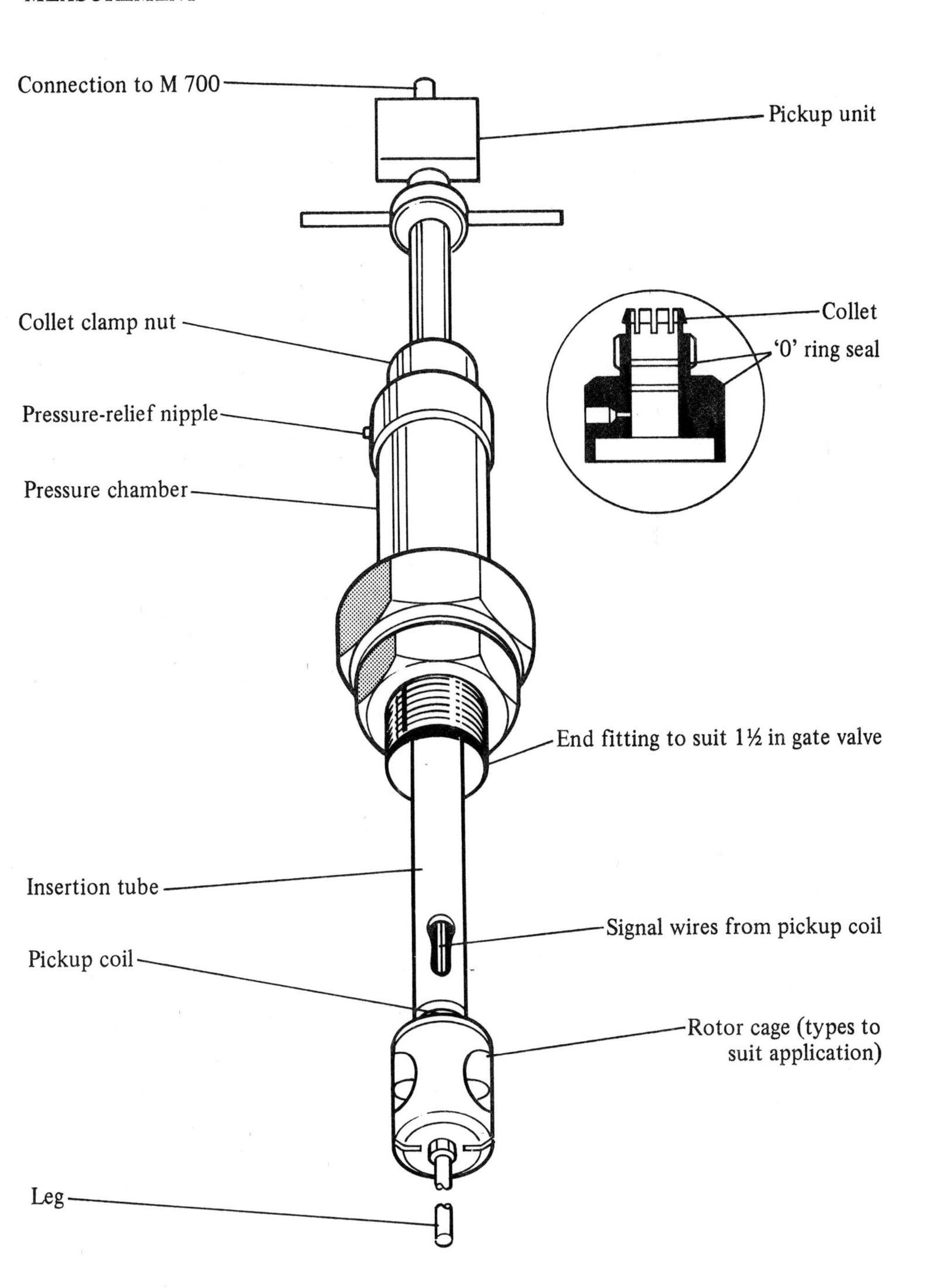

Figure 8:53 Low-pressure insertion meter

mended minimum, which is fixed, however, to ensure the accuracy requirements of ± 1 per cent.

The rangeability is the ratio of the maximum meter capacity to the minimum capacity for specific operating conditions during which the desired accuracy is retained. This figure increases with pressure for turbine meters:

$$\text{rangeability} = RP_F/P_b$$

where R = rangeability at atmospheric pressure
P_F = absolute flowing pressure
P_b = absolute base pressure

Application. The relatively small size, large volume capacity and rangeability of gas turbine meters makes them very suitable for measuring large volume flows. They are compact and light in weight, which makes them easily moveable. They are generally supported by the adjoining pipework and do not need large foundations. Most turbine meters must be mounted horizontally although there are some that may be used in vertical pipework. They can be used for transmission, distribution or industrial applications provided that the flow range is within the operating range of the meter or braking devices or 'switch-over' arrangements are utilized (Section 8.6.2).

In addition to the usual instrumentation, turbine meters can be supplied with correcting devices, etc., identical to those used on other meters (Section 8.6.3).

Insertion type

These instruments have been developed to measure directly the rate and direction of gas flow in individual mains, to validate network analysis models, establish boundary conditions and identify blocked or unconnected mains [36, 37].

The requirements are ease of insertion, portability, robustness and simplicity of operation. The actual rotor cage must be of such a size that a standpipe and valve assembly can be attached to mains down to 100 mm diameter. The meter must give a direct reading (or suitable output signal) in standard volumetric units for a wide range of mains sizes. The calibration should be unaffected by the range of operating densities to be encountered and the meter should be insensitive to small angles of misalignment. The velocity range should be from 0.3 m s^{-1} up to 15 m s^{-1}. Figure 8:53 shows the meter developed by the British Gas Corporation South Eastern Region [36].

8.5.5 Anemometers

There are several types of anemometer, the majority of which are designed to measure air velocity from air diffusers or grilles, with a particular type incorporated in a type of gas meter.

Gas-meter type

The rotary inferential gas meter is designed specifically as a secondary meter for use in industrial and commercial applications. The meter consists of a cast-iron housing within which an aluminium anemometer fan rotates on a vertical shaft which in turn transmits the drive to a counter mechanism that integrates the flow (Figure 8:54). These meters are of rugged construction and are suitable for most non-corrosive

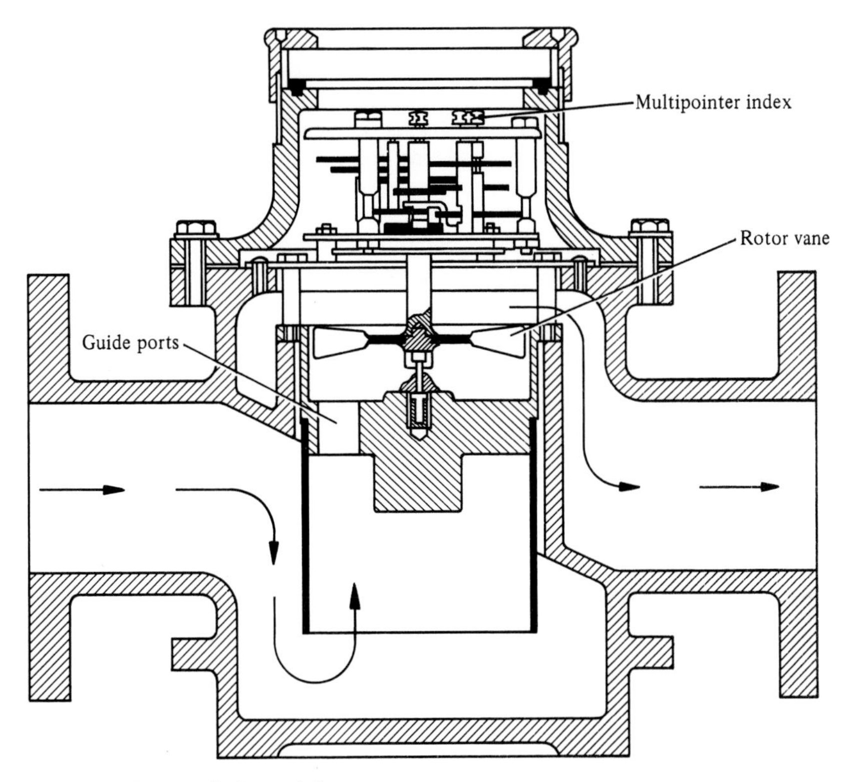

Figure 8:54 Rotary inferential gas meter

gases. They are available in a size range from 4 $m^3 h^{-1}$ to 200 $m^3 h^{-1}$ (141 to 7050 $ft^3 h^{-1}$) at pressures up to 170 kPa (1.7 bar, 24.7 lbf in^{-2}). Accuracy is said to be within ± 2 per cent over a 10 : 1 flow range for a given meter and they must be installed in horizontal pipework.

Air-flow meters

These instruments are designed for measuring air velocity from grilles or diffusers in air-conditioning systems or air movement in rooms and buildings.

Rotating-vane type. The vanes must be light or have a large surface area. The rotor as a whole must be accurately balanced and the bearings must be as friction-free as possible.

The rotor may be of two kinds. In the cup anemometer it consists of three light conical-shaped cups, often made of aluminium, attached to arms carried by a vertical spindle, which in the more sensitive type of instrument runs in jewelled bearings and is geared to a light counting mechanism. The range of the instrument is from 0.13 m s^{-1} to 30 m s^{-1} (0.43 to 98 ft s^{-1}) and it is used for meteorological work [12].

In the air-meter type the rotor consists of a multivane assembly made of aluminium alloy carried on a stainless steel shaft; the other end of the shaft is connected to the counter mechanism; all bearings are enclosed. The instrument measures the air passing in linear metres and should be timed for 100 seconds and the appropriate correction applied to convert the reading to velocity. The velocity range is from 0.3 m s^{-1} to 25 m s^{-1} (1 to 82 ft s^{-1}). The instrument can be designed to measure and/or record the flow rate of air or gases directly, by mounting a capacitance transducer on the outer ring of the anemometer. As the vane assembly rotates a succession of electrical output pulses are developed and fed into the indicator unit, where they operate a conventional meter calibrated directly in m s^{-1} [38].

Hot-wire instruments. When a fluid flows over a heated surface, heat is transferred from the surface and the reduction in temperature is related to the rate of flow [13]. In a hot-wire anemometer, heat is supplied electrically to a wire placed in the flow stream. Figure 8:55 shows the operating principle of the instrument. The temperature

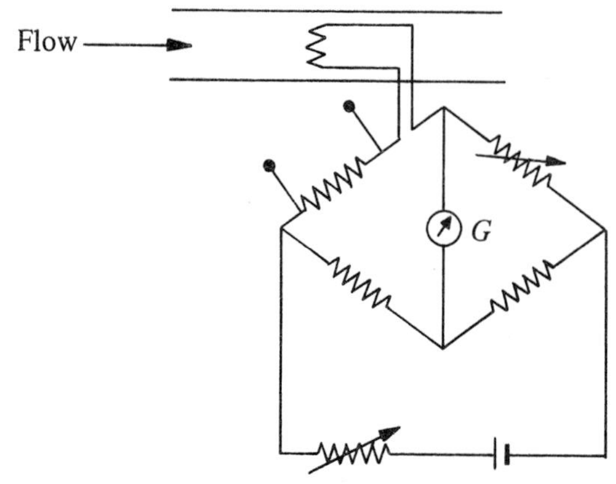

Figure 8:55 Hot-wire anemometer

of the wire is obtained by measurement of its resistance using a Wheatstone bridge. One procedure involves adjusting the current through the wire so that the temperature remains constant and measuring the heating current. In this way the bridge is always balanced. The current is measured by finding the potential difference across a standard resistance in the appropriate arm of the bridge by a potentiometer. It can be shown that the rate of loss of heat from a heated wire is proportional to $(u\rho + b)^{\frac{1}{2}}$ where u is the velocity of flow, ρ is the density and b is a constant. At equilibrium, heat supplied equals the heat lost giving:

$$I^2 R = a\,(u\rho + b)^{\frac{1}{2}}$$

where I is the current, R is the resistance and a is a constant. Hence:

$$u\rho = (R^2 I^4/a^2) - b$$

Thus if the resistance and temperature of the wire are kept constant, a relation is obtained between the velocity of flow and the fourth power of the current. These instruments are available to measure air or gas flows over a wide range. They are particularly useful at low flow rates for laboratory and research applications.

A similar type of instrument [38] which is available for the measurement of air velocities through supply grilles and ceiling diffusers in air-conditioning, ventilating systems, etc., within the range 0 to 5 m s^{-1} (0 to 16 ft s^{-1}) is the Davimeter. The instrument consists of a small hand-held probe made from stainless steel which contains two thermocouple sensing elements. The couples are heated by alternating currents. A change in flow causes a change in temperature of the couples, resulting in a change in the d.c. output. A third thermocouple is put in the d.c. meter circuit identical to the other two but is unheated. A change in ambient temperature will develop voltages in all the couples, but the transient effects in heated and unheated elements are equal and opposite so compensating for changes in ambient temperature. The instrument can measure flows up to 30 m s^{-1} (98 ft s^{-1}). The scale is non-linear and has widely spaced divisions at the lowest velocities, which makes it very suitable for measuring air movement between 0.05 and 2.5 m s^{-1} (0.16 and 8.2 ft s^{-1}) and particularly from 0.05 to 0.5 m s^{-1} (0.16 to 1.6 ft s^{-1}).

8.5.6 Pressure-differential methods

Constant area, variable head [39]

Basic considerations. The application of Newton's laws to a fluid (Chapter 2) shows that a change in velocity, either speed or direction, is associated with a change in pressure. Together with continuity considerations and a knowledge of the equation of state of the fluid, this differential pressure ΔP can be used to infer the flow V.

A pressure-differential flow meter is a device which is put into the fluid path to cause a restriction in some way. This then causes an increase in velocity which is accompanied by a change in pressure, from which the flow can be inferred. The essential features of every device are:

1 A primary element in the flow line to provide the restriction
2 A secondary element which is responsive to the change in pressure, together with a unit to convert the differential pressure into a flow rate

An important distinction must be made between devices where the whole fluid mass is accelerated, and which respond to flow, and devices responding to local velocity only, in which a traverse must be made to find the flow.

There are two main methods of generating the differential pressure:

1 By a change of speed (Bernoulli head)
2 By a change of direction (curvature head)

Most devices have an associated permanent pressure loss, which is an important economic consideration on running costs. The pressure loss can be used to classify the devices but it is better to use loss/differential.

For one-dimensional steady flow of an inviscid fluid:

$$(\mathrm{d}P/\rho) + u\mathrm{d}u = 0 \tag{8.1}$$

where P is pressure, u is velocity and ρ is density. Continuity considerations give:

$$(\mathrm{d}P/\rho) + (\mathrm{d}A/A) + (\mathrm{d}u/u) = 0 \tag{8.2}$$

where A is flow area.

Before equations 8.1 and 8.2 can be used, an equation of state, relating P and ρ, is required. It is normal to assume that:

$$\text{for an incompressible fluid } \rho = \text{constant} \tag{8.3}$$
$$\text{for a compressible fluid } P/\rho^{\gamma} = \text{constant} \tag{8.3a}$$

If equations 8.1, 8.2 and 8.3 are applied to a typical constriction, assuming that the geometric area is equal to the flow area:

$$\text{volume rate of flow} = V = A_t E \sqrt{(2\Delta P/\rho)} \tag{8.4}$$
$$\text{mass rate of flow} = m = \rho V = A_t E \sqrt{(2\Delta P \rho)} \tag{8.4a}$$

where $E = (1 - r_a)^{-\frac{1}{2}}$, r_a = area ratio and A_t = throat area.

Equations 8.4 and 8.4a show that both V and m are proportional to $\sqrt{(\Delta P)}$. But $\Delta P = \rho g h$ and so $V = A_t E \sqrt{(2gh)}$. Thus if the differential pressure is measured as an actual head of the flowing fluid the meter is truly volumetric. If ΔP is measured in any other way then ρ must also be known to obtain either the volume or mass flow.

To account for all the simplifying assumptions and make equations 8.4 and 8.4a applicable to a real situation, two empirical coefficients are introduced, the discharge coefficient C_d and an expansibility factor ε to give:

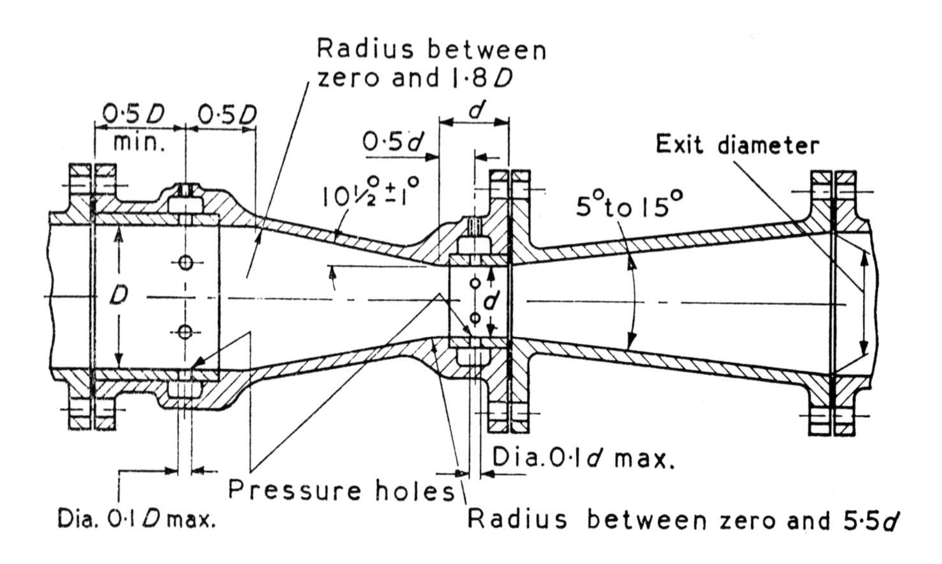

Figure 8:56 Classical venturi [51]

$$m = C_d \, \varepsilon A_t E \sqrt{(2 \Delta P \rho)}$$

where
$$\varepsilon^2 = \left(\frac{\gamma}{\gamma - 1}\right) r_p^{2/\gamma} \left(\frac{1 - r_p^{(\gamma-1)/\gamma}}{1 - r_p}\right) \left(\frac{1 - r_a^2}{1 - r_a^2 \, r_p^{2/\gamma}}\right)$$

r_p = pressure ratio = $1 - (\Delta P/P)$
γ = ratio of specific heats

The discharge coefficient C_d is an empirical coefficient to correct for an inadequate theory. The factors affecting C_d depend basically on the geometry of the device and

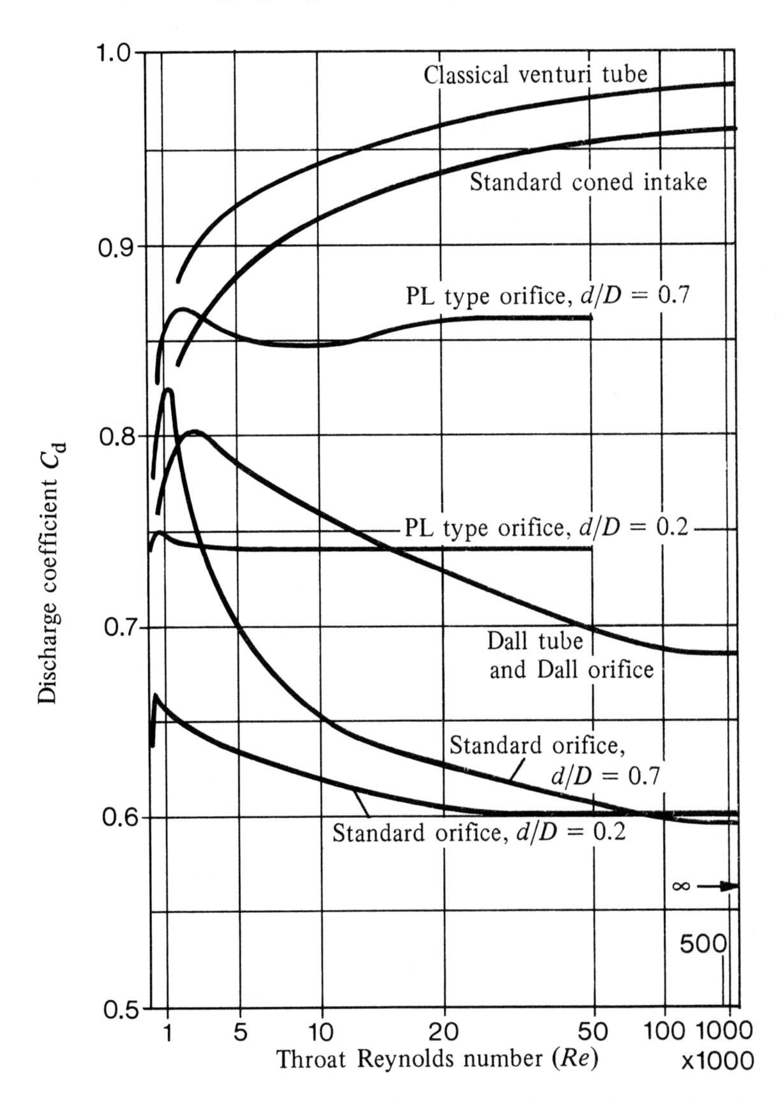

Figure 8:57 Variation of discharge coefficient with Reynolds number for various pressure-differential devices

the invalidity of the four assumptions of one-dimensional, steady flow of an inviscid incompressible fluid used to obtain equation 8.4a. They are dealt with more fully in reference [39] and Chapter 2.

The venturi meter [39]. The geometry of a classical venturi is shown in Figure 8:56. C_d is practically constant at 0.988 when (Re) at the throat is greater than 3×10^5. Figure 8:57 shows the variation of C_d with (Re) for a number of pressure-differential devices. The fact that $C_d \approx 1$ indicates that the theory is reasonable. The geometry ensures no curvature effects at the tappings and almost one-dimensional flow in the throat due to the acceleration.

Care must be taken in the construction, particularly on the blending radius between the upstream cone and the throat which may cause separation effects and on the throat tapping which is in the highest velocity region.

The device has a low pressure loss (Figure 8:58) but at the expense of large size due to the downstream diffuser.

The size of the classical venturi can be reduced by making a nozzle venturi, in

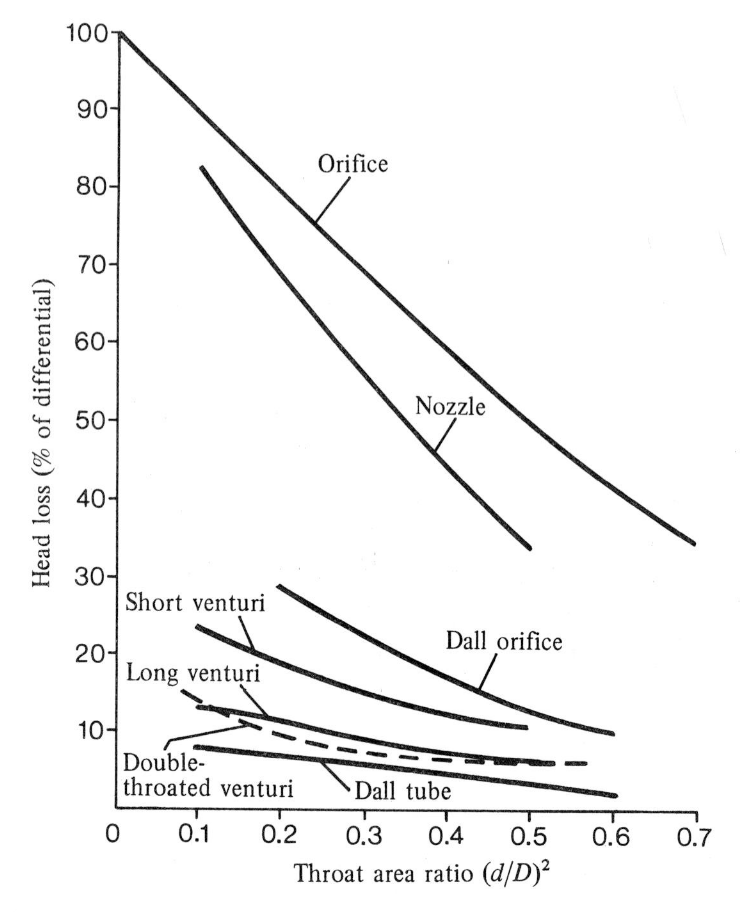

Figure 8:58 Head losses of various pressure-differential devices

which the upstream cone is replaced by a nozzle. However, this is harder to machine than a cone and the upstream tap is in a region of flow curvature. The performance is very similar to the classical venturi.

The size can be still further reduced by discarding the diffuser. The resulting device is known as a nozzle. However, the losses are increased. The performance of a nozzle is similar to that of a nozzle venturi.

The Dall tube is another modification to the venturi which makes use of the streamline curvature at the throat tap to greatly enhance the differential. The double-throat venturi is another device which is very similar to the Dall tube.

The orifice plate [39]. An example of this device is shown in Figure 8:59. It is much simpler and smaller than the venturi and the nozzle. The discharge coefficient is about 0.60 and remains practically constant above $(Re) = 5 \times 10^5$ as seen in Figure 8:57. The main reason for the value of $C_d = 0.6$ is that the stream contracts downstream of the orifice and the flow area is only 0.6 of the actual orifice area.

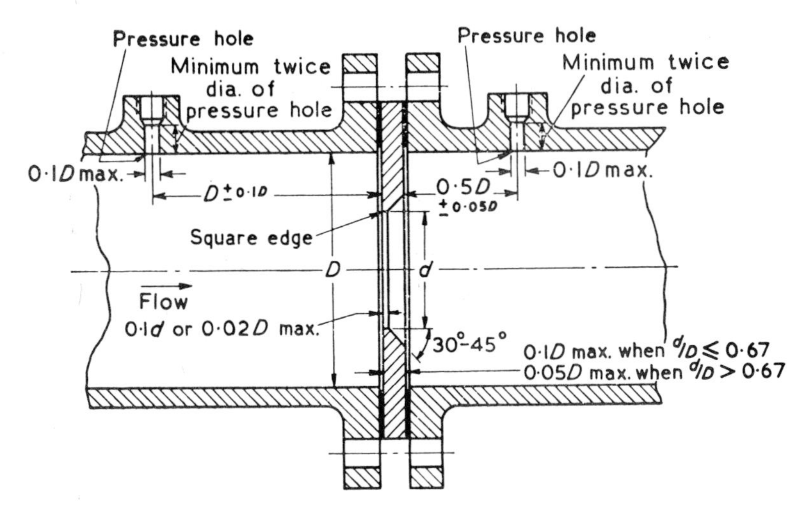

Figure 8:59 Orifice plate with D and $\frac{1}{2}D$ tappings [51]

The most important geometric parameter of the orifice is its sharp edge, which fixes the separation point and determines the contraction. It is much easier to reproduce and maintain geometric similarity than the venturi devices and it is much less expensive. The pressure loss, for the same area ratio is greater than that of a nozzle, but for the same differential pressure they have equal losses. Compressibility effects are smaller than they are with the venturi.

Many tapping positions can be used. They do not alter the fundamental nature of orifice flow but can add local curvature effects or have practical advantages. With vena-contracta taps the upstream tap is one diameter upstream and the downstream tap is at the vena contracta, i.e. minimum pressure. There are no curvature effects, but the position of the vena contracta varies with r_a and gives practical difficulties when interchanging orifices. D and $\frac{1}{2}D$ taps are a compromise vena-contracta tap with the downstream tap at $\frac{1}{2}D$ which is an average vena-contracta position. A practical disadvantage of both these types of tap is that they are located in the pipe wall, remote from the orifice and it is very difficult to supply a combined orifice and tap system. Flange taps and corner taps overcome this and allow a simple orifice carrier and taps to be supplied for insertion between flanges. Flange taps are usually 25.4 mm from the orifice, which makes C_d a function of diameter and brings in an added correction. Corner taps maintain similarity and avoid this problem, making use of curvature effects to enhance ΔP and reduce C_d particularly at high r_a.

The orifice discussed is mounted concentrically within the pipeline. Variants such as the eccentric, segmental and annular types exist to handle dirty fluids. In these, however, the pipe wall becomes part of the throat section with consequent difficulties of reproduction. Another orifice device is the conical entrance (PL) type which is designed to maintain C_d constant to much lower (Re) than normal types.

Orifice meters are extremely important to the British gas industry. All natural gas is currently bought and the largest gas sales made using these meters [40]. It is very difficult to quote accuracies in relation to orifice meters when so much depends on the complete installation and conditions of flow [41, 42]. Reference will be made to the design standards in Section 8.6.

Pitot tubes [12, 43]. The Pitot-static tube is a very useful device for making temporary measurements of flow. It actually measures the velocity at a point, but by traversing the pipe or duct and measuring the velocity at several points it is possible to obtain the average velocity and hence the volume flowing.

The principle of operation depends on the fact that if a tube is placed with its open end facing into a stream of fluid, then the impinging fluid will be brought to rest and its kinetic energy converted into pressure energy. Thus the pressure built up in the tube will be greater than that in the free stream by the 'impact pressure'. This increase will depend on the square of the velocity of the stream. The difference is measured between the pressure in the tube and the static pressure of the stream. The static pressure is measured by a tapping in the pipe wall or by tappings in the Pitot-static tube itself. The pressure difference obtained is a measure of the 'impact pressure' and therefore of the velocity of the fluid at that point.

$$\text{gas velocity} = u = \sqrt{[2(P_t - P_s)/\rho]}$$

where ρ = density
P_t = total pressure
P_s = static pressure

A number of Pitot devices have been developed [5] and the standard type was developed by the National Physical Laboratory as a combined type tube with a hemispherical end as shown in Figure 8:60. The tip must be truly hemispherical and the impact hole truly cylindrical, free from burrs and central with respect to the hemisphere. The tubes are small enough compared with the pipe to incur negligible losses, but must not be so small that vibration becomes a problem.

Pitot tubes are essentially exploratory devices and are rarely used permanently in industry. They can be used to monitor gas flows in the manner described in Section 8.5.5 for insertion turbines. They are of more interest to industrial gas engineers for velocity and direction measurements in combustion studies and heat-transfer work in industrial furnaces. A great deal of work has been carried out by the International Flame Research Foundation on their use for this purpose [5]. A commercially avail-

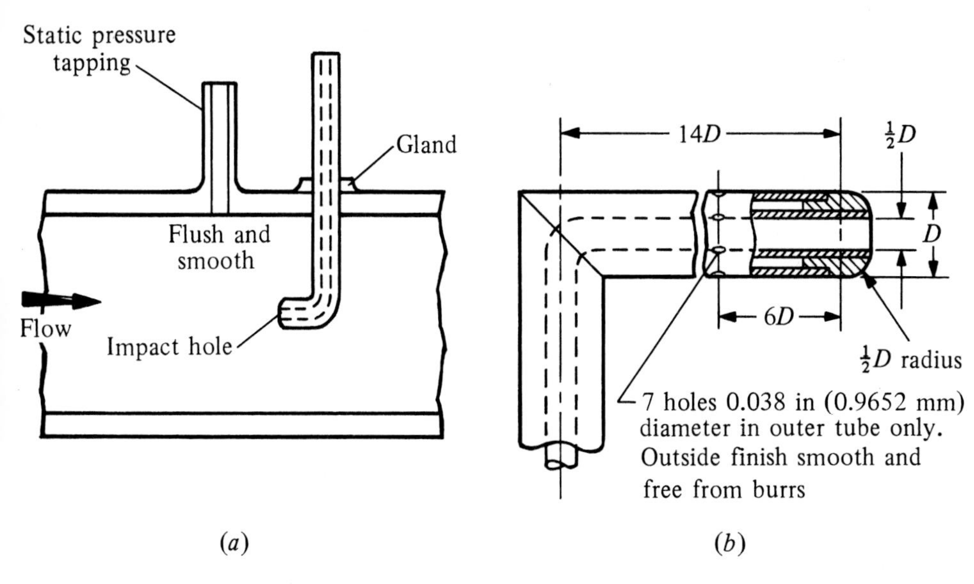

Figure 8:60 Pitot tubes [12]
(*a*) Single-hole Pitot tube
(*b*) Pitot-static tube: NPL hemispherical head

able instrument is the 5-hole Pitot tube [19] which can be used where the gas flow pattern is complex as it usually is in furnaces with recirculation and turbulent zones, which means that the flow direction is unknown. With the 5-hole Pitot if the probe orientation relative to the furnace is known both the magnitude and direction of the gas velocity can be determined. Figure 8:61 shows the construction of this instrument.

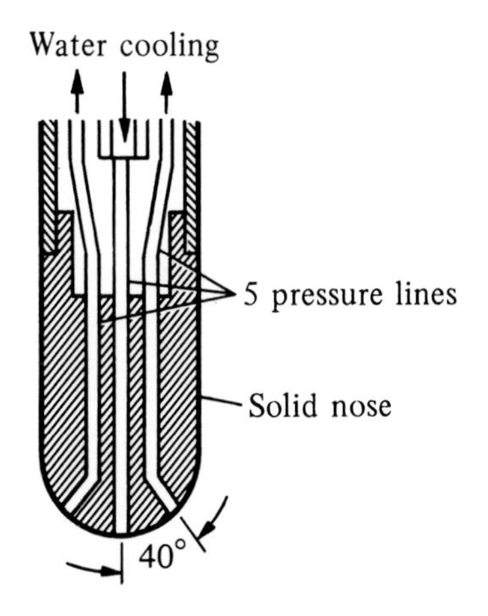

Figure 8:61 Five-hole water-cooled Pitot head [21]

Rotameters

In this class of instrument the principle of operation is that the differential pressure is maintained constant and the magnitude of the variation of the cross-sectional area of the flow is a measure of the flow rate.

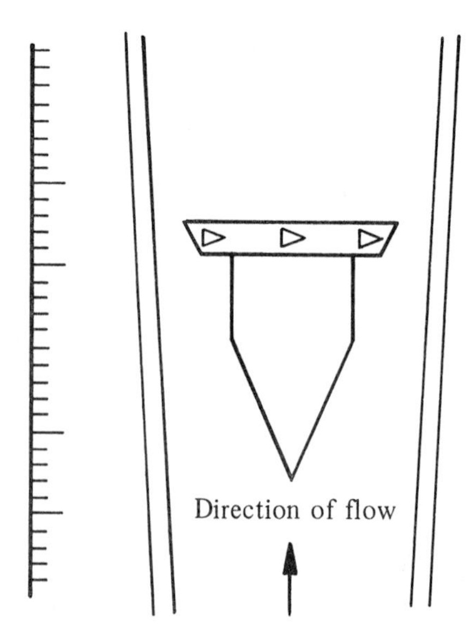

Figure 8:62 Variable-area flow meter

Figure 8:62 shows the principle of a variable-area flow meter a typical example of which is the Rotameter. In this simple device a vertical tube with a bore tapered to be wider at the top contains a free piston often referred to as a 'float' which has the form of a spinning top [10]. The flow through the instrument is vertically upwards and as the rate of flow increases the piston rises in the tube until a point of equilibrium is reached where the upward thrust on the piston due to the differential pressure generated by the metered fluid passing through the annular space balances the apparent weight of the piston. For tubulent-flow conditions the displacement of the piston from its zero position is directly proportional to the rate of flow.

It can be shown that:

$$V = hC_d\pi D_f \sqrt{[8V_f g(\rho_f - \rho_m)/\pi D_f^2 \rho_m]} \tan \tfrac{1}{2}\theta$$

where V = rate of flow (volume units)
h = displacement from zero position
C_d = discharge coefficient
D_f = diameter of piston (float)
θ = angle of taper
V_f = volume of piston
ρ_f = density of float material
ρ_m = density of metered fluid

Therefore : $V \propto h$

For a small value of θ a practically equally spaced scale will be obtained at the lower end. However, there is an opening out of the scale and other small variations in the spacings, due to the variation of C_d with (Re). A number of float shapes have been developed that minimize the effects of viscous forces and produce a more constant flow coefficient for a very large range of flows.

The materials used in the construction of a rotameter depend upon the duty required. The flow range depends upon the tube size, the piston material, the permissible pressure drop and the nature of the metered fluid. They are very useful for metering corrosive liquids and gases. Accuracies as high as ± 0.5 per cent can be obtained over a 10 : 1 range of flow.

Glass tube rotameters are available in sizes up to about 4 kg s^{-1} (8.8 lb s^{-1}) for liquids, with corresponding ranges for air flow being up to about 0.05 m^3 s^{-1} (1.8 ft^3 s^{-1}). Small-diameter tubes can withstand higher pressures than larger sizes. The upper limit for measuring gases with a 100 mm diameter tube is about 400 kPa (4 bar, 58 lbf in^{-2}). For safety, armoured-glass protection can be provided. For use on liquids and gases at higher pressures, metal-tube rotameters are available either with a sight glass and indicator fitted above the tube or a device such as a magnetic follower to indicate the flow. They can also be made as recording or integrating flow meters.

The normal glass-tube indicating rotameters are extremely useful in laboratory, development and industrial applications as an easily installed means of monitoring flow rate of gas and or air supply to burners, etc. They are supplied by manufacturers calibrated for their intended duty and if used for metering fluids of different densities they must be corrected. The accuracy of registration is less affected by flow conditions near to the meter than is the case with other flowmeters.

8.5.7 Novel methods

Work has been carried out at the London Research Station of the British Gas Corporation [44] into the requirements of the industry for improved capacity and flexibility in the measurement of large gas flows. This has led to the development of two methods which are capable of measuring gas velocity directly. They may be usable for field checking of meters and one could possibly serve as a primary standard of flow measurement.

Time of flight

The gas flow rate in a pipeline is measured by timing a pulse travelling at the same velocity as the gas between two detection points. A pulse of organic vapour is fluoresced under ultraviolet light at the detection points and the signals received by photomultiplier tubes which trigger three times at a preselected level of response. The timers measure the time taken for the pulse to pass the first detector, the time taken for the pulse to pass the second detector and the time taken for the pulse to travel from the first to the second detector. Knowing the cross-sectional area of the pipe and the density of the gas, and assuming a fully developed velocity profile, the gas flow rate can be computed.

The work has shown that it should be possible to achieve accuracies close to the theoretically possible value of ± 0.24 per cent. The method is independent of pressure and pipe size and causes the minimum of disturbance to the gas flow. Figure 8:63 shows the arrangement of the laboratory test rig.

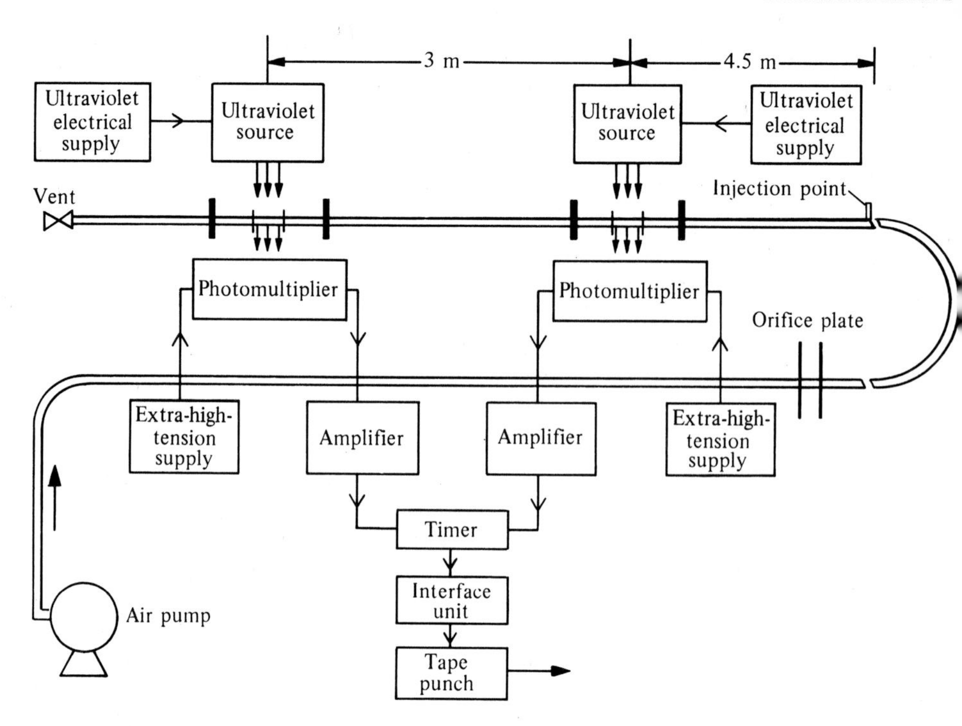

Figure 8:63 Time-of-flight measurement of gas flow rate

Laser Doppler velocity meter

This method is based on the principle of measuring the Doppler shift of laser radiation scattered from particles moving within the gas stream. The laser beam is split into two components which are focused by an optical system to intersect in the gas

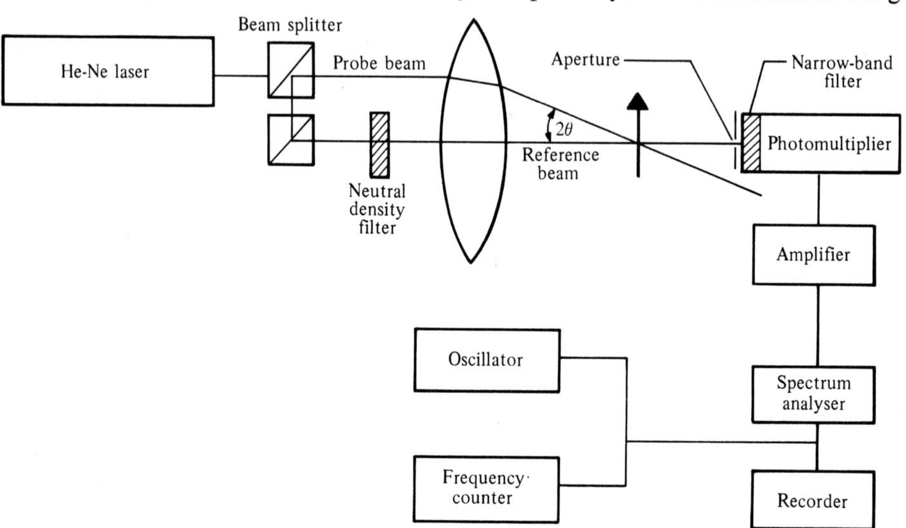

Figure 8:64 Laser Doppler velocity meter

stream as shown in Figure 8:64. Minute particles moving within the gas, either in the form of naturally occurring dust or droplets deliberately introduced into the gas, cause the laser beam to be scattered in all directions. The velocity of these particles causes the scattered light to undergo a Doppler frequency shift directly proportional to the particle velocity given by

$$\Delta f = (2u/\lambda) \sin\theta$$

where Δf = frequency shift
u = velocity of particles in the fluid
2θ = angle between reference and probe beams
λ = wavelength of laser radiation

Particles are selected which are small enough to follow the gas flow with an accuracy of better than 0.1 per cent. A photodetector converts the optical data into electronic signals which can be processed to give a measurement of the point velocity. The optical system has been designed to allow velocity profiles across a pipe to be obtained. This will enable the total volume flow to be calculated. The technique is such that the device needs no calibration, gives an output frequency signal which is linear with velocity over a wide range of velocities from 0.1 mm s^{-1} to three times the velocity of sound and has a high potential accuracy and reproducibility, and it does not disturb the flow. The potential of the laser Doppler velocity meter may be realized in its use as a primary standard against which other metering devices can be calibrated. Optical anemometer systems are now becoming widely used in fluid mechanics and combustion research.

8.6 METERING LARGE INDUSTRIAL AND COMMERCIAL LOADS [41, 42]

In recent years there has been a substantial increase in the number of large measurement systems (volumetric flow rate $> 300\ m^3\ h^{-1}$) installed and also an increase in measurement at high pressure.

8.6.1 Statutory requirements and standards

A fundamental requirement of all metering systems is that they should be within the legal requirements for safety and accuracy. The safety aspect is covered in the Gas Safety Regulations 1972 [45] which are summarized in Section 9.7. The accuracy aspect is governed by the Gas Act 1972 [46] and the Gas (Meter) Regulations 1974 [47] which state:

1 '... no meter shall be used for the purpose of ascertaining the quantity of gas supplied to any person unless it is stamped either by, or on the authority of, a meter examiner ... or in such other manner as may be authorised ...' (Gas Act, section 30 (1))
2 '... where gas is supplied through a meter, the register of the meter shall be prima facie evidence of the quantity of gas supplied' (Gas Act, Schedule 4, paragraph 10 (1))
3 'The meter ... will register such quantity ... as does not differ from the actual

quantity . . . passing through the meter by more than 2 per cent of that actual quantity' (Gas (Meter) Regulations, regulation 4)

The above requirements apply to gas supplied under the terms of published tariffs. However, the following statements are included in the Gas Act 1972:

1 '. . . the Corporation may enter into a special agreement with any consumer for the supply of gas to him on such terms as may be specified in the agreement' (Section 25 (6))
2 Item (2) above 'shall not have effect in relation to the supply of gas to a person under any agreement made with the Corporation and providing for the quantity of gas supplied to him to be ascertained otherwise than by means of a duly stamped meter' (Schedule 4, paragraph 10 (2))

It can be seen that under the Gas Act 1972 the British Gas Corporation may enter into a special agreement with a customer in which case the meter register or index reading no longer has to be taken as the prima facie evidence of the quantity supplied. It follows that, subject to the terms of such a special agreement, the volume of gas consumed may be corrected for the effects of pressure and temperature by means either of fixed factors or of automatic correctors. Special-agreement contracts are individually negotiated for all interruptible loads and for many large loads on a firm basis and they often involve the recording of various parameters, sometimes with remote indication, and the installation of correctors. Some of the included parameters may be as follows:

1 The maximum hourly or daily demand rate
2 The total demand each day during a charging period
3 The hourly demand, to permit the use of tariff terms for charging for differently priced blocks of gas through a single meter
4 The period of interruption and its proof
5 Gas pressure and/or temperature at the meter

For the purpose of metering-system design there is a division in the recommendations based upon the metering pressure.

Where the metering pressure does not exceed 7.5 kPa (75 mbar, 30 inH_2O) reference should be made both to BS CP 331, *Installation of Pipes and Meters for Town Gas: Part 2 Low Pressure Metering* [29] and the relevant sections of IGE Communication 750, *Recommendations for the Design of Governor Installations to Consumers' Premises* [48], which deals with the governor requirements.

For higher-pressure metering installations where the inlet pressure to the metering installation exceeds 7.5 kPa but does not exceed 7 MPa (70 bar, 1000 lbf in^{-2}), reference should be made to *Recommendations on Gas Measurement Practice IGE/GM/1: 1974 Installation of High-Pressure Gas Meters for Inlet Pressures Exceeding 75 mbar* [30]. These recommendations deal with the installation of gas meters in industrial and commercial premises and apply to the measurement of all gaseous fuels distributed by a gas undertaking, where the gas temperature at the meter is in the range −5 °C to 40 °C. For other gases or for use outside these temperature limits the manufacturers of individual items of equipment should be consulted. The recommendations are confined to the following meter types:

1 Positive-displacement diaphragm
2 Rotary displacement
3 Turbine
4 Orifice

Where the inlet pressure to the installation exceeds 700 kPa (7 bar, 100 lbf in^{-2}), reference should also be made to the *Recommendations on Transmission and Distribution Practice IGE/TD/9 Offtakes and Pressure-regulating Installations for Inlet Pressures between 7 and 70 bar* [49].

8.6.2 Flow-meter systems

System specification
The function of a flow system, together with the operating conditions, defines the meter's design specification. It will include the following:

1 Performance targets:
 (*a*) Continuity of measurement
 (*b*) Security of gas supply being monitored
 (*c*) Flow rate accuracy
 (*d*) Integrated flow accuracy
2 Maximum flow capacity
3 Operating load profile
4 Upstream and downstream control equipment characteristics
5 Maximum allowable pressure drop
6 Operating gas pressure and temperature ranges
7 Gas composition limits
8 Read-out requirements, including time data for interruptible loads
9 Available power supplies and their specification
10 Space available

Continuity of measurement is important for charging but from 1 hour to 3 days a year downtime may be required even for scheduled maintenance. In industrial and commercial applications space is also usually at a premium. It is therefore essential that the design of the meter system and any associated pressure-reduction equipment should be integrated. Figure 8:65 shows a layout of typical installation.

Flow meters
The three types considered are:

1 The rotary displacement (RD) meter
2 The turbine meter
3 The orifice meter

RD and turbine meters have been approved by the Department of the Environment for charging purposes, at both low and high pressures [50].

The RD meter has been used extensively for low-pressure measurement. This meter type has a high turndown ratio in the order of 30 : 1 and performance accuracies are quoted as ±1 per cent abs. High-pressure operation should only result in a minor shift of the low-pressure error curve. Flow-straightening lengths are required

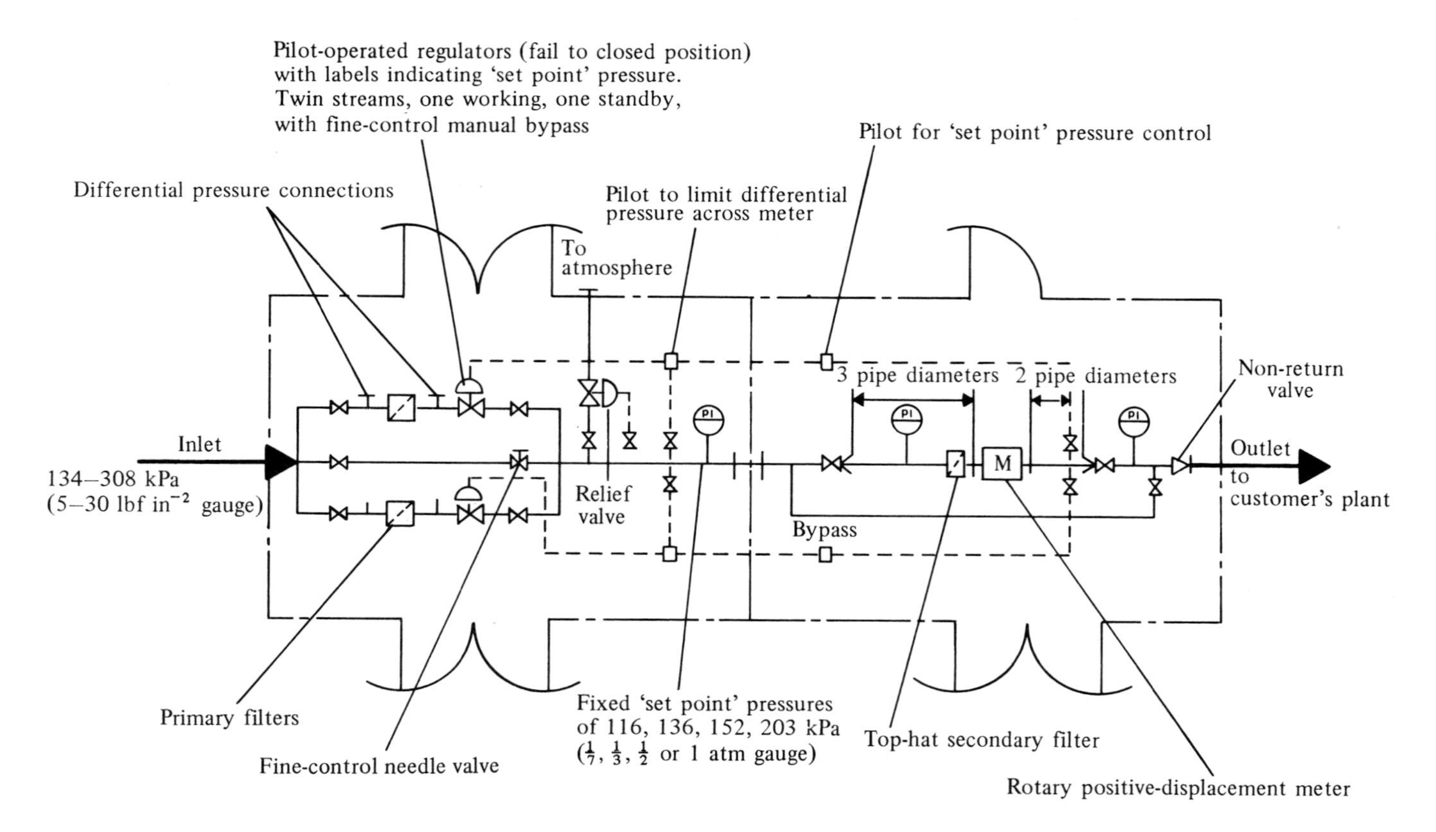

Figure 8:65 Typical governing/metering installation to supply an industrial or commercial customer, utilizing 'fixed-factor' method and rotary-displacement meter, supplied from 300 kPa (3 bar, 30 lbf in^{-2}) system

upstream and downstream, but they are much shorter than those required for orifice plates and are not normally required at low pressures.

If not correctly installed this meter type may give rise to:

1 Jamming—this can occur if dust or debris is trapped between the impellers and the case, causing the meter to stop and the gas flow to be restricted. A filter on the inlet will prevent this and an automatic meter bypass will ensure continuity of supply.
2 Pulsations and impeller inertia—the impeller motion may cause pressure pulsations at the meter outlet. With a rapidly changing load, the time required for the impellers to speed up or slow down may affect local control equipment such as governors and pilot supplies.
3 Pipe resonance—acoustic resonance will result if the pulsation frequency generated by the impellers approaches the resonant frequency of the pipework. Careful pipework design is needed to eliminate this possibility.

The turbine meter is relatively new. Its accuracy is quoted as ± 1 per cent abs. and ± 1 per cent fsd and is dependent upon an established velocity profile being achieved at the meter inlet.

The turndown increases proportionally with the square root of the density ratio. At 2 MPa (20 bar), the turndown ratio may be as high as 100 : 1 compared with 15 : 1 at millibars gauge working pressure. The meter creates no problems with security of gas supply. Line debris may damage or even stop the meter, but the meter will still pass gas.

Problems may be caused by:

1 Blade damage or change in friction, since there is no outward indication if the meter is slowed down or damaged
2 Flow pulsations, which will cause the meter to read high, depending on the pulsations' form factor
3 Swirl, which produces an error dependent on the swirl angle

These problems can be dealt with by regular inspection and spin tests and by the use of flow straighteners and profiles.

The orifice meter has not been generally approved in the UK. The Organisation Internationale de Métrologie Légale (OIML) is ready to legalize the use of orifice meters on the Continent and they are already approved in the USA.

The primary element (Section 8.5.6) is a simple device. It has no moving parts and it will not stop the flow of gas. The accuracy of the primary element varies with the operating conditions, but ± 0.6 to ± 0.75 per cent abs. can be achieved. Flow rangeability is restricted by the ability to measure accurately the differential pressures generated and also by changes in the orifice meter discharge coefficient due to a varying Reynolds number. Using a suitable carrier, the orifice plate and therefore the r_a ratio can be changed under normal operating conditions. A suitable secondary instrumentation system can give a flow rangeability in excess of 40 : 1. All the necessary design data for orifice plates are available in BS 1042 [51], ISO/R541 [52] and AGA Report No. 3 [53]. The performance of this type of meter can be influenced by:

1 Erosion, which can produce changes in edge sharpness and pipe-wall roughness and may give errors of up to 5 per cent

2 Swirl or pulsating flow can be created by governors, compressors, bends, etc., which can result in an error of 15 per cent at a swirl angle of 40°
3 Collection of dust or debris within the system immediately before the orifice plate

A filter and flow straightener can reduce the possibility of these errors occurring but regular checking of the plate for sharpness is necessary.

8.6.3 Secondary instrumentation

Of the three considered, only the orifice meter requires a secondary instrument to derive a flow read-out signal. To reduce the measured volume to the equivalent base volume, secondary instrumentation can be used to obtain automatic correction of the output from all three meter types. The instruments available can be put into four classes:

1 Differential-pressure/pressure transmitters
2 Temperature transmitters
3 Density meters
4 Correctors

Differential-pressure/pressure transmitters
Design improvements have been made on these instruments and an improved long-term stability of ±0.1 per cent fsd a year and a reduction in pressure and temperature coefficients to ±0.1 per cent fsd per 70 bar and ±0.2 per cent fsd per °C has been claimed. Calibration at working pressure is necessary and temperature-controlled enclosures may be required on some applications. Nominal accuracy remains at ±0.5 per cent fsd but by special selection, transmitters can now be obtained with claimed accuracies of ±0.2 per cent fsd.

Temperature transmitters
The platinum resistance thermometer is the standard adopted with nickel and the usual thermocouple junctions used in industrial instruments. These are covered in Section 8.3.3.

Density meters
Three types of density meter are available: the spinner, the vibrating element and the buoyant beam. Accuracies vary between ±0.2 per cent abs and ±0.5 per cent fsd. Gas sampling errors, the influence of dust on the sensing elements and in some cases the high cost of associated secondary equipment, will decide a density meter's acceptability.

Correctors
Automatic correctors can achieve pressure and/or temperature correction and can correct for compressibility either by the inclusion of a fixed factor or by actual correction. Correctors all require an output drive from the meter and this may be by means of direct mechanical linkage or by indirect electromechanical (or electrical) means.

Automatic pressure/temperature correctors are available in two basic types: mechanical and electronic. Their claimed accuracies are ±1 per cent abs and ±0.25 per cent fsd respectively. Mechanical types may be step or continuous integrators, whereas electronic correctors are continuous integrators.

Automatic pressure/temperature/compressibility correctors are also available in two basic types: the trapped sample and the density cell. The trapped-sample type operates by trapping a specific mass of gas in a flexible container and measuring its deflection when subjected to gas line pressure and temperature. A correction accuracy of ± 1 per cent abs is possible provided the line gas characteristics do not differ significantly from those of the trapped sample. The density-cell type utilizes a density transducer mounted in a small gas-sampling line and has a claimed accuracy of within 0.2 per cent fsd.

The reader is referred to references [54], [59], [60], [61] and [63] which cover these instruments in more detail. The reader is also referred to references [55], [56], [57], [58], [62], [64] and [65] in which further information on meter types, meter systems, design and primary equipment for bulk-metering systems is covered.

REFERENCES

1 Connor, N. E., 'Gas quality measuring devices', IGE Seminar on Gas Measurement, University of Salford (1969)
2 Jones, M. W., 'Current work in Gas Standards Branch', IGE Manchester and District Section (1972)
3 Roberts, G. F. I., 'Why measurement is necessary for control', IGE Seminar on Gas Measurement, University of Warwick (1971)
4 Jones, E. B., *Instrument Technology*, Vol. 2, London, Butterworth (1956)
5 Chedaille, J. and Braud, Y., *Measurement in Flames*, London, E. Arnold (1972)
6 'Gas Analysers for use on the district', *BGC Watson House Bulletin*, **33,** No. 249 (November/December 1969)
7 Clifford, R. E., 'Combined detector tube for Draeger gas analyser', *BGC Watson House Bulletin*, **34,** No. 255 (November/December 1970)
8 British Gas Corporation, *Industrial Gas Controls Handbook* (1969)
9 Dingle, M. G. W., 'Temperature measurement' in *Process Engineering Technique Evaluation*, London, Morgan-Grampian, with Process Engineering, (1969)
10 'Measurement and Control Topics No. 2 and 4' (Temperature Measurement, No. 2; Flow Measurement, No. 4), *Hartman & Braun News* (Spring 1974)
11 British Standards Institution, *British Standard 1041: Section 2.1: 1969, Code for Temperature Measurement: Expansion Thermometers: Liquid-in-Glass Expansion Thermometers* (1969)
12 Jones, E. B., *Instrument Technology*, Vol. 1, 2nd ed., London, Butterworth (1965)
13 British Standards Institution, *British Standard 1041 : Part 7 : 1964, Code for Temperature Measurement: Temperature/Time Indicators* (1964)
14 Probert, S. D., Marsden, J. P. and Holmes, T. W., *Experimental Method and Measurement*, London, Heinemann (1971)
15 British Standards Institution, *British Standard 1041 : Part 3 : 1969, Code for Temperature Measurement: Industrial Resistance Thermometry* (1969)
16 British Standards Institution, *British Standard 1041 : Part 4 : 1966, Code for Temperature Measurement: Thermocouples* (1966)
17 Gray, W. A. and Müller, R., *Engineering Calculations in Radiative Heat Transfer*, Oxford, Pergamon Press (1974)

18 British Standards Institution, *British Standard 1041 : Part 5 : 1972, Code for Temperature Measurement: Radiation Pyrometers* (1972)
19 Atkinson, P. G., *Gas Council Res. Commun.* GC33 (1956); *Trans. Inst. Gas Eng.*, **106** (1956), 441
20 Atkinson, P. G. and Hargreaves, J. R., *Gas Council Res. Commun.*, GC57 (1958); *Trans. Inst. Gas Eng.*, **108** (1958), 835
21 Dingle, M. G. W., 'Instruments for the fuel technologist', *Ind. Process Heat.*, *11*, No. 11 (November 1971), 6
22 'National physical laboratory micro-manometer', *J. Sci. Instrum.*, *42* (1956), 677
23 Alexander, A., 'Diaphragm meters', IGE Seminar on Gas Measurement, University of Salford (1969)
24 British Standards Institution, *British Standard 4161 : Part 1 : 1967, Specification for Gas Meters: Meters of Plate Construction up to 1000 cubic feet per hour rating* (1967)
25 British Standards Institution, *British Standard 4161 : Part 2 : 1967, Specification for Gas Meters: Meters of Plate Construction above 1000 cubic feet (28 cubic metres) per hour rating* (1967)
26 British Standards Institution, *British Standard 4161 : Part 3 : 1968, Specification for Gas Meters: Unit Construction Meter of 6 cubic metres (212 cubic feet) per hour rating* (1968)
27 British Standards Institution, *British Standard 4161 : Part 4 : 1970, Specification for Gas Meters: Plate Constructed Positive Displacement Diaphragm Meters for a Pressure of 350 mbar (5 lbf/in²) and up to 170 cubic metres (6000 cubic feet) per hour rating* (1970)
28 British Standards Institution, *British Standard 4161 : Part 5 : 1968, Specification for Gas Meters, Positive Displacement Diaphragm Meters for Ratings above 6 m³/h and Pressures up to 7 bar* (1968)
29 British Standards Institution, *CP331 : Part 2 : 1974, Code of Practice for Installation of Pipes and Meters for Town Gas: Low Pressure Metering* (1974)
30 Installation of high-pressure gas meters for inlet pressures exceeding 75 mbar, *Inst. Gas Eng. Commun.*, 927 (1974)
31 Gilbert, A. T., 'Gas meters', *Watson House Bulletin*, **33**, No. 246 (May/June 1969)
32 British Standards Institution, *British Standard 1179 : 1967 Glossary of Terms Used in the Gas Industry* (1967)
33 Thompson, R. J. S., 'Rotary displacement meters: basic principles, theory and practice', *Inst. Gas Eng. Commun.*, 884 (1972); abstract in *J. Inst. Gas Eng.*, **12** (1972), 272
34 Thompson, R. J. S., 'Rotary displacement meters', IGE Seminar on Gas Measurement, University of Salford (1969)
35 Schneider, H. J., 'Gas turbine meters', IGE Seminar on Gas Measurement, University of Salford (1969)
36 Alder, C., Kemp, B. J. and Lucas, T. A., 'The development of portable insertion turbine meters and their application in medium and low-pressure distribution systems', *Inst. Gas Eng. Commun.*, 887 (1972); abstract in *J. Inst. Gas Eng.*, **12** (1972), 273
37 Howard, G. J. S. and Edgar, A. T., 'The use of insertion-type turbine meters for gas flow in the Eastern Gas Board', *Inst. Gas Eng. Commun.*, 888 (1972); abstract in *J. Inst. Gas Eng.*, **12** (1972), 274

38 *Air Flow Measuring Equipment*, Air Flow Developments Ltd
39 Zanker, K. J., 'Pressure differential devices for flow measurement', IGE Seminar on Gas Measurement, University of Salford (1969)
40 Wood, P. R. and Wilkinson, N. A., 'Orifice plate metering', *Inst. Gas Eng. Commun.*, 885 (1972); abstract in *J. Inst. Gas Eng.*, **12** (1972), 273
41 Griffiths, A. and Newcombe, J., 'Large-volume gas measurement', *Inst. Gas Eng. Commun.*, 834 (1970); *J. Inst. Gas Eng.*, **11** (1971), 23
42 Stokes, B. A., 'Metering large industrial and commercial loads', Midland JGA (February 1972)
43 British Standards Institution, *BS 1042 : Part 2A : 1973, Methods for the Measurement of Fluid Flow in Pipes: Pitot Tubes, Class A Accuracy* (1973)
44 James, H, *et al.*, 'Novel methods of gas flow measurement', *Gas Council Res. Commun.*, GC205 (1972); abstract in *J. Inst. Gas Eng.*, **12** (1972), 279
45 Gas Safety Regulations 1972, Statutory Instrument No. 1178, HMSO (1972)
46 Gas Act 1972, HMSO (1972)
47 Gas (Meter) Regulations 1974, Statutory Instrument No. 848, HMSO (1974)
48 'Recommendations for the design of governor installations to consumers' premises', (IGE/TD/5), *Inst. Gas Eng. Commun.*, 750 (1967)
49 'Recommendations on transmission and distribution practice: offtakes and pressure regulating installations for inlet pressures between 7 and 70 bar, 1972' (IGE/TD/9), *Inst. Gas Eng. Commun.*, 866 (1972)
50 British Standards Institution, *British Standard 4161 : Part 6 : 1971, Specification for Gas Meters: Rotary Displacement and Turbine Meters for Gas Pressures up to 100 bar* (1971)
51 British Standards Institution, *British Standard 1042: Part 1: 1964, Methods for the Measurement of Fluid Flow in Pipes: Orifice Plates, Nozzles and Venturi Tubes* (1964)
52 International Organization for Standardization, ISO/R541 *Measurement of Fluid Flow by means of Orifice Plates and Nozzles*, 1st edition (1967)
53 AGA Gas Measurement Committee Report No. 3, *Orifice Metering of Natural Gas* (1969)
54 British Standards Institution: *BS 4161 : Part 7 : 1973, Specification for Gas Meters: Mechanical Volume Correctors* (1973)
55 Smith, J. W., 'The orifice meter system', IGE Seminar on Gas Measurement, University of Warwick (1971)
56 Thompson, R. J. S., 'Positive displacement meter systems in customers' premises', IGE Seminar on Gas Measurement, University of Warwick (1971)
57 Crompton, G., 'Turbine meters in customers' premises', IGE Seminar on Gas Measurement, University of Warwick (1971)
58 Whittle, K., 'Large volume metering installations for industrial and commercial consumers', Manchester JGA (1972)
59 Newcombe, J., 'Secondary instruments for flowmeters', IGE Seminar on Gas Measurement, University of Salford (1969)
60 Lucas, T. A., 'Volume correctors and load profile devices', IGE Seminar on Gas Measurement, University of Salford (1969)
61 Lucas, T. A., 'Secondary instruments for bulk metering systems', IGE Seminar on Gas Measurement, University of Warwick (1971)
62 Smith, J. F., 'Commissioning, proving and maintenance of bulk metering systems', IGE Seminar on Gas Measurement, University of Warwick (1971)

63 Newcombe, J., Archbold, T. and Jepson, P., 'Errors in measuring gas flows at high pressure: recent developments in correcting methods', *Gas Council Res. Commun.*, GC204 (1972); abstract in *J. Inst. Gas Eng.*, **12** (1972), 279
64 Newcombe, J. and Smith J. W., 'High pressure gas metering system development in Great Britain', Symposium on Flow: Its Measurement and Control in Science and Industry, Pittsburgh, USA (1971)
65 Sutherland, V., 'Primary equipment for bulk metering systems', IGE Seminar on Gas Measurement, University of Warwick (1971)

9

Safety and controls

9.1 INTRODUCTION AND BACKGROUND

There is a long tradition of safety consciousness in the utilization of gas promoted by the supply industry and the equipment and component manufacturers.

In the late 1930s the Industrial Gas Centres Committee was set up to promote exchange of information on all aspects of industrial gas utilization including developments in safety equipment. This function was continued after nationalization by the Gas Council and Area Gas Boards (now the British Gas Corporation). Perhaps the most important document issued in this field by the Gas Council was the *Industrial Gas Controls Handbook* [1] published in 1969. Frequent reference to this publication will be made during the course of this chapter. Although still of considerable general value some of the content has been superseded [2] by later documents as indicated elsewhere in this chapter.

In addition to its independent activities the Gas Council has cooperated with other organizations with interests in the safe operation of industrial gas-fired equipment. Reference [3] may be quoted as an example of this cooperation. Prepared by the Factory Inspectorate it dealt with possible hazards arising from fuel usage or the generation of flammable vapours in gas-heated ovens and furnaces.

In the 1960s the availability of large quantities of natural gas enabled the industry to expand into large, new, unfamiliar markets.

In 1966 standards for automatic gas burners [4] and in 1970 the interim codes of practice for large gas and dual-fuel burners [5] were published by the Gas Council to provide guidance, at an early date, on the basic essential requirements for reliable and safe operation of gas-fired plant in these new fields. The standards for automatic burners were revised [6] in 1970 in collaboration with the Society of British Gas Industries (SBGI) representing the non-nationalized sector of the British gas industry.

The standards and codes of practice referred to above embody the basic principles of safe operation which are common to all industrial gas-fired plant whether manually or automatically controlled. Fitzsimons and Hancock [2] have summarized these as follows:

1 Good standards of engineering
2 Good ventilation and/or purging of combustion spaces
3 The use of good-quality gas-supply shutoff valves
4 The use wherever possible of efficient flame safeguard equipment coupled to good automatic safety shutoff valves
5 The use of correct operating procedures
6 The design of pilot and main burners to give rapid and reliable ignition, cross-ignition and stability
7 The controlled release of gas during the ignition phase of burner operation in accordance with criteria based on acceptable energy release in the event of ignition delay
8 The incorporation of fail-safe principles in the design of control systems

Adoption of all the principles listed above should ensure safe plant operation of gas-fired plant, but 'safe' is a relative term. As Hill [7] has pointed out there are two extremes: if no protective devices are fitted to the plant the operator may consider the plant to be unsafe and never use it which renders it safe; or at the other extreme so many protective devices may be incorporated into the design that the cost of the plant is prohibitive and is never sold and so again it could be considered safe. Clearly manufacturers design plant to be sold and customers invest in plant which will be useful and so between these extremes must lie an economic, acceptably safe mid-course. In considering safety it is necessary to assess the possible hazards, the level of risk and the time at risk [7]. Based on this assessment and economic and practical considerations technical standards may be determined with respect to safety.

Whilst the British gas industry has maintained high standards of safety to date there is a continuing need for research and development into safety aspects, the testing and evaluation of safety systems and components, the drafting of guidelines for safe operation of various types of plant and for the monitoring of current safety procedures. In view of this, the brief review of safety standards and controls presented in this chapter, representing as it does the situation in 1975, will in some respects become quickly out of date as safety procedures and equipment develop.

9.2 TECHNICAL STANDARDS AND STATUTORY REQUIREMENTS

This section lists and briefly summarizes technical standards and legislation relating to industrial gas usage. More detailed treatment of the more important aspects is presented later in this chapter.

Standards for Automatic Gas Burners, Forced and Induced Draught [5, 7]
These apply to automatic burner systems of thermal input > 44 kW (150 000 Btu h^{-1}). 'Automatic' so far as the standards are concerned describes a burner system which (when starting up from the completely shutdown condition) having sensed the start-gas flame actuates the main gas valve without manual intervention. The standards lay down specific requirements for safety shutoff valves, pre-purge, maximum permissible energy release during the start-gas phase, main flame establishment, flame safeguards and safety shutdown.

Midlands Research Station of British Gas operates a certification scheme whereby proprietary safety shutoff valves and programming control units are tested against the standards. In the 1970 edition of the standards [6] two safety shutoff valves in series are required in the gas supply to the main burner for rates in excess of 600 kW (2×10^6 Btu h^{-1}). This requirement has recently [8] been extended to apply to all new installations covered by the standards and in so doing has introduced the concept of various 'classes' of valve.

Interim Codes of Practice for Large Gas and Dual-Fuel Burners [5]
These were published in 1970 and 1971 in four parts:

Part 1. Single gas burner installations
Part 2. Multiple gas burner installations
Part 3. Single burner dual-fuel installations
Part 4. Multiple burner dual-fuel installations

They were published with a view to supplementing the standards for automatic gas burners and relate to plant normally operating below 750 °C and of total rating > 3 MW (102 therm h^{-1}). The codes relate primarily to the essential safety aspects of engineering, startup and operation and are not intended to provide a complete specification for burners and their control equipment. At the time of writing the codes of practice are under review in collaboration with the burner- and component-manufacturing sector of the gas industry.

Purging Procedures for Industrial Gas Installations [9]
This lays down procedures for purging industrial pipework and meter installations of all sizes, purging being either the displacement of air or inert gas by a fuel gas, the displacement of a fuel gas by air or inert gas, two-stage purging using inert gas or finally the displacement of one fuel gas by another.

Procedures are described for calculating purge volumes, for purging meters, for taking pipework in and out of service, for direct, inert gas, air and slug purging, for venting and for vent gas testing. The procedures are under review.

Soundness Testing Procedures for Non-Domestic Gas Installations [10]
This document lays down recommended procedures for soundness testing of installations in non-domestic premises with meters of badged capacity exceeding 400 ft^3 h^{-1} (11.3 m^3 h^{-1}), i.e. D4 size. The soundness procedures are for use after the installation of new pipework or appliances, after the alteration or replacement of existing pipework or where there is a known or suspected gas leak. Attention is drawn to the Gas Safety Regulations 1972 [11] (see below) which prescribe circumstances in which soundness testing must, by law, be carried out. The procedure involves the same 'gauge-watching' technique employed for domestic tests modified only to take into account the diversity of non-domestic premises. The test breaks down into the following phases:

1 Estimation of the system volume
2 Establishment of the test pressure
3 Selection of the test gauge (this affects the sensitivity of the test)
4 Determination of the permitted leak
5 Determination of the test period
6 Carrying out the test

The test pressure for new installations should be 5 kPa (50 mbar, 20 inH_2O) or $1\frac{1}{2}$ times the working pressure or the maximum pressure likely to occur under fault conditions whichever is the higher. For existing installations the test pressure should be at least the normal working pressure.

Overheat Protection for Steam Tube Ovens [12]
This provides a specification for manufacturers and users of steam tube bakers' ovens to enable the plant to be exempted from some of the provisions of the Factories Act 1961. The major requirement is the provision of a manual-reset overheat cutoff.

Non-Return Valves for Oxy-Gas Glassworking Burners [13]
The application of non-return valves in the use of oxygen is covered in respect of all types of hand torches and fixed burners used in glassworking operations.

Evaporating and Other Ovens [3]
Important recommendations include the provision of unambiguous lighting-up and shutdown instructions, the fitting of flame safeguards wherever practicable, the provision of flue down-draught-diverters and terminals, ventilation and purging requirements for flammable vapours, the provision of pressure reliefs and the recommendation that safety shutoff valves and automatic burners should conform with the standards for automatic burners [6].

Technical Notes on the Design of Flues for Larger Gas Boilers [14]
These relate to flues for boilers having an input rating of 50 kW (170 600 Btu h^{-1}) and above. Methods are presented for the determination of chimney height and flue cross-sectional areas for natural and forced draught together with recommendations on materials and methods of construction.

Combustion and Ventilation Air, Guidance Notes for Boiler Installations [15]
In this document guidelines are laid down for ventilation requirements for plant of rating 2 million Btu h^{-1} (586 kW) and over. For air supply by natural ventilation for installations between 2 million and 3.5 million Btu h^{-1} (586–1025 kW) output, and area of 7 ft^2 (0.65 m^2) at low level and 3.5 ft^2 (0.325 m^2) at high level is required. For installations in excess of 3.5 million Btu h^{-1} (1025 kW) output, 2.0 ft^2 for every 1 million Btu h^{-1} (0.634 m^2 per 1000 kW) should be provided at low level and half this amount at high level. For air supply by mechanical ventilation to provide minimum combustion, dilution and ventilation requirements, air volumes range from 735 ft^3 min^{-1} (1.2 m^3 s^{-1}) for a forced-draught burner with direct flue connection to 900 ft^3 min^{-1} (1.45 m^3 s^{-1}) per million Btu h^{-1} (1000 kW) for an atmospheric burner with a draught diverter in the flue. In no case should the air supply be less than 2000 ft^3 min^{-1} (0.94 m^3 s^{-1}).

British Standards Institution Codes of Practice CP 332 [16] *and 331* [17]
CP 331 specifies the installation of pipes and meters for town gas. CP 332 covers boilers in the range 150 000 Btu h^{-1} (44 kW) to 2 million Btu h^{-1} (586 kW) and recommends, for natural ventilation, that openings directly to the outside air of free area 1 in^2 for every 2000 Btu h^{-1} (1100 mm^2 for every 1 kW) output should be provided at low level and half this area at high level. The recommendation for mechanical ventilation is that the air supply should not be less than 1000 ft^3 min^{-1} for every 1 million Btu h^{-1} boiler output (1 m^3 s^{-1} for every 620 kW).

Boiler Changeover, Technical Notes on Boiler Changeover to Gas Firing [18]
This provides an outline of technical procedures to be followed and checks to be made in the changeover of boilers above 250 000 Btu h^{-1} (73 kW) rating (see Chapter 14).

The Gas Act 1972 [19, 20]
This Act is the primary legislation under which British Gas operates. The most obvious change brought about by the Act was the creation of the British Gas Corporation out of the twelve Area Gas Boards and the Gas Council.

The part of the Act relating specifically to industrial utilization is Schedule 4, paragraph 18. This requires the provision of equipment to prevent the ingress of air or other extraneous gases into British Gas mains (usually a non-return valve). It also requires the provision of equipment to prevent boosters, compressors, etc., from causing reduced pressure in mains and services (usually low-presssure cutoff switches). Non-return valves and low-pressure cutoff devices are described in Sections 9.4.7 and 9.4.6, respectively.

The Gas Safety Regulations 1972 [11, 21]
These regulations, which came into effect on 1 December 1972, cover service pipes, meters, installation pipes and appliances and the removal, replacement and maintenance of gas appliances and fittings. Their main provisions are outlined in Section 9.7. Application of recommendations of the technical standards documents reviewed in this section will assist in compliance with the Regulations.

9.3 START-UP OF GAS-FIRED PLANT

A fundamental principle of safe start-up is to ensure that the quantity of fuel accumulating in the combustion chamber of an appliance, especially before or during an ignition attempt, is well below the quantity which if ignited would produce a sufficiently high pressure rise to damage the appliance.

Clearly there is a period during the ignition procedure in which gas is admitted to the combustion chamber with no absolute guarantee that ignition will occur immediately and smoothly. Consequently a reliable ignition source must be used and the admission of gas during this critical phase must be carefully controlled [5, 6]. In setting standards for permissible energy release, the relationship between the energy release, the resulting pressure rise and the strength of the appliance must be quantified.

Section 9.3.1 considers the effects of energy input, venting through ports and flues and the presence of dilute gas–air mixtures on the pressures developed by explosions of stoichiometric gas–air mixtures. Section 9.3.2 considers the possibility of gas accumulation before start-up and of delayed ignition and indicates alternative approaches to setting safe limits with regard to energy input during start-up. An additional approach to limiting the pressure rise is to provide deliberately weak parts of the structure which will vent at an early stage of development of the pressure rise and reduce the maximum pressure to a safe value. Such pressure reliefs are considered in Section 9.3.3.

9.3.1 Pressure rise resulting from ignition of gas–air mixtures in enclosures

Aris, Hancock and Moppett [22] have measured the pressure rise due to the combustion of small volumes of gas–air mixture in an enclosure. The work was restricted to small stoichiometric pockets of natural-gas–air mixture confined in polythene bags in a steel vessel of $0.14\,m^3$ ($5\,ft^3$) volume.

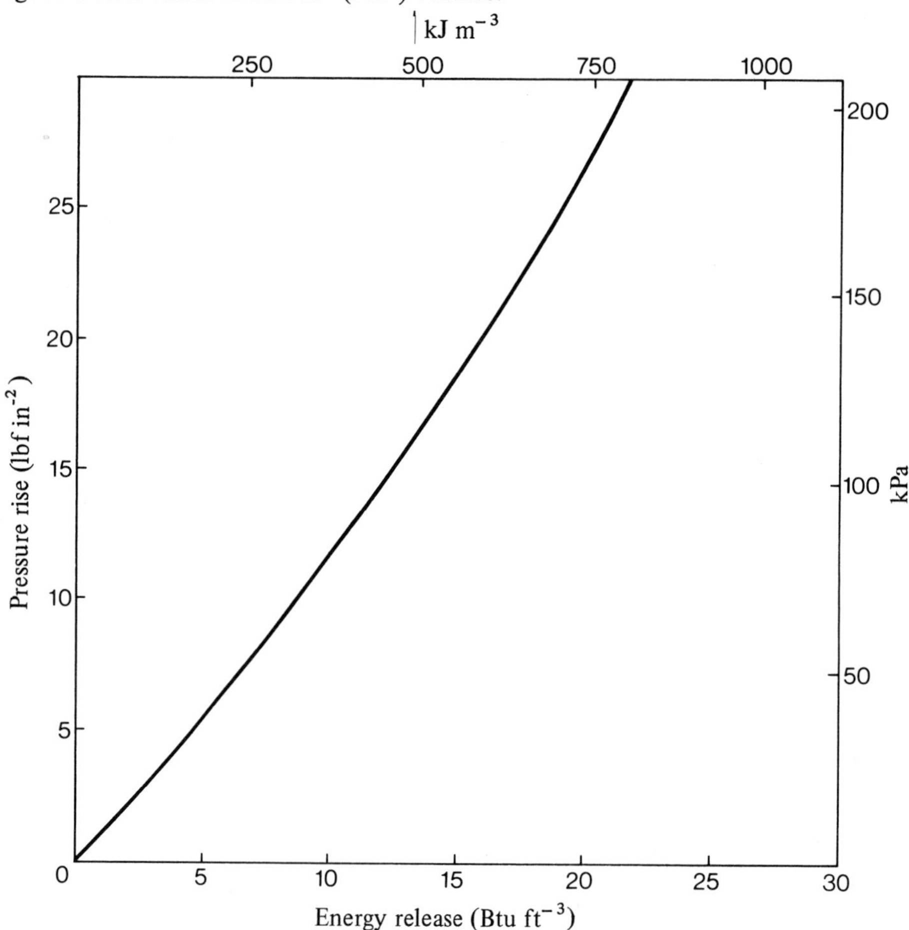

Figure 9:1 Pressure rise due to the combustion of stoichiometric gas–air mixture [22]

The relationship between pressure rise in the closed vessel and energy release is shown in Figure 9:1 in which it will be seen that the relationship is approximately linear (and that energy release in Btu ft^{-3} is numerically equal to pressure rise in lbf in^{-2}) up to a pressure rise of about 5 lbf in^{-2} (34.5 kPa).

A more comprehensive correlation is given by:

$$\frac{P}{\text{lbf in}^{-2}} = 1.013 \frac{E}{\text{Btu ft}^{-3}} + 0.106 \left(\frac{E}{\text{Btu ft}^{-3}}\right)^2 \tag{9.1}$$

where P is the pressure rise and E is the energy release per unit volume of combustion chamber. In SI units this is:

$$\frac{P}{\text{kPa}} = 0.1875\frac{E}{\text{kJ m}^{-3}} + 7.9 \times 10^{-5}\left(\frac{E}{\text{kJ m}^{-3}}\right)^2 \qquad (9.1\text{a})$$

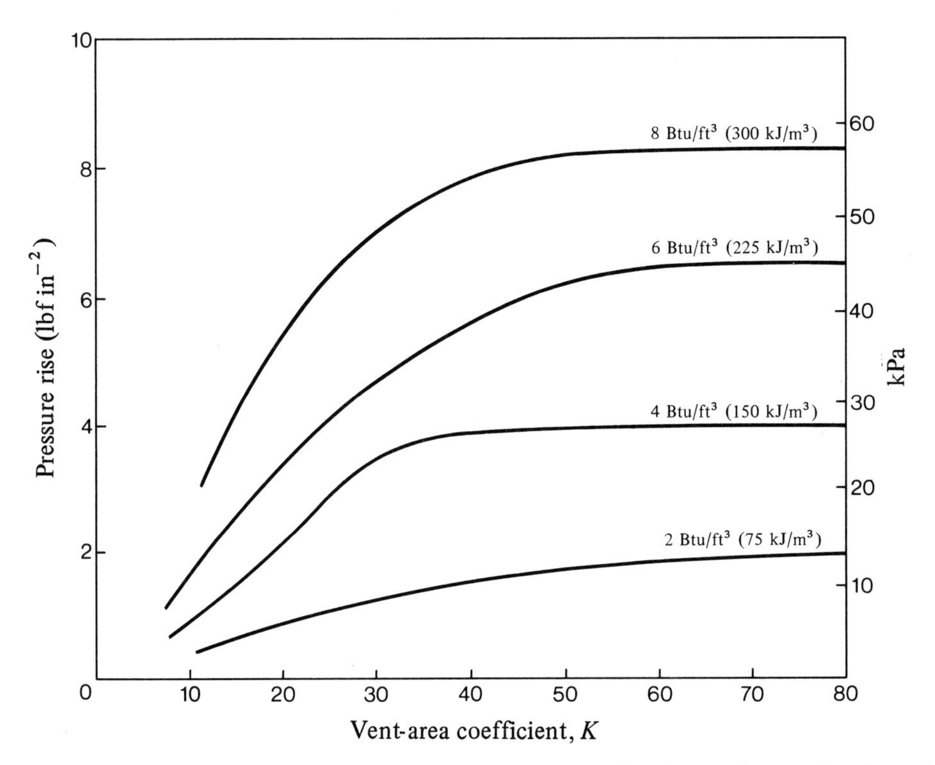

Figure 9:2 Effect of venting via orifices on the pressure rise due to the combustion of stoichiometric gas–air mixture [22]

The effect of venting through simple orifices is shown in Figure 9:2. The vent area coefficient K is defined as the area of the face of the vessel in which the orifice is located divided by the area of the orifice. Figure 9:2 shows that for a given energy release the pressure rise increases as K increases until at a value of about $K = 50$ the pressure becomes that of an unvented enclosure.

Measurements were also made of the effects of venting through flue pipes of various diameters and lengths [22].

During an attempt to restart a burner after a period of operation a situation could arise in which a pocket of stoichiometric mixture may be ignited when surrounded by a mixture of gas and air below the lower explosive limit in either a hot or cold environment. Figure 9:3 shows the effects on the pressure rise generated by the combustion of a 1 Btu ft^{-3} (38 kJ m^{-3}) pocket of stoichiometric mixture when surrounded by very lean mixtures at ambient temperature, 70 °C and 120 °C. Clearly mixtures richer than about 1 per cent natural gas, i.e. only a fifth of the concentration at the lower flammability limit (see Section 4.5) contribute a significant amount of energy to a flame front propagating out of a zone of stoichiometric mixture. Additionally an increase in initial temperature of an ostensibly incombustible gas–air mixture increases its contribution to the heat release and therefore to the pressure rise.

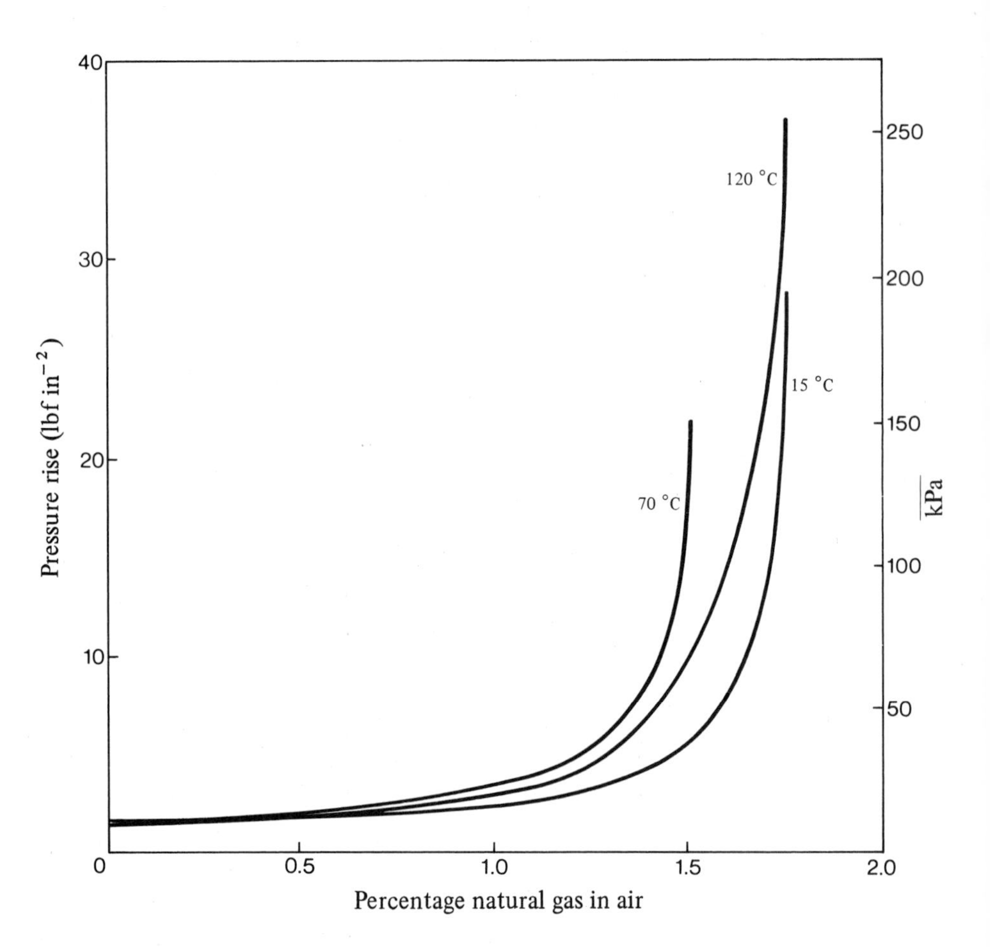

Figure 9:3 Pressure rise in a closed vessel due to combustion of a pocket of stoichiometric gas–air mixture (1 Btu ft^{-3}) surrounded by sub-LEL mixture at ambient and elevated temperatures [22]

9.3.2 System design to prevent unacceptable pressure rise on start-up

The data presented in Section 9.3.1 provide relationships between the energy release and the pressure rise. Thus, provided the appliance strength and the venting effects of air inlets and flue outlets, etc., are known, the maximum amount of gas which may may be allowed to accumulate can be calculated.

Regarding the strength of practical plant there is very little published information available. Even ostensibly strong equipment (e.g. shell boilers) may have relatively weak components such as inspection doors which could become missiles in the event of quite low pressure rises.

All precautions should be taken to avoid pressure rises and, additionally where possible, pressure reliefs should be incorporated into the design to limit the pressures developed (Section 9.3.3). However, this may not always be possible on existing plant. Investigations by Cubbage and Simmonds [23, 24] on box ovens have indicated that pressure rises above about 13.8 kPa (138 mbar, 2 lbf in^{-2}) cause progressively increasing structural damage. Obviously the pressure rises must not exceed relatively low values which for typical appliances, and in the absence of more detailed information, and for relatively small structures may be taken as 13.8 kPa (138 mbar, 2 lbf in^{-2}).

All appliances will have some degree of venting due to air inlets and flue outlets. If these openings may be regarded as simple orifices their vent area coefficient may be calculated and the venting effects estimated from Figure 9:2. For example, if $K = 15$ then the permissible energy release to give a 13.8 kPa (2 lbf in^{-2}) pressure rise would be doubled from 75 to 150 kJ m^{-3} (2 to 4 Btu ft^{-3}). Aris *et al.* [22] have tabulated vent area coefficients for a variety of commercial equipment and their figures suggest that useful venting would occur in the case of most sectional boilers and small high-temperature furnaces but not in the case of air heaters. However, in most installations the flue outlet is fitted with a flue pipe usually including bends and a terminal and consequently without detailed knowledge of a particular installation venting through a flue cannot safely be allowed for. Thus to ensure that a pressure rise of 13.8 kPa (2 lbf in^{-2}) is not exceeded the quantity of gas allowed to accumulate must not exceed that which would release 75 kJ per cubic metre of appliance volume (2 Btu per cubic foot).

Gas accumulation at start-up can occur in two ways, either due to leaking valves or due to an ignition delay.

The risk of leakage is minimized by the correct choice and installation of safety shutoff valves and by adherence to correct operating procedures. With manually controlled systems valve-position checks should be incorporated in the start-up procedure. Low-pressure cutoff valves and valve-position checking systems (Section 9.4.5) provide some protection in this respect. In automatic burner systems, certificated safety shutoff valves are employed (in double-block arrangement on the supply to the main burner [6]) whilst in the case of large burners [5] safety shutoff systems are used in which valve closure is proved either by pressure-proving or position-proving (see Section 9.4).

Referring to delayed ignition there is inevitably a short period during start-up in which gas is allowed to flow into the combustion chamber and at the end of which the start-gas flame must be proved or the gas supply shut off. Any time lag between gas starting to flow to the burner and ignition will allow flammable mixture to accumulate. Clearly good burner head and igniter design will minimize any such delay but for safety the start-gas flow rate and the length of the trial-for-ignition period must be controlled. Two approaches are possible: (*a*) it can be assumed that all the gas entering burns and can contribute to an explosion; or (*b*) it can be assumed that gas and air entering are rapidly mixed in such proportions that the majority of the gas–air mixture in the appliance is so weak as to be incombustible.

The total energy input criterion. This criterion assumes that all the gas entering during an ignition attempt is capable of taking part in explosive combustion, i.e. it behaves in the same way as the discrete pockets of mixture considered in Section 9.3.1. Thus for a maximum pressure rise of 13.8 kPa (138 mbar, 2 lbf in^{-2}) (Section 9.3.2) the energy release during the ignition period should not be greater than 75 kJ per cubic metre (2 Btu per cubic foot) of heating-chamber volume. This criterion was adopted for the first edition of *Standards for Automatic Gas Burners* [4] which states:

> 7.5. The energy release during the pilot flame ignition period shall not be more than 2 Btu per cubic foot of heating-chamber volume.
> 7.6. Where the heating chamber volume is not known, as may be the case for packaged burners for general application, the energy release during the pilot flame ignition period shall not be more than 2 Btu per 100 000 Btu h^{-1} of maximum main flame rate.

Requirement 7.6. is based on a firing intensity of 100 000 Btu ft^{-3} h^{-1}.

In practice the total energy input is controlled by limiting the gas rate and the trial-for-ignition period. The latter is specified as between 2 and 5 seconds for automatic burners and this means that very low start-gas rates may be required to conform with the criterion.

As Aris *et al.* [22] have pointed out, for a firing intensity of 100 000 Btu h^{-1} ft^{-3} (2.93 MJ h^{-1} m^{-3}) and a 5-second trial time the start-gas rate should be $\ngtr$ 1.4 per cent of the maximum burner rating. This presents few problems on large burners with separate pilots but for small burners it is difficult to provide stable low-rated start-gas flames. Due to this problem with small burners the second approach has been adopted as an alternative to the total energy input criterion.

The dilution approach. If, during start-up the gas entering the appliance is mixed with sufficient excess air to dilute it rapidly to well below the lower flammability limit it will not contribute fully to the energy released from a delayed ignition. Consequently a greater total quantity of gas can be admitted without the pressure rise exceeding 13.8 kPa (138 mbar, 2 lbf in^{-2}) [22]. At the time of writing, current standards allow a start-gas rate of up to 10 per cent of the stoichiometric rate corresponding to the proved air flow in some circumstances.

The use of start-up conditions in which up to 10 per cent of the main gas rate together with the full air rate is used provides two advantages. First, considerably larger pilot burners can be used to overcome the stability difficulties referred to above and second, on nozzle-mixing burners capable of running with gross excess air (Section 7.5), direct ignition of the main flame becomes feasible which in turn eliminates the possible hazard of delayed cross-ignition. Aris *et al.* [22] have measured the volume of the flammable mixture zone at the burner head for a variety of burner configurations and based on these values and the considerations above, the second edition of *Standards for Automatic Gas Burners* [6] incorporates, as an alternative to the requirements based on the total energy input criterion, requirements based on this approach as follows:

> 8.6.1 (i) The start-gas rate shall not exceed 25 per cent of the stoichiometric gas rate for the proved air flow at the time of ignition.
> (ii) Notwithstanding paragraph 8.6.1 (i), no start-gas rate shall exceed 60 kW (205 000 Btu h^{-1}).

(iii) The volume of the zone of combustible mixture existing at the point of mixing, defined as that volume at or above the concentration limit of 5 per cent gas in air, shall not exceed 0.5 per cent of the heating-chamber volume, or, where the heating-chamber volume is unknown 5 cm^3 per kilowatt of maximum burner rating (8.95 in^3 per 100 000 Btu h^{-1}).

9.3.3 Pressure reliefs

As previously indicated, all precautions should be taken in the design, installation and operation of plant to avoid unacceptable pressure rises. Nevertheless it is good practice to incorporate in the design an adequately sized pressure relief as an additional safeguard.

Pressure reliefs consist of deliberately weak sections of the plant which it is desired to protect, so that should an explosion occur the venting produced limits the pressure rise to a safe value. The design of pressure reliefs for gas-fired box ovens and similar relatively weak structures was investigated by Simmonds and Cubbage [23, 24] as long ago as 1955 whilst for structures of intermediate strength, for example, boilers and furnaces, Cubbage and Marshall [24, 25] have provided some more recent information.

A further type of pressure protection may be distinguished. This may be described as a bursting disc and, since it is designed primarily to protect pressure vessels, lies outside the scope of this book. For further information the reader is referred to references [1] and [26].

The effectiveness of pressure reliefs is a consequence of the fact that the flame front in an explosion moves relatively slowly through the gas–air mixture. Thus the time taken to reach the peak pressure is significant, varying from tens to a few hundred milliseconds, depending upon gas type and concentration and the configuration of the confining structure [27]. Consequently there is time available to allow some of the combustion products together with some unburnt mixture to escape if an opening is available. Pressure relief consists of providing sufficiently large openings to allow the gases to vent quickly and safely.

Generally, of course, it is not feasible to have large openings in the walls of heating plant but relief can be provided by constructing lightweight or weak sections which will open or give way at an early stage of an explosion. Clearly the burning velocity of the mixture (see Section 4.3) exerts an important influence on the rate of pressure increase; the higher the burning velocity the steeper the pressure rise and therefore the shorter the time interval to reach a particular pressure. Thus a shorter time interval is available to operate the pressure relief. Additionally because of inertia a further finite time interval is required to move the relief sufficiently to allow an appreciable outflow of gases. Until this has occurred the pressure will continue to increase at about the same rate as if the enclosure were unvented. Thus an effective relief must have as low a breaking or relieving pressure as possible and be as light as possible. In addition the larger the vent area the sooner will the outflow rate equal the rate of production of gases by the explosion and the lower will be the maximum pressure developed [27]. It is obviously incorrect to assume that a relief designed to open or break at a particular pressure will limit the maximum pressure developed to that value. A satisfactory relief must have a breaking pressure or weight and area such that it will operate at a very much lower pressure than that which will cause damage to the enclosure to ensure that the maximum time is available for venting the combustion products.

Pressure reliefs for box and conveyor ovens [23, 24]
In addition to the possibility of fuel-gas explosions, drying ovens may experience explosions caused by flammable gases and vapours produced in drying or curing processes. Clearly, adequate ventilation should keep concentrations below the explosive limit and recommendations on ventilation rates are available [3].

As previously indicated the severity of explosions is related to the burning velocity of the flammable mixture and, since the burning velocities of town-gas–air mixtures are higher than those of mixtures of any industrial solvent vapours with air, Cubbage and Simmonds [23, 24] conducted the majority of their tests using town-gas–air mixtures.

From tests carried out on box ovens ranging in volume from 8 ft^3 to 98 ft^3 (0.23 m^3 to 2.8 m^3), Cubbage and Simmonds [23] showed that for stoichiometric town-gas–air mixtures completely filling the oven the relationship between the relief weight, vent area and oven volume and the pressure developed was given by:

$$\left(\frac{p_1}{\text{lbf in}^{-2}}\right)\left(\frac{V}{\text{ft}^3}\right)^{\frac{1}{3}} = 1.18K\left(\frac{M}{\text{lb ft}^{-2}}\right) + 1.57 \qquad (9.2)$$

where p_1 = first peak pressure
V = oven volume
K = vent-area coefficient
M = mass per unit area of pressure relief

or more generally [23, 25]:

$$\left(\frac{p_1}{\text{lbf in}^{-2}}\right)\left(\frac{V}{\text{ft}^3}\right)^{\frac{1}{3}} = \frac{S_u}{\text{ft s}^{-1}}\left[0.3K\left(\frac{M}{\text{lb ft}^{-2}}\right) + 0.4\right] \qquad (9.2a)$$

or, in SI units:

$$\left(\frac{p_1}{\text{kPa}}\right)\left(\frac{V}{\text{m}^3}\right)^{\frac{1}{3}} = \frac{S_u}{\text{m s}^{-1}}\left[0.45K\left(\frac{M}{\text{kg m}^{-2}}\right) + 2.6\right] \qquad (9.2b)$$

where S_u = burning velocity.

In addition a second pressure peak was observed, in some cases larger than the first and which in British units was numerically equal to the vent-area coefficient, i.e.:

$$p_2/\text{lbf in}^{-2} = K \qquad (9.3)$$

The second pressure peak is caused by the fact that after the relief has vented, combustion is still proceeding in the oven and products are being produced at an ever increasing rate. Venting then occurs at such a rate that a high pressure drop occurs across the vent, building up pressure in the oven until the combustion wave reaches the walls when the pressure again falls. In practice the second peak pressure is frequently well below the value given by 9.3 since this was obtained with central ignition, cubical ovens and homogeneous gas–air mixture. Choice of the correct position for the pressure relief is important. The oven top is the most obvious choice but cannot often be used in practice since the effect of horizontal shelves is to render the relief ineffective. Cubbage and Simmonds [24] showed that the effect of a single horizontal shelf with a blockage ratio of 30 per cent increased the pressure rise roughly fourfold compared with the empty oven. Since most industrial drying ovens contain shelves or are loaded such as to prevent easy flow towards top vents, vertical vents are required and these are normally most conveniently accommodated in the back of the oven. The only exception would be the case where the load consists of sheets parallel to the door and practically filling the vertical cross-section of the oven. A suitable design of back relief which should occupy the whole of the ovenback is shown in Figure 9:4. The vent is covered by a light-gauge metal mesh firmly attached to the oven. A sheet of rubberized asbestos $\frac{1}{32}$ in (0.8 mm) thick covers the mesh and seals the oven. This is covered by a layer oflow-density insulating material (e.g. mineral wool, 1 in to $1\frac{1}{2}$ in, (25 mm to 40 mm, thick) which is located by inserting it about $\frac{1}{4}$ in (6 mm) into the double lining of the oven. A second layer of rubberized asbestos covers this, the edges of which are held in place by a $\frac{1}{2}$ in (13 mm) metal strip screwed to the outside of the oven. The screws must not pass through the rubberized asbestos sheet.

The reason for using a back relief is to ensure that when the oven is divided horizontally by shelves each subdivision formed is directly vented. However, the presence of shelves still increases the pressures developed due to the promotion of turbulence and its enhancing effect on burning velocity (see Section 4.8). Based on the results of Cubbage and Simmonds, the *Industrial Gas Controls Handbook* [1] recommends that the solid shelf area should not be less than 50 per cent of the total shelf area to avoid developing the maximum pressure. In some cases this means that the oven will not function when fully loaded. If this is so a smaller coverage area may have to be used and the oven lit fully loaded or with the shelves removed.

The spacing of the relief relative to adjacent walls, ceiling and floor has an important effect on the second peak pressure developed and a minor effect on the first peak pressure. It is recommended [1] that an oven should always be spaced at least 15 in (380 mm) from the wall behind it and, if located in a corner, at least 2 ft (600 mm) from the side wall and 2 ft from the rear wall. These values apply to ovens up to 100 ft^3 (2.83 m^3) in volume. For larger ovens they should be increased in the ratio $(V/100)^{\frac{1}{3}}$.

In the case of treble-cased ovens (i.e. indirectly heated), two reliefs which do not connect with each other should be provided. One relief should occupy the back of the working chamber and the other the back of the combustion chamber if its volume is greater than 10 ft^3 (0.28 m^3).

Consideration should be given to the risk of personnel being burned by flames discharged from reliefs and precautions should be taken to prevent personnel entering the area into which reliefs would discharge. Data are available [24] on the extent of flame spread.

Conveyor ovens normally have open ends but the relief area this provides is inadequate to be effective. They are best considered by imagining them divided into a number of roughly cubical parts each provided with a suitable pressure relief [24].

Pressure reliefs for plant of intermediate strength

Cubbage and Marshall [27] have investigated the suitability of a variety of materials available in the form of rigid panels as pressure reliefs for the protection of plant such as furnaces, kilns and boilers which are structurally stronger than the ovens considered in the previous section or which are designed to operate at pressures in excess of 50–75 Pa (0.5–0.75 mbar, 0.2–0.3 inH_2O). Both natural-gas and town-gas–air mixtures have been considered either as a pocket of stoichiometric mixture or as an approximately homogeneous layer.

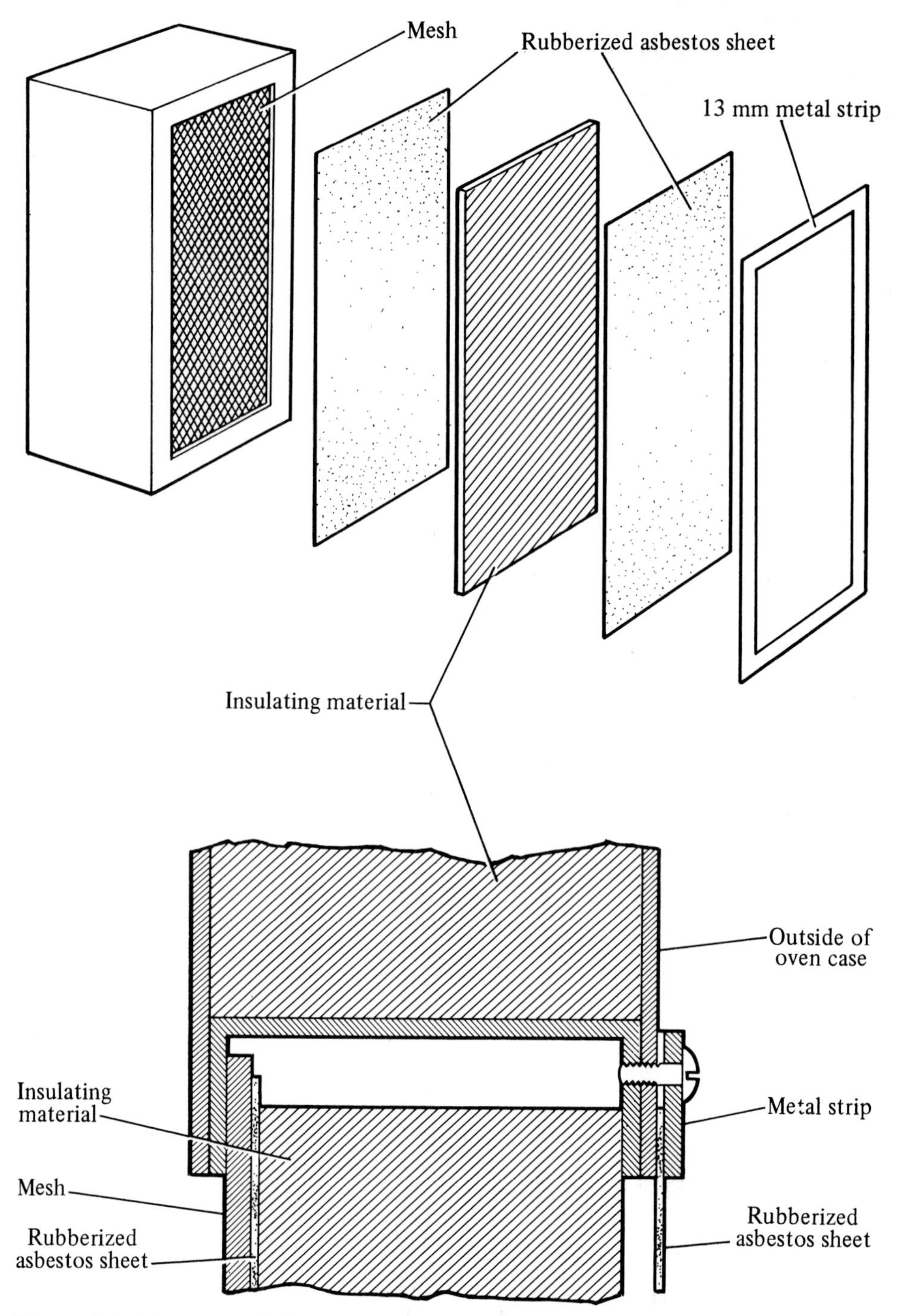

Figure 9:4 Recommended construction of pressure relief for low-temperature ovens, etc. [1]

For the materials considered, breaking pressure was shown to be inversely proportional to the area of the vent cover. Furthermore it is clear from the foregoing that the smaller the area of the vent the less rapid is the venting of the gases. Consequently as the area of a relief is reduced these two factors combine to produce higher pressure rises. Referring to Figure 9:5 (*a*) it is apparent that little pressure reduction results for $K = 9$ because the vent covering has a high breaking pressure and the vent area is small whereas for a larger vent area, $K = 2.94$, the relief is effective in controlling the pressure rise although the vent-cover material is identical in both cases. The weight per unit area of the vent cover also affects the pressure rise since there is a finite time required after the breaking pressure has been reached to move the relief material far enough to allow the outflow of gases to be established. Figure 9:5 (*b*) shows this effect. The heavier vent cover (plasterboard) whilst having the same vent area and breaking pressure as the lighter insulating-board cover is less effective in limiting the pressure rise.

A comprehensive correlation has been devised [25] relating the pressure rise to the characteristics of the enclosure, vent(s) and gas–air mixture ignited. It is:

$$p_{max} = p_v + 500(KM)_{av}\,(S_u^2/V^{\frac{1}{3}})\{1 - \exp[-(E - E_0)/(E + E_0)]\} \quad (9.4)$$

where S_u = burning velocity
p_{max} = maximum pressure generated
p_v = breaking pressure of relief
K = vent area coefficient
M = mass per unit area of vent cover
V = enclosure volume
E = 'energy density' per unit volume of enclosure
E_0 = 'energy density' at which the explosion relief is removed

and $(KM)_{av}^{-1}$ is given by $(KM)_1^{-1} + (KM)_2^{-1} + \ldots$

Pressure vents may be regarded as conduction paths and by electrical analogy (KM) is averaged as resistances in parallel. This is only valid if the breaking pressures p_{v1}, p_{v2}, etc., are approximately equal.

Equation 9.4 is restricted in use to structures having a ratio of maximum to minimum dimensions $\leqslant 3:1$, a vent cladding breaking pressure $(p_v) < 50$ kPa (7 lbf in^{-2}) vent area coefficients between 1 and 10, and values of $KM < 75$ kg m^{-2} (15 lb ft^{-2}). The characteristics of many types of industrial plant fall within these limits.

As Cubbage and Marshall [27] have pointed out, a further consideration which is not widely appreciated is the effect of a plant explosion on the building in which the plant is located. Significant pressures can develop even when the building is vented by windows. With plant occupying only one-fortieth of the building volume it is not impossible, due to the relative strengths of the structures involved, for a pressure rise which would do little damage to the plant to severely damage the building, perhaps even destroy it. Thus there is an added incentive in designing pressure reliefs to minimize the pressures generated to protect the building in addition to the plant.

Investigations of the effects on buildings of plant explosions have been conducted and the reader is referred to reference [25] for details. It is worth pointing out, however, that the conditions necessary for maximum pressure development very rarely apply in practice (e.g. complete filling of the structure with stoichiometric mixture) and few incidents result in severe structural damage; the only significant damage is generally to windows and doors (which act as pressure reliefs) and occasionally to non-load-bearing walls [25].

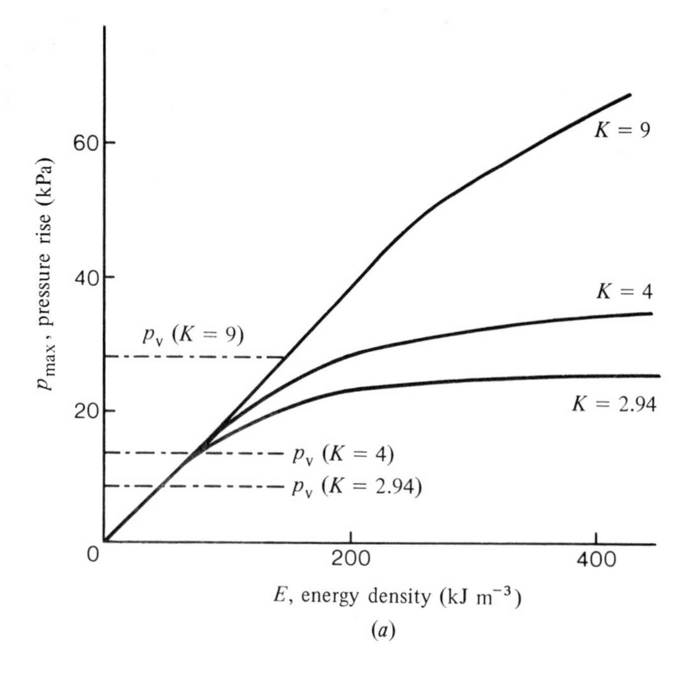

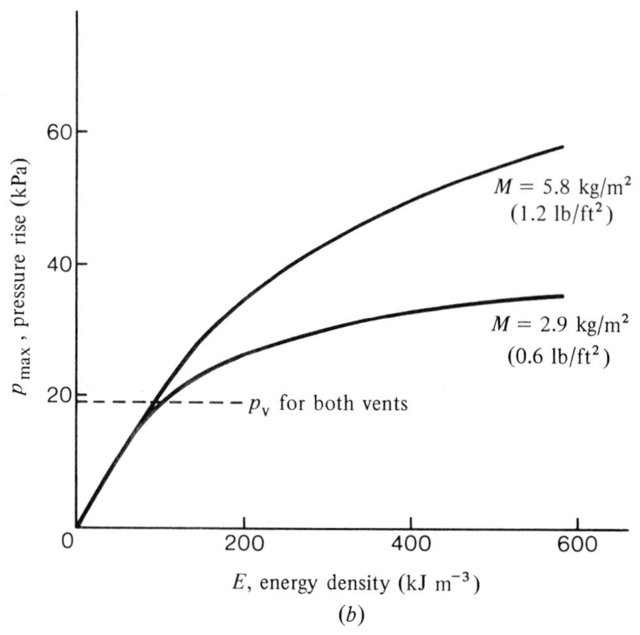

Figure 9:5 Size and weight of vents
(*a*) Pressures generated on ignition of pockets of stoichiometric town-gas–air mixture versus energy density for various values of vent-area coefficient [27]
(*b*) Effect of the weight of the relief cover on pressures generated [27]

9.4 GAS AND AIR FLOW CONTROLS AND SUPPLY SAFEGUARDS

A detailed and comprehensive review of all items of safety and control equipment applied to industrial gas-fired plant is outside the scope of this book. However, in Sections 9.4 to 9.6 the main features are briefly presented. For a comprehensive treatment the reader is referred to the *Industrial Gas Controls Handbook* [1] on which these sections are largely based and the quoted references.

9.4.1 Governors

Gas governors may provide constant outlet pressure or constant volumetric flow. In principle, however, all governors are pressure governors; volumetric governors simply maintain a constant differential pressure across the orifice. Constant-pressure governors are designed to reduce a higher and possibly varying upstream pressure to a lower and constant outlet pressure over a range of throughputs.

Low-pressure governors

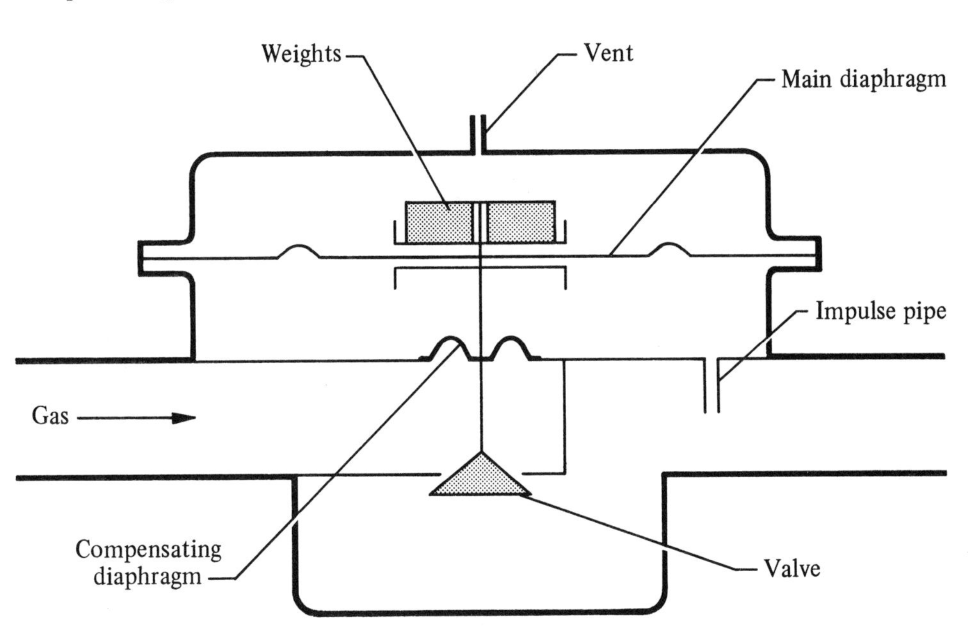

Figure 9:6 Low-pressure governor

These are normally suitable for inlet pressures up to 7.5 kPa (75 mbar, 30 inH_2O). Referring to Figure 9:6 any increase in outlet gas pressure raises the main diaphragm partially closing the valve and tending to restore the outlet pressure to its initial value. Weight-loaded (as shown) or spring-loaded types (especially for higher pressures) are in use. Most modern governors are compensated, i.e. they incorporate a subsidiary diaphragm which has the same effective area as the valve and thereby compensates for the effect of varying inlet pressure on the valve.

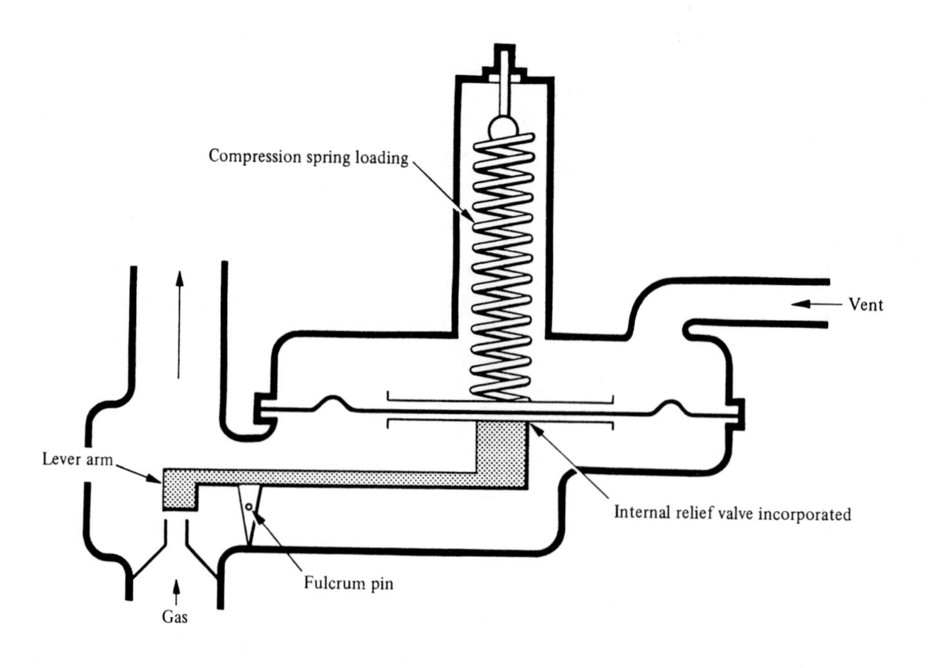

Figure 9:7 Medium- and high-pressure governor

Medium- and high-pressure governors
These are normally lever-operated (Figure 9:7). The comparatively low outlet pressure acts on the large diaphragm area and this, coupled with the mechanical advantage of the lever provides a high closing force on the valve. High-pressure governors or regulators incorporate over-pressure relief arrangements which may be mercury-sealed or more commonly make use of a spring-loaded relief valve.

Zero- or atmospheric-pressure governors
These governors which are designed to produce essentially atmospheric outlet pressure are used in conjunction with self-proportioning air-blast and machine premix burner systems and are considered in Section 7.4.1.

Further information
Further information on the principles of pressure governing is given in references [28–31]. Recommendations for the design of governor installations to consumers' premises are given in reference [32].

9.4.2 Manually operated valves

Plug valves combine simplicity with low pressure loss. They are rapid-closing valves and are widely used as isolation valves although when fitted with a graduated quad-

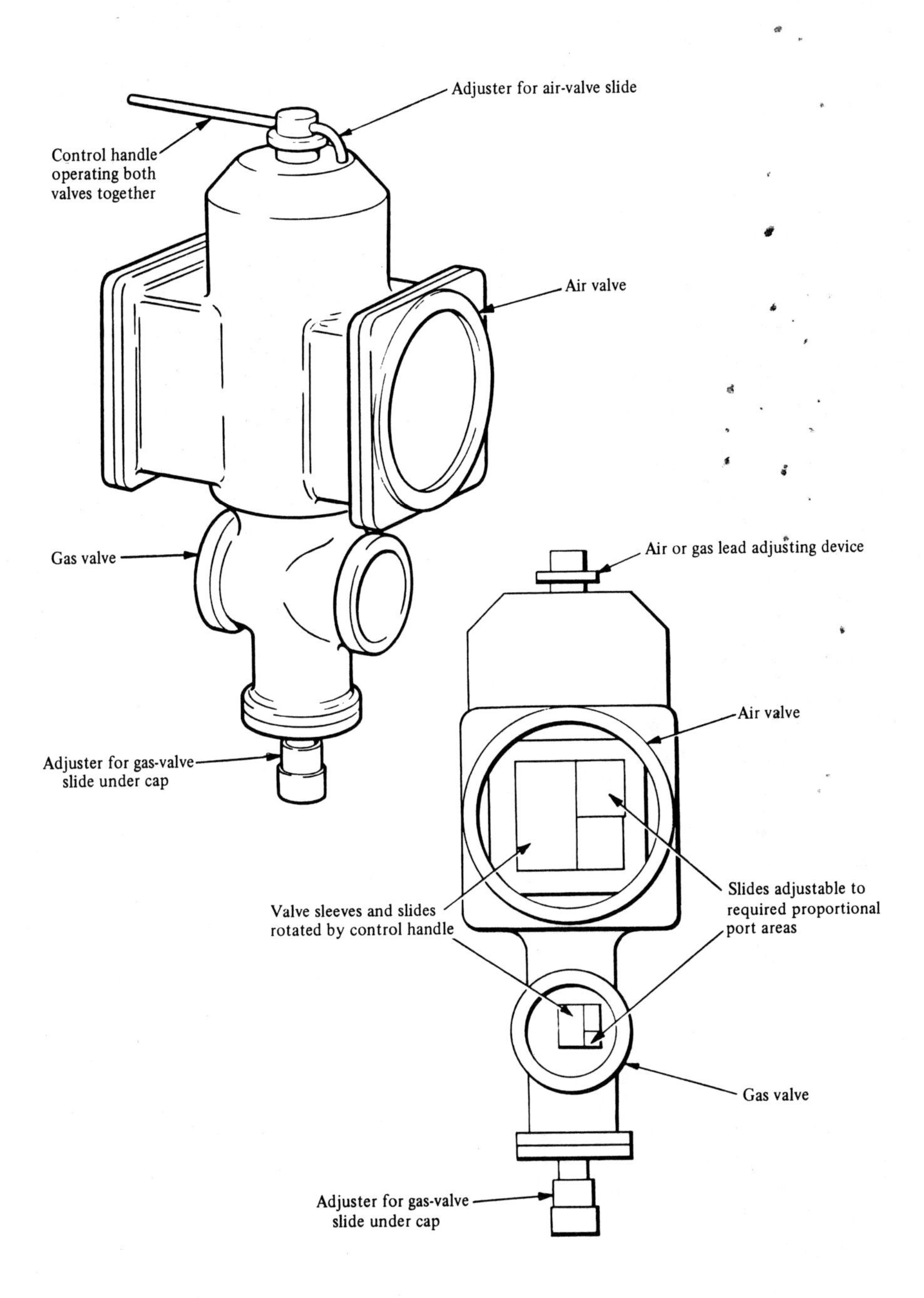
Adjuster for air-valve slide
Control handle operating both valves together
Air valve
Gas valve
Air or gas lead adjusting device
Adjuster for gas-valve slide under cap
Air valve
Slides adjustable to required proportional port areas
Valve sleeves and slides rotated by control handle
Gas valve
Adjuster for gas-valve slide under cap

Figure 9:8 Adjustable-port valve

rant they are commonly used as manual flow-control valves. Capacities of plug valves of various sizes are listed in reference [1]. Ball valves are finding increasing use for similar purposes. *Butterfly valves* are commonly used to control low-pressure gas and air flow. They provide large port openings and consequently very low pressure drops.

Where positive shutoff is required, butterfly valves incorporating rubber seals are available for working pressures up to 1.4 MPa (14 bar, 200 $\mathrm{lbf\,in^{-2}}$).

In *diaphragm valves*, closure is effected by depressing a synthetic rubber diaphragm until it contacts the valve seat. Further compression ensures closure since grit, etc., will embed in the diaphragm material. Diaphragm valves are commonly used as flow-setting valves due to their precise progressive adjustment. However multiturn valves should not be used as burner shutoff gas valves which should be of the quick-opening and closing type with position indication.

The use of linked gas and air valves to control air–gas ratio is considered in Section 7.5.2. The valves employed may be of two types: *adjustable-port valves* of matched (usually linear) flow characteristics in which the air–gas port-area ratio remains constant over the throughput range or a valve of any characteristic coupled with an *adjustable-flow valve* the characteristic of which may be adjusted to match the other valve. Figure 9:8 shows an adjustable-port valve which consists essentially of two mechanically linked plug valves on a common spindle. Rotation of the spindle varies throughput whilst adjustment of the valve slides alters the relative port areas by altering the heights of the rectangular gas and air ports.

9.4.3 Relay valves

Relay valves are used to control gas or air flows using a small actuating valve, which may be a solenoid valve or a thermostat in the weep line. Figure 9:9 shows the principle of operation. A small bleed orifice allows gas to pass from the underside of the diaphragm and thence to a small weep pipe in which is located the actuating (on-off) valve.

The weep may be terminated adjacent to the burner, be returned to the main gas supply downstream of the relay valve or, in the case of air supplies, to atmosphere. If the weep valve is open the pressure above the diaphragm is lower than the inlet pressure which acts on the underside of the diaphragm and opens the main valve. When the weep valve closes, pressure builds up above the diaphragm until the valve closes. The valve opening and closing times are functions of the bleed orifice and weep sizes and the volume above the diaphragm up to the weep valve. Relay valves open and close relatively slowly and have low closing forces. Hence they are not suitable for safety shutoff purposes. In the design shown an adjustable bypass valve is incorporated to permit high-low operation. Alternatively high-low operation is provided by an adjustable stop which prevents the valve from seating fully. Weep relay butterfly valves are also available and are used in particular for high air-flow rates. For further information on relay valves see references [1] and [33].

Figure 9:9 Relay valve
(*a*) On-off operation
(*b*) High-low operation
(*c*) Modulating operation

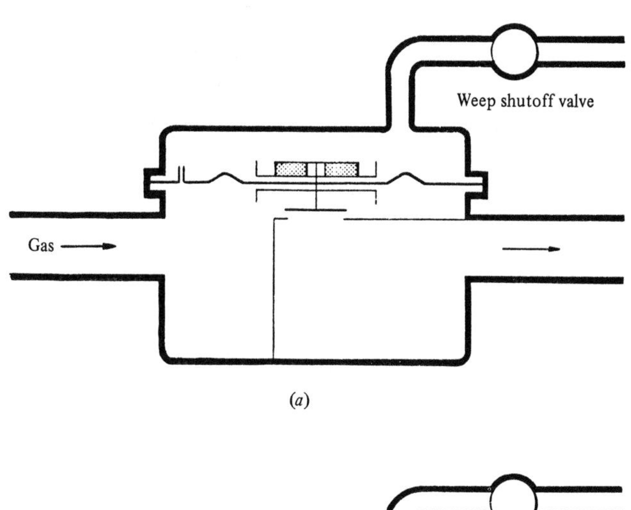
Weep shutoff valve
Gas
(a)

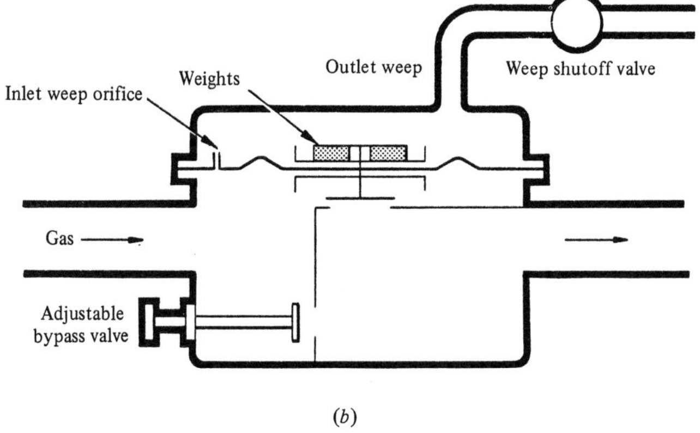
Outlet weep
Weep shutoff valve
Weights
Inlet weep orifice
Gas
Adjustable
bypass valve
(b)

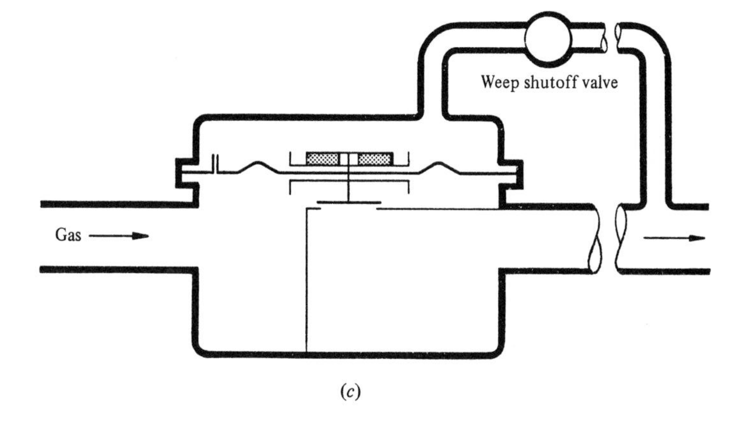
Weep shutoff valve
Gas
(c)

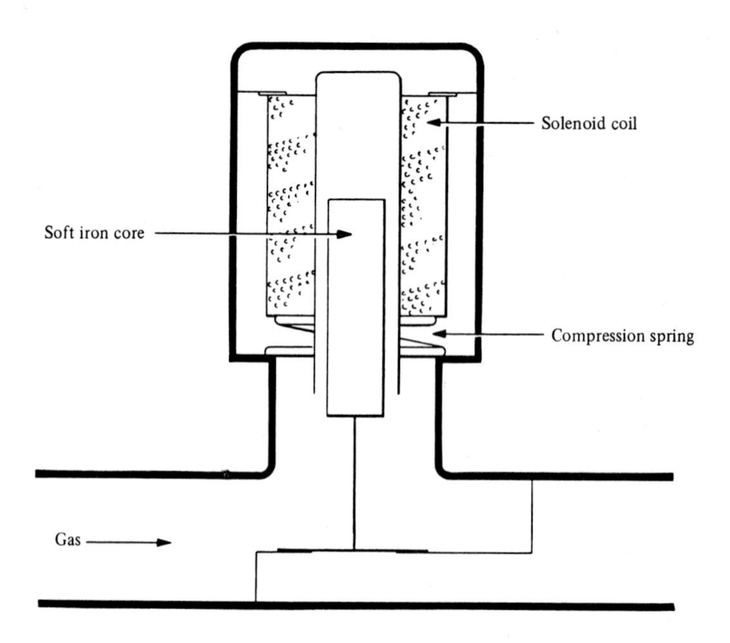

Figure 9:10 Solenoid valve

9.4.4 Solenoid valves and safety shutoff valves

Figure 9:10 shows a simple direct-acting solenoid valve used for electrically operated flow control. The electrical component may be an a.c. solenoid, a rectified a.c. solenoid or a d.c. solenoid. The gas valve component may be direct acting as shown, in which case de-energizing the solenoid causes rapid closing under the action of a spring. Manually reset types are available and also valves which open on being de-energized. Indirect-acting solenoid valves are simply relay valves with a small integral weep solenoid valve. These can, if sufficiently slow acting, avoid the possibility of pressure-wave formation resulting from the snap action of the direct types.

Solenoid valves are widely used for control purposes and as safety shutoff valves. Most shutoff valves are of the disc-on-seat type, either the valve disc or the seat being faced with a soft material. Thus, if grit particles are present they will embed satisfactorily provided the closing force is sufficient. For simple solenoid valves this force increases with decreasing valve size. From an examination of proprietary valves, early work by Atkinson and Moppett [34] showed that whilst the mean seat force for a $\frac{3}{8}$ in BSP valve was as high as 2.5 lbf in^{-2} (17.2 kPa) the value for a 1 in valve was only 0.18 lbf in^{-2} (1.24 kPa). On the basis that solenoid valves of $\frac{3}{8}$ in size and below were known to give reliable leak-free performance, Atkinson and Moppett [34] recommended that port forces of 2 lbf per square inch (14 kN per square metre) of valve port area should be the minimum required for all sizes of valve and this has been adopted as a requirement in *Standards for Automatic Gas Burners* [6] (see Section 9.8). Improvements in design have allowed solenoid valves to meet this requirement over a wide range of sizes.

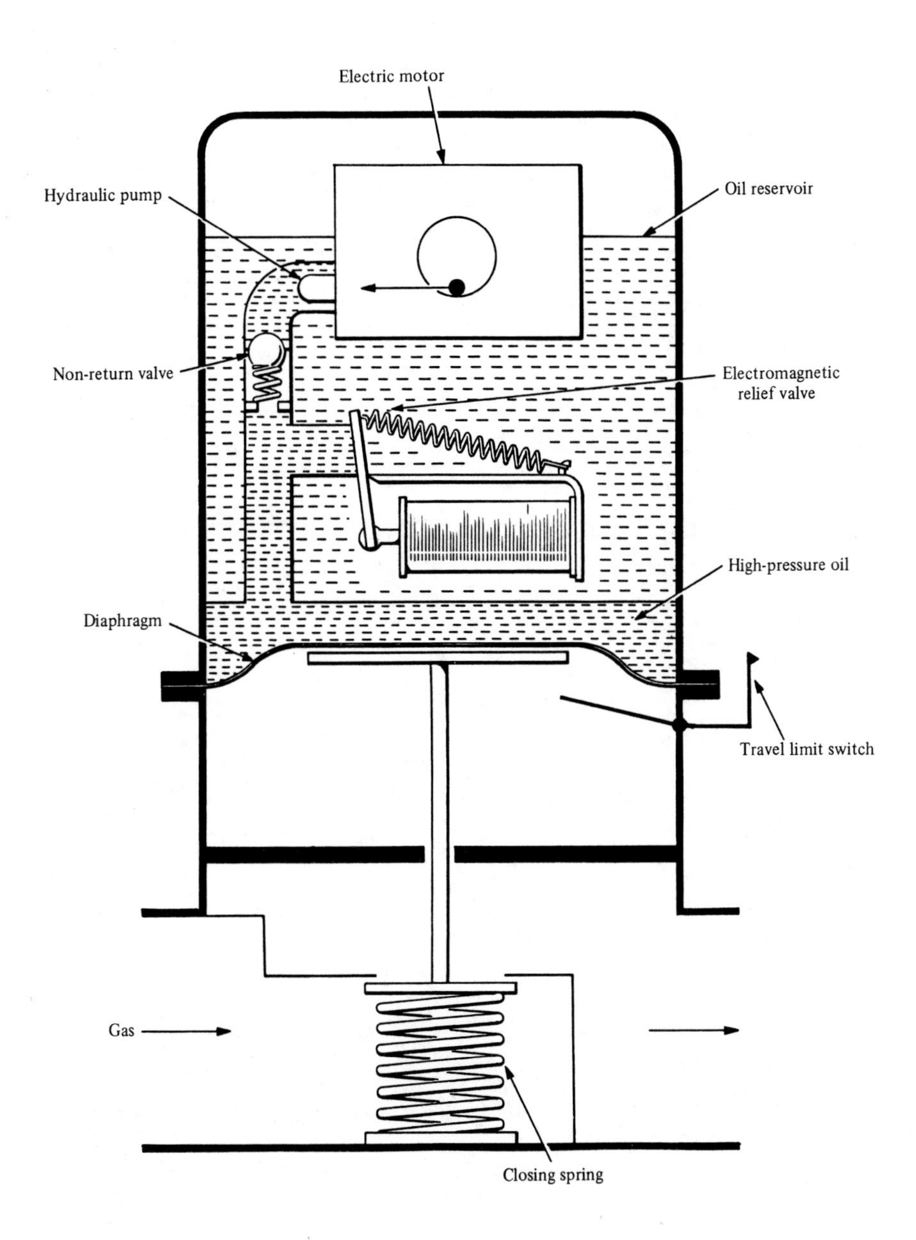

Figure 9:11 Hydramotor valve

An alternative to the conventional solenoid valve using an electro-hydraulic system is shown in Figure 9:11 (Maclaren-ITT Hydramotor). When electrical power is supplied to the valve the electromagnetic relief valve closes and the motor starts pumping oil through the check valve. This acts on the diaphragm and the valve opens against the force exerted by a strong spring below the valve until the travel limit switch is tripped and stops the motor.

The relief valve remains energized until power is cut off to the valve when it opens allowing oil to escape from the high-pressure region above the diaphragm to the reservoir assisted by the spring which seats the valve firmly. Alternatively electromechanical actuators may be used. Soft valve surfaces are employed in all disc-on-seat valves to ensure positive shutoff in the presence of grit, etc. British Gas Corporation's Midlands Research Station operates a certification scheme under which safety shutoff valves are tested to the requirements of *Standards for Automatic Gas Burners* [6].

9.4.5 Safety shutoff systems

As indicated above, reliable safety shutoff valves are now available covering a wide range of sizes. However, there still remains a small chance of maloperation the consequences of which tend to be more serious as the plant increases in size. Fortunately, in the case of large plant it becomes economically more feasible to use more sophisticated equipment to improve the degree of security. Any means of providing safety shutoff which involves more than the use of a single safety shutoff valve may be described as a safety shutoff system.

The simplest means of decreasing the probability of shutoff failure is to use two valves in series. If the probability of one of the valves failing is P then the probability of both valves failing together is nearly equal to P^2 giving a considerable improvement in security. Two valves in series are specified in *Standards for Automatic Gas Burners* [6, 8]. However, if no checking is carried out, this system is vulnerable to both valves failing in succession [35]. To eliminate the effects of systematic manufacturing errors there is a case to be made for using valves from different batches or operating on different principles in multi-valve systems.

The commonest cause of small leakage is a dirty or damaged valve surface preventing gas-tight closure. The addition of a normally open vent valve located in a branch between the two safety shutoff valves very much reduces leakage reaching the combustion chamber. This system is termed *double block and vent*. The vent pipe sizing must of course be adequate and this is specified in the interim codes of practice [5].

The other possible major improvement is the provision of an automatic check on the state of the valves, i.e. whether they are open or closed. This can be arranged to 'lock out' the burner and prevent any ignition attempt if the valves are not properly closed.

Two basic methods are available for checking valve closure: *valve-position proving* and *pressure proving*.

Valve-position proving

These systems check by means of limit switches that the valves are physically open or closed and can easily be incorporated into the overall burner logic to check the valves before and after every operation. For valves to be suitable for position proving they need to possess 'overtravel', i.e. the sealing element must travel beyond the 'just-closed' position. Plug, ball and guillotine valves possess overtravel but unfortunately the commonly used disc-on-seat valves do not, to any practical degree. The interim codes of practice [5] require the use of a double block and vent system with position checking using valves with mechanical overtravel as an alternative to pressure-proving systems on the supply to the main burner(s). If small leakage occurs this will be safely vented through the vent valve which is proved mechanically open. Details of double block and vent systems of this kind are given by Hutt *et al.* [35].

Pressure proving

Pressure-proving systems check that the safety shutoff valves are actually closed by testing for leak-tightness. They consist basically of two block valves usually in conjunction with a normally open vent valve and one or more pressure switches to check the pressure locked in between the valves. Checking may be achieved using gas at line pressure or at higher pressure, using nitrogen or by applying a partial vacuum. Details of these various methods, together with a method of calculating the detectable leakage level for a given system are given by Hutt *et al.* [35]. Only the line-pressure system is described here as a typical method. In this case the pressures used to perform the test sequence are line pressure and atmospheric pressure. If the enclosure between the shutoff valves is initially at atmospheric pressure then monitoring this with a pressure switch enables the upstream valve sealing to be checked. If the enclosure is initially at line pressure then monitoring the enclosure pressure checks the sealing of the downstream valve. Only if both checks are satisfactory can it be concluded that neither block valve is leaking.

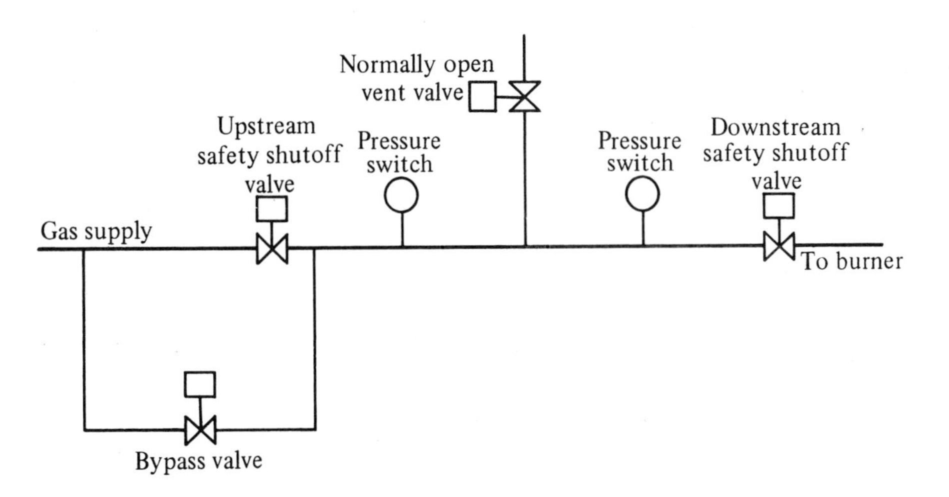

Figure 9:12 Layout of line-pressure-proving system [33]

Figure 9:12 shows a line-pressure proving system. The system comprises two block valves and a normally open vent valve and two pressure switches, one set to monitor atmospheric pressure and the other the line pressure. The upstream block valve is bypassed by a small safety shutoff valve. The test sequence before light-up is as follows.

The vent valve is closed and atmospheric pressure is monitored for an appropriate period. Then, assuming no detectable pressure rise has occurred due to upstream valve leakage, the bypass valve is energized allowing the enclosure to reach line pressure and then closed. The line pressure is then monitored using the other pressure switch for a further period. If both parts of the test are satisfactory lightup can proceed. On burner shutdown the test sequence is reversed.

Only systems incorporating position-proving or pressure-proving are suitable for very large loads and these are specified as alternatives in the interim codes of practice [5].

9.4.6 Low-pressure cutoff devices

Low-pressure cutoff valves

These valves are used to ensure that should the gas pressure fall below a predetermined value (above atmospheric) the valve closes. For gas to be restored to the burners three conditions must be satisfied, an adequate gas pressure, a manual reset of the valve and all downstream valves closed. Low-pressure cutoff valves have been

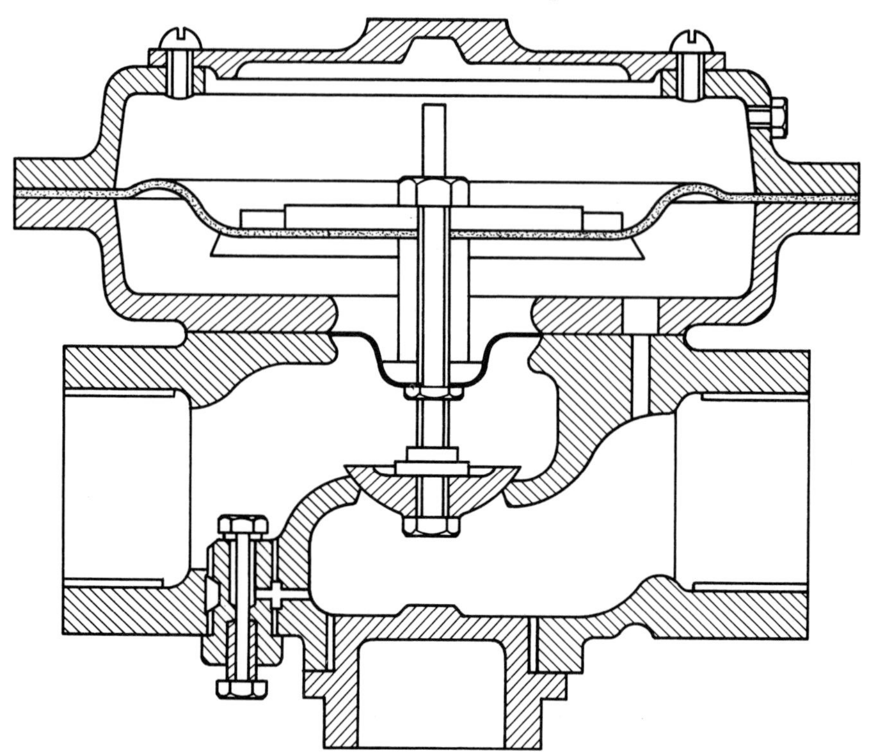

Figure 9:13 Low-pressure cutoff valve

used for many years in manual systems and have some points of similarity with the pressure-proving safety shutoff systems described above. Figure 9:13 illustrates a typical low-pressure cutoff valve. When the valve is in the closed position after a pressure failure, restoration of inlet pressure alone will not open the valve since this is compensated by the auxiliary diaphragm. A manual reset is required and this is achieved by pressing the spring-loaded bypass button. If all the downstream valves are closed pressure builds up on the valve outlet until there is sufficient force on the diaphragm to open the valve. The bypass weep rate is very small so that opening will not occur if there is significant leakage through the burner valves. The reopening time obviously increases as the downstream pipework volume increases and 30 metres (100 ft) of pipe is considered to be a practical maximum [1].

Low-pressure cutoff valves provide protection against some of the circumstances which may allow unburned gas to enter a combustion chamber. They must not, however, be regarded as a substitute for a flame safeguard system [1]. The trend has been towards their replacement by pressure switches used in conjunction with safety shutoff valves.

Low-pressure cutoff switches and anti-suction valves

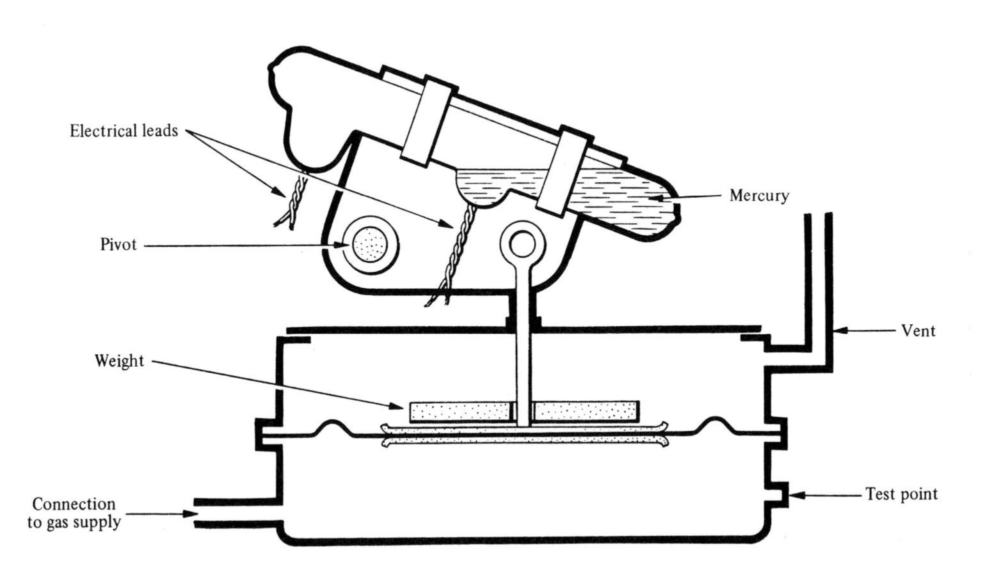

Figure 9:14 Low-pressure cutoff switch

Low-pressure cutoff switches are used to prevent gas (or air) compressors or boosters from causing reduced or negative pressures on the inlet side of the compressor which might lead to damage to the meter, interference with other appliances or other consumers' appliances and drawing in air at other appliances and to provide protection against the effects of low supply pressure. Paragraph 18 of the Fourth Schedule to the Gas Act 1972 [19] (see Section 9.2) requires the installation of an anti-suction device on the inlet to compressors and this normally takes the form of a low-pressure cutoff switch (Figure 9:14). Low-pressure cutoff switches should be used in such a way that no automatic restart can occur when the pressure is restored. One means of achieving this is to use the switch in conjunction with a compressor motor starter switch incorporating a 'no-volt release'.

9.4.7 Non-return valves

Non-return valves are:

1 Inserted in the gas supply to pressure-air or oxy-gas burner systems of both premix and nozzle-mixing classes to prevent air or oxygen (or any other extraneous gas) entering BGC's supply system in contravention of paragraph 18 of the Fourth Schedule to the Gas Act 1972 [19] (see Section 9.2) which requires the consumer to provide a device effectually to prevent the admission of compressed air or 'extraneous gas' into the service pipe or mains.
2 Fitted in the air line to prevent the admission of gas into the air supply and fan especially where the fan is higher than the injector.

Early non-return valves were simple hinged-flap valves but because of their poor performance at low reverse pressures they have been superseded by the diaphragm type shown in Figure 9:15. In the event of return pressure the diaphragms seat and prevent return flow past the valve. If the return pressure is increased the complete metal valve seats against its spring-loading providing an extra seat to withstand high back-pressures. At the time of writing improved designs of flap valves are coming back into use.

Note that if flame safeguard equipment is fitted using a safety shutoff valve conforming to *Standards for Automatic Gas Burners* [6] this can meet the requirements of a non-return valve. Non-return valves for oxy-gas glassworking burners are covered by reference [13].

9.4.8 Air-flow failure devices

These are normally employed to cut off the gas supply should the combustion air supply or a recirculation fan fail in circumstances where this would lead to flame extinction, excessive combustion-chamber temperatures or to a hazardous accumulation of flammable vapour. Mechanical types have been used but normally electrical types in conjunction with safety shutoff valves are used. Devices sensing air flow rather than air pressure are preferable and these normally take the form of differential-pressure switches.

9.4.9 Warning notices

In addition to fitting automatic protective devices it is essential to provide notices requiring the use of the correct operating procedure and indicating any hazards. Reference [1] gives examples of appropriately worded warning notices for gas compressors, gas engines and air-blast systems.

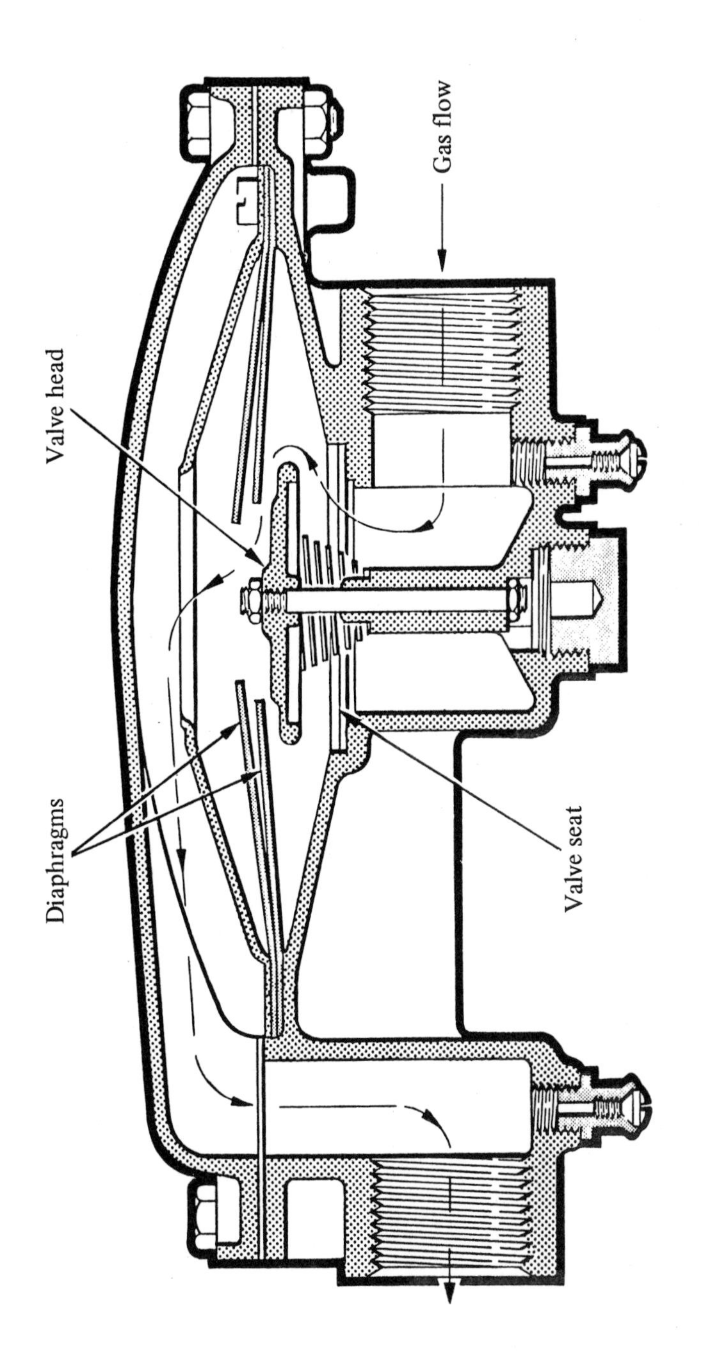

Figure 9:15 Non-return valve

9.4.10 Flame traps or arresters

In industrial gas practice there are situations in which flammable gas–air mixtures may be present in pipework for short periods of time, during purging and venting procedures, etc. Additionally in machine-premix burner systems (Section 7.6) flammable mixtures may be distributed over significant distances to individual burners in, for example, working flame and atmosphere generation applications. If lightback occurs it is possible for the flames to propagate through the pipework to the mixing plant. If the mixture strength and run-up length (Section 4.10) permit, the flame may develop into a detonation propagating at thousands of metres per second. Detonations are described in Section 4.10. The use of systems in which flammable mixtures are conveyed in pipes is not advocated but where such systems exist it is necessary to prevent the flame travelling back through the system and this is the function of the flame trap. Flame traps, in principle, consist of an assembly of narrow passages or apertures through which gas can flow but which are too small for a flame to pass through, i.e. the passages are less than the quenching diameter (Section 4.6). Note however that normally quoted quenching diameters relate to flame propagation at normal burning velocities and give little guide to the aperture size in a flame trap which may be required to stop a detonation with a velocity a thousand times greater than this. Since a flame trap must have a low pressure drop in normal flow it generally consists of a large number of small passages in parallel in an expansion housing. As Cubbage [36] has demonstrated, the expansion housing serves the equally important purpose of causing detonations to degenerate temporarily exposing the flame-trap element to a much slower combustion wave.

A variety of constructional methods have been employed including drilled plates, gauzes, sintered metal and gravel pots. However, the most satisfactory design (manufactured by Amal Ltd) consists of alternate layers of crimped and flat metal ribbon coiled around a former to produce a cellular structure with triangular passages. Although flame traps can be designed to arrest detonations it is nonetheless advisable to site the traps as near the burners as possible so as to avoid exposure to unnecessarily severe conditions. Good sealing between the arrester and its housing is essential since very small gaps may render the device useless.

If a flame is arrested but not extinguished the fuel supply must be cut off before the arrester becomes sufficiently hot to ignite the mixture on the upstream side. A system based on a thermocouple located on the downstream side of the element is commercially available and is described in reference [1].

If a flame is arrested and extinguished while mixture is flowing, unignited mixture will continue to be discharged. If this cannot be guaranteed to be ignited then a flame safeguard device should be fitted to the burner.

For further information on flame traps the reader is referred to references [1, 36, 37, 38]. Their application in purging procedures is detailed in reference [9] (see Section 9.2).

9.5 FLAME SAFEGUARDS

The majority of potential hazards in industrial gas equipment are associated with lighting up. Consequently in addition to providing protection against flame failure a satisfactory flame-safeguard system must also protect against incorrect lighting procedures. The simpler systems described in this section based on thermal expansion or thermoelectric effects are complete in themselves and in principle are capable of satisfying these two requirements. The other more sophisticated electronic devices are in effect flame detectors or sensors which detect the presence or absence of a flame by responding to flame radiation or electrical conductivity and are coupled to appropriate ancillary equipment to control the gas supply. The major advantage of the electronic types is their speed of operation. However, devices based on all these properties are in current use although the trend is strongly towards electronic devices.

Reference [1] defines the indispensable characteristics of flame safeguard systems as follows.

Every flame-safeguard system shall:

1 Ensure that the correct lighting procedure appropriate to the type of burner and appliance is followed
2 Entirely prevent gas from being supplied to the main burner until a pilot flame is established or prevent the full gas rate from being supplied to the main burner until a flame at a lower rate subject to an appropriate trial-for-ignition period (Section 9.6) is established
3 Be free from any inherent weakness in design which could give rise to failure to danger provided each component is fitted correctly
4 Stop all gas flow to the burners on flame failure and then require manual resetting

(Thus the use of an unprotected pilot in any installation fitted with a flame safeguard is unsatisfactory for industrial use.)

5 When detecting a pilot flame be actuated only by that part of the flame which will always ignite the main flame, i.e. the flame detector shall not be actuated either by a flame which cannot directly light the main flame or by a flame-simulating condition. This is dependent on correct application as well as on correct design
6 With the exception of thermal expansion and thermoelectric devices be provided with a safe-start check to prevent energizing of the gas valve(s) and where applicable any electrical ignition device should a 'flame-on' condition be present prior to ignition. Thermoelectric types should ensure manual isolation of the main gas until the pilot is established
7 Be mechanically and electrically satisfactory and readily serviced

For a flame-safeguard system to be satisfactory for industrial use it must comply with *all* of the above requirements. In addition there are a number of desirable rather than essential features such as tolerance to changes in gas characteristics, supply voltages, etc.

9.5.1 Thermal devices

Heat-operated flame safeguards are in common use in domestic gas utilization and may be either expansion-activated as in bimetal strips and rods, liquid-expansion and vapour-pressure types or thermoelectrically activated as in thermoelectric valves and thermocouple-relay systems. The advantages of thermal types are that they are relatively cheap and generally do not require a mains electricity supply. They are

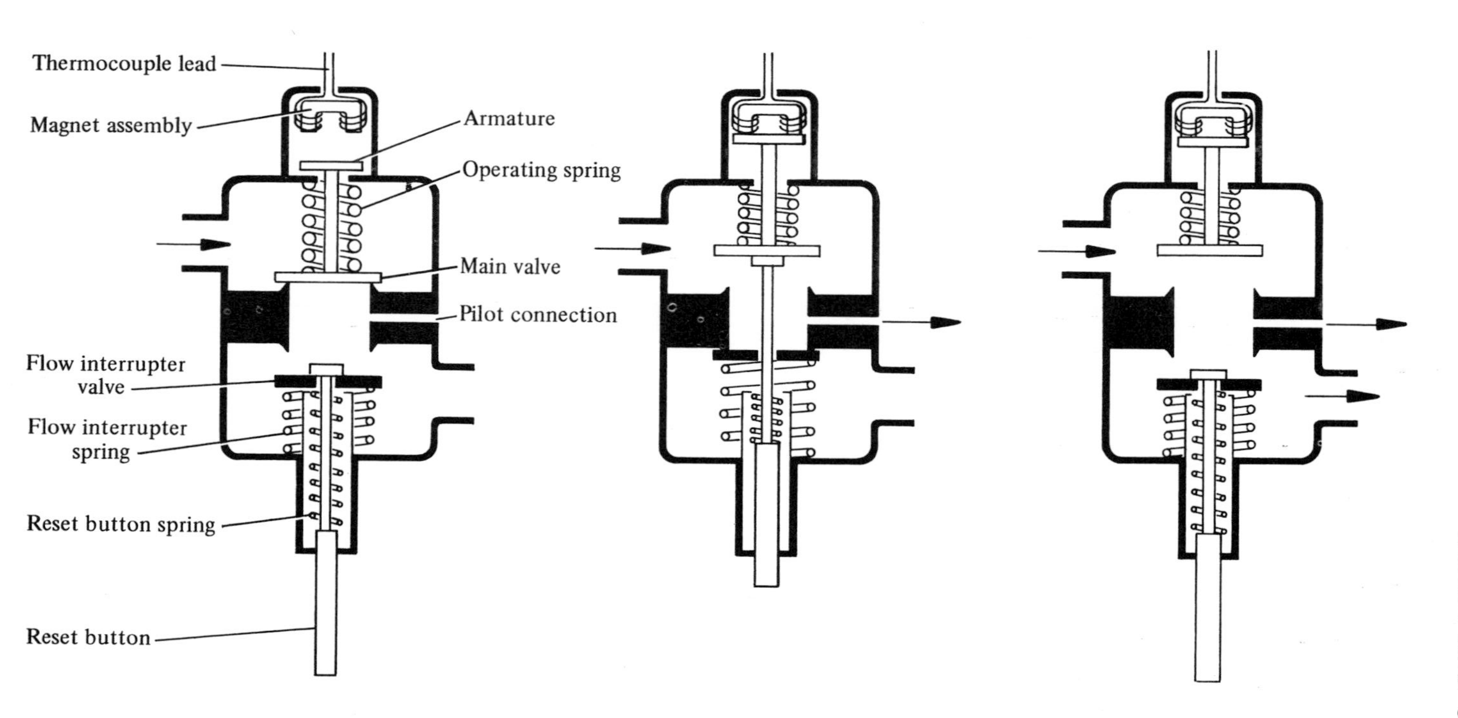

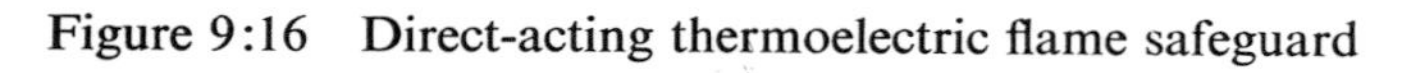
Figure 9:16 Direct-acting thermoelectric flame safeguard

restricted to applications in which the gas input is small.

Few expansion types satisfy the requirements outlined above and they have only rarely been used in industrial gas applications; thermoelectric types on the other hand are in widespread use.

Figure 9:16 shows the most commonly used thermoelectric device, the direct-acting gas-valve type. Pressing the push button A causes valve D to close sealing the exit to the main burner. Continued movement of A raises the main valve F admitting gas to the pilot burner which may be lit. Completion of the push-button movement presses armature H against the electromagnet which, when the thermocouple is sufficiently hot, is energized and retains the main gas valve open. When the push button is released, spring B opens valve D allowing gas to pass to the main burner. If the pilot flame is extinguished cooling the thermocouple de-energizes the electromagnet and the main gas valve F closes. Switching types are in use in which the thermocouple operates an electromagnetic switch rather than a gas valve. Since this type requires mains electricity supplies and the use of safety shutoff valves it possesses only a marginal cost advantage over electronic types and for this reason has not been widely adopted.

Thermal devices suffer from two serious disadvantages. They are very slow to react to loss of flame, often requiring 20 seconds or more to shut off the gas supply after flame failure. Secondly they may be actuated by heat from sources other than the flame, e.g. from hot combustion-chamber refractories.

9.5.2 Ionization detectors

These devices use a flame electrode for flame sensing together with an electronic amplifier. They possess a number of important advantages over the types described above; they detect only the flame not heat, they have a very fast response and they are suitable for higher temperatures.

As outlined in Chapter 4 the flame processes involve the presence of large numbers of free ions and electrons and these may be attracted to appropriately charged electrodes to produce a small current flow ($\approx$ 0.01 mA). One of the electrodes is earthed and this usually takes the form of a metallic burner head whilst the other, the flame electrode, is a thin rod of heat-resistant metal.

Early systems involved applying a d.c. potential across the electrodes, the presence of a conductive path indicating the presence of the flame. However, this system is susceptible to dust or condensate which produces a conductive path to earth leading to a spurious flame signal and so the flame-conduction method has been largely superseded by the flame-rectification technique. If an alternating voltage is applied between two electrodes of different areas (the burner head forming the larger electrode) more ions will strike the burner on the half-cycle when it is at a negative potential than will strike the smaller flame electrode when it in turn is at a negative potential. Thus as a result of the low mobility of the ions compared with the electrons the result is a small rectified current which is amplified in a simple d.c. amplifier to operate a flame relay. The flame electrode must be positioned such that the detected pilot is always of sufficient magnitude to ignite the main burner smoothly. To satisfy this requirement the flame electrode should only contact the pilot flame just inside the outer edge of the main flame under all conditions of draught, throughput and gas–air ratio likely to occur. Care must also be taken to prevent ignition sparks arcing onto the flame electrode. Often the area of the burner head is sufficient to provide an adequate earthed electrode area (area ratio normally > 4 : 1). If it is not, the area may be extended by short extension rods, fins or a wire spiral. Flame-rectification detection is inherently very sensitive but practical systems are designed so that only stable, well-aerated flames can be detected and unstable flames which lift intermittently are likely to cause shutdown [39].

9.5.3 Radiation detectors

Flames radiate over a wide range of wavelengths from the ultraviolet through the visible range to the infrared. Gas flames, unlike solid and liquid fuel flames, radiate only weakly in the visible range and radiation detectors for gas flames are sensitive to either ultraviolet or infrared.

Infrared detectors

Infrared detectors are based on detector cells the resistance of which varies in accordance with the radiation falling on them. The current passed by the cell is amplified before being used to operate a relay. The wavelengths over which infrared detectors operate lie between 1 and 3 μm. Since all hot bodies radiate in the infrared, simple resistive change detectors are unable to discriminate flames from hot surroundings. However, whilst radiation from refractories, etc. is fairly constant in level, that from flames is characterized by having a modulation or flicker superimposed on the steady output. Thus infrared detectors are designed to accept only an alternating signal of preset bandwidth (commonly 10–30 Hz) to avoid flame simulation. Despite this, infrared detectors can in some cases be affected by the flicker caused by air flowing over heated surfaces (air modulation). In addition, infrared detectors may be saturated by steady radiation at high background temperatures which desensitizes the detector and leads to nuisance shutdown. Infrared detectors can, however, be successfully applied in situations in which background temperatures are relatively low (< 800 °C).

Ultraviolet detectors

These are inherently more discriminatory than infrared types since flames (and electric sparks) emit ultraviolet radiation whilst hot refractories do not to any significant degree. The ultraviolet emission spectrum of a flame consists of a relatively uniform level on which are superimposed peaks due to the presence of the OH and CH radicals at 310 to 325 nm and about 430 nm respectively (for natural-gas–air flames) [40]. For heated backgrounds to emit appreciably in these wavelengths the temperature would need to be in excess of 1650 °C (for a perfect black body). Sunlight is, of course, a strong source of ultraviolet radiation. If a detector is installed correctly it should never be exposed to direct sunlight but this possibility cannot be completely discounted. However, normal window glass filters out all wavelengths below about 330 nm [40] and since daylight will almost inevitably have to pass through glass before reaching the detector the likelihood of flame simulation by daylight is remote.

Early ultraviolet detectors were based on a Geiger-Müller tube in which a d.c. potential is applied between the two electrodes (the cathode being a graphite deposit on the inside of the tube). If ultraviolet radiation of a sufficient energy level strikes the photosensitive cathode when there is an adequate potential difference between the electrodes, electrons are emitted and accelerated towards the anode. These collide with gas molecules in the tube ionizing them and producing further electrons. This 'avalanche' causes a significant d.c. current to flow in the amplifier circuit and this will continue as long as there is a sufficient potential across the electrodes. Thus the circuitry associated with the Geiger-Müller tube must be designed to reduce the potential to zero periodically to ensure that the flame signal derives from a continued supply of radiation.

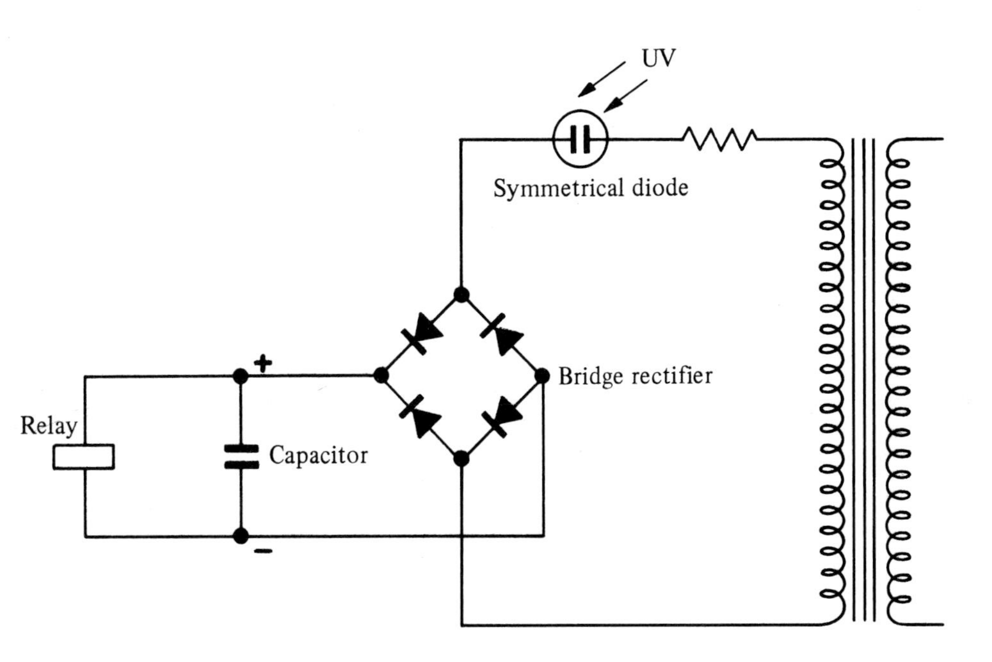

Figure 9:17 Symmetrical-diode ultraviolet flame-detector circuit

A simpler and cheaper alternative which is now in widespread use is the symmetrical diode detector tube a schematic circuit for which is shown in Figure 9:17. The gas-filled quartz-glass tube contains two identical photosensitive electrodes to which an alternating potential below the striking potential of the tube is applied. Radiation initiates the emission of electrons from the negatively charged electrode which, if the potential is sufficient, causes an avalanche discharge. The resulting current may be as high as 100 μA. Since an alternating voltage is applied, the potential at each electrode passes through zero twice per cycle automatically quenching the discharge. Also since the cell does not conduct over the complete cycle, the system can discriminate against the complete alternating signal which would occur for example if the electrodes shorted together [39].

Ultraviolet detectors are suitable for flames in high-temperature environments but must themselves be kept below 60–100 °C depending on type, which may necessitate air cooling in some circumstances. Since ultraviolet radiation is emitted by electrical sparks the detector should be sited so that it cannot 'see' ignition sparks unless the system is such that radiation from sparks is used to prove their presence before gas is admitted.

Although ultraviolet radiation emitted from flames is relatively intense it will be somewhat attenuated before it reaches the detector both by the normal inverse square law and as a result of absorption by gases. This varies approximately linearly with absorptive path length and differs significantly for different gases. Paraffin hydrocarbons, air and combustion products are virtually non-absorbing whilst aromatics are semi-opaque to ultraviolet, even in trace quantities, as are odorants. Thus ultraviolet absorption is due mainly to the trace components and town gas which contains many trace compounds is highly absorbent up to 280 nm. Natural gas (with odorant) generally presents less of an absorption problem than town gas [40]. However, as a general principle, detectors should not be sighted through burner gasways for reliable detection.

9.5.4 Application of flame safeguards

Thermal types of flame safeguard are restricted to applications involving small gas rates and where the risk of explosion due to their slow response is small. The general trend away from multiburner installations towards single forced-draught (often packaged) burners has contributed to the replacement of thermal types by electronic types. For other applications the choice rests mainly between flame-rectification and ultraviolet systems. Flame-rectification systems are suitable for low-temperature applications where probe overheating is not a problem. As the electrode is situated in the flame it is unaffected by adjacent flames in multiburner installations. Ultraviolet detection is more suitable for high-temperature applications and for burners which are not amenable to the application of ionization systems.

Standards for Automatic Gas Burners [6] require a flame safeguard to be fitted incorporating a safe-start check, i.e. if during this check which precedes the ignition sequence a fault or flame-simulating condition is present then the system will shut down and lock out in such a way that restart cannot be accomplished without a manual reset.

The maximum time elapsing between flame failure and de-energization of the safety shutoff valve is specified at 1 second.

For larger burners on low-temperature plant covered by the interim codes of practice [5] two main flame detectors were recommended for cases where a very high degree of operational reliability is required. Indication of loss of flame by one detector gives an alarm condition whilst simultaneous loss by both detectors leads to lockout. For burners firing continuously for periods greater than one week self-checking flame detectors are required by [5] and in other circumstances are desirable but optional. These commonly use a mechanical shutter in front of the detector tube which closes at frequent intervals to check that the flame signal is lost when the flame cannot be 'seen'. With intermittent pilot operation (i.e. using a pilot which is ignited prior to the main flame and shut off simultaneously with it) separate flame safeguards are essential for the pilot and main burner, the main flame detector being positioned so that it cannot detect the pilot flame.

9.6 IGNITION SYSTEMS

9.6.1 Pilot and start-gas flames

A start-gas flame may be defined as a flame established at the start-gas rate which may be either a pilot flame or the main flame at a reduced gas rate (see Section 9.3.2). A pilot flame is a subsidiary flame used for the ignition of a main flame. A separate pilot burner may be employed or it may be integral with the main burner but provided with a separate gas supply.

Pilot flames are required where it is desirable to provide a flame of limited thermal input to ignite a main burner either because of the hazards of lighting a main flame directly or for reasons of process intermittency [1]. They must be designed and positioned such that the pilot flame will provide immediate and smooth ignition of the main flame under any operative thermal input of both pilot and main flames and under any changes in gas characteristics, pressure and draught as may occur.

All pilots should be subjected to a pilot-turndown test as outlined in reference [1]. The importance of carrying out such tests with great care cannot be over-emphasized.

Care must be taken to avoid gas starvation on pilot burners when main burners are turned on and to effect this the gas supply may need to be taken from upstream of the main governor and separately governed.

Pilot-flame ratings are considered in Sections 9.3.2 and 9.8.

9.6.2 Lighting torches

A lighting torch is a portable burner supplied through a flexible tube arranged for manual ignition of fixed burners. Since flexible pipes are a possible source of leakage, a valve should be provided upstream of the flexible tube which should be kept as short as possible. For town gas a simple orifice flame port is adequate but for natural gas a flame-retention device will be necessary. Anything which might assist the torch being left inside the appliance should be avoided to prevent indiscriminate entry of gas. The gas flow to the torch must be restricted to that required for reliable ignition and in any case must not be more than 7.3 kW (25 000 Btu h^{-1}) or 3 per cent of the main-burner rating whichever is the greater [41].

Ignition and sighting holes must be provided to give easy access for ignition and a clear view of the flame together with a prominently displayed notice indicating the lighting procedure. Consideration must always be given to the potential hazards associated with lighting torches and the desirability of replacing this method of ignition by a properly controlled system incorporating a flame safeguard.

9.6.3 Electrical ignition

This may be by means of an electrical spark or using a hot wire. The general principles of both these ignition methods are considered in Section 4.7. The location of hot-wire igniters with respect to the flame is critical and in general they are restricted in use to standard appliances using natural-draught burners. The short life resulting from incorrect application prevents their more general adoption [1].

Spark ignition requires a minimum potential of between 5 kV and 10 kV depending on the ignition device used. These voltages may be obtained using mains transformers, electronic pulse systems or piezoelectric devices.

Mains-transformer ignition

Conventional step-up transformers are employed, normally resin-encapsulated. Transformers with a single secondary tapping should preferably be used. Any unused tappings on other types must be protected. The transformer should be mounted as close to the burner as possible but not where it will be overheated and all high-tension leads must be kept as short as possible both for safety reasons and to prevent electrical interference, especially with flame detectors. An adequate earth return must always be provided.

Electronic-pulse ignition

In this system a high-speed pulse of electrical energy is fed to a pulse transformer, i.e. a car ignition coil, to produce a high-voltage spark [41, 42]. The pulse is obtained from a solid-state switching circuit controlled by a thyristor or silicon-controlled rectifier which triggers periodically to release the energy stored in a capacitor into the pulse transformer. The main advantage over mains transformers is that up to 100 pulse transformers may be fed from a single pulse generator producing appreciable reductions in cost in multiburner installations. In addition cheap single- to seven-channel devices are available from the domestic-cooker market. The spark is of very short duration and thus the power input is much lower than for mains systems. For this reason their application is rather more critical than for mains-transformer systems.

Piezoelectric ignition

Piezoelectric devices make use of the property of certain polycrystalline ceramics (e.g. lead zirconate/lead titanate) which can produce high voltages when mechanically deflected in compression, tension or torsion. They are normally rubber encapsulated and arranged to be compressed by a manually operated lever and cam system. Typically, piezoelectric generators develop a peak voltage of 20 kV for a maximum stress of 70 MPa (700 bar).

Although in common use on domestic gas appliances this system has had little application in industrial utilization.

A final spark-ignition method which is worthy of mention is one employing a spark which tracks over the surface of a refractory semiconductor such as silicon carbide. Commonly known as surface-discharge ignition and used in such applications as jet engines, this system provides ignition despite heavy contamination with soot, oil or water but requires the use of high spark energies around 10 J [42].

9.7 THE GAS SAFETY REGULATIONS 1972 [11, 21]

These Regulations came into force on 1 December 1972 under the provisions of section 67 of the Gas Act 1948 which were continued under paragraph 16 of Schedule 7 of the Gas Act 1972. Part I is primarily concerned with definitions. Part II deals with the installation of service pipes, including restrictions on service pipe runs, requirements regarding corrosion protection, service pipe falls and condensate provision, service governor venting (for inlet pressures greater than 30 inH_2O (7.5 kPa, 75 mbar)) and the provision of a readily accessible service valve for services of 2 in (50 mm) bore or greater or where there is a special risk or where the service supplies more than one primary meter. In addition Part II requires that the installer of a service pipe, fitting, valve or governor shall test the installation for gas tightness to ensure that the installation conforms with the Regulations, apply any necessary protective coating to the joints after testing for gas tightness, purge the system and, if the service is not to be used immediately, to cap or plug it off at the meter control (i.e. the valve on the inlet side of the primary meter).

Part III deals with the installation of meters. Regulation 17 (1) restricts the installation of primary meters and primary meter valves, bypasses and governors to employees of the British Gas Corporation or its contractors. Both primary and secondary meter installations are covered by this Part. The installer's employer is required to ensure that the provisions of the Part are complied with. In the case of secondary meters and ancillaries on premises forming part of a factory the occupier is required to ensure compliance. Regulation 18 requires good workmanship and materials whilst regulation 19 specifies positioning to ensure accessibility and freedom from accidental damage.

Meter controls are specified in regulation 20 and regulation 22 requires that in the case of buildings having two or more floors above the ground floor, meters shall not be installed on or under a stairway or elsewhere which provides the only means of escape in case of fire. In the case of other buildings meter installations on or under stairways providing the only fire escape are required to be of fire-resistant construction or housed in an enclosure of at least half-an-hour fire resistance, the doors of which are fitted with automatic self-closing devices or the service pipe to incorporate a thermal cutoff near the meter (regulation 22 (2)). Other requirements of this Part include the provision of a permanent notice at the meter indicating the consumer's course of action in the event of an escape (regulation 23), the use of a temporary electrical connection while a meter is being installed (regulation 29) and requirements regarding the installation of meter governors (regulations 30, 31 and 32). As with service pipes the onus is placed on the installer of the meter or ancillaries to test it for soundness, ensure that it conforms with the requirements of the Regulations and purge it and its associated pipework.

Part IV is concerned with installation pipes and fittings. The installer must comply with the provisions of this Part and in the case of premises forming part of a factory the occupier is required to ensure compliance. Regulation 35 requires that all installation pipes and fittings are of sound construction and materials, of adequate strength and size to secure safety and soundly installed and jointed by a competent person. Restrictions are placed on pipe runs in cavity walls and under foundations and footings (regulation 37). Sleeving is required for pipes passing through walls and solid floors (regulation 37) and corrosion protection or the use of corrosion-resistant materials is required for pipes exposed to corrosion risk (regulation 38). Adequate support and condensate-removal provisions are specified (regulations 39 and 40). For non-domestic installations having a service pipe of more than 2 in (50 mm) diameter a readily accessible and conspicuously sited valve is required on the incoming pipe to each floor in a building having two or more floors or, where the floor of a building has self-contained areas to which gas is supplied, on the incoming pipe to each area (regulation 42). This regulation also requires the provision of a line diagram of the installation system to be fixed as near as possible to the primary meter in the case of such non-domestic installations. As with service pipes the installer is required to test for soundness, apply any necessary protective coating to the joints, purge and cap off if it is not to be used immediately (regulation 43).

Part V governs the installation of appliances. Regulation 44 requires the installer to comply with the following provisions, his employer to ensure compliance and in the case of premises forming part of a factory the occupier to ensure that the provisions are complied with. All gas appliances must be installed by competent persons and no gas appliance shall be installed unless the appliance and fittings, etc., the means of removal of combustion products, the availability of combustion air, the ventilation to the room and the general conditions of installations are such that the appliance can be used safely (regulation 45).

Appliances, fittings, flues and means of ventilation must conform to the appropriate building regulations (regulation 46). The installer is required to test the connection of every appliance for soundness and to examine and adjust the appliance, fittings, flue, ventilation means, etc., to ensure that they conform with the provisions of Part V, that the heat input and operating pressure are as recommended by the manufacturers, that safety controls are in working order and that any means of combustion-product removal and ventilation and of supply of combustion air are in safe working order (regulation 46).

Part VI, entitled 'Use of Gas', requires that no person shall use or permit a gas appliance to be used if at any time he knows or has reason to suspect:

1 That there is insufficient supply of air available for the appliance for proper combustion at the point of combustion
2 That the removal of the products of combustion from the appliance is not safely being carried out
3 That the room or internal space in which the appliance is situated is not adequately ventilated
4 That gas is escaping
5 That the appliance, fittings, etc., is so faulty or maladjusted that it cannot be used without constituting a danger to person or property (regulation 47)

Regulation 48 requires that if a consumer knows or has reason to suspect that gas is escaping in the premises he must shut off the supply. Where gas continues to escape in any premises after the supply has been shut off the consumer must as soon as possible notify the Corporation. The supply must not be re-opened until all necessary steps have been taken to prevent the gas escaping again.

Part VII deals with the removal, disconnection, alteration, replacement and maintenance of gas fittings, etc. A temporary electrical connection while pipes, fittings, meters, etc., are disconnected (regulation 49) is required. A person who disconnects a gas fitting or any part of the gas-supply system is required to seal it off, cap it or plug it at every outlet of every pipe to which it is connected with the appropriate pipe fitting (regulation 50). Regulation 51 requires that no alteration be made (whether the system was installed before or after the introduction of the Regulations) such that as a result there would be a contravention of any provision of Parts II to V. On every replacement of a gas fitting or of any part of the gas supply the provisions of Parts II to V apply to its replacement as they apply to its installation—except that when replacing a meter that is installed on or under a stairway, etc., which forms the only means of fire escape in a building having two or more floors above the ground floor, the replacement meter may be placed in the former position if it is installed to comply with regulation 22 (2) relating to buildings with less than two floors above ground floor. Alterations and replacements must be tested for soundness and it must be verified that there is no contravention or failure to comply with the provisions of Parts II to V by the person carrying out the work. Finally the Corporation is required to keep in proper working order all service valves (regulation 52).

Part VIII states that any person offending against these Regulations shall be liable on summary conviction to a fine not exceeding £100.

9.8 STANDARDS FOR AUTOMATIC BURNERS, FORCED AND INDUCED DRAUGHT [6]

These standards relate to automatic burners using forced or mechanically induced draught in which, when starting, gas is automatically turned on to the main burner (see Section 9.2) and in which the thermal input is greater than 44 kW (150 000 Btu h^{-1}). The standards lay down basic minimum requirements of safety. For some specific applications and for larger burners [5] more stringent requirements may be necessary.

Section 2 specifies the requirements regarding safety shutoff valves. Safety shutoff valves are required to maintain a tight shutoff at all forward-flow pressure differentials up to one-and-a-half times their rated working gas pressure which should not be less than 5 kPa (50 mbar, 20.1 inH_2O) for class I valves and 2.5 kPa for class II. In the case of air-blast injector systems it is possible to obtain a suction in the gas line of up to 1.75 times the air pressure [43] thus the forward differential pressure across the valve would be this plus the gas pressure. Where air-blast injectors are used, therefore, the valve must be suitable for twice the air pressure plus the gas pressure.

A convenient way of specifying the closing force of a safety shutoff valve is by specifying a reverse-flow pressure differential (see Section 9.4.4). The standards specify either 15 kPa (150 mbar, 2.18 lbf in^{-2}) for valve sizes up to 50 mm (2 in BSP) and 10 kPa (100 mbar, 1.45 lbf in^{-2}) for valve sizes over 50 mm for class I valves (5 kPa for class II) or twice the maximum air-supply pressure whichever is the greater. In addition to specifying an adequate closing force implementation of these requirements may obviate the need for a separate non-return valve (see Section 9.4.7).

Two safety shutoff valves are required in series in the gas supply to the main burner as a minimum, i.e. the simplest safety shutoff system (Section 9.4.5). Note, however, that if the circumstances warrant by virtue of large thermal input, location or other factors, a multiple-valve system incorporating a closure-checking system will be required. It is obligatory for burners rated in excess of 3 MW (102 therms per hour).

The maximum safety shutoff valve closing time is specified at 1 second. Thus, since the maximum time for the flame safeguard to de-energize the safety shutoff valve is specified also at 1 second, the maximum total shutoff time from loss of flame is 2 seconds.

Section 5 requires the use of a device (Section 9.4.8) to prove the air supply which must be proved throughout the purge before any ignition attempt and during operation. A prepurge is specified in section 6 to take place immediately before any ignition attempt. The minimum purge period is 30 seconds at full combustion air rate or proportionally longer periods at lower air rates. Note that these rates may not be adequate for removal of solvent vapours, for example, and reference [3] should be consulted. The choice of 30 seconds prepurge period is based on the fact that if a stoichiometric gas–air mixture is diluted with 10 times its own volume of air its composition is well below the lower explosive limit and will not contribute appreciably to an explosion (Section 9.3.1). Since the maximum total shutdown time is 2 seconds then a 20-second purge would give the required dilution and a 30-second purge an adequate margin of error.

A postpurge is optional. Arguments may be put for and against the use of a postpurge. In its favour is the fact that upon flame failure unburned gas will have passed into the appliance and some ignition source may cause on explosion before the prepurge of the next ignition attempt. Against its use is the argument that a possible cause of flame failure is an over-rich mixture leading to blowoff and any subsequent addition of air would increase the explosion hazard. Considering the fact that it takes a finite time for the combustion air to cease after the fan has been switched off there is no overwhelming case for or against its adoption.

Section 7 requires that burners be fitted with a flame safeguard which incorporates a safe-start check (see Section 9.5.4) and recommends but does not require the use of continuous self-checking flame safeguards.

Section 8 covers the important subject of start-gas flame establishment and requires that the start-gas rate conforms either to the total energy input criterion or to the dilution criterion which are considered in Section 9.3.2 of this chapter. The length of the start-gas period is specified as $3\frac{1}{2} \pm 1\frac{1}{2}$ seconds ignition period plus $\nless$ 5 seconds start-gas flame proving period during which it is established that the flame is stable on its own.

Section 9 limits the main flame establishing period to $3\frac{1}{2} \pm 1\frac{1}{2}$ seconds at the end of which the pilot flame is normally extinguished and supervision of the main flame alone begins. If the pilot flame is required to coexist with the main flame then separate unambiguous detection of the main flame is necessary.

Although the standards apply strictly only to automatic burners as defined, their requirements embody good practice with regard to safety, the principles of which can be applied with advantage to manual and semi-automatic burner systems.

REFERENCES

1 Gas Council, *Industrial Gas Controls Handbook* (Gas Council Report No. 763/69) (1969)

2 Fitzsimons, W. A. and Hancock, R. A., *Inst. Gas Eng. Commun.*, 924 (1973); abstract in *J. Inst. Gas Eng.*, **13** (1973), 269

3 Department of Employment, HM Factory Inspectorate, *Evaporating and Other Ovens* (Health and Safety at Work, Booklet 46), London, HMSO (1971)

4 Gas Council, *Standards for Automatic Gas Burners, Forced- and Induced-Draught* (Report No. 762/66, U051, U9) (1966)

5 Gas Council, *Interim Codes of Practice for Large Gas and Dual-fuel Burners* (Reports Nos. 764/70, 766/70, 767/71, 768/71) (1970 and 1971); Amendment No. 1 (April 1974). New code in preparation

6 Gas Council, *Standards for Automatic Gas Burners, Forced- and Induced-Draught*, second edition (Report No. 765/70, U0515, U9, U96) (1970)

7 Hill, K. B., 'Technical standards on industrial process heating plant', Midland J.G.A. (1974)

8 *Gas Eng. Manage.*, **15** (January 1975), 19

9 Gas Council, *Purging Procedures for Industrial Gas Installations*, second edition (Report No. IM/2, U763) (1975)

10 British Gas Corporation, *Soundness Testing Procedures for Non-Domestic Gas Installations* (Report No. IM/5, U06, U9) (June 1974)

11 The Gas Safety Regulations 1972, Statutory Instrument No. 1178, London, HMSO (1972)

12 Gas Council, *Overheat Protection for Steam Tube Ovens* (Report No. 770/71) (1972)

13 Gas Council, *Non-Return Valves for Oxy-Gas Glassworking Burners* (Report No. IM/1) (1972)
14 Gas Council, *Flues; Technical Notes on the Design of Flues for Larger Gas Boilers* (1971)
15 British Gas Corporation, *Combustion and Ventilation Air, Guidance Notes for Boiler Installations in Excess of 2×10^6 Btu h^{-1} (586 kW) Output* (Publication DCM/2/73) (1973)
16 British Standards Institution, *CP332, Selection and Installation of Town Gas Space Heating*, 4 parts (1961-70)
17 British Standards Institution, *CP331, Code of Practice for Installation of Pipes and Meters for Town Gas*, 3 parts (1973-4)
18 Gas Council, *Boiler Changeover: Technical Notes on Boiler Changeover to Gas Firing*, 2nd edition (1971)
19 Gas Act 1972, London, HMSO (1972)
20 Watts, R., 'The Gas Act 1972', *Watson House Bulletin*, **37,** No. 271 (1973)
21 British Gas Corporation, *A Guide to the Gas Safety Regulations* (April 1973)
22 Aris, P. F., Hancock, R. A. and Moppett, D. J., *Gas Council Res. Commun.*, GC166; *J. Inst. Gas Eng.*, **10** (1970), 97
23 Cubbage, P. A. and Simmonds, W. A., *Gas Council Res. Commun.*, GC23 (1955); *Trans. Inst. Gas Eng.*, **105** (1955–6), 470
24 Cubbage, P. A. and Simmonds, W. A., *Gas Council Res. Commun.*, GC43 (1957); *Trans. Inst. Gas Eng.*, 107 (1957–8), 503
25 Cubbage, P. A. and Marshall, M. R., *Inst. Gas Eng. Commun.*, 926 (1973); abstract in *J. Inst. Gas Eng.*, **13** (1973), 269
26 Lake, G. F. and Inglis, N. P., *Proc. Inst. Mech. Eng.*, **142** (4) (1940), 365
27 Cubbage, P. A. and Marshall, M. R., 'Explosion relief protection for industrial plant of intermediate strength', Symposium on Chemical Process Hazards, Institution of Chemical Engineers Symposium Series, No. 39 (1974)
28 Benton, W. E., *Trans. British Junior Gas Assoc.*, **24** (1933–4), 179
29 Hadow, H. J. and Pinkess, L. H., *Trans. Inst. Gas Eng.*, **86** (1936–7), 579
30 Parkinson, B. R., *Governors and Governing*, London, Walter King (1947)
31 British Standards Institution, *British Standard 3554, Specification for Gas Governors*, 2 parts (1971–2)
32 Institution of Gas Engineers, 'Recommendations for the design of governor installations to consumers' premises', IGE TD/5 *Inst. Gas Eng. Commun.*, 750 (1967)
33 British Standards Institution, *British Standard 1963 : 1969, Specification for Pressure Operated Relay Valves for Gas Burning Appliances* (1969)
34 Atkinson, P. G. and Moppett, D. J., *Gas Council Res. Commun.*, GC135 (1966); *J. Inst. Gas Eng.*, **7** (1967), 369
35 Hutt, S. H., Moppett, D. J. and Stein, K., *Gas Council Res. Commun.*, GC190 (1971)
36 Cubbage, P. A., *Gas Council Res. Commun.*, GC63 (1959); *Trans. Inst. Gas Eng.*, **109** (1959–60), 543
37 Ministry of Labour, *Guide to the Use of Flame Arresters and Explosion Reliefs* (Safety, Health and Welfare Booklet No. 34), London, HMSO (1965)
38 Gas Council, *The Design of Flame Traps for Use with Town-Gas–Air Mixtures* (Industrial Gas Development Committee Report No. 731/60) (1960)

39 Hopkins, H. F., 'Flame sensors for large boilers', *Watson House Bulletin*, **34** (November–December 1970)
40 Marshall, M. R. and Ward, R. G., BGC Midlands Research Station Ext. Rept. 222 (1973)
41 Arnold, J. and Atkinson, P. G., Gas Council Midlands Research Station Ext. Rept. No. ER105
42 Sayers, J. F., Tewari, G. P., Wilson, J. R. and Jessen, P. F., *Gas Council Res. Commun.*, GC171 (1970); *J. Inst. Gas Eng.*, **11** (1971), 322
43 Francis, W. E., *Gas Council Res. Commun.*, GC101 (1964); *J. Inst. Gas Eng.*, **4** (1964), 399

10

Aerodynamics

10.1 INTRODUCTION

In this chapter the basic principles of Chapter 2 are applied to some flow situations which are of common occurrence in combustion systems.

10.1.1 Entrainment by the use of injectors

In many burners the air–gas mixture is produced by causing a jet of one gas to share its momentum with the other, as in the Bunsen burner. The basic theory applicable is given in Section 2.2.5 (Example 3).

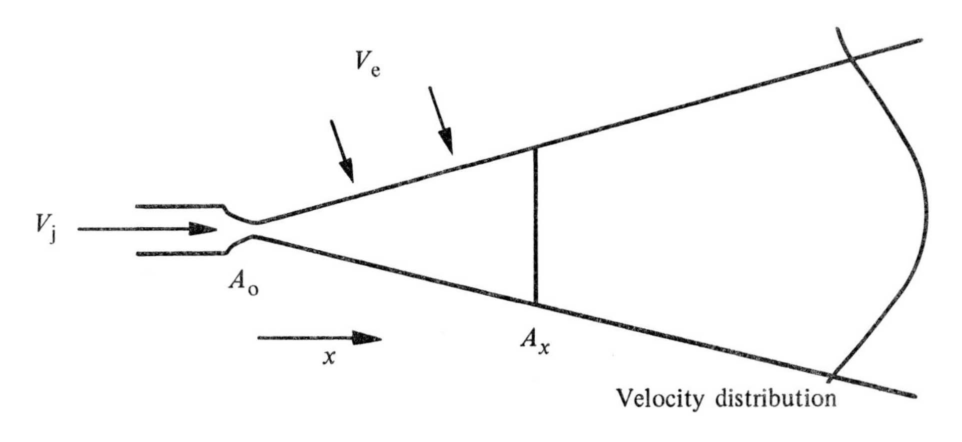

Figure 10:1 Entrainment by an unconfined jet

Consider the free jet shown in Figure 10:1. The entrained gas diffuses into the jet gas to form a mixture and, as x increases the volume V_e of entrained gas increases. Since the momentum passing through any cross-section remains constant it follows that the velocity decreases and the cross-sectional area A_x of the mixture increases.

Also the Bernoulli head decreases in the x-direction. In a burner a certain head is required in order to overcome the resistance in mixture tube and burner ports. Hence a 'throat' of appropriate area must be placed at a suitable position so as to limit the entrainment and produce a mixture having sufficient head to carry it through the burner ports.

The velocity distribution across the mixture stream is of the form shown in Figure 10:1, but in investigating the process theoretically it will be assumed to be constant over the cross-section.

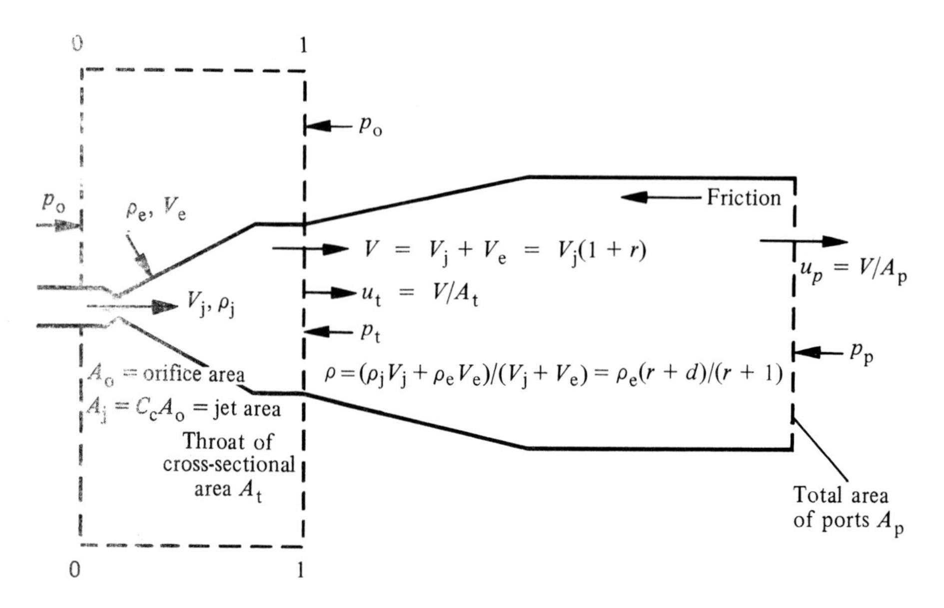

Figure 10:2 Entrainment by a confined jet: simple Bunsen-type burner

Figure 10:2 shows the form of a simple burner. A mass balance over the section 00–11 gives the total mass flow rate as: $\rho_j V_j + \rho_e V_e$ and hence the mixture density $\rho = (\rho_j V_j + \rho_e V_e)/(V_j + V_e)$. Writing $V_e = rV_j$ and $\rho_j = d\rho_e$ this becomes:

$$\rho = \rho_e (r + d)/(r + 1) \tag{10.1}$$

Next apply the momentum theorem over the region 00–11, neglecting any frictional force in the throat:

$$p_o A_t + (\rho_j V_j^2/A_j) = p_t A_t + (\rho V^2/A_t) \tag{10.2}$$

where A_j, the cross-sectional area of the gas jet is:

$$C_c \times \text{orifice area}$$

Bernoulli's equation may be applied to the flow between throat exit and burner-port exit. The discharge coefficient of the burner ports is denoted by C_{dp} (see Section 10.5) and there is a friction head loss in the mixture tube which for convenience is expressed as a fraction of the nominal kinetic head at the burner ports. Hence:

$$(p_t/\rho) + \tfrac{1}{2}u_t^2 = (p_p/\rho) + \tfrac{1}{2}(u_p^2/C_{dp}^2) + \tfrac{1}{2}C_F\,u_p^2$$

Multiplying by ρA_t and rearranging gives:

$$p_t\,A_t + \tfrac{1}{2}(\rho V^2/A_t) - p_p\,A_t - \tfrac{1}{2}\rho A_t\,u_p^2\,(C_{dp}^{-2} + C_F) = 0 \tag{10.3}$$

u_p is the nominal velocity in the burner port, i.e. V/A_p, and the term $\tfrac{1}{2}(u_p^2/C_{dp}^2)$ takes account of the kinetic head of the issuing gas and the loss of head in the burner ports (see Section 10.5). Rewriting 10.2:

$$p_t\,A_t + (\rho V^2/A_t) - p_o A_t - (\rho_j V_j^2/A_j) = 0 \tag{10.2a}$$

Subtracting to eliminate p_t:

$$(p_o - p_p)\,A_t - \tfrac{1}{2}\,(\rho V^2/A_t) + (\rho_j\,V_j^2/A_j) - \tfrac{1}{2}\rho A_t\,u_p^2\,(C_{dp}^{-2} + C_F) = 0$$

Substituting for ρ, V and u_p and multiplying through by $A_t/\rho_e V_j^2$ finally yields the convenient form:

$$[(p_o - p_p)\,A_t^2/\rho_e\,V_j^2] - \tfrac{1}{2}(r + 1)\,(r + d)\,[(C_{dp}^{-2} + C_F)\,(A_t/A_p)^2 + 1] + d\,(A_t/A_j) = 0 \tag{10.4}$$

in which the terms are dimensionless.

Equation 10.4 shows that if $p_o = p_p$ then r is independent of V_j, which means that there is an inherent self-proportioning action giving a constant air–gas ratio for all throughputs. This situation occurs whenever the entrained air and the flame are at atmospheric pressure. However, in practice r is not entirely independent of throughput since C_{dp} and C_F are both functions of the Reynolds number.

For the case where $p_o - p_p = 0$, equation 10.4 becomes:

$$(r + 1)\,(r + d) = 2d\,(A_t/A_j)/[1 + K\,(A_t/A_p)^2] \tag{10.5}$$

where $K = C_{dp}^{-2} + C_F$. From equation 10.5 the value of r resulting from given dimensions and given K can be found. It is clear that the maximum value of r will be obtained when the factor $B = (A_t/A_j)/[1 + K(A_t/A_p)^2]$ is a maximum, and the optimum size of throat is that which maximizes B.

Differentiating with respect to A_t gives:

$$dB/dA_t = \{A_j^{-1}\,[1 + K(A_t/A_p)^2] - 2KA_j^{-1}\,(A_t/A_p)^2\}/[1 + K(A_t/A_p)^2]^2 \tag{10.6}$$

It will be found that d^2B/dA_t^2 is negative so that B is a maximum when the quantity in curly brackets in equation 10.6 is set equal to zero, i.e. when:

$$1 - K(A_t/A_p)^2 = 0 \tag{10.7}$$

or

$$A_t/A_p = K^{-\frac{1}{2}} \tag{10.8}$$

Typically $C_{dp} \approx 0.65$ and C_F may be of the order of 2. If $C_F = 0$, $K^{-\frac{1}{2}} = C_{dp}$ and if $C_F \approx 2$, $K^{-\frac{1}{2}} = 0.45$, so the optimum throat area usually lies between 0.45 and 0.65 of the burner port area.

Substituting from equation 10.8 into 10.5 gives

$$(r + 1)(r + d) = A_t d/A_j \tag{10.9}$$

for a burner with optimally designed throat. For typical values of r and d it is easily shown that $(r + 1)(r + d) \approx (r + \sqrt{d})^2$ whence 10.9 can be rearranged as:

$$r \approx \sqrt{d}[\sqrt{(A_t/A_j)} - 1] \tag{10.10}$$

This is Prigg's formula which is frequently used in burner design.

If the optimal value of A_t/A_p is put into equation 10.3 it will be found that $p_t = p_p$, i.e. for the optimum design of throat the gauge pressure at throat exit is zero. It is also clear that the optimum design maximizes the Bernoulli head at the throat exit.

Combining equations 10.8 and 10.9 gives:

$$A_j/A_p = C_{dp}d/(r + 1)(r + d)\sqrt{(1 + C_F C_{dp}^2)} \tag{10.11}$$

10.1.2

In some entrainment systems, for example those used in tunnel burners, the entrained gas approaches the throat through an inlet pipe, in which case the momentum of the entrained gas entering the throat ought to be included in equation 10.2, and this might be expected to give an increased value of r. For example, in the arrangement

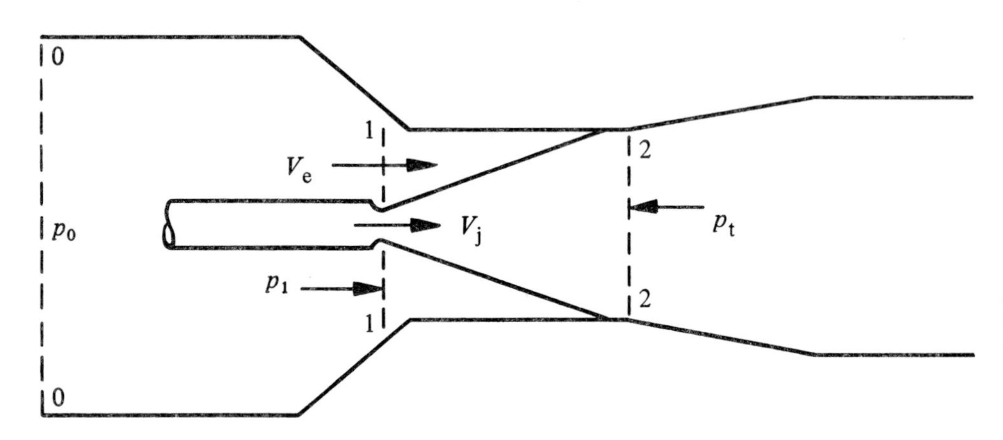

Figure 10:3 Entrainment by a confined jet: entrained air contributes axial momentum

shown in Figure 10:3 Bernoulli's equation ought to be applied between 0 and 1 to give p_1, and the momentum equation over 11–22 which would give an inlet momentum of $p_1 A_t + (\rho_e V_e^2/A_t) + (\rho_j V_j^2/A_j)$ which replaces the left side of equation

10.2. However, in such a case there will be an entrance head loss at 11 of approximately $V_e^2/2A_t^2$ together with an increased frictional loss in the throat, and these nullify the effect of the increased inlet momentum so that equation 10.2 still properly applies.

Reference [1] contains a fuller account of the topics dealt with in this section. Note that in this paper the author expresses friction losses as a function of the throat kinetic head.

10.1.3 Recirculation in furnaces and radiant tubes

Here the stream of combustion products leaving the burner is used to entrain gases which have already circulated round the furnace or tube. This has the effect of increasing heat transfer rates and producing a more even temperature. Figure 10:4 illustrates these applications.

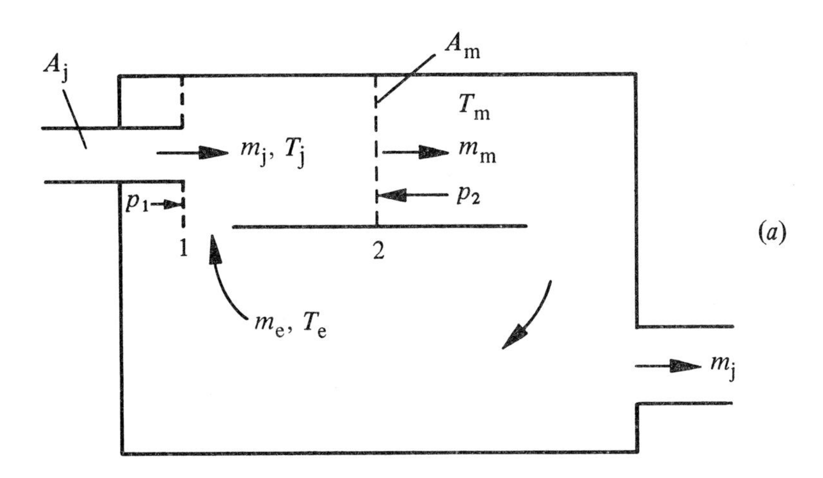

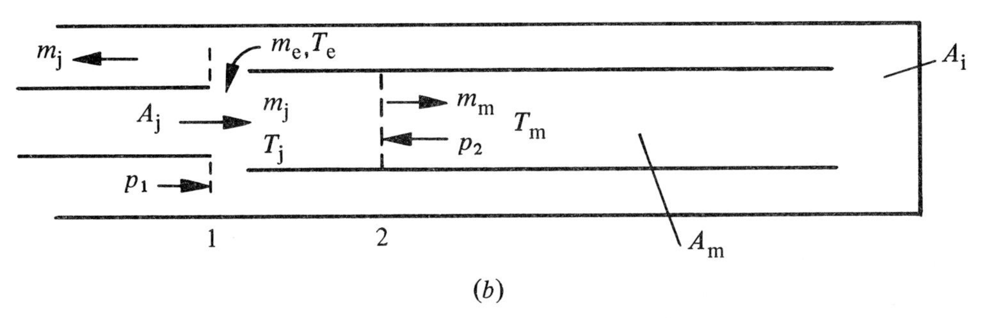

(*b*)

Figure 10:4 (*a*) Recirculation in a furnace due to entrainment by the jet of combustion gases

(*b*) Recirculation in a radiant tube due to entrainment by the jet of combustion gases

The relevant momentum equation is 2.29 which refers to Figure 2:16. Suffixes j and e refer to the burner products and the recirculated gases, respectively. The pressure difference $p_2 - p_1$ is then equal to all the pressure losses experienced by the gases in making their circuit. These consist principally of friction and bend losses, and those due to changes in cross-section, and all may conveniently be expressed in the form $\frac{1}{2}C_F u_m{}^2$. In addition to these losses there are pressure changes due to temperature changes and offtakes.

10.1.4 Pressure rise due to fall in gas temperatures

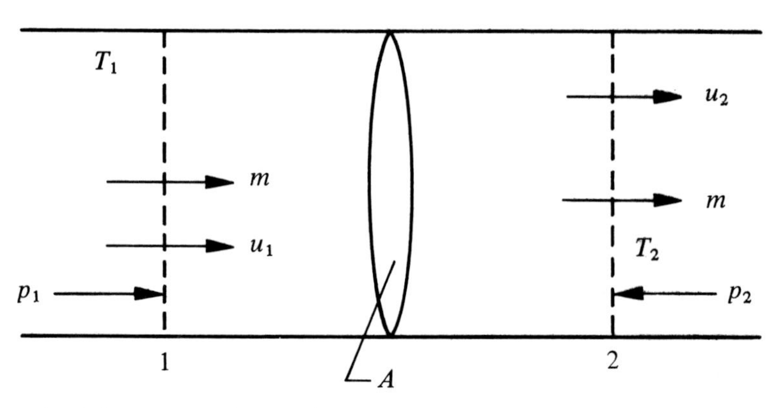

Figure 10:5 Momentum change due to temperature change in flowing gas

In Figure 10:5 the momentum theorem is applied to the region 1–2 where the mass flow rate is m and the absolute temperature changes from T_1 to T_2:

$$p_1 A + m u_1 = p_2 A + m u_2 = p_2 A + m u_1 (T_2/T_1)$$

whence:

$$p_1 - p_2 = m u_1 A^{-1} (T_2 T_1{}^{-1} - 1) \tag{10.12}$$

which shows that there is a pressure recovery in the direction of falling temperature.

10.1.5 Pressure rise due to an offtake

Figure 10:6 shows part of a flow system where a fraction ζ of the flow is taken out of the system. The momentum theorem applied to the section 1–2 gives:

$$p_1 A + m u = p_2 A + m(1 - \zeta) u (1 - \zeta)$$

or:

$$p_2 - p_1 = m u \zeta (2 - \zeta)/A \tag{10.13}$$

In the case where the offtake is the flue of a furnace or radiant tube in which recirculation is occurring it is clear that $m = m_m$, while $\zeta m = m_j$.

The above principles are applied to the case of recirculation in furnaces in reference [2] where it is deduced that the recirculation ratio $r_r = m_m/m_j$ is given by:

$$C_F = \left(\frac{2}{r_r{}^2}\right)\left(\frac{T_j}{T_m}\right)\left(\frac{A_m}{A_j}\right) - 2 + \beta\frac{2(r_r - 1)^2}{r_r{}^2}\left(\frac{T_e}{T_m}\right)\left(\frac{A_m}{A_e}\right) \tag{10.14}$$

where the meaning of the symbols is shown in Figure 10:4 (*a*), and where β takes the value 0 or 1 according as the recirculated gases enter the mixing section perpendicular or parallel to the burner jet respectively.

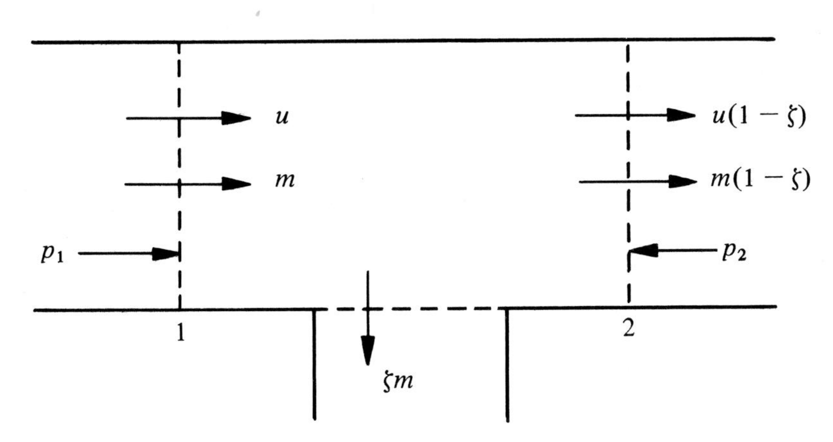

Figure 10:6 Momentum change due to change in direction of fluid flow at an offtake

Reference [3] applies these methods to the recirculation in a single-ended radiant tube, and arrives at equation 10.14 again, with $\beta = 0$. It is clear that for a given outer tube cross-section A_i there will be an optimum value for A_m to minimize the total value of C_F. By assuming reasonable values for friction and bend losses it is deduced that the optimum value of A_i/A_m is 1.73, which gives:

$$r_{r,\,max}{}^2 = 0.167\; T_j A_i / T_m\, A_j \qquad (10.15)$$

10.1.6 Bar burners

These are dealt with in Section 7.4 and in reference [4], which applies the methods of Sections 10.1.1, 2 and 5 to their design. It is clear that the pressure increase along the burner bar due to the progressive removal of the gas through the burner ports is equivalent, in effect, to a reduction in the loss coefficient C_F and equation 10.11 may be put in the form:

$$A_p/A_j = (r + 1)\,(r + d)\sqrt{[F(1 + C_F)]}/dC_{dp} \qquad (10.16)$$

where the factor F is less than unity, and takes into account the pressure recovery along the bar. The determination of F is given in reference [4] together with a graph of F as a function of burner geometry.

10.2 CHIMNEYS AND FLUES

The purpose of a chimney is to provide the necessary pressure gradient to drive combustion products through the furnace, and to disperse these products into the atmosphere. Usually the draught is produced by buoyancy, i.e. natural draught, but may be augmented by a fan.

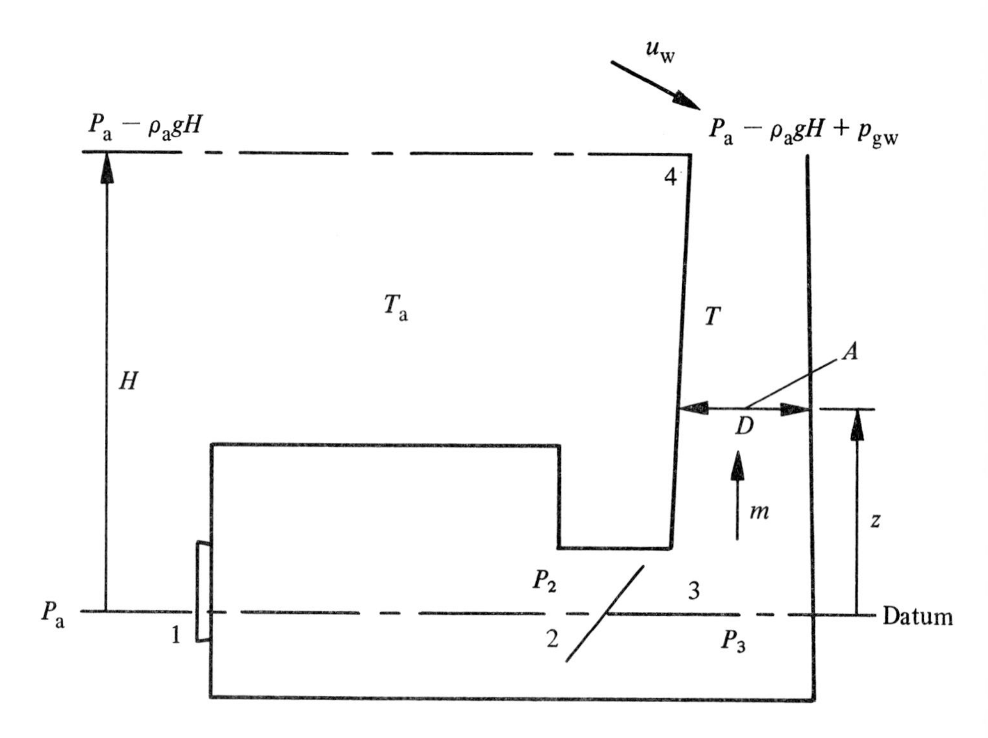

Figure 10:7 Pressures in a furnace chimney

Figure 10:7 shows a chimney producing natural draught. p_{gw} is a wind pressure caused by any component of wind velocity parallel to the chimney axis. Pressure differences in the chimney are small compared with the average pressure, so the effect of pressure change on density may be ignored and only the effect of temperature T need be considered. Bernoulli's equation 2.15 may be applied to the flow in the chimney, and by using equation 2.42 to account for friction in the chimney, and writing the mass flow rate up the chimney as $m = \rho u A$ equation 2.15 becomes:

$$(m/A)\mathrm{d}u + \mathrm{d}P + g\rho \mathrm{d}z + (4fmu/2DA)\mathrm{d}z = 0 \tag{10.17}$$

Integrating this between the base (4) and top (3) of the chimney gives:

$$m\int_0^H A^{-1}\,\mathrm{d}u + (P_4 - P_3) + g\int_0^H \rho \mathrm{d}z + 2fm\int_0^H (u/DA)\mathrm{d}z = 0 \tag{10.18}$$

If D, A and T are known as functions of z then ρ and u may be expressed in terms of z since $u = m/\rho A$ and $\rho = \rho_a T_a/T$. The integrals in equation 10.18 may then be evaluated.

If, for example, T, A and D are constant (insulated cylindrical chimney) then ρ

and u are constant, so equation 10.18 becomes:

$$P_4 - P_3 + g\rho H + (2fmHu/DA) = 0$$

or
$$P_4 - P_3 + (g\rho_a T_a H/T) + (2fHm^2T/DA^2T_a\rho_a) = 0 \tag{10.19}$$

By forming the derivative dm/dT from equation 10.19 and setting this equal to zero it is easily shown that m attains a maximum value at a certain value of T, for a given value of $(P_4 - P_3)$. Also, since $P_4 = P_a - \rho_a gH + p_{gw}$ (see Figure 10:7), equation 10.19 becomes:

$$P_a - P_3 = \rho_a gH(1 - T_a T^{-1}) - (2fHm^2T/DA^2T_a\rho_a) - p_{gw} \tag{10.20}$$

This is the draught available to overcome resistances in the furnace and to provide the combustion products with the necessary kinetic energy, and it is readily shown that this is maximized at a certain value of T. Differentiating equation 10.20 with respect to T:

$$\frac{d}{dT}(P_a - P_3) = \frac{\rho_a gH\,T_a}{T^2} - \frac{4fHm^2}{2DA^2\,T_a\rho_a} = 0$$

for maximum $P_a - P_3$, since clearly $d^2(P_a - P_3)/dT^2$ is negative. Hence for maximum draught:

$$T = \frac{\rho_a\,T_a\,A}{m}\left(\frac{Dg}{2f}\right)^{\frac{1}{2}} \tag{10.21}$$

10.3 TURBO-PUMPS

These are machines in which energy is transferred to a continuously flowing fluid in a duct by the dynamic action of moving blades. Such machines are often called compressors, pumps, boosters, or fans according to the magnitude of the pressure increase generated.

The path of the fluid through the machine may be mainly radial (centrifugal pump) or axial (axial-flow or propeller-type pumps). The former will usually produce a greater pressure lift than the latter type.

The essential mechanism by which these machines operate is that the torque supplied to the blades by the driving motor increases the angular momentum of the fluid passing over the blades according to equation 2.22.

Figure 10:8 shows the basic arrangement of the two types of machine. In the centrifugal machine of Figure 10:8 (*a*) there may be inlet guide vanes to induce prerotation of the gas. The purpose of the outlet diffuser guide vanes is to convert most of the kinetic head $u^2/2g$ into pressure head by gradually reducing the velocity. In some simpler types of machine these vanes may be absent.

In the axial machine of Figure 10:8 (*b*) the stator vanes serve to guide the fluid onto the rotor blades without eddy loss, and to act as diffusers at outlet. It is easy in this machine to have several compressor stages in series.

The familiar axial-flow fan used for ventilation systems, where only a very small pressure lift is required, is essentially a single-stage axial compressor, but usually the

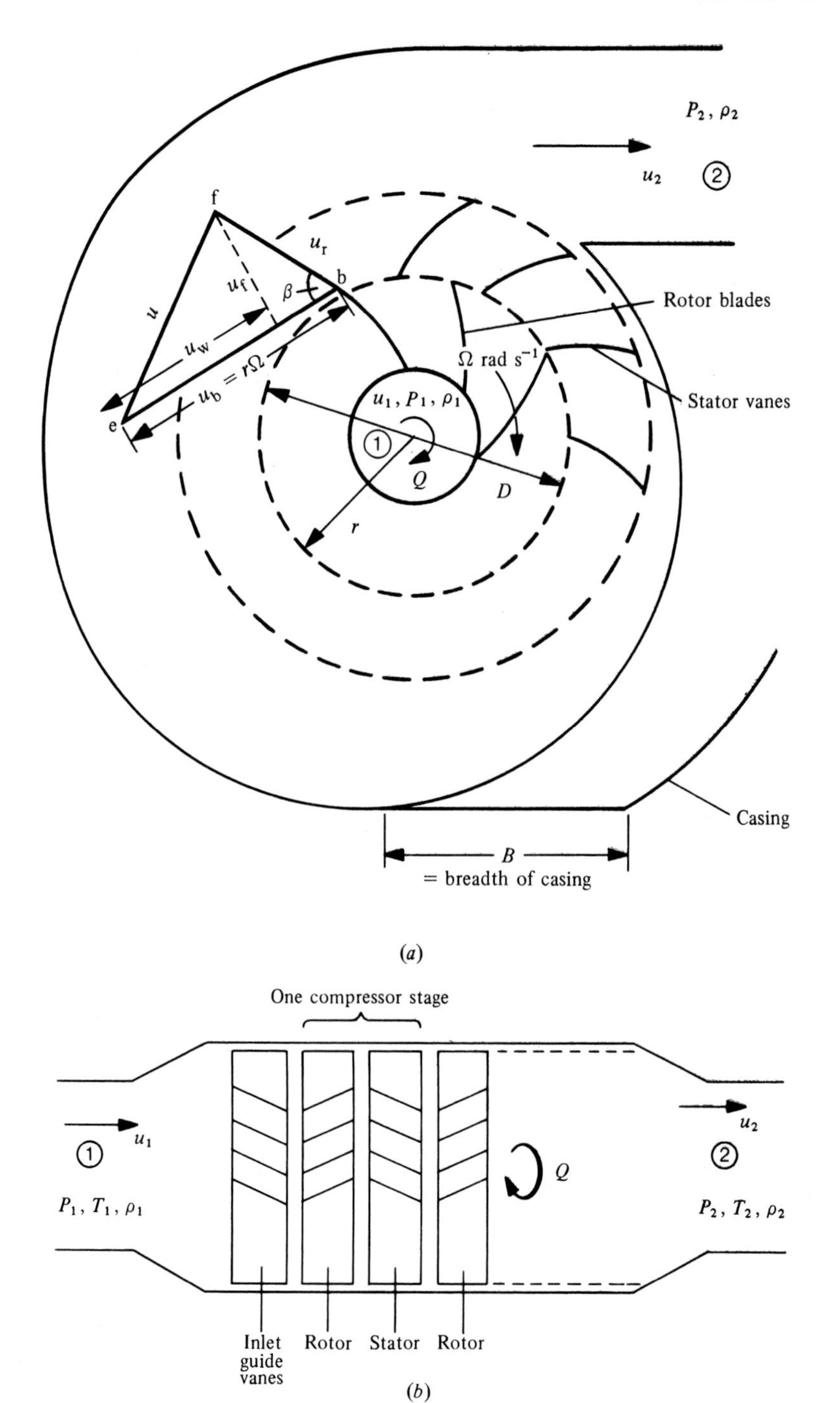

Figure 10:8 (*a*) Radial-flow centrifugal fan
(*b*) Axial-flow fan

number of rotor blades is smaller and often stator vanes are not used.

The change in fluid momentum is found from vector velocity triangles at the inlet and outlet of the rotor blades. In Figure 10:8 (*a*) the outlet triangle is shown, and the lettering used facilitates its construction. Thus, the lines $\overleftarrow{be}$, $\overleftarrow{fb}$, $\overleftarrow{fe}$ represent the velocities, *b*lade relative to *e*arth, *f*luid relative to *b*lade, and *f*luid relative to *e*arth, respectively. The radial component of u is u_f, the velocity of flow, and clearly the volume flow rate of fluid $= V = u_f \pi DB$. Hence for a given blade angle β and rotor dimensions, and V, the triangle ebf can be drawn. If the inlet fluid is assumed to have no whirl, i.e. to enter radially, then the change in moment of momentum per unit weight of fluid is $u_w r/g$ and this equals the shaft torque Q per unit weight of fluid. Then the work done per unit weight of fluid is:

$$w = Q\Omega = u_w \Omega\, r/g = u_w u_b/g = u_b\,(u_b - u_f \cot\beta)/g \qquad (10.22)$$

The energy equation 2.18 applied between inlet (1) and outlet (2) gives, for adiabatic conditions (i.e. no heat losses from machine):

$$w = (I_{s2} - I_{s1})/g \qquad (10.23)$$

Where I_s is the specific stagnation enthalpy $c_p T + \frac{1}{2}u^2$ (see equation 2.31). Equation 10.23 is true for any compressor, but for a machine producing a small pressure rise so that density changes can be ignored Bernoulli's equation 2.15 can be applied between inlet and outlet. In this case $\int F ds$ is the useful mechanical work done on unit weight of fluid, which is $w - w_L$ where w_L is the work lost per unit weight in friction in the casing. Hence

$$w - w_L = E_{M2} - E_{M1} = H \qquad (10.24)$$

where $E_{M2} = (P_2/\omega) + (u_2{}^2/2g)$ and similarly for E_{M1}.

Since the purpose of a pump is to increase the pressure in the fluid it is useful to obtain an expression for the pressure rise from equations 10.22 and 10.23 or 10.24, but this can only be done if the losses in the casing are known. These are due mainly to friction losses in the blades, eddy losses at blade tips, and losses due to secondary circulation of the fluid in the spaces between the moving blades. These can only be found by experiment, but before dealing with actual performance in the next section it is instructive to derive something about the theoretical performance. From equations 10.22 and 10.24:

$$H = u_b\, g^{-1}\,(u_b - u_f \cot\beta) - w_L$$

or

$$gH = \pi^2 D^2 N^2 - (\pi DNV/\pi DB)\cot\beta - w_L g \qquad (10.25)$$

where N is the frequency of rotation of the machine in revolutions per second.

Clearly, if $\beta < 90°$ H decreases with increasing V and for $\beta > 90°$ H increases with flow, in the ideal case where w_L is zero. For stable operation the head/flow characteristic should have a negative slope so that backward-sloping blades are usually used, although for mechanical reasons radial blades may be used in high-speed machines. Furthermore the velocity triangle shows that u is small when $\beta < 90°$, so that there is a smaller energy loss in the diffuser vanes for backward-sloping rotor blades.

Also, by dividing through equation 10.25 by D^2N^2 and by taking B as being proportional to D, as it would be for geometrically similar machines, it is clear that gH/N^2D^2 is a function of V/ND^3.

Finally, equation 10.23 shows that a useful dimensionless form of the performance of the machine would be $(I_{s2} - I_{s1})/I_{s1}$ and if it is assumed that the inlet and outlet kinetic heads are small, this becomes $(I_{s2}/I_{s1}) - 1 = (T_2/T_1) - 1$, since $I_s = c_pT$.

In the ideal case where the flow in the machine is isentropic, i.e. reversible adiabatic, and if the gas laws are obeyed by the fluid, then $P_2v_2^{\gamma} = P_1v_1^{\gamma}$, and:

$$(I_{s2}/I_{s1}) - 1 = (P_2/P_1)^{(\gamma-1)/\gamma} - 1 \qquad (10.26)$$

10.3.1 Efficiency: performance curves for compressors

The efficiency of a machine may be defined in many ways. For low-pressure-rise machines a suitable measure is:

$$\eta = H/w = 1 - w_L w^{-1} \quad \text{(from equation 10.24)} \qquad (10.27)$$

For compressors the isentropic efficiency is used. This is the ratio:

$$\eta = [(P_2/P_1)^{(\gamma-1)/\gamma} - 1]/[(P_2/P_1)^{(n-1)/n} - 1] \qquad (10.27a)$$

where for the actual compressor $P_1v_1^n = P_2v_2^n$. This efficiency is the ratio of work ideally required for isentropic compression from P_1 to P_2 to actual work required.

Dimensionless groups

For low-pressure-rise pumps the main dependent variables are the net energy transfer gH, the power supplied W, and the efficiency η, and these will be a function of the independent variables V, N, D, ρ, μ and shape factors which will be expressed as length ratios in the machine.

Similarly for compressors, the dependent variables are ΔI_s or P_2/P_1, W and η and the independent variables are m, γ, N, D, μ_1, ρ_1, a_1 where the viscosity, density and sound velocity are taken at inlet conditions.

These can be arranged to give the following convenient forms.

For incompressible flow:

$$gH/(ND)^2,\ \eta,\ W/\rho N^3D^5 \text{ are functions of } (V/ND^3,\ \rho ND^2/\mu,\ \text{shape}) \qquad (10.28)$$

For compressible flow:

$$\Delta I_s/(ND)^2,\ \eta,\ W/\rho_1N^3D^5 \text{ are functions of } (m_1/\rho_1ND^3,\ \rho_1ND^2/\mu_1,\ ND/a_1,\ \gamma) \qquad (10.28a)$$

or, alternatively, if the gas laws hold:

$$P_2/P_1,\ \eta,\ \Delta T/T_1 \text{ are functions of } (m\sqrt{(RT_1)}/D^2P_1,\ ND/\sqrt{(RT_1)},\ \gamma,\ \rho_1ND^2/\mu_1) \qquad (10.28b)$$

In these expressions the group $\rho ND^2/\mu$ is the Reynolds number and is not usually very important since the conditions in the machine are usually very turbulent. The group ND/a_1 or $ND/\sqrt{(RT_1)}$ is the blade Mach number.

The forms of the performance curves are shown in Figure 10:9. In 10:9 (*b*) the vertical portions of the characteristics indicate that the flow is choked, i.e. that sonic velocity occurs somewhere in the machine.

Specific speed

From Figure 10:9 (*a*) it will be seen that any point on the graph of H/N^2D^2 versus V/ND^3 corresponds to a particular efficiency for a given type of machine. Thus it is possible to write $H/N^2D^2 = f_1(\eta)$ and $V/ND^3 = f_2(\eta)$. Then, eliminating D between these gives a parameter which is independent of machine size and is therefore a measure of efficiency for a given type of machine:

$$V^{\frac{1}{2}}N/H^{\frac{3}{4}} = f(\eta) = N_s$$

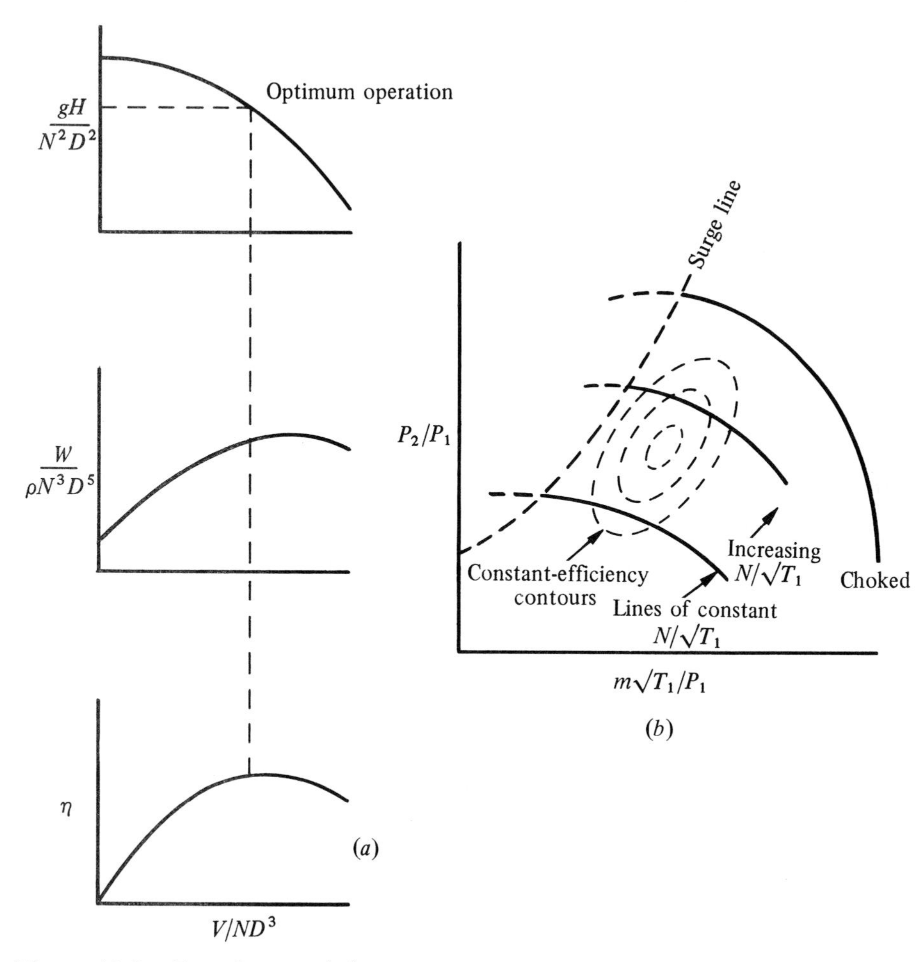

Figure 10:9 Fan characteristics

which is inappropriately called the specific speed. Then for any type of centrifugal machine there is a range of values of N_s corresponding to its best efficiency range. Since N_s is specified by the duty required from the machine this enables the best type of machine to be chosen.

Fan scaling laws
The following relationships enable one to calculate changes in performance resulting from changes in size and operating conditions, for fans of the same geometric shape at the same efficiency point.

$$V_2/V_1 = N_2/N_1; \qquad H_2/H_1 = (N_2/N_1)^2; \qquad W_2/W_1 = (N_2/N_1)^3$$

for constant D and ρ.

$$V_2/V_1 = (D_2/D_1)^3; \qquad H_2/H_1 = (D_2/D_1)^2; \qquad W_2/W_1 = (D_2/D_1)^5$$

for constant N and ρ.

W and H vary directly as ρ for constant V and N.

A pump or compressor should not be used in the region, shown dotted in Figure 10:9 (*b*), to the left of the surge point since the flow may be unstable there. This phenomenon may be simply explained by reference to Figure 10:10 showing a pump

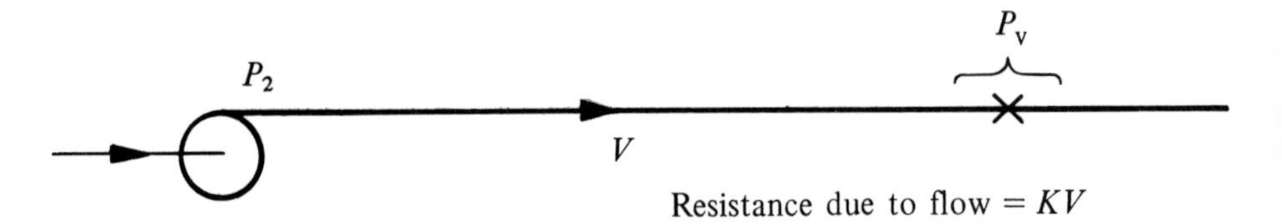

Figure 10:10 Pump feeding into pipeline

and pipeline. The equation of motion for the fluid in the pipe is approximately:

$$P_2 - P_v - KV = \rho(\mathrm{d}V/\mathrm{d}t)$$

where P_v is the pressure drop across the valve. In the surge-point region P_2 is given approximately by:

$$P_2 = P_{\text{no flow}} + bV$$

where b is the slope of the P–V curve in the surge-point region. These combine to give:

$$(\mathrm{d}V/\mathrm{d}t) + (K - b)V/\rho = (P_{\text{no flow}} - P_v)/\rho \qquad (10.29)$$

Since the solution of this includes a term $V = \mathrm{e}^{-(K-b)t}$; this will represent an unstable system if $(K - b)$ is negative, but in practice compressible-flow instability occurs more readily than this would indicate.

Instability may also occur between two compressors used in parallel if slight variations in the sharing of the total flow can cause one of the machines to operate in its surge region.

Another phenomenon in a compressor which has a similar effect to surge is that of blade stall where separation of the boundary layer occurs in a blade, causing a maldistribution of the fluid between the blades, and this moves round the rotor system causing mechanical vibration.

For a given machine of constant size, the performance is often plotted as H against V or P_2/P_1 against m for different values of N. By plotting the head-loss flow characteristic for the rest of the flow system on the pump characteristic diagram the size and speed of machine can be chosen so that it will normally operate at maximum efficiency without going into the surge region.

10.4 AERODYNAMICS WITHIN A FURNACE

The aim of the furnace designer is to ensure that the required flow of combustion gases through the furnace can be maintained and that an even temperature distribution and good heat transfer to the furnace charge will occur.

It is not possible to apply the basic principles of fluids very easily to this problem since the streamline pattern is not known, and the designer often relies on empirical data obtained from measurements on actual furnaces or models.

10.4.1 Total flow through the system

The basic equation involved in determining the total flow is that the pressure lost in friction and eddies in the system is equal to the pressure provided by the chimney, i.e. $P_2 - P_a$ in Figure 10:7.

The head losses may be calculated directly, using the methods of Section 2.4 for those parts of the system where the flow is through regular cross-sections and where a reasonably accurate average velocity in the cross-section is known, as for example in the case of the flow of combustion gases through the fire tubes of a boiler.

10.4.2 Flow pattern within the furnace

The complex flow pattern in a furnace is brought about by the jet action of streams of combustion products, by buoyancy forces due to the temperature gradients inevitably associated with heat transfer, and by the pressure gradient resulting from the chimney pull, together with the geometry of the furnace and its contents.

The application of the principles of fluid mechanics in a simple form can predict some general features of the flow, but for detailed knowledge studies on actual furnaces or models are necessary.

The main effect of the jet of combustion gases is to entrain gas from the main body of the furnace chamber and thus to set up a circulation in the furnace as mentioned in Section 10.1.3 and Figure 10:4 (*a*). The circulation still occurs if the dividing wall between the jet and the body of the furnace is absent, and this has the desirable effect of increasing the rate of convective heat transfer to the surfaces in the furnace and of producing a more even temperature. However, the jet can also produce isolated or stagnant eddies at sharp corners in a furnace, thus reducing the useful circulation.

The recirculation of combustion products to the flame has the effect of increasing the flame length in the case of diffusion flames, since it is clear that the combustion air will be diluted and hence diffusion rates reduced. Furthermore, if the jet of gas expands to meet the surface of the combustion chamber, a smaller area will be offered over which diffusion can occur. For these reasons an excess of air will be needed.

The effect of buoyancy is to produce circulation in the vertical plane which is the basis of natural convection. The flow of gas over a surface which is receiving heat from the gas should ideally be downward, and it is desirable to have the surfaces which are receiving heat well above the floor of the furnace to avoid the formation of a stagnant layer of cooler gases at the bottom of the furnace.

It is also an aid to even temperature distribution to arrange for the hot gas to originate at the top of the furnace chamber since gas flowing downwards over cooler surfaces will tend to distribute itself evenly between parallel channels.

10.4.3 Aerodynamics of jets

A turbulent free jet entrains fluid equivalent to the nozzle fluid mass flow for every three nozzle diameters distance along the axis:

$$\rho_e V_e / \rho_j V_j = 0.32 \, (\rho_e / \rho_j)^{\frac{1}{2}} \, (x / D_j) - 1 \quad (10.30)$$

where D_j is the nozzle diameter (Figure 10:1).

Turbulent mixing depends on velocity gradients so that when the jet is surrounded by a coaxial stream of another gas the rate of mixing is a minimum when the two streams have the same velocity. Hence rapid mixing is promoted when the jet gas is given a high swirl velocity.

If the jet is confined, and there is an ample supply of the secondary gas, the entrainment continues until the jet expands to reach the duct walls, as dealt with in Section 10.1.1 and Figure 10:2. If the supply of secondary gas is restricted there will be a recirculation eddy of the jet fluid back towards the nozzle. Interference between adjacent jets can have the same effect of causing recirculation in the inner flames.

When a bluff body is placed in the path of a jet of gas a boundary-layer gradient is set up at the surface of the bluff body which promotes stability by providing a region where the flame speed can match the flow velocity. Also there is a positive (adverse) pressure gradient downstream of the bluff body where the jet re-expands, and this promotes recirculation in the form of a vortex in the wake of the bluff body. This further assists stability by supplying heat back to the base of the flame.

Swirl

A high tangential velocity of the jet leaving the nozzle (strong swirl) increases the rate of mixing with secondary fluid, and also produces a radial pressure gradient (Section 2.2.2). This causes an inwards recirculation of the gas back to the base of the jet axis, thus achieving a similar result to the bluff body.

References [5] and [6] give a more detailed account of the topics of Section 10.4.

10.5 FLOW THROUGH CONSTRICTIONS

Changes in cross-section occur in most flow systems. These give rise to pressure changes and energy losses which are frequently the major losses in the system.

Common examples are the orifice plate used for flow measurement (Section 8.5.6) and the changes in cross-section occurring in a furnace at flue entrances and where the load occupies part of the furnace space. Figure 10:11 shows a number of configurations, of which (*a*) and (*b*) are flow-measuring devices. In each case section 1 is the minimum cross-section of the fluid stream, called the vena contracta, which occurs at or just after the minimum duct cross-section, or throat. The ratio A_1/A_t is called the coefficient of contraction, C_c, and its value may range from 1 in (*a*) to about 0.62 for (*b*).

If the density does not change significantly over the distance 0–1–2 the continuity equation 2.12 is:

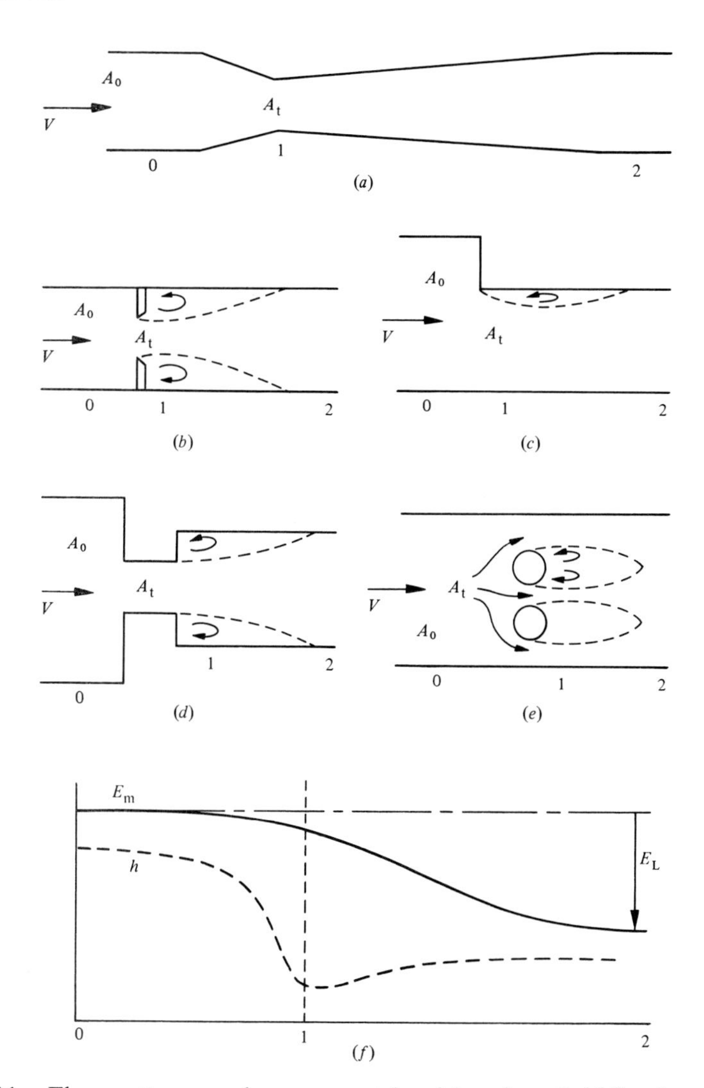

Figure 10:11 Flow patterns and consequent head loss in a fluid flowing through a change in cross-sectional area

$$V = u_0 A_0 = u_1 A_1 = u_1 C_c A_t = u_2 A_2 \tag{10.31}$$

and if there is no degradation of energy to heat due to friction and eddies over 0–1, Bernoulli equation 2.16 is:

$$E_m = (u_0^2/2g) + h_0 = (u_1^2/2g) + h_1 \tag{10.32}$$

where $h = (P/\omega) + z$ is the pressure head relative to a horizontal datum. Note that from equation 2.8, $h_0 - h_1$ is what would be recorded by a manometer connected between 0 and 1.

Equations 10.31 and 10.32 applied to the section 0–1 yield:

$$h_0 - h_1 = \frac{V^2}{2g\, C_c^2\, A_t^2}\left[1 - \left(\frac{C_c\, A_t}{A_0}\right)^2\right] \tag{10.33}$$

In practice there will be a small loss of head between 0 and 1 which means that $h_0 - h_1$ will be larger than indicated by equation 10.33. This is taken care of by replacing C_c^2 in the denominator by C_d^2, where C_d is the coefficient of discharge which has to be determined experimentally. The equation is then in the form applied to flow-measuring instruments.

If the enlargement after the throat is in the form of a gradual taper as in Figure 10:11 (*a*), boundary-layer separation is avoided and almost all of the pressure head is recovered so that h_2 is only a little less than h_0. But in all the other cases shown considerable dissipation of mechanical energy occurs by eddy formation. The maximum head loss which can occur is $u_1^2/2g$ for the case where $A_2 \gg A_1$ (i.e. the sudden enlargement discussed in Section 2.4.3). For the cases shown in Figure 10:11 (*b*)–(*e*) the loss would be of the order of 0.4 to 0.7 $u_1^2/2g$. The head loss, E_L, is shown in Figure 10:11 (*f*).

10.6 FURTHER READING

Dixon, S. L., *Fluid Mechanics: Thermodynamics of Turbomachinery*, Oxford, Pergamon Press, 1966

Ferguson, T. B., *The Centrifugal Compressor Stage*, London, Butterworths, 1963

REFERENCES

1 Francis, W. E., *Gas Council Res. Commun.* GC101; *J. Inst. Gas Eng.*, **4** (1964), 399
2 Francis, W. E., *Gas Council Res. Commun.* GC34; *Trans. Inst. Gas Eng.*, **106** (1956), 483
3 Bridgens, J. L. and Francis, W. E., 'The development of a single-ended gas-fired radiant tube', 1st Symposium on Industrial Uses of Town Gas, 26 June 1963, Institute of Fuel and Institution of Gas Engineers
4 Goodwin, C. J., Hoggarth, M. L. and Reay, D., *Gas Council Res. Commun.* GC165 (1969); *J. Inst. Gas Eng.*, **10** (1970), 324
5 Thring, M. W., *The Science of Flames and Furnaces*, 2nd edition, London, Chapman & Hall, 1962
6 Beér, J. M. and Chigier, N. A., *Combustion Aerodynamics*, London, Applied Science Publishers, 1972

11

Methods of heat recovery

11.1 INTRODUCTION

This chapter will discuss the methods of heat recovery available to and practised by the furnace-using and heat-process industries with which an industrial gas engineer must be familiar.

The subject is really one aspect of the much wider study of furnace design and heat utilization. In broad terms, the plant for any heat process should be made to high thermal standards and equipped with the maximum amount of instrumentation that can be economically justified.

A list of items which must be considered in any thermal project design is given below. It is not in order of priority and it will be seen that in many cases the items overlap one another.

1 Method of process heating—batch or continuous
2 Arrangement and distribution of the load
3 Method of charging and handling material
4 Mode of heat transfer—radiative and/or convective and/or conductive
5 Length of operating period
6 Methods of heat recovery
7 Type of burners and control systems and fuel used
8 Velocity of furnace gases and their flow pattern
9 Process plant atmosphere
10 Method of construction and materials employed
11 Capital costs

These items will be discussed more fully in Chapter 13. They are listed here in order to emphasize the interdependence of the present topic and the rest of the list. The engineer must also have an understanding of the heat transfer factors involved (Chapter 3) and the necessary sections of Chapter 4 where the subject of heat balances is discussed.

Heat balances can be constructed to include the whole process or they can be produced in stages or sections [1]. Their basis can be purely thermal or can include costing which will greatly enhance the value to the designer, operating staff and management. The results can be presented in graphical form by means of the Sankey diagram. This consists of a two-dimensional map of the parts of the system on which is superimposed a stream representing the flow of energy. The width of the stream at any given point is proportional to the number of units of energy flowing past that point (Figure 11:1).

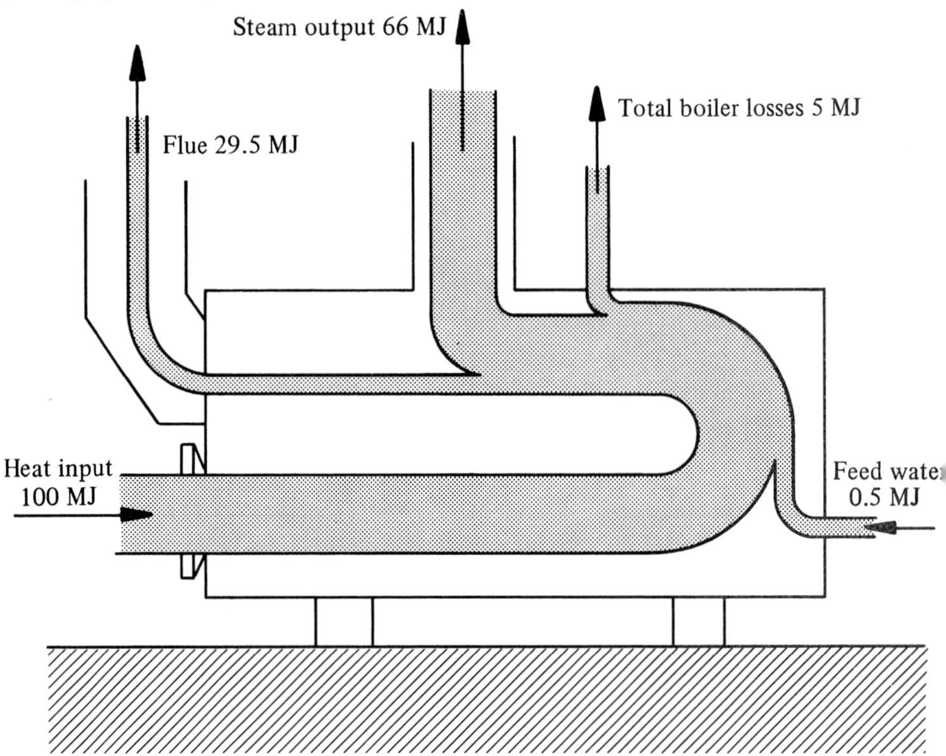

Figure 11:1 Simple diagrammatic heat flow or Sankey diagram

Thermal efficiency is another item which must be discussed before starting to study heat-recovery methods. If 'thermal efficiency' is defined as the ratio of useful energy obtained from a unit of fuel to the energy initially contained in that unit, it does not mean very much when applied to an industrial process. Here, the object is to produce a given item at minimum cost and the use of equipment and procedures designed to achieve maximum thermal efficiency may not result in minimum cost. The engineer is always faced with a compromise situation between technical excellence and economics.

A more useful concept to consider for the determination of equipment and procedures likely to give the lowest production cost is that of 'heating cost'. The 'heating cost' includes the cost of fuel, the cost of furnace maintenance and depreciation and the cost of output spoiled by improper heating. This illustrates why an extremely well insulated furnace fitted with a recuperator and any of the other heat-recovery devices available might give the highest possible thermal efficiency, but if the fuel savings attributable to each heat-saving device do not economically justify the annual cost of owning that device, then a design giving a lower thermal efficiency may have to be selected. Another factor which must be considered is the dependence of the total operation on any particular piece of optional equipment. This can be illustrated with the rejection by a designer of an economically attractive heat-saving device on the grounds that a plant failure would result in a furnace shutdown which could negate the annual return of the device in a few hours downtime. In other words, care must be taken to ensure that increased utilization of waste heat sources does not interfere with the prime function of the particular process, i.e. the production. However, the present energy situation with the emphasis on saving must be considered with the overall economics of production.

11.2 FORMS OF HEAT WASTE

A unit of fuel burned in an industrial furnace is distributed as follows:

1 A part which increases the temperature and/or changes the state of the material being processed
2 A part which may not undergo combustion
3 A part which maintains the furnace interior at the same temperature as the material being processed
4 A part which leaves the furnace as high-temperature combustion products
5 Radiation and other losses, such as heat in bogeys or solid waste

The designer's job is to maximize the first of these items and keep the losses to a minimum. Items (2), (3) and (5) can be minimized by careful design. However, once the operating temperature has been selected, this fixes the quantity of heat energy which leaves the furnace in the combustion products (4). There is no way to 'design out' or minimize the energy contained in these hot gases. The only way to prevent the total loss of this energy is to recover it for some useful purpose.

11.3 RECOVERY OF HEAT WASTE

The energy leaving the furnace in combustion products, or some part of it, can be utilized in one of two main ways:

1 It can be returned to the furnace
2 It can be used for some function independent of the furnace

11.3.1 Energy returned to the furnace

Energy can be returned to the furnace in a number of ways, the most common are to preheat the combustion air and/or preheat the stock.

The value of air preheating is shown in Figure 11:2.

In addition to increasing the thermal efficiency of the process, air preheating raises the flame temperature. For natural gas 600 °C air preheat will increase the flame temperature by about 220 °C to 2170 °C. The higher temperature may be advantageous or essential to the process.

The theoretical volume of waste gases produced by burning 1 m^3 of dry methane is 10.52 m^3 of which 2.0 m^3 is water vapour. Taking the mean heat capacity per unit volume of the combustion products as 1.407 kJ m^{-3} °C^{-1} their total heat capacity becomes $10.52 \times 1.407 = 14.80$ kJ °C^{-1} per cubic metre of gas burnt (0.221 Btu °F^{-1} per cubic foot of gas burnt).

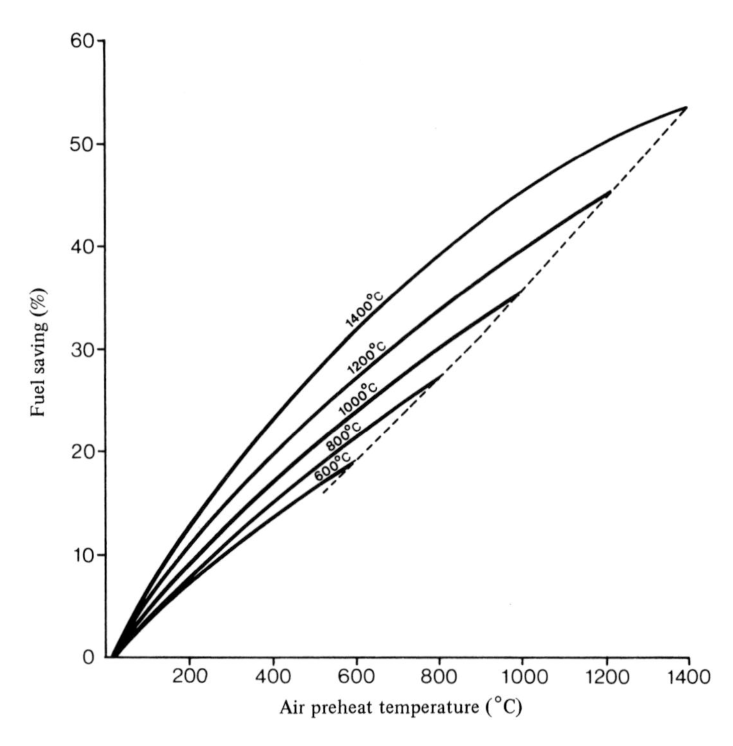

Figure 11:2 Variation of fuel saving with air preheat temperature for a number of furnace outlet temperatures

The amount of air required for combustion of 1 m^3 of methane is 9.52 m^3. Taking the mean heat capacity per unit volume of air as 1.227 kJ m^{-3} °C^{-1} its thermal capacity is $9.52 \times 1.227 = 11.68$ kJ °C^{-1} per cubic metre of gas burnt (0.174 Btu °F^{-1} per cubic foot of gas burnt).

If the theoretical amount of air is used for combustion, the maximum heat recovery from the combustion products is $11.68/14.8 \approx 80$ per cent.

In practice some excess air is used to ensure complete combustion, which increases the amount of heat theoretically recoverable. However, it is important not to reduce the temperature of the combustion products below their dew point. This eliminates the latent heat of water from the heat available approximating to 10 per cent of the total heat of combustion of the methane. Therefore the maximum heat recoverable from the products remains around 80 per cent.

11.3.2 Independent energy recovery

Energy recovery independent from the furnace may be carried out by steam generation and/or hot-water heating by using waste-heat boilers. Other uses for waste heat include low-temperature heating applications, for example, in drying plant and for space heating. In cases where very large amounts of hot gases are available, the possibility of power generation can be investigated, but the substantial capital and operating costs of a power-generating system—and, in Britain, the complicated tariffs operated by the Central Electricity Generating Board for such installations—preclude their feasibility in many cases (see Chapter 14).

The heat-recovery equipment used to carry out the abovementioned functions generally falls into one of the following classifications:

1 Regenerators
2 Recuperators
3 Waste-heat boilers or water heaters
4 Preheating and cooling stock
5 Self-recuperative burners

Regenerators [1, 2, 3]

Regenerative furnaces have a limited application and regenerator design is complicated by the many variables involved.

A regenerator consists of an arrangement of passages in a solid, the refractory matrix, through which hot waste gases and then cold combustion air are passed alternately. The heat given up to the refractory from the hot gases is stored and subsequently passes to the combustion air which is thus preheated before entering the furnace.

From thermal considerations, the governing equation at the gas–refractory interface is:

$$h\,A_r\,(t_g - t_r) = \lambda\,(\delta t_r/\delta x_r)_s$$

heat transfer from gas to surface = heat conducted into refractory

h = heat-transfer coefficient
A_r = the area of refractory of temperature t_r swept by gases of temperature t_g
λ = thermal conductivity of the refractory
$(\delta t_r/\delta x_r)_s$ = temperature gradient in the refractory at the heat-transfer surface

The problem of design including the choice of materials is complex and is very much influenced by the reversal interval and conditions of use. The thermal conductivity and the thermal capacity per unit volume $c\rho$ (c is the specific heat capacity, ρ is the density) of the refractory should be high. For high efficiency of heat transfer h and A_r and $t_g - t_r$ should be as high as possible. This means that the mass flow of gases through the ducts in the refractory (chequer brickwork) should be as high as possible consistent with a satisfactory pressure drop, because h is a direct function of the gas velocity. The thinner the refractory walls for a given mass of matrix, the greater the heat-transfer area. The temperature difference between the gas and the matrix for a given h and A_r will be increased as λ, c and the mass of the brickwork are increased.

As indicated, the conditions of use have a big influence on the extent to which the desirable thermal design criteria discussed can be achieved in practice. For example, channels should not be liable to blockage by dust or slag which may mean compromising on larger-sized channels. This will prolong the regenerator life, but reduce its efficiency.

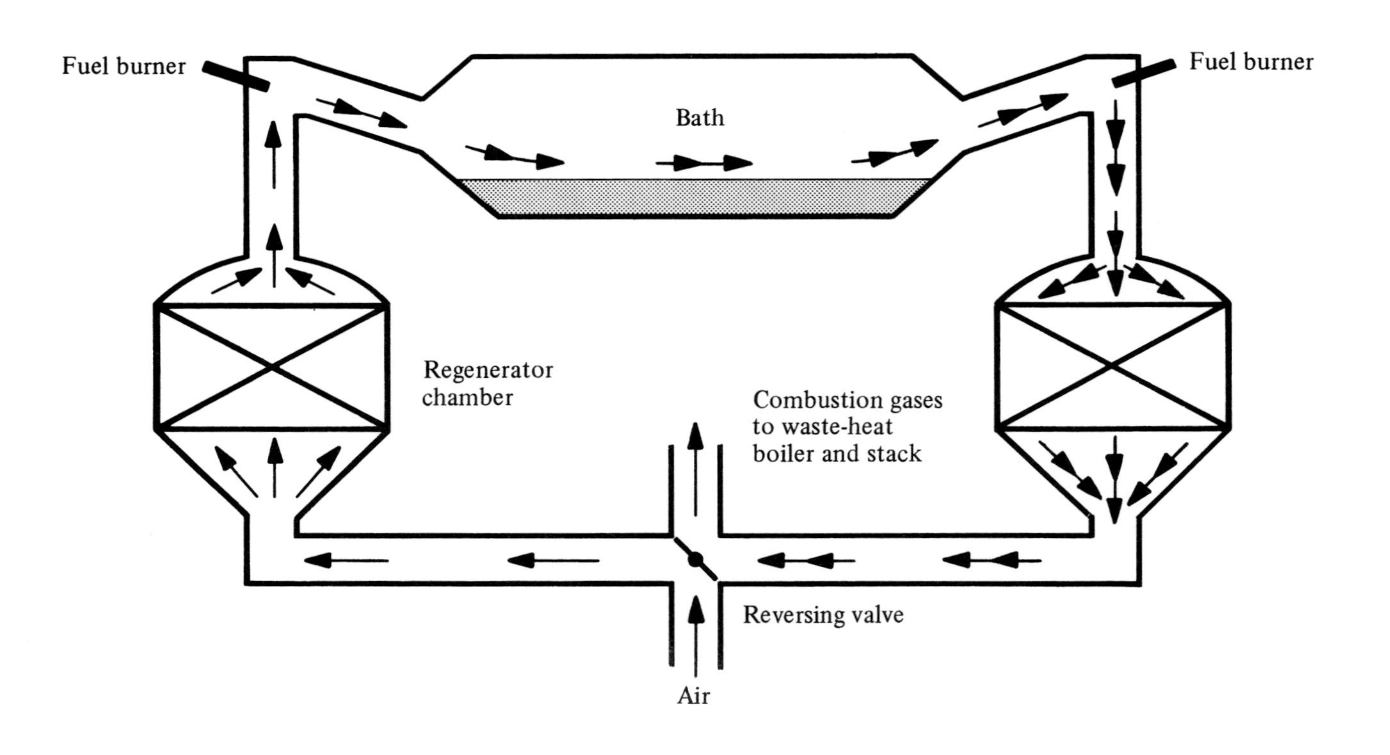

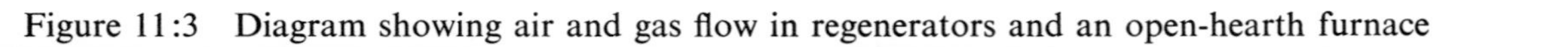

Figure 11:3 Diagram showing air and gas flow in regenerators and an open-hearth furnace

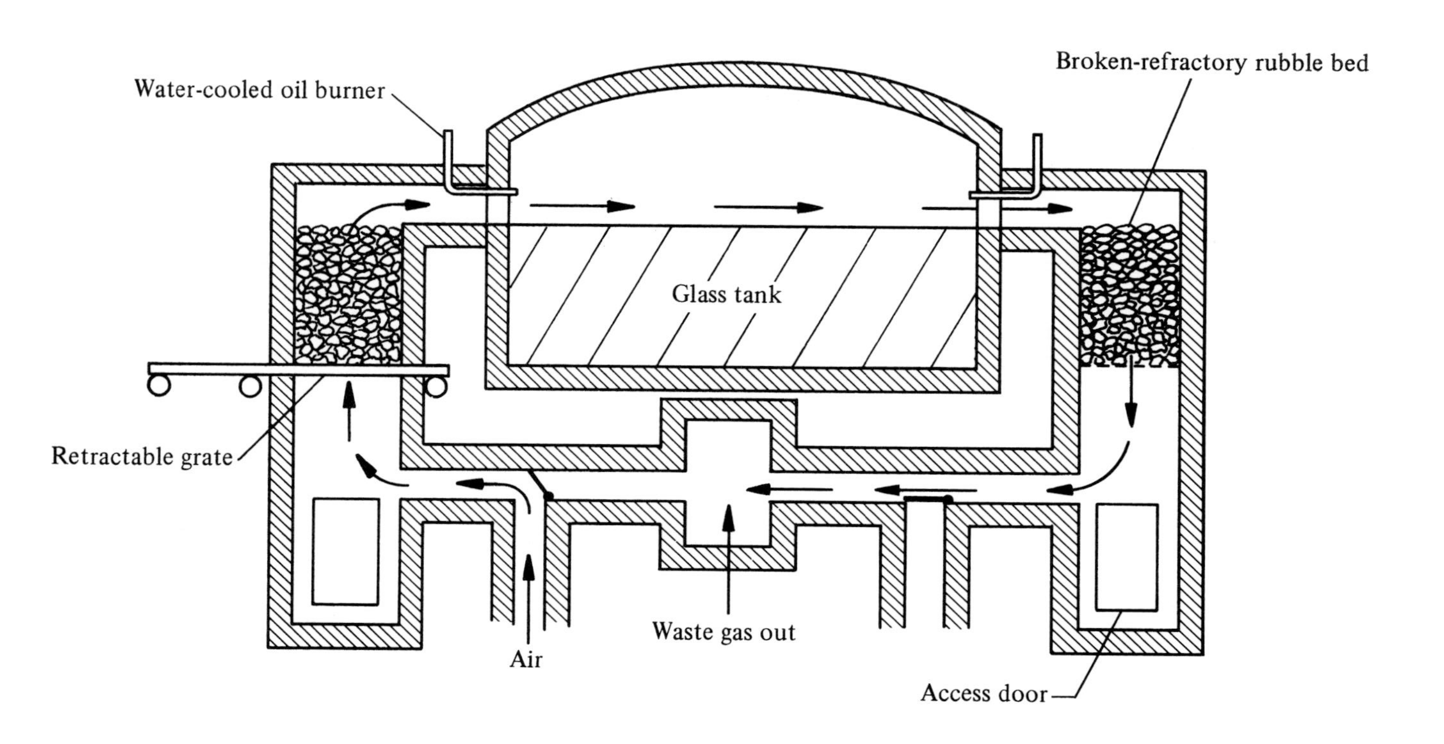

Figure 11:4 Rubble-bed regenerator

Designs are available with the gases flowing vertically or horizontally. Another variation is the use of a rubble bed instead of refractory channels for the heat exchange section. This system gives greater surface area with the use of less refractory material but the pressure drop across the bed is greater. Figures 11:3 and 11:4 show two typical regenerator arrangements. The common feature of the regenerators discussed is the cyclical nature of their operation. Continuous operation has been achieved by the Ljungstrom metal-plate air heater, which is based on the principle of a slowly rotating drum permeable to gas flow in an axial direction, but not radially. The combustion products pass through one half, with the air passing in the opposite direction through the other half. Rotation of the drum carries the heat in the plates from one section to the other past seals which prevent excessive mixing of the two streams. Another type of continuous regenerator is the pebble air heater in which the combustion products pass upwards through a stream of small pebbles moving downwards. The pebbles pass through a comparatively narrow neck into a lower chamber through which a stream of air is flowing. The cooled pebbles from the base are returned by an elevator to the top of the system.

These two systems are truly counter-current and theoretically can raise air temperatures to near the combustion products' temperature. However, both have rather complex mechanical flow systems and are difficult to seal against high pressure differences. The Ljungstrom type is normally built with metal plates which limits the temperature of operation. Regenerative systems have been used in steel forge furnaces in which sub-stoichiometric combustion is used to minimize scaling and decarburization. The partially burned combustion products are passed to a regenerator chamber in which combustion is completed using additional air. On reversal the primary air supplied to the burner is preheated by passage through the regenerator. Similarly two-stage combustion systems using recuperative burners have been developed. Further information on these applications is given in Chapter 7.

Recuperators [1, 2, 3]

A recuperator is a device in which heat is transferred continuously from hot combustion products to combustion air thus preheating it. The heat transferred is limited by the thermal resistances of the two gas films which are large compared to the resistance of the partition separating the gases.

The heat exchange is usually carried out in one of the following ways:

1 Parallel flow
2 Counterflow
3 Cross-flow

Figure 11:5 illustrates a simple concentric-tube system. The controlling equation defining the heat transfer between the flowing streams in a counter-flow system with the hot products flowing in the inner tube is:

$$Q = UA\,(\Delta t_2 - \Delta t_1)/\ln\,(\Delta t_2/\Delta t_1)$$

A is the surface area available for heat transfer and it is obvious that for a high rate of heat transfer, A should be as high as possible, i.e. a large number of small-diameter tubes is better than one large one. U is the overall heat-transfer coefficient and is defined by:

$$\frac{1}{U} = \frac{1}{h_g} + \frac{1}{h_a} + \frac{d_w}{\lambda_w}$$

h is the gas-to-surface heat-transfer coefficient, the subscripts g and a indicating the gas and air sides of the dividing wall respectively. d_w and λ_w are the thickness and thermal conductivity of the dividing wall. h_g and h_a should be as high as possible consistent with the pressure-drop restrictions. The tube should be thin walled and have a high conductivity.

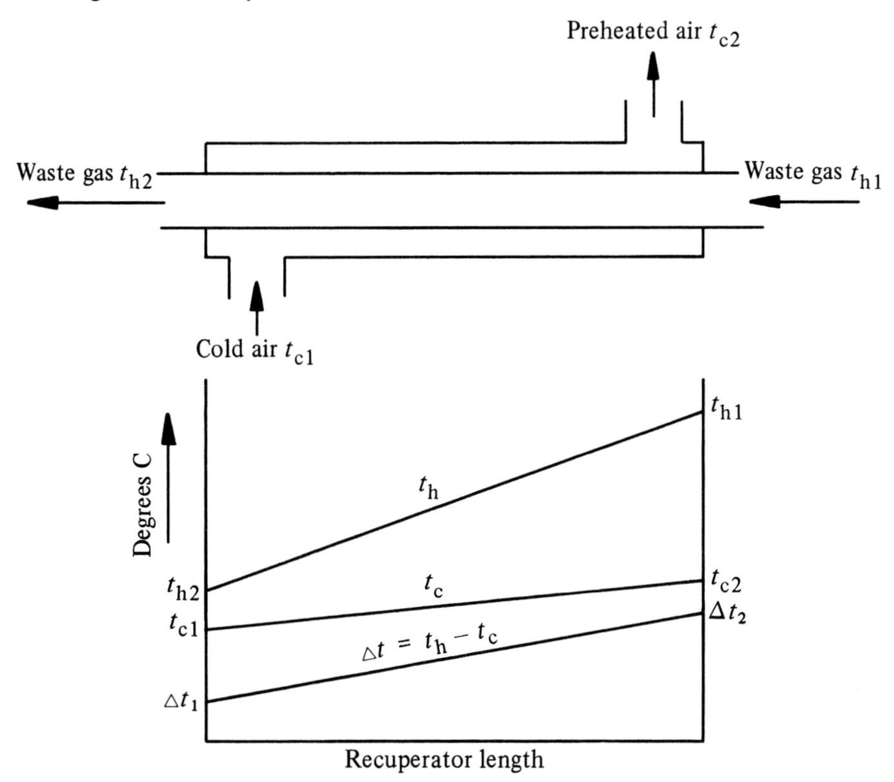

Figure 11:5 Characteristics of a counter-flow recuperator

The heat-resisting steels at present available are able to cope with preheat temperatures of about 900 °C. For higher temperatures refractory recuperators must be used. As well as having thin walls and high conductivity, in refractory recuperators the porosity of the tubes must be low, the thermal expansion levels should be acceptable to reduce breakage in service and the material should be spall resistant.

The designs of recuperator available are the steel-plate type, the tubular type, the fused or needle-tube type and the refractory sections type. Figure 11:6 shows a Newton vertical needle-tube recuperator with details of the needle elements.

Waste-heat boilers [2]

The design of waste-heat boilers has developed to complement improvements in furnaces. For example, waste heat is rejected in narrower ranges of volume and

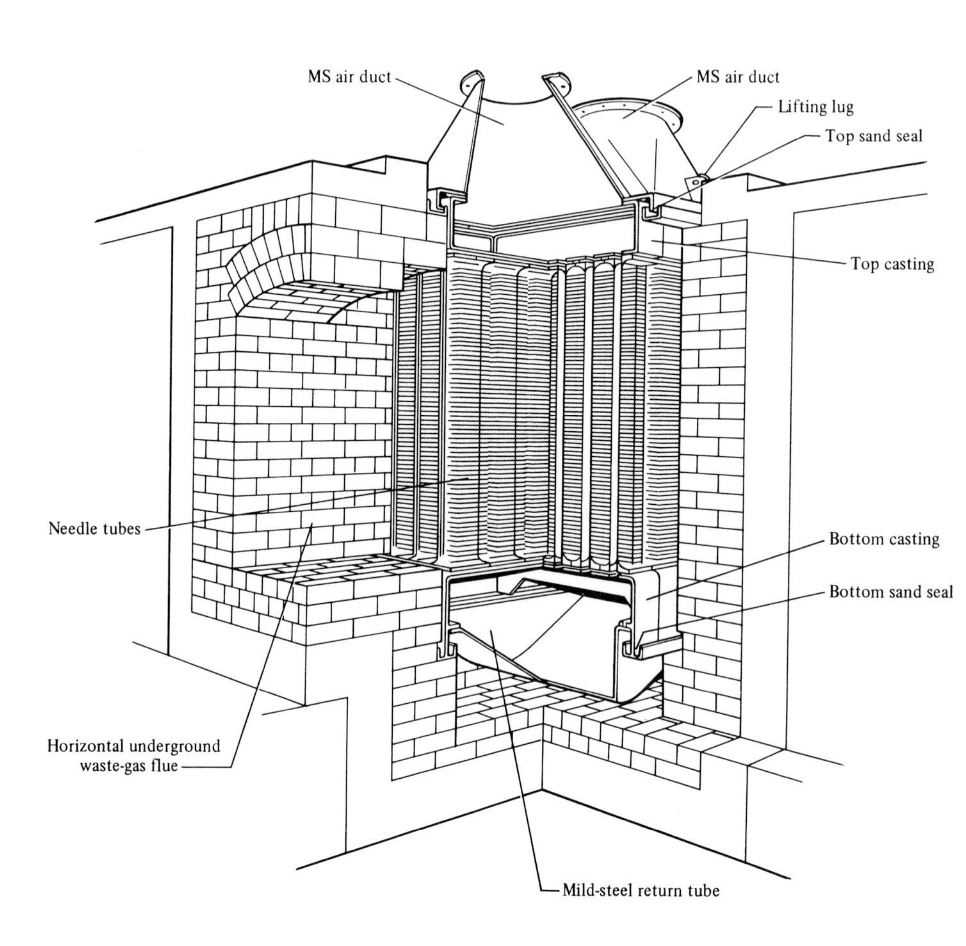

Figure 11:6 Newton vertical needle-tube recuperator

temperature than previously. In some applications, waste-heat boilers form an integral part of multi-stage heat recovery, draught control and gas cleaning. The waste-heat boiler on, say, a high-duty metallurgical furnace provides the only economical and efficient means of reducing the waste-gas temperature to a point suitable for induced fan draught control or for gas cleaning by electrostatic precipitation.

British antipollution legislation, in particular the Clean Air Act, means that the boiler must be capable of on-load cleaning and be operated automatically. Design has also been influenced by the inclusion of some furnace cooling systems in the boiler circuit.

Shell boilers. The basic design of the heating surface in a waste-heat boiler differs fundamentally from a fuel-fired unit because the radiant-heat factor is small. The heat absorption, particularly in the lower temperature ranges, depends largely on convection. Therefore, the gas velocity must be high to give good heat transfer. For straight tubes with a high length/diameter ratio mean velocities should be about $20\ m\ s^{-1}$ ($65.6\ ft\ s^{-1}$).

The shell boiler has largely dominated the field of waste-heat recovery by its simplicity, compactness and avoidance of casing. Apart from the basic factors of design for heat transfer under the temperature limitations of furnace operation, the waste-heat boiler must be capable of functioning for long periods of furnace operation. The importance of on-load cleaning is emphasized as dirty tubes will result in difficulties in furnace draught control leading to reduced output and lower steam yield. The amount of cleaning required is obviously dependent on the furnace duty.

Water-tube boilers. The demand for higher steam pressures has brought the water-tube boiler into use for waste-heat recovery. The design factors important in water-tube boilers are tube diameters, tube formations and layout of tube nests. These factors affect draught loss, heat transfer and accessibility for cleaning. Therefore, some compromise must be made in the design of a water-tube boiler for a waste-heat role because of these factors. If in a water-tube waste-heat boiler the tubes were of constant cross-section and unchanging layout, the progressive reduction in gas velocity, through cooling, would result in the outlet end doing very little work. This is compensated for in waste-heat boilers by progressively changing the tubes from an in-line layout at the inlet to a close-pitched diamond formation at the outlet. The reduction of tube diameters may mean pump-assisted circulation, the rate being dependent on inlet gas temperature and water temperature.

Preheating and cooling stock

In discussing forms of heat waste, it has been stated that the quantity of energy which leaves the furnace in the combustion products is fixed by the operating temperature, that it is not possible to 'design out' or minimize the energy in these gases and also that the energy can be returned to the furnace in some way.

This section deals with a special case of returning energy to the furnace. If the material to be processed has an effectively constant specific heat capacity, so that it can be heated and cooled at a constant rate, then a counter-flow heating process will give the best heat-utilization efficiency. Examples of such materials are found particularly in the ceramic industries and in certain metallurgical applications. In Chapter 13 a number of specific industrial applications of natural gas involving counter-flow furnaces will be discussed.

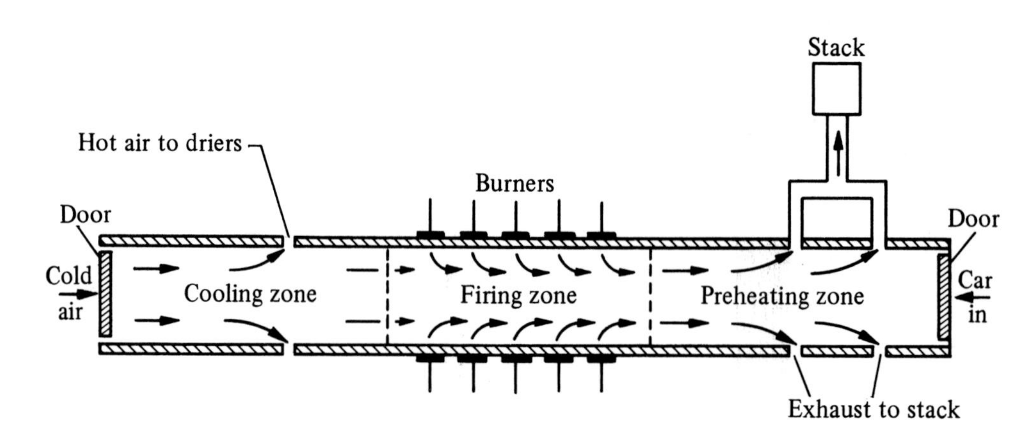

Figure 11:7 Car tunnel kiln

For the purposes of this chapter, a tunnel kiln, Figure 11:7, as used by the ceramic industry, will be described to illustrate the method of operation. The ware to be fired is progressed through the long tunnel construction on cars, in counter-flow with the combustion gases, passing an approximately central firing zone and then continuing through the tunnel in counter-flow with incoming combustion air, so becoming cooled. The temperature profile undergone by any particular ceramic must be precise in all three regions since the thermal treatment is the main determining factor in the quality of the fired material. The design of the heating and exhaust system must be carefully carried out to suit the required firing specifications and duty. Usually heating is supplied by multiple burners distributed over the length of the firing zone. The exhaust system may involve recirculation, natural or forced draught, designed to give controlled cooling conditions.

Self-recuperative burners [4, 5]

Regenerators, recuperators and waste-heat boilers all involve independent units and are mainly confined to large-scale process plant because of the cost and size of the equipment required. However, it is often the smaller furnaces and kilns which are the least efficient. A typical batch forge furnace has an efficiency of about 10 per cent with some 70 per cent of the heat supplied being lost in the flue gases. The recovery of this heat by air preheating in a recuperator would depend as indicated on the size of the operation. Load recuperation is another possibility, but is usually only practised in large continuous furnaces.

The MRS design of recuperative burner is shown in Figure 7:29 Chapter 7. It consists essentially of a high-velocity nozzle-mixing burner surrounded by a counter-flow heat exchanger which supplies hot combustion air to the burner nozzle. The heat exchanger is constructed from a series of concentric tubes which define annular air and flue-gas passages and provide heat-exchange surfaces. All recuperator components are made from an appropriate heat-resisting steel and the burner can operate at high process temperatures since all the metallic components in contact with the flue products are cooled by the combustion air. The refractory burner quarl is designed to protect the front face of the recuperator from direct furnace radiation.

In operation, the combustion air enters the burner at a manifold, which totally encloses the flue annuli, Figure 7:29. It then passes forward along the air annulus, thus keeping the external surfaces of the recuperator cool. Flue products are extracted in a counter-flow direction around the outside of the air annulus and thus preheat the air. At the front of the recuperator the air flow doubles back and enters the burner nozzle. The gas and preheated air mix at the nozzle, on the face of which the flame stabilizes and combustion is essentially complete before the gases pass into the furnace chamber. Exit velocities from the burner tunnel exceed 50 $m\ s^{-1}$ (164 $ft\ s^{-1}$). Normally all the combustion products are extracted through the recuperator by using an air-driven eductor mounted on the burner flue. The furnace pressure is maintained at the desired level by controlling the amount of eductor air.

The design procedure for the recuperator section is as described. It allows the maximum air preheat to be determined for a given pressure differential across the recuperator and for any specified material surface temperature. The calculation takes into account the change in properties of the gases with temperature and in particular the effect, due to preheat, of increased air volume on the air-pressure requirement. This increase, which can be as much as threefold, must also be taken into consideration in the design of the burner nozzle. The nozzles used give a stable flame, are capable of a turndown ratio of about 10 : 1 and can operate if necessary with over 100 per cent excess air.

A range of burners is available which covers thermal inputs from 100 kW to 900 kW. These burners can give air preheats of up to 600 °C, they are compact in that the integral recuperator occupies only one-tenth of the volume of a conventional flue-stack recuperator and they can achieve fuel savings of up to 50 per cent.

In many applications of recuperative burners, such as intermittent kilns, installation can be simply replacement of existing burners. There is, however, the possibility of such a high degree of recirculation being obtained that it may be possible to reduce the number of burners required. In these cases physical and mathematical models are very useful aids in determining the best burner and chamber configuration and in predicting the furnace performance (Chapter 12).

REFERENCES

1 Thring, M. W., *The Science of Flames and Furnaces*, 2nd edition, London, Chapman & Hall (1962)
2 Institute of Fuel, *Waste Heat Recovery*, London, Chapman & Hall (1963)
3 Fitzgerald, F. and Lakin, J. R., *J. Inst. Fuel*, **45** (1972), 555
4 Harrison, W. P., Oeppen, B. and Sourbutts, S., *Gas Council Res. Commun.*, GC164 (1969); *J. Inst. Gas Eng.*, **10** (1970), 538
5 Bryan, D. J., Masters, J. and Webb, R. J., *Inst. Gas. Eng. Commun.*, 952 (1974)

12

Mathematical models

12.1 INTRODUCTION

This chapter will look at some of the ways in which the theoretical topics of the earlier chapters may be applied to the design of the kind of industrial equipment described in later chapters.

For the technologist, the basic principles outlined in previous chapters serve two main purposes. First, they are one of the main sources of inspiration in the invention of new designs of equipment. Obvious examples of this are the concept of high-velocity burners for forced-convection heating and the use of the momentum principle in promoting recirculation in furnaces. In this role of synthesis the basic principles are used almost unconsciously and intuitively so that it is very difficult to systematize the process of invention.

The second use of the basic principles is in the analysis of proposed designs of equipment in an attempt to predict performance accurately. The ultimate aim of the designer is to replace the costly process of actual testing of hardware by calculations on paper.

This process of describing the behaviour of an engineering system by equations adequately representing the basic flow, heat and mass transfer, and combustion processes, etc., occurring in the system is called mathematical modelling. Designers have always made some use of calculations in particular cases, but only recently has it been possible to carry out the complex numerical computations which arise from detailed mathematical models of complete processes.

Of course the prudent designer wishes to see some validation of a prediction from mathematical modelling and confidence will only be gained as more actual performances can be compared with predictions. For this reason it is very useful to apply mathematical modelling to well established designs such as simple tube heaters for tanks, and shell boilers (see Section 12.5).

Most design work builds on existing designs and consists of limited extrapolation from existing data and experience, and this process is aided very much by having a reasonable mathematical model on which to base the extrapolations or modifications.

The successful use of mathematical modelling requires accurate data. Some data, such as emissivity of surfaces or gases, and reaction rates in flames, are fundamental and only obtainable by experiment. Others, such as heat transfer to complex shapes like finned tubes is also found experimentally, but may also be amenable to mathematical modelling, and most experimenters try to support their findings by a theoretical analysis.

As well as mathematical models, use is made of physical models and analogues. Spalding [1] defines these as connoting the practice of predicting the likely results of one experiment (the prototype) by interpreting the results of another experiment (the model), and these are used to obtain data or to validate mathematical models in cases where prototype experiments are too difficult or costly. Some relevant basic principles are given in Sections 2.3 and 3.5 and the subject is treated more comprehensively in Section 12.8.

12.2 PRINCIPLES OF MATHEMATICAL MODELLING IN COMBUSTION SYSTEMS

There are essentially two distinct stages in the formulation of a mathematical model. These are:

1 A simplification of the actual system to give a tractable problem, for example, the replacement of complicated surface shapes by simpler ones for which heat transfer data are available.
2 The formulation of equations adequately describing the combustion, flow and heat transfer in the simplified system. Since this is a continuum the equations will be differential equations, but these will have to be solved as finite-difference equations in a computer so that it may be better to divide the system into a finite number of zones and to write down the system of finite-difference heat, momentum, and mass balances for each zone.

The equations may be steady-state or unsteady-state equations, the latter being considerably more difficult to deal with since they involve the extra dimension of time. In cases where unsteady-state modelling is necessary, as for example in batch processes or in investigating automatic-control problems, it may be necessary to make very large simplifications in the model in order to make it usable.

It is only fairly recently that complete mathematical modelling of the interdependent flow, combustion and heat-transfer processes has been attempted. In most applications of immediate use to designers it is possible to make assumptions about the flow pattern and/or combustion pattern, based on experience or on physical model studies. For example, in the case of the rapid billet-heating furnace, combustion is complete at the tunnel burner and the furnace geometry is such as to give a stream of gas of uniform temperature and velocity, so the problem is essentially that of determining the heat-transfer coefficient from this gas to the walls and billet, and the unsteady heat flux into the billet. (This example is dealt with in more detail in Section 12.4.)

This separation of the flow and combustion processes from the heat-transfer process is possible because the flow and combustion depend very much on each other, but not greatly on the heat transfer.

Patanker and Spalding [2] classify mathematical models in order of complexity as follows:

Zero-dimensional models. In zero-dimensional models uniform gas temperature and uniform wall temperature are assumed. This implies a 'well-stirred' furnace and may be an appropriate model if recirculation is included. The steady-state model consists of algebraic equations only. The unsteady-state model may include ordinary differential equations, and this relative simplicity makes it particularly useful for unsteady-state modelling. Examples are given in Section 12.4.

One-dimensional model. This is appropriate where the temperatures of wall and gas vary in one direction only and is a reasonable model for the long furnace or tube in which variations across the tube are small. If radiative heat transfer along the tube is ignored the steady-state model is a first-order ordinary differential equation, while the unsteady-state model is a partial differential equation. Examples are given in Section 12.5.

Two-dimensional model. Here variations in temperature, etc., occur in two directions and the commonest example is that of an axisymmetric furnace. The mathematical model consists of partial differential equations which are relatively easy to solve numerically by explicit 'marching integration' if there is no recirculation in the furnace, but require longer iterative methods if there is axial recirculation. General methods for dealing with this type of problem are described by Gosman *et al.* [3].

Three-dimensional model. This is the most general case where variations occur in all directions which is the case in most real systems. Computer solution of the resulting partial differential equations requires very large storage and long computation time and is hardly practicable at present. It is the present author's feeling that new mathematical techniques and possibly new analogue methods are needed in this problem.

12.3 RADIATIVE TRANSFER BETWEEN GASES AND SURFACES AND THE GENERAL HEAT BALANCE

In most reasonably large furnaces radiation is the principal mode of heat transfer and its correct representation often forms a large part of the mathematical model.

The basic methods have been outlined in Section 3.4 and detailed treatments are given by Hottel and his coworkers [4, 5, 6, 7]. More specialized work, including treatment of luminous radiation in furnaces is given in references [8, 9, 10, 11].

The book by Gray and Müller [12] is an excellent introduction to the subject as it includes many worked numerical examples.

This section gives a brief outline of the general method and deals with a few special cases of frequent occurrence.

For convenience, we rewrite equation 3.38:

$$B_i = \varepsilon_i E_i + \rho_i H_i \tag{12.1}$$

where E_i is the black-body radiant intensity, and equation 3.39a:

$$H_i = \varepsilon_{gi} E_g + \sum_k (1 - \alpha_{gki}) B_k F_{ik} \tag{12.2}$$

or, using Hottel's notation:

$$H_i = \frac{\overrightarrow{gs_i}}{A_i} E_g + \sum_k \frac{\overrightarrow{s_k s_i}}{A_i} B_k \tag{12.2a}$$

The quantities $\overrightarrow{gs_i}$ and $\overrightarrow{s_k s_i}$ are called the direct radiant exchange areas from gas to surface i and from surface k to surface i including the effect of gas absorption in between.

By reference to equations 3.36 and 3.37 it will be seen that the quantity $\overline{s_1 s_2}$ is defined by equation 3.36 with the additional term $(1 - \alpha_{g12})$ included in the integral to give the heat flux actually arriving at surface 2 from surface 1.

If the surfaces are grey, equation 3.41 is the most convenient form for the net radiant energy leaving a surface:

$$Q_i = \varepsilon_i (E_i - B_i) A_i/(1 - \varepsilon_i) \tag{12.3}$$

and, for the net radiant energy leaving the gas, equation 3.42 is:

$$Q_g = E_g \sum_i \varepsilon_{gi} A_i - \sum_i \sum_k \alpha_{gki} B_k F_{ki} A_k \tag{12.4}$$

or:

$$Q_g = E_g \sum_i \overrightarrow{gs_i} - \sum_k \overrightarrow{s_k g} B_k \tag{12.4a}$$

12.3.1 Gas emissivity, absorptance and transmittance

The radiation emitted from or absorbed by a 'grey' gas is proportional to the number of particles or molecules encountered, which implies that the change in intensity of a beam passing through the gas is proportional to the intensity and to the number of molecules encountered, i.e.:

$$\mathrm{d}I = -KI\mathrm{d}L = -\chi pI\,\mathrm{d}L$$

where p is the partial pressure of the gas. This integrates to give:

$$I = I_0 \mathrm{e}^{-KL} \tag{12.5}$$

if K is constant, or:

$$I = I_0 \mathrm{e}^{-\int K \mathrm{d}L} \tag{12.5a}$$

if K is a function of L. I/I_0 is the transmittance τ_g. The fraction absorbed is:

$$\alpha_g\,(\text{grey}) = \varepsilon_g\,(\text{grey}) = 1 - \tau_g = 1 - \mathrm{e}^{-KL} \tag{12.6}$$

This law is obeyed by luminous gases where the radiation is due to incandescent particles. For large particles of diameter $> 5\lambda/\pi$, where λ is the wavelength of the radiation, $K = na$ where n is the number of particles per unit volume and a is the surface area of each particle.

For small particles of diameter $< 0.6\,\lambda/\nu\pi$ where ν is the refractive index of the particle, $K = Ff/\lambda$, where f is the fraction of the total volume occupied by the particles and F depends on their nature. Data for natural-gas flames are given in reference [13].

However, equation 12.6 is not obeyed by a real gas with its main absorption in a few discrete wavebands, because if a beam of radiant energy containing the whole range of wavelengths (as would be emitted by a surface) passes through a real gas, the energy in those wavebands is absorbed relatively quickly leaving the rest of the energy to be absorbed more slowly.

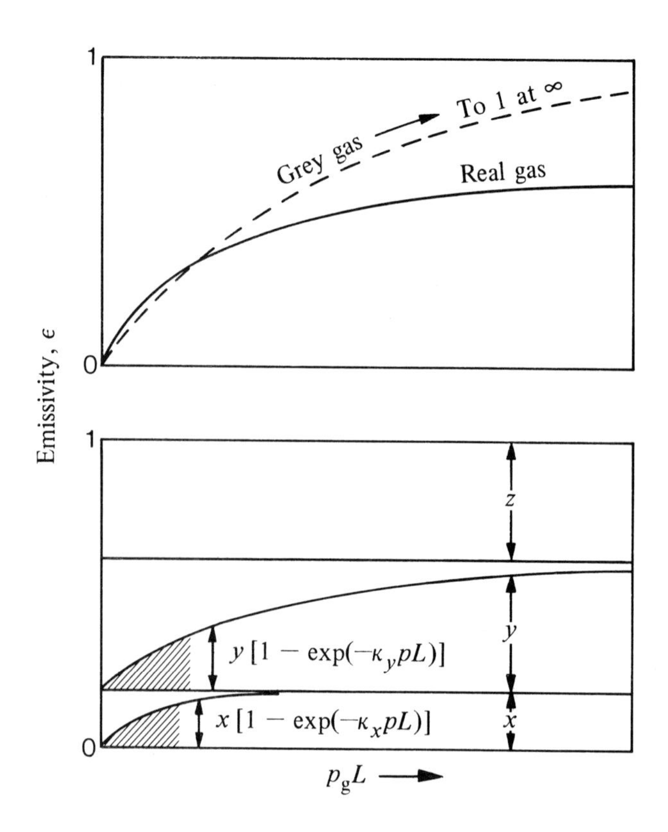

Figure 12:1 Representation of a real gas by a mixture of grey gases (From *Heat Transmission* by William H. McAdams. Copyright 1933, 1942, 1954 William H. McAdams. Used with permission of McGraw-Hill Book Company)

Figure 12:1 shows the difference between a real and a 'grey' gas. The curve for the real gas can be accurately represented by a mixture of grey gases, each having a different value of K:

$$\varepsilon_g(\text{real}) \approx \sum b_i\,(1 - e^{-K_i L}) \tag{12.7}$$

where the fractions b depend on the temperature of the gas. It is usually found that sufficient accuracy over a given path length can be achieved by two or three grey gases, one of which is usually clear, i.e. has a value $K = 0$.

Similarly, the absorbtance can be represented by a *different* mixture of the same gases, and in this case the fractions depend on the temperature from which the radiation comes.

This method requires the storage of less data than would a complete account of the real gas properties. The topic is covered in detail in Chapter 6 of reference [7].

12.3.2 Direct exchange areas and path lengths

It can be shown that for a small volume of gas the appropriate path length is 4 × gas volume/radiating surface area. Also, for small values of KL, equation 12.6 gives $\varepsilon_g = KL$, so that the total radiation leaving a small volume of gas in all directions is

$$Q_g = 4KVE_g \tag{12.8}$$

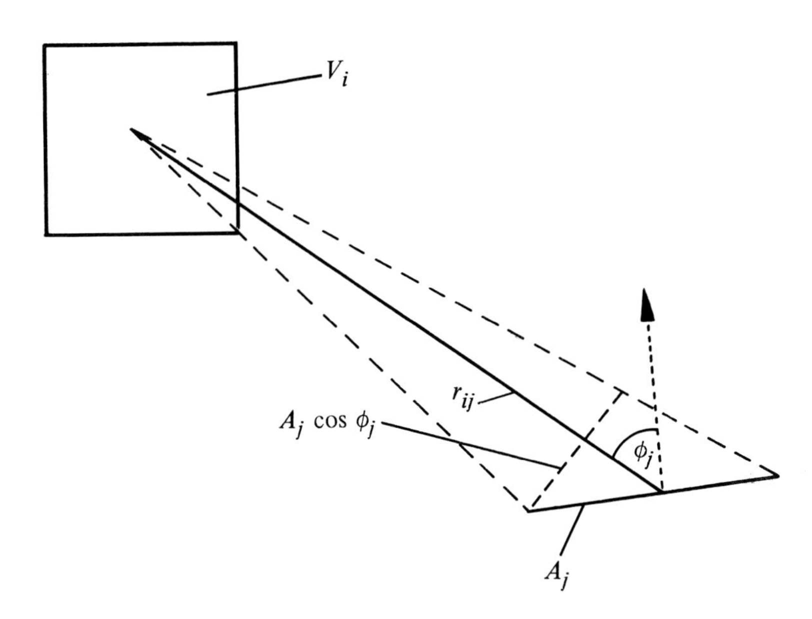

Figure 12:2 Direct exchange of radiation between a surface element and a gas volume element

Referring to Figure 12:2 where V_i is taken to be small, the proportion going towards A_j of the radiation leaving V_i is $A_j \cos\theta_j / 4\pi\, r_{ij}^2$ so that, using 12.8:

$$Q_{i \to j} = \left(4\, K_i\, V_i\, A_j \frac{\cos\, \phi_j}{4\pi\, r_{ij}^2}\, \tau_{ij} \right) E_{gi} \tag{12.9}$$

is the energy actually reaching A_j from V_i so that the quantity enclosed in brackets is $\overrightarrow{g_i\, s_j}$. Equation 12.9 can be summed over the whole area A and volume V to give $\overrightarrow{gs}$ for these larger units:

$$\overrightarrow{gs} = \sum_A \sum_V K_i\, A_j \cos\phi_j\, \tau_{ij} / \pi r_{ij}^2 \tag{12.9a}$$

The other direct exchange areas can be found similarly. Hottel [7] has applied equation 12.9 to a variety of shapes to a *completely surrounding surface* and shows that in this case a path length of 3.5 × volume/surface can be used to find ε_g and α_g from gas emissivity charts for use in equation 12.2.

Equation 12.9a may have to be used when radiation to different portions of surface is to be calculated, or when the gas temperature is not uniform (see Section 12.3.6 on the zone method of analysis).

12.3.3

From a knowledge of the view factors or direct exchange areas for given surface configurations, those for others may be deduced by geometrical means in two-dimensional systems. We make use of the definitions:

$$\overline{s_i s_j} = \overline{s_j s_i} \quad \text{(equation 3.37a)} \tag{12.10}$$

and

$$F_{i1} + F_{i2} + F_{i3} + \ldots + F_{in} = 1 \tag{12.11}$$

or, multiplying both sides by A_i:

$$\overline{s_i s_1} + \overline{s_i s_2} + \overline{s_i s_3} + \ldots + \overline{s_i s_n} = A_i \tag{12.12}$$

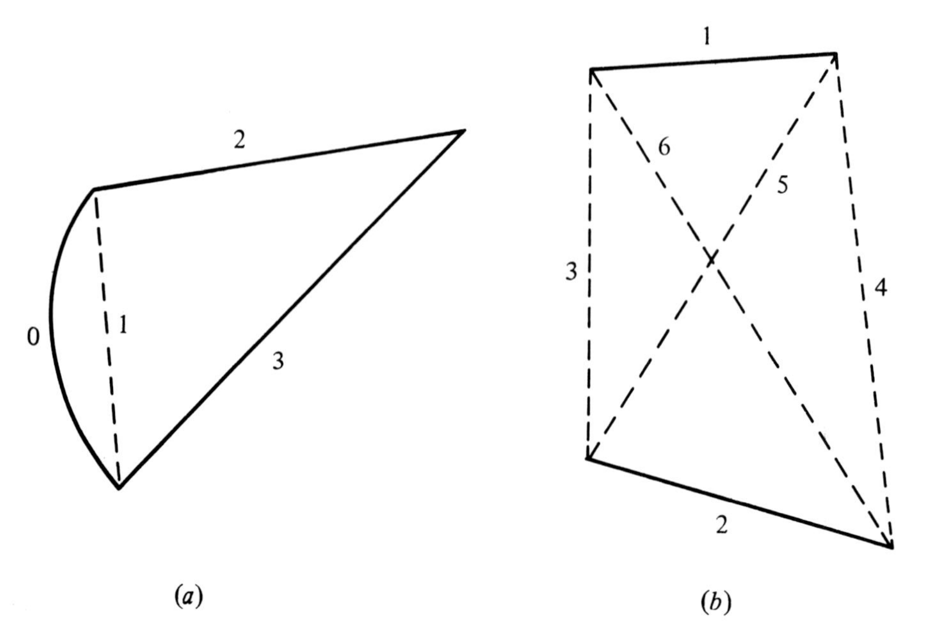

Figure 12:3 Direct exchange areas for two-dimensional configurations

As an example, consider the enclosure in Figure 12:3(*a*). For the enclosure 01, equation 12.2 gives, using simplified symbols:

$$\overline{00} + \overline{01} = A_0, \qquad \overline{11} + \overline{10} = A_1$$

Also $\overline{11} = 0$ and, from equation 12.11, $\overline{01} = \overline{10}$ which gives:

$$\overline{00} = \overline{s_0 s_0} = A_0 - A_1$$

Also, for the enclosure 123 we have, $\overline{11} = \overline{22} = \overline{33} = 0$, since they are all plane, and equations 12.12 and 12.10 give:

$$\begin{aligned} \overline{12} + \overline{13} &= A_1 \\ \overline{13} + \overline{23} &= A_3 \\ \overline{12} + \overline{23} &= A_2 \end{aligned}$$

from which:

$$\overline{12} = \overline{21} = \tfrac{1}{2}(A_1 + A_2 - A_3) \tag{12.13}$$

and similarly for the other exchange areas.

This method can be extended to enclosures having any numbers of sides, by dividing them into triangular areas.

Figure 12:3(*b*) shows two areas 1 and 2. To find $\overline{s_1 s_2}$ join the ends by the imaginary plane areas 3, 4, 5, 6 (or 'uncrossed and crossed strings'). Then by applying the above method to the enclosure 1234 and to the triangular enclosures, the general result will emerge that:

$$\overline{s_1 s_2} = \tfrac{1}{2}[(A_5 + A_6) - (A_3 + A_4)] \tag{12.14}$$

These methods are used for finding the exchange areas for surfaces inside furnaces, for example between radiant tubes, and between radiant tubes and objects being heated.

In the next two sections radiant energy interchange will be calculated for two common configurations.

12.3.4 Single-temperature gas enclosed by a single-temperature grey surface

This is the simplest possible configuration in a zero-order model, and is shown in Figure 12:4. Equation 12.2 is:

$$H_1 = \varepsilon_g E_g + (1 - \alpha_{g11}) B_1 \quad \text{since } F_{11} = 1$$

i.e. the surface 'sees itself' in all directions. Then equation 12.1 becomes:

$$B_1 = \varepsilon_1 E_1 + (1 - \varepsilon_1)\,[\varepsilon_g E_g + (1 - \alpha_{g11})\,B_1]$$

i.e.

$$B_1 = \frac{\varepsilon_1 E_1 + \varepsilon_g (1 - \varepsilon_1)\,E_g}{\varepsilon_1 + \alpha_{g1} - \varepsilon_1\,\alpha_{g1}}$$

and using this in equation 12.4 yields

$$\frac{Q_g}{A_1} = \varepsilon_g E_g - \alpha_{g1}\left[\frac{\varepsilon_1 E_1 + \varepsilon_g (1 - \varepsilon_1)\,E_g}{\varepsilon_1 + \alpha_{g1} - \varepsilon_1 \alpha_{g1}}\right]$$

or

$$\frac{Q_g}{\sigma A_1} = \frac{\varepsilon_1\,\varepsilon_g\,T_g^4 - \varepsilon_1\,\alpha_{g1}\,T_1^4}{\varepsilon_1 + \alpha_{g1} - \varepsilon_1\,\alpha_{g1}} \tag{12.15}$$

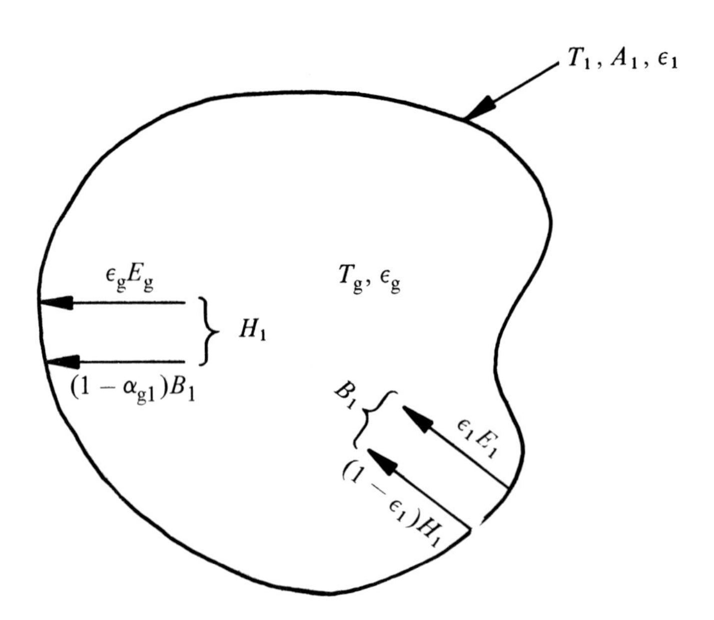

Figure 12:4 Radiant interchange between a gas and an enclosure

Alternatively, if the above expression for B_1 is used in equation 12.3 then exactly the same expression as 12.15 will be obtained for the quantity $-Q_1/\sigma A_1$. This is as it should be since the energy leaving the gas must all be absorbed by the wall.

Equation 12.15 can be written as:

$$Q_g = -Q_1 = \sigma(\overrightarrow{GS_1}\,T_g^4 - \overrightarrow{S_1G}\,T_1^4) \tag{12.16}$$

Hottel terms $\overrightarrow{GS}$ and $\overrightarrow{SG}$ the *total radiant exchange areas* and the arrow denotes that they are not equal when the gas is non-grey. But if the gas is grey, i.e. $\varepsilon_g = \alpha_{g1}$, which would be true only if $T_g = T_1$, then equation 12.15 becomes:

$$\begin{aligned} Q_g &= -Q_1 \\ &= \sigma A_1 (T_g^4 - T_1^4)/(\varepsilon_g^{-1} + \varepsilon_1^{-1} - 1) \\ &= \overline{GS}\,(T_g^4 - T_1^4) \end{aligned} \tag{12.17}$$

i.e. $\overrightarrow{GS} = \overrightarrow{SG}$ for a completely grey enclosure.

Note that for a black surface, equation 12.15 becomes:

$$Q_g = \sigma A_1\,\varepsilon_g\,(T_g^4 - T_1^4) \tag{12.17a}$$

12.3.5 Single-temperature gas in an enclosure consisting of two grey surfaces at different tempe:atures

This corresponds to the practical case of a 'well-stirred' furnace with walls and load at different temperatures, and the rapid billet-heating furnace in Section 12.4 is a good example which will be worked out numerically in this section. The system is shown in Figure 12:5 and consists essentially of an inner surface which is completely

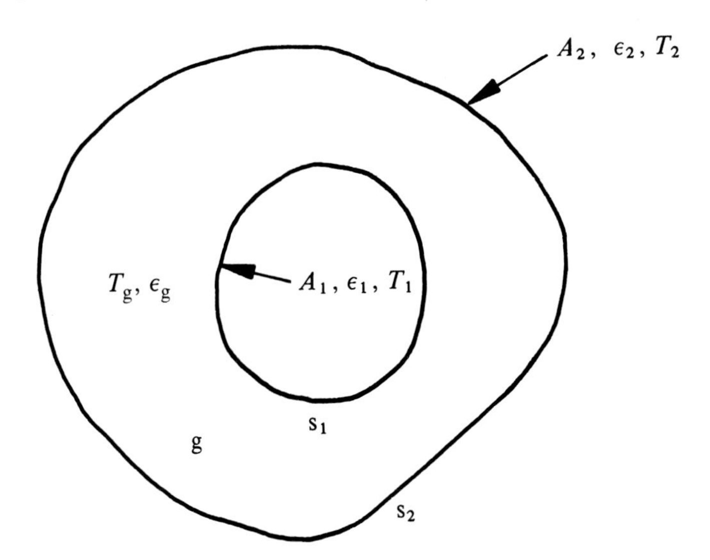

Figure 12:5 Radiant interchange between gas, furnace wall and load

enclosed by an outer with the radiating gas in between. The following data will be assumed: $\varepsilon_1 = 0.6$, $\varepsilon_2 = 0.8$, $\varepsilon_g = 0.1$, $\alpha_g = 0.2$ for both surfaces and $A_2 = 2A$, $A_1 = A$.

Since the temperatures of surfaces and gas are not known at this stage, assumed values of α_g and ε_g have to be used, and then more accurate values can be chosen by the methods in Sections 3.4.4 or 12.3.1.

The view factors

If the inner surface s_1 is wholly convex then $F_{11} = 0$. Since, by definition, $F_{11} + F_{12} + F_{13} + \ldots = 1$, in this case $F_{11} + F_{12} = 1$ hence $F_{12} = 1$.

Then, by equation 3.37, $F_{21} = F_{12}\, A_1/A_2$. Therefore $F_{21} = 0.5$ and, since $F_{21} + F_{22} = 1$, $F_{22} = 0.5$.

Equations 12.2 are then:

$$H_1 = 0.1E_g + 0.8B_2$$
$$H_2 = 0.1E_g + 0.8 \times 0.5B_2 + 0.8 \times 0.5B_1$$

Using these in equations 12.1 gives:

$$B_1 = 0.6E_1 + 0.4(0.1E_g + 0.8B_2)$$
$$B_2 = 0.8E_2 + 0.2(0.1E_g + 0.4B_2 + 0.4B_1)$$

For computation solution a matrix representation would be used:

$$\begin{bmatrix} 1 & -0.32 \\ -0.08 & 0.92 \end{bmatrix} \begin{bmatrix} B_1 \\ B_2 \end{bmatrix} = \begin{bmatrix} 0.6\,E_1 + 0.04\,E_g \\ 0.8\,E_2 + 0.02\,E_g \end{bmatrix} = \begin{bmatrix} C_1 \\ C_2 \end{bmatrix}$$

This is easily solved by Cramer's rule which states that if $[A]\,[B] = [C]$ then $B_i = \det[A_i]/\det[A]$, where $[A_i]$ is formed from $[A]$ by replacing column i by the column $[C]$. Hence

$$B_1 = \frac{\begin{vmatrix} C_1 & -0.32 \\ C_2 & 0.92 \end{vmatrix}}{\begin{vmatrix} 1 & -0.32 \\ -0.08 & 0.92 \end{vmatrix}} = \frac{0.92\,C_1 + 0.32\,C_2}{0.894}$$

i.e.

$$\left.\begin{aligned} B_1 &= 0.617E_1 + 0.285E_2 + 0.048E_g \\ B_2 &= 0.054E_1 + 0.895E_2 + 0.026E_g \end{aligned}\right\} \tag{12.18}$$

These are used in equations 12.3 and 12.4 The former gives:

$$\begin{aligned} Q_1 &= (0.575E_1 - 0.43E_2 - 0.072E_g)A \\ Q_2 &= (0.847E_2 - 0.43E_1 - 0.208E_g)A \end{aligned}$$

Equation 12.4 is:

$$Q_g = \varepsilon_g(A_1 + A_2)E_g - \alpha_{g11}\,B_1F_{11}A_1 - \alpha_{g21}\,B_2\,F_{21}\,A_2 - \alpha_{g12}\,B_1F_{12}A_1 - \alpha_{g22}\,B_2\,F_{22}\,A_2$$

which yields

$$Q_g = (0.280E_g - 0.417E_2 - 0.145E_1)$$

However, it is often more convenient to find the individual total radiant exchange areas separately. This is done by setting all but one of the values of E in equation 12.18 equal to zero, and equations 12.3 and 12.4 then give the energy interchange due to the one surface or gas remaining.

Thus, setting E_2 and $E_g = 0$ in equation 12.18, equation 12.3 gives $-Q_{21}$ (i.e. the heat entering 2, which must have come from 1)

$$= \left(\frac{0.8}{1 - 0.8}\right) \times 0.054\,E_1 \times 2A = 0.43\,AE_1 = \overrightarrow{S_1S_2}\,E_1$$

and from equation 12.4:

$$\begin{aligned} -Q_{g1} &= (0.2 \times 0.054 \times 0.5 \times 2A + 0.2 \times 0.617 \times A \\ &\qquad + 0.2 \times 0.054 \times 0.5 \times 2A)E_1 \\ &= 0.145AE_1 = \overrightarrow{S_1G}E_1 \end{aligned}$$

As a check, equation 12.3 gives:

$$\begin{aligned} Q_1 &= 0.6(E_1 - 0.617E_1)A_1/(1 - 0.6) \\ &= 0.575AE_1 \end{aligned}$$

which is equal to $(\overrightarrow{S_1G} + \overrightarrow{S_1S_2})E_1$ as it should be, since $Q_1 = Q_{1\to g} + Q_{1\to 2}$.

Similarly, setting E_1 and $E_g = 0$ and E_1 and $E_2 = 0$ gives the following values:

$$\begin{aligned} \overrightarrow{S_2S_1} &= 0.43A, & \overrightarrow{S_2G} &= 0.417A \\ \overrightarrow{GS}_1 &= 0.072A, & \overrightarrow{GS_2} &= 0.208A \end{aligned}$$

Hence the radiant interchanges are:

$$\left.\begin{aligned} Q_{1\rightleftarrows 2} &= 0.43(E_1 - E_2)A \\ Q_{g\rightleftarrows 1} &= 0.072AE_g - 0.145AE_2 \\ Q_{g\rightleftarrows 2} &= 0.208AE_g - 0.417AE_2 \end{aligned}\right\} \qquad (12.19)$$

If the same calculations are done using a grey gas for which $\alpha_g = \varepsilon_g = 0.1$ it will be found that

$$\begin{aligned} \overrightarrow{S_1S_2} = \overrightarrow{S_2S_1} &= 0.495A \\ \overrightarrow{S_1G} = \overrightarrow{GS_1} &= 0.075A \\ \overrightarrow{S_2G} = \overrightarrow{GS_2} &= 0.224A \end{aligned}$$

12.3.6 The zone method

This is used where substantial variations in temperature occur in different parts of the heat-transfer system. The furnace volume and surface area are subdivided into conveniently shaped zones assumed to have uniform properties, such as temperature, composition, emissivity and velocity, and an energy balance is written for each zone.

Figure 12:6 shows such a zoned enclosure where the dotted lines separate well stirred zones of different uniform properties. Referring to radiant heat exchange

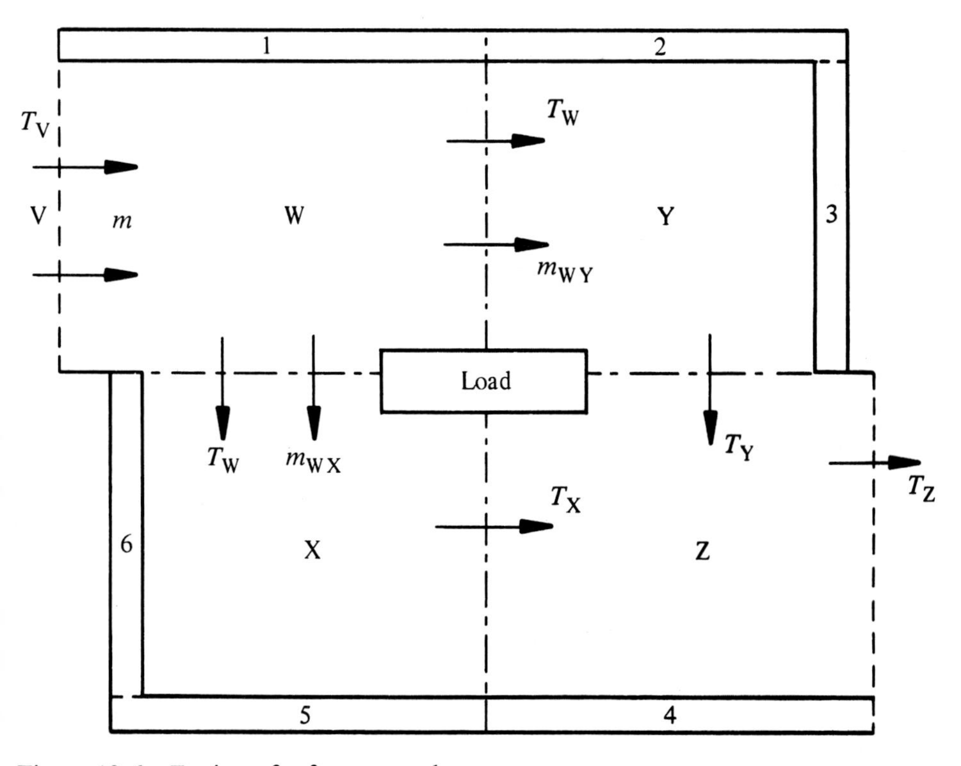

Figure 12:6 Zoning of a furnace enclosure

alone one method is to apply equations 12.1 to 12.4a to the whole enclosure. This will require prior computation of direct exchange area such as $\overrightarrow{g_w s_4}$ and $\overrightarrow{s_5 g_2}$ and $\overrightarrow{s_2 s_8}$ by the methods of Section 12.3.2.

An alternative method which does not require these computations to be made separately applies the radiant-interchange equations to each zone individually and includes the dotted zone-demarcation areas as if they were actual transparent surfaces.

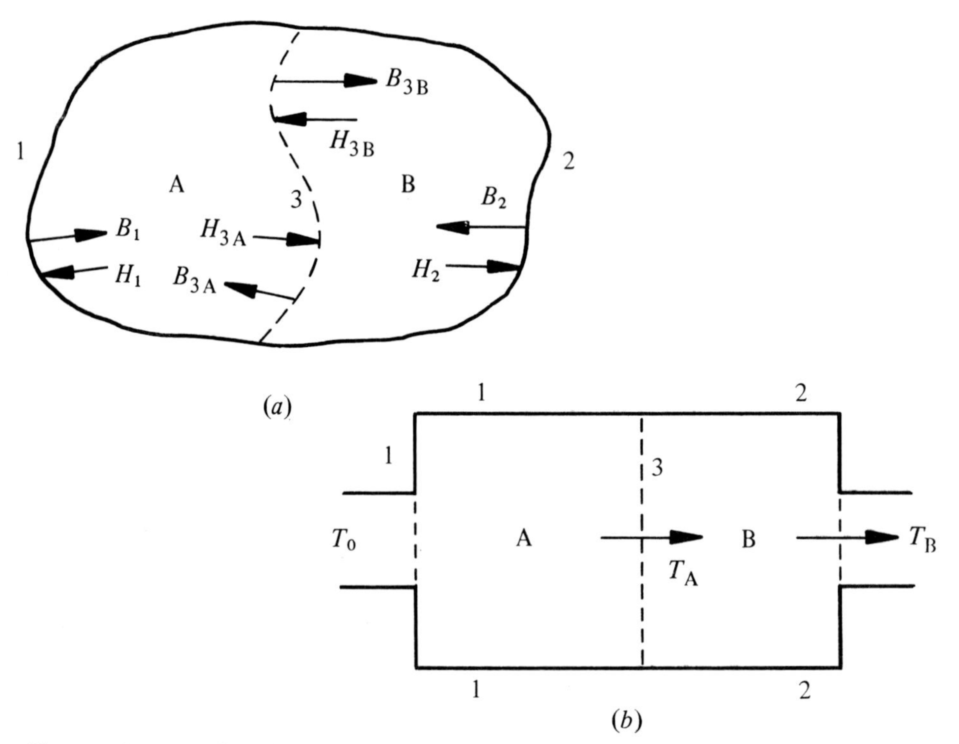

Figure 12:7 Radiant interchange between zones

To explain this in more detail, consider the simple two-zone system of Figure 12:7 (a). Equation 12.2 gives:

$$H_1 = \varepsilon_{A1} E_A + (1 - \alpha_{A11}) B_1 F_{11} + (1 - \alpha_{A31}) B_{3A} F_{13}$$
$$H_{3A} = \varepsilon_{A3} E_A + (1 - \alpha_{A13}) B_1 F_{31} + (1 - \alpha_{A33}) B_{3A} F_{3A\ 3A}$$

and a similar pair of equations for H_2 and H_{3B}.

Then equation 12.1 gives:

$$B_1 = \varepsilon_1 E_1 + (1 - \varepsilon_1) H_1$$
$$B_2 = \varepsilon_2 E_2 + (1 - \varepsilon_2) H_2$$
$$B_{3A} = \tau H_{3B} + \rho H_{3A} = H_{3B}$$
$$B_{3B} = \tau H_{3A} + \rho H_{3B} = H_{3A}$$

because the dotted zone 'walls' have zero emissivity and reflectance, but unity transmittance.

Thus there are eight equations and eight unknowns, namely B_1, B_2, B_{3A}, B_{3B}, H_1, H_2, H_{3A} and H_{3B} which can therefore be determined and used in equations 12.3 and 12.4 to yield Q_1, Q_2, Q_A and Q_B.

As a numerical example, consider the system shown in Figure 12:7(*b*):

$$F_{11} = F_{22} = 0.5, \quad F_{33} = 0, \quad F_{13} = F_{23} = 0.5$$
$$F_{31} = F_{32} = 1, \quad \varepsilon_A = \alpha_A = 0.2, \quad \varepsilon_B = \alpha_B = 0.3, \quad \varepsilon_1 = \varepsilon_2 = 0.7$$

(1) $H_1 = 0.2E_A + 0.8 \times 0.5B_1 + 0.5 \times 0.8B_{3A}$
(2) $H_{3A} = 0.2E_A + 0.8B_1$
(3) $H_2 = 0.3E_B + 0.7 \times 0.5B_2 + 0.7 \times 0.5B_{3B}$
(4) $H_{3B} = 0.3E_B + 0.7B_2$

Therefore:

(5) $B_1 = 0.7E_1 + 0.3(0.2E_A + 0.4\,B_1 + 0.4B_{3A})$
(6) $B_2 = 0.7E_2 + 0.3(0.3E_B + 0.35B_2 + 0.35B_{3B})$
(7) $B_{3A} = 0.3E_B + 0.7B_2$
(8) $B_{3B} = 0.2E_A + 0.8B_1$

Using (7) and (8) in (5) and (6) enables the latter to be solved for B_1 and B_2, and by setting all but one of the values of E to zero the total interchange areas can be calculated.

For example, retaining only E_A yields:

$$B_1 = 0.071E_A, \quad B_2 = 0.0305E_A$$
$$B_{3A} = 0.0214E_A, \quad B_{3B} = 0.257E_A$$

Using these values in equation 12.3 gives:

$$Q_{A\to 1} = -\,Q_1 = 0.166A_1E_A = \overrightarrow{G_AS_1}E_A$$
$$Q_{A\to 2} = -\,Q_2 = 0.071A_1E_A = \overrightarrow{G_AS_2}E_A$$

and from equation 12.4:

$$Q_{A\to B} = -Q_B = (0.0092A_2 + 0.077A_3)E_A = \overrightarrow{G_AG_B}E_A$$

All the other total exchange areas may be similarly calculated.

12.3.7 The complete heat balance

This can be written for each zone surface and each portion of gas. For a gas zone the statement is:

(A) (enthalpy of gas entering) + (heat radiated to the gas)
+ (heat convected to the gas) + (heat liberated by combustion in zone)
− (enthalpy leaving) = (change in zone enthalpy)

This statement would also be true for any material passing through a zone. For example in a continuous strip annealing furnace it would apply to the metal moving through the zone, except, of course, for the combustion term.

For a stationary surface:
(B) (heat radiated to surface) + (heat convected to surface)
− (heat lost from the material) = (change of enthalpy of the material)

For example, statement A for the gas zone W in Figure 12:6 is:

$$mc_p(T_V - T_W) + (\overrightarrow{S_1G_W}E_1 + \overrightarrow{S_2G_W}E_2 + \ldots + \overrightarrow{S_{10}G_W}E_{10} + \overrightarrow{G_XG_W}E_X + \overrightarrow{G_YG_W}E_Y + \overrightarrow{G_ZG_W}E_Z) - (\overrightarrow{G_WS_1} + \overrightarrow{G_WS_2} + \ldots + \overrightarrow{G_WG_X} + \ldots)E_W + Q(\text{combustion}) + h_{W1}(T_1 - T_W)A_1 + h_{W7}(T_7 - T_W)A_7 = \mathrm{d}(Mc_pT_W)/\mathrm{d}t$$

where M is the mass of gas in the zone and m the mass flow rate of gas through the zone.

Statement B for surface 1 is:

$$\sum_1^{10} \overrightarrow{S_jS_1}E_j + \sum \overrightarrow{G_iS_1}E_i - \sum_1^{10} \overrightarrow{S_1S_j}E_1 - \sum \overrightarrow{S_1G_i}E_1$$

$$+ h_{W1}(T_W - T_1)A_1 - (\text{rate of heat conduction from 1 to 2}) - (\text{rate of heat loss from back of 1}) = \mathrm{d}(\text{heat stored in 1})/\mathrm{d}t \qquad (12.21)$$

Thus a full heat balance on a solid material will involve the unsteady-state equations dealt with in Section 3.2.1, and a numerical solution may involve zoning the solid. This is discussed further in Section 12.4.4.

The solutions of 12.20 and 12.21 for every zone and surface can in principle enable all temperatures to be calculated.

12.3.8 Convective heat-transfer coefficients

These are dealt with in Section 3.3. A great deal of experimental data exists, and correlations are given in such forms as

$$(Nu) = K(Re)^n (Pr)^m \qquad (12.22)$$

and

$$j = (St)(Pr)^{2/3} = K_1(Re)^n \qquad (12.23)$$

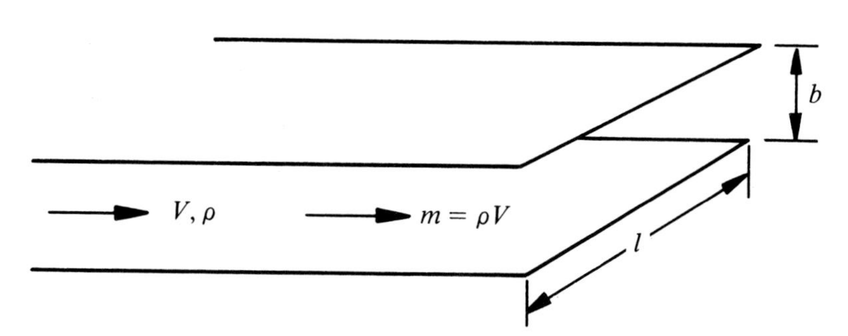

Figure 12:8 Convective heat transfer between gas and surface

For example, 3.19a can be written as:

$$hd/k = 0.023(Re)^{0.8}(Pr)^{0.4} \tag{12.24}$$

and is a reliable equation for flow through ducts of simple cross-sections such as circles and squares. d is the hydraulic mean depth. For other shapes correlations derived from model experiments have been used and will be quoted in later sections.

For combustion gases at furnace temperatures $(Pr)^{0.4} \approx 0.85$.

Referring to Figure 12:8

$$(Re) = (V/bl)(d\rho/\mu) = (m/l)(d/b\mu) \tag{12.25}$$

For example, if the mass flow rate of burner gases is m_g and there is a recirculation ratio of r_r, then $m = r_r m_g$, and if the duct is narrow relative to its length ($b < l$) then $d \approx 2b$, so that equation 12.24 becomes:

$$h = 0.01\,(k/b)(2r_r\,m_g/l\mu)^{0.8} \tag{12.26}$$

Kays and London [14] give a great many correlations for convective heat transfer and friction coefficients for flows through a variety of commonly encountered cross-sections.

12.4 THE 'WELL-STIRRED' ZERO-ORDER MODEL

A uniform temperature is promoted by recirculation of the gases in the furnace, either by the jet action of the burner, or by means of a fan in low-temperature applications. The effect is to eliminate very high gas temperatures close to the burner and to increase convective heat transfer because of the high velocity, although this will be opposed by a reduction in gas radiation due to the temperature reduction.

Furthermore, reducing the size of the furnace space increases the convective heat transfer (equation 12.26) and reduces radiative heat transfer from the gas by reducing the emissivity, as well as probably reducing the cost of the furnace structure.

Clearly there is an optimization problem in maximizing the heat-transfer rate to a given furnace load, or in minimizing the cost of heat transfer, whichever is the more appropriate criterion. It is not possible to derive optimal dimensions analytically since heat transfer is highly nonlinear with respect to temperature, flow rate and dimensions so that the designer is obliged to try various sets of conditions in order to find the best. Graphical presentation of heat-transfer coefficient in terms of design parameters such as gas temperature T_g, linear size b, gas flow per unit length of furnace m_g/l and recirculation ratio r_r can be of great assistance to the designer. Figure 12:9 is such a graph, in which h_c has been calculated from equation 12.26 and h_r, defined as $\sigma\varepsilon_g(T_g^4 - T_s^4)/(T_g - T_s) = \sigma\varepsilon_g(T_g^2 + T_s^2)(T_g + T_s)$, has been calculated for a path length of b in a flue gas corresponding to stoichiometric combustion of methane.

It will be seen that the smallest total heat transfer occurs in the region where $h_r = h_c$ and this might well define a furnace dimension which ought to be avoided.

Two types of furnace are used. In the batch-type furnace, the cold work is loaded,

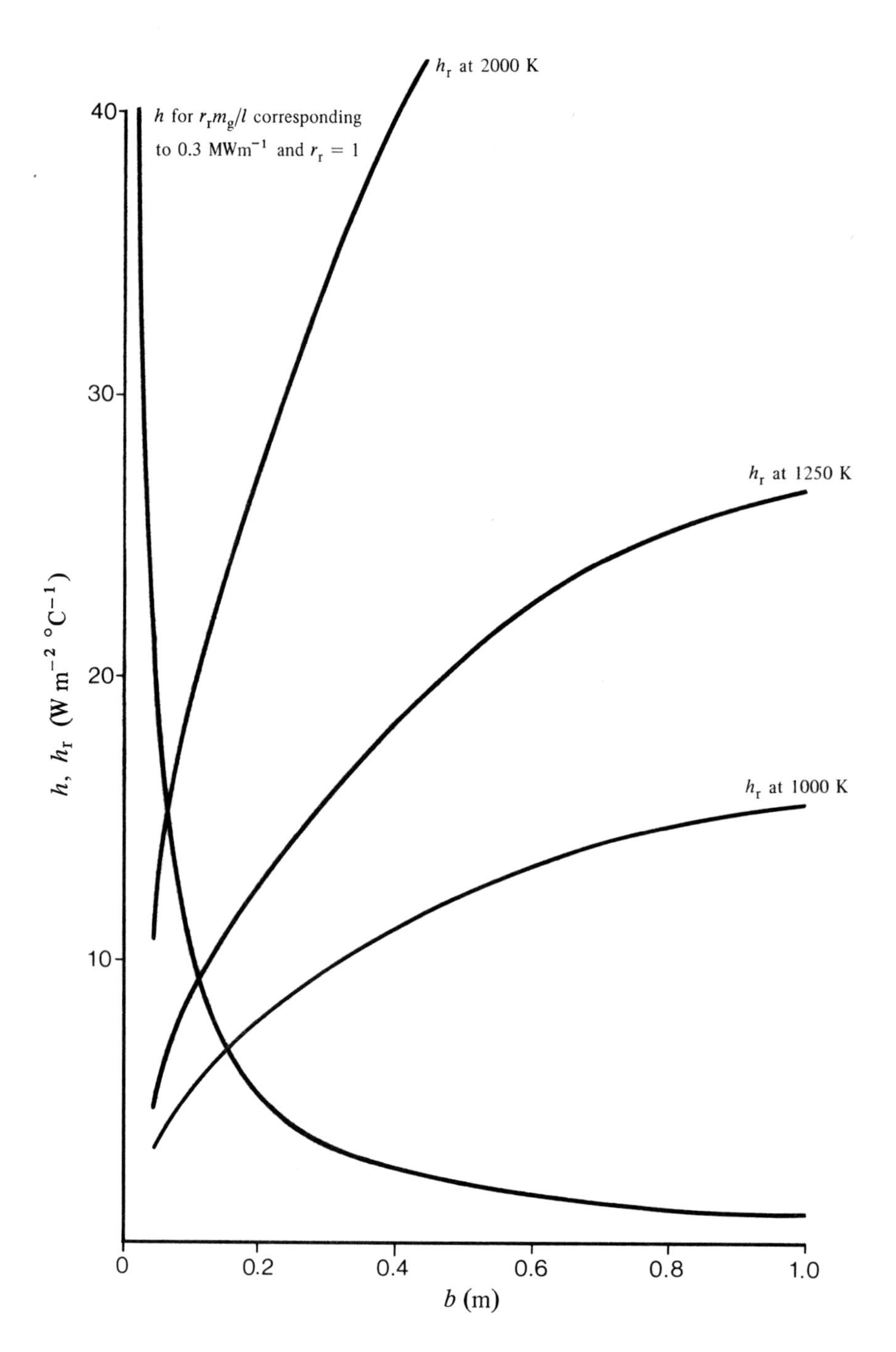

Figure 12:9 Dependence of convective and radiative heat-transfer coefficients on furnace dimension

heated to the required temperature, taken out and replaced by the next load, and this is clearly an unsteady-state process.

In the continuous type, material flows at a steady rate through the furnace and is heated to the desired exit temperature. The material may be a fluid flowing in a pipe, as in an oil or air heater, or it may be a solid such as a continuous metal strip or line of billets. This is normally a steady-state process.

There is one important difference in the design problem presented by the two cases. In the continuous furnace, if the throughput rate is decided, the designer has to determine the required heat-transfer area for the best heat-transfer coefficients he can contrive, whereas in the batch furnace the heat transfer area is fixed by the shape of the load and the designer has to try to maximize the heat-transfer coefficients, and it is this which then decides the throughput. Sometimes the rate of heat penetration into the load may be the limiting factor.

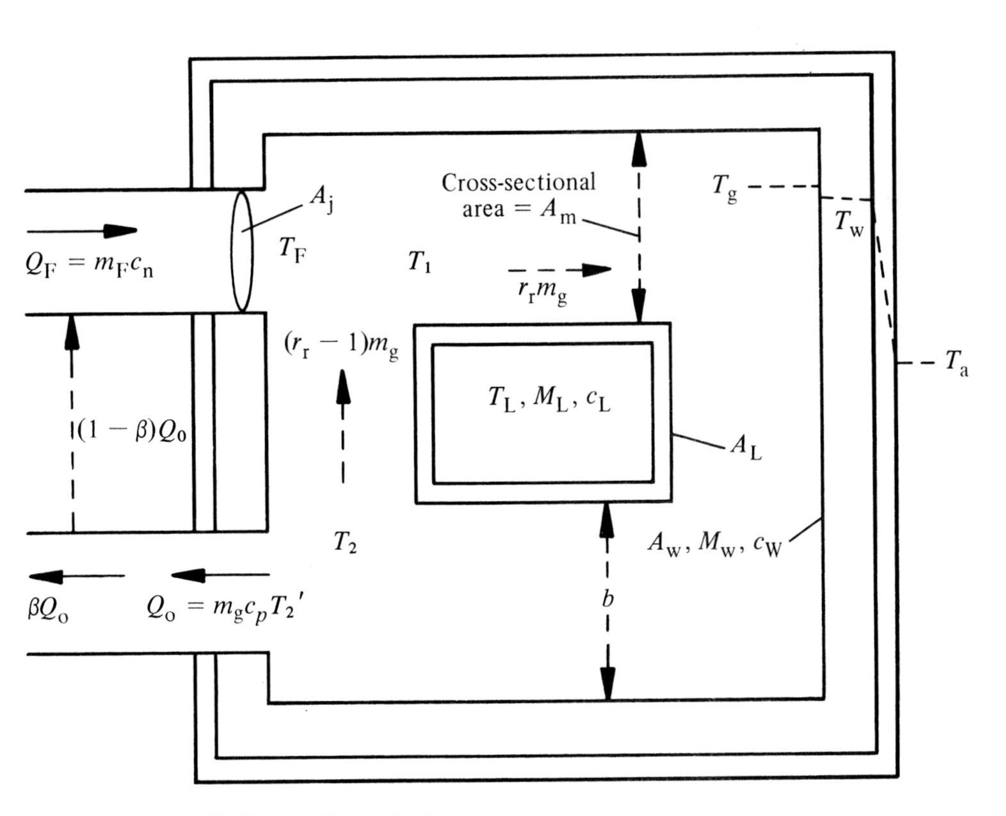

Figure 12:10 'Well-stirred' batch furnace

12.4.1 General approach to the design: batch furnace

Figure 12:10 shows the essential features of a batch furnace in which the load has a relatively thin section so that a uniform temperature T_L can be assumed. The refractory wall is assumed to have fairly high conductivity so that it can be assumed to have a fairly uniform average temperature T_W and it is backed with an insulating layer in which the major temperature gradient occurs. A fraction $(1 - \beta)$ of the heat in the flue gases is recovered, for example in a self-recuperative burner.

The recirculation ratio r_r is given by Francis [15, 16] as:

$$C_F = (2/r_r^2)(T_F/T_g)(A_m/A_j) - 2 \tag{12.27}$$

where $C_F u^2/2g$ is the total pressure loss in the flow circuit in which u is the average velocity, T_F is the temperature of the gases leaving the burners, and A_m and A_j are defined in Figure 12:10.

A heat balance on the whole system gives:

$$Q_g = Q_F - \beta Q_o = m_F c_n - \beta m_g c_p T_2' = m_F c_n - \beta m_g I_2 \tag{12.28}$$

where Q_g is the total rate of heat transfer from the gas, c_n is the net calorific value of the fuel per unit mass, and c_p is the mean specific heat capacity of the product gases.

A heat balance at the point where the burner gas and recirculated gas mix is:

$$m_F c_n + (1 - \beta) m_g I_2 + (r_r - 1) m_g I_2 = r_r m_g I_1 \tag{12.29}$$

For heat-transfer calculations an average furnace gas temperature T_g may be defined on the basis of average heat content:

$$I_g = \tfrac{1}{2}(I_1 + I_2) \tag{12.30}$$

Equations 12.29 and 12.30 assume complete combustion at mixing point 1. If combustion occurs beyond this point it does not affect I_g, provided the gas is completely burnt by the time it has reached the flue exit.

Eliminating I_1 between 12.28, 12.29 and 12.30 yields:

$$m_F c_n + (2r_r - \beta) m_g I_2 - 2m_g I_g = 0 \tag{12.31}$$

and

$$Q_g = 2r_r (m_F c_n - \beta m_g I_g)/(2r_r - \beta) \tag{12.32}$$

Heat-transfer balances on gas, load and wall yield:

$$Q_g = \overrightarrow{GW}\,\sigma T_g^4 - \overrightarrow{WG}\,\sigma T_w^4 + \overrightarrow{GL}\,\sigma T_g^4 - \overrightarrow{LG}\,\sigma T_L^4 + h_L A_L (T_g - T_L) + h_w A_w (T_g - T_w) \tag{12.33}$$

$$-Q_L = \overrightarrow{GL}\,\sigma T_g^4 - \overrightarrow{LG}\,\sigma T_L^4 + \overrightarrow{WL}\,\sigma T_w^4 - \overrightarrow{LW}\,\sigma T_L^4 + h_L A_L (T_g - T_L) = M_L c_L (\mathrm{d}T_L/\mathrm{d}t) \tag{12.34}$$

$$-Q_w = \overrightarrow{GW}\,\sigma T_g^4 - \overrightarrow{WG}\,\sigma T_w^4 + \overrightarrow{LW}\,\sigma T_L^4 - \overrightarrow{WL}\,\sigma T_w^4 - h_{wa} A_w (T_w - T_a) + h_w A_w (T_g - T_w) = M_w c_w (\mathrm{d}T_w/\mathrm{d}t) \tag{12.35}$$

For any given values of T_w and T_L, T_g may be found from equations 12.32 and 12.33, and this value can then be used in equations 12.34 and 12.35 to find the rate of heating of the load and wall.

A complete procedure might be shown in Figure 12:11.

1 Assume β and furnace dimensions A_m, A_j, d.
2 Assume a value for m_F and hence m_g.
3 Assume values for T_w and T_L.
4 Assume T_g.
5 Calculate r_r from equation 12.27. The calculation of T_F requires an assumption about the completeness of combustion at burner exit.
6 Calculate (Re) and hence h_w and h_L.
7 Calculate $\overrightarrow{GW}$, $\overrightarrow{WG}$, $\overrightarrow{GL}$, $\overrightarrow{LG}$, $\overrightarrow{WL}$ and $\overrightarrow{LW}$ from the geometry and T_g.
8 Calculate T_g from equations 12.32 and 12.33.
9 Hence I_2 from equation 12.31. If I_2 is too high (i.e. too much waste heat) then there is a heat-transfer 'bottleneck'. Can the heat transfer coefficients be increased?
If not *reduce* m_F i.e. go to 2.
If I_2 is unnecessarily low m_F may be increased.
Otherwise:
10 Recalculate the heat-transfer coefficients using the new value of T_g, i.e. go to 4.
11 Calculate dT_L/dt from equation 12.34.
12 Calculate dT_w/dt from equation 12.35. (Are there limitations to be imposed on dT_L/dt and dT_w/dt for reasons of thermal stress?) If the load has been put into a preheated furnace, dT_w/dt should not be positive. If it is then m_F is too high. Therefore reduce m_F i.e. go to 2.
Otherwise:
13 Calculate T_L after a time Δt.
14 Calculate T_w after a time Δt.
15 Go to 3.

Figure 12:11 Design procedure for a batch furnace

12.4.2 Continuous furnace

For the continuous type of furnace, running under steady conditions, the right-hand side of equation 12.34 is replaced by $m_L c_L(T_{L2} - T_{L1})$, where m_L is the mass flow rate of material through the furnace, and T_{L1} and T_{L2} are the entry and exit temperatures of the material. Also, the right-hand side of equation 12.35 becomes zero.

Furnace efficiency can be defined as:

$$\eta = Q_g/m_F\, c_n = 1 - \beta\,(m_g/m_F)\,(I_2/c_n) \qquad (12.36)$$

(This, of course, includes the heat losses through the walls.)

The design procedure might then be as follows:

1 For a given throughput Q_g is estimated.
2 Assuming η and β use equation 12.36 to calculated m_F (and thereby m_g) and T_2
3 Hence T_g and r_r from equations 12.27 and 12.31
4 Calculate heat-transfer coefficients and total exchange areas

5 From equations 12.34 and 12.35 calculate T_w and A_L, probably by assuming $T_L = \frac{1}{2}(T_{L1} + T_{L2})$

This assumption is only valid if the change $T_{L2} - T_{L1}$ is small relative to the difference between furnace and load temperatures, as for example in continuous heating of liquid. Otherwise zoning of the load is required, and this is more properly dealt with in the next section on the long furnace. Continuous furnaces are rarely 'zero order'.

12.4.3 Rate of cooling during shutdown periods

During shutdown, the furnace contains no load or radiating gas so that equation 12.35 becomes, if it is assumed that there are no flue losses:

$$(M_w c_w / h_{wa} A_w)(dT_w/dt) + T_w = T_a \tag{12.37}$$

the solution of which, for a constant T_a, is:

$$(T_w - T_a) = (T_{w1} - T_a)(1 - e^{-t/\tau}) \tag{12.38}$$

where T_{w1} is the wall temperature at the beginning of the shutdown period, and $\tau = M_w c_w / h_{wa} A_w$ is the *time constant* for the wall.

It follows that the total heat lost during a shutdown is:

$$Q_w(\text{shutdown}) = M_w c_w (T_{w1} - T_a) e^{-t/\tau} \tag{12.39}$$

12.4.4 Rate of heat penetration into a solid

The assumption of uniform temperature in load and walls made in Section 12.4.1 is at best an approximation and is quite inaccurate for thick sections of low conductivity.

Unsteady heat conduction into a solid is represented by equation 3.11, the one-dimensional form of which is equation 3.11a:

$$\alpha(\partial^2 T/\partial x^2) - (\partial T/\partial t) = 0 \tag{12.40}$$

The two-dimensional form, in polar coordinates, is:

$$\alpha[(\partial^2 T/\partial r^2) + r^{-1}(\partial T/\partial r)] - (\partial T/\partial t) = 0 \tag{12.41}$$

where α is the thermal diffusivity, $\lambda/\rho c$.

Equations 12.40 and 12.41 must be solved by numerical methods for all but the simplest boundary conditions.

Lucas *et al.* [17, 18] use finite-difference forms for the partial derivatives and favour an implicit method of solution which is stable for large time intervals.

An alternative method might be to use a 'lumped-parameter' model of the solid, that is, one in which the solid is divided into zones and the resistance to heat flow is assumed to occur at the zone boundaries. This will give a set of ordinary differential equations. For example, Figure 12:12 represents a furnace wall. For section 1:

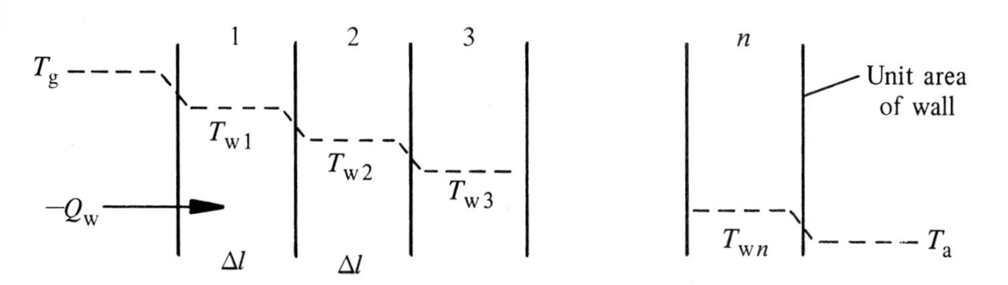

Figure 12:12 'Lumped parameter' approximation to unsteady heat flow in a solid

$$\left.\begin{aligned} & -Q_w - (\lambda_1/\Delta l)(T_{w1} - T_{w2}) = \rho c_1 \Delta l \,(\mathrm{d}T_{w1}/\mathrm{d}t) \\ & \text{(where } Q_w \text{ can be calculated from equation 12.35) and, for section 2:} \\ & (\lambda_1/\Delta l)(T_{w1} - T_{w2}) - (\lambda_2/\Delta l)(T_{w2} - T_{w3}) = \rho c_2 \Delta l \,(\mathrm{d}T_{w2}/\mathrm{d}t) \end{aligned}\right\} \quad (12.42)$$

and so on.

This approach is often favoured by control engineers who are used to handling ordinary differential equations.

12.4.5 Rapid billet-heating furnace

Experimental work on both batch and continuous rapid billet-heating furnaces is described by Lawrence and Spittle [19] and the mathematical modelling of the batch furnace is given in references [17, 18, 20, 21].

Rapid heating of billets is desirable to minimize hydrogen or sulphur contamination of nonferrous billets and oxidation of ferrous billets. The thermal conductivity of nonferrous billets is so high that the billet temperature is practically uniform, but this is not so for ferrous billets where heat penetration rate into the metal may be the limiting factor in heat transfer.

In Lawrence and Spittle's experiments on a batch furnace, the billets were of circular cross-section, and the furnace, also of circular section, was fired by several tangential type-C tunnel burners. Furnace diameters of 7 in, 10 in and 13 in (178 mm, 254 mm and 330 mm) were used with billets having a range of diameters from $3\frac{1}{2}$ in to $10\frac{1}{2}$ in (89 mm to 267 mm). Heat inputs ranged between 1 and 5 therms per hour per foot length of furnace (0.1 and 0.5 MW per metre length).

Rates of rise of billet surface and core temperatures were measured for aluminium and copper billets. No other temperatures were measured but simple heat balances for gas, billet and wall, similar in principle to equations 12.33, 12.34 and 12.35, were set up and enabled the three quantities T_w, T_g and h to be calculated, assuming that $h_L = h_w = h$.

The emissivities of the aluminium and copper were 0.15 and 0.6 respectively. This resulted in marked differences in the relative importance of different modes of heat transfer as summarized in Figure 12:13 which gives the extremes of the range.

Since in most cases convection is considerably greater than gas radiation, the efficiency increases as the convection increases, as mentioned in Section 12.4 and Figure 12:9. The second result given for copper shows that the lowest efficiency tends to occur when gas radiation and convection are equally important.

Lawrence and Spittle [19] outline a design procedure for this type of furnace on

Billet diameter (in)	Furnace diameter (in)	Gas rate (Mw/m)	Efficiency (%)	Percentage heat transfer received by billet			h	
				Wall radiation	Gas radiation	Convection	(Btu h^{-1} ft^{-2} °F^{-1})	(W m^{-2} °C^{-1})
Aluminium								
$3\frac{1}{2}$	10	0.13	6.7	53.9	18.9	27.2	1.3	7.4
$3\frac{1}{2}$	10	0.43	6.3	50.8	6.3	42.9	6.4	36
$8\frac{1}{2}$	10	0.13	22.3	39.8	3.8	56.4	4.7	27
$8\frac{1}{2}$	10	0.43	18.8	31.4	1.5	67.1	14.6	83
Copper								
5	10	0.16	10.5	38.9	41.8	19.3	1.8	10
5	10	0.33	9.6	49.6	22.1	28.3	5.0	28
5	10	0.38	9.8	51.1	18.8	30.1	6.3	36
7	10	0.2	15.3	45.3	20.4	34.6	4.9	28
7	10	0.38	13.9	47.6	12.5	39.9	9.1	52

Figure 12:13 Heat transfer in a billet-heating furnace

the lines of that given in Section 12.4.1, but rather more specific in that they suggest a maximum heating rate which can usefully be used.

The continuous furnace in Lawrence and Spittle's experiment was of a pusher type, of circular cross-section with tangential burners in the forcing zone and with the gases then flowing counter-current to the billets in the preheating zone. This type of furnace gives higher throughputs and is desirable for materials of low thermal conductivity since the lower heating rates will produce smaller temperature gradients. Very much higher efficiencies, in the range 27 to 50 per cent, were obtained with the continuous furnace.

The mathematical model [17] which was subsequently developed uses the methods of Section 12.4.1. The gas emissivity is based on a 'one grey one clear gas' model (see Section 12.3) as defined by equation 4-102 of reference [4], in which the real-gas emissivity is accurately represented over two path lengths (see also [22]).

The convective heat-transfer coefficients h_L and h_w were determined from a mass-transfer analogue (Section 12.8.2) which showed the correlations:

$$\left.\begin{aligned} j_L &= 0.525(Re)_L^{-0.5} \\ j_w &= 1.13(Re)_w^{-0.5} \end{aligned}\right\} \qquad (12.43)$$

where j is the Colburn j factor and the values of (Re) were based on the diameters of billet and wall respectively.

Flow-visualization studies using a water model were used to determine r_r.

Initially it was assumed that the wall temperature was constant during the heating of a billet. The predictions of the mathematical model were tested against the actual performance of the furnace described in [19], and reasonable agreement was found in most cases once correct values had been chosen for billet emissivities and wall losses.

The flow-visualization studies had shown that in the single top flue furnace some of the gases from the upper burner pass straight out of the flue without participating in the recirculation, thus resulting in values of r_r less than 1. Further flow-visualization studies of two-flue and self-recuperative burner designs showed values of r_r up to about 2.8 and the mathematical model was used to predict the performance of these designs. This showed, for example, that the self-recuperative burner design would heat a billet twice as quickly as the original top flue design.

Further applications in [18] include the effect of variations in wall temperature and show that by using a dense cast inner refractory backed with a refractory brick, the temperature variations can be confined to the dense material.

This paper also presents the heating time (and hence efficiency) as a function of the furnace diameter for a given billet size (of diameter 159 mm). It shows that the efficiency is a minimum at a particular furnace diameter (250 mm) and that equal efficiencies can be obtained, for example, with 39 per cent convection, 3 per cent gas radiation, 58 per cent wall radiation, or with 5.5 per cent convection, 24 per cent gas radiation, 70.5 per cent wall radiation. These two cases correspond respectively to furnace diameters of 172 mm and 760 mm.

The model is also used to predict the effect of a temporary stoppage or shutdown.

12.4.6 Low-temperature oven with recirculation

In this type of appliance a fan is often used to promote circulation, and the mode of heat transfer is almost entirely by forced convection. In this case an analytical solution of the batch-heating problem is possible if the conductivity of the load is such as to give a uniform load temperature.

Eliminating Q_g between equations 12.32 and 12.33 yields an expression for T_g' (i.e. I_g/c_p) in terms of T_L' and this can then be substituted into equation 12.34 to give:

$$\tau_L(\mathrm{d}T_L'/\mathrm{d}t) + T_L = Q_F/m_g c_p \tag{12.44}$$

where τ_L (the load time constant) is given by:

$$\tau_L = \left(\frac{M_L\, c_L}{m_g\, c_p}\right)\left(\frac{2r_r - 1}{2r_r}\right)\left[1 + \left(\frac{2r_r}{2r_r - 1}\right)\left(\frac{m_g\, c_p}{h_L\, A_L}\right)\right] \tag{12.45}$$

assuming no heat losses through the wall and $\beta = 1$, i.e. no waste-heat recovery. h_L will, of course, be dependent on r_r, as for example in equation 12.26.

The solution of equation 12.44, assuming a constant fuel heat input Q_F is:

$$T_L' = Q_F\,(1 - \mathrm{e}^{-t/\tau_L})/m_g\, c_p \tag{12.46}$$

assuming that $T_L' = 0$ when first placed in the oven.

It has been assumed in the above treatment that the oven walls do not change in temperature.

For many types of load it will be difficult to evaluate h_L, for example, in the case of randomly packed components of irregular shape, where the velocity of the gases over the load may be uncertain.

The final value of T_L' is $Q_F/m_g\, c_p$, which is the adiabatic combustion products temperature, which in this type of application would be kept at a low value by having a large m_g, i.e. a large amount of excess air.

12.4.7 High-temperature muffle furnace

In this case the load is heated only by radiation from the muffle walls, since it may be assumed that the muffle is filled with nonradiating gas (e.g. air). The muffle walls are, of course, heated from the outside.

If again it is assumed that there is rapid heat penetration into the load, and also that load and walls are 'grey', equation 12.34 becomes:

$$\overline{\mathrm{WL}}\,\sigma\,(T_w^{\,4} - T_L^{\,4}) = M_L\, c_L\,(\mathrm{d}T_L/\mathrm{d}t) \tag{12.47}$$

and equation 12.35 becomes:

$$h_w\, A_w\,(T_g - T_w) - \overline{\mathrm{WL}}\sigma\,(T_w^{\,4} - T_L^{\,4}) = M_w\, c_w\,(\mathrm{d}T_w/\mathrm{d}t) \tag{12.48}$$

where the gas is on the outside of the muffle wall.

In practice the heat transfer to the muffle from the combustion chamber will be very little affected by T_w, so the first term of equation 12.48 can be replaced by a constant Q_g.

This pair of simultaneous equations would have to be solved by numerical methods, but in the case where the wall temperature is assumed to remain constant, equation 12.47 can be integrated directly to give:

$$t = (M_L c_L/\sigma \overline{WL} T_w^{\ 3}) \left[\tfrac{1}{2}\arctan(T_L T_w^{\ -1}) + \tfrac{1}{4}\ln \{(1 + T_L T_w^{\ -1})/(1 - T_L T_w^{\ -1})\}\right]_{T_{L1}}^{T_{L2}} \tag{12.49}$$

The radiant interchange area $\overline{WL}$ is found by the method used in section 12.3.5, but, of course, with $\varepsilon_g = \alpha_g = 0$.

For example, in the case where $A_w = 2A_L$, $\varepsilon_L = 0.5$, $\varepsilon_w = 0.8$, and the load cannot 'see itself', then $F_{LL} = 0$, $F_{Lw} = 0.5$, $F_{ww} = 0.5$, and it will be found that $\overline{WL} = 0.47A_L$.

As a numerical example, suppose these data related to a load of flat vitreous enamelled sheet plus perretts (supports) weighing 180 kg and having an area of 9.3 m^2. The mean specific heat capacity is 0.2 Btu/1b °F = 840 J/kg °C. The load at 293 K is inserted into the muffle whose walls are at 1273 K. How long will it take for the load to reach 800 °C (1093 K)?

Putting the relevant values into equation 12.49 gives:

$$t = \left(\frac{180 \times 840}{5.7 \times 10^{-8} \times 0.47 \times 9.3 \times (1273)^3}\right) \times$$

$$\left(\tfrac{1}{2}\arctan\frac{1093}{1273} - \tfrac{1}{2}\arctan\frac{293}{1273} + \tfrac{1}{4}\ln\frac{1 + (1093/1273)}{1 + (293/1273)} + \tfrac{1}{4}\ln\frac{1 - (1093/1273)}{1 - (293/1273)}\right) \text{ seconds}$$

$$= (0.29 \times 10^3)(0.242 + 0.103 + 0.425) \text{ seconds}$$
$$= 223 \text{ seconds}$$
$$= 3.72 \text{ minutes}$$

which is a perfectly acceptable figure for this duty.

The heat required is:

$$180 \times 840 \times 800/223 \text{ J s}^{-1} = 540 \text{ kW (18.4 therm/h)}$$

The actual requirement would be rather less than this because the complete time per load including loading and unloading would be about 300 seconds.

Thus the application would require a constant heat input of

$$540 \times 223/300 \text{ kW} = 400 \text{ kW (13.7 therm/h)}$$

An estimate of the effect of the cooling of the wall can now be made by using a heat balance on the wall over the estimated heating cycle.

Suppose the wall weighs 2000 kg and has a specific heat capacity of 1000 J kg^{-1} $°C^{-1}$ (0.24 Btu/lb °F). Then equation 12.48 gives, for the load heating period already calculated:

$$2000 \times 1000 \times \Delta T_w = 223(400\text{–}540) \times 1000$$

whence $\Delta T_w = -15.6$ °C.

So at the end of the heating period the wall temperature falls to 1257.4 K. The heating time can now be recalculated from equation 12.49 using the average wall temperature over the heating period, $T_w = 1265.2$ K, to obtain a better estimate of the heating time, and the iteration may be repeated if necessary until the procedure converges to a constant value.

12.5 ONE-DIMENSIONAL 'LONG-FURNACE' MODEL

This is appropriate in equipment where the combustion products travel through a duct which is long compared with its diameter so that gas temperature is not uniform in the direction of flow, although it is assumed uniform over any cross-section.

Internal recirculation may occur within the furnace and also external recirculation may be used whereby some of the gases from a cooler section in the furnace are led back to a hotter section. Both types of recirculation increase convective heat transfer by increasing the gas velocity, but the external recirculation also has the effect of reducing the hot-end temperature and the temperature gradient along the furnace.

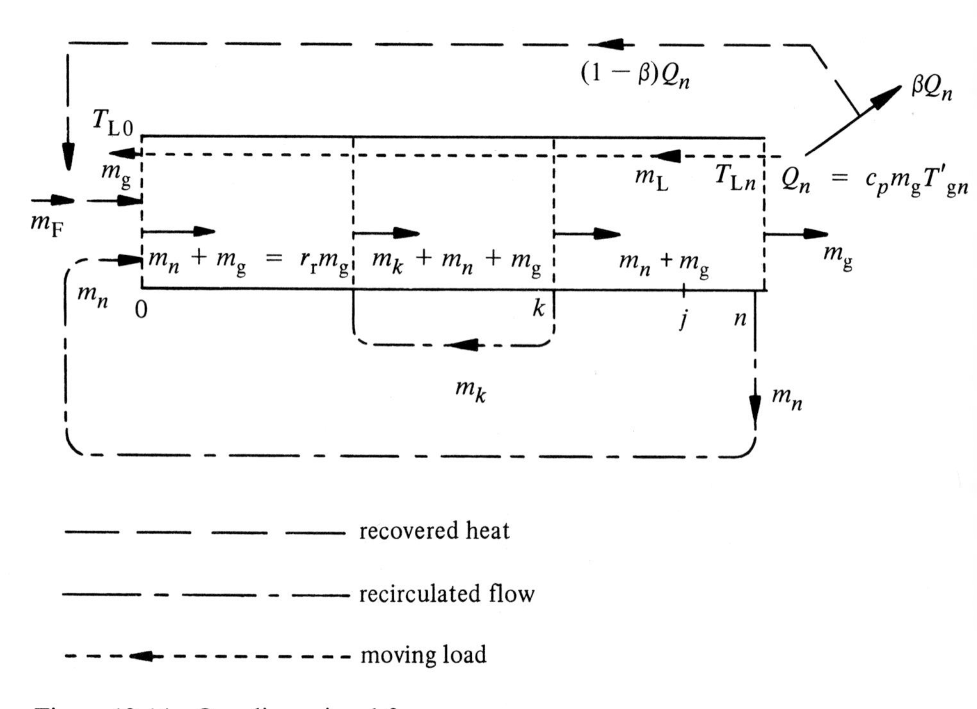

Figure 12:14 One-dimensional furnace

Figure 12:14 shows a rather general system of this kind in which a continuous stream of material is moving counter-current to the furnace gases. However, in one important type of application the useful heat passes through the furnace wall to a fluid as in the case of the shell boiler and the immersion-tube heater. Another example

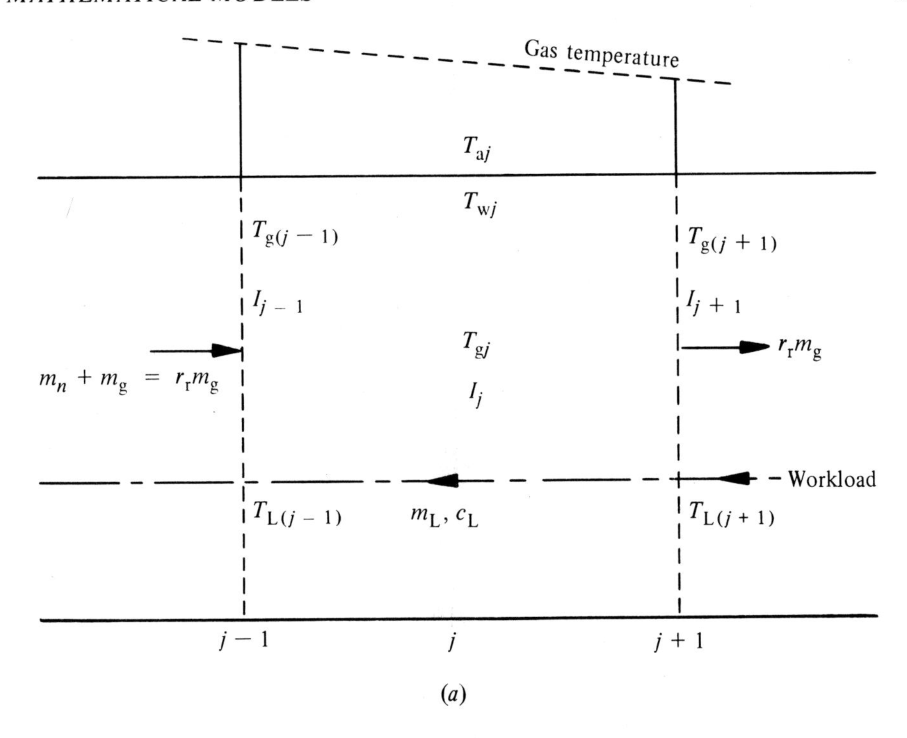

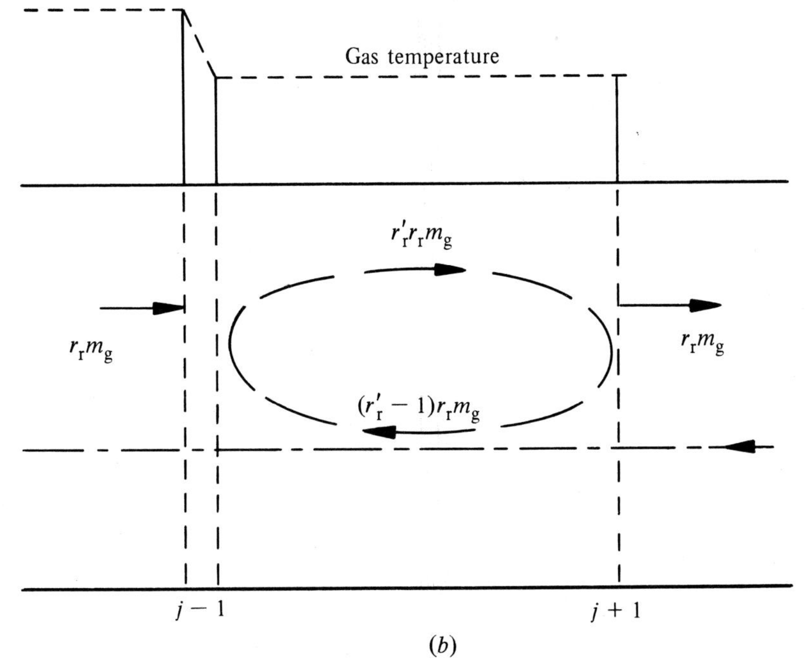

Figure 12:15 (*a*) 'Plug-flow' zone
(*b*) 'Well-stirred' zone

is the single-ended radiant tube in which the tube is bent back on itself so that flue and burner are at the same end.

The term 'plug-flow' model is used to describe one in which there is no internal recirculation, and Figure 12:15(*a*) shows a short length of such a furnace in which the gas temperature changes gradually over the whole furnace length.

If recirculation occurs within the section as is shown in Figure 12:15(*b*) then the temperature changes more abruptly at the place where the recirculation begins, and the section then approximates to a 'well-stirred' model. In this case the average temperature in the section approximates to the exit temperature from the section.

The heat-balance equations 12.50 to 12.53 are the same for both models since internal recirculation cannot affect heat or mass conservation, but the heat-transfer coefficients are affected by the amount of internal recirculation.

Both forms of recirculation will affect the rate of combustion (see Section 12.8).

A heat balance on the gas in section j is:

$$r_r m_g (I_{j-1} - I_{j+1}) - \Delta Q_F - \Delta Q_g = 0 \tag{12.50}$$

where ΔQ_F is the heat produced by combustion in the section and ΔQ_g is the net heat leaving the gas by radiation and convection in the section.

A heat balance on the load in the section is:

$$m_L c (T_{L(j+1)} - T_{L(j-1)}) - \Delta Q_L = 0 \tag{12.51}$$

and for the wall, where h_{wa} may be radiative as well as convective:

$$h_{wa} A_w (T_{wj} - T_{aj}) - \Delta Q_w = 0 \tag{12.52}$$

A heat balance on the gas at the entrance to the furnace (section 0) is:

$$r_r m_g I_0 = m_g(r_r - 1) I_n + \varphi m_F c_n + m_g (1 - \beta) I_n \tag{12.53}$$

where φ is the fraction of the fuel burnt up to section 0.

An efficiency may be defined as in equation 12.36, with I_n replacing I_2.

The heat transfer rates ΔQ_g, ΔQ_L and ΔQ_w are of course calculated as in equations 12.33 to 12.35 but without the unsteady-state terms. For the 'plug-flow' model, the convective portion would be calculated using the mid-section temperatures T_{gj}, T_{Lj} and T_{wj}, and these can reasonably be taken as the average of T_{j+1} and T_{j-1}. This can also be done for the radiative heat transfer as long as radiant interchange between zones can be ignored, and this may be the case if the zone length is relatively large compared with its cross-sectional dimension. For the 'well-stirred' model element the average gas temperature in the section may be taken as $T_{g(j+1)}$.

The design procedure is clearly going to be more complex than that for the 'well-stirred' model, and the following points can only hope to provide a broad outline.

1 The duty of the furnace will be specified, and this, together with an assumed efficiency and heat recovery factor, will determine the fuel input m_F and the exit gas temperature T_{gn}' from equation 12.36. In some applications, as for example the immersion-tube heating of liquid in a tank, both the heating-up rate and the working rate will need to be considered.

2 If there is a limitation on the maximum temperature of the furnace wall T_{w0}, this can be used to decide the recirculation ratio r_r. This may occur when the heat transfer is through the wall, and where there is a low value of h_{wa} (equation 12.52) as in the radiant tube. In this case, the use of T_{w0} (max) in equation 12.52 for the first furnace zone 0–2 fixes ΔQ_w and hence T_{g0}, and hence r_r from equation 12.53.
3 With this knowledge of hot and cold end temperatures and gas flow rate, average heat-transfer coefficients can be estimated for assumed furnace cross-sections and thus an estimate made of total heat-transfer area and hence required furnace length. For example, Trinks [23] states that a logarithmic mean temperature difference as defined in equation 3.23 of Section 3.3.5 has been used for this purpose.

Having thus arrived at an estimate of the main dimensions these can be used in equations 12.50 to 12.52, which can be solved, using a small number of zones in the first instance, for the zone temperatures. Probably the most convenient technique would be to try different zone areas until the correct boundary conditions for the load and gas are arrived at. The use of a computer is obviously very desirable, although a relatively straightforward iterative procedure using a hand-calculator is possible.

For example, starting with zone 012 (i.e. $j = 1$) the known values of T_{g0} and T_{L0} can be used instead of T_{g1} and T_{L1} in equation 12.52 which can then be solved directly for T_{w1}.

This first estimate of T_{w1} is then used in equations 12.50 and 12.51, again replacing T_{g1} and T_{L1} by T_{g0} and T_{L0}, to give estimated values of T_{g2} and T_{L2}.

These values then give a closer estimate of T_{g1} and T_{L1} which can be used in equations 12.50 and 12.51 for a still closer estimate of T_{g2} and T_{L1}. This explicit scheme could, of course, be used instead of the more usual implicit computer method.

If the principal mode of heat transfer in the furnace is either convective or radiative alone, then the calculation becomes much simpler, and mathematical integration may be possible as in the case of the muffle furnace dealt with in Section 12.4.7.

An example of this type is dealt with next.

12.5.1 Continuous radiant-tube furnace

In this type of furnace, shown schematically in Figure 12:16, the load is heated almost entirely by radiation from radiant tubes above and below the load. The furnace may be filled with an inert atmosphere having practically no emissivity and only a small amount of heat is transferred by natural convection.

In calculating the radiant interchange it is very convenient to use the zone method to replace the radiant tubes by an equivalent grey surface at the same temperature as the tubes.

Figure 12:17 shows an enlarged view of the tubes with the refractory wall behind and the imaginary plane in front of them. The refractory can reasonably be regarded as 'adiabatic', since almost no heat is lost by conduction through it or convection from it, and, as no heat is generated within it, the only radiation it emits is that which it has reflected from the tubes or absorbed and re-emitted. Since it clearly does not matter which of these two mechanisms occurs, the refractory can be assumed to have zero emissivity and unit reflectivity. This is quite an accurate and very convenient assumption in furnace calculations where radiation predominates.

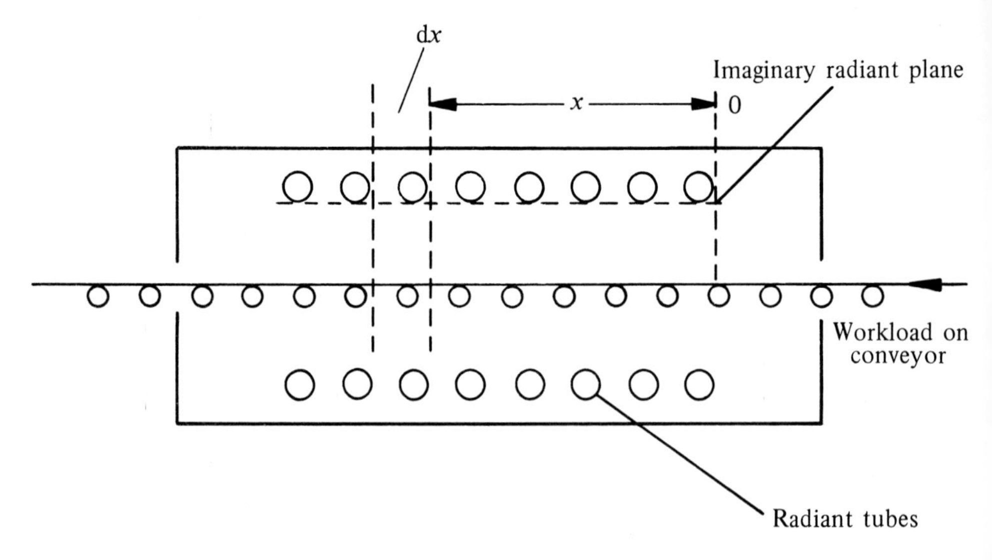

Figure 12:16 Continuous radiant furnace

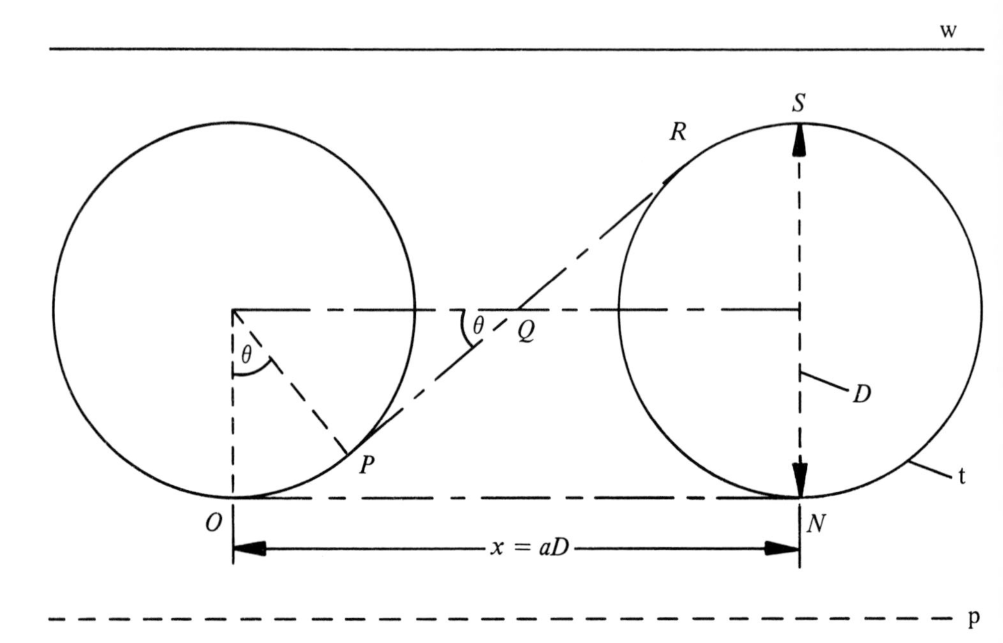

Figure 12:17 Radiant interchange between tubes, furnace wall and load. Equivalent radiant plane

Using equations 3.38 and 3.39:

$$B_w = H_w$$
$$B_t = \varepsilon_t E_t + \rho_t H_t$$
$$H_w = B_t F_{wt}$$
$$H_t = B_w F_{tw} + B_t F_{tt}$$

and

$$H_p = B_w F_{pw} + B_t F_{pt}$$

Solving for B_w and B_t gives:

$$H_p = \{(F_{pt} + F_{pw} F_{wt})\varepsilon_t/[1 - \rho_t (F_{tw} F_{wt} + F_{tt})]\} E_t \qquad (12.54)$$
$$= \varepsilon_p E_t$$

Since $Q_p = H_p A_p = \varepsilon_p A_p \sigma T_t^4$ the quantity in curly brackets in equation 12.54 is the emissivity of the plane ε_p.

The various view factors are related as follows:

$$\left.\begin{aligned}
& F_{tt} + F_{tw} + F_{tp} = 1 \\
\text{Therefore:}\quad & F_{tw} = F_{tp} \quad \text{(symmetry)} \\
& \quad = \tfrac{1}{2}(1 - F_{tt}) \\
& F_{wt} = F_{tw} A_t/A_w = F_{tw}\,\pi D/x = F_{tw}\,\pi/a \\
& F_{wt} + F_{wp} = 1 \\
\text{Therefore:}\quad & F_{wp} = 1 - F_{wt} \\
\text{and, by symmetry,}\quad & F_{pw} = F_{wp} \\
& F_{pt} = F_{wt}
\end{aligned}\right\} \qquad (12.55)$$

The value of F_{tt} is found by the method of Section 12.3.3. From equation 12.14 it follows that:

$$F_{tt} = \tfrac{1}{2}(OPQRS - ON)\,\pi D \qquad (12.56)$$

where $ON = x = aD$ = tube-centre spacing and, by geometry, $PR = \sqrt{(x^2 - D^2)}$ and $OP = \frac{1}{2}D\theta$ where $\theta = \arcsin(D/x)$. Hence

$$F_{tt} = 2[\sqrt{(a^2 - 1)} - a + \theta]/\pi \qquad (12.57)$$

Values of ε_p as a function of tube spacing for different values of tube emissivity are tabulated in Figure 12:18.

a	ε_p		
	$\varepsilon_t = 1$	$\varepsilon_t = 0.8$	$\varepsilon_t = 0.6$
1	1	0.93	0.83
1.5	0.97	0.87	0.74
2	0.88	0.78	0.64
3	0.72	0.61	0.49
4	0.59	0.50	0.39
8	0.34	0.28	0.22

Figure 12:18 ε_p as a function of tube spacing for various values of ε_t

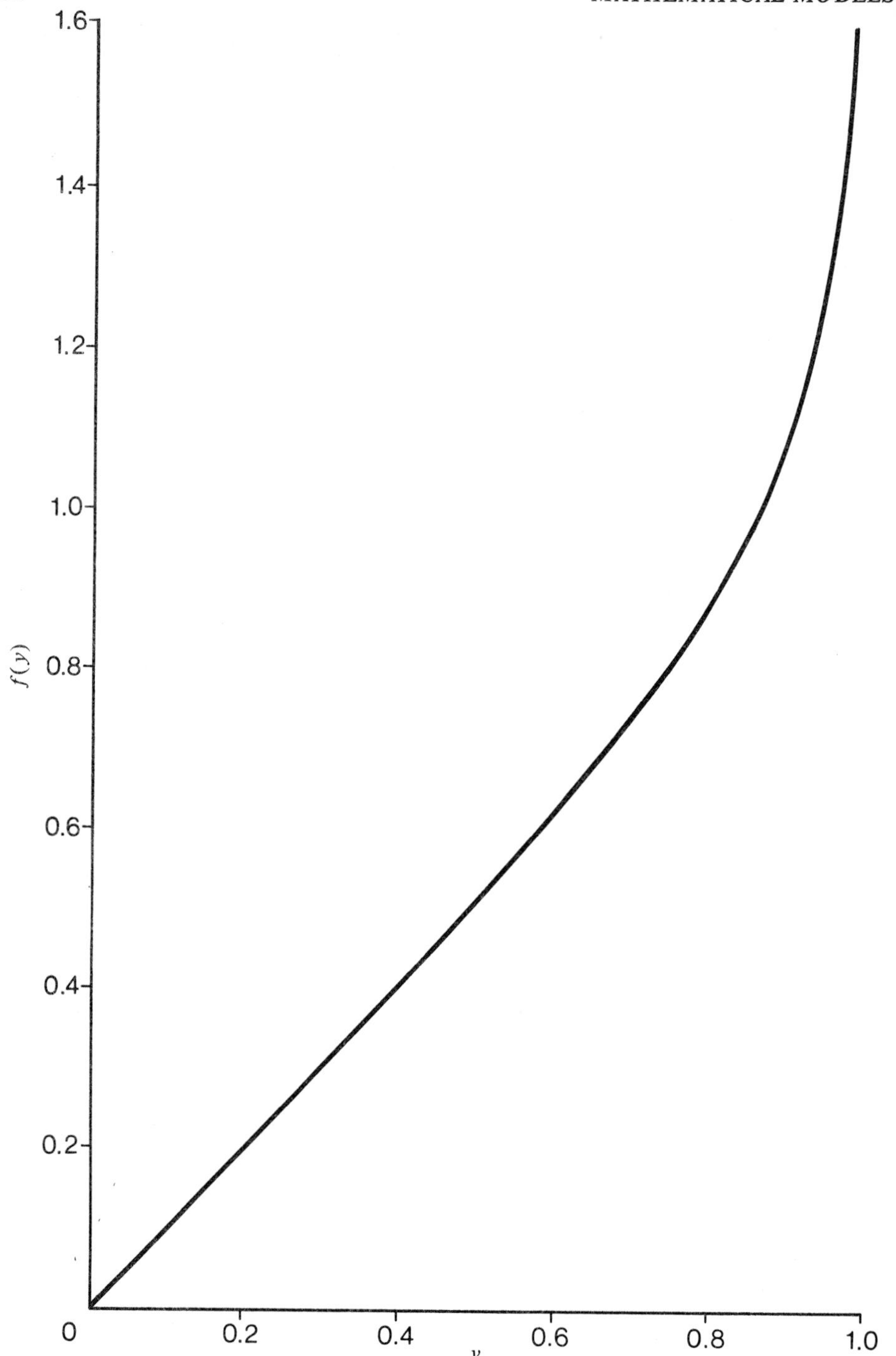

Figure 12:19 Area required for load to achieve a temperature of $y \times$ radiant tube temperature

The rate of heat transfer to the load depends on the geometry of the load and the furnace interior and on load emissivities, which may vary with load temperature.

As a first approximation it can be assumed that the load and radiant plane are parallel planes, sufficiently close together so that any area on the load has a unity view factor to the radiant plane (i.e. $F_{Lp} \approx 1$ for any part of the load). For such a configuration it is easy to show by the methods of Section 12.3 (i.e. 3.4.3) that the radiation entering a portion of load of area ΔA_L is given by:

$$-\Delta Q_L = \sigma \varepsilon_F \Delta A_L (T_p^4 - T_L^4) \quad (12.58)$$

where:

$$\varepsilon_F = 1/(\varepsilon_p^{-1} + \varepsilon_L^{-1} - 1) \quad (12.59)$$

A heat balance on the load passing through the region dx in Figure 12:16 is:

$$mc\,(\mathrm{d}T/\mathrm{d}x) = \varepsilon_F \sigma b (T_p^4 - T_L^4)$$

where m is the mass of load passing through per second, c is the specific heat capacity and b is the breadth of the load.

This may be integrated over a length x to give:

$$bx = (mc/\varepsilon_F \sigma T_p^3)\,[f(r_{Tx}) - f(r_{TO})] \quad (12.60)$$

where $f(y) = \frac{1}{2}\arctan y + \frac{1}{4}\ln[(1+y)/(1-y)]$, $r_{Tx} = T_L/T_p$ at the position x and $r_{TO} = T_{LO}/T_p$ at the entry to the radiant section.

The function f has been plotted in Figure 12:19 and can now be used in a numerical example.

The required duty is to heat 4.5 kg s^{-1} of metal of specific heat capacity 830 J kg^{-1} °C^{-1} (0.2 Btu lb^{-1} °F^{-1}) from 300 K to 1173 K, the radiant tubes being at 1273 K. The width of the load is 3 m.

What is the length of the radiant plane and how many radiant tubes are required?

Assuming no losses, the heat required is:

$$4.5 \times 830 \times (1173 - 300) = 3.26 \text{ MW} = 112 \text{ therms per hour}$$

The value of $r_{TO} = 300/1273 = 0.235$, and $r_{Tx} = 1173/1273 = 0.92$, so from Figure 12:19, $f(r_{Tx}) - f(r_{TO}) = 0.96$. Using this in equation 12.60 gives:

$$x = \frac{4.5 \times 830}{\varepsilon_F \times 5.7 \times 10^{-8} \times 1273^3 \times 3} \times 0{\cdot}96 \text{ m}$$

$$= \frac{10.2}{\varepsilon_F} \text{ m}$$

ε_F can now be calculated for any value of load emissivity ε_L, tube emissivity ε_t and (tube spacing/tube diameter) a, using equation 12.59 and the tabulated data for ε_p. The results are tabulated in Figure 12:20 for two load emissivities. The length of the radiant section is $\frac{1}{2}x$ since heating is from both sides. The number of tubes assumes them to be 3 metres long and 150 mm diameter. In practice one might use two $1\frac{1}{2}$ m tubes from opposite sides of the furnace.

If the load is not flat but can 'see itself', then an equivalent plane emissivity can be calculated as for the tubes.

a	ε_p	ε_F	x (m)	Length of radiant section (m)	Number of tubes 3 m long, 0.15 m diameter	Average Q per m^2 of tube (kW)
$\varepsilon_L = 0.5$		$\varepsilon_t = 0.8$				
1.5	0.868	0.465	21.9	10.95	100	23
2	0.775	0.437	23.3	11.65	80	28.8
3	0.61	0.38	26.8	13.4	62	37.2
8	0.28	0.22	46.4	23.2	40	57.7
$\varepsilon_L = 0.8$		$\varepsilon_t = 0.8$				
1.5	0.868	0.713	14.3	7.15	66	34.9
2	0.775	0.65	15.7	7.85	54	42.7
3	0.61	0.53	19.2	9.6	44	52.4
8	0.28	0.26	39.2	19.6	34	67.8

Figure 12:20 Design parameters for a radiant-tube furnace with various tube spacings and load emissivities

The tabulated furnace dimensions will be underestimates since there are furnace losses to be accounted for, and the radiant interchange between load and tubes will be less than assumed because the view factor between load and radiant plane is less than one, and the conveyor system will reduce radiation falling onto the load.

However, the results bring out the following points:

1. The load emissivity has a pronounced effect on the furnace length required.
2. There may be an opportunity for optimizing the tube spacing since an increase in spacing, and hence furnace length, is accompanied by a reduction in the number of tubes required, since they then radiate more effectively.
3. The limiting factor in the design may be the maximum heat input into the tubes. This may be decided by the maximum size of commercial burner which can be accommodated in the tube. Other relevant factors are the reduction in tube efficiency concomitant with increased input, and the increase in internal temperatures which are likely to reduce tube life. Chatwin *et al.* [24] give experimental data on these aspects of tube performance. In practice, a radiant flux from the tube of 45 kW/m^2 (14 500 Btu/hr ft^2) is not unusual, and over twice this amount has been used in applications where it has been possible to remove heat at this rate from the tube.

If tubes are too closely spaced ($a \lessapprox 2$) local 'hot spots' can develop where the tubes face each other and the back wall, because of the low tube conductivity in a circumferential direction. Uneven heating at the load plane is likely if the tubes are closer than about $\frac{1}{2}$ m to the load, and if a is too large.

It is important that safe tube temperatures should not be exceeded and this requires close control of the gas input to the radiant tubes. The tabulated radiant output per unit tube surface area is an average, being higher at the cool end than at the hot end of the furnace. An exact calculation of the heat radiated from each part of the radiant plane would necessitate zoning the radiant section of the furnace to obtain the radiant interchange areas between each part of the radiant plane and load, but an approximation is that:

$$\frac{Q_t \text{ (cold end)}}{Q_t \text{ (hot end)}} \approx \frac{T_p^4 - T_L^4 \text{ (cold)}}{T_p^4 - T_L^4 \text{ (hot)}}$$

which has a value of about 3.5 in the above example.

The increased radiant flux requirement at the cold end may be met by closer spacing of the tubes at that end, and in this case the analysis would involve using radiant plane zones of different emissivities.

It is very difficult to make an accurate estimate of the heat transfer by natural convection which will occur mainly to the underside of the load.

Assuming that the gas temperature is midway between that of the tubes and the average of the load, then the magnitude of the Grashof number (see Section 3.3.2) is about 3.2×10^6, which from equation 3.17 gives $h = 0.47\,(3.2 \times 10^{-6})^{0.25}\, k/d$.

Hence $h \approx 1$ Btu/ft² h °F, and this will result in a convective heat transfer from the lower tubes of about 1000 Btu/ft² h (3.13 kW/m²) which is about 5 per cent of the total radiant heat transfer. This could be increased considerably by the addition of forced convection using fans to circulate the atmosphere.

12.5.2 Immersion-tube heating of aqueous solutions

Experimental data on this application are found in references [25–8], the last of these giving a systematic design procedure. No recirculation is used, and since the tube is water cooled its temperature can be considered uniform. The liquid-side heat-transfer

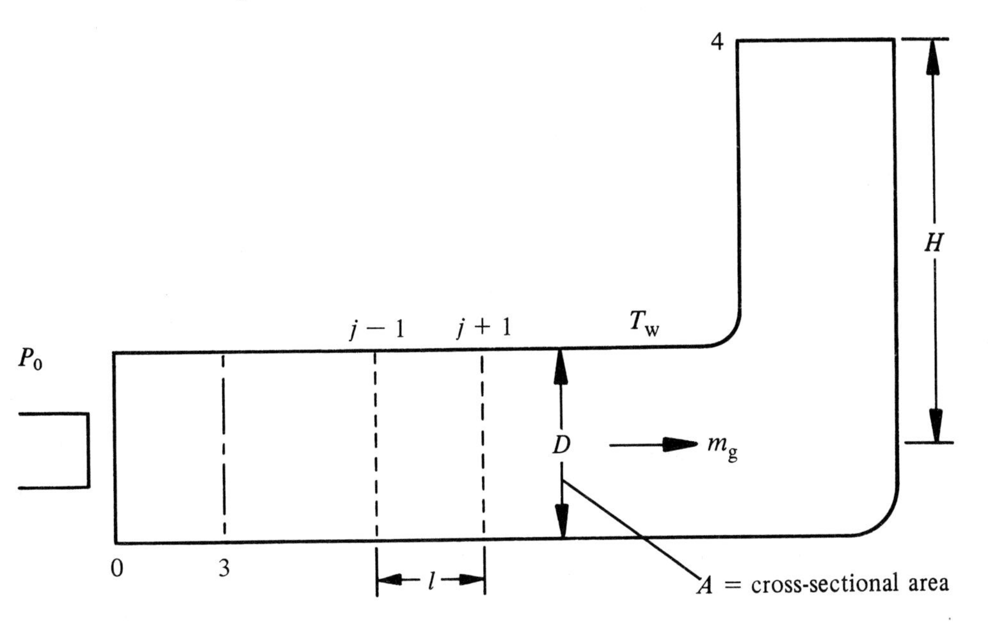

Figure 12:21 Immersion tube heater

coefficient is large compared with the gas side so that only the latter has to be calculated.

The system is shown in Figure 12:21. The flow of gases, and hence the heat transfer, is governed by the chimney height and temperature, so that heat-transfer calculations are needed for various proportions of excess air and exit-gas temperature, and for each of these the required chimney height may be calculated.

Radiant interchange along the tube may be neglected and an assumption needs to be made about the progress of combustion along the tube. This depends on the degree of primary aeration and the type of burner. It is simplest to assume complete combustion close to the tube entrance and that the 'plug-flow' model applies.

A steady-state heat balance on the gas in the tube element l (equation 12.50) is:

$$\begin{aligned} m_g(I_{j-1} - I_{j+1}) - h\pi Dl\,(T_{gj} - T_{wj}) \\ - \tfrac{1}{2}\sigma\pi Dl\,(\varepsilon_w + 1)\,(\varepsilon_g T_{gj}^4 - \alpha_{gw} T_w^4) = 0 \end{aligned} \tag{12.61}$$

where equation 3.42 has been used to evaluate the radiant heat transfer. Equation 12.15 could equally well be used.

Equation 12.24 might be used to predict the convective heat-transfer coefficient, but would not be very accurate in the following numerical example where the Reynolds number is low (< 5000). Instead, data are taken from [14] in which (Nu) is plotted as a function of distance from tube entrance for both laminar and turbulent conditions.

Gas emissivity data have been taken from [22] using a path length equal to the tube diameter. The same reference also supplies data on other gas properties.

Data

Tube diameter	= 0.1 m (4 in)
Heat input	= 33.5 kW (115 000 Btu/h)
Excess air	= 20%
Tube temperature	= 100 °C throughout (373 K)
Tube emissivity	= 1

T_w^4 is everywhere small compared with T_g^4 so that the term $\alpha_{gw} T_w^4$ can safely be ignored.

The required flow of methane is:

$$0.625 \times 10^{-3}\ \text{kg/s} = 0.87 \times 10^{-3}\ \text{m}^3/\text{s}$$

and the resulting products flow:

$$m_g = 13.2 \times 10^{-3}\ \text{kg/s}$$

The flame temperature is 2039 K.

The Reynolds number ranges from 2600 at the tube entry temperatures to about 6000 at the likely tube-exit temperature. Data on convective heat transfer are scarce in this near-transition region. For both laminar and turbulent flows (Nu) is greater at the tube-entry region, i.e. for small values of D/L, than in the region of fully developed flow, and (Nu) also increases as (Re) increases. The effect of these factors in the present case is that (Nu) will have a value of about 20 at the tube entry, falling to about 10 at the middle of the tube and rising again to about 15 at the tube exit.

For tube sections 1 m long, equation 12.61 becomes:

$$I_{j+1} = I_{j-1} - 23.6h(T_{gj} - T_{wj}) - 1.35\varepsilon_g (T_{gi}/1000)^4 \quad (12.62)$$

where I is in MJ/kg and h in MJ/m^2 s K.

The next step is to apply equation 12.62 to the first tube section 0–2, using an assumed value for T_{g1}. One could assume this to be T_{g0}, but convergence will obviously be more rapid the more accurate the first guess is, so take a value of $T_{g1} = 1600$ K. Then $h_1 = 20 \times 0.11 \times 10^{-5} = 22 \times 10^{-6}$, $\varepsilon_g = 0.058$, $T_{g1} - T_w = 1227$ $I_2 = 2.37 - 23.6 \times 22 \times 10^{-6} \times 1227 - 1.35 \times 0.058 \times (1.6)^4 = 2.37 - 0.64 - 0.51 = 1.22$. Therefore $T_{g2} = 1290$ K, so that the next approximation to $T_{g1} = \frac{1}{2}(2039 + 1290) = 1665$ K.

Then $h = 20 \times 0.114 \times 10^{-5} = 23 \times 10^{-6}$, $\varepsilon_g = 0.055$, $T_{g1} - T_w = 1292$, $I_2 = 2.37 - 0.70 - 0.57 = 1.1$. Therefore $T_{g2} = 1200$ K, so that the next approximation to T_{g1} is 1620 K.

The next iteration converges to $T_2 = 1240$ K, $T_1 = 1640$ K.

The next section can be dealt with in the same way and the continued process gives the temperature changes along the tube. The results are tabulated in Figure 12:22.

L (m)	0	1	2	3	4	5
T_g (K)	2039	1240	950	820	740	690
η based on gross input						
Calculated	0	0.48	0.62	0.68	0.72	—
Experimental [28]	0	—	0.62	0.65	0.68	—

Figure 12:22 Temperature and efficiency at various points in an immersion-tube heater

Direct comparison of the calculated efficiency and the experimental figures quoted by Robertshaw is not possible since the latter are averaged and do not refer to a particular thermal input or excess air. Nevertheless it is clear that there is reasonable agreement and it might be inferred that the convective heat-transfer coefficient at the cooler end of the tube has been overestimated.

Chimney height requirement. To calculate this it is necessary to apply the chimney-draught equation 10.17:

$$(m_g/A)(u_4 - u_3) + (P_4 - P_3) + \rho_4 gH + (4fm_g/2DA)\sum u_j l + \Delta P_b = 0 \quad (12.63)$$

where ΔP_b represents pressure losses due to bends, etc. (the suffixes refer to the numbers on Figure 12:21 which agree with those on Figure 10:7). The temperature in the vertical part of the flue is assumed uniform, so ρ_4 is constant in this section.

From equation 2.7:

$$P_4 = P_0 - \rho_a gH \quad (12.64)$$

where ρ_a is the density of the ambient air.

To find the value of P_3 the momentum equation 10.2 has to be applied over the section 0–2 and to do this accurately requires knowledge of the gas flows and velocities through the burner. If, for example, it is assumed that 40 per cent of the total gases issue from the burner where they entrain the remainder of the air and that the velocity of the gases leaving the burner is the same as the velocity at 3 then equation 10.2 becomes:

$$P_0A + 0.4\,mu_3 = P_3A + mu_3 \tag{12.65}$$

assuming that the secondary air contributes no axial momentum.

Equations 12.64 and 12.65 are used to eliminate P_3 and P_4 from 12.63. Then substitute $u_j = V_g T_j / A T_s$ and $m_g = \rho_g V_g$ where ρ_g and V_g are, respectively, the density and volume flow rate at the standard temperature T_s. Then dividing through by $m_g V_g / A^2$, and noting that $\rho_g = \rho_a$ almost exactly yields the convenient form:

$$\frac{T_4 - 0.4T_3}{T_s} - \frac{A^2 g}{V_g^2}\left(1 - \frac{T_a}{T_4}\right)H + 2\frac{f}{D}\sum \frac{T_j}{T_s}\, l + \tfrac{1}{2}\frac{T_4}{T_s} = 0 \tag{12.66}$$

where the bend loss has been taken as $\frac{1}{2}m_g u_4/A$, which is one kinetic head.

Equation 12.66 can be evaluated for any given flue temperature, flow rate and pipe length to give the required flue height H. For example, suppose $T_4 = 690$ K; total tube length = 8 m; length of horizontal tube = 5 m; $f = 0.007$; and the other data are as used previously. Then equation 12.66 becomes:

$$\frac{690 - 0.4 \times 2039}{288} - 4.86\left(1 - \frac{288}{690}\right)H +$$

$$\frac{2 \times 0.007}{0.1}\left[\frac{1640 + 1095 + 885 + 780 + 715 + (3 \times 690)}{288}\right] + \tfrac{1}{2}\frac{690}{288} = 0$$

whence $H = 1.5$ m.

This is a realistic estimate since [28] quotes a primary flue height of 5 ft (1.52 m) as typical.

12.5.3 Shell boiler

This appliance is described in Section 14.2.1. Many industrial boilers of this type previously fired by oil now use natural gas and this may result in a change in the heat-transfer distribution in the boiler because of the change in flame luminosity. For this reason detailed mathematical modelling has been carried out and is described in references [21, 29, 30].

The model used for the heat transfer in the boiler fire tube is essentially one-dimensional and plug flow and therefore uses equations 12.50, 12.52, 12.33, 12.34 and 12.35, but the following special features have been included:

1 In the entrance section of the tube the jet of combustion gases causes internal recirculation in a length of about 1.3 diameters so that this portion of the fire tube was represented by a series of well-stirred elements.

2 In some of the computations combustion was assumed to be complete in the well-stirred region: in others, experimentally obtained information on the progress of combustion was incorporated into the model.
3 Radiation from luminous carbon particles is included in the model.
4 Because there will in practice be scale and soot on the outer and inner tube surfaces, the heat transfer to the water includes the conductivity of the tube as well as the water-side heat-transfer coefficient which was assumed to be constant at 5.7 $kW/m^2\ K$ (1000 $Btu/h\ ft^2\ °F$).

The nonluminous radiation was computed on the basis of a 'one grey, one clear gas' model (Section 12.3.1 and reference [22]).

The convective heat-transfer coefficient was difficult to evaluate correctly. It is known that equation 12.24 which relates to fully developed flow in a tube considerably underestimates the convective heat transfer in the entry region of a tube. Experimental work by other workers, on heat transfer in the region of a sudden expansion and from a jet expanding into a tube showed that in these circumstances the heat-transfer coefficient was increased to as much as four times that given by equation 12.24.

Even with the use of these data it was found that the calculated heat flow from both oil and gas flames to the fire tube was an underestimate, so heat-transfer coefficients were determined using mass-transfer models and also from the data of Wu [31] who has made direct measurement of heat transfer to a cylindrical water cooled furnace by both gas and oil flames. This showed up to a tenfold increase in the heat-transfer coefficient over that predicted by equation 12.24. A comparison of the model calculations with experimental results showed also that axial radiative heat transfer is considerable and that for the case of a long flame (zero swirl) the most accurate model requires a well-stirred region for three-quarters of the length, while for a burner with a swirl number of 0.84 only about one-eighth of the fire-tube length is well-stirred.

The main purpose of the model was to predict gas temperatures in the reversal chamber, and tube-end temperatures. The reversal chamber is treated as a well-stirred region within which convective heat transfer is given by equation 12.24 in which (Re) and (Nu) are based on reversal-chamber diameter and it is necessary to consider cases where the chamber walls are water cooled or refractory backed.

Convective heat transfer in the tube ends is given by:

$$(Nu) = 0.3\,(Re)^{0.8}\,(Pr)^{0.33}\,e^{0.318D/x} \tag{12.67}$$

where D = tube diameter and x = distance along tube.

Accurate results from the model can only be obtained when correct values for the thermal resistance of end plate and chamber walls has been found experimentally, but once this has been done the model can predict the effect of changing the type of flame, fuel rate and excess air.

Results from the model computations showed that the gas temperature at the fire-tube exit is about 15 per cent higher for natural-gas firing compared with oil firing, but that the tube-end temperatures are only about 6 per cent higher for natural-gas firing. Model predictions are found to agree very closely with experimental results, particularly for the case of gas firing.

Lockwood [32] uses a simplified model of a packaged boiler in order to determine the design parameters which produce a given boiler output at the least total cost. His model boiler comprises a fan-driven burner, a well-stirred combustion chamber in which radiative heat transfer only occurs, followed by tubes in which only convection takes place, followed by an induced-draught fan. Heat transfer is increased as combustion-chamber size increases and as tube frictional resistance increases and this frictional resistance determines the fan size required. Hence capital and running costs can be expressed in terms of combustor and tube diameters and in terms of the temperatures of the gases leaving these two parts of the boiler. A computer program is used to find the optimal design.

12.5.4 Coiled-tube air heater

In this heater the burner fires into a central cylindrical chamber comprising a refractory wall lined with parallel air tubes in an axial direction. The gases then pass into an outer annulus in which the air tubes form a multistart helical coil. The air passes counter-current to the combustion gases in both sections.

The mathematical model is described in [33] and uses a well-stirred central combustion chamber, followed by a plug-flow annular section. The emissivity of the refractory-backed tubes is found by the method of Section 12.5.1 and standard data are used for heat transfer by convection both inside and outside the tubes. The model has been used to predict the length of the tube required to achieve various efficiencies for a given duty. The quoted results, the salient features of which are tabulated in Figure 12:23, show the possibilities of optimizing the design by the correct choice of tube lengths in the two zones. In each case the duty was 1.185 kg/s of air at 650 °C.

Tube surface area (m^2) (high-temperature zone)	19.9	10.9	8.42	7.24	6.52
Overall heat-transfer coefficient (W/m^2 °C)	34.8	41.9	44.8	41.9	41.2
Tube surface area (m^2) (low-temperature zone)	31.7	18.2	13.35	11.5	10.35
Overall heat-transfer coefficient (W/m^2 °C)	22.9	27.6	32.8	37.3	40.0
Efficiency (%)	85.4	73.1	64.0	57.0	51.2

Figure 12:23 Tube lengths for various efficiencies in a coiled-tube air heater

The tubes in the inner combustor zone appear to be more effective in transferring heat, presumably because of the increased radiative contribution, so that, like the case of the packaged boiler and as mentioned in Section 12.4, there should be an optimal distribution of the two modes of heat transfer in a given space.

12.6 TWO- AND THREE-DIMENSIONAL MODELS

This section gives a necessarily brief account of a few published model studies of various furnaces. There is a need by furnace designers of methods for predicting heat-flux distribution in furnaces and the effect thereon of such variables as changes in type of fuel, in length and luminosity of flame and the number and positioning of burners. Even simplified models can provide this type of comparative information and often a mathematical model which is easy to use can be of more benefit to the designer than a more accurate one requiring lengthy computation.

Radiation is the main heat-transfer mode in most industrial furnaces, and the majority of the models are essentially radiative heat-transfer models which either ignore convection or represent it by a simple empirical equation. The zone method of analysis is usually used and the assessment of luminous radiation is often important.

In most of the models, combustion and flow pattern are input data, so that the effect of changes in these factors is readily found from the model computation.

All the models use, in various forms, equations 12.20 and 12.21 and the methods of Section 12.3.7.

12.6.1 Gas-fired pusher-reheating furnace [34]

This is essentially a long continuous furnace fired by longitudinal burners which in the case studied in reference [34] were at one end of the furnace only. The model was used to support the results of experiments on two pilot-scale furnaces. One of these was completely rectangular with two symmetrically placed burners at one end; the other was of a conventional design having a sloping roof at one end, and asymmetrically placed burners at the other. For the first an essentially one-dimensional model was used with a well-stirred burner region followed by a plug-flow section. The second required a two-dimensional model having upper and lower zones, the former containing the burners. For this case the flow pattern was obtained using a cold-flow visualization technique. The convective heat-transfer coefficient was taken as $h = 1 + 2.7\, m_g/A$, where A is the furnace cross-sectional area [23], and a finite-difference computation scheme for the heat conduction into the billets was included. Only non-luminous flames were considered and the computed and experimental results were in good agreement except in the region where combustion is occurring.

12.6.2 Reheating furnace [35]

Reference [35] describes the modelling of the general type of pusher-reheating furnace which may have several sets of longitudinally firing burners placed along the furnace, which thus consists essentially of a number of simple furnaces as described in Section 12.6.1 placed in series. A two-dimensional model is used, and convective heat transfer is ignored. A simple grey gas is assumed and the flame is represented by a set of elliptical equal-temperature and K (equation 12.5) contours.

The computed billet temperatures agree well with measurements on production furnaces.

12.6.3 Rotary kilns [36]

In a rotary kiln, feed material (usually in the form of granular solids) tumbles steadily down the sloping kiln counter-current to the hot-gas flow. Because gas velocities are low, convective heat transfer is small and is neglected in Pearce's model, but heat transfer to the charge is complex because of the way the hot particles on the outside of the burden are continually being mixed with the cooler interior particles.

The model is essentially of a one-dimensional plug-flow furnace with a well-mixed zone at the burner end occupying a length of about three kiln diameters, together with a mixing model for the moving burden. In this the particles are assumed to be carried through a certain angle by the rotating kiln before slumping to a new position during which a new sector of the burden surface is exposed to radiation from gas and walls. A further complication is the fact that in many applications a heat of reaction is involved, as for example in the endothermic decomposition of limestone and dolomite.

12.6.4 Modelling of an experimental furnace [37]

Reference [37] compares the calculated temperatures and heat-flux distributions with those measured in an experimental furnace which is water cooled. Although the actual furnace is square in cross-section, a cylindrical model was used which allowed an axisymmetric zoning system to be used with the furnace divided into annular coaxial gas zones. The refractory wall is partly covered by water tubes so that effective view factors for wall and tubes have to be calculated.

The nonluminous radiation is modelled by a 'clear plus grey gases' model and luminous radiation is assessed using equation 12.6. Flow and combustion patterns are assumed from experimental data.

12.6.5 Water-tube boiler [38]

Modern water-tube boilers use very high radiative heat fluxes of $> 500\ \mathrm{kW/m^2}$, and there is a danger of tube failure because of local overheating due to non-nucleate boiling. There is therefore a need for predictive methods of calculation of heat flux. The model study was of a boiler converted from coal to oil firing and having eight opposed burners on each side wall. Convection is negligible and flame and wall emissivities were obtained by comparing readings from optical and total-radiation pyrometers on the actual furnace (see [5] for details of this method).

12.6.6 Petroleum heater

Thorneycroft and Thring [39] have used a three-dimensional model of an actual heater which uses three oil burners in a rectangular furnace (38 ft $\times$ 15 ft $\times$ 13 ft; 11.6 m $\times$ 4.6 m $\times$ 4.0 m) forming part of a reboiler unit. Cubical zones of side 4.55 ft (1.387 m) were used, and a 'grey' gas was assumed. This assumption is stated to be quite accurate if the refractory emissivity is > 0.4. Convective heat transfer is included in the model. Reasonable agreement between model calculations and actual performance is obtained; for example for an input of 40×10^6 Btu/h (11.7 MW) actual heat transferred to the oil was 22.6×10^6 Btu/h (6.62 MW) compared with 20.2×10^6 Btu/h (5.92 MW) calculated.

12.6.7 General furnace study

This is a very comprehensive three-dimensional model developed by Yardley and Patrick [40] which is used to investigate the influence of furnace design variables such as type of flame, furnace size, disposition of the load and load temperature, on furnace performance. The model uses 40 surface zones, 20 gas zones and 4 flame zones, and assumes black walls and load and a grey gas. A single value for h is used throughout.

The input variables used were:

1 Type of flame, i.e. emissivity, heat-release pattern, recirculation and air–gas ratio.

2 Size of furnace.
3 Area of load relative to furnace area.
4 Load temperature (280 °C and 980 °C).
5 Heat input per unit load area, which was taken as 100 000 Btu/h ft^2 (315 kW/m^2) in all cases.

From a study of the results the following main conclusions emerge:

1 Thermal efficiency is little affected by type of flame.
2 Thermal efficiency is always increased by increasing the furnace size, the increase being rather more marked for luminous flames.
3 To spread the load over the total surface area of a furnace, rather than use a larger furnace with the same total load area covering the floor only, decreases the efficiency considerably.

12.7 FLOW AND COMBUSTION

In a furnace, combustion-chamber flow and combustion are interdependent. For example, the flow pattern in the absence of a flame is different from that when the flame is present, and the rate of combustion, particularly in turbulent diffusion flames, is governed by the rate of mixing of fuel and air which is dependent on the flow pattern.

Also, radiative heat transfer depends on flame size and luminosity, while convective transfer depends on flow velocity.

All these processes and interrelations are complex and most of our present knowledge comes from experiments, mostly on rather idealized model systems, with some supporting work done on actual furnaces.

12.7.1 The effect of recirculation of combustion gases on the rate of combustion

The rate of the combustion reaction increases with temperature and with the concentrations of the reacting molecules. The recirculation of hot combustion products back to the earlier part of the combustion region increases the temperature there but also reduces the concentrations of the reacting gases, and these two opposing tendencies can therefore cause different effects according to which is the more important in the particular flame considered.

It is known that the rate of heat release in a hydrocarbon combustion approximates to a second-order rate equation with an activation energy E_a of about 42 kcal/mol (176 J/mol) ([41] and Section 4.1.3). Hence:

$$Q_{\text{comb}} = (K\varphi_f\varphi_a/T^2)e^{-E_a/RT} \tag{12.68}$$

where φ_f and φ_a are the local mole or volume fractions of fuel and oxidizer, $K = k_1 c_n P^2/R^2$ and Q_{comb} = rate of heat release per unit volume.

Table 4.1 shows that an increase in temperature in the region of $T = 300$ K has a very much larger effect on the term $e^{-E_a/RT}$ than it does at higher temperatures, so that recirculation of hot products to the base of the flame always increases reaction rate, and this is the basis of flame stabilization (see Chapter 4). Not only is the temperature increased in this case, but E_a is probably reduced by introducing active molecular species.

Apart from this special case the effect of recirculation depends on whether the flame is premixed or is a diffusion flame, and on the amount of excess air present.

A diffusion flame is controlled by the rate of mixing of fuel and air, i.e. by the magnitude of $\varphi_f \varphi_a$, and in this type of flame the predominant effect of recirculation is usually to cause dilution and hence to lengthen the flame.

In a premixed flame recirculation increases the rate of reaction, the more so as the excess air increases. For near-stoichiometric mixtures there is an optimum recirculation which maximizes the rate of heat release.

12.7.2 Premixed flames

Longwell and Weiss [41] measured combustion rates in a perfectly stirred reactor and found a maximum of 3×10^8 Btu/ft^3 h (3.1 MW/m^3) for a stoichiometric mixture. This occurred when the uniform composition in the reactor contained 80 per cent combustion products and 20 per cent fuel–air mixture, which corresponds to a recirculation ratio of 5.

A mathematical model of essentially the same system has been studied by Hedley and Jackson [42] for a wide range of recirculation levels and percentage burnaway. The basic feature of this type of mathematical model, which is of use in studying chemical reactors generally, is shown in Figure 12:24 where a heat and material

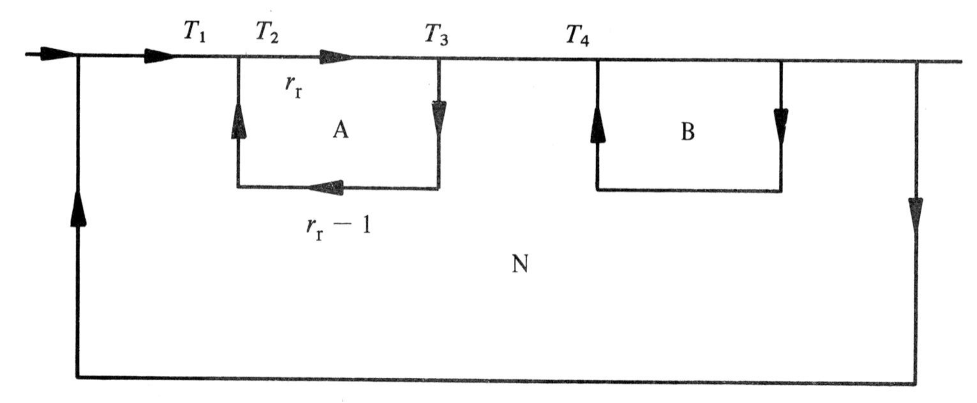

Figure 12:24 Model of reactor with well-stirred and plug-flow zones

balance is carried out over each recirculation zone. For example, for the zone A, assuming constant specific heat capacity:

$$\left.\begin{aligned}(r_r - 1)T_3 + T_1 &= r_r T_2 \\ (r_r - 1)\varphi_{f3} + \varphi_{f1} &= r_r \varphi_{f2} \\ (r_r - 1)\varphi_{a3} + \varphi_{a1} &= r_r \varphi_{a2}\end{aligned}\right\} \tag{12.69}$$

Since the average time spent by gases in the zone is given by (zone volume)/(volume flow rate), equation 12.68 can be numerically integrated over the region 2–3 to give the change in temperature $T_2 - T_3$ and the corresponding changes in fuel and oxidant concentrations. Hedley and Jackson consider only one recirculation zone between outlet and inlet, shown as N in Figure 12:24. Figure 12:25 summarizes some of their results which show that about 60 per cent recirculation (i.e. $(r_r - 1)/r_r = 0.6$ or $r_r = 2.5$) gives the highest rate of combustion which is about 10^4 times that with no recirculation. The model assumes adiabatic conditions, and this increase of combustion intensity would be unobtainable in practice. Nevertheless, the combustion intensity quoted by Longwell [41] is of the order of 10^3 times that obtained in normal combustion chambers. Perhaps the best example of a high-intensity combustion

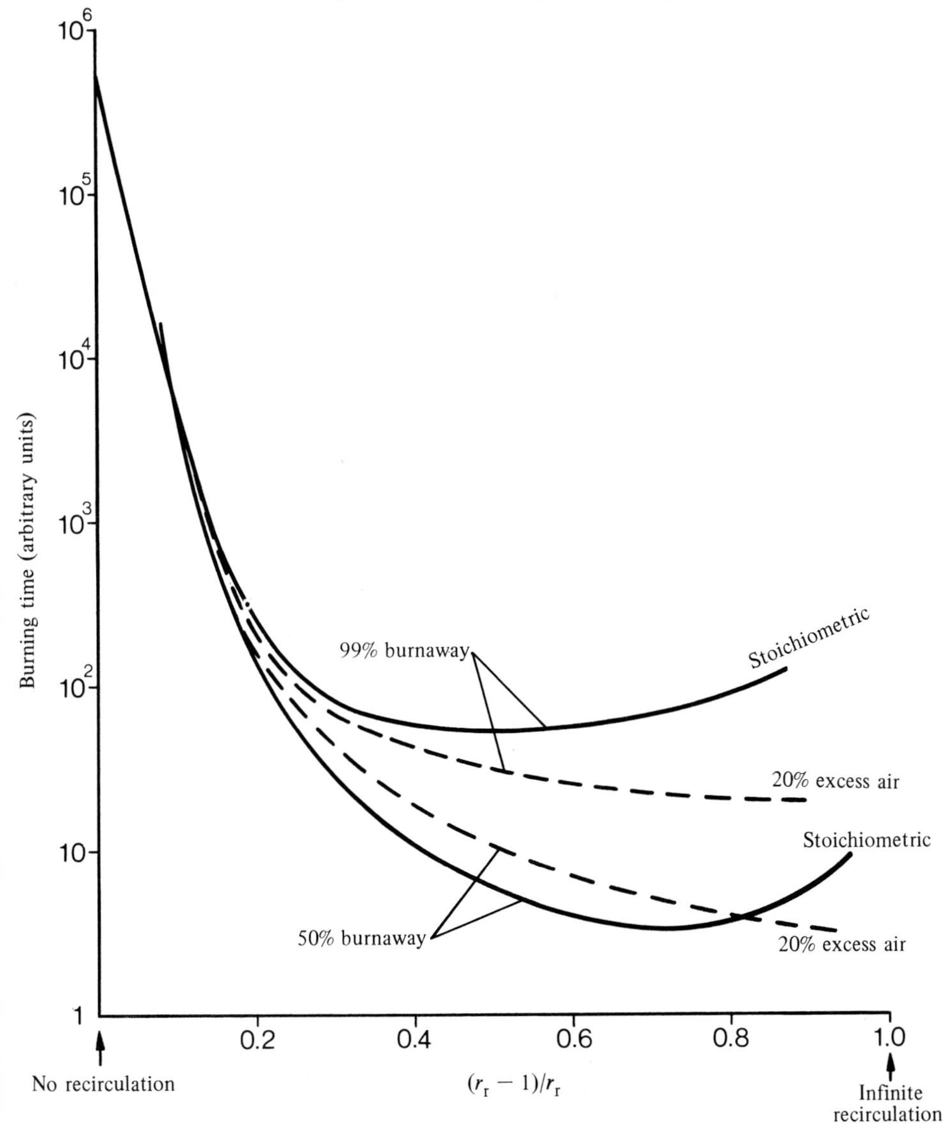

Figure 12:25 Computed effect on rate of combination of recirculation in a premixed combustion chamber

system using this principle is the tunnel burner in which combustion intensities of the order of 2.5×10^7 Btu/ft^3 h (0.26 MW/m^3) are achieved.

The effect of recirculation is discussed by Thring [43] who quotes the work of Spalding and Bragg as showing that the ideal combustion chamber for maximum intensity at a high efficiency, i.e. complete combustion at the exit, will comprise well-stirred zones followed by plug-flow zones. The effect of this is shown in Figure 12:26 (see also [44]).

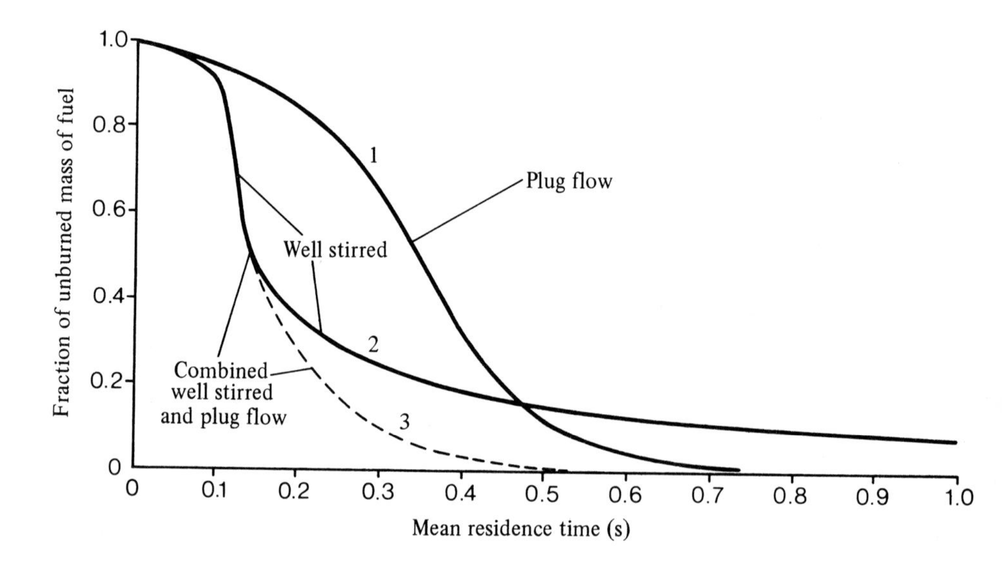

Figure 12:26 Fraction of unburned mass of fuel or combustion efficiency plotted as a function of mean residence time calculated for stirred and plug-flow combustors. Curve 1: plug-flow combustor; curve 2: stirred combustor with complete 'segregation'; curve 3: stirred combustor with complete micromixing

12.7.3 Diffusion flames

Turbulent diffusion flames are widely used in large furnaces where large volumes of radiating gas are required. They are simple to produce and enable the designer to control the progress of combustion, although they have the disadvantage that some of the combustion space is wasted in accommodating the air and the mixing process.

The free turbulent diffusion flame is dealt with in Section 4.9.2 where several expressions are given for flame length. If such a flame is confined so that the expanding jet meets the confining walls then recirculation occurs of the combustion products towards the burner nozzle, and the flame length is usually increased. Beér [44] gives the formula:

$$L_{\text{confined}}/L_{\text{free}} = (0.066/\theta) + 0.95 \qquad (12.70)$$

where θ is the Thring–Newby parameter (Section 12.9.1).

Beér and Chigier [45] also state that for a confined jet:

$$r_r = (0.47/\theta) + 0.5 \qquad (12.71)$$

and these combine to give:

$$L_{\text{confined}}/L_{\text{free}} = 0.14(r_r - 0.5) + 0.95 \qquad (12.72)$$

These equations refer to an axisymmetrically confined jet. Conditions in a combustion chamber would normally be rather different so that recirculation might occur on one side of the flame only.

Some published work seems to show that a reduction in flame length can occur by recirculation, and the effect must be influenced by the temperature of the recirculating gases (cf. the difference between a shell boiler and a muffle combustion chamber) and by the amount and entry position of the combustion air.

12.7.4 Swirl

The rate of entrainment and mixing of ambient gas by a moving gas stream increases with the relative velocity, and hence with the shear stress between the two streams. For this reason a swirling jet produces a combustible mixture in a shorter length than does a nonswirling jet.

Also, the reduced pressure at the centre of a rotating mass of gas (Section 2.2.2) causes a recirculation to occur, in the centre of the swirling jet, of hot combustion products to the base of the flame. These products are at flame temperature since they have not been involved in any furnace heat-transfer process, and they do not directly dilute the combustion air. So the effect of this recirculation is to increase the speed of combustion and hence to stabilize and shorten the flame.

There may also be a recirculation zone on the outside of the flame.

These characteristics of a swirl burner require a swirl number > 0.6 (Section 4.8.1). Diffusion flames with lower values of S may not be stable. The shape of the nozzle exit also has an effect on the flame shape. For example, a divergent nozzle exit on a swirl burner increases the size of the recirculation zone and reduces the critical S at which recirculation occurs.

Beér [44] describes two basically different forms of swirling flame. Type 1 has medium swirl in which the axial fuel jet is strong enough to pierce the recirculation eddy to give a long, silent, luminous flame. Type 2 is of high swirl, and the flame is short, nonluminous and noisy.

Syred and Beér [46] give a comprehensive account of swirl in combustion.

12.7.5 Pollutants from combustion

There is increasing concern about pollution, and combustion processes are known to contribute hydrocarbons and oxides of nitrogen.

Hydrocarbons are not likely to be produced in gas-fired furnace combustion chambers provided there is a reasonable amount of excess air. Nitric oxide (NO) may be produced and this may be oxidized to nitrogen dioxide (NO_2).

The formation of NO is favoured by high temperature, long residence time at an elevated temperature, and high concentrations of O_2 and N_2. Thus the following conditions will favour a reduction in NO formation:

1 Low excess air
2 Two-stage combustion, preferably with intercooling
3 High heat-transfer rates from combustion gases
4 Recirculation of cooler products into the flame
5 Injection of water or steam into the flame. This is not particularly practicable for furnaces but may be a promising method in the case of IC engine combustion

The combustion of gas produces less NO than does oil or coal. Careful design of the burner and combustion system to reduce residence time at a high temperature can reduce NO formation, for example the use of a divergent nozzle in a swirl burner is beneficial by increasing recirculation. Bartok *et al.* [47] is a useful summary of this topic.

12.8 MATHEMATICAL MODELLING OF THE FLOW AND COMBUSTION PROCESS

Progress in the accurate modelling of this very complex process is hampered by the lack of basic understanding of turbulence and by the large computer size required. But experimental work on large-scale combustion is also very difficult and expensive so that there is a vast benefit to be had if satisfactory mathematical models can be perfected.

12.8.1

The most comprehensive type of model is one firmly based on the momentum, energy and matter conservation equations. These were introduced in Chapter 2 for an element of a streamtube, but for the present purpose it is necessary to use more general cartesian coordinates. Polar coordinates will be more useful for axisymmetric problems.

To outline the method, consider the two-dimensional finite element ABCD in Figure 12:27:

1 Steady-state conservation of matter gives:

$$(\rho_{21} u_{21} - \rho_{23} u_{23})\Delta Y + (\rho_{12} v_{12} - \rho_{32} v_{32})\Delta X = 0 \qquad (12.73)$$

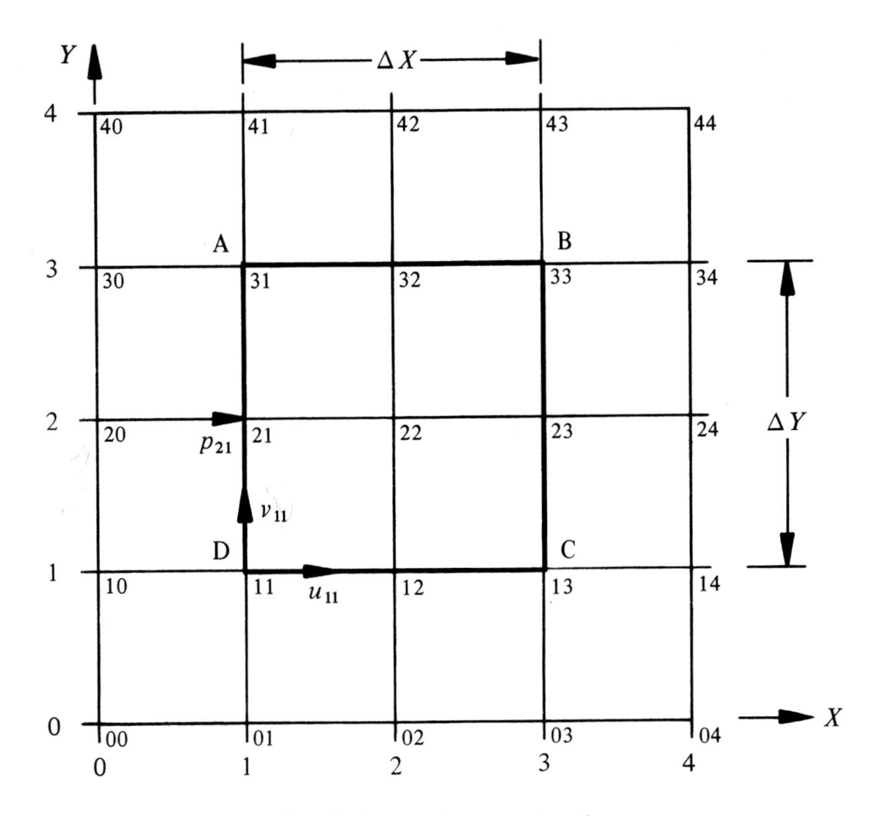

Figure 12:27 Two-dimensional element in a combustion system

2 Steady-state conservation of fuel is:

$$(\varphi_{f21} u_{21} - \varphi_{f23} u_{23})\Delta Y + (\varphi_{f12} v_{12} - \varphi_{f32} v_{32})\Delta X$$

$$+ \left[D_{21} \frac{(\varphi_{f20} - \varphi_{f22})}{\Delta X} - D_{23} \frac{(\varphi_{f22} - \varphi_{f24})}{\Delta X}\right] \Delta Y$$

$$+ \left[D_{12} \frac{(\varphi_{f02} - \varphi_{f22})}{\Delta Y} - D_{32} \frac{(\varphi_{f22} - \varphi_{f42})}{\Delta Y}\right] \Delta X$$

$$+ \frac{Q_{comb} \Delta X \Delta Y}{c_n} = 0 \qquad (12.74)$$

For laminar flow, equation 12.74 is correct as it stands, provided of course that ΔX and ΔY are small enough to ensure that the concentrations and velocities are reasonable averages and that the concentration differences in the diffusion terms adequately represent the concentration gradients. But for turbulent flow the product terms have to be written as products of steady plus fluctuating components.

Using the notation of Section 4.8:

$$\varphi u = (\bar{\varphi} + \varphi')(\bar{u} + u') = \bar{\varphi}\bar{u} + \bar{\varphi}u' + \varphi'\bar{u} + \varphi'u'$$

which gives, when averaged over a period of time:

$$\overline{\varphi u} = \bar{\varphi}\bar{u} + \overline{\varphi' u'} \tag{12.75}$$

The last term represents the effect of turbulent mixing which is likely to be much greater than the molecular diffusion. At present its magnitude can only be determined by experiment, and it will be replaced by an eddy diffusivity term as defined in equation 3.47. Thus the first term of equation 12.74 becomes:

$$\varphi_{f21}\, u_{21} = \overline{\varphi_{f21}\, \bar{u}_{21}} + D_{E21}\,(\varphi_{f20} - \varphi_{f22})/\Delta X \tag{12.76}$$

and it will be seen that the eddy and molecular diffusivity terms can be combined.

The last term of equation 12.74 relates the rate of disappearance of the fuel in the element to the rate of heat production by its combustion. Q_{comb} will be given by an expression such as equation 12.68.

Similar equations to 12.74 can be written for oxidant and combustion products.

3 Steady-state conservation of momentum in the X-direction is:

$$(p_{21} - p_{23})\Delta Y + (\rho_{21}\, u_{21}{}^2 - \rho_{23}\, u_{23}{}^2)\Delta Y + (\rho_{12}\, v_{12}\, u_{12} - \rho_{32}\, v_{32}\, u_{32})\Delta X + \left[\frac{\mu_{12}\,(u_{02} - u_{22})}{\Delta Y} - \frac{\mu_{32}\,(u_{22} - u_{42})}{\Delta Y}\right]\Delta X = 0 \tag{12.77}$$

There is a similar equation for the Y-direction which includes the weight term $\rho_{22}\, g\Delta X\Delta Y$. These equations are finite-difference forms of the Navier–Stokes equations.

Introducing mean and fluctuating velocity components leads to:

$$\left.\begin{aligned} \rho u^2 &= \rho\bar{u}^2 + \rho\overline{u'u'} \\ \rho uv &= \rho\bar{u}\bar{v} + \rho\overline{u'v'} \end{aligned}\right\} \tag{12.78}$$

The term $\rho\overline{u'u'}$ is a normal stress or pressure and can be denoted as:

$$\rho\overline{u'u'} = p_{\text{T}} = \mu_{\text{TN}}\,(\partial u/\partial x) \tag{12.79}$$

The term $\rho\overline{u'v'}$ is a shear stress and can be denoted as

$$\rho\overline{u'v'} = \tau_{\text{T}} = \mu_{\text{T}}\left(\frac{\partial u}{\partial y} + \frac{\partial v}{\partial x}\right) \tag{12.80}$$

by analogy with molecular viscosity.

The substitution of 12.79 and 12.80 in 12.77 replaces molecular viscosity by an effective viscosity which is the sum of the molecular and turbulent viscosities.

4 The energy equation accounts for the rate at which kinetic energy and enthalpy flow into the element, the rate at which work is done on the element by shear stresses

and against gravity, the rate at which heat is produced in the element by combustion, and the rate at which it leaves by radiation, conduction and convection.

The net enthalpy entering per second is

$$(\rho_{21} u_{21} I_{21} - \rho_{23} u_{23} I_{23})\Delta Y + (\rho_{12} v_{12} I_{12} - \rho_{32} v_{32} I_{32})\Delta X \tag{12.81}$$

The mechanical energy entering per second, in the X-direction for example, is of the form $\frac{1}{2}\rho u\,(u^2 + v^2)\Delta Y$ which, upon expansion and after rejecting terms which have a zero time average (such as $\bar{u}^2 u'$ and $\bar{u}\bar{v}v'$), yields:

$$\{\rho\bar{u}[\tfrac{1}{2}(\bar{u}^2 + \bar{v}^2) + \tfrac{1}{2}(\bar{u}'^2 + \bar{v}'^2)] + \rho(\overline{\bar{u}u'^2} + \overline{\bar{v}u'v'})\}\Delta Y \tag{12.82}$$

= bulk KE + turbulence KE + work done by turbulent pressure + work done by turbulent shear stresses.

Using equations 12.79 and 12.80 this may be expressed in terms of the effective viscosities and the velocity gradients. For example, for the side AD these terms become:

$$\left\{\frac{\rho_{21}\,\bar{u}_{21}\,(\bar{u}_{21}{}^2 + \bar{v}_{21}{}^2)}{2} + \frac{\mu_{TN}\,\bar{u}_{21}\,(\bar{u}_{22} - \bar{u}_{20})}{\Delta X} + \frac{\rho_{21}\,\bar{u}_{21}\,(\bar{u}'^2 + \bar{v}'^2)}{2}\right.$$

$$\left. + \mu_T\,\bar{v}_{21}\left[\frac{(\bar{u}_{31} - \bar{u}_{11})}{\Delta Y} + \frac{(\bar{v}_{22} - \bar{v}_{20})}{\Delta X}\right]\right\}\Delta Y \tag{12.83}$$

with similar terms for the other sides.

The rate of doing work against gravity is:

$$\rho_{22}\, g\, v_{22}\, \Delta X \Delta Y \tag{12.84}$$

The rate of heat production in the element is given by equation 12.68 as in the last term of equation 12.74:

$$Q_{\text{comb}}\, \Delta X \Delta Y \tag{12.85}$$

The rate at which heat leaves by conduction or convection, through side AD for example, is:

$$k_{21}\,(T_{22} - T_{20})\,\Delta Y/\Delta X \tag{12.86}$$

and

$$h_{21}\,(T_{22} - T_{20})\Delta Y \tag{12.87}$$

In the former case, where the heat transfer is to an adjacent fluid element, the transport coefficient k is an experimentally determined function of the turbulent fluctuations.

The radiation leaving the element is of the form:

$$\sum\overrightarrow{\mathrm{G}_{22}\,\mathrm{S}_i}\,T_{22}{}^4 - \sum\overrightarrow{\mathrm{S}_i\mathrm{G}_{22}}\,T_i{}^4 - \sum\overrightarrow{\mathrm{G}_j\mathrm{G}_{22}}\,T_j{}^4 \tag{12.88}$$

Thus the energy equation is of the form:

$$(12.81 + 12.83 + 12.84 + 12.85 - 12.86 - 12.87 - 12.88) = 0. \quad (12.89)$$

5 Finally, the equation of state can be applied at each point.

The usual procedure, which is followed in references [3, 45, 48] is to use these conservation expressions in the form of partial differential equations as the starting point. This carries with it the danger of inaccurate finite-difference representation of the derivatives and this is known to cause spurious results.

Some of the terms, particularly those relating to mechanical work, may be negligible in particular cases, but the full application to a three-dimensional case is a major computational task if reasonably small elements are used.

The model may also be used in an inverse manner to calculate the various turbulent transport coefficients from experimental results.

12.8.2 Models based on a combination of well-stirred and plug-flow zones

In a plug-flow zone all particles have a residence time t_r equal to (zone volume)/(flow rate), while in a well-stirred zone there is a distribution of residence times with an average of t_r. Residence-time distribution can be determined experimentally by injecting an easily measured tracer into a region and measuring its rate of removal from the zone, thus giving an experimentally based model comprising well-stirred and plug-flow zones. See Beér [44, 49].

12.9 PHYSICAL MODELLING AND ANALOGUES

In a model the process used is recognizably the same as in the prototype. For example, a model flow system will be another flow system, scaled down and possibly distorted, and probably using a different fluid.

An analogue uses a system which may be physically quite different from the prototype but which is governed by the same equations and which is easier and cheaper to handle experimentally than is the prototype.

12.9.1 Physical modelling

A model will behave in the same way as the real system if it has the same geometrical shape and if the various kinds of energy transfer and force in the model bear the same ratio to each other as they do in the prototype. This means that the ratios of the terms in each of the governing equations must be the same for model and prototype.

This topic was introduced in Sections 2.3, 3.3 and 3.5 and will be extended here by considering the equations derived in Section 12.7.1.

For example, the ratios:

$$\frac{\text{momentum or heat or mass transferred by flow}}{\text{momentum or heat or mass transferred by diffusion}}$$

may be found.

For example, in the case of momentum, this ratio is (inertia force)/(viscous force), and from equation 12.77 this is of the form:

$$\rho u^2 \Delta Y \Delta Y / \mu u \Delta X = \rho u l / \mu = (Re) \tag{12.90}$$

Clearly, the values chosen for ΔX and ΔY are arbitrary and cannot affect the processes taking place in the system so l can represent any characteristic dimension in the system. This particular group is called the Reynolds number and its value characterizes the degree of turbulence and also the flow pattern, provided buoyancy forces are relatively small.

For heat, the ratio is, from equation 12.89 (i.e. 12.81/12.86):

$$\rho u c_p T l / k T = c_p \rho u l / \lambda \text{ or } c_p \rho u / h = R_2 \tag{12.91}$$

The inverse of this is the Stanton number.

For matter transport the ratio is from equation 12.74:

$$\varphi u l / \varphi D = u l / D = R_3 \tag{12.92}$$

The ratios of these dimensionless groups are a measure of the similarity in the shapes of the velocity, temperature or concentration gradients at a boundary and are used when considering transport processes taking place through those gradients. Thus the ratio:

$$R_2/(Re) = (c_p \rho u l / \lambda)(\mu / \rho l u) = c_p \mu / \lambda = (Pr) \tag{12.93}$$

characterizes the similarity of heat transfer and momentum transfer.

$$R_3/(Re) = (\rho l u / \mu)(D / u l) = \rho D / \mu = (Sc)^{-1} \tag{12.94}$$

(Schmidt number)$^{-1}$, characterizes the similarity of mass transfer and momentum transfer.

$$R_2/R_3 = (c_p \rho u l / \lambda)(D / \mu l) = c_p \rho D / \lambda \tag{12.95}$$

characterizes heat and mass transfer similarity.

If buoyancy forces are relatively large, as in the case of natural convection or in chimneys, the ratios (buoyancy force)/(inertia force) and (buoyancy force)/(viscous force) will be important parameters.

The first of these is the Archimedes number $(Ar) = l \Delta T g / u^2 T_g$ and the second gives rise to the Grashof number $(Gr) = l^3 \gamma g \Delta T \rho^2 / \mu^2$.

It is quite impossible to achieve complete similarity between model and prototype, i.e. equal values of dimensionless groups in model and prototype, because no fluid will have just those properties which permit all the dimensionless groups to be matched simultaneously. Furthermore it is practically impossible to simulate the combustion process or radiative heat transfer in a model.

In practice, therefore, 'partial modelling' is used, in which only the dimensionless groups which control the particular process under consideration are satisfied. For example, if (Re) in the prototype is high so that flow is very turbulent, then groups which relate to molecular transport processes need not be satisfied if interest lies in the bulk flow process. But if one is attempting to model transport processes at a surface, equality of groups such as (*Pr*) and (*Sc*) must be satisfied.

Many models are built to investigate flow patterns in combustion systems so that the groups involving heat and matter transport processes can be ignored. It is quite easy to simulate correctly the Reynolds number since cold air and water have kinematic viscosities of about 1/12 and 1/120 of hot combustion products, so that a 1/12-scale cold-air model using the same velocities as in the prototype, or a 1/12-scale water model using velocities 1/10 of that in the prototype would be correct.

But in many furnace combustion chambers the flow pattern is largely determined by the momentum interchange between the jet of combustion gases and the combustion air. In this case it is important to model the jet momentum correctly, and provided the value of (*Re*) in the model is high enough to ensure turbulence, exact equality of (*Re*) in model and prototype is not essential.

Thring and Newby [50] considered that in this type of system where a jet was causing recirculation the important parameter for similarity is the ratio:

$$\theta_1{}^2 = \frac{\text{momentum of fluids in fully expanded jet}}{\text{momentum of jet fluid at nozzle}}$$

and reference to Sections 2.2.5 and 10.1.1 will support this.

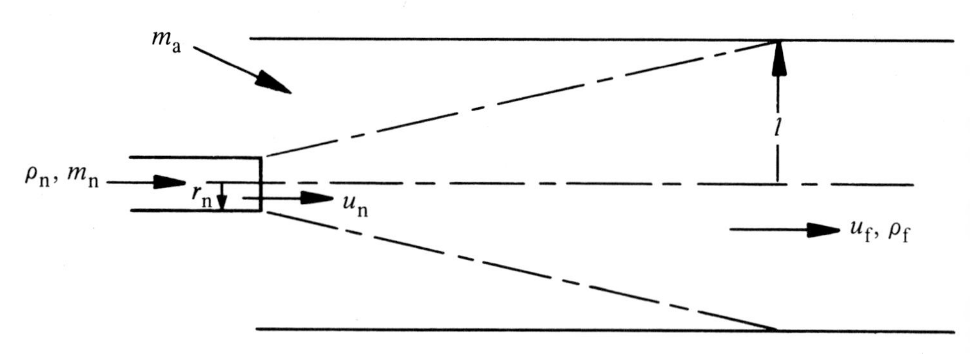

Figure 12:28 Enclosed jet

Referring to Figure 12:28:

$$\theta_1{}^2 = (m_n + m_a)u_f/m_n u_n, \qquad u_f \rho_f \pi l^2 = m_n + m_a, \qquad u_n \rho_n \pi r_n{}^2 = m_n$$

Eliminating u_n and u_f gives:

$$\theta_1 = [(m_n + m_a)/m_n]\,(r_n/l)\sqrt{(\rho_n/\rho_f)} \tag{12.96}$$

For an isothermal system:

$$\theta_1 = \theta = [(m_n + m_a)/m_n]\,(r_n/l) \tag{12.97}$$

The maximum value of θ and θ_1 is unity which would occur in a free jet. A value of

$\vartheta < 1$ means that some of the jet momentum is dissipated by the friction force between recirculating gases and walls.

If θ_1 is the value of the Thring–Newby parameter in a furnace then in order to make $\theta = \theta_1$ in the isothermal model it is necessary to have:

$$r_{nm} = r_n (l_m/l)\sqrt{(\rho_n/\rho_f)} \tag{12.98}$$

where the suffix m refers to the model. Since $\rho_n/\rho_f > 1$ this means that the model nozzle has to be larger than the geometrical scaling-down would give, i.e. the model has to be distorted.

A similar parameter, which was defined by Craya and Curtet [51], includes the momentum of the entrained stream. This parameter is

$$(Ct) = \frac{1 + (m_a/m_n)}{[(m_a i_n/m_n i_a) - (m_a/m_n) - \frac{1}{2}]^{\frac{1}{2}}} \tag{12.99}$$

where i_n and i_a are the momenta of the nozzle and entrained fluids respectively.

The most commonly used cold flow models are usually constructed of transparent sheets. Flow visualization may be achieved by injecting short pulses of dye into different regions, or by using small visible particles. Mixing may be studied by introducing an easily measured tracer into one of the streams.

When buoyancy is important, as in large combustion spaces such as occur in glass tanks and open-hearth furnaces, its effect may be partially simulated in the cold model by using a fluid of different density for the jet of combustion gases. For example, an inverted model may be used in which the flame gases are simulated by a salt solution with water as the combustion air. If an air model is used then the jet may be preheated.

References [1, 52, 53, 54] deal with modelling principles and practice and the last of these gives an account of the application of model results to the design of industrial flames.

12.9.2 Analogues

Various types of analogue have been used for investigating heat flow.

Conductivity analogues

The equations for one-dimensional heat conduction through a solid (3.4, 3.9, 3.10) are of the same form as those relating current flow to potential difference, where Q and T would represent current and voltage, respectively. A number of thermal resistances in series and parallel can therefore be represented by a similar arrangement of electrical resistors.

The case of two-dimensional conduction is governed by two terms of equation 3.12 (the Laplace equation), which also describes the distribution of electrical potential over a conducting surface. A special paper coated with an electrical conductor is available. This is cut out to the shape of the cross-section through which the heat conduction is to be studied and boundary conditions are applied as electrical potentials corresponding to the temperature boundary conditions in the heat-conduction problem. The resulting lines of equal potential can be plotted using a probe and voltmeter. Since these correspond to the isothermals, the heat-flux directions can be plotted as a set of orthogonal lines.

Convection analogues

The analogy between heat transfer and mass transfer which is discussed in Sections 3.3.6 and 3.5 has been used for determining heat-transfer coefficients in forced-convection types of furnace and in other heat exchangers.

The value of the method lies in the fact that mass transfer can easily be measured, over small areas of surface if necessary, so that local heat-transfer coefficients can be deduced. This is valuable in finding which regions in a heat exchanger or furnace are contributing least to the transfer of heat, and this kind of information enables the designer to optimize heat-exchanger design.

The method is based on the Chilton Colburn analogy which equates the two j factors:

$$(h/\rho c_p u)\,(Pr)^{2/3} = (h_D/u)\,(Sc)^{2/3} \qquad (12.100)$$

where h_D is the mass transfer coefficient defined analagously to h as in equation 3.50.

Thus if there is geometric and flow similarity (i.e. Reynolds number similarity) between a scale model and a prototype:

$$h = c_p u \rho\,(h_{Dm}/u_m)\,[(Sc)_m/(Pr)]^{2/3} \qquad (12.101)$$

where the suffix m refers to the model. (Re) similarity requires that $ul\rho/\mu = u_m l_m \rho_m/\mu_m$ so that equation 12.101 can be written as:

$$h = (l_m/l)\,(\mu/\mu_m)\,[(Sc)_m/(Pr)]^{2/3}\,\rho_m c_p\,h_{Dm} \qquad (12.102)$$

In the *naphthalene sublimation* technique, the heat transfer surfaces of the model are coated with naphthalene which is transferred into the air flowing in the model. The amount of naphthalene removed can be found by direct weighing or by measuring changes in thickness. Local rates of mass transfer can be found by making the heat-transfer surfaces in the form of removable panels coated with naphthalene. In calculating h_D the concentration of naphthalene in the airstream has to be known, and this is easily found in the case of a well-stirred furnace from the total weight of naphthalene lost divided by the volume of air passed. If the furnace is not well-stirred then local concentrations in the airstream would have to be found by sampling and analysis.

In the *electrolytic mass-transfer method* the model is usually made of plastic with heat-transfer surfaces which have been made electrically conductive by electrodeposition of Ni, or which are formed from removable metal panels. A suitable electrolyte is passed through the model and a potential difference applied between electrolyte and model surfaces. If there is a large concentration of an indifferent strong electrolyte, ion movement is due solely to convective diffusion. If the applied potential is made large enough the surface concentration of the migrating ion becomes negligible so that the current reaches a limiting value of

$$I_0 = AnFh_D C_b \qquad (12.103)$$

where A = electrode area, n = valency change in the reaction, F = Faraday's constant, C_b = ion bulk concentration.

Thus a measurement of the limiting current determines h_D and thereby the corresponding h from equation 12.102.

Several electrochemical reactions have been used, such as the reduction of the ferricyanide ion and the oxidation of the ferrocyanide ion, both in alkaline solution, and the deposition of copper from $CuSO_4$ solution.

Both the naphthalene and electrolytic methods have been applied to the rapid billet-heating furnace [55, 56]. The results are in close agreement and the resulting heat-transfer correlations are given in equation 12.43.

Also, using the mass-momentum transfer analogy (equation 3.25), the naphthalene technique has been used to evaluate the shear stress in a burner port. Since this is a measure of the local velocity gradient it has considerable influence on flame stability (see Section 4.4.2).

A particularly valuable feature of mass-transfer analogue methods is the ease with which local coefficients can be found; this kind of knowledge could prove very useful in developing more efficient heat exchangers.

12.10 STUDIES OF PARTICULAR APPLICATIONS

This section gives a brief account of some published studies of flow and combustion in particular furnaces, using mathematical or physical modelling.

12.10.1 Mathematical modelling of combustion-chamber performance [57]

This describes work at IJmuiden in which measurements were made of heat flux, gas temperature and velocity, residence times (by means of an He tracer) and NO concentrations, on a full-scale water-cooled experimental furnace, 6 m long, using various types of burner, for the purpose of testing the validity of various mathematical modelling techniques, and of providing empirical data for such models, particularly effective viscosities and diffusivities for concentric swirling jets.

Typical results showed that heat flux is concentrated at the burner end for strongly swirling jets (type 1 and type 2 flames) and at the opposite end for nonswirling flames, irrespective of the fuel, and that there are very wide variations in the levels of NO produced in different parts of the flames.

It is concluded that no model is capable of predicting NO_x levels, and that in the short term the prediction of combustion-chamber performance is best achieved by a combination of physical modelling and radiation zone analysis, while in the long term the Spalding type of mathematical model, if supported by adequate turbulence data, may prove superior.

12.10.2 Aerodynamics of the rotary cement kiln [58]

Reference [58] describes a water and an air model of the kiln together with comparative results from actual kilns.

The cement kiln approximates to an ideal axisymmetric enclosed jet system, but differs from this on account of its very large size which makes buoyancy effects substantial, and the various shapes of the firing hood and kiln shell.

(Re) is of the order 10^6 and the nozzle momentum parameters (Thring–Newby, Craya–Curtet and Becker) show that circulation is low. Partial modelling was used with values of (Re) between 10^4 and 10^5, but using correct values of (Ct) and (Th) (Becker). Buoyancy was not modelled. Reasonably good agreement was obtained, and the authors favour the use of (Ct) as the similarity parameter for nozzles.

12.10.3 Aerodynamic and heat transfer characteristics of proprietary burners fired with natural gas in a cylindrical furnace [59]

Two types of similar dual-fuel burner were used in a model firetube and measurements made of heat flux to various parts of the walls, gas composition, temperature and velocities. Two mixing parameters are defined, characterizing the macro or bulk mixing process, and the micro or molecular mixing process (which permits combustion to proceed), and the burners are compared on this basis. 'Unmixedness' is also defined as indicating how the combustion lags behind the macro mixing process. The measured gas temperatures and velocities are used in a mathematical heat-transfer model to predict the heat fluxes.

12.10.4 Predictions of a convective heating system [60]

This deals with the mathematical modelling of the flow of a high-velocity impinging jet of hot combustion products over a surface, using a finite-difference solution of the two-dimensional energy, vorticity, and stream-function equations. This uses the methods of [3] and assumes the heat transfer to be entirely convective. The model was not entirely satisfactory since it did not converge well and it proved difficult to compute the effective viscosities. The necessary heat-transfer coefficients were obtained from a mass-transfer analogue.

12.10.5

Hulse and Rigg [61] describe 1/12-scale cold-air and water models of a two-drum marine boiler. There was good agreement between the models which showed up regions where overheating would occur, and where space was being wasted. A better arrangement of burner positions was evolved which would be expected to increase efficiency by 4 per cent.

12.10.6

D. Anson [62] discusses general modelling experience of large boilers, and emphasizes the difficulty of modelling jet momentum. The Thring–Newby correction is not easily achieved since it may take up too much space or cause interference between adjacent burners. A recommended alternative is to use a gauze to provide flow resistance at the combustion front.

12.10.7

Fitzgerald and Robertson [63] discuss the isothermal cold modelling of open-hearth, arc, and heat treatment furnaces generally. Recirculation is usually important so that the corrected Thring–Newby parameter should be used.

12.10.8

Wingfield [64] used a cold model to study mixing in double concentric-jet burners. Using N_2O as a tracer he found that optimum speed of mixing occurred at a value of (external diameter of primary nozzle)/(internal diameter of primary nozzle) between 2 and 3.

REFERENCES

1 Spalding, D. B., 'The art of partial modelling', 9th Symposium on Combustion
2 Patanker, S. V. and Spalding, D. B., 'Mathematical models of fluid flow and heat transfer in furnaces: a review', 4th Symposium on Flames and Industry; *J. Inst. Fuel*, **46,** 279–83 (1973)
3 Gosman, A. D., Pun, W. M., Runchal, A. K. and Spalding, D. B., *Heat and Mass Transfer in Recirculating Flows.* New York: Academic Press (1969)
4 McAdams, W. H., *Heat Transmission*, 3rd edition. New York: McGraw-Hill (1954)
5 Hottel, H. C. and Cohen, E. S., 'Radiant heat exchange in a gas filled enclosure', *A.I.Ch.E. Journal*, **4** (1), 3 (March 1958)
6 Hottel, H. C., 'Radiative transfer in combustion chambers', *J. Inst. Fuel*, **34,** 220 (1961)
7 Hottel, H. C. and Sarofim, A. F., *Radiative Transfer.* New York: McGraw-Hill (1967)
8 Johnson, T. R. and Beér, J. M., 'The zone method of analysis of radiant heat transfer: a model for luminous radiation', 4th Symposium on Flames and Industry; *J. Inst. Fuel*, **46,** 301–9 (1973)
9 Thring, M. W. *et al.*, 'Prediction of the emissivity of hydrocarbon flames', International Heat Transfer Conference 1961–2
10 Beér, J. M. and Howarth, C. R., 'Radiation from flames in furnaces', 12th Symposium on Combustion
11 Ross, G., 'Calculation of non-uniform radiant flux to tube banks', *The Chemical Engineer*, No. 214, CE272 (December 1967)
12 Gray, W. A. and Müller, R., *Engineering Calculations in Radiative Heat Transfer.* Oxford: Pergamon Press (1974)
13 Newall, A. J. and Lowes, T. M., 'Luminous emissivities of natural gas diffusion flames', *Inst. Gas Eng. Commun.*, 862 (1971)
14 Kays, W. M. and London, A. L., *Compact Heat Exchangers*, 2nd edition. New York: McGraw-Hill (1964)
15 Francis, W. E., 'Forced recirculation in industrial gas appliances', *Gas Council Res. Commun.* GC34; *Trans. Inst. Gas Eng.*, **106,** 483 (1956–7)
16 Francis, W. E. and Roughton, H. E., 'Calculation of flue gas recirculation and heat transfer in heaters using high velocity gas burners', International Heat Transfer Conference 1961
17 Lucas, D. M., Masters, J. and Toth, H. E., 'Prediction of the performance of rapid heating furnaces', *Gas Council Res. Commun.* GC151 (1968); *J. Inst. Gas Eng.*, **9,** 397 (1969)
18 Lucas, D. M. and Barber, A. J., 'Transient thermal response of a rapid-heating furnace', *J. Iron & Steel Inst.*, **209,** 790 (1971)

19 Lawrence, M. N. and Spittle, J., 'Forced-convection techniques for gas-fired rapid billet heating', *Gas Council Res. Commun.* GC108; *J. Inst. Gas Eng.*, **5,** 515 (1965)

20 Davies, R. M., Lucas, D. M., Masters, J. and Toth, H. E., 'Use of models for predicting the performance of rapid heating furnaces', *Iron & Steel Inst. Publ.* 123

21 Davies, R. M., Lucas, D. M. and Toth, H. E., 'The prediction of heat transfer and thermal performance in gas fired processes', *Gas Council Res. Commun.* GC184 (1971)

22 Davies, R. M. and Toth, H. E., 'Equilibrium compositions, heat contents and physical properties of natural gas—air combustion products in SI units', Gas Council Midlands Research Station, Report N187

23 Trinks, W. and Mawhinney, M. H., *Industrial Furnaces*, Vol. 1, 5th edition. New York: Wiley (1961)

24 Chatwin, G. R., Francis, W. E., Harrison, W. P. and Lawrence, M. N., 'A recuperative version of a single-ended radiant tube', *Gas Council Res. Commun.* GC124; *J. Inst. Gas Eng.*, **6,** 207 (1966)

25 Patrick, E. A. K. and Thornton, E., 'Immersion tube heating of aqueous liquids', *Gas Council Res. Commun.* GC47; *Trans. Inst. Gas Eng.*, 108, 360 (1958–9)

26 Tailby, S. R. and Clutterbuck, E. K., 'Heat transfer in horizontal town-gas immersion tubes', *Trans. Inst. Chem. Eng.*, **42,** T64 (1964)

27 Tailby, S. R. and Ashton, M. D., 'Heat transfer from town gas flames to water jacketed horizontal tubes', *Trans. Inst. Chem. Eng.*, 36, 1 (1958)

28 Thornton, E. and Robertshaw, G. W., 'Thermal design of gas-fired tubular immersion heaters', *J. Inst. Fuel*, **38,** 307 (1965)

29 Lucas, D. M. and Toth, H. E., 'Calculation of heat transfer in the fire tubes of shell boilers', *J. Inst. Fuel*, **45,** 521 (1972)

30 Lucas, D. M. and Lockett, A. A., 'Mathematical modelling of heat flux and temperature distribution in shell boilers', *J. Inst. Fuel*, **47** (1974)

31 Wu, H. L., 'Comparison of the performance of natural gas and oil flames in a cylindrical furnace', *J. Inst. Fuel*, **42,** 316 (1969)

32 Lockwood, F. C., 'Optimum design of the packaged boiler', *J. Inst. Fuel*, **42,** 150 (1969)

33 Lucas, D. M. and Toth, H. E., 'Calculating the performance of a gas fired coiled tube air heater', *Br. Chem. Eng.*

34 Fitzgerald, F. and Sheridan, A. T., 4th Symposium on Flames and Industry; *J. Inst. Fuel*

35 Salter, F. M. and Costick, J. A., 4th Symposium on Flames and Industry; *J. Inst. Fuel*

36 Pearce, K. W., 'A heat transfer model for rotary kilns', 4th Symposium on Flames and Industry; *J. Inst. Fuel*, **46,** 363–71 (1973)

37 Johnson, T. R., Lowes, T. M. and Beér, J. M., 4th Symposium on Flames and Industry; *J. Inst. Fuel*

38 Anson, D., Godridge, A. M. and Hammond, E. G., 4th Symposium on Flames and Industry; *J. Inst. Fuel*

39 Thorneycroft, W. T. and Thring, M. W., 'A theoretical method for predicting heat transfer distribution in a petroleum heater', *Trans. Inst. Chem. Eng.*, **38,** 63 (1960)

40 Yardley, B. E. and Patrick, E. A. K., 'A theoretical study of furnace performance', Institute of Fuel and Institution of Gas Engineers, Conference on Industrial Uses of Town Gas (1963)

41 Longwell, J. P. and Weiss, M. W., 'High temperature reaction rates in hydrocarbon combustion', *Ind. & Eng. Chem.*, **47,** 1635 (1955)
42 Hedley, A. B. and Jackson, E. W., 'The effect of recirculation on reaction rates in a homogeneous combustion system', *Trans. Inst. Chem. Eng.*, **44,** T85 (1966)
43 Thring, M. W., *The Science of Flames and Furnaces*, 2nd edition. London: Chapman & Hall (1962)
44 Beér, J. M., 'Recent advances in the technology of furnace flames', *J. Inst. Fuel*, **45,** 370 (1972)
45 Beér, J. M. and Chigier, N. A., *Combustion Aerodynamics*. London: Applied Science Publishers (1972)
46 Syred, N. and Beér, J. M., 'Combustion in swirling flames: a review', *Combust. Flame*, **23,** 143–201 (1974)
47 Bartok, W., Crawford, A. R. and Skopp, A., 'Control of NO_x emissions from stationary sources', *Chem. Eng. Prog.*, **67,** No. 2, 64 (February 1971)
48 Zuber, I. and Konečný, V., 'Mathematical model of combustion chambers for technical applications', *J. Inst. Fuel*, **46,** 285–94 (1973)
49 Beér, J. M. and Lee, K. B., 'Effect of residence time distribution on performance and efficiency of combustors', 10th Symposium on Combustion
50 Thring, M. W. and Newby, M. P., 'Combustion length of enclosed turbulent jet flames', 4th Symposium on Combustion (1963)
51 Craya, A. and Curtet, R., *C.R. Acad. Sci. Paris*, **241**, 621 (1955). Derivation also given in Hottel, H. C. and Sarofim, A. F., *Int. J. Heat & Mass Transfer*, **8,** 1153–69 (1965)
52 Beér, J. M., 'The significance of modelling', *J. Inst. Fuel*, **39,** 466 (1966)
53 Thring, M. W., 'The construction of models in which more than one process is similar to the original', *Trans. Inst. Chem. Eng.*, **26,** 91 (1948)
54 Chigier, N. A., 'Application of model results to design of industrial flames', 4th Symposium on Flames and Industry; *J. Inst. Fuel*, **46,** 271–8 (1973)
55 Lucas, D. M., Davies, W. A. and Gay, B., 'Evaluation of local and average convective heat transfer coefficients in a furnace using an electrolytic mass transfer model', *J. Inst. Fuel*, **48,** 31–7, (1975)
56 Davies, R. M. and Lucas, D. M., 'Measurement of the local shear stress in burner ports and other short ducts by a mass transfer technique', *Int. J. Heat & Mass Transfer*, **13,** 1013–27 (1970)
57 Lowes, T. M., Heap, M. P., Michelfelder, S. and Pai, B. R., 'Mathematical modelling of combustion chamber performance', 4th Symposium on Flames and Industry; *J. Inst. Fuel*, **46,** 343–51 (1973)
58 Moles, F. D., Watson, D. and Lain, P. B., 'The aerodynamics of the rotary cement kiln', 4th Symposium on Flames and Industry; *J. Inst. Fuel*, **46,** 353–62 (1973)
59 Rhines, J. M., 3rd Members Conference of the International Flame Research Foundation, IJmuiden (1974)
60 McBride, R. N., 4th Symposium on Flames and Industry; *J. Inst. Fuel*
61 Hulse, C. and Rigg, C., 'Aerodynamics and heat transfer in an oil-fired marine boiler: a model study', *J, Inst. Fuel*, **41,** 243 (1968)
62 Anson, D., 'Modelling experience with large boilers', *J. Inst. Fuel*, **40,** 20–5 (1967)
63 Fitzgerald, F. and Robertson, A. D., 'Isothermal fluid modelling of steelmaking plant', *J. Inst. Fuel*, **40,** 7–19 (1967)
64 Wingfield, G. J., 'Mixing and recirculation patterns from double concentric jet burners using an isothermal model', *J. Inst. Fuel*, **40,** 456–64 (1967)

13
Furnaces

13.1 GAS-FIRED FURNACES: TYPES, TERMINOLOGY AND TRENDS

The general aim of this chapter is to introduce the terminology of furnaces, to consider briefly the particular requirements of the more important industrial processes for which gas-fired furnaces are used and to illustrate how these requirements are met in modern designs.

Many furnaces, particularly in the smaller size range, have been designed largely on the basis of experience and empiricism, but recently there has been a growing trend towards the application of scientific principles and the use of mathematical modelling techniques. This aspect is considered in Chapter 12 and is complementary to the practical descriptions presented in this chapter.

In general terms, furnace design involves considerations of:

1 The liberation of heat by combustion.
2 The transmission of heat to the charge.
3 Heat losses through the furnace structure.
4 Heat losses in the combustion products.
5 The aerodynamics of the furnace gases.
6 Furnace construction and materials.

These aspects are considered individually elsewhere in this book, combined in Chapter 12 and applied to industrial processes in this chapter.

New design techniques have been applied to gas-fired furnaces since the 1950s resulting in furnaces which differ in a number of significant respects from traditional designs. Traditional furnaces tend to have inherited the large empty spaces in the working chamber required on solid-fuel or producer-gas firing to ensure complete combustion and to avoid flame impingement on the stock. Large furnace dimensions favour gas radiation since this is a function of beam length, and convection plays only a very small part in the heat-transfer process. The presence of ash in the combustion products, in addition to having possible harmful effects on the stock, limits the choice of furnace lining materials to dense refractories which are resistant to slag attack. These two factors result in large furnaces of high thermal capacity which require lengthy heating-up periods, are wasteful in fuel usage and are difficult to control. Gas firing allows, in many applications, the use of hot face insulation which dramatically reduces heating-up periods and thermal storage. The recent development of ceramic-fibre refractories is a further important step in this direction (see Chapter 6).

Although structures of low thermal capacity have been used in gas-fired furnaces for many years, furnace dimensions tended to retain the generous proportions inherited from earlier types. Natural-draught firing was commonly employed and for good temperature uniformity this usually necessitated the use of quite large numbers of small individual burners. The trend away from this approach has been very marked in recent years. Forced circulation using hot-gas fans has been introduced with a view to improving temperature uniformity and convective heat transfer but, more important, the development of high-intensity tunnel burners has provided a convenient means of promoting jet-driven hot-gas recirculation. When applied to furnaces of the conventional kind the main effects are the promotion of hot-gas circulation providing well-mixed combustion products leading to good temperature uniformity and the avoidance of stratification. Convective heat transfer is increased but the overall improvement in heat-transfer rate is generally small. A major advantage, however, is that the large number of small burners previously necessary for temperature uniformity may be replaced by a few (often only one) strategically placed burners thereby simplifying furnace construction and permitting the economic use of more sophisticated control equipment.

This line of development may be taken a stage further. As indicated in Chapter 7, premixed and tunnel-mixing burners are available in which combustion is completed inside a compact refractory tunnel from which emerges a stream of high-velocity gases at near flame temperature. It is possible therefore to disregard any considerations of combustion in the design of the furnace. In other words the designer can concentrate on the problem of heat transfer from a rapidly moving stream of hot gases to the stock. By reducing furnace dimensions to a minimum, very high convective heat-transfer rates can be obtained resulting in very short heating times. Additionally the momentum of the gas stream may be used to entrain and recirculate combustion products further increasing convective heat transfer and reducing temperature differentials. This approach has been used to considerable effect in the development of rapid-heating furnaces for the heating of steel and non-ferrous metals to forging temperatures and for localized heating applications (see Sections 13.5.2 and 13.5.3). The very high forced convective heat-transfer rates obtained by this method are of course not appropriate to all industrial heating processes; ceramics firing, for example, could not be done in this way. However, the use of forced-draught burner systems to produce well-stirred furnace gases and to minimize the number of burners is of wide applicability and many examples will be quoted during the course of this chapter.

Many of the industrial processes to be considered involve the heat processing of metals. There is a growing demand for heating processes in which the effects of the furnace gases on the metal surface are minimized or eliminated, e.g. bright-annealing and scale-reduced reheating. Indirectly fired furnaces using prepared furnace atmospheres are becoming more widely used in the heat treatment of both ferrous and non-ferrous metals. With a very few exceptions, controlled atmospheres produced by the combustion or reforming of natural gas provide suitable conditions for bright heat treatment and for carburizing. Section 13.2 deals with the production, properties and uses of atmospheres derived from gas. In high-temperature processes, e.g. steel-billet heating for forging purposes, it is desirable to minimize oxidation and decarburization but it is difficult to adopt indirect firing using a prepared atmosphere. Two approaches to this problem have been adopted in gas-fired furnace design: the use of rapid-heating techniques to minimize the time spent by the stock at high temperature and the use of two or more stages of combustion in which the stock is exposed to partially burned combustion products in the high-temperature zone, the potential heat content of the gases being recovered by completion of combustion elsewhere in the furnace. These methods are reviewed in Section 13.5.

As indicated in Section 4.2.7, a furnace operating at, say, 1200 K (927°C) loses about 47 per cent of its total heat input in the heat content of its combustion products. This figure relates to stoichiometric combustion and is higher for combustion with excess air. A substantial proportion of this heat loss may be recovered by air preheating or, in the case of continuous furnaces, load preheating. Until the increases in fuel prices of the early 1970s it was often argued that fuel savings resulting from waste-heat recovery did not generally justify the capital cost of the equipment required. Heat recovery has therefore been practised mainly on large-scale, high-temperature plant where the increase in flame temperature resulting from the use of preheated air was perhaps as large a motive for its adoption as fuel saving. This situation has changed; there is a growing awareness of the need for waste-heat recovery. The use of recuperative burners (Section 7.6.3) in a wide variety of furnace types is likely to increase considerably in the next few years. Some examples of their use are presented in this chapter.

Increasing labour charges and the continuing growth of automation in industry favour the development of increased automation in furnace operation. The trend is towards continuous furnaces, with fully automatic stock conveyance and burner control.

Most of the foregoing remarks have been directed towards furnaces specifically designed for gas firing. Since natural gas first became available in large quantities in Britain, gas usage has expanded rapidly into new areas of industry, in many cases replacing oil or coal in furnaces originally designed for those fuels. In changeovers of this kind the adaptability of gas as a fuel has allowed it to perform at least as well and in many cases much better than the fuel it replaces. Had the furnace been designed for gas firing the performance would in many cases be better still. Large furnaces represent substantial capital investment, however, and it will be necessary to wait for the next generation of furnaces to realize fully the potential of gas firing in several fields of application.

Following the brief introduction to trends in gas-fired furnace design presented above the remainder of this section is devoted to an attempt at classifying the major furnace types.

Gas-fired furnaces cover an enormous range of types and any attempt at categorization inevitably will contain oversimplifications and omissions. Two means of basing a broad classification are immediately apparent:

1 By method of firing: directly fired or indirectly fired.
2 By the method of handling the load in the furnace: intermittent (or batch) or continuous.

Furnace terminology differs widely depending upon the industry in which the furnace is used and often disguises common features. To take a simple example, gas-fired intermittent furnaces used in pottery firing have a great deal in common with furnaces

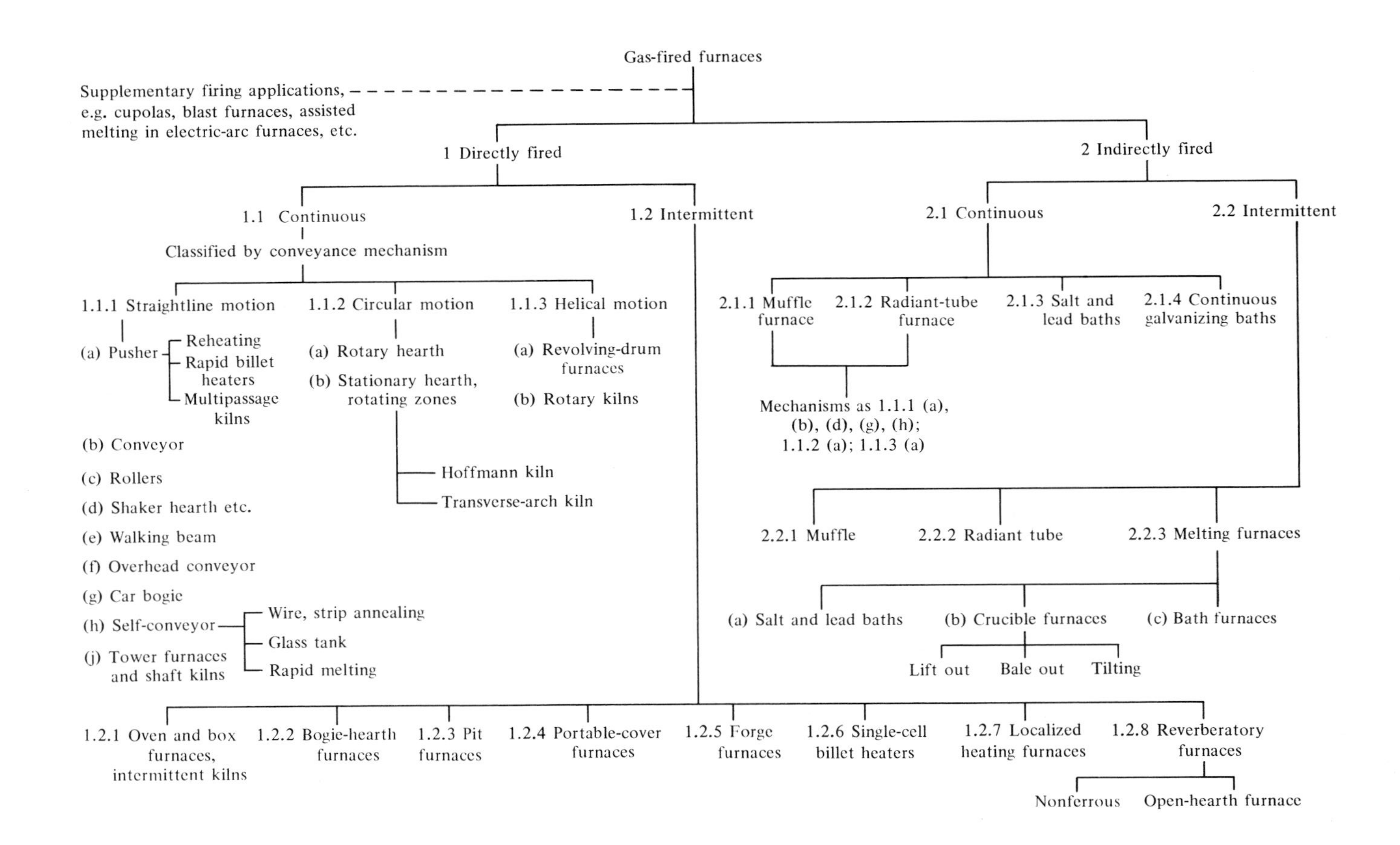

Figure 13:1 Classification of gas-fired furnaces

designed for metal heat treatment but are termed kilns. On the other hand some furnaces are specific to a particular industry, for example the annular continuous kiln used in ceramics firing, in which the load remains stationary and the heating zones progress, has no parallel in metallurgical furnace types.

Figure 13:1 shows a breakdown of furnace types based on classifications (1) and (2) above.

1 Directly fired furnaces

Directly fired furnaces make up the bulk of industrial-furnace types although in the metallurgical field indirectly fired furnaces using protective atmospheres are increasingly being used. Semi-muffle furnaces are sometimes employed in which the roof of the muffle is omitted, the muffle sides serving to protect the stock from flame impingement rather than from the effects of the gases. In a way the crucible furnaces used in nonferrous metal melting (Figure 13:16) are semi-muffle furnaces, the crucible being the muffle but some direct heating occurring at the metal surface. They are regarded, however, as indirectly fired for the purposes of this classification.

1.1 Directly fired, continuous furnaces

These are most conveniently classified on the basis of the conveyance mechanism employed [1] which may involve:

1.1.1 Straight-line motion.
1.1.2 Circular motion.
1.1.3 Helical motion.

For a comprehensive treatment of continuous-furnace conveyance systems the reader is referred to reference [1].

1.1.1 (a) Pusher furnace. In this commonly used system the stock is moved through the furnace by pushing along the hearth. It is the most widely used type for reheating steel to rolling or forging temperatures. A major drawback is the possibility of pile-ups occurring but this is reduced by inclining the hearth downwards in the direction of motion so as to reduce the pusher force required. In most steel-reheating pusher furnaces the stock slides on skid rails in order to avoid excessive hearth wear and reduce the pusher force required.

In general, pusher furnaces are used for heating relatively large regular shapes (cuboids, cylinders, etc.) to high temperatures at which the simplicity of the conveyance method offers particular advantages. The principle has in recent years been applied to new designs of continuous pottery kilns as replacements for car-bogie tunnel kilns. Such kilns include multipassage kilns and the Trent kiln (Figure 13:42(*c*) and 13:43) in which the ware is loaded on flat tiles or 'bats' standing on refractory bricks which are pushed through the furnace by a hydraulic ram. The system offers many advantages over the cumbersome conventional car-bogie, in particular, the much reduced deadload-to-payload ratio. A comparison of the two systems is presented in Section 13.6.2.

1.1.1 (b) Conveyor furnaces. Conveyors for continuous furnaces cover a wide variety of types including chain conveyors, chain-link belts, solid metal belts and wire-mesh belts [1].

Conveyors may remain in the furnace while going and returning, or may return outside the furnace. Generally, in the interests of fuel economy, the former is preferable. Also, in furnaces with controlled atmospheres the former method simplifies sealing. Wear on conveyor moving-parts increases with temperature whilst their ability to carry loads without stretching decreases. Conveyors are generally limited to relatively light stock and fairly low temperatures.

Systems in which the chain conveyor remains relatively cold have advantages in terms of both maintenance and fuel economy. The conveyors travel entirely outside the furnace or alternatively in grooves in the hearth. In both cases they are equipped with attachments which carry the stock through the furnace (see Figure 13:2(*a*)). Woven mesh belts of stainless steel or high nickel-chrome alloys are widely used for transporting light articles through furnaces. The flexibility of the material prevents breakdown due to repeated heating and cooling. A major advantage is the low heat storage but woven-mesh conveyors are limited to about 800 °C maximum working temperature. They are often used in continuous lehrs, (glass-annealing furnaces) and in heat-treatment furnaces for small metal components, often loaded on trays.

1.1.1 (*c*) *Roller furnaces.* Roller systems in their simplest form consist of a series of heat-resistant steel round bars with bearings and drive arrangements outside the furnace. The rollers may be driven by individual motors, by worms or by chains. An advantage from the fuel-economy standpoint is that, unlike conveyors, rollers extract no heat from the furnace (except a small amount by conduction). On the other hand, for high furnace temperatures, metallic rollers may need to be air- or water-cooled which may increase fuel consumption appreciably. Uncooled metallic rollers need to revolve fast enough to minimize the temperature difference between top and bottom, otherwise warping results. For higher temperatures refractory rollers may be used. Firebrick rollers used with water-cooled shafts are prevented from spalling by interspersing firebrick discs with metallic conducting rings. Silicon carbide is a suitable material for sleeving water-cooled rollers. Roller diameters range from 12.7 mm ($\frac{1}{2}$ in) up to 305 mm (12 in).

1.1.1 (*d*) *Shaker and vibrating hearths.* Shaker hearths are used to transport small flat articles, like knife blades and washers, through furnaces by a system based on using inertia and friction. The hearth moves forward with an acceleration inadequate to cause the charge to slide on the hearth. The hearth is then stopped suddenly and the charge slides forward, the hearth then being retracted slowly. Shaker hearths are not suitable for large stock or stock that rolls. Since the hearth is of metal this limits the working temperature. Vibrating hearths are inclined downwards at a small angle. As the hearth vibrates the charge is thrown up at right angles to the hearth but falls vertically thus moving forward slightly.

1.1.1 (*e*) *Walking-beam furnaces.* A walking-beam furnace comprises a slotted stationary hearth with one or (usually) more moving beams having a lifting and reciprocating movement, lifting, carrying forward and depositing the stock again on the hearth (Figure 13:2(*b*)). Scale deposition in the hearth slots presents some difficulties but walking-beam furnaces are quite widely used for large, heavy stock.

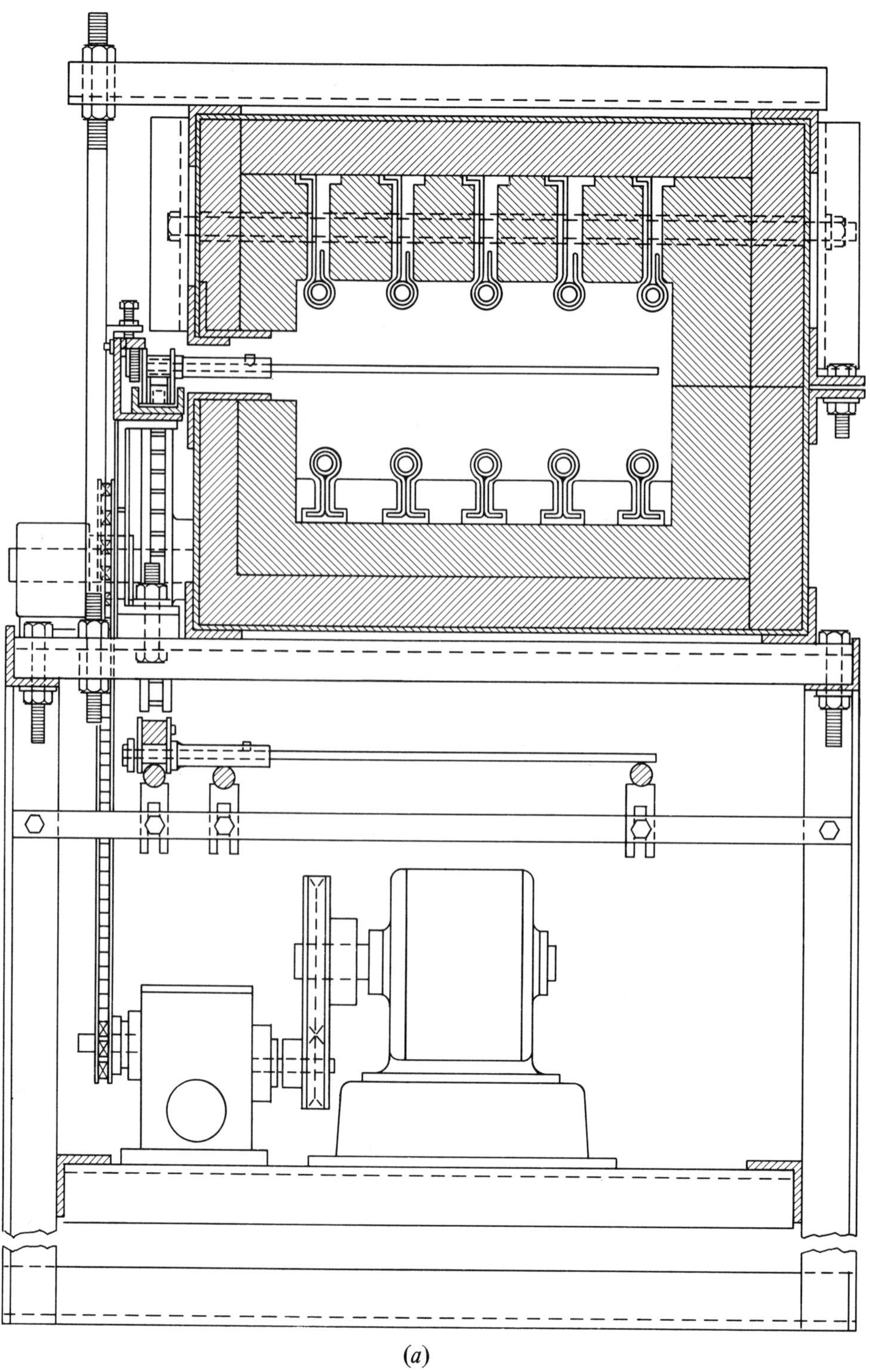

(a)

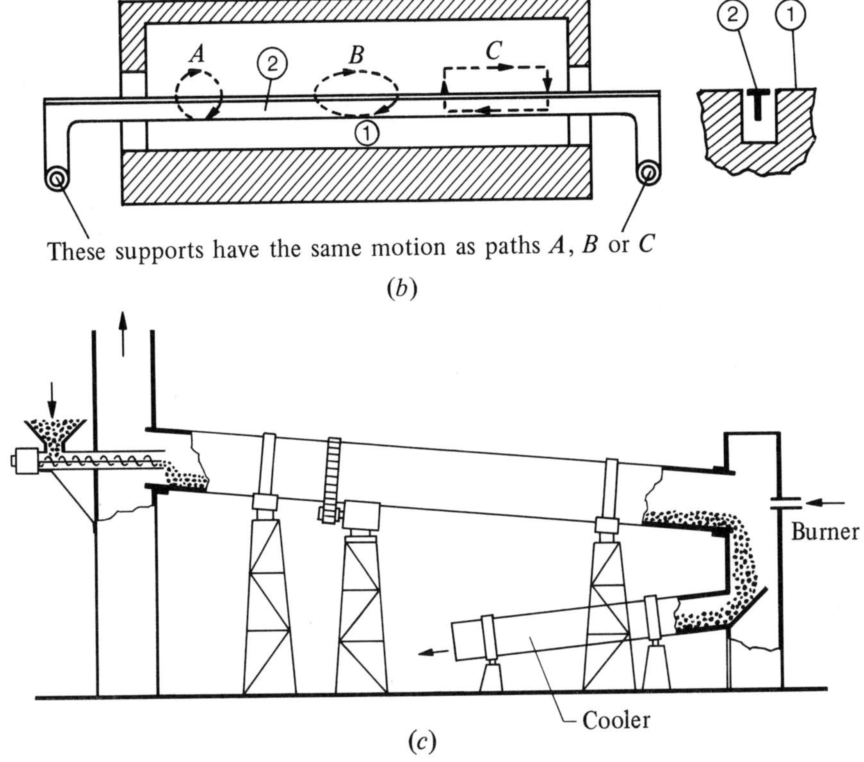

Figure 13:2 Directly fired continuous furnaces
(*a*) Cold-chain conveyor furnace [1]
(*b*) Walking-beam furnace: diagram of mechanism [1]
(*c*) Rotary kiln

1.1.1 (f) Overhead-conveyor furnaces. Some objects are particularly suited to being transported suspended from an overhead monorail conveyor. This system is commonly employed in low-temperature continuous ovens (see Section 15.2) but is not commonly used in furnace work. Two reasons for its lack of popularity are the extra complexity of the split furnace arch and the difficulty of sealing the slot from excessive hot-gas leakage. Overhead conveyors are best suited to transporting light objects at low temperatures (up to 850 °C) although Trinks [1] has reported systems operating up to 1300 °C.

1.1.1 (g) Car-bogie kilns. In car-bogie tunnel kilns the conveyance consists of a train of heavy, wheeled refractory cars running on rails in a sand-sealed, usually air-cooled, duct (Figure 13:42(*a*) and (*b*)). The system is widely used in ceramics firing. Comparative ease of loading and unloading makes it suitable for large and heavy objects and kilns with elaborate furniture. Car-bogie tunnel kilns are large (typically 30–90 m, 100–300 ft long) and incorporate extensive waste-heat recovery arrangements. They are considered in more detail in Section 13.6.2. More modern designs of continuous kiln have abandoned the heavy and cumbersome car-bogie for simple pusher systems.

1.1.1 (h) Self-conveyor furnaces. For loads like wire, strip and chains the load may conveniently act as its own conveyor, being pulled through from the outlet end. Wire patenting (Section 13.4.1) is almost universally carried out in this way. Furnaces may be horizontal or vertical, the latter being commonly used for strip annealing in which the work makes several up-and-down traverses guided by rollers before leaving the furnace.

1.1.1 (j) Tower furnaces. In packed-tower melting furnaces the charge of ingots or scrap is fed into a vertical shaft as a randomly packed bed and, as melting proceeds, travels by gravity down the shaft counter-current to the flow of combustion products from burners at the base of the shaft (Figure 13:22). This is an inherently efficient melting technique and efficiencies as high as 70 per cent have been obtained for aluminium melting. The method suffers from some practical drawbacks, however, as outlined in Section 13.3.1. Vertical-shaft kilns used for lime manufacture, etc., also fit into this category.

1.1.2 Circular motion

1.1.2 (a) Rotary-hearth furnaces. The simplest arrangement is a flat rotating hearth with one door serving for both loading and unloading access. The design is compact compared with straightline systems and since the charge does not move relative to the hearth the system is particularly suitable for objects like tall containers which could not safely be pushed through a furnace.

1.1.2 (b) Annular kilns. These differ from other continuous furnaces reviewed in this section in a very fundamental way; the charge remains stationary and the furnace zones rotate. These kilns, which may be longitudinal (Figure 13:45) or transverse arch, are widely used in the heavy-clay industry.

1.1.3 Helical motion

The conveyor is an inclined drum which rotates about its axis, continually tending to carry the material up the side then allowing it to fall vertically back. The material travels from the top to the bottom of the kiln in a zigzag fashion, being tumbled over repeatedly in the process. Drum furnaces are used in the heat treatment of small metal components but find their main use as rotary kilns in cement manufacture, calcining processes, etc. (Figure 13:2(*c*)).

1.2 Directly fired intermittent furnaces

This classification covers the largest and, in principle, the simplest, group of furnaces. They are subdivided partly on the basis of the charging and discharging arrangements employed and partly on their function. The most basic method of charging is employed in oven and box furnaces and in intermittent kilns (type 1.2.1) where the charge is placed on the furnace hearth, heated, cooled to a handleable temperature and removed.

Oven furnaces differ from box furnaces in being smaller and having the hearth at waist level for easy charging. Hand placing and removal is the general rule for ceramics and for fairly light metal objects. Heavy metal objects may be placed using fork trucks

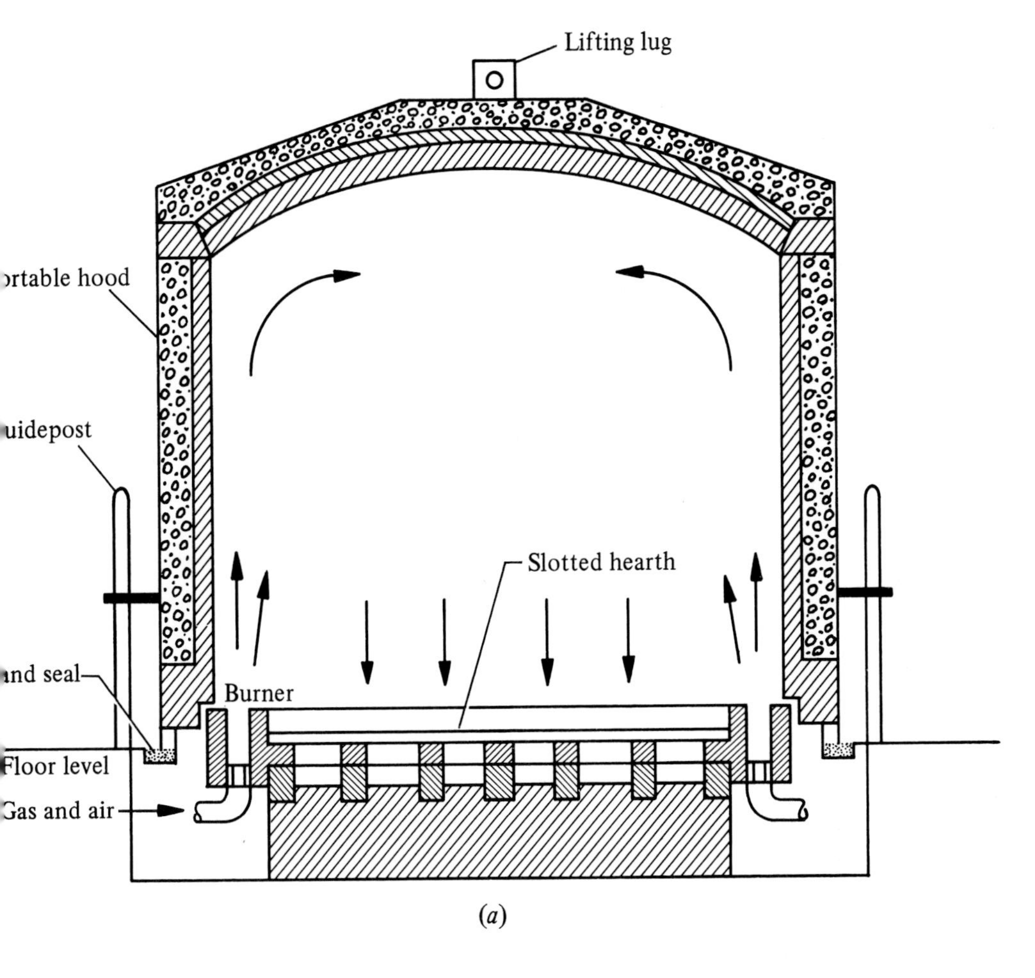

Figure 13:3 Directly fired intermittent furnaces
(*a*) Portable-cover furnace
(*b*) Bogie-hearth furnace
(*c*) Slot forge furnace [3]

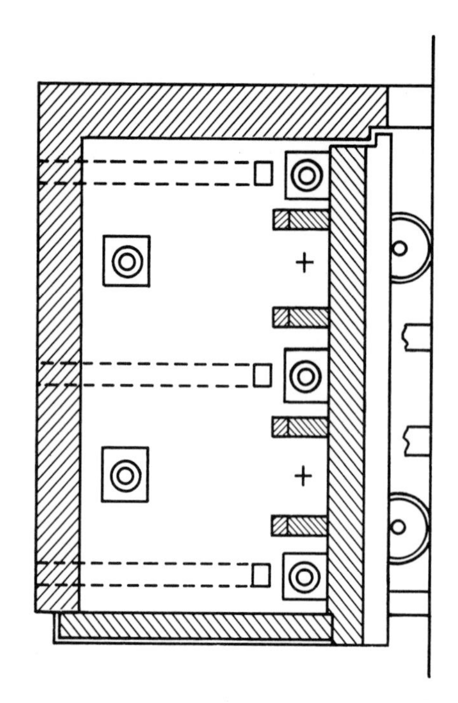

(b)

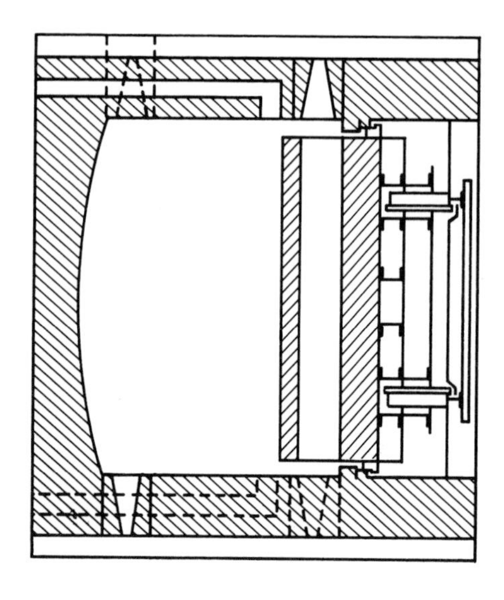

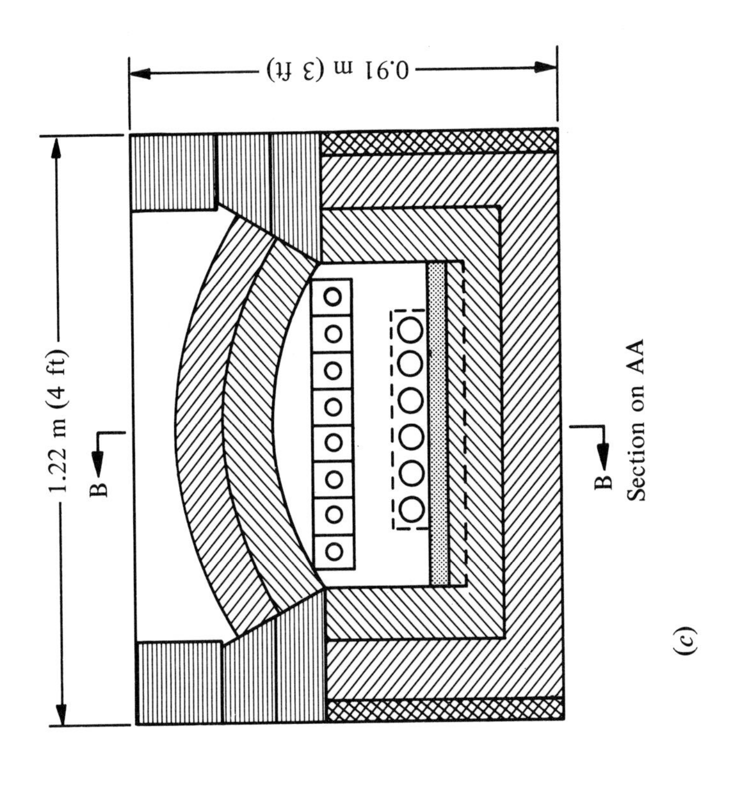
0.91 m (3 ft)
1.22 m (4 ft)
B
B
Section on AA

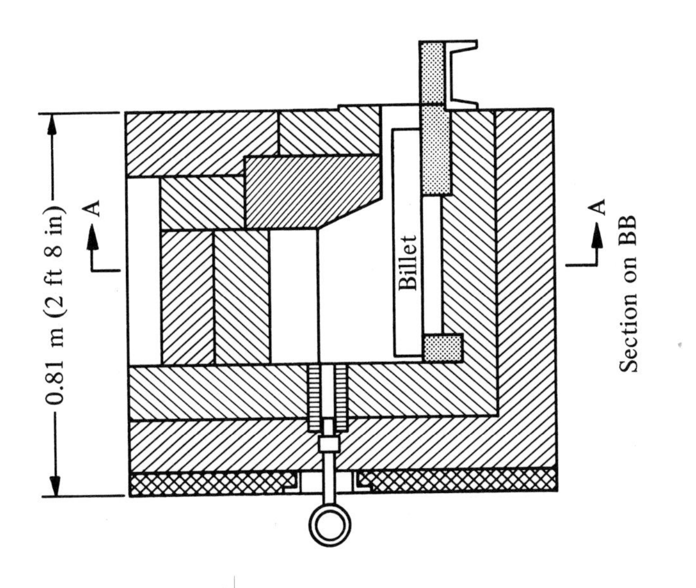
0.81 m (2 ft 8 in)
A
A
Billet
Section on BB

(*c*)

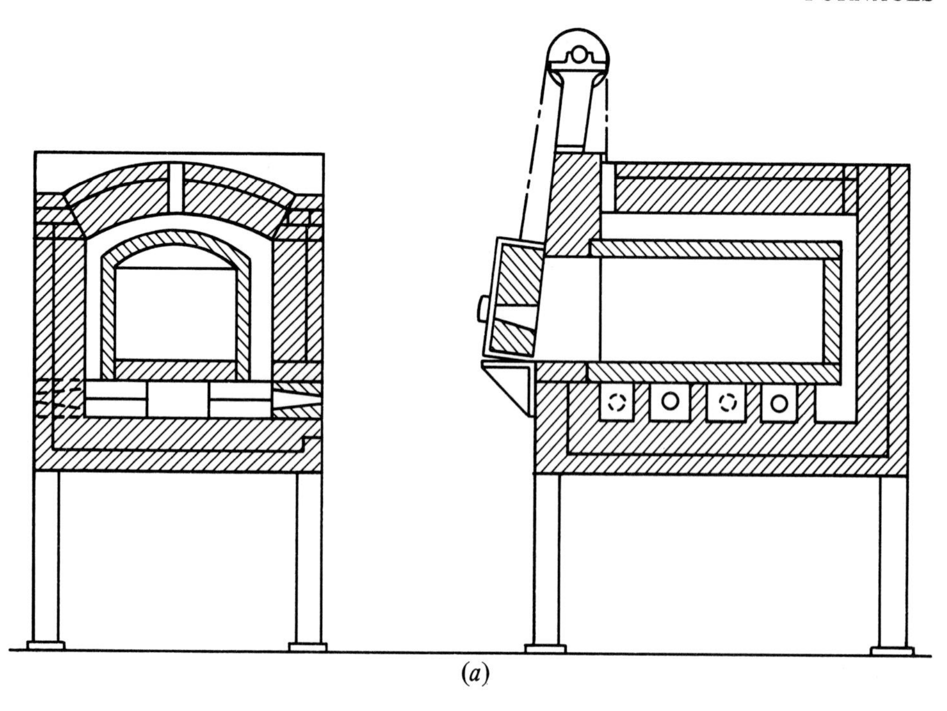

(*a*)

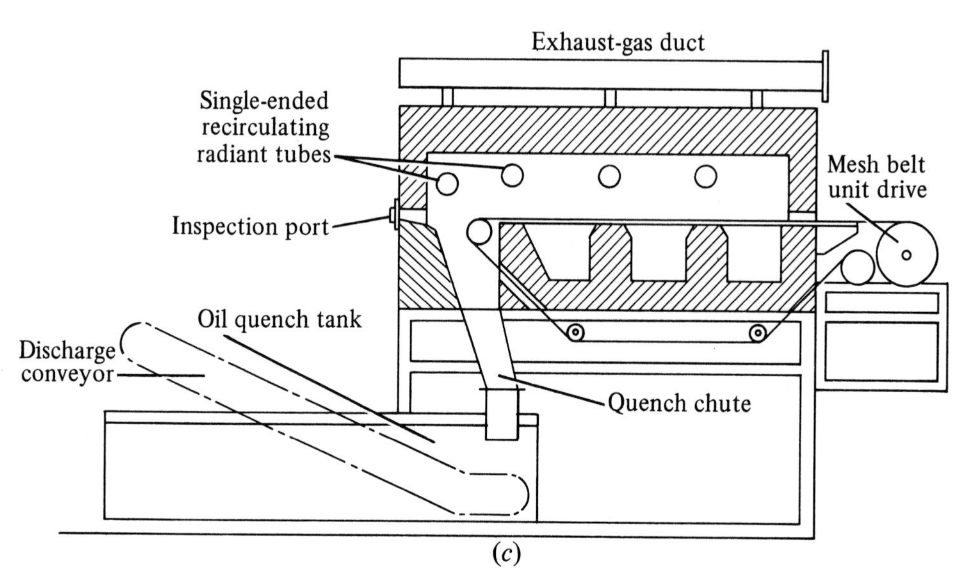

(*c*)

Figure 13:4 Indirectly fired furnaces
(*a*) Small underfired muffle furnace
(*b*) Radiant-tube portable-cover furnace [5]
(*c*) Radiant-tube continuous-conveyor furnace

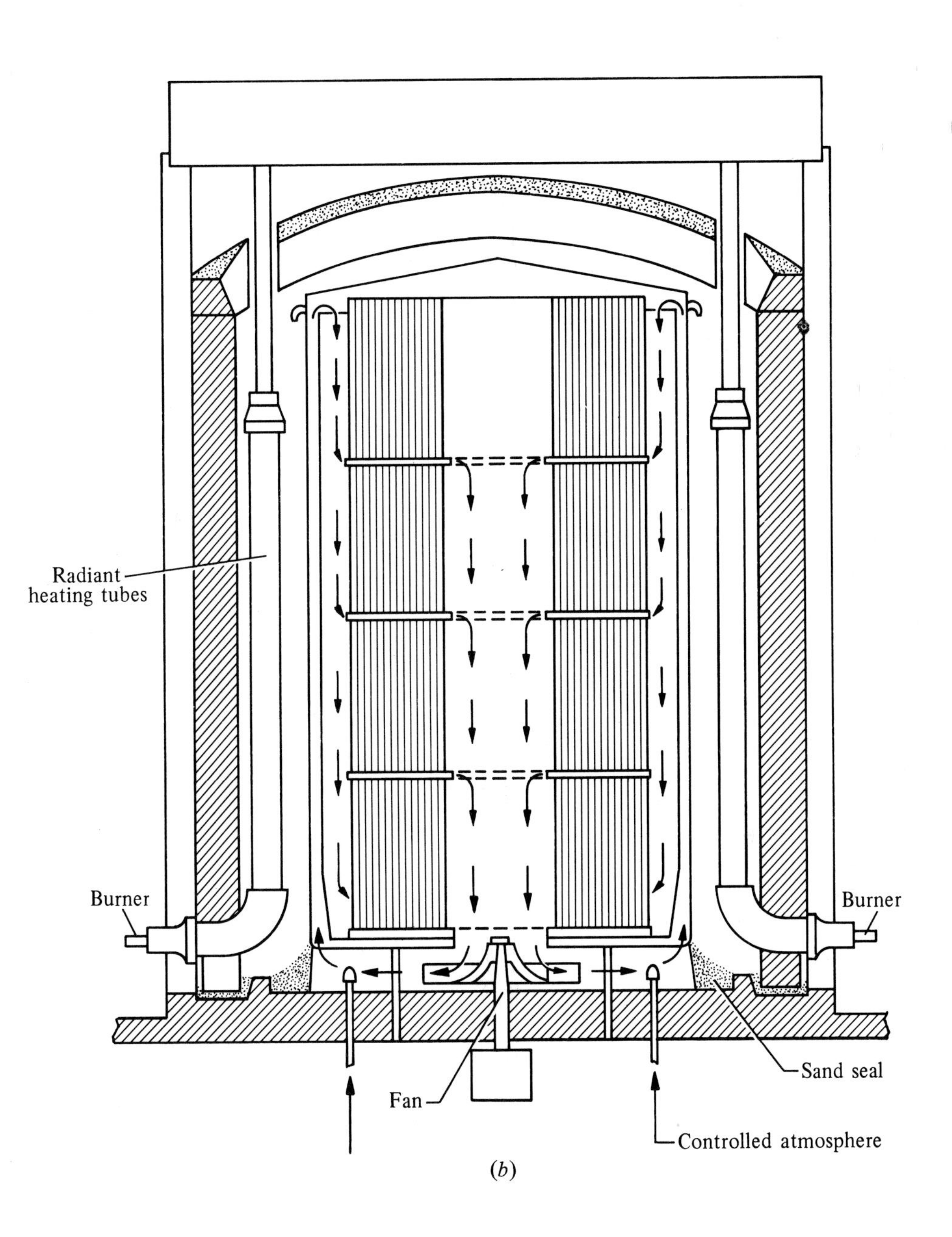
Radiant heating tubes
Burner
Burner
Sand seal
Fan
Controlled atmosphere

(*b*)

or porter-bars or in some instances specially designed furnace chargers and manipulators. Heavy objects like steel ingots are very cumbersome to handle even with charging machines and for this reason furnaces for their heat treatment are commonly constructed with removable roofs, i.e. pit furnaces (type 1.2.3) and are loaded using an overhead crane.

A similar principle is used in portable-cover furnaces (type 1.2.4) (Figure 13:3(*a*)) which differ in that the whole furnace structure except the hearth is removed for easy placing of the stock. Portable-cover furnaces are not restricted to heavy loads and since the hearth does not move they are suitable for tall stacks of material and are finding increasing use in ceramics firing. The cover is built as lightly as possible and the use of two furnace bases to one cover improves fuel economy and reduces capital costs. Bogie-hearth furnaces (type 1.2.2) owe their existence also to easy loading and unloading. In these, the hearth is withdrawn on rails, loaded and wheeled back into the furnace, the moving parts being protected from the hot gases by longitudinal sand seals. Figure 13:3(*b*) illustrates a typical bogie-hearth furnace suitable for heavy pieces. Better temperature uniformity is achieved in slotted-hearth designs but these are less suitable for very heavy loads.

Forge furnaces (type 1.2.5) are small-scale high-temperature furnaces used to bring steel bars or small billets quickly to forging temperatures. If the ends of a number of bars are to be heated simultaneously, as in drop-forging, the furnace contains a horizontal slot through which the bars enter (and which acts as the flue) and the furnace is termed a slot forge furnace (Figure 13:3(*c*)). Simple forge furnaces as shown are inefficient and their rates of heating are less than ideal. As indicated earlier in this chapter great improvements have been achieved in heating rates by the use of high-velocity tunnel burners and furnace chambers tailored to the stock to promote high convective heat-transfer rates and a high degree of recirculation. This has led to the development of rapid billet heaters (type 1.2.6) and a variety of localized heating furnaces (type 1.2.7), e.g. plate- and bar-end heaters.

The final group of intermittent directly fired furnaces considered are reverberatory furnaces (type 1.2.8). These are normally metal-melting furnaces (the tank furnace used in glass manufacture is a reverberatory furnace, but a continuous one) and the term strictly relates to furnaces in which the flame is deflected onto the hearth by an arched or sloping roof. The term is used generically, however, for melting furnaces in which the combustion products are in contact with the surface of a pool of molten metal.

Bulk melting of nonferrous metals is done in reverberatory furnaces and the open-hearth steel furnace can be classified as reverberatory.

2 Indirectly fired furnaces

Indirect furnaces can be subdivided on the basis of the type of solid partition used to separate the heating gases from the material to be heated. This partition may be a refractory (or in some cases metal) muffle (2.2.1, 2.1.1) (Figure 13:4(*a*)), a radiant tube (2.1.2, 2.2.2) (Figure 13:4(*b*) and (*c*)) or the refractory or metal wall of a crucible or bath furnace (2.1.3, 2.2.3) (Figure 13:16).

The main aim of the muffle is to protect the charge from the effects of the products of combustion and it is normally used in conjunction with a separately generated controlled atmosphere. Muffles are also used sometimes to minimize temperature inequalities. The heating capacity of an indirectly fired furnace is less than that of

Industry or furnace type	Operating temperature (°C)	Natural-gas consumption (m^3 (st) t^{-1})	Heat consumption (MJ kg^{-1})
Iron and steel industry			
Melting furnaces			
OH furnaces	1700	135–172	4.75–6.00
OH furnace for steel castings	1700	160–295	5.55–10.30
Reheating furnaces			
Pusher furnaces	1150–1350	37–55	1.25–1.85
Rotary-hearth furnaces	1150–1280	50–62	1.65–2.05
Roller hearth furnaces for:			
Steel tubes	1050–1100	25–50	0.80–1.65
Plates	1050–1100	42–55	1.45–1.85
Packs	1050–1100	20–50	0.65–1.65
Walking-beam furnaces for:			
Plates	1050–1100	37–50	1.25–1.65
Packs	1050–1100	15–52	0.50–1.85
Bogie-hearth furnace for:			
Large forgings, first heat	1250–1300	62–122	2.05–4.35
Subsequent heats	1250–1300	25–50	0.80–1.65
Die-forge furnaces:			
Chamber furnace for:			
Bar end heating	1200–1300	75–245	2.45–8.65
Piece heating	1200–1300	72–245	6.00–8.65
Short pusher furnaces ($l:b < 4:1$)	1200–1300	97–197	3.5–6.80
Longer pusher furnaces ($l:b > 6:1$)	1200–1300	65–122	2.25–4.35
Intermittent furnaces	1200–1300	97–197	3.5–6.80
Rivet and bolt heating furnaces	1250–1300	97–122	3.5–4.35
Heat-treatment furnaces			
Bogie hearth large annealers (castings)	800–900	50–97	1.65–4.35
Continuous furnaces for forged pieces:			
Normalizing, hardening	800–1000	55–70	1.85–2.45
Tempering	450–700	37–55	1.25–1.85
Continuous furnaces for steel strip	600–1050	17–50	0.60–1.65
Coiled strip annealers (l-o)	600–900	17–40	0.60–1.45
Lift-off annealers	650–850	50–60	1.65–2.05
Pack annealers	700–850	75–97	2.45–3.50
Heavy plate normalizers	800–920	40–60	1.45–2.05
Tube and bar annealers	900–950	32–75	1.05–2.45
Wire patenting furnaces	800–900	40–97	1.45–3.50

Figure 13:5 Approximate performance characteristics of heat-using processes

Industry or furnace type	Operating temperature (°C)	Natural-gas consumption (m^3 (st) t^{-1})	Heat consumption (MJ kg^{-1})
Tempering furnaces			
White-heart elevator annealer	1050	197–245	6.60–8.25
Black-heart elevator annealer	950	50–87	2.45–3.10
White-heart continuous annealer	1050	122–147	4.35–5.15
Black-heart continuous annealer (ferritic)	950	82–97	2.90–3.50
Black-heart continuous annealer (pearlitic)	950	75–975	2.45–3.50
Enamelling furnaces			
Continuous reversing furnaces: Throughput (kg/h)			
for sheet 500		82–117	2.90–4.10
1000	900–950	50–75	1.65–2.45
3000		22–35	0.75–1.25
for castings 500		72–102	2.45–3.50
1000	700–800	42–65	1.45–2.25
3000		22–32	0.75–1.05
Chamber furnaces:			
for sheet	900–950	97–245	3.50–8.65
for castings	700–800	97–210	3.50–7.20
Galvanizing furnaces	445–460	20–50	0.65–1.65
Drying stoves			
For steel casting moulds (recirculation)			
For grey-iron casting moulds (recirculation)	500–600	3–4	0.105–0.135
For grey-iron casting moulds (recirculation)		0.5–1.0	0.015–0.035
For grey-iron casting moulds (no recirculation)	200–250	1.5–3.0	0.050–0.105
Nonferrous metal industry			
Copper refiners (barrel furnaces):			
For particulate scrap	1420	87–107	3.10–3.70
For wire scrap		122–172	4.35–6.00
Pusher furnace (copper)	800–900	25–40	0.80–1.45
Roller hearth (Cu-brass tube)	800–900	15–30	0.50–1.05
Brass melting	1000	122–147	4.35–5.15
Brass reheating	700–800	25–50	0.80–1.65
Aluminium reverberatory melting	750	122–147	4.35–5.15
Aluminium crucible melting	750	122–147	4.35–6.00
Aluminium radiant tube roller hearth	500	17–22	0.60–0.75
Bunt metal radiant tube roller hearth	550–750	10–30	0.35–1.05

Industry or furnace type	Operating temperature (°C)	Natural-gas consumption (m^3 (st) t^{-1})	Heat consumption (MJ kg^{-1})
Glass industry			
Pot furnace, up to 6 pots:			
Melting		540–712	18.50–24.70
Holding		295–392	10.30–13.60
Tank furnace, up to 20 tonnes capacity:			
Tank firing		417–690	14.40–23.90
Feeder heating		65–87	2.25–3.10
Tank furnace, over 20 tonnes capacity:			
Tank firing		295–447	10.30–16.10
Feeder heating		50–75	1.65–2.45
Lehr coolers	500	10–15	0.35–0.50
Ceramics industry			
Batch kiln, sanitary ware	1200–1300	245–885	8.65–30.50
Batch kiln, drainpipes	1200–1300	245–392	8.65–13.60
Tunnel kiln, sanitary ware	1200–1350	222–295	7.60–10.30
Tunnel kiln, fine ceramics	1270–1370	60–72	2.05–2.45
Tunnel kiln, fireclay ware	1100–1200	172–227	5.75–7.85

Figure 13:5 Approximate performance characteristics of heat-using processes

an equivalent directly fired furnace because of the absence of combustion products which are hotter than the furnace wall and the effect of the muffle as a radiation shield which reduces wall radiation by about one-half [1]. For temperatures up to about 750 °C, cast-iron or calorized-steel muffles are reasonably durable. High-chromium and nickel alloys may be used for higher temperatures but in general high-temperature muffles are of refractory materials. If fireclay is used, the large temperature differential across the muffle wall reduces the heating capacity, increases the temperature of the waste gases and presents cracking problems. Thin-walled silicon-carbide muffles overcome most of these difficulties but, increasingly, the greater flexibility offered by radiant tubes is leading to their widespread adoption. Radiant tubes are considered in Section 7.10.

Indirectly fired furnaces need to be operated at a slight positive pressure and, to minimize atmosphere requirements and because some atmospheres are toxic and combustible, they need to be gastight. To reduce the escape of atmosphere gases vestibules are frequently provided, often with double doors. The need for atmosphere sealing limits the type of conveyance used in continuous indirect furnaces. An alternative method of indirect heating is by conduction using a bath of molten salt or metal (usually lead) (types 2.1.3 and 2.2.3 (*a*)). Very high heating rates and good temperature uniformity are possible by this method. The bath may be used purely for heat transfer or may impart some change to the metal, e.g. cyaniding (Section 13.2.4) or galvanizing (Section 13.3.1).

Crucible furnaces make up the final group in this category. These furnaces are used primarily for nonferrous-metal melting and are described in Section 13.3.1.

Figure 13:5 lists the temperature ranges and typical heat consumptions of a wide range of heat-using processes including many referred to in this section.

13.2 FURNACE AND PROTECTIVE ATMOSPHERES

13.2.1 Introduction

An important aspect of the heat treatment of materials is the effect of the furnace atmosphere on the stock being heated. In most cases the need is to minimize or eliminate the undesirable effects of the furnace gases, e.g. oxidation or decarburization, but there are processes in which the interaction between the gases and the stock is the *raison d'être* of the process, e.g. the carburizing of steels.

In directly fired processes only a limited control of furnace atmosphere is possible and although this is adequate for some low-temperature heat-treatment processes, e.g. stress-relieving of low-carbon steels, an increasing number of heat-treatment processes are carried out in indirectly fired furnaces. In these the stock is separated from the heating gases by a muffle or radiant tubes and the working chamber is fed with a separately generated prepared atmosphere.

Protective atmospheres find their widest application in metallurgical heat-treatment processes but are also used for protective and fire-prevention purposes in, for example, the paint, pharmaceuticals and rubber industries and for purging. Coverage in this section will be limited to their application in the metallurgical field.

13.2.2 Effects of gases on metals

The terms 'oxidizing', 'neutral' and 'reducing' are often applied to furnace atmospheres. A common misconception is that if a furnace atmosphere contains no free oxygen it is necessarily neutral in its effects. This is far from the truth since carbon dioxide and water vapour are both oxidizing agents and furthermore a furnace atmosphere which may be neutral to one material may be highly oxidizing to another. Additionally the temperature of the process exerts an important influence on the neutrality or otherwise of the furnace atmosphere. Consequently the terms oxidizing, neutral and reducing when applied to furnace gases produced from combustion with excess air, stoichiometric air and sub-stoichiometric air respectively, should not be taken as implying that these effects will occur.

The consideration of metal–gas reactions occurring in the furnace chamber is essentially one of the chemical equilibria existing at the temperature and pressure of the process.

The main reaction to be considered is oxidation and this may be caused by CO_2 and H_2O as well as free oxygen, i.e.

(1) $R + CO_2 \rightleftharpoons RO + CO$

(2) $R + H_2O \rightleftharpoons RO + H_2$

(3) $R + O \rightleftharpoons RO$

The equilibrium constants K are given by equation 4.28 of Chapter 4, i.e. for reaction (1):

$$K_1 = p_{RO}\, p_{CO} / p_R\, p_{CO_2} \qquad (13.1)$$

where p denotes the partial pressure.

The vapour pressures of the metal and its oxides are negligibly small and 13.1 may be rewritten:

$$K_1 = p_{CO}/p_{CO_2} \tag{13.2}$$

Consider the reaction between iron and carbon dioxide:

(4) $$Fe + CO_2 \rightleftharpoons FeO + CO$$

Figure 13:6 shows the variation of the equilibrium constant with temperature.

Temperature (°C)	450	500	550	600	650	700	750	800
$K = p_{CO}/p_{CO_2}$	0.87	0.952	1.02	1.18	1.36	1.52	1.72	1.89
Temperature (°C)	850	900	950	1000	1050	1100	1150	
$K = p_{CO}/p_{CO_2}$	2.05	2.20	2.38	2.53	2.67	2.85	2.99	

Figure 13:6 Variation of equilibrium constant with temperature for the reaction $Fe + CO_2 \rightleftharpoons FeO + CO$ [6]

As a simple illustration of the use of the data consider a furnace atmosphere at 950 °C containing 4 per cent CO_2 and 8 per cent CO, the remainder being N_2. The furnace pressure is 1 atmosphere.

From Figure 13:6 the equilibrium constant for the oxidation reaction at 950 °C is 2.38. p_{CO}/p_{CO_2} for the gas is $0.08 \times 1/0.04 \times 1 = 2.00$.

Since this is below the value of the equilibrium constant the reaction would proceed to the right towards equilibrium, i.e. the iron would oxidize. If the temperature were 800 °C the reaction would proceed to the left since the equilibrium constant (1.89) is less than the ratio of partial pressures.

The concept of dissociation pressure is useful in indicating the oxidation tendencies of metals. If a metal oxide is heated in a pure inert gas, e.g. argon, it will tend to dissociate i.e. $RO \rightleftharpoons R + O$ and dissociation will proceed until the partial pressure of the oxygen in the inert gas reaches a value defined as the dissociation pressure of that oxide. The system is now in equilibrium and further dissociation would require either an increase in temperature or a change of atmosphere to reduce the oxygen content. If the oxygen were continually removed eventually complete dissociation would occur. However, the dissociation pressures of most metal oxides are so low at feasible practical temperatures that this method could not be used commercially for oxide reduction [7]. If, on the other hand, the atmosphere contains an easily oxidized gas, hydrogen for example, the partial pressure of the oxygen above the metal and oxide may be reduced below its equilibrium value and complete reduction may occur.

Clearly the dissociation pressure of an oxide has a strong influence on the composition of a prepared atmosphere intended to prevent oxidation. Metals whose oxides have high dissociation pressures (e.g. copper, 4.4×10^{-1} atm (44.6 kPa) at 2000 °C) are resistant to oxidation and undemanding in atmosphere requirements. Metals whose oxides have low dissociation pressures (e.g. chromium, 1.0×10^{-33} atm (1.01×10^{-28} Pa) at 2000 °C) require atmospheres of very high purity.

The tendency of a metal to oxidize is related to the free energy of formation of its

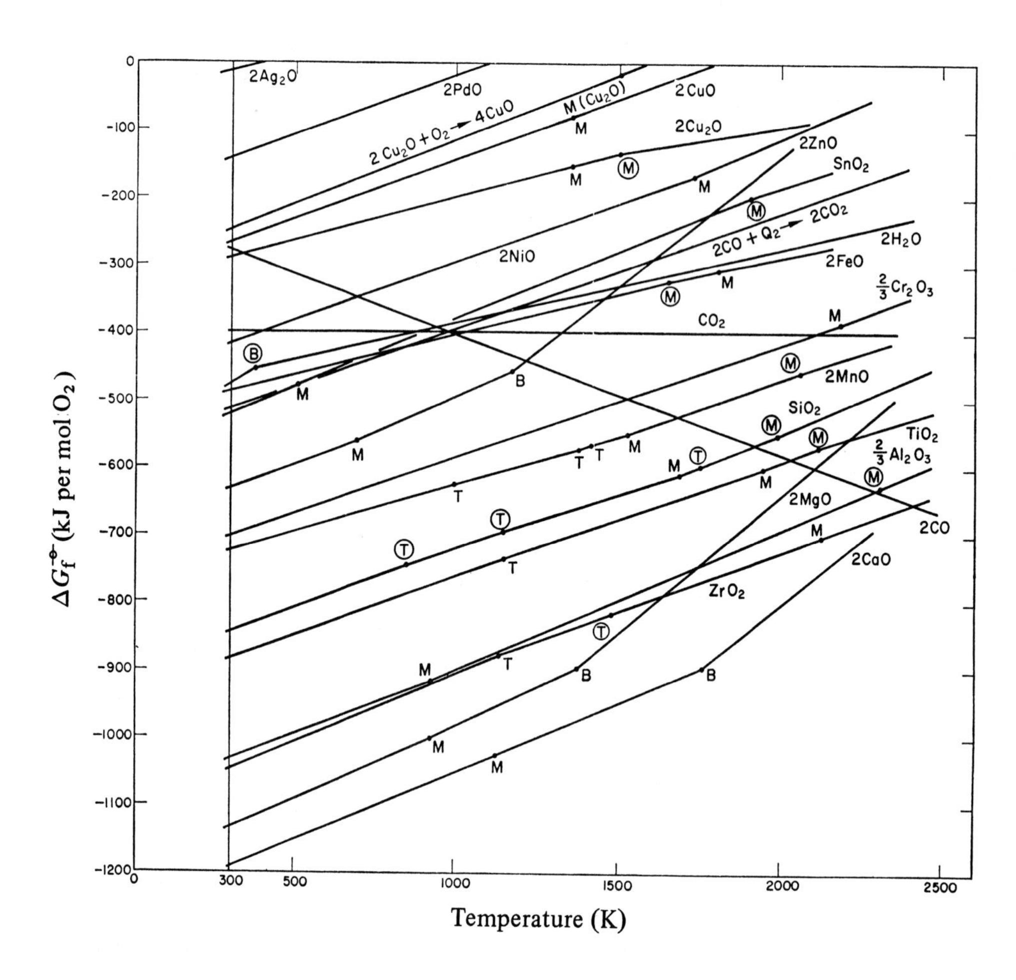

Figure 13:7 Simplified Ellingham diagram for oxide formation [9]. M, B and T denote the melting, boiling and transition points for the elements. Circled they refer to the oxides

oxide. As outlined in Chapter 4 the equilibrium constant is linked with free energy ($\Delta G^{\ominus}$):

$$\Delta G^{\ominus} = -RT\ln K \qquad (4.29)$$

In high-temperature processes it is necessary to take account of the variation of $\Delta G^{\ominus}$ with temperature and this is conveniently presented in the form of Ellingham diagrams [8] in which free-energy data for many reactions of one type (e.g. metal oxidation) are plotted against temperature providing, in Ellingham's words 'a ground-plan of metallurgical possibilities'. A simplified Ellingham diagram is shown in Figure 13:7.

A complete discussion is outside the scope of this book but Figure 13:7 illustrates a number of general points. First the greater the negative value of the free energy of the oxide the more thermodynamically stable is the oxide and the greater the difficulty of preventing oxidation [10]. With the important exception of CO the lines generally have a positive slope indicating that the oxides become less stable at elevated temperatures.

Carbon monoxide is the only oxide to increase in stability as the temperature increases, its line moving under those of the other oxides indicating that almost any metal oxide is reducible by CO at a sufficiently high temperature.

As an example [9], up to 1680 K, MnO is stable in the presence of carbon. At 1680 K the MnO and CO lines intersect and $\Delta G^{\ominus}$ for the reaction $MnO + C \rightarrow Mn + CO$ is zero, i.e. $\Delta G_f^{\ominus}(CO) - \Delta G_f^{\ominus}(MnO) = 0$. Thus from equation 4.29, $K = 1$, and, at a slightly higher temperature, liquid manganese starts to form. It is on this ability of CO to reduce metals at modest temperatures that the production of iron and many other metals is based.

In the case of aluminium, Figure 13:7 indicates that reduction will be possible above about 2300 K where the Al_2O_3 and CO lines intersect. Since the cost of producing such temperatures is prohibitive, electrolytic methods are employed in aluminium extraction. As Warn [9] has indicated, a free-energy change of 600 kJ per mol O_2 can be overcome by as little as 1.6 volts applied potential allowing reduction to take place at more modest temperatures.

For further information on the applications of Ellingham diagrams the reader is referred to books by Warn [9] and Ives [11].

Furnace gases and most prepared atmospheres consist of multicomponent gas mixtures and so in practice it is necessary to consider not only the reactions which may take place between the gases and the solids but between the gases themselves. In addition, the metals being treated, in particular nickel and chromium and to a lesser extent, iron, may act as catalysts on the reactions between the constituents of the furnace atmosphere. The inter-gas reactions are of course temperature-dependent and it is necessary to consider the atmosphere conditions at all stages of a heat-treatment process including cooling.

Fortunately, as Fleming [10] has pointed out, the kinetics of most of the reactions are relatively slow even at the high temperatures involved and some latitude exists in the composition of prepared atmospheres.

With the exception of a few metallurgical processes which require the provision of inert gases from cylinders or the use of a vacuum, the vast majority of industrial protective atmospheres are derived from the combustion or reforming of fuel gases and contain some or all of the following constituents: nitrogen, carbon dioxide, carbon monoxide, hydrogen, water vapour and hydrocarbons. The atmospheres of direct gas-fired furnaces will contain these constituents and in addition sulphur dioxide will be present in the case of oil or coal firing.

Industrially the most significant reactions are those between these gases and the constituents of steel, i.e. ferrite (α-Fe) and cementite (Fe_3C) and these are considered below.

The following gas–metal and inter-gas reactions may be involved in the heat processing of ferrous metals:

Oxidation–reduction

(4) $Fe + CO_2 \rightleftharpoons FeO + CO$

(5) $Fe + H_2O \rightleftharpoons FeO + H_2$

Decarburizing–carburizing

(6) $Fe_3C + CO_2 \rightleftharpoons 3Fe + 2CO$

(7) $Fe_3C + H_2O \rightleftharpoons 3Fe + CO + H_2$

(8) $3Fe + CH_4 \rightleftharpoons Fe_3C + 2H_2$

Producer-gas reaction

(9) $C + CO_2 \rightleftharpoons 2CO$

Water–gas reaction

(10) $CO + H_2O \rightleftharpoons CO_2 + H_2$

Thermal decomposition

(11) $CH_4 \rightleftharpoons C + 2H_2$

Others

(12) $H_2O + C \rightleftharpoons CO + H_2$

(13) $H_2O + CH_4 \rightleftharpoons CO + 3H_2$

(14) $CO_2 + CH_4 \rightleftharpoons 2CO + 2H_2$

(15) $C + O_2 \rightleftharpoons CO_2$

The gas–metal reactions are oxidation–reduction and decarburization–carburization. Equilibrium considerations for reaction (4) have already been made and it is interesting to compare these with the decarburization reaction (6). Again considering a furnace atmosphere containing 4 per cent CO_2 and 8 per cent CO at 950 °C:

$$p_{CO}^2/p_{CO_2} = (0.08 \times 1)^2/(0.04 \times 1) = 0.16$$

Temperature (°C)	450	500	550	600	650	700	750	800	850
$K = p_{CO}^2/p_{CO_2}$	0.00045	0.0037	0.0186	0.0794	0.347	1.01	2.98	10.0	17.2

Temperature (°C)	900	950	1000	1050	1100	1150
$K = p_{CO}^2/p_{CO_2}$	37.6	74.1	143.0	462.0	759.0	822.0

Figure 13:8 Variation of equilibrium constant with temperature for the reaction $Fe_3C + CO_2 \rightleftharpoons 3Fe + 2CO$* [6]

*The decarburizing reaction (6) is more accurately expressed as $C_{Fe} + CO_2 \rightleftharpoons 2CO$. The solid reactant is in solution in the parent metal. At any temperature, for every carbon concentration there is a different equilibrium ratio of p_{CO}^2/p_{CO_2} and this is shown in Figure 13:10.

From Figure 13:8 the equilibrium constant at 950 °C is 74.1 and at 800 °C is 10.0 (assuming the iron to be completely saturated with carbon) and so in both cases the reaction would proceed to the right. Thus although the atmosphere under consideration is not oxidizing at 800 °C it is strongly decarburizing.

Figure 13:8 indicates the sharp rise in decarburizing power of CO_2 with temperature. To avoid decarburization at quite modest temperatures it is clearly necessary to employ very low CO_2 concentrations.

To summarize, reactions (4) to (7) involving reactions with CO_2 and H_2O cause oxidation and decarburization in steel. In general, it is necessary to keep water-vapour and carbon-dioxide levels to a minimum in heat-treatment processes. Whether the process is one of oxidation or decarburization CO and H_2 are reaction products and clearly an increase in their concentration will inhibit these reactions. It is possible, as Figure 13:9(*c*) indicates, to produce an atmosphere which is nonscaling but decarburizing, but nondecarburizing atmospheres must be nonscaling at all practical temperatures. For high-carbon steels (Figure 13:10) and high temperatures (Figure 13:9(*c*)) the CO/CO_2 ratio will need to be high primarily to prevent decarburization.

Figure 13:11 summarizes in general terms the effects of atmosphere constituents on ferrous and nonferrous metals.

13.2.3 Practical controlled atmospheres

Industrial controlled atmospheres are generated from two main sources, single relatively pure gases in cylinders and fuel gases. As an alternative to the use of a prepared atmosphere a vacuum is sometimes employed particularly for copper annealing (see Section 13.2.4). Of the gases drawn from cylinders pure hydrogen is used where extreme reducing conditions are required and argon for processes in which nitrogen is insufficiently inert. Ammonia is, however, the only cylinder gas which is widely used. This may be used straight in carbonitriding processes or more commonly cracked to produce its constituent elements:

(15) $$NH_3 \rightleftharpoons \tfrac{1}{2} N_2 + \tfrac{3}{2} H_2$$

Cracking is carried out over a catalyst at a temperature around 520 °C and the atmosphere produced used direct in processes requiring a strongly reducing atmosphere or burned and dried to reduce the hydrogen content or ultimately to produce nearly pure nitrogen. The heat release from the combustion process may be used to promote the cracking process using a regenerative burner [7]. Where atmospheres are required in large quantities, in general the cost and inconvenience of obtaining the gases in cylinders make it necessary to generate the atmosphere on site from a fuel gas.

Atmospheres derived from fuel gases may be termed endothermic or exothermic, the terms refer to the production processes, the former being produced by catalytic reforming of the gas with a small amount of air (the mixture being well beyond the rich limit of flammability) in an externally heated retort and the latter by combustion of the fuel with a controlled amount of air ranging from above-stoichiometric down to just sufficient to maintain combustion. These atmospheres have a wide range of composition which can be further extended and controlled by selective purification after generation if required. The processes are flexible and controllable and the gases

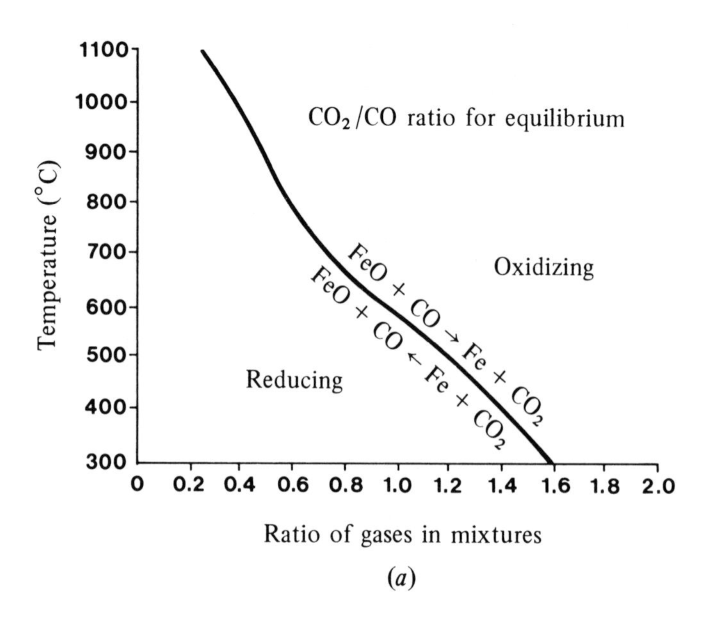
1100
1000
900
800
700
600
500
400
300
Temperature (°C)
0
0.2
0.4
0.6
0.8
1.0
1.2
1.4
1.6
1.8
2.0
Ratio of gases in mixtures
CO_2/CO ratio for equilibrium
Oxidizing
Reducing
FeO + CO → Fe + CO_2
FeO + CO ← Fe + CO_2

(*a*)

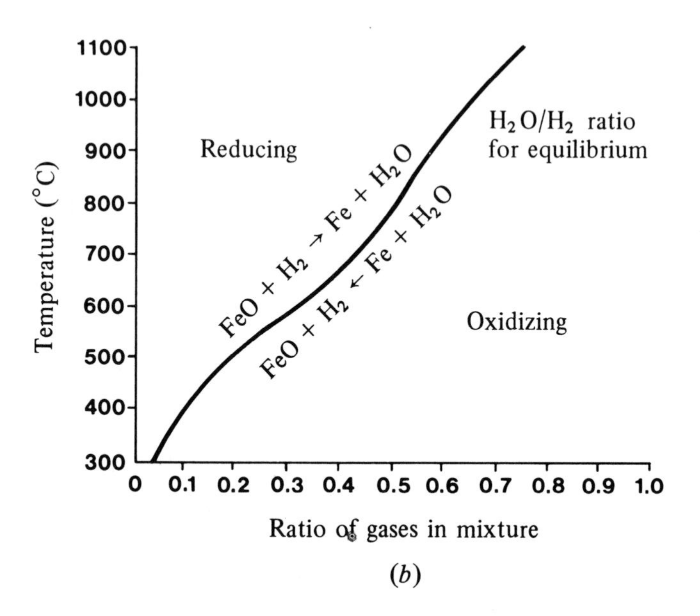
1100
1000
900
800
700
600
500
400
300
Temperature (°C)
0
0.1
0.2
0.3
0.4
0.5
0.6
0.7
0.8
0.9
1.0
Ratio of gases in mixture
Reducing
H_2O/H_2 ratio
for equilibrium
Oxidizing
FeO + H_2 → Fe + H_2O
FeO + H_2 ← Fe + H_2O

(*b*)

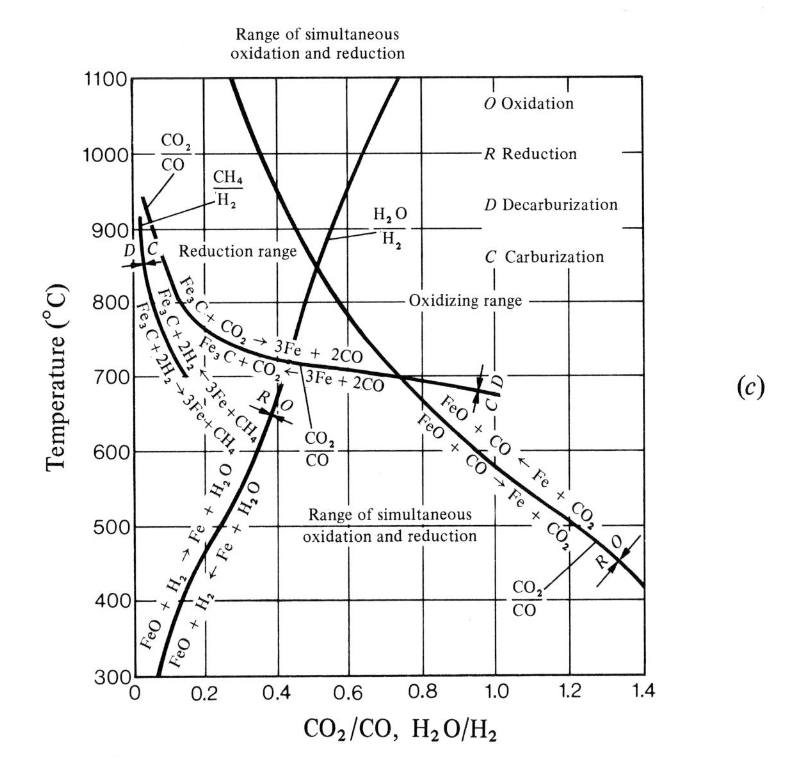

Figure 13:9 Equilibrium ratios for some oxidation and decarburization reactions [10, 12]

(*a*) Reaction (4)
(*b*) Reaction (5)
(*c*) Reactions (4), (5), (6) and (8)

obtained range from active atmospheres containing free oxygen and highly reducing and carburizing atmospheres to inert atmospheres comprising almost pure nitrogen.

Production and properties of exothermic atmospheres

Exothermic atmosphere generators consist essentially of a burner, combustion chamber and atmosphere cooler. Premixed burners are used almost exclusively and, although air-blast burners have been used, the most common arrangement is a machine-premix system (Section 7.7) which combines relatively accurate gas–air ratio control with a high mixture pressure to overcome the appreciable system resistance. Tests by Cubbage and Ling [13] have indicated that nozzle-mixing burners, because of their poor mixing characteristics, produce nonequilibrium combustion products and are therefore unsuitable for the generation of controlled atmospheres except for purging or fire-extinguishing purposes. Fleming [10], however, quotes one example of a very large nozzle-mixing burner (305 $m^3\ h^{-1}$; 11 000 $ft^3\ h^{-1}$) being used successfully for the generation of lean exothermic atmospheres close to stoichiometric composition.

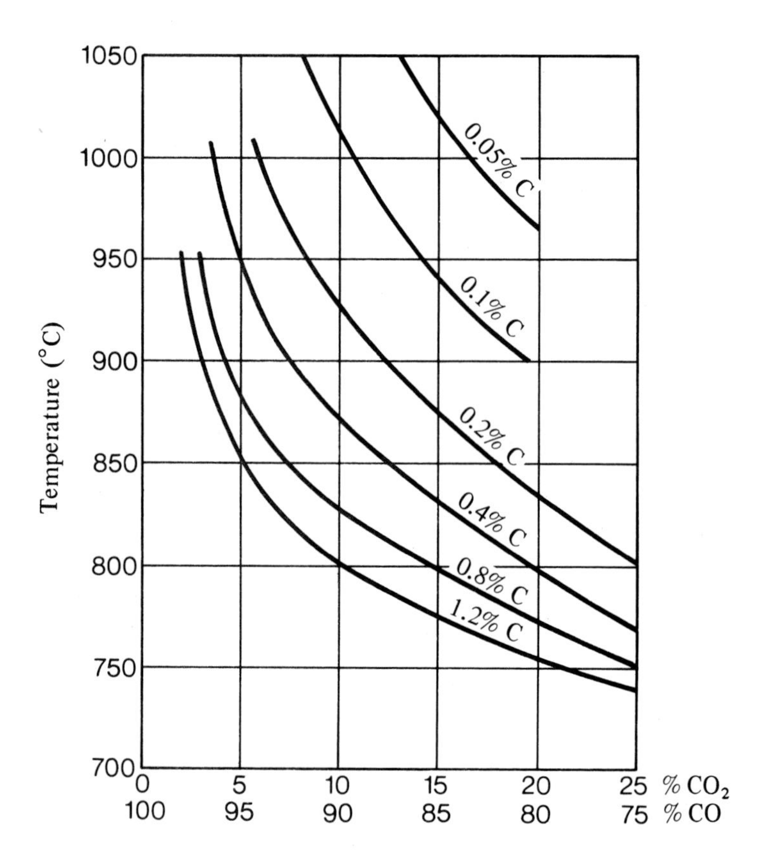

Figure 13:10 Variation with temperature of the equilibrium value of the CO_2/CO ratio in the reaction $C_{Fe} + CO_2 \rightleftharpoons 2CO$ for various concentrations of carbon [7]

Figure 13:12 shows in diagrammatic form a typical exothermic atmosphere generator.

Exothermic atmospheres may be classified as rich or lean, the limit of the former being dictated by the upper flammability limit. Lean exothermic atmospheres are produced by combustion around stoichiometric proportions, generally slightly below to produce a gas consisting of CO_2, N_2 and H_2O with a small excess of combustibles, but for a few specialized processes combustion with a very slight excess of air is used (see Section 13.2.4). The air–gas ratios on various fuels are typically [10]:

Natural gas:	lean 8.5–10.0,	rich 6.0– 7.5
Town gas:	lean 3.5– 4.0,	rich 1.7– 2.5
Propane:	lean 20.0–24.0,	rich 12.0–16.0

The intermediate atmosphere ranges are rarely used.

The control of lean exothermic atmospheres derived from town gas has been studied by Cubbage and Whiteside [14] who concluded that it was enough to control

Gas		Ferrous					Nonferrous		
		Inert	Oxidizing	Reducing	Decar-burizing	Carburizing	Inert	Oxidizing	Reducing
Elements									
Argon	A	X					X		
Helium	He	X					X		
Nitrogen	N_2	X					X		
Oxygen	O_2		X					X	
Hydrogen	H_2			X					X
In chemical combination									
Water			X		X		X		
CO_2			X		X		X		
CO				X		X			X
CH_4				X		X			X
As mixtures									
$N_2 + H_2$				X					X
$N_2 + H_2 + H_2O$			X or	X	X				X
$N_2 + H_2 + H_2O + CO_2 + CO$			X or	X	X or	X			X
$N_2 + H_2 + CO$				X		X			X

Figure 13:11 Effects of atmosphere constituents on hot ferrous and nonferrous metals [10]

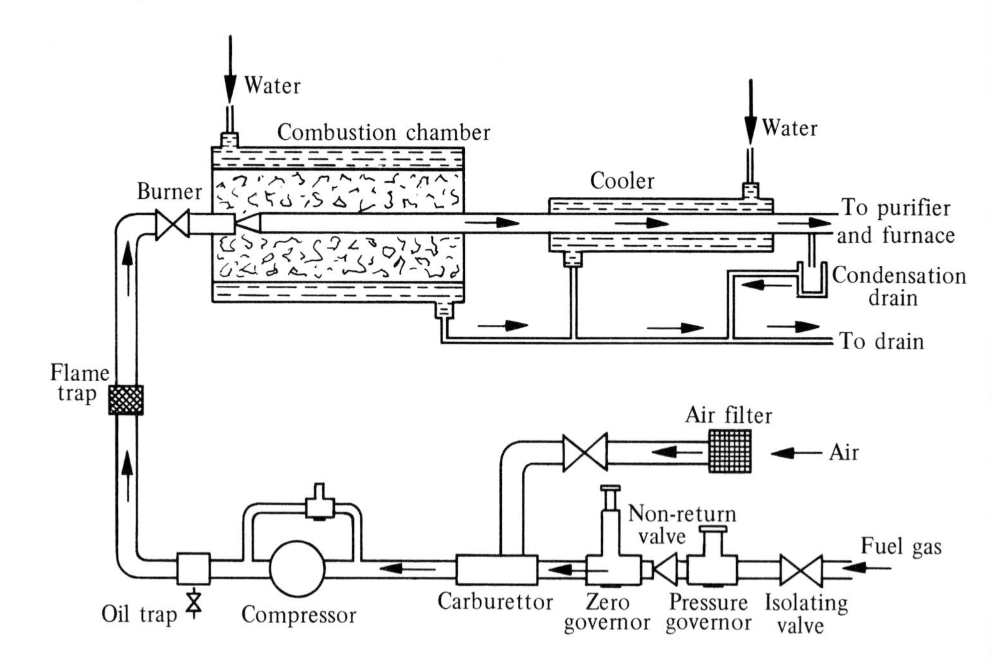

Figure 13:12 Exothermic atmosphere generator [10]

the Wobbe index of the gas–air mixture or of the gas (e.g. hydrogen content constant to within ±0.4 to ±0.7 per cent for a change of one gas group—see Section 4.2.2). The constancy of composition of natural gas and LPG reduces the problem to one of simple flow control.

The method adopted for cooling the combustion products in the generator has a significant effect on the composition of the atmosphere generated and is of particular importance for lean exothermic atmospheres close to stoichiometric composition. In addition to containing N_2, CO_2 and H_2O, equilibrium considerations dictate that at flame temperature stoichiometric combustion products also contain small but significant amounts of CO, H_2 and O_2 together with a variety of atomic species. Equilibrium compositions for stoichiometric natural gas combustion products are tabulated in Figure 4:8 which indicate that not until the temperature has fallen to about 1000 °C are the amounts of these constituents negligible. If the combustion products are cooled slowly, equilibrium composition is maintained throughout the cooling process and the cooled atmosphere consists of N_2, CO_2 and H_2O only. If on the other hand the combustion products are rapidly cooled, for example by a direct water quench, the prepared atmosphere retains the same proportions of the trace constituents as at flame temperature. Where the composition is critical, Cubbage and Ling [13] have suggested the best practical cooling method is by recirculation cooling down to 1200–1300 °C, by which temperature the concentration of H_2 plus O_2 plus CO is around 0.1 to 0.2 per cent, followed by a rapid quench.

Turning to rich exothermic atmospheres, the analysis varies but (for natural gas) is typically 15 per cent H_2, 10 per cent CO, 5 per cent CO_2 and 1 per cent CH_4 (dry analysis). Rich exothermic gas may be used 'raw' or alternatively 'stripped', i.e. after CO_2 and water vapour have been removed. CO_2 and water vapour may be removed by three alternative methods: compression over water, the use of molecular sieves or by the use of organic solvents, e.g. ethanolamines. Detailed descriptions of atmosphere-stripping processes have been given by Moppett [15]. Stripped rich exothermic gas is extremely reducing and may be used as an alternative to endothermic gases. Near-pure nitrogen may be produced by CO_2 and water removal from lean exothermic atmospheres.

Production and properties of endothermic atmospheres

Endothermic atmospheres are produced by the catalysed reforming of fuel gases with a small amount of air. Air and gas from a machine-premix system are fed into an externally heated nickel-chrome steel or refractory retort containing nickel-impregnated refractory as a catalyst. The generator temperature is normally between 1000 °C and 1150 °C, the higher temperatures being required with town gas. The retort is generally heated by tangential-firing air-blast burners or in some instances by natural-draught burners [10].

Normal air–gas ratios on various fuel gases are: natural gas 2.4 : 1, town gas 0.5 : 1, propane 7.5 : 1, butane 9.8 : 1. As Fleming [10] has pointed out, little deviation from these ratios is generally required since the aim is to produce exact conversion to H_2, CO and N_2, which can be written approximately for methane as:

$$CH_4 + \tfrac{1}{2}O_2 + 2N_2 \rightarrow CO + 2H_2 + 2N_2 \tag{16}$$

A typical composition using natural gas (reactor temperature = 1050 °C) is [16]: 0.2 per cent CO_2, 19.8 per cent CO, 39.4 per cent H_2, 0.2 per cent CH_4, balance N_2, dew point 2 °C.

Carbon is deposited on the catalyst and this needs to be burned off periodically by reducing the retort temperature below 700 °C and passing air through the retort. The interval between burning-off runs depends largely on the dew point (see below) at which the plant has been running, e.g. for −1 °C to −5° C the interval is about three weeks. After burning off, the nickel catalyst is partially oxidized and needs to be reduced to metallic nickel by 'conditioning', i.e. passing a mixture with the correct air–gas ratio at the rated throughput for a few hours.

In addition to CO, H_2 and N_2, endothermic gases contain traces of H_2O, CO_2 and hydrocarbons. The quantities of CO_2 and H_2O are very sensitive to changes in air–gas ratio and since even a small increase of these decarburizing agents may be significant, accurate proportioning is required and fuel gases of constant chemical composition are advantageous. In this connection natural gas is a particularly suitable feedstock. Commercial-grade LPG, containing varying amounts of unsaturated hydrocarbons, and town gas are less satisfactory in this respect. Since the dew point of endothermic gas is an accurate and easily measured index of the concentration of decarburizing agents, control is usually based on this property. Alternatively the oxygen potential of the atmosphere may be measured using a solid electrolytic cell [10, 17].

For gas-carburizing purposes it is usually necessary to enrich endothermic gas by the addition of a few per cent of propane or natural gas to increase its carbon potential. This is defined as percentage carbon content of the steel that would be in equilibrium with the atmosphere at the temperature of the heat-treatment process.

13.2.4 Applications of controlled atmospheres

Ferrous metals

Low-carbon steel. In general decarburization is unimportant and the main concern is to prevent oxidation, i.e. reactions (4) and (5) are of prime interest. From Figure 13:9 it is evident that, for a given equilibrium ratio of H_2O/H_2 at heat-treatment temperature, oxidation may occur on cooling and that fast cooling is desirable. The reaction slows as the temperature decreases, however, and some latitude is permissible in the H_2O content. At normal heat-treatment temperatures the equilibrium CO_2/CO ratio is between 0.4 and 0.6. This ratio can be provided by a rich exothermic gas containing around 10 per cent CO and 5 per cent CO_2 and an H_2O/H_2 ratio below equilibrium.

The scaling of mild steel (as distinct from oxide-tarnishing) at the high temperatures involved in preheating the metal for forging or rolling operations requires separate consideration and is dealt with in Section 13.5. Many low-temperature stress-relieving processes are carried out on mild-steel fabrications in directly-fired furnaces using stoichiometric combustion, the limited degree of oxidation occurring being regarded as acceptable.

For some applications it is desirable to oxidize mild-steel surfaces lightly and uniformly to provide partial protection against corrosion. This 'blueing' process is carried out in air or steam at between 400 and 600 °C often during cooling from a previous heat-treatment process.

Carbon steels. Similar considerations apply regarding the prevention of oxidation as with mild steel although the temperatures involved are normally higher. The suppression of decarburization, however, becomes the main requirement (reactions (6) and (7)). As indicated in Section 13.2.2 an atmosphere which is nonoxidizing may be quite strongly decarburizing and much lower CO_2 and H_2O levels must be used. For short, low-temperature heat-treatment processes rich exothermic atmospheres may be used accepting some small carbon loss but at high temperatures the choice is limited to endothermic or stripped rich exothermic gases.

Alloy steels. Low-alloy steels present few special difficulties in preventing oxidation and decarburization. If the combined concentrations of chromium, manganese and silicon total less than about 1.5 per cent then similar atmospheres may be used to those for low-carbon steels [7]. High-alloy steels on the other hand are very prone to oxidation and may require bright heat treatment in dissociated ammonia. Alloy steels, particularly nickel chromium types, are especially susceptible to attack by sulphur compounds, the attack being much more serious if the sulphur oxides are present in a reducing atmosphere. A comprehensive treatment of the complex subject of atmosphere requirements for alloy steels is outside the scope of this book and the reader is referred to reference [7] for a full account.

Carburizing. Carburizing is carried out on low-carbon steels to increase the carbon content of the surface to combine the advantages of a hard surface and a tough core. A hard surface layer of a few hundredths of a millimetre is all that is required in most instances since wear beyond this usually renders the article useless. Three methods of carburizing may be used employing a solid, liquid or gaseous carbon-containing medium respectively.

The oldest method, pack carburizing, involves packing the workpieces in heat-resistant steel boxes containing charcoal together with promoters (e.g. barium carbonate or potassium cyanide) followed by slow heating in a directly fired furnace to about 900 °C. Although the medium is solid, the reactions are gaseous. The box initially contains air and carbon and as the temperature increases these combine to form CO which reacts with the ferrite surface, the cycle of reactions being [18]:

$$C + O_2 \rightleftharpoons CO_2 \quad (15)$$
$$C + CO_2 \rightleftharpoons 2CO \quad (9)$$
$$3Fe + 2CO \rightleftharpoons Fe_3C + CO_2 \quad (6)$$
$$CO_2 + C \rightleftharpoons 2CO \quad (9)$$

Time and temperature variation allow virtually any case-depth to be obtained. Case-depths required are generally in the range 0.8 to 3 mm requiring between two and 10 hours to attain. The exact role of the promoters is not fully understood although the generally accepted view is that they provide a constant supply of CO_2 by slow dissociation compensating for the loss of gas from the container through leakage [7]. Pack carburizing has a number of disadvantages: it is labour-intensive, slow, difficult to control and leads to carbon deposition.

A logical development is to eliminate the carbonaceous materials and simply use the promoters in the form of a bath of molten salts. This process, known as cyaniding or cyanide case-hardening, may use a variety of proprietary salt mixtures but all contain sodium cyanide in proportions ranging from 30 to 95 per cent. Cyaniding offers a number of advantages: uniformity of heating is ensured by the salt fluidity, short immersion periods (15 minutes to 2 hours) are possible and the product remains clean. Depth and case-carbon content are generally limited to about 0.5 mm and 0.8 per cent respectively. The main disadvantages of cyanide carburizing are, however, the extreme toxicity of the salts used and the difficulty of control over the salt-bath composition.

The third method, gas carburizing, is now the most widely used, being rapid, clean and controllable. Either endothermic gas or stripped rich exothermic gas may be used as a carrier gas, the carbon potential being increased by the controlled addition of hydrocarbons, either propane or, increasingly, natural gas. The latter is preferable since, besides being cheaper, its rate of decomposition allows the maximum carburizing rate without sooting [10]. The two reactions $CH_4 \rightleftharpoons C + 2H_2$ and $CH_4 \rightleftharpoons C_{Fe} + 2H_2$ have different equilibrium constants and carburizing is possible without sooting. Higher hydrocarbons have a greater carburizing potential but more likelihood of soot deposition.

The diffusion coefficient of carbon in iron increases sharply with temperature, increasing about four times between 925 and 1000 °C, and this has led to some carburizing plant being operated around 1050 °C to speed up the process. High carburizing temperatures save time only in the carburizing period. The heating-up time remains the same and it is necessary to balance the time saving against the increased cost of running and maintaining the furnace.

Carbonitriding, in which both carbon and nitrogen enter the steel surface, also uses endothermic gas with the addition of controlled amounts of ammonia. The process is carried out between 850 and 925 °C.

Nonferrous metals
The bright heat treatment of nonferrous metals and their alloys in general presents less problems than the treatment of carbon steels since the question of balanced oxidation–reduction reactions is less critical [7]. Two broad groups of nonferrous metals may be distinguished: those forming oxides with high dissociation pressures permitting the use of exothermic atmospheres (e.g. nickel, copper, cobalt, molybdenum) and those forming oxides of low dissociation pressures with which oxidation will take place even in an endothermic atmosphere and which require a pure atmosphere such as pure hydrogen or dissociated ammonia. This group includes chromium and manganese. In addition to oxidation almost all nonferrous metals are prone to staining and intergranular attack by sulphur gases.

Copper. Copper is easily bright annealed, oxidation being prevented by the presence of very small amounts of reducing gases. Apart from free oxygen and some sulphur-containing gases, combustion products are inert to copper. The reaction:

$$2Cu + CO_2 \rightleftharpoons Cu_2O + CO$$

has, at temperatures in the annealing range (400–600 °C), an equilibrium constant p_{CO}/p_{CO_2} as low as 10^{-13}. Thus lean exothermic atmospheres slightly on the rich side of stoichiometric are satisfactory for pure-copper heat treatment.

Hydrogen is inert to pure copper but may cause embrittlement in contact with copper containing copper oxide in solution. Hydrogen is capable of diffusing into the metal where it reacts with the oxide inclusions to produce water vapour at a sufficient pressure to produce blisters and embrittlement. Copper of this type (high-conductivity or tough-pitch copper) must be heat treated in a lean exothermic atmosphere containing a controlled minimum of hydrogen ($<$ 0.5–1.0 per cent). Wet CO also causes embrittlement, the water–gas reaction liberating sufficient H_2 to cause embrittlement.

Hydrogen sulphide attacks copper and complete absence of this gas is a prerequisite of bright annealing. Sulphur dioxide and organic sulphur compounds on the other hand are practically inert to copper. Annealing of copper wire is normally carried out under vacuum conditions. Even under moderate vacuum conditions the concentration of active constituents is very low, for example if the pressure in a vessel containing air is reduced to 10^{-3} mmHg (0.133 Pa) the oxygen content is reduced to 1.3 ppm [19]. This order of vacuum is easily obtained using mechanical pumps. There are two main reasons for vacuum annealing being used for copper wire:

1 The surface of the wire is often contaminated with die lubricants having low vapour pressures and these are removed more easily when pressure is reduced.
2 Vacuum annealing prevents successive coils of wire tending to stick together so that on unwinding the wire breaks or kinks. Sticking of this kind is caused by the presence of a layer of copper oxide which normally reduces to the metal in prepared atmospheres giving rise to local welding. This difficulty is overcome in vacuum annealing.

Copper alloys. The atmosphere requirements for copper alloys containing nickel, silver, or up to about 15 per cent zinc or tin are similar to those for pure copper. Brasses containing more than about 15 per cent zinc are difficult to heat-treat without

Atmosphere	Source	Approximate composition % (DP = dew point)	Processes	Metals	Advantages	Disadvantages
(1) Rich exothermic	Partial combustion of fuel gases	9–12 CO 11–15 H_2 7– 5 CO_2 2– 3 H_2O 1– 2 CH_4 Bal N_2	Normalizing, copper brazing and sintering	Ferrous metals	Cheap	Can cause decarburization and sooting
(2) Lean exothermic	As (1)	0– 3 CO 0– 4 H_2 10–13 CO_2 2– 3 H_2O Bal N_2	Bright and clean annealing	Copper Nickel Brasses Aluminium	Cheap; nonexplosive	Sulphur removal (town gas) sometimes necessary; Traces of O_2 near full combustion
(3) Rich stripped exothermic	As (1) but carbon dioxide and water removed	10–13 CO 12–15 H_2 Bal N_2	Substitute for (6) for most purposes		Relatively cheap; nondecarburizing	Can cause sooting; increased plant size and sometimes increased maintenance over (1)
(4) Lean stripped exothermic	As (3)	0– 3 CO 0– 4 H_2 Bal N_2	Annealing; carrier gas or ca bon restoration	Ferrous metals	As (3); also non-explosive and nondecarburizing	As (3); traces of O_2 near full combustion
(5) Modified stripped exothermic	As (1) but CO_2 removed, water gas shift, CO_2 and H_2O removed	3–12 H_2 Bal N_2	Long-cycle annealing	Low-carbon and mild steel	Nonsooting; cheaper than (8)	
(6) Endothermic	Catalytic reaction of fuel gas and air	20–25 CO 30–45 H_2 0.5–1.0 CH_4 −15 to +15 °C DP	Hardening; brazing and sintering; carrier gas for carburizing and carbonitriding	Ferrous materials	Nondecarburizing	Explosive; can cause sooting; good maintenance and control required
(7) Cracked ammonia	Catalytic cracking of ammonia	75 H_2 25 N_2 −15 to −70 °C DP*	Bright and clean annealing; sintering and brazing	Stainless and other steels; certain copper and nickel alloys	Pure; cheaper than (10)	Explosive; can cause nitriding effects
(8) Partially burnt ammonia	Catalytic oxidation of ammonia	0.5–25 H_2 Bal N_2 +5 to −70 °C DP	As (7)	As (7)	Pure; cheaper and less explosive than (7)	
Fully burnt ammonia	As (8)	0 to 0.5 H_2 Bal N_2 +5 to −70 °C DP	Special annealing, hardening, and sintering	Special nonferrous metals and steels	Nonexplosive; purer than (4)	
(9) Hydrogen† (purer grades)	Electrolysis of water Diffusion purification of (7) In large cylinders	>99.99 H_2	As (7) but additional applications where nitriding must be avoided		Very pure; no nitriding effects; cylinders require little attention and capital	Expensive; explosive
(10) Nitrogen† (purer grades)	In large cylinders as liquid	>99.99 N_2	As (4) and (9) but generally more exacting applications		Pure; nonexplosive; inert to most metals; little attention and capital needed	Expensive
(11) Argon (purer grades)	In large cylinders as liquid	>99.99 A	Bright annealing and other heat treatments on special components	Titanium; special steels; Nimonic alloys	Nonexplosive; inert; little attention and capital needed	Very expensive
(12) Steam	Electric boiler	>99.9 H_2O	Tempering and blueing; bright-annealing	Steel; copper	Very cheap; non-explosive; little attention needed	Condensate difficulties

*Variable according to grade and type of raw material, consumption rate, etc.

†Mixtures of these gases can be used instead of (7), (8), and (9

Figure 13:13 Composition and uses of prepared atmospheres [12]

discoloration. Brasses oxidize readily in gases containing CO_2 and H_2O, forming a film mainly of zinc oxide which provides some protection against further oxidation up to 600 °C. Brasses are also susceptible to staining by controlled atmospheres and lubricants containing sulphur. The major difficulty with brass is the significant loss by volatilization which may occur at temperatures above 450 °C. Volatilization is a function of the density of the controlled atmosphere and is particularly serious *in vacuo*.

Other metals and alloys. The requirements for nickel are similar to those for copper but nickel is slightly more resistant than copper to oxidation and attack by sulphur gases. Nickel alloys with metals which, like nickel, have oxides of high dissociation pressure can be bright annealed in exothermic atmospheres. On the other hand alloys containing chromium, for example, require very pure atmospheres for bright heat treatment.

Figure 13:13 summarizes the composition and uses of protective atmospheres in ferrous and nonferrous heat treatment.

Finally, it should be mentioned that a safety code relating to the design and operation of atmosphere generators and associated plant is currently in preparation by the British Gas Corporation.

13.3 MELTING FURNACES

13.3.1 Melting nonferrous metals

The thermal properties of nonferrous metals and alloys are listed in Figure 1:29.

Metallurgical aspects

In general, metals and alloys with low melting points, such as lead, tin and type metals, present few problems in melting. On the other hand metallurgical problems such as gas absorption and oxidation in the case of aluminium and copper, and volatilization and bath attack in the case of zinc, have an important bearing on the design and operation of melting equipment.

Aluminium and its alloys. Melting aluminium presents two main problems. First, it is highly reactive to oxidizing agents, the oxidation rate increasing sharply above the melting point (Figure 13:14). If the oxide film remains undisturbed further reaction is minimized but this is rarely possible in practice. Agitation of the bath surface has the added disadvantage of mixing the surface oxide skin into the molten metal and, since the metal and oxide have nearly the same density, oxide particles are likely to be carried into the castings. As indicated in Section 13.2 the oxide is extremely stable and the reaction is irreversible at normal temperatures.

Moisture on the charge, tools, fluxes, etc., also causes rapid oxidation sometimes with explosive results. Oxides forming on the surface of the molten bath may be removed by surface-cleaning fluxes which contain constituents with low melting points that react exothermically on the bath surface [20]. The oxides separate to form a dry powdery dross which is removed by skimming. Oxidation presents particular problems in the melting of light scrap due to the high ratio of surface area to volume and the difficulty of separating the filaments of molten metal and oxide.

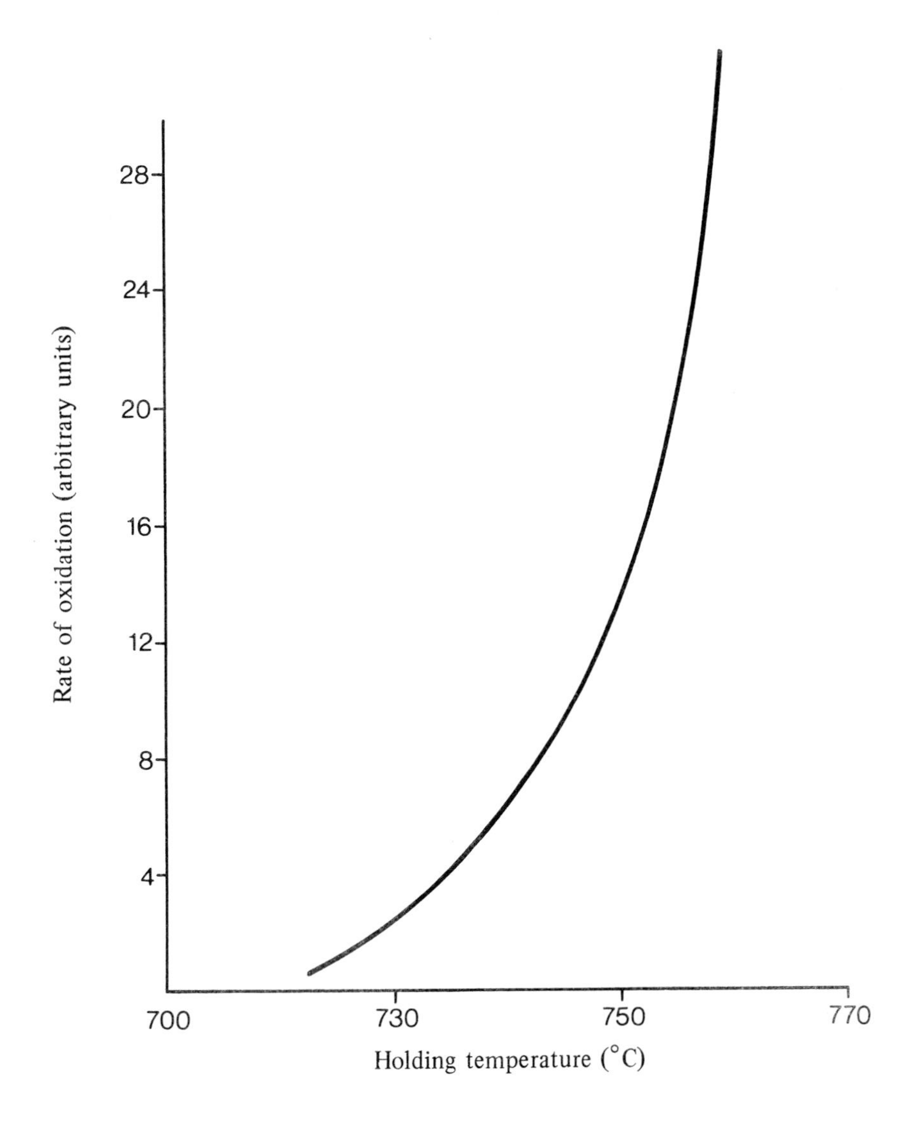

Figure 13:14 Oxidation rate of aluminium

The other major problem is hydrogen pickup. Hydrogen, derived from moisture and dirt on the charge and from the combustion products, dissolves readily in molten aluminium but as Figure 13:15 indicates, the solubility reduces drastically on solidification. Thus any hydrogen in excess of the solid solubility comes out of solution and results in 'pinhole porosity'. Avoidance of wet, dirty or oily charge materials, excessive exposure of the melt to combustion products or excessive stirring reduce hydrogen pickup but in practice it is often necessary to 'degas'. Degassing agents include gases (chlorine or chlorine–nitrogen mixtures) bubbled through the melt or pelleted solids,

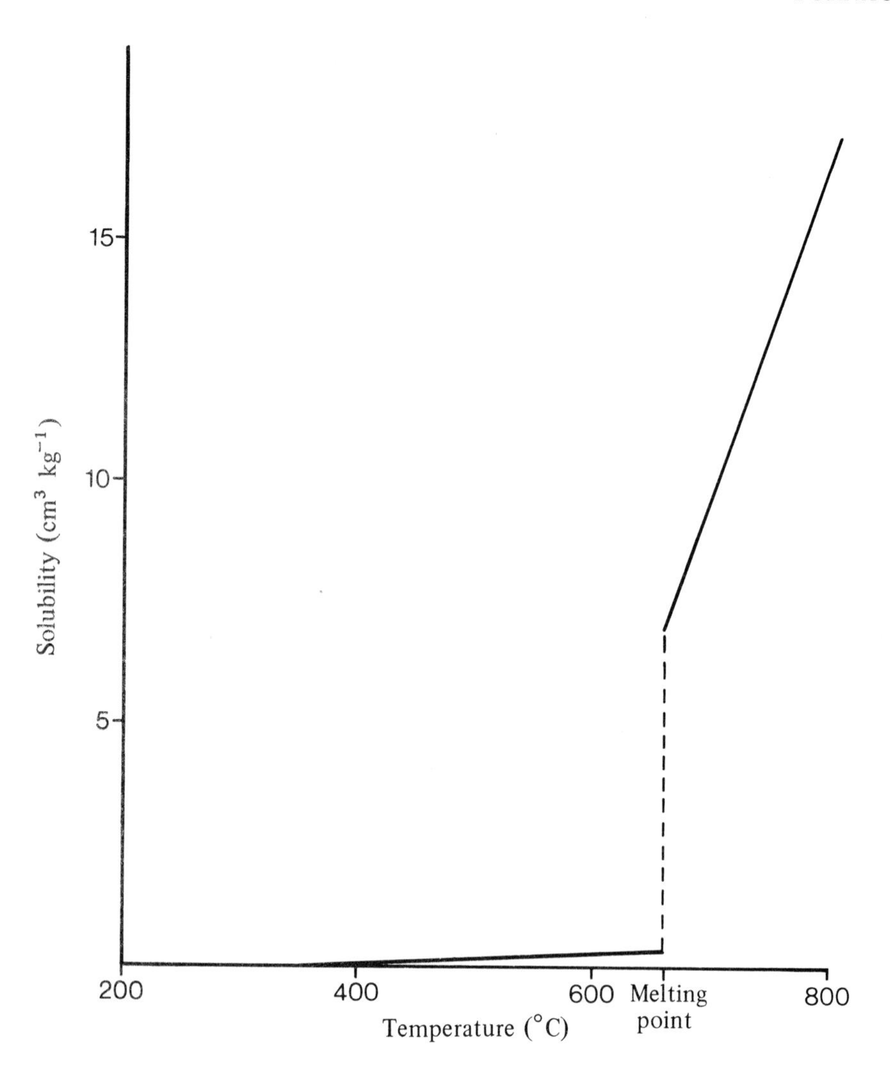

Figure 13:15 Solubility of hydrogen in aluminium

e.g. hexachloroethane [20]. The normally acceptable hydrogen content lies within the range 0.1 to 1.0 cm^3 per 100 g.

Copper and its alloys. Problems of oxidation and gas pickup occur with copper. Oxygen is soluble in the metal to an extent but when the solubility limit is reached an oxide Cu_2O is produced which also is soluble in the metal [21]. An excess of Cu_2O makes the metal unworkable. Hydrogen pickup also occurs as for aluminium but on solidification it is expelled as water vapour if dissolved oxygen is present. Sulphur pickup may also occur if SO_2 is present in the furnace atmosphere; the sulphide is formed but as the metal cools the SO_2 is regenerated and produces porosity.

In copper alloys containing zinc, volatilization losses may present problems.

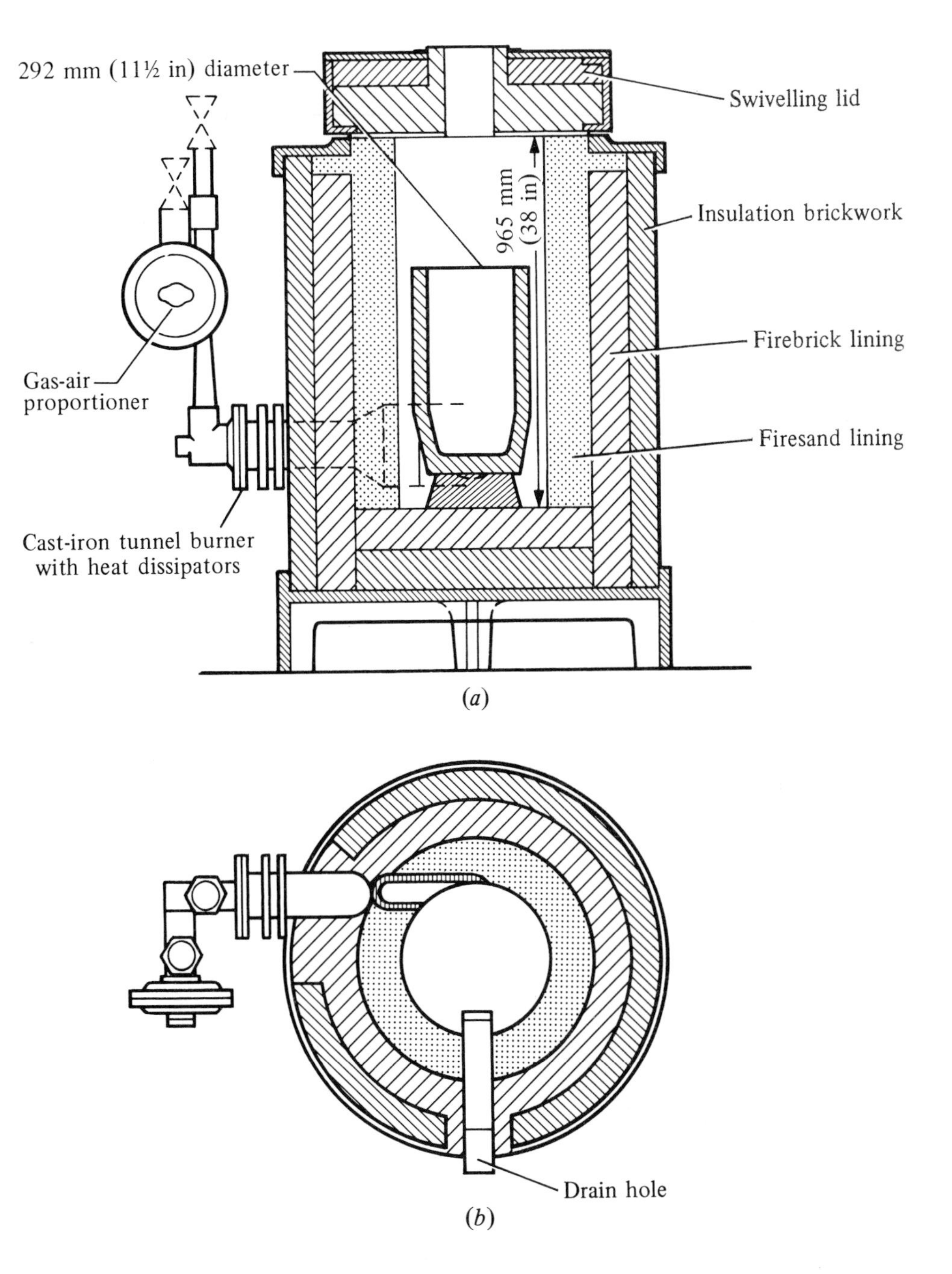

Figure 13:16 Tangentially fired crucible furnace
(*a*) Sectional elevation
(*b*) Sectional plan

Zinc. When zinc alloys are melted for casting purposes melting usually takes place in refractory crucibles upon which the zinc has little effect. The other major zinc-melting application, however, is for galvanizing and this is done in large welded steel baths. Galvanizing presents the major difficulty that zinc attack dissolves the steel of the bath limiting its life. The rate of attack becomes rapid at temperatures in excess of 450 °C (melting point of zinc is 419 °C). At 520 °C, for example, the pot wall thickness is reduced by 0.025 mm per hour [22]. Attack increases also where there is excessive metal agitation and particularly in the presence of metal hot spots. The attack results in dross which sinks to the bottom of the bath. Since its thermal conductivity is much less than that of zinc its presence exacerbates attack in bottom-heated installations by increasing temperature gradients. Gas-fired galvanizing settings have a long history of development aimed at minimizing these effects and thereby increasing bath life [22, 23, 24].

Furnace types

Crucible furnaces. Crucible furnaces are employed for melting metals for which direct-flame methods are unsuitable, e.g. magnesium alloys and some types of brass, and also for relatively small-scale melting of other metals. Melting in crucible furnaces is less efficient than direct firing although the recent use of self-recuperative burners [25] has led to significant improvements in this respect. Metal losses and gas pickup are low. Air-blast burners firing tangentially into the base of the annulus between the lining and the crucible constitute the normal firing method (Figure 13:16).

Nozzle-mixing and packaged burners are also used, the latter in particular being used for the melting of zinc die-casting alloys and aluminium [26]. Crucibles may be of refractory or metal construction. It is particularly important in the latter case to avoid localized heating and burners of high radiant output and low forward momentum, e.g. surface-combustion panels, flat-flame burners and radiant-cup burners have found application. Crucible furnaces are generally classified on the basis of the means of removal of the molten metal. In pit-type or lift-out crucible furnaces the crucible is removed by manual or mechanical tongs for pouring. Sizes range from around 5 kg (11 lb) capacity (for use in schools, etc.) up to about 550 kg (1200 lb) for copper and 180 kg (400 lb) for aluminium.

Tilting crucible furnaces eliminate the need for crucible removal and are suitable therefore for larger capacities. The whole furnace tilts about an axis which may be on the centre-line or preferably at the pouring lip to avoid the necessity for moving the moulds as pouring progresses. Capacities range from 140 to 1400 kg (300 to 3000 lb) for copper and up to 900 kg (2000 lb) for aluminium.

Bale-out furnaces are mainly used for die-casting and may be melting/holding furnaces or simply holding furnaces supplied from a bulk melter. Removal of the metal is by hand ladling or less commonly by submerged pumps. A limited degree of load preheating is practised in what are termed ‘supercharge preheating’ furnaces. Crucible furnaces lend themselves particularly well to self-recuperative burner firing, the burner momentum being used to produce recirculation in the annular space. In a series of trials reported in [25] on a 180 kg (400 lb) capacity aluminium crucible furnace, using an MRS recuperative burner showed fuel savings of 27.8 per cent and 61.5 per cent over a proprietary nozzle-mixing burner and a packaged burner respectively. In terms of thermal input the figures were 6.5 MJ kg^{-1} (2730 Btu lb^{-1}) for the recuperative burner, 8.8 (3780) for the nozzle-mixing burner and 16.7 (7110) for the packaged burner (all for a cold start melt to 720 °C).

Bath furnaces. This group includes steel or cast-iron bath furnaces for a wide variety of uses, including galvanizing, type-metal melting and lead heat treatment. The low-temperature examples present few problems, lead and tin baths often have simple underfiring arrangements using natural draught. Type metals, while having low melting points and heat capacities (see Figure 1:29), are usually melted using air-blast burners or by immersion tubes (see Chapter 15) because of the rapid recovery rate required in this process.

The metal attack referred to on page 570 precludes the simple use of immersion tubes for heating galvanizing baths, although systems have been described by Higgs [22] in which immersion tubes are used to heat an intermediate heat-transfer medium, either molten salt or lead, which in turn heats the zinc.

Temperature uniformity is of prime importance and until recently this has been achieved to some degree by using large numbers of small burners in a variety of configurations. Such systems have been reviewed by Higgs [22], Lupton [23] and Hall [24]. The most recent techniques, however, are based on jet-driven recirculation fired by a single high-velocity burner. Figure 13:17 shows a horizontal cross-section through a galvanizing setting designed along these lines. Combustion products from the high-velocity tunnel burner flow along two low-level heating ducts along the bath sides, are discharged through ports and pass upward between the setting and the bath, over the baffle and down over the bath sides to the collecting duct and thence to the mixing duct where a proportion is recirculated. The high mass flow rate and low temperature head make convective heat transfer predominate supplemented by low-temperature radiation from the baffle walls. The maximum heat transfer rate obtained was 22 kW m^{-2} (7000 Btu h^{-1} ft^{-2}). Results obtained on a prototype of similar design showed a maximum temperature differential of 20 °C on the bath sides and a temperature head of about 200 °C. In addition to providing temperature uniformity the greatly reduced number of burners allows the economic application of automatic controls and safety equipment.

Reverberatory furnaces Reverberatory furnaces are generally used to melt large quantities of metal to supply holding furnaces or used to remelt scrap on a large scale. Capacities range from 1 tonne to 50 tonnes and in a few instances up to 90 tonnes. Reverberatory furnaces are directly fired and, unlike crucible melting, the charge is unprotected from the combustion products. Two basic types may be distinguished, wet-hearth and dry-hearth. In the former the metal is charged into an already molten bath and heat transfer is to the surface of the melt primarily by radiation from the roof and side walls.

In dry-hearth furnaces, on the other hand, the charge is placed on a sloping hearth and is enveloped in the combustion products. The solid charge melts and drains down the sloping hearth, gaining some superheat in the process and collects in the holding chamber. Because of the much larger exposed heat-transfer area, dry-hearth furnaces have higher melting rates but, for the same reason, have higher melting losses. An additional major drawback of dry-hearth designs is the necessity for keeping the hearth free from dross buildup. Wet-hearth designs are essential for the melting of light scrap, swarf, etc., and some incorporate a subsidiary charging well. Burners for reverberatory furnaces are normally conventional nozzle-mixing designs.

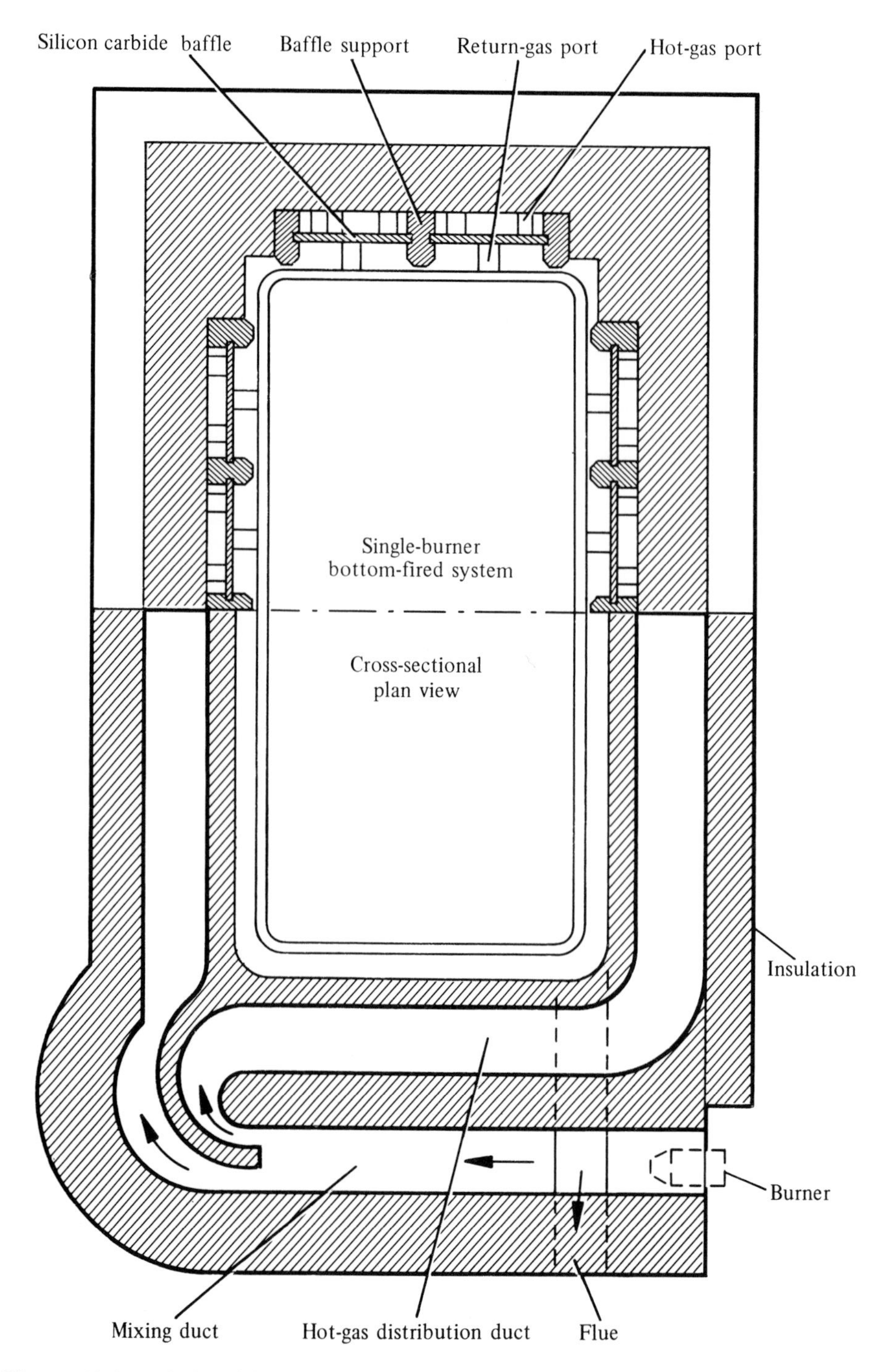

Figure 13:17 Galvanizing setting: single-burner recirculating firing [24]

Although the thermal efficiency of reverberatory furnaces is better than that of crucible furnaces it is still of a fairly low order (typically 25–40 per cent) and there is considerable scope for the application of waste-heat recovery by air preheating or load preheating.

A variant is the rotary reverberatory furnace which partially rotates, the hot walls of the furnace being continually in contact with the molten metal. This furnace melts swarf with some success due to the mixing action resulting from rotation [21].

Metal losses in reverberatory melting are generally high but are so dependent on the type of charge, holding period and the skill of the operators, that it is difficult to quote reliable figures. Rice [27], for example, in describing the changeover of an oil-fired 9 tonne dry-hearth furnace melting aluminium scrap quotes a metal loss of 1.3 per cent on gas firing whilst Potter [28] quotes a 13.8 per cent loss on a 12 tonne furnace fed with baled light aluminium scrap. The only common factor is that these two figures represent an appreciable improvement over oil firing. Figure 13:18 shows average melting losses for Al–Si alloys using various furnace types.

	All ingot (%)	All scrap (%)
Large reverberatory	1.3	3.6
Small reverberatory	1.0	3.9
Crucible, bale-out	2.4	2.2
Crucible, tilting	1.4	2.0

Figure 13:18 Metal losses: effects of furnace type [29]

Davies, Lucas and Toth [30] have described the application of mathematical modelling and isothermal scale modelling to reverberatory furnaces. Flow-visualization studies have indicated that despite the use of low-momentum burners there is a surprisingly large amount of recirculation and it is reasonable to apply the well-stirred furnace model (Section 12.4). The performance of a furnace 4.1 m long $\times$ 2 m $\times$ 1 m has been calculated for an assumed heat input of 2.17 MW (7.4×10^6 Btu h^{-1}) resulting in a firing intensity of 0.25 MW m^{-3} (25 000 Btu h^{-1} ft^{-3}). Predicted heat-transfer rates for oil and gas firing at the same (20 per cent) excess-air level are shown in Figure 13:19(*a*) which shows that the higher emissivity of oil flames increases heat transfer, but not by a very significant amount. If more realistic excess-air levels are taken the difference in heat-transfer rates is eliminated as shown in Figure 13:19(*b*). Thus the virtues of firing in such applications with luminous flames are by no means as great as they are often supposed. High-momentum burners with good mixing and short flames give improved results.

Rapid melting furnaces. As indicated above, crucible and reverberatory furnaces leave much to be desired in terms of thermal efficiency. In addition they are essentially intermittent furnaces and it is common practice to store molten metal, in many cases hundreds of kilograms, when only a few kilograms are required for each casting. Associated with this procedure are a number of disadvantages: heat losses from the metal not only raise the temperature of the working environment but have to be replaced, and metal losses and contamination increase with prolonged storage of molten metal.

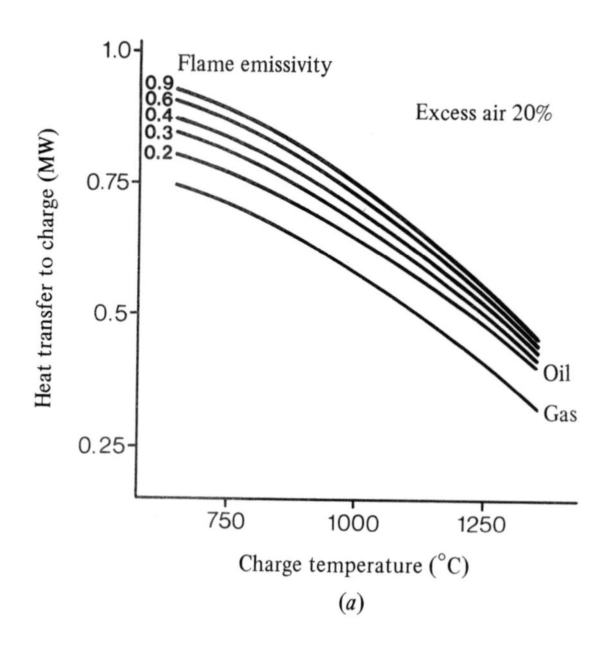

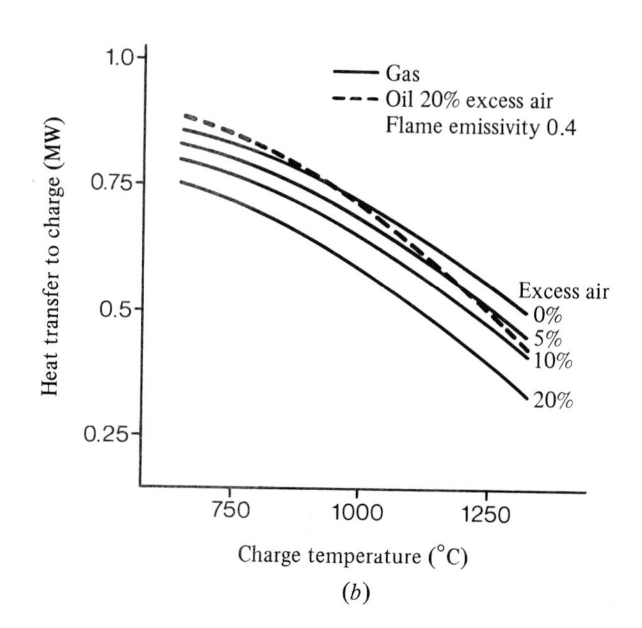

Figure 13:19 Reverberatory furnace: comparison of oil and gas firing [30]
(*a*) Excess air: 20%
(*b*) Excess air: 20% for oil and 0%, 5%, 10% and 20% for gas

Lawrie and Ayres [31] and later Watson and Glen [32] have recommended the replacement of the conventional processes by a method in which only the quantity of metal immediately required is melted rapidly. This approach is attractive in principle: drastic reduction of the time the metal is molten decreases oxidation and gas pickup, the capital cost of a small rapid melting furnace is much lower than a conventional one and the operating costs are likely to be lower because of lower metal losses and removal of the necessity for degassing. An additional attractive feature is that a system of this type lends itself to integration into mechanized casting systems.

An early furnace designed with this aim was the atmosphere-controlled rapid melting furnace (ACRM) developed by the North Eastern Gas Board (Figure 13:20 (*a*)). One set of burners provides gas-rich flames over the melting hearth and another set, over-aerated, fire across the roof. Air jets between the combustion chamber and the preheat zone provide sufficient air to complete combustion of the gas at this point. Metal is charged into the preheating section through which combustion products pass en route to the flue.

The charge is preheated then pushed onto the sloping melting hearth where it melts and immediately runs out of the furnace into a ladle. Watson and Glen [32] reported that using a heat input of 280 kW (9.5 therm h^{-1}) a continuous melting rate of 4 kg min^{-1} (8.8 lb min^{-1}) of aluminium alloy was obtained at a thermal efficiency of 39.4 per cent with a metal loss of 1.7 per cent. Although this is a higher efficiency than obtained with conventional methods, the preheating arrangements are rather rudimentary and, as Watson and Glen's work showed, much higher efficiencies can be obtained using a packed tower forming both the preheat and melting zones.

Figure 13:20(*b*) shows the simple prototype packed tower used by Watson and Glen. This consisted of a simple insulated refractory shaft fired by a single high-velocity tunnel burner, the combustion products passing up the shaft down which the charge was fed as a randomly packed bed, the bed depth being kept roughly constant by continuous feeding as the molten metal ran out. Heat transfer was primarily by convection and extremely high efficiencies (up to 70 per cent), heat-transfer rates and melting rates (5 kg min^{-1} (11 lb min^{-1})) were obtained. Metal losses obtained lay between 1 and 3 per cent for aluminium and gas pickup was less than obtained with conventional reverberatory furnaces. Glen [33] has described a comprehensive design procedure for packed-tower furnaces of this kind.

The high efficiency, compact size and low cost of these furnaces are very attractive features and a variety of production prototype melters have been developed from Watson and Glen's design. Considerable differences in design have emerged from the simple laboratory arrangement stemming from the difficulties of maintaining a constant superheated metal offtake temperature, difficulties in obtaining a continuous uniform metal flow from the furnace and problems resulting from dross buildup. Development has taken two main directions:

1 Towards a small-scale rapid melter incorporating an intermittent automatic delivery device to feed automatic die-casting machines.
2 Towards a large-scale ($1\frac{1}{2}$ t h^{-1}) bulk aluminium melter.

Figure 13:20(*c*) shows a furnace of the former type, the Temgas Autocaster developed by North Thames Gas. Departures from Watson and Glen's basic design include the provision of a hearth to retard the flow of molten metal in order to provide it with

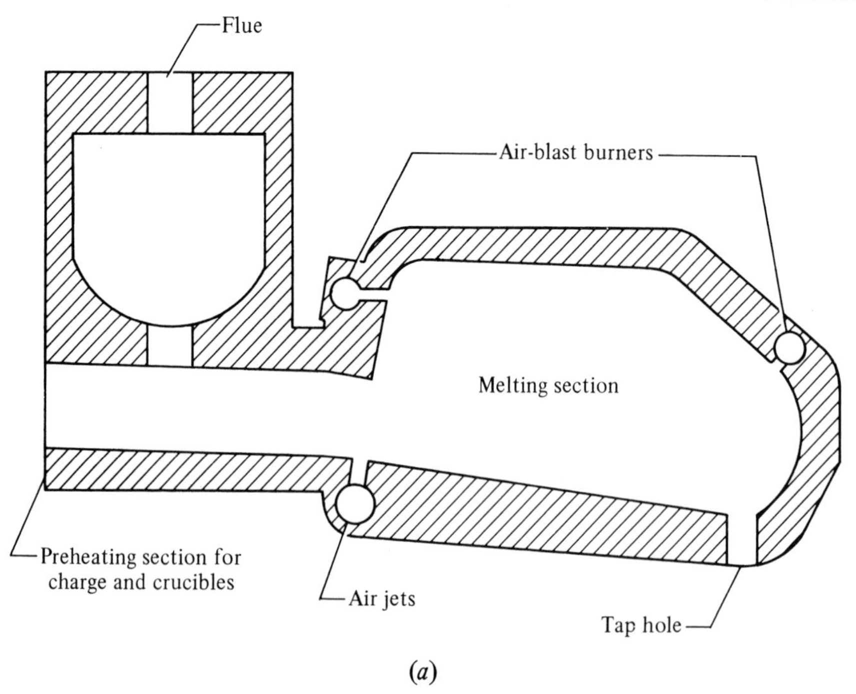
Flue
Air-blast burners
Melting section
Preheating section for charge and crucibles
Air jets
Tap hole

(a)

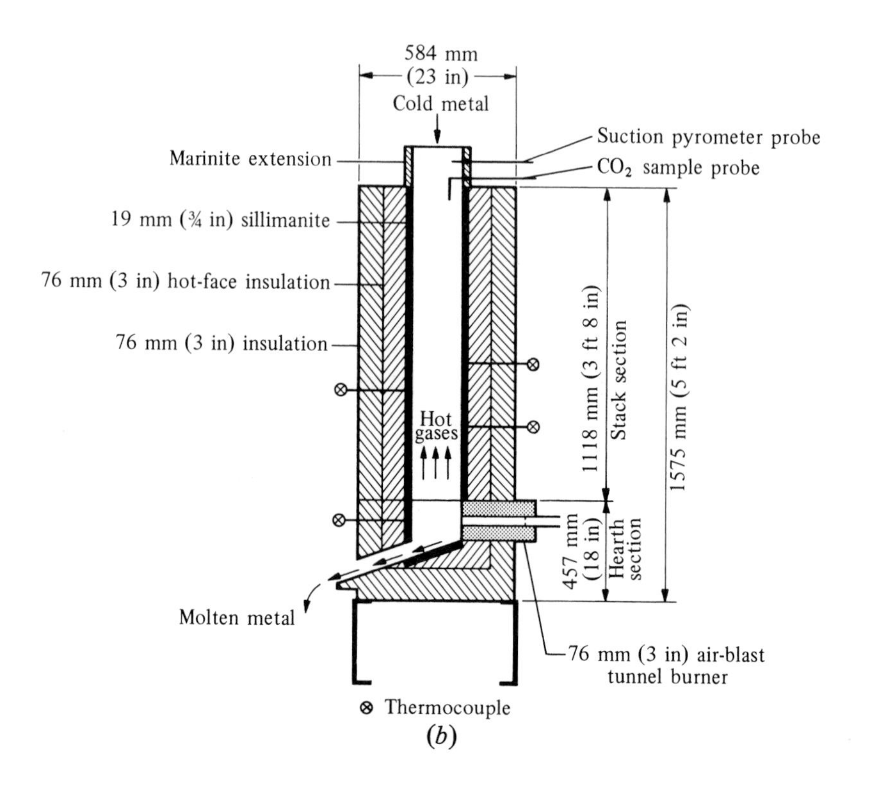
584 mm (23 in)
Cold metal
Suction pyrometer probe
CO_2 sample probe
Marinite extension
19 mm (¾ in) sillimanite
76 mm (3 in) hot-face insulation
76 mm (3 in) insulation
Hot gases
1118 mm (3 ft 8 in)
Stack section
1575 mm (5 ft 2 in)
457 mm (18 in)
Hearth section
Molten metal
76 mm (3 in) air-blast tunnel burner
⊗ Thermocouple

(b)

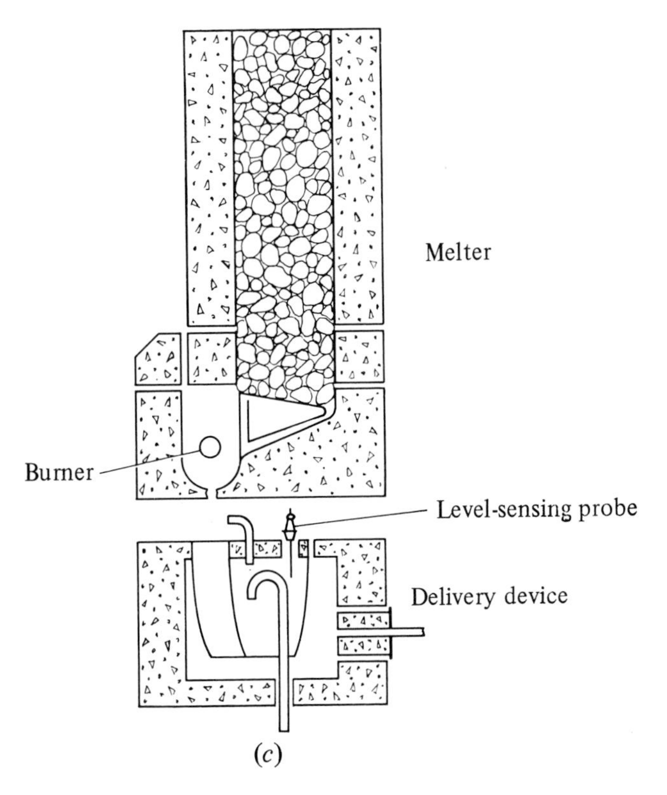

Figure 13:20 Gas-fired rapid-melting furnaces
(*a*) ACRM
(*b*) Laboratory prototype packed-tower furnace [32]
(*c*) Temgas Autocaster

some superheat and the addition of a small holding furnace incorporating the delivery device which is a pressure-syphon system using a partitioned crucible. From a cold start the furnace provides the first molten metal in 2 minutes and conversely on turning off the burner the flow of molten metal ceases within 30 seconds.

The departures from Watson and Glen's prototype are much greater in the bulk melter shown in Figure 13:21. The Hi-melt furnace was developed from a simple packed-tower prototype incorporating a slotted silicon-carbide grate which proved unsatisfactory because of the rapid buildup of aluminium oxide above the grate which reduced the melting rate by over 50 per cent and caused aluminium fires by producing excessive combustion-chamber temperatures. Even with a 1 per cent oxide-formation rate a melting rate of $1\frac{1}{2}$ t per hour requires the removal of the considerable quantity of 15 kg h^{-1} of oxide from the furnace. Figure 13:21 indicates how this was achieved in the Hi-melt furnace. Melting did not take place in the tower section. The ingots, preheated almost to melting point, were pushed quickly onto the melting hearth by two automatically controlled hydraulic pushers where they melted almost instantaneously. Dross buildup was removed from the inclined dry hearth via a large door. Metal losses averaged as little as 0.7 per cent. Firing was accomplished by four nozzle-mixing burners of total rating 820 kW (2.8 million Btu h^{-1}) firing across

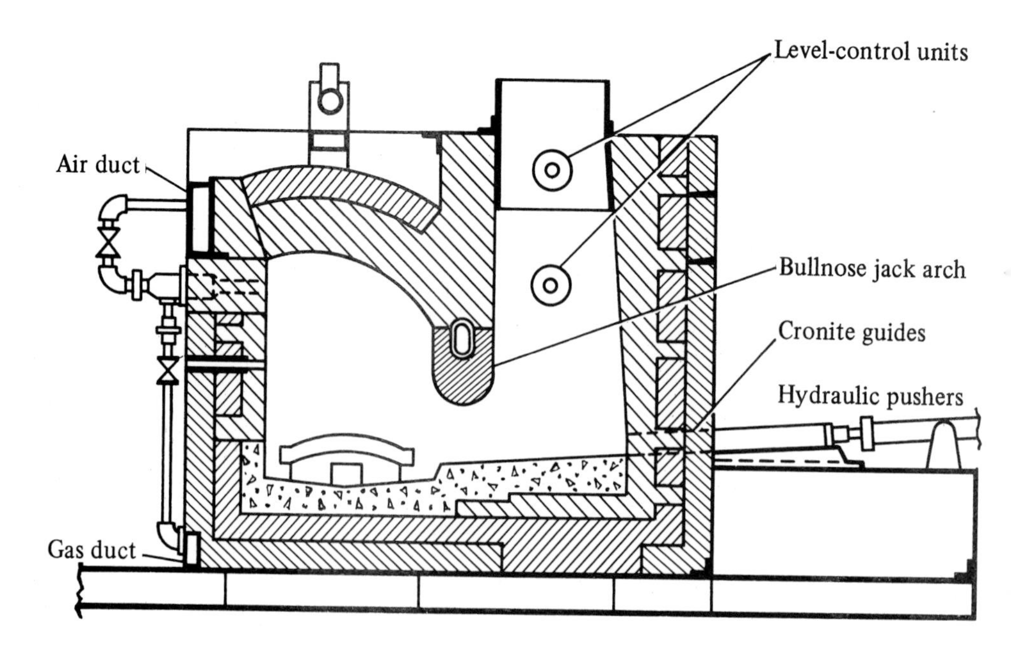

Figure 13:21 Gas Council/Scottish Gas Hi-Melt tower furnace [34]

the arch, the bullnose preventing direct flame impingement on the charge. The melting furnace operated in conjunction with a 454 kg (1000 lb) capacity holding furnace heated by a single 152 mm (6 in) radiant tube. Load-transducer cells fitted in the support legs of the holding furnace had overriding control over the temperature controller setting the control valves to a minimum. Thus melting ceased when the holding furnace was filled to capacity and automatically restarted after metal removal.

Despite their early promise, rapid melting furnaces have not proved very successful outside laboratory conditions. Major problems have been associated with dross buildup and its removal and the charge jamming in the tower.

Figure 13:22 shows the general features of a wet-hearth aluminium scrap recovery furnace designed by one of the authors [35]. Like the Hi-melt furnace this comprises a preheating tower and reverberatory melting hearth but differs in that the tower is dimensioned to accommodate single bales of compressed scrap one above the other and the reverberatory section is a wet-hearth design. Light aluminium scrap, e.g. swarf, collapsible tubes, etc., is particularly difficult to melt without incurring high metal losses because of the large surface area and the difficulty of separating the thin films of molten metal and oxide. In the design shown, the bales are heated almost to melting point in the shaft and melt, under compression from the bales above, in the molten pool. The bales measure 400 mm × 400 mm × 75–100 mm thick (16 in × 16 in × 3–4 in), 50 mm (2 in) less all round than the internal dimensions of the tower. Refractory guide rails locate the bales in the centre of the tower. Heating is by

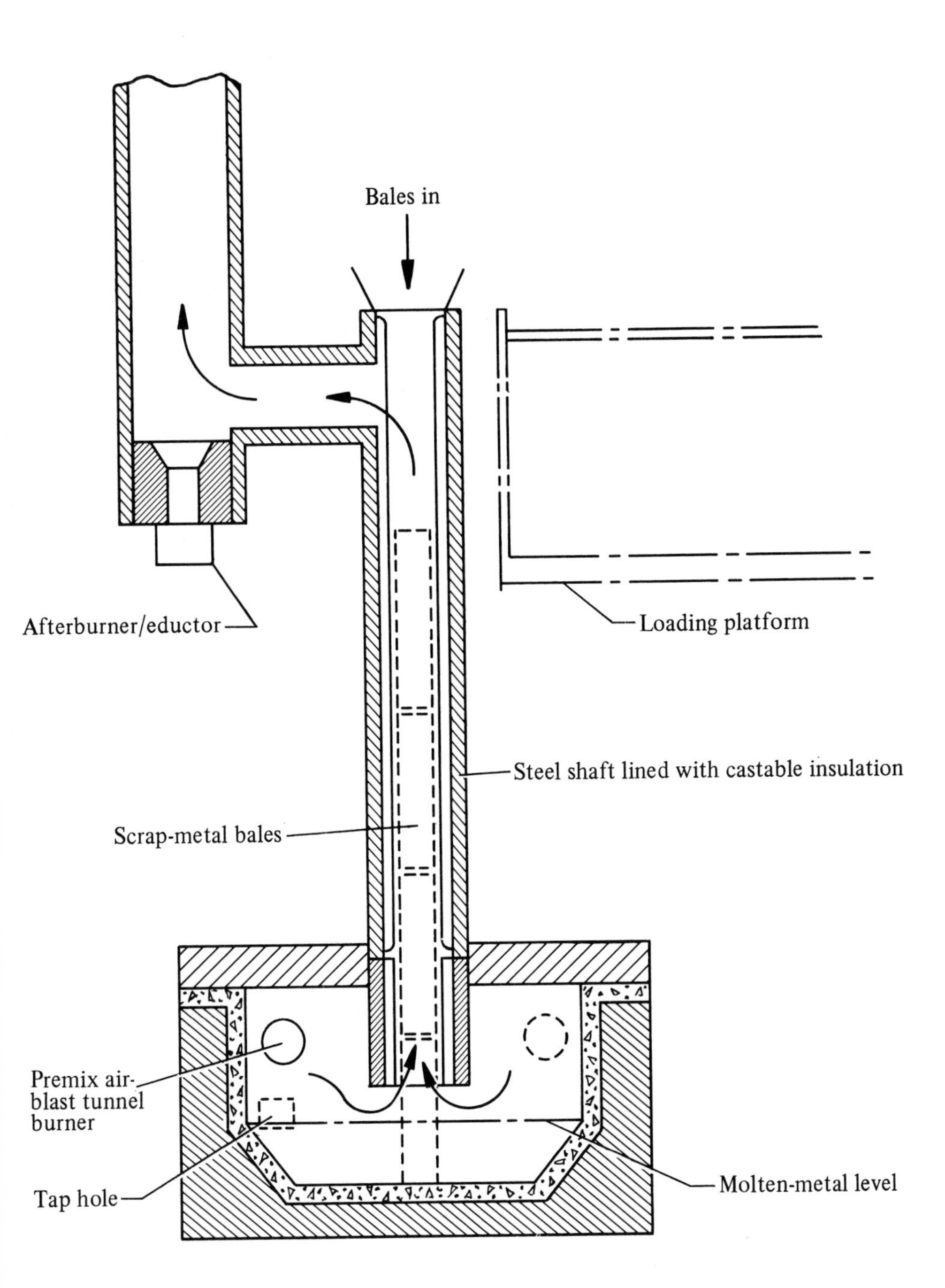

Figure 13:22 Furnace for recovering aluminium from scrap [35]

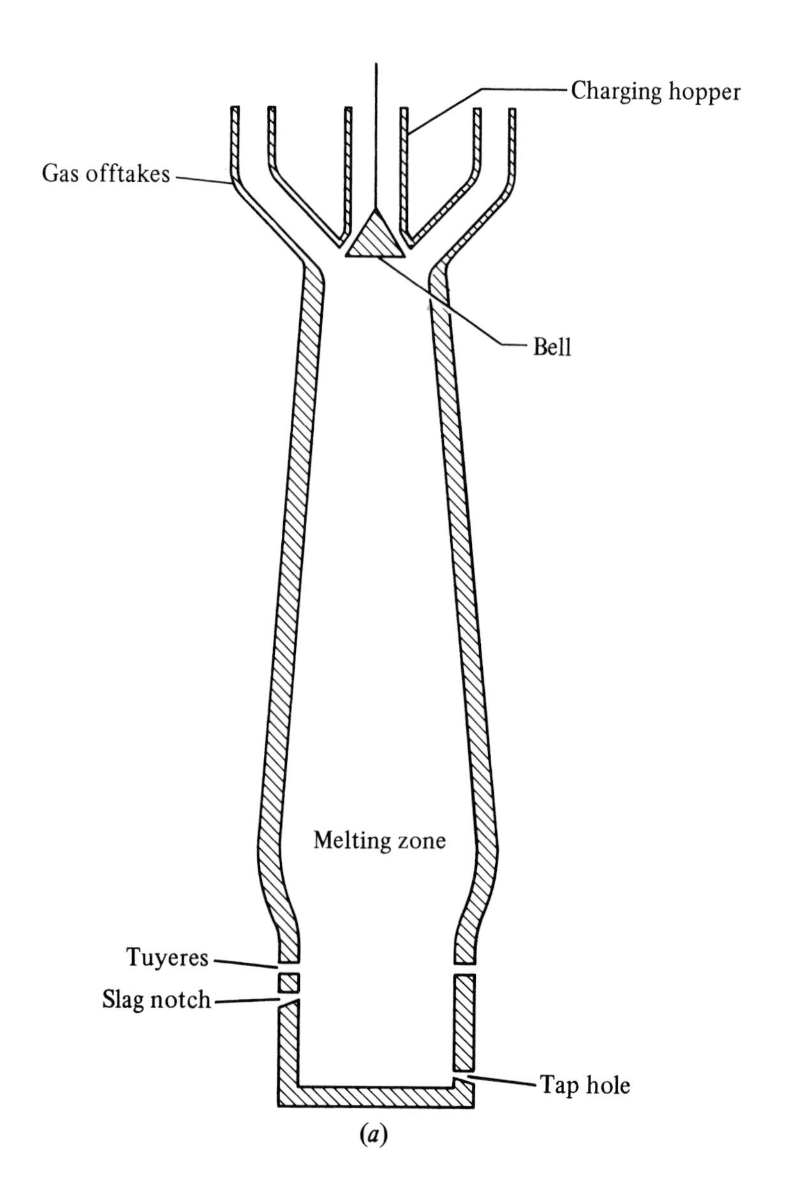

Figure 13:23 Iron-melting furnaces
(*a*) Blast furnace
(*b*) Cupola, with supplementary gas firing [40]

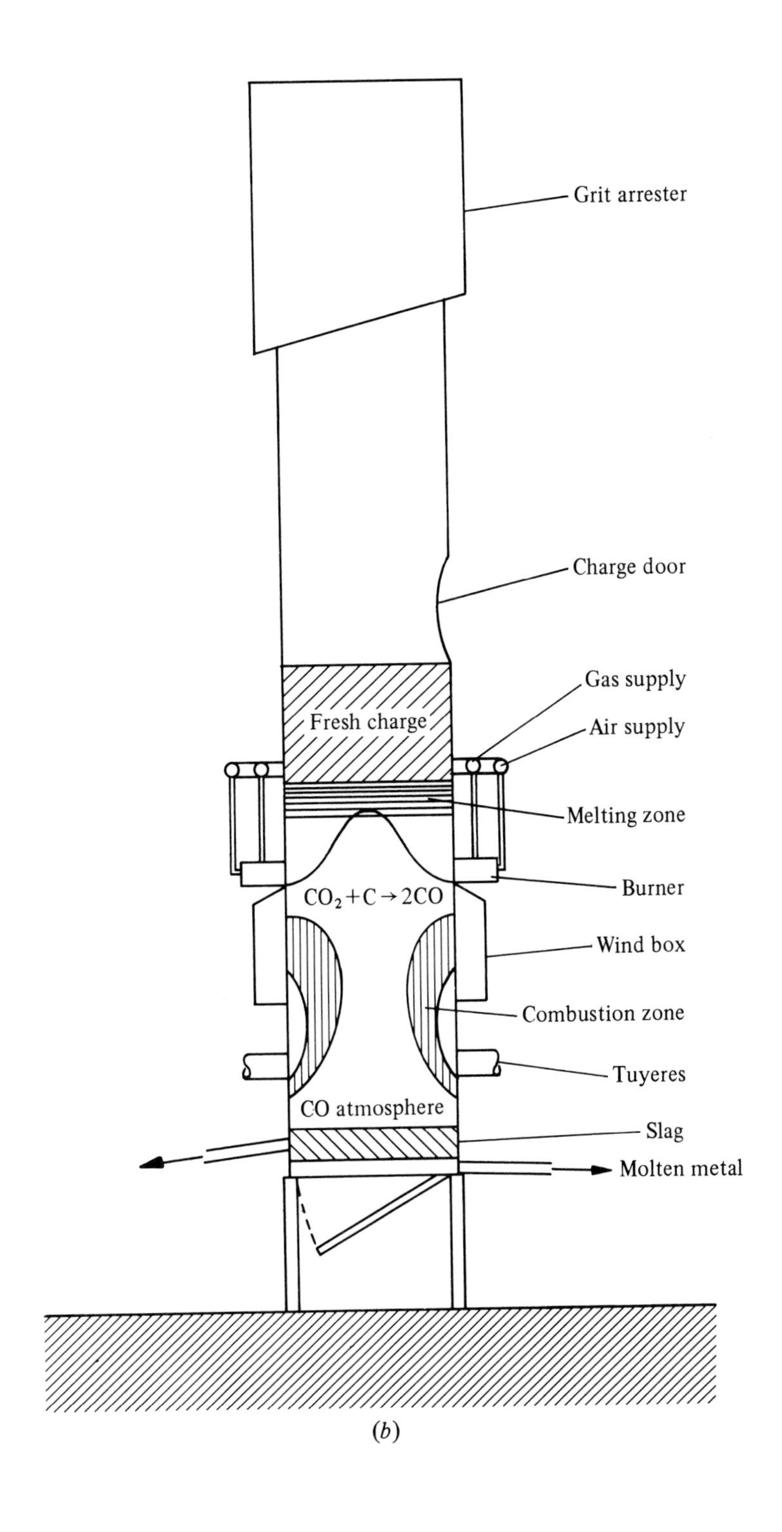
Grit arrester
Charge door
Gas supply
Air supply
Fresh charge
Melting zone
Burner
$CO_2 + C \rightarrow 2CO$
Wind box
Combustion zone
Tuyeres
CO atmosphere
Slag
Molten metal

(*b*)

two premix tunnel burners firing across the bath surface. The wet-hearth design minimizes oxide formation and facilitates its removal by skimming from the molten pool's surface. A nozzle-mixing burner running at high excess-air level firing up the flue adjacent to the shaft charging point produces a negative pressure at this point preventing the discharge of combustion products and serves as an afterburner for smoke resulting from oily or painted scrap. Depending on the type of scrap, output varies from 300 to 500 kg h^{-1} (660 to 1100 lb h^{-1}). Metal losses average 8 per cent on very light scrap, i.e. swarf, toothpaste tubes, etc., down to 0.2 per cent on 8 mm thick strip scrap. Fuel consumption is very low, 2490–3115 MJ t^{-1} (24–30 therm ton^{-1}) compared with 5815 MJ t^{-1} (56 therm ton^{-1}) quoted in [28] for baled-scrap melting in a large reverberatory melter, and the capital-cost and floor-space requirements are small fractions of those for a conventional reverberatory furnace of the same output.

13.3.2 Melting ferrous metals

The thermal properties of ferrous metals are listed in Figure 1:29.

Iron

Blast furnaces (Figure 13:23(*a*)). Iron production in blast furnaces involves thermochemical processes for which external heat is supplied by coke-oven coke and preheated air. Coke in addition to providing heat is a reactant and cannot be dispensed with; however, partial replacement is possible with alternative fuels. This is widely practised in, for example, Russia and Japan where the demand for iron has outstripped blast-furnace and coke-oven capacity. In the UK on the other hand only about one-third of blast furnaces use supplementary fuel injection [36]. Both oil and natural gas have been used as supplementary fuels [36–38]. The use of supplementary fuels is not merely a straightforward replacement on a thermal basis and other factors such as the cooling effect of the injectant need to be considered. Flux [36] quotes maximum replacement rates of about 20 per cent of the charged coke for oil and 12 per cent for coke-oven and natural gases, the replacement ratio for natural gas being about 600 m^3 per tonne of coke replaced (1100 ft^3 per cwt). Where gas is used it is simply injected via the tuyeres, there being no burners as such, and the capital cost is low. Blankley [39] quotes figures of 42–70 m^3 (1500–2500 ft^3) of natural gas per tonne of metal as representative of North American practice.

Ore sintering. Pre-treatment of iron ore is necessary for high blast-furnace outputs and this generally involves sintering. The process consists of mixing fine iron ore with coke breeze followed by just sufficient combustion to cause the ore particles to begin to melt on their surfaces. About 7–10 per cent coke breeze is employed and the mixture is spread on a travelling grate and ignited by means of a bank of gas burners firing downwards from an ignition hood. The heat requirement is about 2900 MJ t^{-1} (28 therm ton^{-1}) of finished sinter [39] of which coke supplies 80 per cent and the ignition system 6 per cent (180 MJ t^{-1}, 1.7 therm ton^{-1}), the balance being supplied by recycled flue dust. In general, sintering is carried out at integrated works and natural gas would generally be in competition with site-produced coke-oven gas.

Cupolas. Cupolas are used mainly for the production of cast iron and resemble small-scale blast furnaces, the main difference being that the charge consists of pig iron and scrap rather than iron ore. The charge is introduced in alternating layers of metal and coke together with small additions of limestone.

Partial replacement of coke by natural gas is commonly practised in the USA and Europe. Allan [40] has described in detail the successful replacement of 45 per cent of the coke used in a 610 mm (24 in) diameter cold-blast cupola by gas. Figure 13:23(*b*) shows the modified cupola incorporating four nozzle-mixing burners firing into 6 in (152.4 mm) bore refractory-lined combustion tubes. The burners are operated on boosted gas at 10 per cent excess air and controlled by a pressure governor backloaded with cupola pressure at burner level (2 kPa $\pm$ 250 Pa, 20 mbar $\pm$ 2.5 mbar, 8 inH_2O $\pm$ 1 inH_2O, gauge). Ultraviolet flame detection is applied to one of the burners. The 45 per cent coke replacement was accomplished by a 25 per cent increase in metal output while maintaining the metal temperature and analysis virtually unchanged. In addition a reduction of about 25 per cent in the grit collected by the grit arrester was recorded. Hirst and Ward [41] have described an installation in which 47 per cent coke replacement was obtained and consider that much higher levels of replacement may be possible. Furthermore it has proved possible to operate with 25 per cent steel whereas normal cupola operation allows a maximum of only 15 per cent.

Steel

Steelmaking is essentially a refining process carried out on a charge comprising pig iron and steel scrap. Carbon, manganese and silicon are removed by oxidation and phosphorus and sulphur by reaction with lime slag. A slag is produced which may be acidic or basic depending on the raw materials used and determines the type of refractory lining employed. A wide variety of steelmaking furnaces exist, some of which have been wholly or partially fired by natural gas as outlined briefly below.

Open-hearth furnaces. An open-hearth furnace is essentially a regenerative, usually cross-fired reverberatory furnace (Figure 13:24(*a*)). The working temperature is limited by the maximum usable temperature of the silica refractories used for the roof construction and is about 1660 °C. Open-hearth furnaces are generally very large, basic furnaces having capacities usually between 10 and 400 t, acid furnaces being rather smaller—up to 100 t [39]. Open-hearth furnaces are being replaced by LD converters and electric-arc furnaces but still represent a formidable fuel usage in Britain (63 PJ (600 million therms) per year in 1969 [36]).

Heavy fuel oil is the major fuel employed for open-hearth furnaces and luminous gas radiation is the major heat-transfer mode. Natural gas is, however, widely used outside Britain, particularly in the USSR and USA. Two main approaches have been adopted, both aimed at increasing flame luminosity. In a system originating in the USSR and currently used in one British steelworks, the gas is partially thermally cracked by burning some of the gas to give a temperature of about 800 °C in a cracker chamber through which the main gas flows. The gas entering the furnace consists of a mixture of raw gas, products of partial combustion and cracked gas [26]. The other approach used mainly in the USA involves mixed oil and gas firing. In some cases the gas, at up to 1 MPa (10 bar, 145 lbf in^{-2}) is used as the atomizing agent replacing steam or air. In other cases gas at pressures up to 350 kPa (3.5 bar, 50 lbf in^{-2}) is simply injected adjacent to the fuel-oil burner.

Increasing luminosity may not be the only satisfactory approach to gas firing of

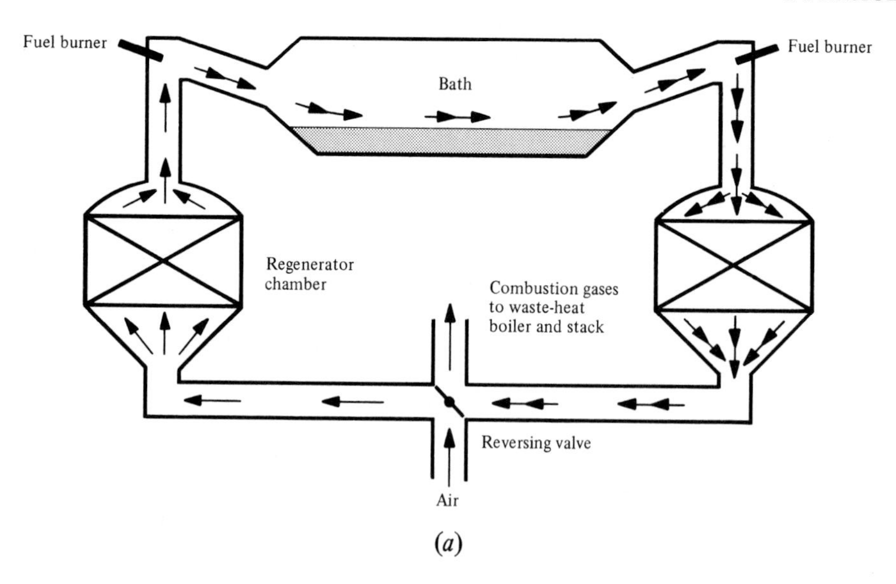

(a)

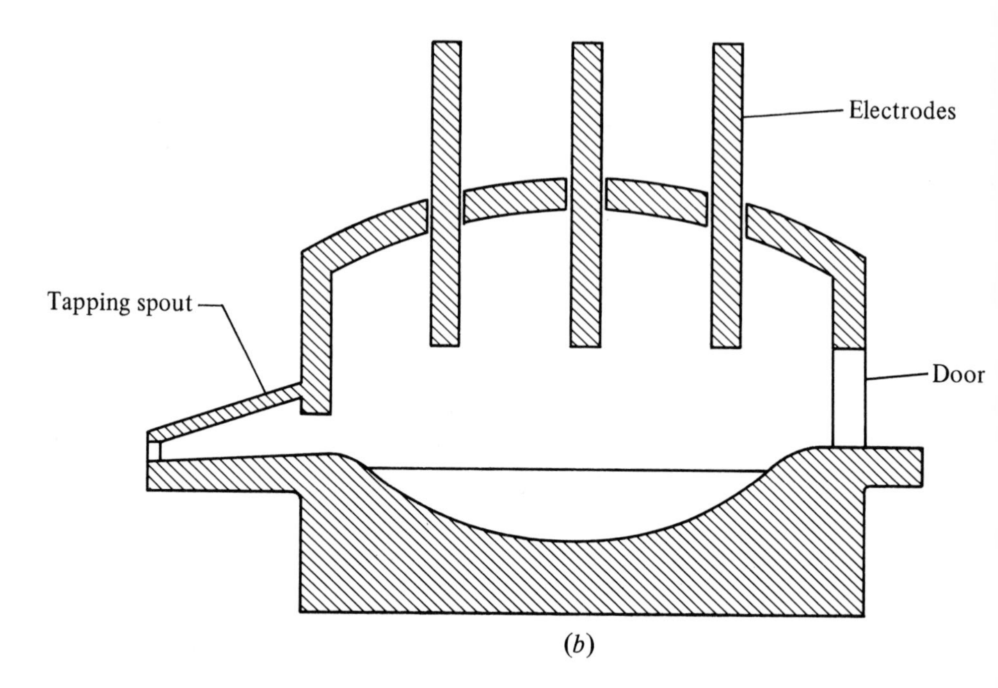

(b)

Figure 13:24 Steelmaking furnaces
(a) Open hearth
(b) Electric arc

open-hearth furnaces and high-momentum burners producing short flames may provide an effective alternative.

Electric-arc furnaces. The essential features of an electric-arc steel-melting furnace are shown in Figure 13:24(*b*). Carbon electrodes are located in the roof of the furnace and these may be lowered or retracted as the melt proceeds. Arc furnaces range in size from around 10 t up to 130 t. The tendency in recent years has been towards much higher power ratings for a given furnace diameter, the power rating increasing with the cube of the furnace diameter rather than with the diameter squared as was the case in the 1950s [36].

Considerable development work has been carried out on assisted melting in electric arc melters using oxygen–gas burners, and to a lesser extent air–gas burners, and these have been successful in overcoming electricity-supply restrictions and increasing production together with reducing energy costs in some instances.

Arnold [42] and Pomfret [43] have provided detailed descriptions of oxygen–town-gas and oxygen–natural-gas burners and their application in assisted melting. For their most effective application the combustion products should be directed at the base of the scrap charge and allowed to permeate through the charge before being exhausted. When melting is under way and the bath has become flat the burner effectiveness, i.e. ratio of electrical power saved to burner input, is much reduced.

The location of the burner is largely a matter of convenience and is to some extent governed by the availability of space around the furnace; wall-mounting, roof-firing and door-firing arrangements have all been employed, the latter having the advantage that some degree of manipulation is possible. Oxidation and decarburization are difficult to avoid when the burner is operated at stoichiometric oxygen–gas ratio because the oxygen–gas flame contains a large proportion of dissociated combustion products including free oxygen. Consequently gas-rich flames with oxygen–gas ratios between 75 and 100 per cent stoichiometric are employed [43].

Air–gas burners have been applied to a limited extent but the reduced heat-transfer rate requires a greater heat input for an equivalent effect and the lower flame temperature restricts their use to the earlier stages of melting.

In general, assisted melting has proved successful particularly in small, under-powered arc furnaces where significant reductions in tap-to-tap times have been achieved. The burner is least effective when the bath is flat and optimum savings are achieved in energy costs if the burner is used for less than 50 per cent of the time.

An integral part of electric-arc steelmaking is the refining period in which carbon and other elements are partially removed. Originally this was done simply by allowing the molten steel to boil for long periods.

Oxygen injection accelerates the refining process and this is widely practised using a consumable tubular lance but has the disadvantage of producing large amounts of fumes of red iron oxide. A combined fuel-fired oxygen lance allows the refining reactions to take place with a greatly reduced visible fume emission but at a slightly slower rate. This method, termed fumeless refining, has been accepted by the Alkali Inspectorate as a suitable means of fume suppression. Pomfret [43] has listed the following recommendations regarding operating conditions.

1 Oxygen requirements. The quantity of oxygen required to decarburize a given charge depends on the charged weight, the initial and final carbon levels and the levels of other oxidized elements. In general, however, no significant increase in

carbon-removal rate can be achieved with oxygen inputs greater than 2.8 m^3 min^{-1} t^{-1} (100 ft^3 min^{-1} ton^{-1}).

2 Oxygen–gas ratio. Excess oxygen is required for refining. Refining takes places with between 20 and 700 per cent excess oxygen although fume concentration increases with increasing oxygen–gas ratio. Pomfret [43] considers that 200 to 300 per cent excess oxygen offers the best conditions for fumeless refining.

3 Burner position. To allow refining to take place the slag layer over the molten metal must be penetrated by the oxygen–fuel flame. Shallow burner angles give large impingement areas but reduced penetrating power and angles close to the vertical are preferred. Flux [36] considers the scope for fumeless refining to be limited to relatively small low-rated furnaces on which time costs are low and in which there is no fume emission during the melting stage so that fume-removal equipment is not required.

Converter processes. These are not fuel-using processes but involve blowing air (in the Bessemer converter) or oxygen (in the LD, Kaldo and Rotor converters) through or onto charged molten pig iron. In modern Bessemer practice, oxygen diluted with steam or carbon dioxide may be used during the refining process. In the LD converter oxygen is blown onto the surface of the molten metal while the converter is in the vertical position. The rate of oxygen usage is around 56 m^3 t^{-1} (2000 ft^3 ton^{-1}) [39]. In the LD-AC or OLP (oxygen, lance, powder) method a two-slag process is used for pig iron of high phosphorus content involving the injection of powdered lime via the oxygen lance. The Kaldo converter is similar to the LD but is inclined at about 20° to the horizontal and rotates. The Rotor converter on the other hand is a horizontal rotating drum. Two oxygen lances are employed one above the metal surface and one below [39].

As indicated above, the converters are not fuel users and in fact could be fuel producers, liberating large quantities of carbon monoxide which has generally been burned off at the mouth of the vessel but is now beginning to be recovered for boiler firing and other ancillary uses. The scrap-consuming capacity of converters is much less than that of the open-hearth furnaces they are replacing. To increase this, scrap preheating using oxygen–gas burners has been used in the USA and this represents a large potential market in Britain. Hansrani and Richard [44] have described trials in which oxygen–fuel lances are used instead of oxygen lances in the LD process with a view to reducing fume emission. The results indicated that fume formation is governed mainly by the charge composition and the oxygen flow rate and hardly at all by fuel injection.

13.3.3 Nonmetals

Glass tanks. There are many similarities between the firing of glass tanks and open-hearth steel furnaces. Both are large-scale, regenerative, usually cross-fired, reverberatory furnaces. The glass tank operates at slightly more congenial temperatures (1450–1590 °C) but the necessity for long periods between repair ('campaigns') and the characteristics of the refractories used demand careful burner application [26]. Figure 13:25(*a*) shows a cross-section and (*b*) a diagrammatic longitudinal section through a typical glass tank. Glass tanks for sheet-, plate- or float-glass processes range in size from 30.5 m to 46 m long (100–150 ft), 7.6 m to 10.6 m (25–35 ft) wide by about 1 m deep and contain between 1000 and 2200 t of molten glass [45]. Glass tanks for container ware are smaller, typically 9.1 m by 6.1 m (30 ft × 20 ft). Still smaller tanks ('unit melters') are commonly end-fired or 'horseshoe-fired'. It is a continuous process, molten glass being drawn from the conditioning end while the raw materials, sand, soda ash, limestone, dolomite, salt cake, cullet (scrap glass) and fining agents, are fed in at the other.

In the first zone (melting zone) the temperature is raised quickly to 1450–1500 °C. It is desirable to have a low oxygen content in the furnace atmosphere without actual reducing conditions since this allows for the dissociation reactions of fining agents such as sodium sulphate which promotes deseeding, i.e. bubble removal [46]. The glass then enters the refining or planing zone of the furnace. The chemical reactions are complete at this point but the glass contains quantities of fine bubbles which in the refining phase are allowed to rise to the surface and be released. At the same time the thermal currents in the bath homogenize the glass. Any unmelted material is prevented from passing to the refining zone by means of a throat or by the use of floating skimmer blocks. In the conditioning zone the glass is cooled to the working temperature, a shadow wall forming a screen against radiation from the refining zone possibly supplemented by cold air from fans [45].

Almost all British glass tanks are oil fired, gas firing being restricted to some smaller container tanks so far, although there is extensive experience outside Britain of natural-gas firing of large-scale tanks. The most common method of oil firing is underport firing (Figure 13:25(*c*)). Burners are sited immediately below the port sill. Despite the fact that the combustion air flow is asymmetric to the oil spray good results may be obtained and the burners are easily accessible for maintenance. For large flat-glass (i.e. sheet, plate or float glass) tanks, throughport burners are generally used in which the burner head is sited near to the port centre-line (Figure 13:25 (*d*)). This location provides more uniform mixing, allowing greater throughput to individual burners but requires extensive water cooling [45]. The oil burners used are commonly of the high-pressure air type (see Section 7.12.3).

The low sulphur content of natural gas presents no great advantage in tank firing since sulphur is present in fairly substantial quantities in the melt where fining agents such as sodium sulphate are used [46]. On the other hand the poorer control possible with oil firing may give rise to local reducing conditions which can affect the glass and refractories even though there is excess oxygen present in the furnace atmosphere as a whole. Additionally in some types of glass manufacture some of the constituents of the oil ash are undesirable [46].

In general, on natural gas, sideport firing is used. Introducing gas in this way allows it to crack before combustion leading to flame luminosity. Figure 13:25(*e*) shows a sideport firing arrangement used on large flat-glass tanks [45]. The jets impinge in the mouth of the port producing a highly turbulent flame. Burner pressures in the range 13.8 to 27.6 kPa (138–276 mbar, 2–4 lbf in^{-2}) are used to provide high gas velocities. In general terms the aim is to produce a highly radiant turbulent flame avoiding direct impingement on the glass surface or on the furnace roof. The burners themselves may be simple single orifices or complex multi-jet arrangements, some including air injection [26]. On tanks specifically designed for oil firing, sideport natural-gas firing sometimes requires extensive structural modifications. The practicability of throughport firing with replacement gas burners is currently under investigation [45].

Before leaving glass tanks mention should be made of the extensive usage of gas in preheating oil-fired tank furnaces after rebuilding or repair allowing the carefully

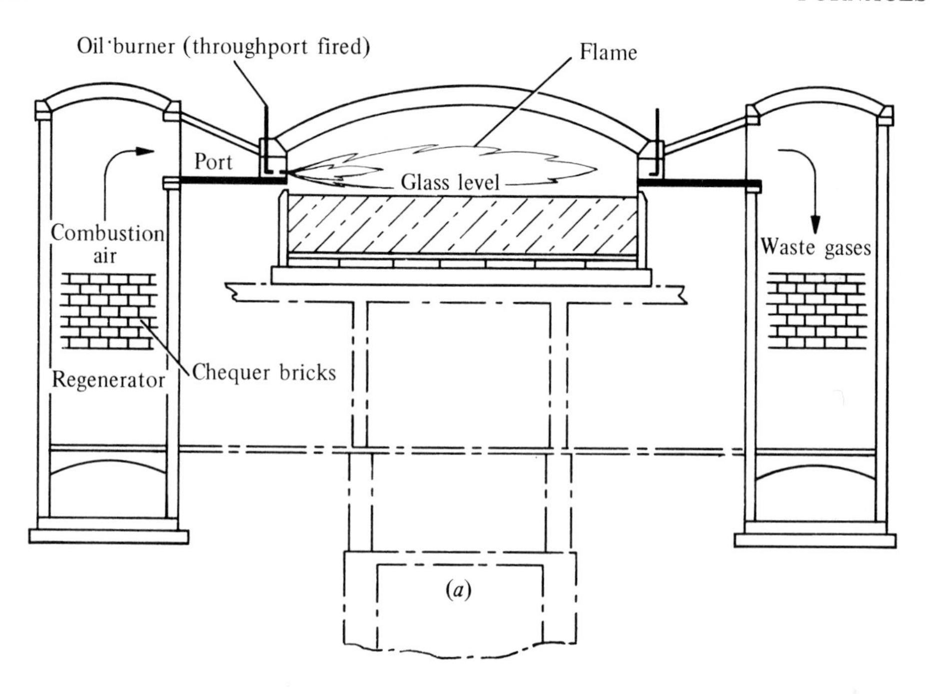
Oil burner (throughport fired)
Flame
Port
Glass level
Combustion air
Waste gases
Regenerator
Chequer bricks
(a)

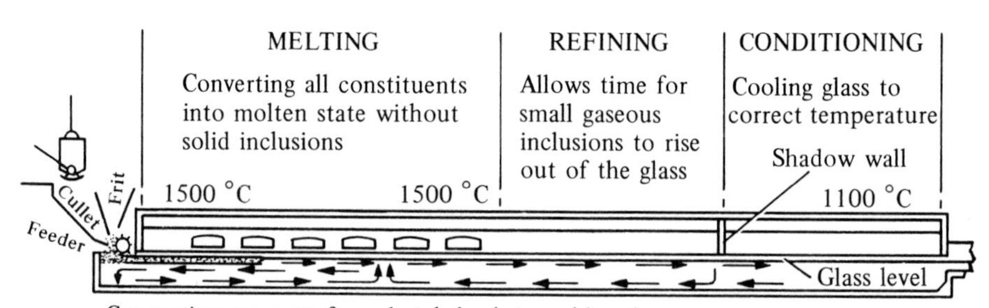
MELTING
Converting all constituents into molten state without solid inclusions
REFINING
Allows time for small gaseous inclusions to rise out of the glass
CONDITIONING
Cooling glass to correct temperature
Shadow wall
Cullet
Frit
Feeder
1500 °C
1500 °C
1100 °C
Glass level
Convection currents force batch back to cold end
(b)

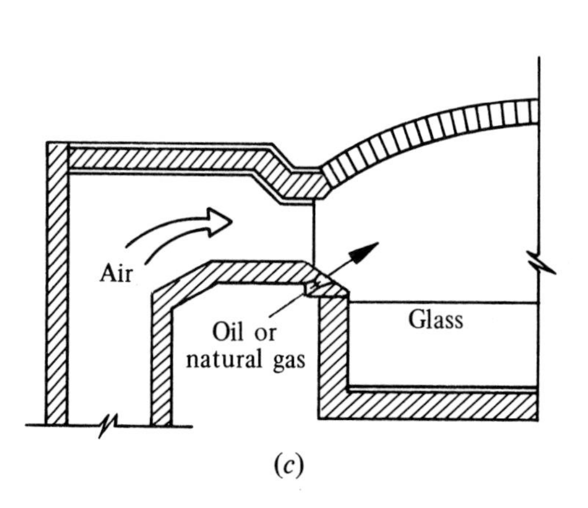
Air
Oil or natural gas
Glass
(c)

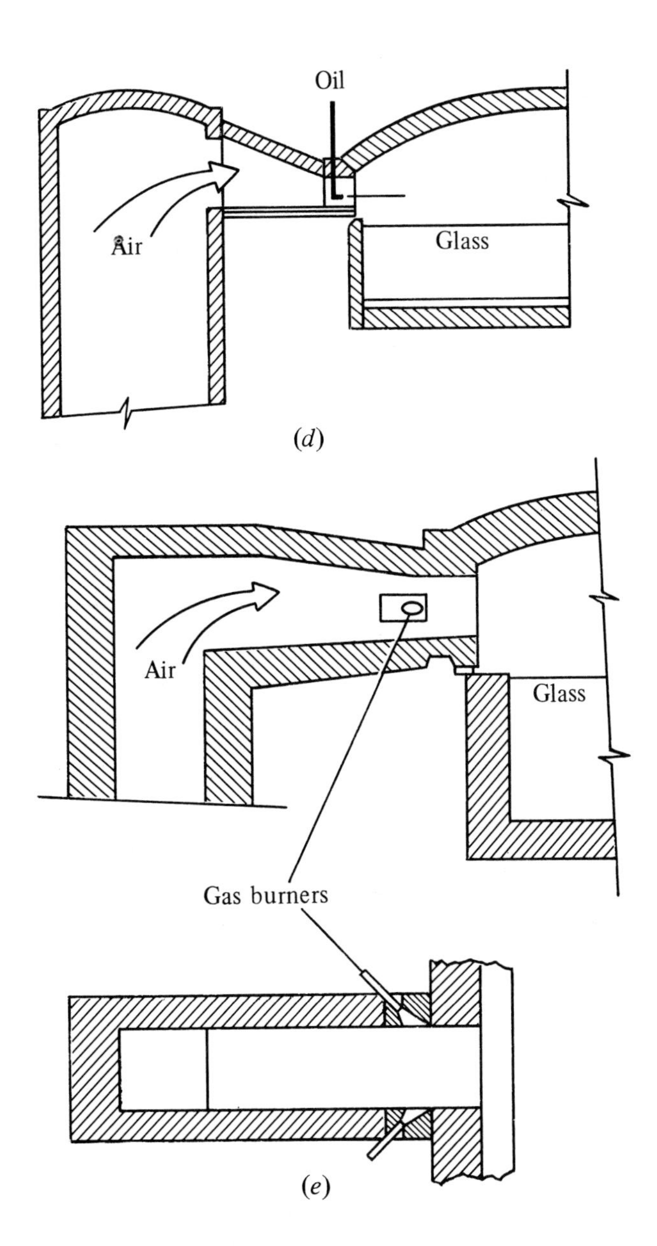

Figure 13:25 Glass tank [45]
(*a*) Cross-section through tank and regenerators
(*b*) Diagrammatic longitudinal section
(*c*) Underport firing
(*d*) Throughport firing
(*e*) Sideport firing

controlled temperature schedule required by the silica-brick structure to be obtained. Multiburner systems were used originally but the modern trend is to use jet-driven recirculation to provide good temperature uniformity [41, 45].

Additionally, gas firing, initially town gas and later natural gas, has had a virtual monopoly of heating the forehearths and feeders associated with glass tanks. Here the glass viscosity, and therefore temperature, needs to be controlled to within very narrow limits for subsequent glass-forming operations. Precise temperature control (in some instances ± 1 °C) is possible with gas firing and this has led to its predominant position.

Details of gas-fired forehearths and feeders have been given by Hooper [86] and Cole [47].

13.4 METAL HEAT-TREATMENT FURNACES

The object of heat-treating metals is to produce mechanical or physical properties suited to their industrial use. Heat treatment in its broadest sense includes any process involving heating and cooling the solid metal by which its properties are altered without intentional alteration of its chemical composition [48]. Carburizing is excluded by this definition and is considered separately in Section 13.2.4. Also, heating preparatory to hot working, e.g. forging and rolling is outside the scope of this definition and is considered in Section 13.5.

13.4.1 Metallurgical requirements and definitions

Ferrous metals

Heat-treatment processes are required to conform to a time-temperature cycle consisting of three periods: heating, holding at the desired temperature ('soaking') and cooling. The heating rate is not very important from a metallurgical point of view but is often limited by the need to avoid distortion. This is particularly so in the case of steel which, before heat treatment, is in a stressed condition due to cold working or to welding. In these instances the heating rate must be slow. The length of the soaking period is governed by the nature of the heat-treatment process and by the thickness of the section being heated in order to achieve temperature uniformity. If the stock is of varying thickness clearly the length of the soak period is governed by the time required to heat the thickest section uniformly. There is an increasing demand for bright heat-treatment processes using indirect firing and controlled atmospheres (see Section 13.2). A complete description of the metallurgy of heat treatment is, of course, outside the scope of this book and the reader is referred to the many excellent texts available, e.g. [48], [49] and [50]. However, it is hoped that the sections which follow will serve to introduce the requirements of the common heat-treatment processes and their terminology.

Iron–carbon equilibrium diagram. This diagram (Figure 13:26) is essentially a map of the temperatures at which the various crystal structures or phases change on very slow heating or cooling in relation to carbon content. For example pure iron or ferrite (also called α-iron) possesses a body-centred cubic crystalline structure and exists, in the absence of carbon, from room temperature up to 910 °C (point G on

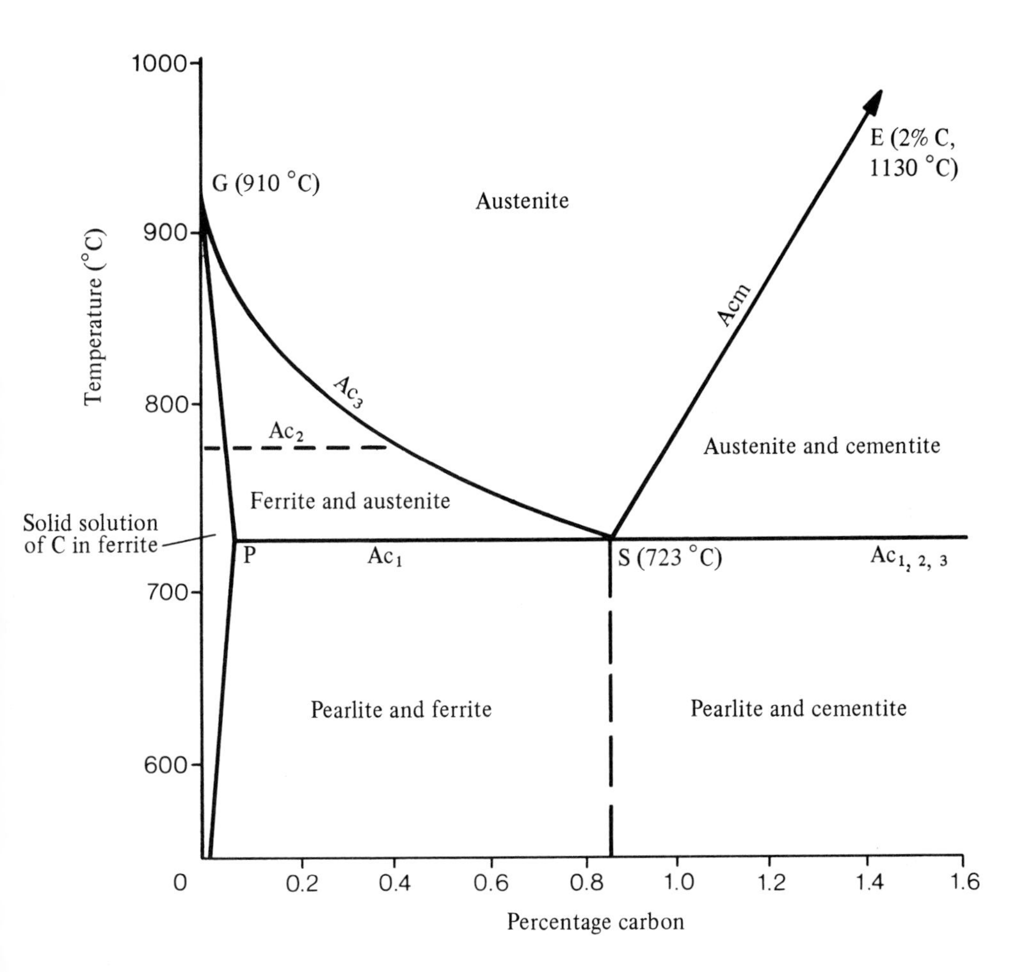

Figure 13:26 Iron–carbon equilibrium diagram

the diagram). As the carbon content increases its upper limit of existence reduces to about 723 °C at about 0.8 per cent C. Above these temperatures it converts to austenite (γ-iron) which has a face-centred cubic structure. The temperatures at which such transformations occur are termed critical temperatures or critical points. Equilibrium diagrams for nonferrous metals provide much of the basic information required regarding alloys and heat-treatment processes. The same may be said for the iron–carbon equilibrium diagram but with some serious reservations.

It must be remembered that equilibrium diagrams represent behaviour under conditions of very slow heating and cooling and delineate areas of composition and temperature in which the various constituents are stable. No provision is made in equilibrium diagrams for the situation where, by rapid cooling, constituents exist on the 'wrong' side of the critical line. In addition steel can exist as martensite, a phase of great importance in the hardening of steels which does not appear at all on the equilibrium diagram. As Bullens [48] has pointed out, the only type of heat treatment which the diagram describes with any degree of accuracy is the process of full annealing. None of the treatments by which hardening and strengthening of steels are obtained by rapid cooling are described by it. Nonetheless the iron–carbon diagram is the starting point for the study of heat-treatment processes and its terminology is common currency in the heat-treatment field.

There are two constituents in cold (annealed or normalized) steel, α-iron (Fe) and cementite (Fe_3C). If the steel is heated to a sufficiently high temperature there is only one constituent, austenite which is a solid solution of cementite in γ-iron. The diagram indicates that austenite does not exist at temperatures below 723 °C but in some alloy steels austenite is stable at room temperature. Austenite can contain various amounts of Fe_3C, the amount increasing with temperature, but the amount is limited and in the field to the right of line SE there are two constituents, austenite and cementite, the point E at 2 per cent C representing the boundary between steels and cast irons.

Point S is of particular interest. A carbon-rich austenite, say, of 1.5 per cent C cooling from 1100 °C remains the same until it reaches 1050 °C at which point the solid-solubility limit is reached and cementite comes out of solution. The carbon-depleted austenite continues to deposit more and more until the temperature approaches 723 °C. At this point iron carbide corresponding to about 0.67 per cent C of the original 1.5 per cent has come out of solution leaving 0.83 per cent [48]. At point S the iron in the austenite changes to the α-form in which the carbon solubility is low (maximum 0.025 per cent C) and so the remaining 0.8 per cent or so comes out of solution forming (under slow-cooling conditions) interleaved plates of cementite and ferrite, this composite structure being termed pearlite.

If on the other hand austenite of a lower carbon content, say, 0.4 per cent, is cooled it remains unchanged down to about 790 °C when the ferrite separates leaving the austenite relatively richer in carbon. The process continues down the line GS until the carbon content of the austenite has increased to 0.83 per cent C and the temperature has fallen to 723 °C, i.e. point S. Further cooling causes the austenite to change to pearlite.

Thus in the first case the steel comprises pearlite and cementite and in the second case pearlite and ferrite.

Point S is the lowest point at which α-iron can exist and is termed the eutectoid as the solid-solution analogy with the eutectic point of liquid–solid transitions. Thus steel of around 0.83 per cent C is commonly termed eutectoid steel, steel with < 0.83 per cent C is hypoeutectoid steel and steel with > 0.83 per cent C is hypereutectoid steel. In commercial steels containing alloying ingredients the eutectoid occurs over a band of temperatures rather than at 723 °C.

Referring again to the diagram, the line GS is designated Ac_3, the line SE, Acm, etc. This terminology requires some explanation. If a steel sample is heated and a time–temperature record is made it will show a break at 723 °C and this distortion of the plot will persist up to the line GS or SE. This 'arrest' indicates that heat is required for the α- to γ-iron transformation and for the solution of carbon in iron. The arrest on heating is designated Ac (French *arrête chauffage*) and further designated Ac_1

etc., to indicate the change involved. In the case of hypereutectoid steels the Acm line denotes the cementite-solution arrest points.

As previously stated the equilibrium diagram relates strictly only to very slow heating and cooling. In practice, heating and, more importantly, cooling generally take place at a much faster rate. This allows austenite to be retained below the temperature indicated on the equilibrium diagram and this in turn transforms to martensite, which is very hard, and pearlite. Thus a steel of, say, 0.45 per cent C quenched from 715 °C would contain ferrite, pearlite and a small amount of martensite. If it were quenched from, say, 750 °C its hardness would increase markedly and the microstructure would indicate that the pearlite had been entirely replaced by martensite which moreover had encroached on the ferrite areas since at sufficiently high temperatures the austenite dissolves the ferrite. Higher temperatures still render the steel wholly austenitic and consequently wholly martensitic after quenching. This process is reversed by heating and subsequent slow cooling. Martensite being brittle and hard does not itself confer very widely useful properties on the steel but it is the constituent which can readily be converted into other microstructures providing a valuable combination of hardness and strength.

Annealing. An annealing process is in general terms one which renders a product soft. Steel annealing may be subdivided into full annealing, i.e. heating the steel above its critical temperature range and cooling it slowly, and subcritical annealing in which the steel is heated to just below its critical temperature. Full annealing results in a structure which is pearlite together with free ferrite or free carbide depending on the carbon content. Full annealing has an additional effect generally described as grain-size refinement. This latter effect improves the ductility of the softened steel or, if it is to be subsequently hardened, improves the toughness of the steel as hardened. It is necessary to exceed Ac_3 to transform the ferrite wholly to austenite. However, in steels without additions for grain-size control the grain starts to coarsen at temperatures just above Ac_3 and the annealing temperature should be kept as little as possible above Ac_3. The necessary slow cooling through the critical range is usually provided by cooling in the furnace. Directly fired furnaces may be employed but increasingly bright annealing is practised using controlled atmospheres (Section 13.2.4).

Subcritical annealing has as its chief aim the removal of internal stress usually in low-carbon steels (up to about 0.25 per cent C) which have been cold-worked. The process temperature is below the Ac_1 line. No major structural changes are promoted but grain distortion is corrected. If the steel is not to be further cold-worked temperatures well below the critical range may be employed, e.g. 550 °C. Stress relief is commonly carried out on welded structures to render the welding stress distribution uniform. Temperatures are generally in the range 500 to 650 °C. British Standard requirements exist for the heat treatment of welded pressure vessels. Typically for straight carbon steels stress relief at 580 to 650 °C for one hour per 25 mm (1 in) thickness is required increasing to 740 °C for some alloy steels (e.g. Cr, Mo) [51, 52].

Normalizing. Normalizing consists in principle of heating steel to above its critical temperature (Ac_3 or Acm) followed by cooling in air. This may of course cover a wide range of cooling conditions. A small single component will cool much more quickly in air than will a number of components piled together leading to rather different properties in the finished product.

Normalizing generally involves rather higher temperatures than are involved in full annealing with a view to obtaining full absorption of excess ferrite and equalization of the carbon content throughout the austenite. To avoid the coarse grain structure which would result from the slow cooling of full annealing from this high temperature the much faster air cooling is adopted.

Spheroidizing. Spheroidizing is essentially a type of tempering. If steels are subjected to prolonged heating close to but below Ac_1 so that no austenite is produced the carbide which is normally present as plates or needles interspersed in the ferrite of pearlite, bainite or martensite tends to contract into spheroids. A spheroidized structure provides machinability and improves uniformity in hardening with some grades of steel.

Hardening. Hardening occurs when steels are heated above Ac_3 to develop an austenitic structure followed by quenching to develop martensite. Two factors determine the hardness after quenching, first the 'hardenability' of the steel, which is a function of carbon content for straight carbon steels but which is strongly influenced by the presence of alloying elements, and second the speed of quenching. Water is used for fast quenching, oil for slower quenching. Hardening generally produces a steel which is too hard and brittle for most engineering purposes and tempering is required after hardening.

Tempering. Tempering or 'drawing' involves heating martensitic steels to below Ac_1, generally in the range 120 to 675 °C, the higher temperatures inducing greater toughness. The changes occurring during tempering are complex but consist mainly of carbide precipitation and agglomeration. Short tempering times are generally undesirable. At least $\frac{1}{2}$ h and preferably 1 to 2 h are required at tempering temperature for most steels. Tempering usually follows hardening before the hardened steel has cooled to room temperature to avoid stress–cracking. The term tempering is also applied to a post-normalizing stress-relief carried out at 650–750 °C on some classes of steel.

Patenting. Patenting is carried out on work-hardened steel wire which requires further cold-drawing. Temperatures may be as high as 1000–1050 °C but the cooling is such that a fine pearlitic structure is obtained. Air cooling may be employed but in steels with high hardenability a lead-bath quench (480–620 °C) is used to control the cooling rate.

Austempering. In this process austenite is transformed to bainite without the intermediate production of martensite. It consists of heating steel above Ac_3 followed by quenching in a hot bath held at a temperature below that of fine-pearlite formation.

Martempering. Martempering is a quench-hardening process in which the final stages of the quench are carried out at a much slower rate than in normal quenching, i.e. in air, to minimize stresses. The steel is quenched first in a molten-salt bath which cools it rapidly to slightly above the M_s temperature (i.e. the temperature at which martensite starts to form on cooling) then cooled in air. The finished product is fully martensitic and may be subsequently tempered.

Nonferrous metals

Aluminium and its alloys. Figure 13:27 shows a generalized phase diagram for age-hardening aluminium alloys upon which are superimposed the various heat-treatment ranges.

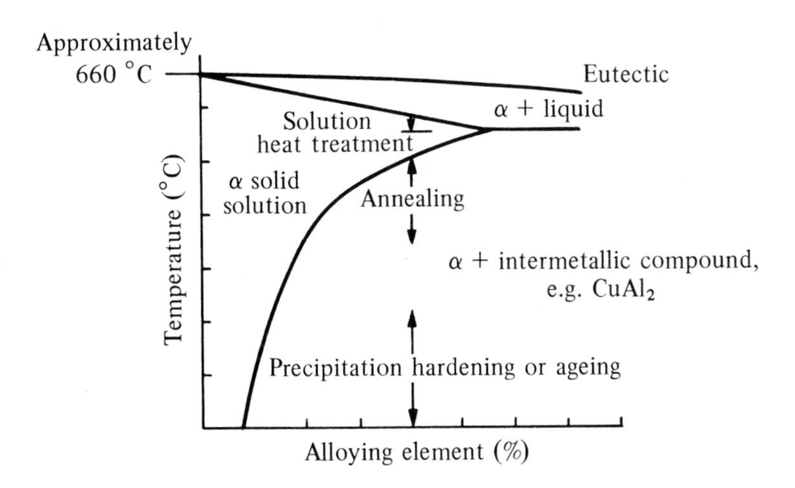

Figure 13:27 Typical phase diagram for heat-treatable (age-hardening) aluminium alloys [53]

The main heat-treatment processes used with aluminium alloys are:

1 Solution treatment. This consists of heating the alloy in the α solid-solution field for sufficiently long to take all the phases into solid solution and to anneal the metal thoroughly.
2 Quenching. If the alloy is cooled rapidly it becomes supersaturated and the alloying elements are precipitated out as an aluminium–alloying-element intermetallic. The size and distribution of the precipitate determine the final properties of the alloy. Quenching is done as rapidly as possible, usually in water.
3 Ageing. This is carried out at low temperatures, always below the α solid-solution boundary. Maximum hardness is obtained with lower ageing temperatures and longer times [53].

Solution-treatment temperatures are taken as high as possible above the α + intermediate compound boundary but at such levels as to avoid the possibility of eutectic melting [53]. Good furnace-temperature uniformity and control are essential for this process. Ageing is carried out in ovens, generally in the temperature range 130 to 240 °C for periods ranging from 1 to 50 hours [5].

Other metals. In the heat treatment of copper and its alloys the major point of interest is the selection of appropriate heat-treatment atmospheres. This topic is covered in Section 13.2.4. For a comprehensive review of heat-treatment require-

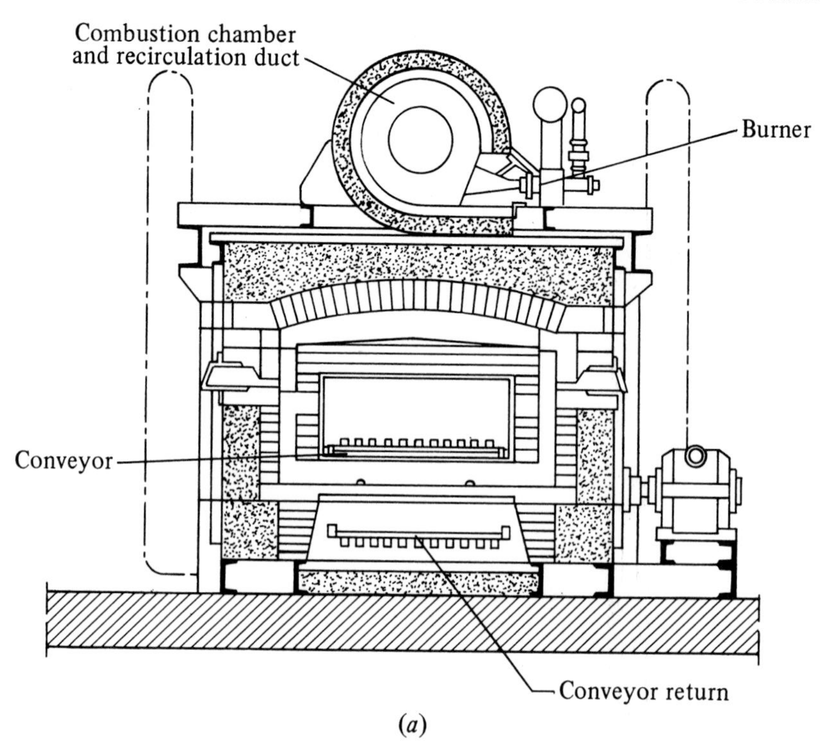
Combustion chamber
and recirculation duct
Burner
Conveyor
Conveyor return

(a)

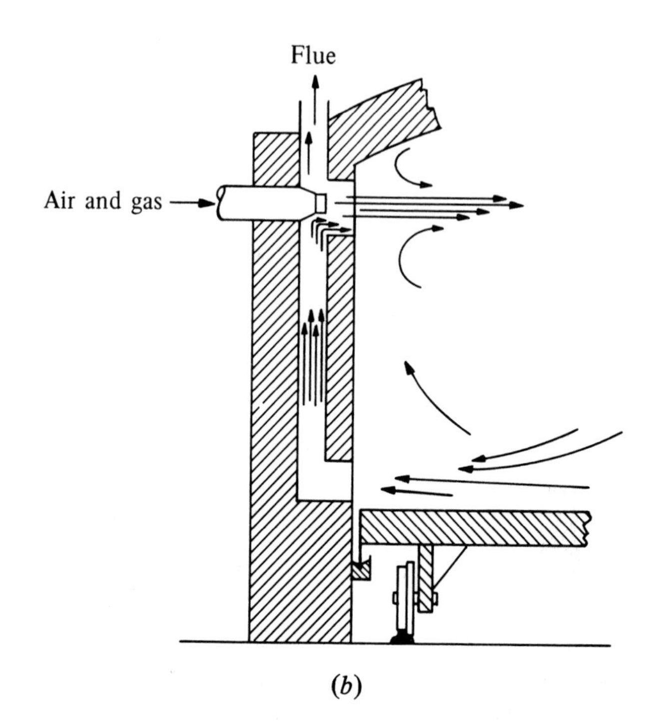
Flue
Air and gas

(b)

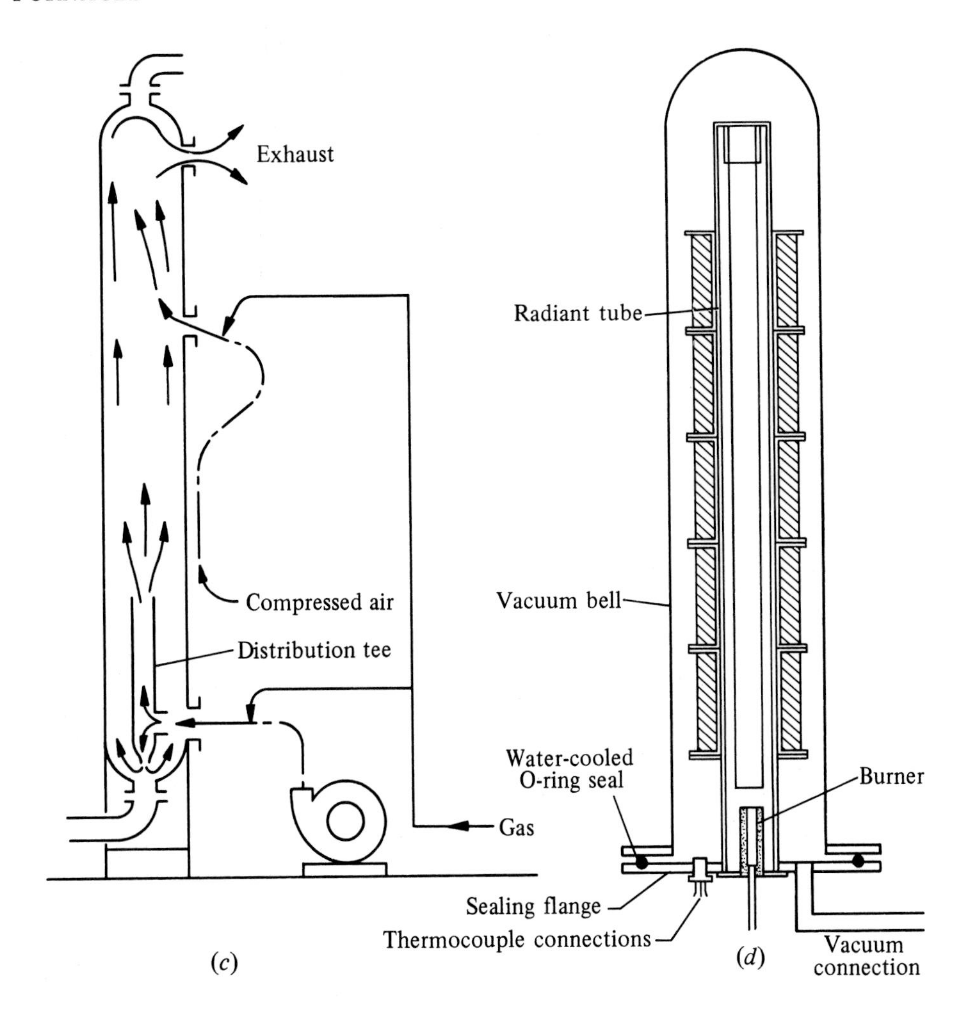

Figure 13:28 Heat treatment of metals
(*a*) Conveyor furnace with external combustion chamber for light alloys
(*b*) Bogie-hearth furnace with jet-driven recirculation [54]
(*c*) On-site stress relief of tall column [41]
(*d*) Gas-fired vacuum furnace for copper wire [19]

ments and processes for titanium-based and nickel-based alloys the reader is referred to [53] which, in addition, deals fully with the heat treatment of stainless steels.

13.4.2 Furnaces for metal heat treatment

As the brief outline of metallurgical requirements presented above indicates, heat-treatment furnaces are required for processes covering temperatures from as low as 130 °C up to around 1000 °C. A wide variety of furnace types including both direct and indirect firing and continuous and batch operation are employed but a universal requirement is for accurately controlled and uniform temperatures. In high-temperature furnaces where wall and gas radiation are the important heat-transfer modes, temperature uniformity (except for 'shadow' areas) is usually quite good. The situation is different in low-temperature furnaces and ovens, in which radiation is weak. Additionally, the materials usually heated in the low-temperature range (e.g. non-ferrous metals) are normally bright and have very low emissivities. In directly fired furnaces this presents particular problems since, if the work chamber is also the combustion chamber, some parts of the charge are in line of sight with the flames, which leads to poor temperature uniformity. It is necessary, therefore, in these low-temperature processes, to promote convective heat transfer by the use of fan- or jet-driven recirculation, often using an external combustion chamber to avoid the stock 'seeing' the flame (Figure 13:28(*a*)). The necessity for such measures is particularly great in the case of light-alloy heat treatment, especially solution treatment for which, as indicated in Section 13.4.1, the temperature tolerances are very small. In the higher temperature range, i.e. in furnaces for heat treatment of steel, jet-driven recirculation is now being applied to provide temperature uniformity in place of the banks of burners previously employed. Figure 13:28(*b*) shows this principle applied to a bogie-hearth stress-relieving furnace. In the example shown, two burners are used firing from diametrically opposite corners, each rated at 732 kW (2.5×10^6 Btu h^{-1}). The furnace dimensions are 5.5 m × 4.9 m × 1.8 m high (18 ft × 16 ft × 6 ft high).

Welded pressure vessels are sometimes so large that stress relief cannot be carried out in a conventional furnace. North Eastern Gas Board in the 1950s pioneered the use of the pressure vessel itself as the furnace in which a high-velocity burner promotes recirculation of combustion products around a temporary heat-distribution system in the vessel. Temperature uniformity of ±10 °C may be obtained and the fuel consumption is, as might be expected, very low, e.g. 520 MJ t^{-1} (5 therm ton^{-1}) at 650 °C [41]. This technique is now very well established and vessels up to 73 m (240 ft) long by 6.7 m (22 ft) diameter have been heat-treated by this system [41]. Figure 13:28(*c*) illustrates the system applied to a 21.3 m (70 ft) high column.

The availability of ceramic-fibre refractories makes possible the construction of extremely light large-scale heat-treatment furnaces. Hirst and Ward [41] have described a bogie-hearth furnace of low thermal capacity measuring 23 m × 7 m wide × 7.1 m high (75 ft × 23 ft × 23.3 ft) for stress relieving at 700 °C ± 15 °C. Heating is carried out by six high-velocity burners, four at hearth level, two in the roof, all firing anticlockwise. An additional feature provided by ceramic-fibre refractories is their applicability to quickly erected lightweight furnace structures for on-site stress relieving.

As outlined in Section 13.2.4, copper wire is normally annealed under vacuum conditions to vaporize the die lubricants and to prevent successive wires sticking together. This is largely carried out in electrically heated furnaces. Burton and Cubbage [19] have applied the single-ended recirculating radiant tube described in Section 7.10 to this process as illustrated in Figure 13:28(*d*). Heat transfer in vacuum annealing presents particular problems. Because of the removal of most of the gases, convection plays little part and, since there is usually little contact between the load and the heating source, conduction also is small. Radiation is the main mechanism but since the process is a low-temperature one (250–450 °C) and the charge has a low emissivity (0.1), heating rates and especially cooling rates are very slow. In the design shown in Figure 13:28(*d*) these problems are overcome by applying heat along the centre-line. The bobbins upon which the wire is wound are threaded onto the central radiant tube. Heat is transferred uniformly to the inside of the tube by forced convection, thence by conduction and radiation to the bobbins which, being of steel, have an emissivity of about 0.7. The combustion air may be used for forced cooling of the tube at the end of the heat-treatment cycle. Heating times to 300 °C are around 30 minutes and cooling times to 70 °C around 300 minutes, both showing considerable improvements over electrical heating. Fuel consumption is low, typically 780–1140 MJ t^{-1} (7.5–11 therm ton^{-1}).

As an alternative to the use of indirect firing and protective atmospheres, molten-lead or salt baths are employed in both batch and continuous heat-treatment processes, the molten medium providing protection from oxidation and decarburization and

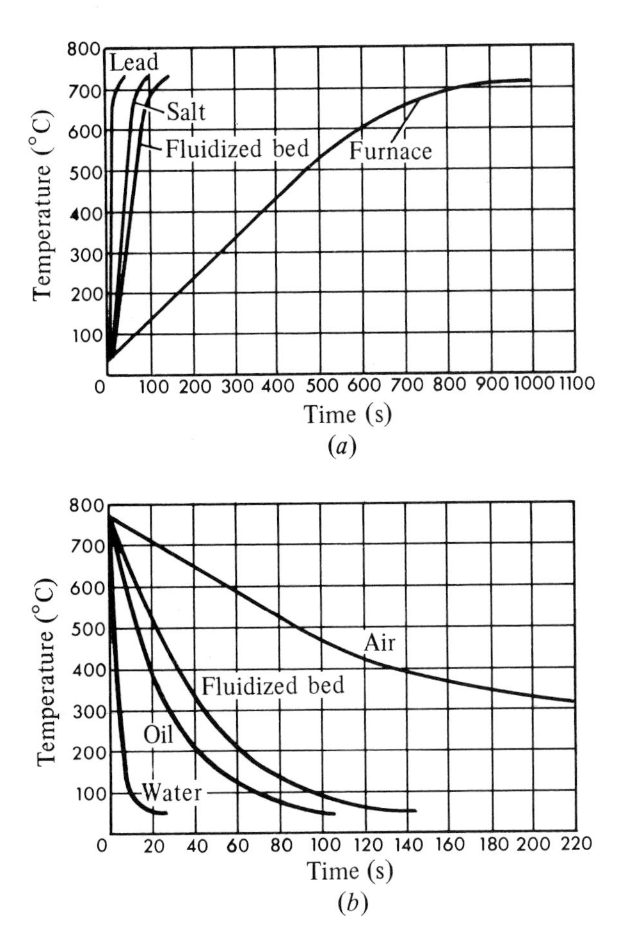

Figure 13:29 Heating and quenching rates for 16 mm ($\frac{5}{8}$ in) diameter steel bars: conventional furnaces, liquid baths and fluidized beds [55]

(*a*) Heating

(*b*) Quenching

fast and uniform heating. Figure 13:29 illustrates the very much faster heating rates of liquid baths compared with conventional furnaces. Among other advantages, it is claimed [5] that salt and lead baths:

1 Facilitate selective heating by immersion of only the part of the metal to be heated. The fast heating rate gives sharp demarcation between the heated and unheated parts of the work.
2 In principle, create less distortion since the work is partially supported by the liquid.

Salt baths possess a number of disadvantages:

1 Molten salts are potentially hazardous and operators must be protected from splashing.
2 Components to be heat-treated must be clean and dry and must be free of cavities containing air or other gases which may lead to explosions.
3 The salt must be prevented from coming into contact with combustibles or reducing agents.
4 The pots have a limited working life.

Lead baths also are used in heat treatment, particularly in continuous wire processing where they are employed for cooling after annealing and also for preheating prior to galvanizing. Lead baths suffer from a number of disadvantages:

1 Since lead is denser than steel, floating of immersed objects may present problems.
2 Lead is easily oxidized and starts to vaporize at around 650 °C.
3 Lead fumes are extremely toxic and collecting hoods and extraction systems are necessary.

Most of the difficulties can be overcome by the use of fluidized beds and these are beginning to be used in heat treatment [55, 56]. Fluidized beds may be designed for both heating and quenching stock. Heating rates may be as high as eight times that of a conventional furnace and temperature control and uniformity are very good. They are slightly slower than molten-salt or metal baths but have few of their disadvantages and the added advantage that a single piece of equipment may be used at room and elevated temperatures. Figure 13:29 compares heating and cooling rates for steel bars in fluidized, lead and salt baths and conventional furnaces. Fluid beds have been used successfully for normalizing, austempering and martempering, usually on a batch basis, but continuous annealing of wire and strip is achieved by passing the material horizontally through close-fitting slots in the side walls of the bed [53].

13.5 FURNACES FOR HOT-WORKING

This section deals with furnaces for heating metals to temperatures suitable for hot-working processes such as forging, rolling and extruding. Hot-working temperatures for steel are generally in the range 1000 to 1300 °C and for aluminium alloys 480 to 500 °C. Figure 13:30 [5] indicates the temperature requirements of a wide variety of hot-working (and heat-treatment) processes.

Material	Operation	Approximate temperature range	
		K	°F
Aluminium alloy	Ageing	395–510	250–460
Aluminium alloy	Annealing	505–685	450–775
Aluminium alloy	Forging	670–755	750–900
Aluminium alloy	Heat-treating	425–810	310–1000
Aluminium	Melting	920–1030	1200–1400
Aluminium alloy	Die-casting	950–1030	1250–1400
Antimony	Melting point	903	1166
Asphalt	Melting	450–505	350–450
Babbitt	Melting	590–700	600–800
Brass	Annealing	590–755	600–900
Brass	Extruding	1030–1060	1400–1450
Brass	Forging	920–1030	1200–1400
Brass, yellow	Melting	1200	1705
Brass, red	Melting	1270	1830
Bread	Baking	420–530	300–500
Brick	Burning	1255–1700	1800–2600
Brick, refractory	Burning	1645–1920	2500–3000
Bronze, 5% aluminium	Melting	1330	1940
Bronze, manganese	Melting	1170	1645
Bronze, phosphor	Melting	1320	1920
Bronze, Tobin	Melting	1160	1625
Cadmium	Melting point	595	610
Cake (food)	Baking	420–450	300–350
Calcium	Melting point	1123	1562
Calender rolls	Heating	420	300
Candy	Cooking	380–420	225–300
Cast iron	Annealing	1145–1200	1600–1700
Cast iron	Malleablizing	1170–1255	1650–1800
Cast iron	Vitreous enamelling	920–975	1200–1300
Cement	Kiln firing	1810–1920	2800–3000
China	Glazing	1090–1310	1500–1900
Clay, refractory	Burning	1480–1700	2200–2600
Cobalt	Melting point	1763	2714
Coffee	Roasting	590–700	600–800
Coke	By-product oven	1270–1770	1830–2730
Cookies	Baking	460–505	375–450
Copper	Annealing	700–920	800–1200
Copper	Refining	1420–1700	2100–2600
Copper	Melting	1420–1530	2100–2300
Copper	Smelting	1420–1700	2100–2600
Cores, sand	Baking	395–560	250–550
Cupronickel, 15%	Melting	1450	2150
Cupronickel, 30%	Melting	1500	2240
Electrotype	Melting	665	740
Enamel	Baking	395–505	250–450

Material	Operation	Approximate temperature range	
		K	°F
Everdur 1010	Melting	1290	1865
Frit	Smelting	1365–1590	2000–2400
Glass	Annealing	700–920	800–1200
Glass	Melting, pot furn	1530–1645	2300–2500
Glass, bottle	Melting, tank furn	1645–1865	2500–2900
Glass, flat	Melting, tank furn	1645–1920	2500–3000
Gold	Melting	1340–1450	1950–2150
Iron	Blast furnace tap	1645–1810	2500–2800
Iron, malleable	Melting	1590–1980	2400–3100
Iron, malleable	Annealing	1090–1200	1500–1700
Japan	Baking	420–505	300–450
Lacquer	Drying	340–365	150–200
Lead	Melting	600–670	620–750
Lead	Blast furnace	1170–1480	1650–2200
Lead	Refining	1255–1365	1800–2000
Magnesium	Ageing	450–480	350–400
Magnesium	Heat-treating	660–675	730–760
Magnesium	Melting	950–975	1250–1300
Magnesium	Superheating	1060–1170	1450–1650
Meat	Smoking	310–340	100–150
Mercury	Melting point	234	38
Molybdenum	Melting point	2898	4757
Monel metal	Annealing	865–1075	1100–1480
Monel metal	Melting	1810	2800
Moulds, foundry	Drying	475–670	400–750
Muntz metal	Melting	1175	1660
Nickel	Annealing	865–1075	1100–1480
Nickel	Melting	1725	2650
Palladium	Melting point	1827	2829
Phosphorus, yellow	Melting point	317	111
Pie	Baking	530	500
Pigment	Calcining	1150	1600
Platinum	Melting	2046	3224
Porcelain	Burning	1700	2600
Potassium	Melting point	336	145
Potato chips	Frying	450–480	350–400
Primer	Baking	420–480	300–400
Silicon	Melting point	1703	2606
Silver	Melting	1225–1310	1750–1900
Sodium	Melting point	371	208
Solder	Melting	480–590	400–600
Steel, sheet	Blue annealing	1030–1140	1400–1600
Steel, sheet	Box annealing	1090–1200	1500–1700
Steel, sheet	Bright annealing	950–1000	1250–1350
Steel, sheet	Job mill heating	1365–1420	2000–2100

Material	Operation	Approximate temperature range	
		K	°F
Steel, sheet	Mill heating	1250–1420	1800–2100
Steel, sheet	Open annealing	1090–1200	1500–1700
Steel, sheet	Tin-plating	615	650
Steel, sheet	Vitreous enamelling	1030–1170	1400–1650
Sheet, tinplate	Box annealing	920–1170	1200–1650
Sheet, tinplate	Hot mill heating	1250–1365	1800–2000
Sheet, tinplate	Lithographing	420	300
Steel	Bessemer converter	1810–1920	2800–3000
Steel	Calorizing	1200	1700
Steel	Carburizing	1090–1200	1500–1700
Steel	Case hardening	1140–1200	1600–1700
Steel	Cyaniding	1030–1250	1400–1800
Steel	Drawing forgings	725	850
Steel	Drop-forging	1475–1590	2200–2400
Steel	Forging	1200–1450	1700–2150
Steel	Form-bending	1140–1250	1600–1800
Steel	Galvanizing	700–760	800–900
Steel	Heat-treating	650–1250	700–1800
Steel	Lead hardening	1030–1250	1400–1800
Steel	Melting, open hearth	1810–1975	2800–3100
Steel	Melting, electric furn	1590–2030	2400–3200
Steel	Normalizing	1170–1310	1650–1900
Steel	Rolling	1530	2300
Steel	Soaking pits	1310–1420	1900–2100
Steel	Upsetting	1365–1530	2000–2300
Steel	Welding	1590–1810	2400–2800
Steel bars	Heating	1310–1480	1900–2200
Steel billets	Heating	1500	2250
Steel blooms	Heating	1500	2250
Steel bolts	Heading	1480–1530	2200–2300
Steel castings	Annealing	975–1150	1300–1600
Steel flanging	Heating	1250–1420	1800–2100
Steel ingots	Heating	1365–1480	2000–2200
Steel nails	Blueing	615	650
Steel pipe	Butt welding	1590–1700	2400–2600
Steel rails	Hot bloom reheating	1310–1400	1900–2050
Steel rivets	Heating	1365–1480	2000–2200
Steel rod	Mill heating	1310–1420	1900–2100
Steel shape	Heating	1310–1480	1900–2200
Steel slabs	Heating	1365–1480	2000–2200
Steel spikes	Heating	1365–1480	2000–2200
Steel springs	Annealing	1090–1170	1500–1650
Steel wire	Annealing	920–1030	1200–1400
Steel wire	Pot annealing	1170	1650
Steel wire	Baking	420–450	300–350

Material	Operation	Approximate temperature range	
		K	°F
Steel alloy, tool	Hardening	1050–1450	1425–2150
Steel alloy, tool	Preheating	920–1090	1200–1500
Steel alloy, tool	Tempering	435–950	325–1250
Steel, carbon	Hardening	1010–1120	1360–1550
Steel, carbon, tool	Hardening	1060–1090	1450–1500
Steel, carbon	Tempering	420–870	300–1100
Steel, carbon, tool	Tempering	420–560	300–550
Steel, high-carbon	Annealing	1030–1090	1400–1500
Steel, high-speed	Hardening	1500–1575	2250–2375
Steel, high-speed	Preheating	1060–1150	1450–1600
Steel, high-speed	Tempering	810–900	1000–1150
Steel, S.A.E.	Annealing	1030–1170	1400–1650
Tin	Melting	530–615	500–650
Type metal	Stereotyping	545–615	525–650
Type metal	Linotyping	560–615	550–650
Type metal	Electrotyping	615–670	650–750
Varnish	Cooking	545–590	520–600
Vitreous enamelling	Firing	1030–1200	1400–1700
Zinc	Melting	700–760	800–900
Zinc alloy	Die-casting	730	850

Figure 13:30 Industrial heating operations [5]

13.5.1 Steel-reheating furnaces

Reheating is required at various stages of the steel-forming process involving the use of both intermittent and continuous furnaces. Intermittent types include pit furnaces (soaking pits), bogie-hearth furnaces and box furnaces fed by charging machines. Continuous furnaces are predominantly of the pusher type but rotary-hearth and walking-beam types are also used. Heavy and medium fuel oil, coke-oven gas, producer gas and natural gas, have all been used in the firing of steelworks reheating furnaces. Natural gas is a relative newcomer to the field in Britain although it is extensively used elsewhere.

Soaking pits

These are used for reheating ingots of up to 20 t weight and form the link between the melting shop and rolling mills. They consist of deep rectangular or cylindrical chambers into which the ingots are placed upright via a removable roof. Loading and unloading are by an overhead stiff-leg crane to avoid swaying. Individual soaking pits hold up to 10 ingots. Scale and slag buildup in the pit bottom present problems and a layer of coke breeze is sometimes spread in the pit to absorb slag and scale. Alternatively scale fusion is allowed to occur and the pit bottom is periodically scraped using a crane attachment. Air preheating by recuperators or regenerators is applied to the larger pit furnaces and in some cases waste-heat boilers are employed. Heat requirements for soaking pits vary from 260 to 1560 MJ t^{-1} (2.5–15 therm ton^{-1}) [5] dependent mainly on the initial and final temperatures of the ingots. Operating temperatures are in the range 1200 to 1300 °C. Gas firing is by large nozzle-mixing burners normally two or four in number. High-velocity burners have recently been applied, reducing the number of burners to two, providing faster and more uniform heating and improving furnace performance by 20 to 30 per cent [26]. Typical furnace charge weights are between 70 and 140 t at an output of around 740 kg h^{-1} per square metre of floor area (150 lb h^{-1} ft^{-2}) [5].

Continuous reheating furnaces
The majority of continuous reheating furnaces for slabs, blooms or billets are of the

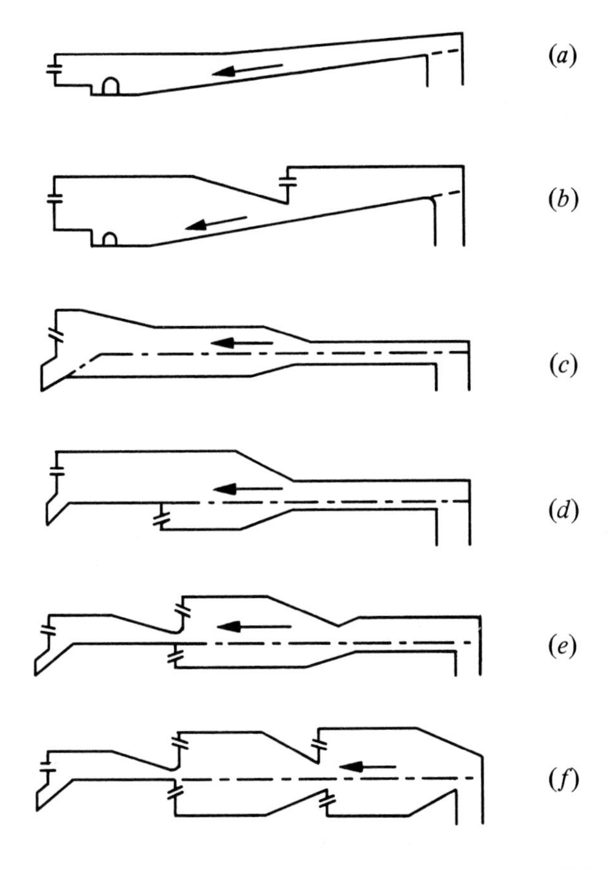

Figure 13:31 Evolution of multizone pusher reheating furnaces [57]
(*a*) Single zone
(*b*) Two zone, top fired
(*c*) Split flame, top fired
(*d*) Two zone, top and bottom fired
(*e*) Three zone
(*f*) Five zone

counter-flow pusher type. Pattison [57] has presented a comprehensive review of the development and design of multizone pusher furnaces to which the reader is referred for details. The following brief review is based mainly on this paper. As shown in Figure 13:31(*a*) the basic single-zone furnace consists of a long low refractory box only slightly wider than the length of the stock. The furnace is fired at the stock-exit end by a single burner or row of burners and the waste-gas ports are located below the stock-entry aperture. Stock discharge may be via an end or side port and may be fully automatic or manual. Furnace dimensions are up to 9–12 m (30–40 ft) wide by 18–30 m (60–100 ft) long. The thermal efficiency is increased by the counter-flow of stock and combustion products, the furnace tapering towards the gas-exit ports to increase convective heat transfer in the preheat zone. If the furnace were sufficiently long the waste gas could be reduced almost to ambient temperature. This is not economically practicable and the furnace length is designed to give a waste-gas temperature around 320 °C and a recuperator is used to recover much of the remaining sensible heat.

The main limitation of the single-zone furnace is that there is an optimum length for a given heating duty and the narrow good-performance range makes it most suitable for a fixed stock size at a constant feed rate. Attempts to increase throughput may result in overheating the stock since all the extra heat must be applied in the region where the stock is at its highest temperature. The two-zone top-fired furnace (Figure 13:31(*b*)) was developed to raise the stock temperature more evenly at higher throughputs. It is also better able to cope with thicker stock sizes. There is some temperature equalization between removal of the stock from the furnace and its arrival at the mill and a 20–30 °C top-to-bottom temperature difference in the stock is satisfactory. Temperature differences in excess of 40 °C are unacceptable. In the split-flame furnace (Figure 13:31(*c*)), acceptable temperature differentials can be obtained with thicker stock since there is some bottom heating but the skid rails required set up another type of temperature inequality which requires 'soaking out'. The top-and-bottom fired design shown in Figure 13:31(*d*) incorporates a solid fore-hearth for soaking-out skid-marks. In the three-zone furnace (Figure 13.31(*e*)) the fore-hearth becomes a soak-zone with its own burners. Development to date is completed in the five-zone furnace (Figure 13:31(*f*)) in which further increases in throughput rates are possible. The term 'zone' in this context refers to a particular furnace volume in which heat transfer to the stock is controlled by a burner or bank of burners.

Single-zone furnaces are generally used for stock sizes up to 100 mm (4 in) with throughputs up to 30 t h^{-1}. For throughputs between 30 and 100 t h^{-1} and stock sizes above 100 mm, three-zone furnaces are widely used, giving way to five-zone furnaces above 100 t h^{-1}.

Figure 13:32 indicates comparative fuel consumptions and efficiencies for a wide range of British and French pusher furnaces fired by coal, producer gas, oil, pitch, coke-oven gas and blast-furnace gas. The figure shows that most of the large furnaces operate at efficiencies between 45 and 60 per cent at a throughput of around 220 kg h^{-1} per square metre of heat-receiving surface (45 lb h^{-1} ft^{-2}). Most of the deviations from this optimum region were caused by furnaces being run at low throughputs due to reduced mill demand.

Hot-gas flow patterns in multizone furnaces are complex and complicate attempts at a mathematical analysis of heat transfer. Water-flow models indicate the presence of stable vortices which, if they occur close to the burner nozzle, improve mixing and

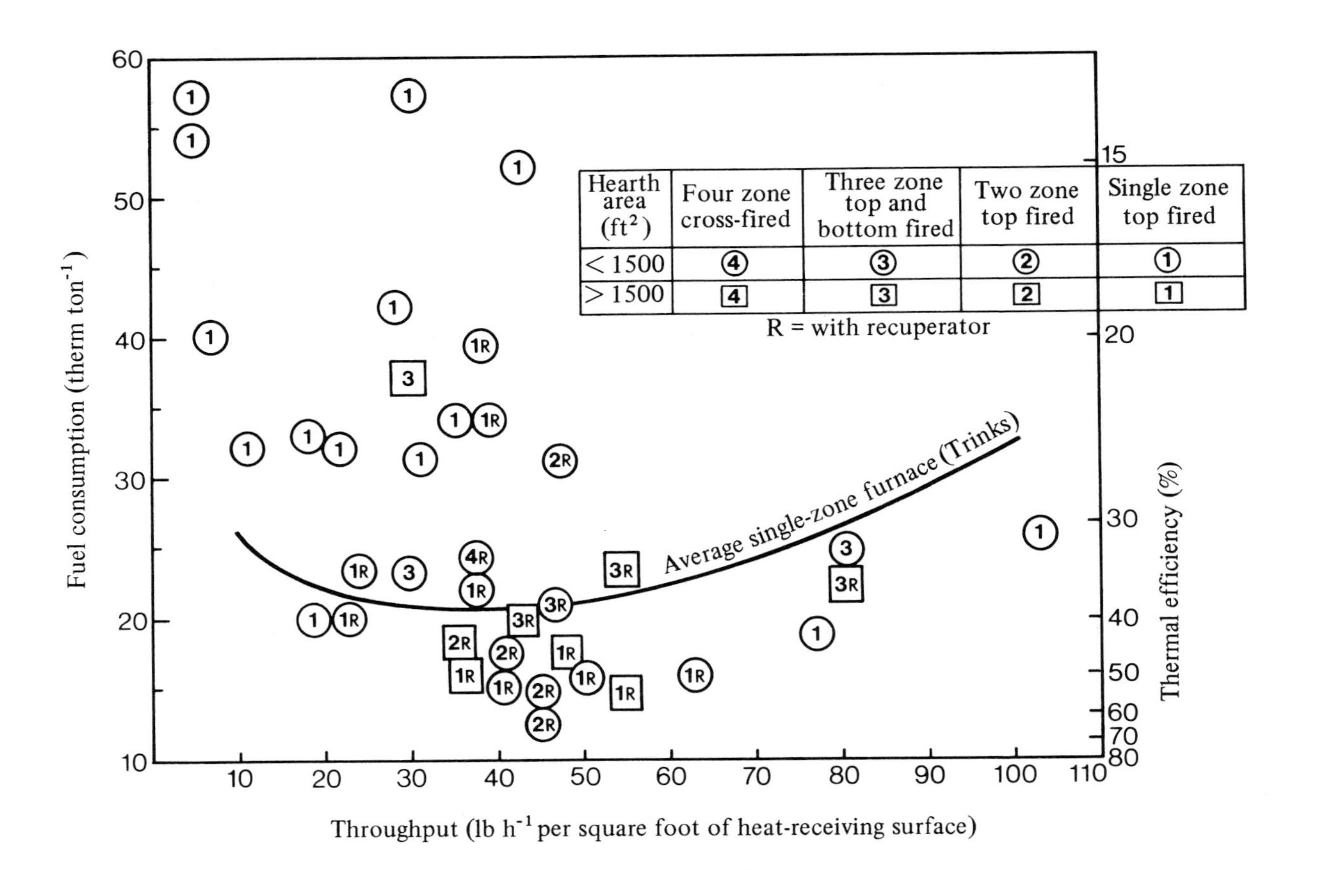

Figure 13:32 Variation of fuel consumption with hearth loading for continuous pusher preheating furnaces (adapted from [57])

contribute to flame stability by preheating the fuel. Unstable vortices may also occur because of an incorrect choice of burner angle and these are a source of uneven heating and poor temperature control. In addition, ingress of air at the discharge end can create a cold recirculating pattern which contributes to temperature non-uniformity in the stock. Reverse firing of the soaking zone is sometimes employed to minimize this problem.

During reheating, around 1 per cent of the stock is converted to scale. This is an expensive loss and scale formation interferes with heat transfer since the diffusivity of scale is only 5 per cent of that of steel. Scale formation can be reduced by minimizing excess air but even though the overall level of excess air may be controlled to a reasonable level the formation of scale will vary in different parts of the furnace because of uneven flow and mixing patterns and because of changes in throughput and firing rates. These variations can be minimized by using premixed gas burners of the radiant-cup or flat-flame types (Sections 7.6 and 7.8) set into the roof. Bosworth [58] has described such a three-zone furnace incorporating 85 roof-mounted radiant burners giving higher throughput rates than conventional three-zone types and having a fuel consumption of 2076 MJ t^{-1} (20 therm ton^{-1}) (yearly average). Dessarts and Fauvel [59] have described the use of roof-mounted tunnel-mixing radiant burners applied to French reheating furnaces producing simple furnace construction, ease of temperature control and improved draught control.

For normal mild-steel billets and blooms a 1 per cent scale loss is regarded as largely unavoidable and not too serious. In the case of high-carbon and alloy steels there is a demand for improving the conditions of the product by reducing scaling and decarburization in the reheating processes. Scaling and decarburization are both dependent on the residence time of the stock in the high-temperature zones of the furnace and on the furnace atmosphere.

As indicated in Section 13.2 the rate of scale formation and decarburization are functions of the CO_2/CO and H_2O/H_2 equilibria. At these temperatures sulphur oxides also exert an important influence. The diffusion rate of iron in iron sulphide is about 80 times that in iron oxide [57] and so scale formation increases markedly if a sulphur-bearing fuel such as oil is used. Processes involving partial gas combustion to provide rich atmospheres in the high-temperature regions of reheating furnaces have been developed and are reviewed in Section 13.5.3. Residence-time reductions can be obtained using notched-hearth, walking-beam or rotary furnaces in which, because of stock separation, more than one face is exposed for heat transfer.

Davies [60] has described a recently installed natural-gas-fired walking-beam reheating furnace for 100 mm (4 in) square by 12 m (40 ft) long billets. Figure 13:33 shows the general features. The chamber dimensions are 12.8 m $\times$ 12.8 m $\times$ 1.5 m high (42 ft $\times$ 42 ft $\times$ 5 ft) and the furnace is divided into three separately controlled zones. 65 roof-mounted flat-flame burners are employed rated at 590 kW (2×10^6 Btu h^{-1}) each in the heating zones and 350 kW (1.2×10^6 Btu h^{-1}) each in the preheat zone. Air supply to the burners is at 310 °C from a tower recuperator 38 m (125 ft) high, the combustion products leaving the recuperator at about 700 °C.

Improvements in heating rates can be achieved by the choice of suitable conveyance arrangements, but much greater increases are possible by the use of forced-convection heating in rapid billet-heating furnaces. This form of heating has been applied to a wide variety of nonferrous billets and to ferrous billets in the smaller size range and has been the subject of considerable research and development work within the gas industry. Rapid billet heating is considered in Section 13.5.2.

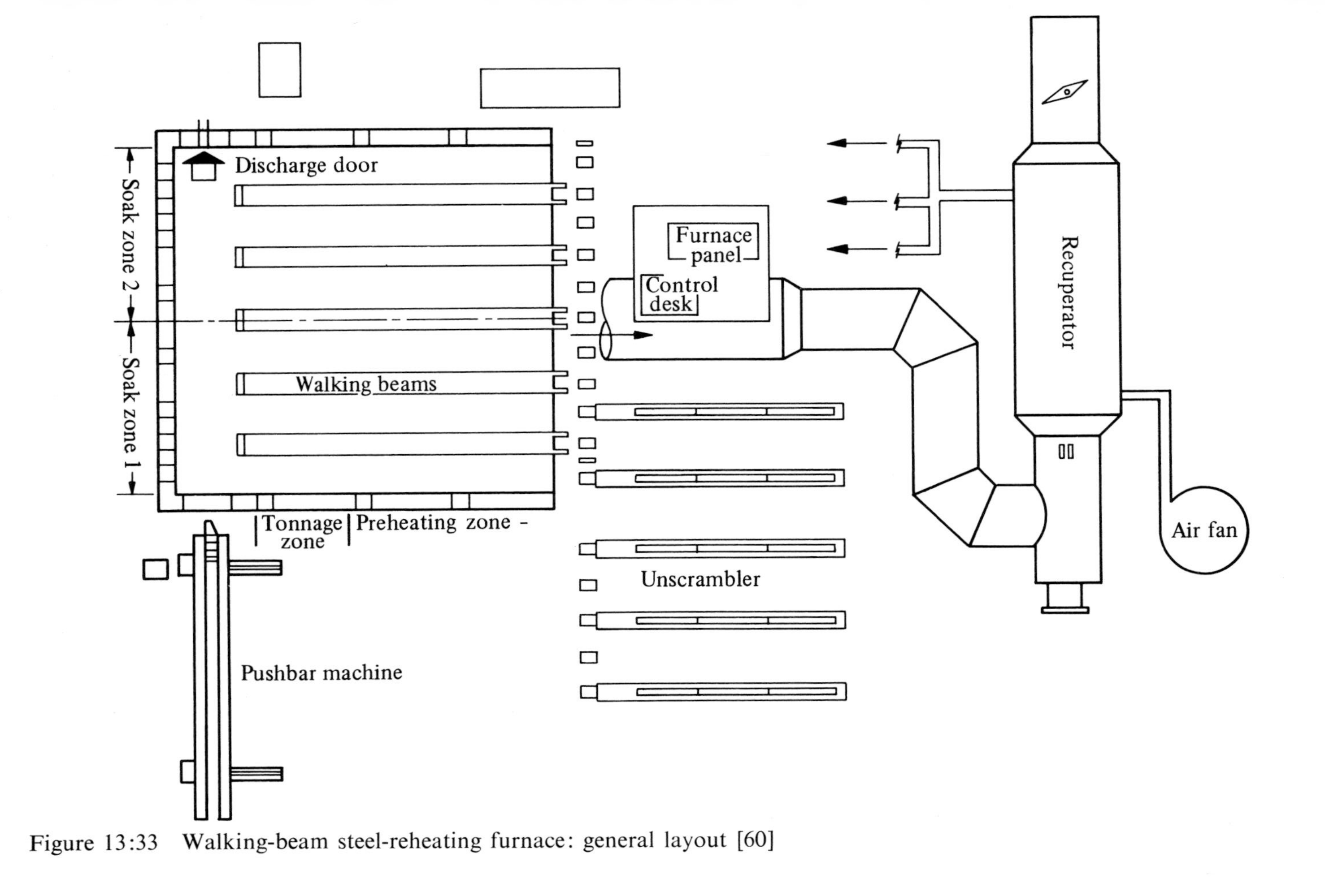

Figure 13:33 Walking-beam steel-reheating furnace: general layout [60]

13.5.2 Rapid billet heating

Introduction

As Masters *et al.* [4] have pointed out, in processes where metal is heated in conventional furnaces for hot-working purposes, the rate and pattern of working is determined by the characteristics of the furnace. These characteristics are far from ideal and the need to develop better furnaces has been recognized by the gas, electricity and LPG industries and by furnace manufacturers, resulting in modern designs of rapid-heating furnaces based on forced convection in fuel-fired designs and direct resistance or induction heating in electrical designs. The aim of this development work has been towards what could be termed 'heating machines' which are suitable for integration into production-line systems.

Before considering rapid billet-heating furnaces it is worthwhile considering the general characteristics of what might be called conventional furnaces. These have large working chambers in order to ensure complete combustion and to promote hot-gas radiation, since this is a function of beam length. The furnace walls are heated mainly by gas radiation and very little by convection and the stock is heated by a combination of wall radiation, gas radiation and hardly at all by convection. The heavy refractory structure associated with a large chamber is slow to heat up and, should a stoppage occur, continues to radiate heat to the stock long after the burners have been turned off. This thermal inertia renders temperature control difficult and, in order to prevent temperature overshoot, furnace-wall temperatures are usually restricted to values little above the maximum temperature the stock can tolerate. Inevitably this slows the heating rate, particularly as the final temperature is approached and results in large quantities of stock being held at risk to obtain the desired output [4]. Lengthy heating times at high temperatures are harmful for all metals and steel in particular. Scaling is undesirable from many points of view; it results in a metal loss, produces a poor surface finish, tends to slag refractories and causes wear on dies, rolls, etc. In addition scale build-up in furnaces is harmful to conveyance mechanisms and often necessitates shutdown for removal. Decarburization also increases with heating time and may have even more serious consequences than scaling. Variations in residence time also occur in practice, particularly in batch furnaces and this is often dependent on furnace operation. As indicated in the previous section, exposing more of the billet surface to heat transfer by the use of walking beams, rotary hearths, etc., produces improvements in heating times but the term rapid billet heating can only be accurately applied to forced-convection furnaces of low thermal capacity.

The development of high-velocity burners in which combustion is completed inside a small refractory tunnel (Section 7.5.2) has freed the furnace designer from the necessity to design a compromise heat-release/heat-transfer chamber and to concentrate on producing a furnace chamber tailored to suit the stock and shaped to develop high convective heat-transfer rates. This implies small gas spaces in the loaded furnace thereby decreasing heat transfer by nonluminous gas radiation but this can be tolerated if it is adequately compensated for by the increase in convection. For metals with low emissivity and low forging-temperature requirements, i.e. nonferrous metals, the advantages of increasing the convective component are obvious but are less so in the case of ferrous metals with emissivities approaching unity and much higher working temperatures. An additional factor is the lower thermal conductivity of steel compared with nonferrous metals. With copper and aluminium the speed of heating is limited only by the rate of heat transfer to the surface and it is almost impossible to set up serious local temperature gradients, thus complete uniformity of heating is not required and the design does not need to be compromised by the inclusion of a soak zone or soak period. In the case of steels, however, it is quite possible to transfer heat to the surface faster than it can be conducted away to the core. Thus the approach to the rapid heating of steels must of necessity be less pure; uniform heating over the surface is important and it may well be necessary to provide a soak zone. The high exit-gas temperatures involved make the use of heat recovery almost obligatory. Self-recuperative burners and continuous counter-flow designs of furnace incorporating load recuperation have both been employed.

Single-cell rapid billet heaters

The earliest rapid billet heaters to be developed were single-cell, intermittent furnaces for nonferrous metals. Figure 13:34(*a*) illustrates a prototype furnace of this type [61]. Premixed high-velocity slot burners (type F_1) of total rating 350 kW (1.2×10^6 Btu h^{-1}) fire above and below the billet to provide rotational flow and hot-gas re-entrainment. The furnace diameter is 254 mm (10 in) and was designed to heat billets up to 229 mm (9 in) in diameter by 711 mm (28 in) long.

Figure 13:34(*b*) shows the large effect on the efficiency obtained of the ratio of billet diameter to furnace diameter. This indicates that the larger the billet size the greater the efficiency although the requirements of burner size and temperature uniformity in general restrict the maximum diameter ratio to about 0.85 even when heating very large billets [61]. Figure 13:34(*b*) also shows a considerable variation in efficiency with furnace type. The high efficiencies obtained with the 330 mm (13 in) type can be attributed to the use of convergent slot burners producing very high exit velocities (Section 7.5.2) avoiding hot-gas short-circuiting and promoting increased recirculation. Extensive flow and heat-transfer model studies have been conducted on single-cell furnaces [30, 62–4] making use of electrolytic and naphthalene-sublimation techniques. These techniques are outlined in Section 12.8.1.

Figure 13:34(*c*) illustrates the effects of burner/flue-port configuration on calculated heating times (for mild steel). Model studies have shown that in the prototype design shown in Figure 13:34(*a*) most of the combustion products from the upper burner passed straight out of the adjacent flue. Significant improvements are obtained if the more symmetrical twin-flue arrangement shown in Figure 13:34(*c*) is adopted. The third arrangement, shown in Figure 13:34(*d*), uses self-recuperative burners which in addition to providing fuel savings by waste-heat recovery also produce a symmetrical flow pattern. The self-recuperative burners have lower exit velocities than the premix burners used in Figure 13:34(*a*) yet reduce the heating times by about half, illustrating the importance of designing the heating chamber as a whole as opposed to simply applying high-velocity burners.

An alternative approach to the tangentially fired designs considered so far is to use direct flame impingement on the stock. Mass-transfer/heat-transfer experiments have been conducted on models of a 250 mm (10 in) twin-flued furnace using tangential and impingement firing [30]. This work showed that convective heat transfer to the billet increases with impingement but convection to the wall decreases. For aluminium billets this has the effect of decreasing the heating time by 10 to 20 per cent since, due to aluminium's low emissivity, most of the heat gained by the billet is by direct convection. Steel billets, in contrast, have high emissivities and gain much of their heat by re-radiation from the walls. Consequently the increase in convective

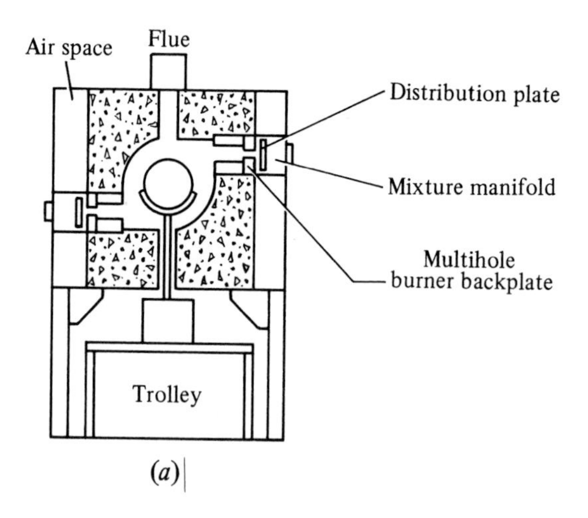

(a)

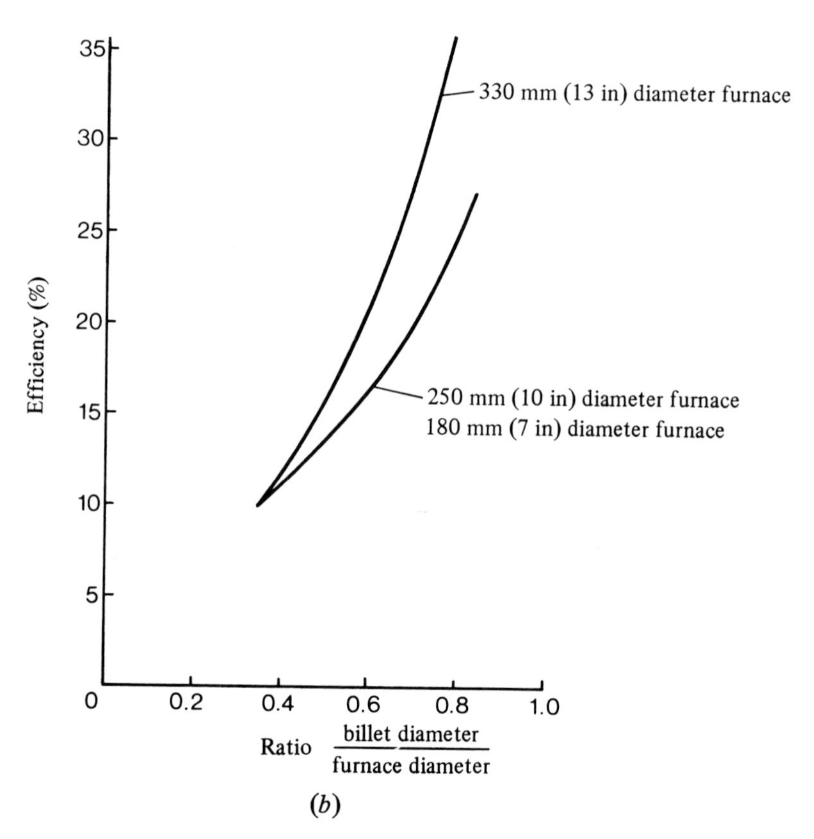

(b)

Figure 13:34 Single-cell rapid billet heating
(a) Prototype single-cell furnace [61]
(b) Variation of efficiency with ratio of billet diameter to furnace diameter [61]

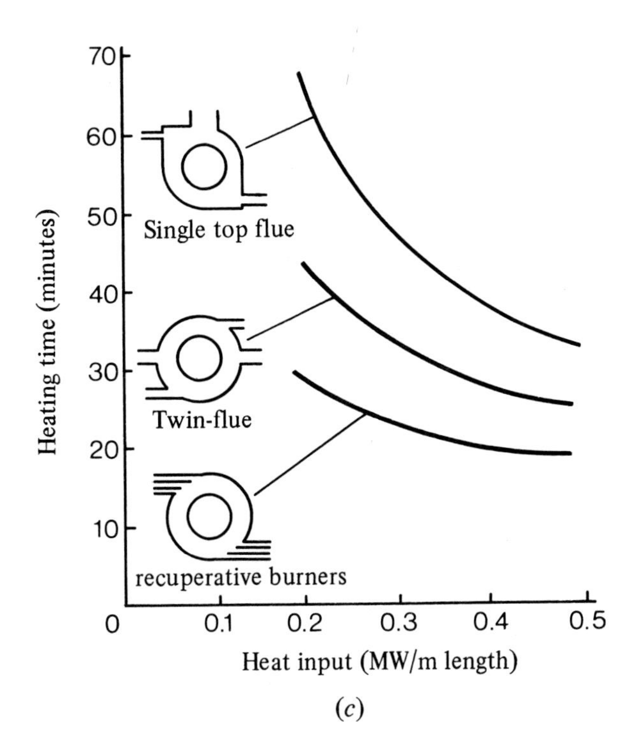

(*c*)

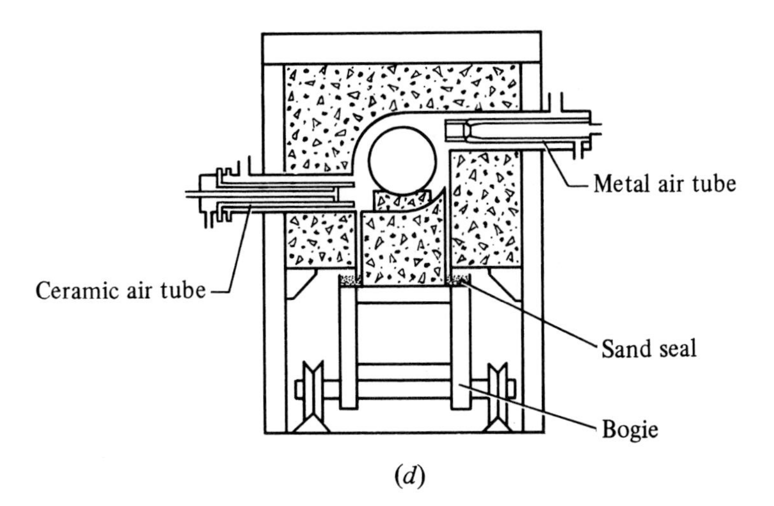

(*d*)

(*c*) Rapid-heating furnaces: theoretical heating times for a 149 mm ($5\frac{7}{8}$ in) mild-steel billet in three 254 mm (10 in) diameter furnaces [30]

(*d*) Recuperative single-cell furnace [4]

heat transfer is obtained at the expense of reduced transfer to the walls and no reduction in heating time results. Direct impingement inevitably leads to uneven heat transfer to the billet surface which may not be serious with nonferrous metals but may cause unacceptable temperature differentials in large steel billets. Techniques have been developed for predicting temperature distributions in billets heated by direct impingement [30]. Calculations indicate that for a round-section steel billet, the maximum surface-temperature difference at the end of the heating period may be as little as 15 °C but much higher differences occur in the early part of the heating schedule. Similar calculations on square and rectangular stock have indicated that impingement heating may assist temperature uniformity since a uniform heat flux causes the temperatures at the corners to be higher than in the middle of the face.

An important advantage of rapid-heating furnaces is their small physical size and low thermal capacity allowing them to be put into operation very quickly from cold and providing ease of control to meet fluctuating demands.

During operation the wall temperature fluctuates widely during each heating cycle since as a cold billet is charged into a hot furnace the walls radiate to the billet and their temperature drops rapidly, slowly recovering during the remainder of the cycle. These changes present problems in the choice of suitable refractory materials. Bubble-alumina castable refractories were used in the earlier experimental prototypes but these start to crumble after a few cycles. Some 30 different refractory materials have been subjected to thermal cycling tests [65], the most suitable being a composite brick comprising a dense cast head mechanically keyed to a lightweight backing brick [4].

Figure 13:34(*d*) shows an experimental furnace fired by self-recuperative burners for billets up to 229 mm (9 in) diameter by 838 mm (33 in) long which are mounted on ceramic supports on a bogie hearth forming the base of the chamber. The furnace diameter is 305 mm (12 in). For 165 mm (6.5 in) billets the efficiencies after a 10-minute heat-up from cold were 17.5, 20.5 and 23.5 per cent for the first, second and fourth cycles respectively, the heating times being 53, 45 and 38 minutes. Higher efficiencies (up to 26 per cent) were obtained with higher ratios of billet diameter to furnace diameter. Surface-to-core temperature differentials (at 1200 °C surface temperature) ranged from 20 to 50 °C depending on billet size and these differentials disappeared within one minute of discharge from the furnace. Using the top burners only was found not to affect the furnace performance. Work on this and earlier prototypes has led to the development of a fully engineered 3630 kg h^{-1} (8000 lb h^{-1}) single-cell unit for steel-billet heating [66]. 25 recuperative burners rated at 103 kW (3.5 therm h^{-1}) each are employed, firing from the top only. The furnace chamber is lined with composite bricks and is designed to accommodate square billets up to 127 mm (5 in) square by 4.9 m (16 ft) long carried on water-cooled skids mounted on an air-cushion pallet. Heating time to 1200 °C is 17–20 minutes. It is intended to install this furnace in series with a conventional pusher reheating furnace which will preheat the billets to 800 °C before entry to the rapid billet heater in order to gain the benefits of rapid heating, i.e. low metal loss and decarburization together with increased output to 3630 kg h^{-1}, at a relatively low capital cost.

Continuous rapid billet heaters

Although rapid heating on a continuous basis may be achieved by transporting the stock on skid rails through a single-cell furnace, load recuperation, and hence higher efficiencies than can be achieved with recuperative burners, can be obtained by drawing the combustion products counter-flow to the billet flow through an extension of the heating cell. An early prototype [67] in which 76 mm (3 in) diameter billets could be heated to 900 °C at a rate of 318 kg h^{-1} (700 lb h^{-1}) in a furnace only 1.2 m (4 ft) long demonstrated the potential of the method. The throughput obtained was comparable to that obtained in a single-cell furnace but continuous operation increased the thermal efficiency by a factor of about three.

Figure 13:35(*a*) shows in diagrammatic form a later prototype furnace for heating 900 kg h^{-1} (about 2000 lb h^{-1}) of 76 mm (3 in) or 102 mm (4 in) billets to 1250 °C. The overall length is 6.1 m (20 ft) and comprises a cylindrical 165 mm (6.5 in) diameter preheat zone and a burner zone of similar dimensions fired by convergent slot burners. The skid rails consist of stainless-steel tubes which could be air- or water-cooled. As mentioned earlier, the design of skid rails for high-temperature pusher furnaces presents difficulties. For water-cooled rails the water flow rate has to be high to prevent boiling and this has a serious cooling effect, reducing furnace efficiency and resulting in temperature inequalities in the billet. Air-cooled skids, on the other hand, operate at higher temperatures but this reduces their mechanical strength and creep resistance. Masters *et al.* [4] have reported the results of trials on this furnace showing that air-cooling resulted in a 40 per cent increase in output compared with water-cooling and much lower temperature differentials in the billet. The thermal efficiency averaged 31 per cent for air-cooled rails and 21 per cent for water-cooling.

Solid silicon-carbide skid rails have also been employed but with little success because attack by molten ferrous scale caused excessive wear. The most successful skid-rail design to emerge from these trials is the indirectly water-cooled rail shown in Figure 13:35(*b*). The outer tube is cooled by radiation to the water-cooled inner tube and the system can be designed so that the outer heat-resistant alloy tube can be operated at a specified temperature (say 1050 °C). If the cooling rate needs increasing, low-pressure air from the combustion-air system can be passed through the annulus. Other advantages claimed for this system are that the cooled inner tube acts as a continuous support for the outer tube preventing creep and that the skid-rail temperature is fairly uniform along its length unlike simple air-cooled tubes.

Masters *et al.* [66] have described the application of the results of these trials to the design of fully engineered units of 1730 kg h^{-1} (about 3800 lb h^{-1}) and 454 kg h^{-1} (1000 lb h^{-1}) capacity for site trials. The former is a twin-chamber unit of 4.6 m (15 ft) effective length suitable for 76 mm (3 in) to 127 mm (5 in) square billets and is fired by five 146 kW (5 therm h^{-1}) high-velocity tunnel-mixing burners giving exit velocities of 73 m s^{-1}. Each burner is fitted with spark ignition and ultraviolet flame detection and air–gas ratio control is by the pressure-divider technique (Section 7.6.2). Provision is made for direct water-cooling of the skid rails in the event of an electricity supply failure.

Scaling and decarburization tests carried out on the 454 kg h^{-1} unit indicated that these effects were very small. Percentage metal losses of 0.42–0.64 were obtained compared with typical forging industry values of 2–5 per cent. Mean carburization depths were 0.178 to 0.229 mm (0.007–0.009 in) for EN45 steels. Masters *et al.* [66] state that the figures are lower than those published for any other heating method including electrical induction furnaces.

Rapid-heating furnaces for long billets have also been developed [66] in which the billets are pushed sideways in the form of a flat bed towards burners firing from an end wall. A 4.5 t h^{-1} unit has been built for heating 2.4 m (8 ft) long billets to 1250 °C

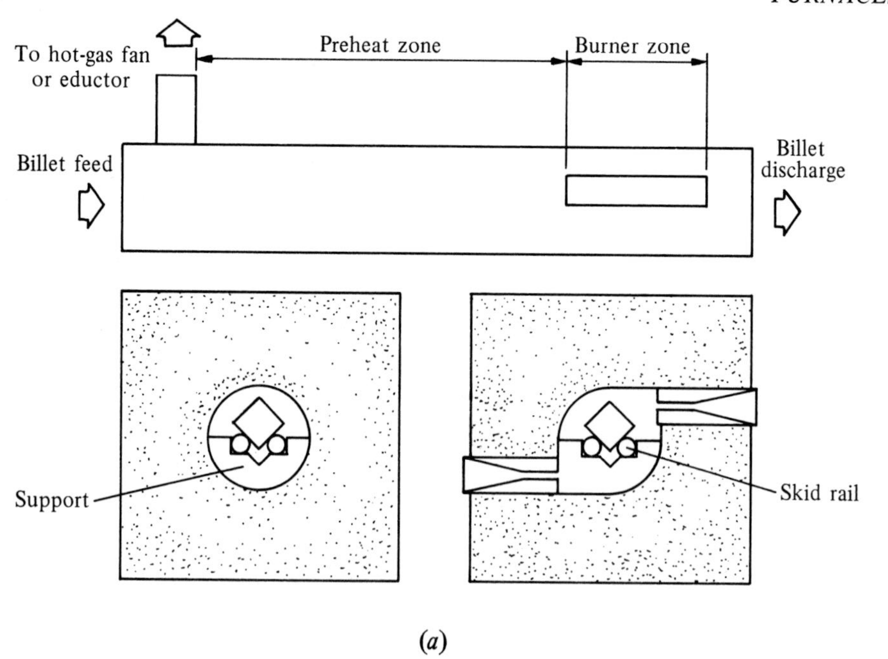

(*a*)

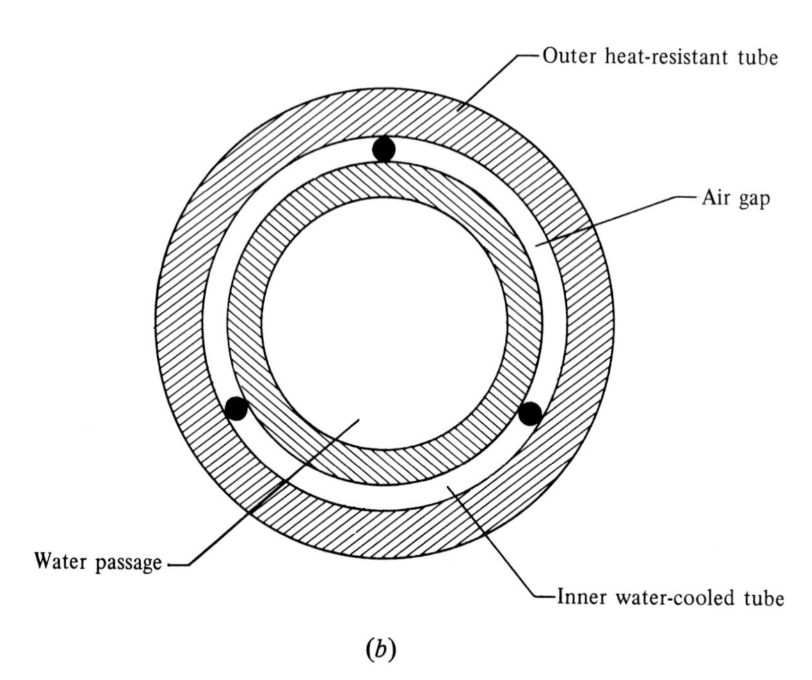

(*b*)

Figure 13:35 Continuous billet-heating furnaces
(*a*) 900 kg h^{-1} prototype furnace [4]
(*b*) Indirectly water-cooled skid rail [4, 68]

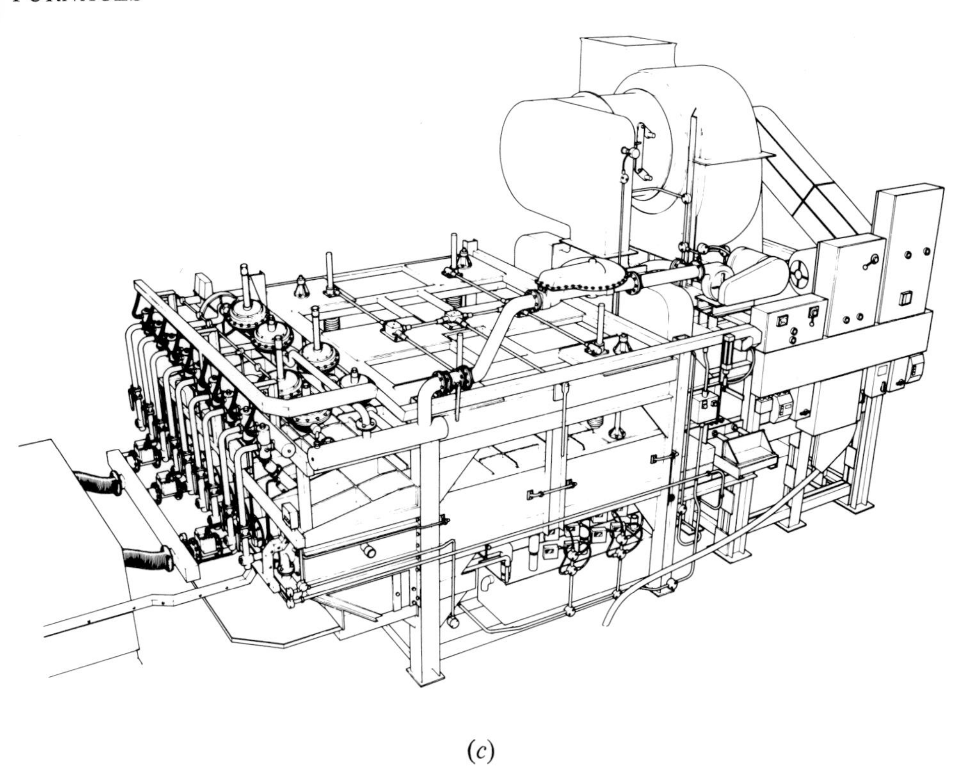

(*c*)

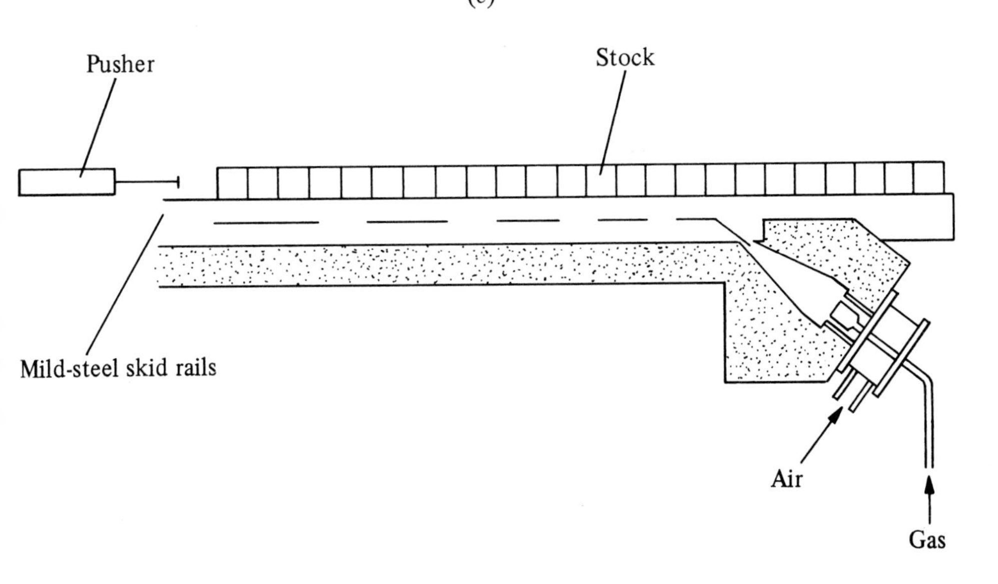

(*d*)

(*c*) Furnace for heating long billets (BGC Midlands Research Station)
(*d*) Warm-cropping heater [66]

for rolling (Figure 13:35(*c*)). The same basic idea has been applied to bar heating prior to cropping. Preheating to 350 °C increases cropping rates and extends the life of the cropping blades. The heater shown in Figure 13:35(*d*) is rated at 6 t h^{-1}, uses fourteen 103 kW (3.5 therm h^{-1}) burners firing below the bars and can deal with bars up to 83 mm (3.25 in) square by 5.5 m (18 ft) long. Bars reach temperature within 10 minutes of light-up and the specific fuel consumption is low (28.6 MJ t^{-1}, 3.5 therm ton^{-1}).

It is clear from the foregoing that rapid billet heaters have now reached a stage of development which allows them to be integrated in automated factory systems; they are in fact becoming closer to machine tools than to furnaces in the traditional sense. Their advantages can be summarized as:

1 Improved and more consistent product quality, in particular less scaling and decarburization.
2 High output from small units of relatively low capital cost, with low floor-space requirements.
3 Less materials at risk in the event of a plant stoppage.
4 Low thermal inertia leading to fast start-up and lowered risk of temperature overshoot.
5 Better working conditions and reduced need for skilled labour.
6 Easy maintenance. Rapid cool-down avoids delays and adaptability to modular construction could reduce downtime.

Masters *et al.* [66] have presented a detailed cost comparison between a wide variety of rapid billet heaters and their conventional counterparts including gas, oil and induction furnaces. Direct furnace costs per tonne (i.e. fuel, maintenance, repayment of capital and cost of water, electricity, etc.) are lower for rapid heating than for any of the other methods. Costs affected by the heating process (e.g. metal loss, die wear, etc.) are less than for conventional furnaces and the same as for electrical methods. If all items of cost are totalled, in all the cases considered the cost per tonne is lowest using gas-fired rapid-heating methods and highest using conventional furnaces.

Localized heating

In many metal-forming processes only the end of the stock, which may be in the form of pins, plates or bars, requires heating to forging temperatures. The slot-forge furnace described in the introduction to this chapter is an example. Work carried out on the development of rapid billet heaters has led to the development of so-called part-bar heaters which provide faster heating rates than conventional designs for this type of application. The MRS part-bar heaters [4, 66] are based on the idea of using the end of the stock as part of the burner tunnel enclosure. The main advantages of this type of furnace are its simplicity and low capital cost and its very low thermal inertia. Very fast starting-up is possible and in most cases stock is delivered at temperature within 5 minutes of ignition. The main design consideration is to minimize the longitudinal temperature gradient which tends to arise because of conduction along the stock, cooling of the combustion products and thickening of the boundary layer in the direction of flow. For this reason part-bar heaters are normally constructed with converging tunnels to accelerate the combustion products as they flow over the stock. Units have been built giving good uniformity over heated lengths of up to 305 mm (12 in) [4].

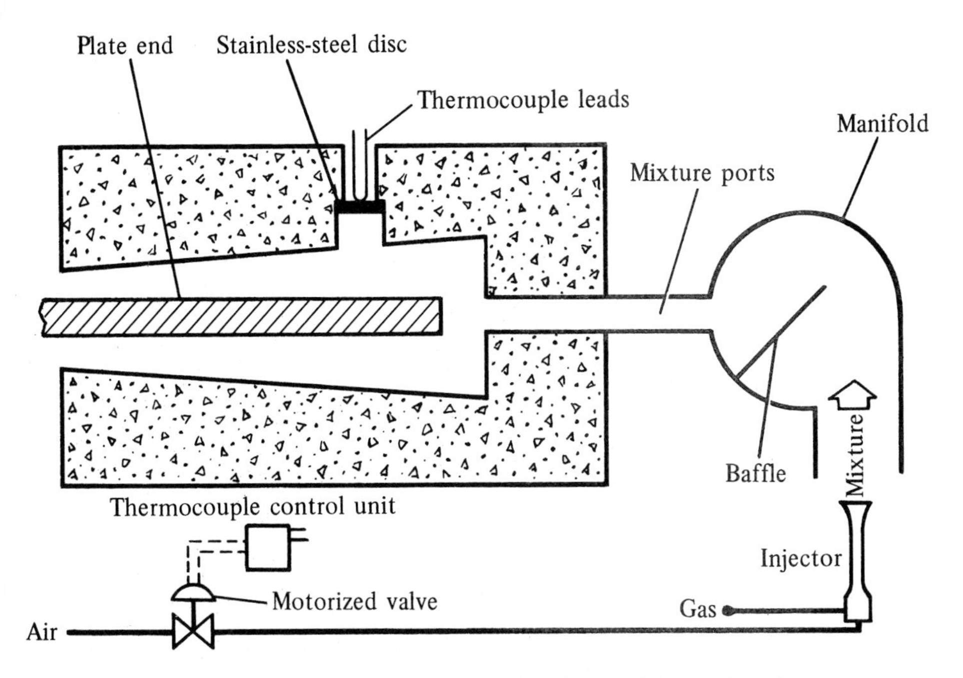

Figure 13:36 Schematic diagram of plate-end heater illustrating the temperature-control system [4]

Figure 13:36 shows a fully engineered plate-end heater designed along these lines. The heating tunnel is 1.2 m long and the furnace is rated at 440 kW (15 therm h^{-1}) being capable of heating 454 kg h^{-1} (1000 lb h^{-1}) of steel to 1150 °C. Up to 229 mm (9 in) of the plate end can be heated at a thermal efficiency of 20 per cent, the heating time being about 5 minutes. The unit is controlled by a thermocouple disc in the tunnel roof which serves as a simple radiation pyrometer giving a reading which is representative of the stock temperature. Because heating schedules are very short, scaling and decarburization are again much lower than for conventional methods.

13.5.3 Minimization of scaling and decarburization in hot-working furnaces

The mechanisms of scaling and decarburization have been considered in Section 13.2.2 and the use of prepared atmospheres to eliminate their effects in heat-treatment processes has been outlined in Section 13.2.4. Their prevention in high-temperature steel-heating furnaces for hot-working purposes is altogether more difficult. Except for some small-scale processes, muffle furnaces are unsuitable and the temperatures involved (1100–1300 °C) are too high for the use of metal radiant tubes. The development of reliable ceramic radiant tubes would allow indirect firing but that stage of development has not yet been reached.

Scaling and decarburization are functions of:

1 Temperature
2 Time
3 Furnace atmosphere

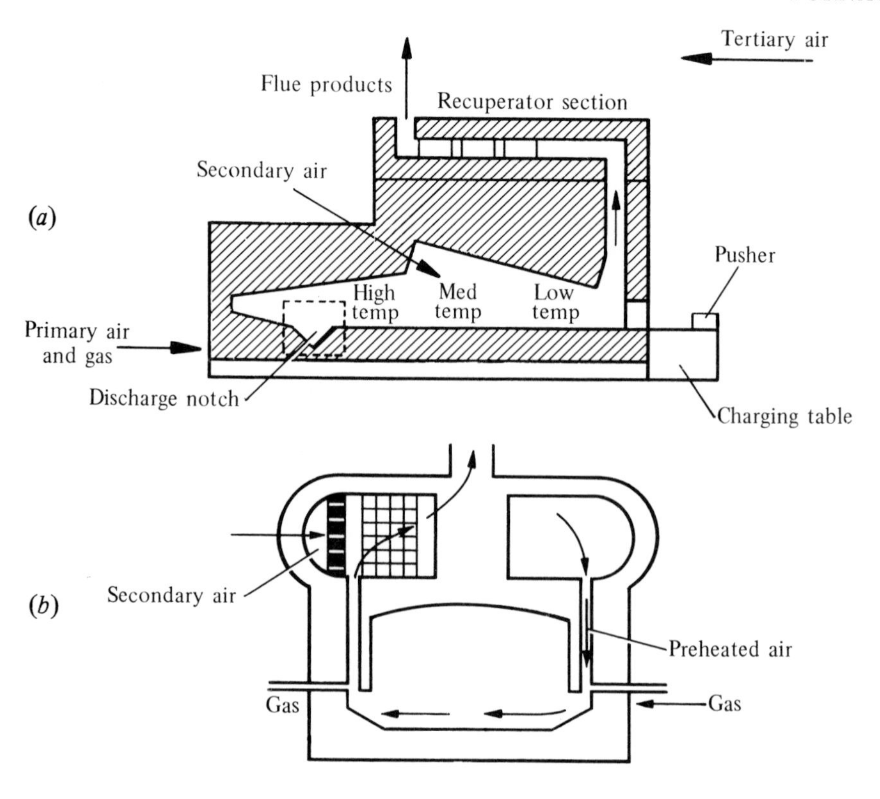

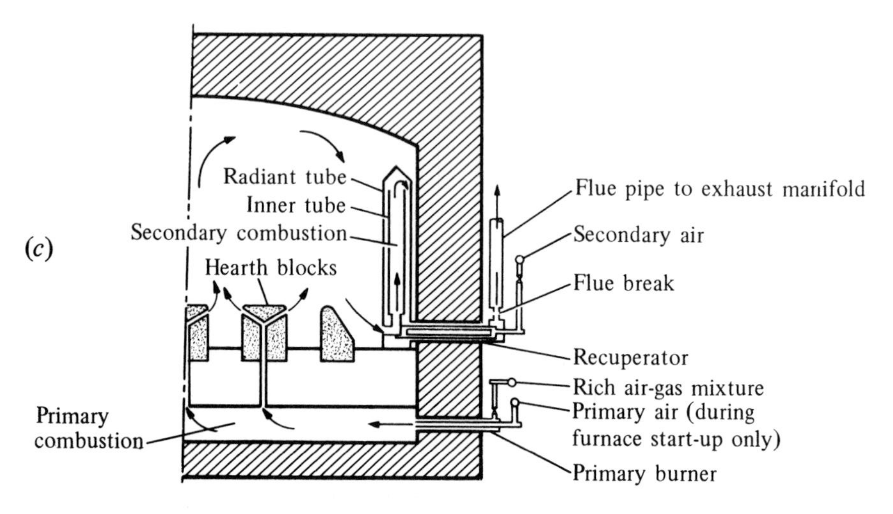

Figure 13:37 Furnaces using partial combustion
(*a*) Protek system [71]
(*b*) Equiverse system
(*c*) EMGB 'SF' system [71]

In considering methods of minimizing their effects, (1) cannot be varied since it is a function of the hot-working process and the steel composition. (2) and (3) are, however, variables and represent possible ways in which reduced scaling and decarburization may be achieved.

Residence-time reduction has been the most widely used method, starting with conveyance mechanisms designed to separate the stock, e.g. walking-beam and rotary-hearth furnaces, and culminating in the development of forced-convection rapid billet heaters. The development of two-stage heating furnaces, in which slow heating up to 750 °C is used, permitting maximum load recuperation, followed by rapid heating up to 1250 °C, has shown improvements in this respect. This idea has taken the form of using two walking-beam mechanisms in the same furnace so that after moving slowly through the preheat section the stock moves rapidly at a wider spacing through the high-temperature section [69]. However, the simplicity of pusher furnaces has led to this principle being used in almost all rapid-heating furnaces. Rapid billet heating will not be considered further in this section, instead attention is directed to the alternative approach, i.e. directly fired furnaces using a reducing atmosphere produced from partial combustion of gas.

As indicated in Section 13.2 equilibrium considerations dictate that much more strongly reducing atmospheres are required to prevent decarburization than oxidation. The heat liberated by combustion of the requisite rich gas–air mixture is such a small proportion of the potential heat content that some means must be found for gaining the remainder if the furnace is not to be hopelessly inefficient. A number of systems have been developed which involve subsequent burning of the rich combustion products either out of contact with the stock, or in contact with the stock in a cooler part of the furnace where scaling and decarburization do not occur.

Figure 13:37 shows three commercial furnaces based on partial combustion. In the Protek system (*a*), which is applicable only to continuous operation, either two-stage or three-stage combustion is used. About 60 per cent of the stoichiometric air requirement is supplied as primary air to the burners which are located at the high-temperature, discharge end. The billet temperature is raised from about 900 to 1200 °C in this zone and the CO/CO_2 ratio is about 2 : 1. Further combustion takes place with additional air in the medium-temperature zone (550–900 °C) giving a CO/CO_2 ratio of about 1.5 : 1 and in the low-temperature zone (20–550 °C), combustion finally being completed in the recuperator section with the addition of tertiary air. The recuperator provides preheating for the primary and secondary air supplies.

The Equiverse system (*b*) incorporates no load preheating and is therefore applicable both to continuous and batch operation. Combustion takes place in the furnace chamber using about 60 per cent of the stoichiometric air requirement, highly preheated from a regenerative system. The products of partial combustion leave the working chamber and combustion is completed with secondary air in one of the regenerator chambers, heating the chequer brickwork. On reversal, primary air flows through the brickwork and becomes preheated, secondary combustion taking place in the other regenerator chamber.

In the SF system (*c*) the products of primary combustion flow through the perforated hearth, thence to open recuperative radiant tubes in which secondary air is provided to complete combustion. This system is capable of operating under appreciably richer combustion conditions than either (*a*) or (*b*) and is much more effective in preventing decarburization [71] and presumably more effective against scaling. The metal radiant tubes used in the SF system, however, impose an upper limit of about 1100 °C which places the system outside the range of most hot-working steel processes. Johnson [70] has compared the effects on decarburization of partial combustion of town gas and natural gas and has concluded that, for a given proportion of the stoichiometric air requirement, natural gas is much more effective in reducing decarburization than town gas. This point is clearly illustrated in Figure 13:38 which shows variations in decarburization of a straight carbon steel using the two fuels at 1150 °C and 1200 °C [70].

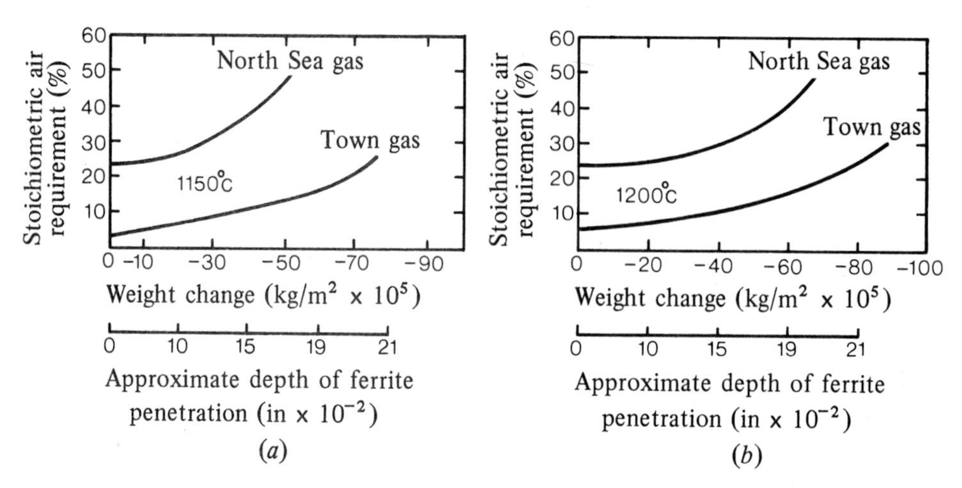

Figure 13:38 Variations in decarburization of 1.2 per cent carbon steel held at (*a*) 1150 °C (*b*) 1200 °C for 33 minutes

Other systems of scale and decarburization reduction have been summarized by Johnson [70], including a Czech technique [72] involving tangential firing in which it is claimed that centrifugal forces cause the oxidizing components to migrate towards the furnace walls leaving the reducing constituents around the stock and a French method [73] using air-deficient firing in loose refractory beds to achieve rich combustion without the necessity for high air-preheat temperatures. For scale-reduced heating, the most promising system is at present the self-recuperative burner with provision for burning the rich combustion products, an example of which is the burner described in Section 7.6.3 and illustrated in Figure 7:32. A particular advantage of this type of burner is its suitability for converting existing furnaces. Virtually scale-free heating is obtained up to 1250 °C with natural gas at between 50 and 60 per cent stoichiometric air requirement, the air-preheat temperature being in the range 500–700 °C. Whilst 50–60 per cent stoichiometric air is suitable for scale-free heating, the burner system must be capable of operating satisfactorily on 25–30 per cent stoichiometric air if a large reduction in decarburization is to be obtained. This presents much more difficult problems, particularly of flame stability and involves preheating both gas and air and possibly the use of oxygen-enriched air [70].

13.6 CERAMICS KILNS

13.6.1 Process requirements

The term 'ceramic' implies a material formed from clay or a similar substance in the plastic state then dried and fired at a sufficiently high temperature to provide the necessary strength. Ceramic products are generally divided into:

1 Fine ceramics, e.g. pottery and special ceramics.
2 Heavy clay products, e.g. bricks, drainpipes and refractories.

There has been a long history of gas firing in group (1) where the technical advantages of town gas outweighed unfavourable fuel-price differentials. Most of the pottery kilns described in this section were designed specifically for gas firing.

In contrast, gas is a comparative newcomer to the heavy-clay industry but has made substantial inroads into the market following the availability of large quantities of natural gas. Consequently most of the kilns described in this section for heavy-clay firing were initially designed for solid-fuel firing and have, in many cases, been changed over to oil and then to gas firing.

Composition of clay
With the exception of some special ceramics and refractories the basic raw material of ceramic products is clay which is used in a wide variety of forms, both naturally occurring and artificially blended. The major constituents are, however, kaolin, feldspar, silica, lime, magnesia and iron.

Kaolin ($Al_2O_3.2SiO_2.2H_2O$). This represents the theoretical decomposition product of feldspar and is termed 'pure clay' or 'clay-substance' [74]. Pure kaolin does not occur naturally. All clays contain impurities. The terms 'clay', 'clayware', etc. do not imply pure clay but a mixture of clay with other materials. The clay-substance provides clays with their characteristic plasticity which increases with reducing particle size. Thus bentonite, an ultra-plastic clay, consists of particles of around 0.001 mm in size whilst the much less plastic china clay ('kaolin') has a particle size of about 0.01 mm.

When pure clay is heated the following changes take place. At 400–500 °C the chemically combined water is evaporated and meta-kaolinite ($Al_2O_3.2SiO_2$) is formed. Between 800 and 910 °C meta-kaolinite decomposes and alumina (Al_2O_3) is formed. Above 910 °C sillimanite ($Al_2O_3.SiO_2$) is formed followed by mullite ($3Al_2O_3.2SiO_2$) which starts to form at 975 °C. These changes are accompanied by large increases in shrinkage, densification and vitrification as the clay particles draw together [74].

Clays have no definite melting temperature, the melting point being arbitrarily defined as the temperature at which a clay cone slumps such that its apex leans over and touches the base. A pure-clay cone slumps in this way at about 1770 °C (for a temperature rise of 100 °C per hour). (See Section 8.3.5.)

The firing shrinkage experienced by a clay is dependent on the grain size of the clay-substance and of the impurities, on the chemical composition of the impurities (fluxes such as feldspar, lime, etc., increase shrinkage whilst alumina reduces shrinkage) and on the furnace atmosphere.

Feldspar. Clay is a decomposition product of feldspar and most clays therefore still contain undecomposed feldspar. The main feldspars are potassium feldspar, $K_2O.Al_2O_3.6SiO_2$ (orthoclase or microline) and sodium feldspar, $Na_2O.Al_2O_3.6SiO_2$

(albite). Feldspar is the most important ceramic flux, the average fusion temperature being about 1230–1250 °C.

Silica. Silica is an important constituent of clays and may be present as quartz or sand or added in the form of ground calcined flint. The crystal transformations of silica together with their expansion characteristics are considered in Chapter 6. They have an important bearing on ceramics firing since they place restrictions on heating and cooling rates to avoid cracking and distortion.

Lime and magnesia. Lime, CaO, is used as a flux, being introduced in the form of chalk or dolomite. Brick clays are often rich in finely divided chalk. Chalk is a strong flux, the addition of 1 per cent of chalk has the same fluxing effect as 10 per cent feldspar above 1100 °C [74].

Magnesia (MgO) is present in small quantities in many clays as magnesite or dolomite and acts as a flux in a similar way to lime.

Iron. Iron compounds are present in almost all clays and act as fluxes, reducing the vitrification temperature particularly when present with other fluxes. Their more important effects, however, are on the colour of the finished product. In oxidizing atmosphere iron compounds yield colours ranging from cream to browns and reds. In reducing atmospheres grey or blue colours are produced.

Drying

After forming by a wide variety of processes, including throwing, casting, pressing, jolleying, extruding, etc., the ware may be dried down to around 5 per cent moisture before firing. Depending upon the free-moisture content of the clay article, which in turn is governed by the forming process, it may be necessary to use a separate drier before the article enters the kiln. In other cases, e.g. semi-dry and stiff-plastic pressed bricks and tiles, the ware can in general go straight into the kiln, drying taking place during the initial stage of the firing schedule.

Firing

Firing takes a variety of forms dependent on the ware. Pottery is normally fired at least twice. The first firing is termed 'biscuit' or 'bisque' firing, at up to 1150–1250 °C for earthenware and up to 1250–1410 °C for porcelain, after which glaze is applied and the glaze or 'glost' firing takes place at temperatures in the range 1040–1150 °C. Additional firings at 700–850 °C may be needed for fusing 'on-glaze' decoration. Unglazed products like bricks are of course fired only once but in addition some glazed products may also be 'once-fired'. Sanitary ware, in particular, is commonly once-fired, the glaze being formulated to possess the same maturation temperature as the clay body. A rather more rudimentary form of once-firing is the salt glazing used in the manufacture of glazed stoneware drainpipes in which common salt is introduced into the kiln atmosphere during the later stages of firing.

The firing of raw products can be divided into three broad periods:

1 Water removal (free and combined). Most of the free water is removed by the time the temperature reaches 200 °C. The rate of temperature rise during this early 'water-smoking' period must be relatively low as rapid production of steam may damage the ware. It is good practice to use high excess-air levels during this period.

The removal of the combined water commences around 400 °C and is completed at about 800 °C.

2 Recrystallization, and carbon, sulphur and organic-binder burnout. These occur in the temperature range 300–900 °C and also require oxidizing conditions.

3 Vitrification. This commences at temperatures around 900 °C and continues up to the final firing temperature which is typically 1150–1250 °C for earthenware bisque and up to 1410 °C for porcelain.

Compared with studies of the effects of clay composition and firing temperatures on the properties of the finished ware the effects of kiln atmospheres have been somewhat neglected except in so far as they affect coloration. Throughout the temperature ranges in which chemical reactions occur, where vitrification and densification take place and where physical properties and colour are determined, the kiln atmosphere reacts to a considerable extent with the ware to determine its finished properties.

Gibeaut and Rice [75] have studied the effects of varying kiln atmospheres on biscuit firing of porcelains, on glost firing and on brick firing. A 60 °C reduction in the temperature required to produce a fixed value of shrinkage (which is a measure of the degree of maturation) was obtained for the combustion products from 5 per cent excess gas compared with pure air. This effect is attributed to the water vapour present which increases the vitrification rate by significantly affecting the viscosity of the melt especially at lower temperatures. The addition of water vapour to an air atmosphere reduced the maturation temperature by 100 °C between the dry atmosphere and 19 per cent water vapour content but further additions have little effect.

Similar improvements were obtained in glost firing and in reduction firing of bricks. In the firing of blue engineering bricks a reducing atmosphere is required towards the end of the heating schedule. Similarly some types of multicoloured facing bricks require reducing conditions during part of the firing. It is extremely difficult to avoid smoke emission from coal- or oil-fired kilns producing these types of product. Gas firing eliminates this difficulty and offers other important advantages. The main advantages of gas for ceramics firing may be summarized as follows [76]:

1 Cleanliness.
2 Controllability leading to uniform heating and controlled heating schedules.
3 Elimination of the need for storage space and handling.
4 Reduced maintenance.
5 Avoidance of pumping and oil-heating facilities.
6 Elimination of fume and smoke emission permitting the use of shorter and cheaper stacks.
7 Negligible sulphur content eliminating problems of acid attack on stacks, exhaust fans, etc., acid smut emission and scumming of the product. Gas firing eliminates the need for saggars (i.e. protective refractory boxes) and muffle firing in some instances.
8 In many cases, increased kiln output and capacity.
9 Reduced manpower requirements.
10 Reduced kiln refractory maintenance.
11 Reduced fuel requirements per tonne of fired ware.
12 Increased kiln productivity by improved product quality and virtual elimination of rejects caused by poorly controlled firing.
13 Reduced salt requirements for salt-glazed products facilitating compliance with

the requirements of antipollution legislation, such as the English Alkali, etc. Works Regulation Act 1906.

14 Improved working conditions.
15 Ease of application of waste-heat recovery, e.g. recuperative burners in intermittent kilns and air and load recuperation in continuous kilns.

13.6.2 Pottery kilns

The main developments in firing methods for pottery since the 1950s have been:

1 A trend towards once-firing certain types of glazed ware, e.g. vitreous sanitary ware.
2 A changeover in the type of fuel used. Coal in intermittent kilns has given way to town and natural gas, LPG, electricity and oil.
3 The replacement of saggar- and muffle-furnace firing by open-firing methods using refined fuels and yielding higher efficiencies.
4 The introduction of new kiln types, e.g. modern car tunnel kilns, multipassage kilns and continuous kilns of small cross-section. Also the development of modern intermittent kilns fired by gas, LPG, electricity and oil.
5 The attainment of faster firing cycles, e.g. by firing a stream of single articles.

Intermittent pottery kilns

The pottery industry adopted the car-bogie tunnel kiln in large quantities in the 1950s as an efficient replacement for the old solid-fuel-fired intermittent bottle kilns. Large tunnel kilns must, however, be run with nearly maximum payloads to ensure efficient operation. Market conditions may make this impossible and so the flexibility of intermittent kilns has led to their reintroduction in a convenient and efficient form.

Gas-fired intermittent kilns may be of the box, bogie-hearth or portable-cover types. Figure 13:39 shows a typical portable-cover intermittent kiln fired by vertical tunnel burners along the side walls and flueing centrally through the perforated hearth. Portable-cover furnaces possess two main advantages: the ware is not disturbed after placing and one cover may be used for two bases thereby shortening heating and cooling times and making a small contribution to fuel efficiency.

To obtain consistent ware quality it is vital that each piece of ware has a similar thermal history. Thus an essential feature of any satisfactory firing system is even temperature distribution. Forced recirculation using high-velocity burners is particularly useful in providing temperature uniformity in pottery kilns since the ware is normally loaded in a fairly elaborate stack such that the central pieces are shaded from direct radiation by their neighbours and by the kiln furniture. Compared with continuous kilns, in which extensive load preheating is carried out, intermittent kilns are inefficient devices unless air recuperation is practised. A relatively cheap and convenient means of combining the virtues of forced recirculation and recuperation is the use of self-recuperative burners. Bryan, Masters and Webb [25] have described the application of the MRS recuperative burner (Section 7.6.3) to existing single- and twin-truck intermittent kilns (Figure 13:40).

The single-truck kiln of 4.5 m^3 (160 ft^3) capacity is fitted with a single 290 kW (10 therm h^{-1}) recuperative burner in place of two banks of four premix air-blast burners. The control system provides fully automatic start-up and operation and proportional temperature control is fitted, based on a cam system to control the time–

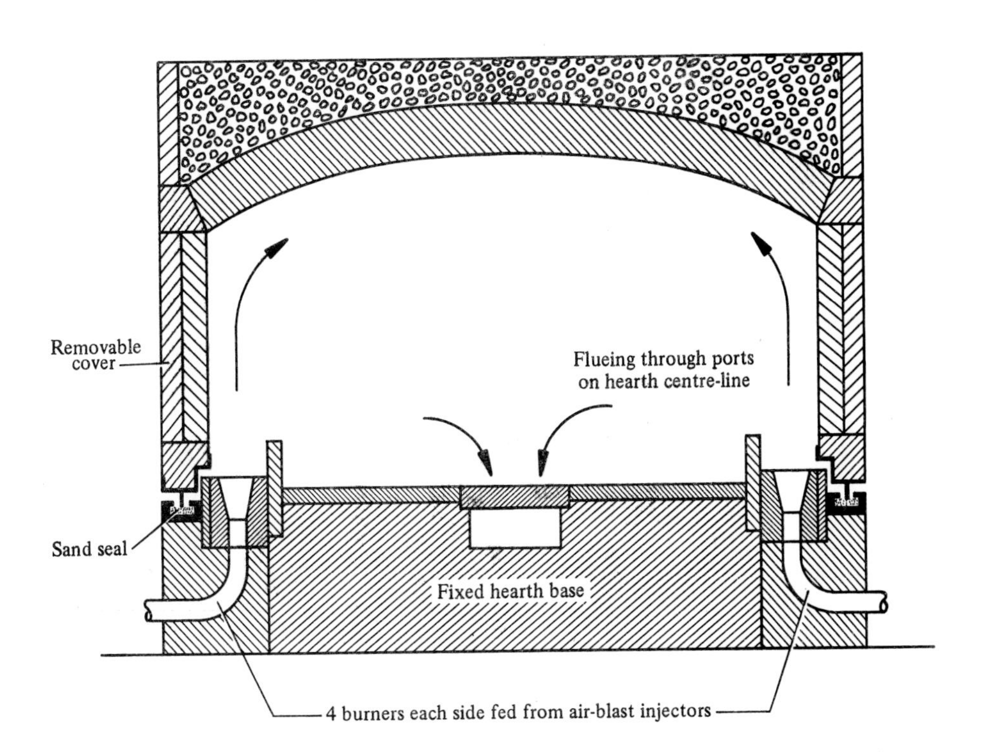

Figure 13:39 Portable-cover pottery kiln [77]

temperature programme. Ultraviolet flame detection is incorporated and the controls are arranged such that at the end of the predetermined heating cycle the burner switches itself off and the air-flow controls open to provide rapid cooling. Comparative performance figures are shown in Figure 13:41.

Additional advantages are that due to the high mass flow rate the drying period of the cycle is reduced appreciably and as a result of the greater turndown ratio the firing cycle is optimized. Even greater fuel savings have been achieved on an 18.4 m^3 (650 ft^3) twin-truck kiln fired by two 440 kW (15 therm h^{-1}) recuperative burners [25] the kiln turn-round time being reduced from 50 to 35 hours.

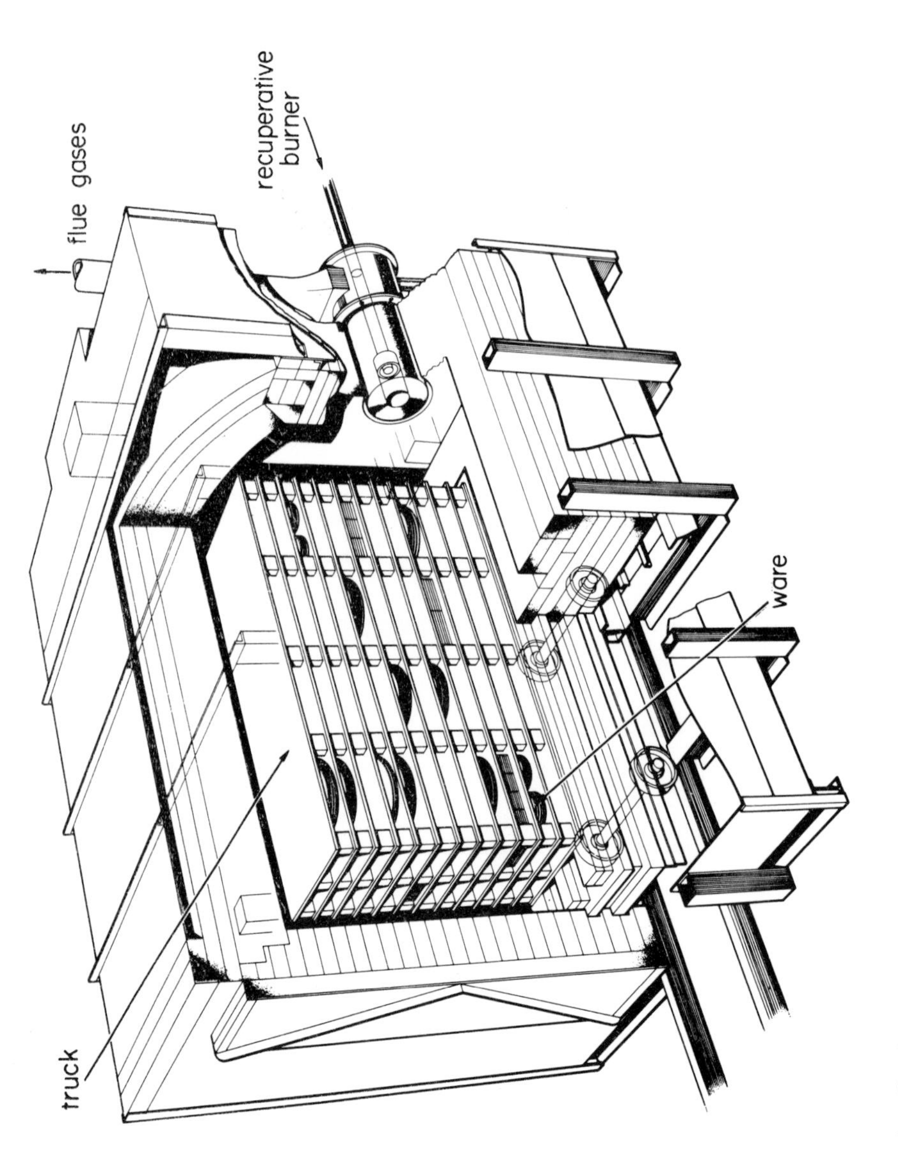

Figure 13:40 Application of self-recuperative burner to single-truck intermittent kiln [25]

	Nonrecuperative	Recuperative
Burner type	Premix air blast	MRS
Rating	60 kW	290 kW
Number	8	1
Fuel consumption over one cycle	17.9 GJ (170 therm)	12.1 GJ (115 therm)
Fuel saving	—	32.3%

Figure 13:41 Fuel saving by using a recuperative burner in a single-truck intermittent pottery kiln [25]

Continuous pottery kilns
The vast majority of continuous kilns used at present in pottery firing are car-bogie tunnel kilns. These may be either muffle or open fired. With the latter the original practice was to place the ware in saggars to protect it from direct flame impingement and from the effects of the combustion products. Open placing has been rendered possible by gas firing and is increasingly being adopted in tunnel kilns for biscuit firing and for some glost firing. Open firing provides improved heat transfer to the ware and the absence of saggars reduces the deadload-to-payload ratio. Comprehensive waste-heat recovery arrangements are made in tunnel kilns. Cold air is blown or drawn into the kiln at the ware-discharge end counter-flow to the car travel and cools the ware. This air may be used for drying purposes and/or flows to the burner zones as preheated combustion air. The combustion products flow towards the ware-inlet end preheating the ware.

Figure 13:42(*a*) shows a diagrammatic longitudinal section of a typical gas-fired car-bogie tunnel kiln and Figure 13:42(*b*) a cross-section through the burner zone. Tunnel kilns have lengths typically between 30 m and 100 m (100–350 ft) and placing cross-sections up to about 1.4 m wide by 1.2 m high (4 ft 6 in by 4 ft). The kiln illustrated has an overall length of 76 m (250 ft) and is divided into 50 bays, firing being confined to nine central sections. Each section has a burner each side firing at car-floor level and two sections at the exit end of the firing zone each have a further two burners firing at high level. Despite their great advantages over the inefficient batch kilns they replaced, tunnel kilns possess a number of important disadvantages:

1 High deadload-to-payload ratio. This results from the use of heavy trucks and extensive kiln furniture. For sanitary ware, Davis and Manuel [79] quote a ratio of 10 : 1 as being typical.
2 Long firing cycle. This results from the use of a kiln of large cross-section since a considerable time is required for the ware in the centre of the load to reach a soaked condition. This situation is of course much worse in the case of muffle tunnel kilns in which heat transfer is almost solely by wall radiation.
3 Cold centre-bottom zone. Since the conveyance method does not permit centre-bottom heating, the cold base resulting from this further increases the firing cycle.
4 Unevenly fired ware. This is due to masking of one piece of ware by another. Again the effect is much more serious in muffle tunnel kilns.
5 Inflexibility in operation.

As a result of these limitations, alternative designs for continuous kilns have been

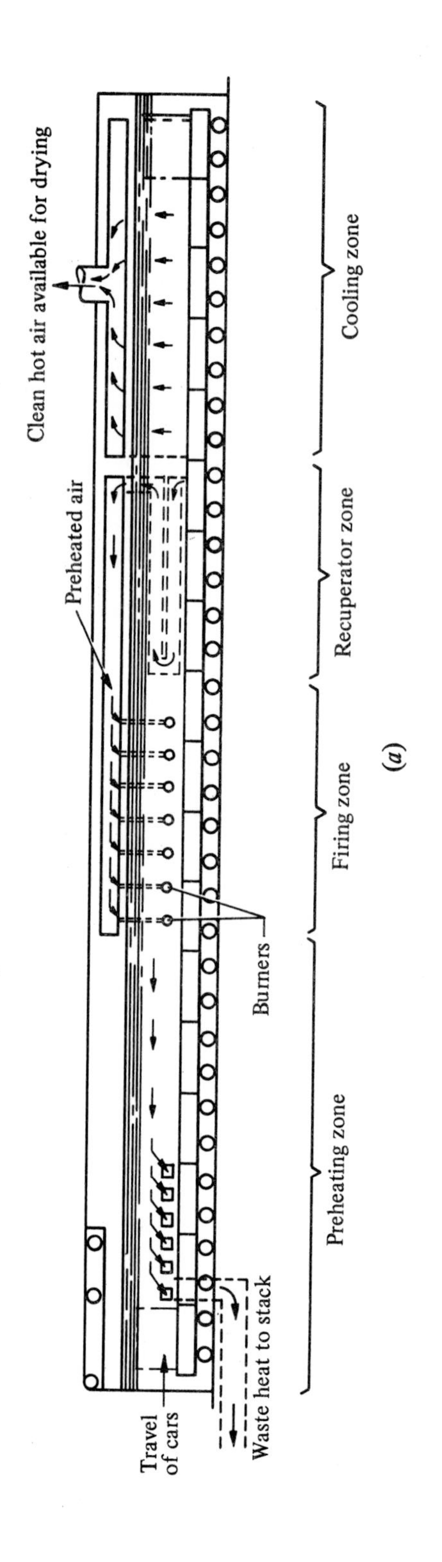

Figure 13:42 Continuous kilns
(*a*) Car-bogie tunnel kiln: longitudinal section [78]

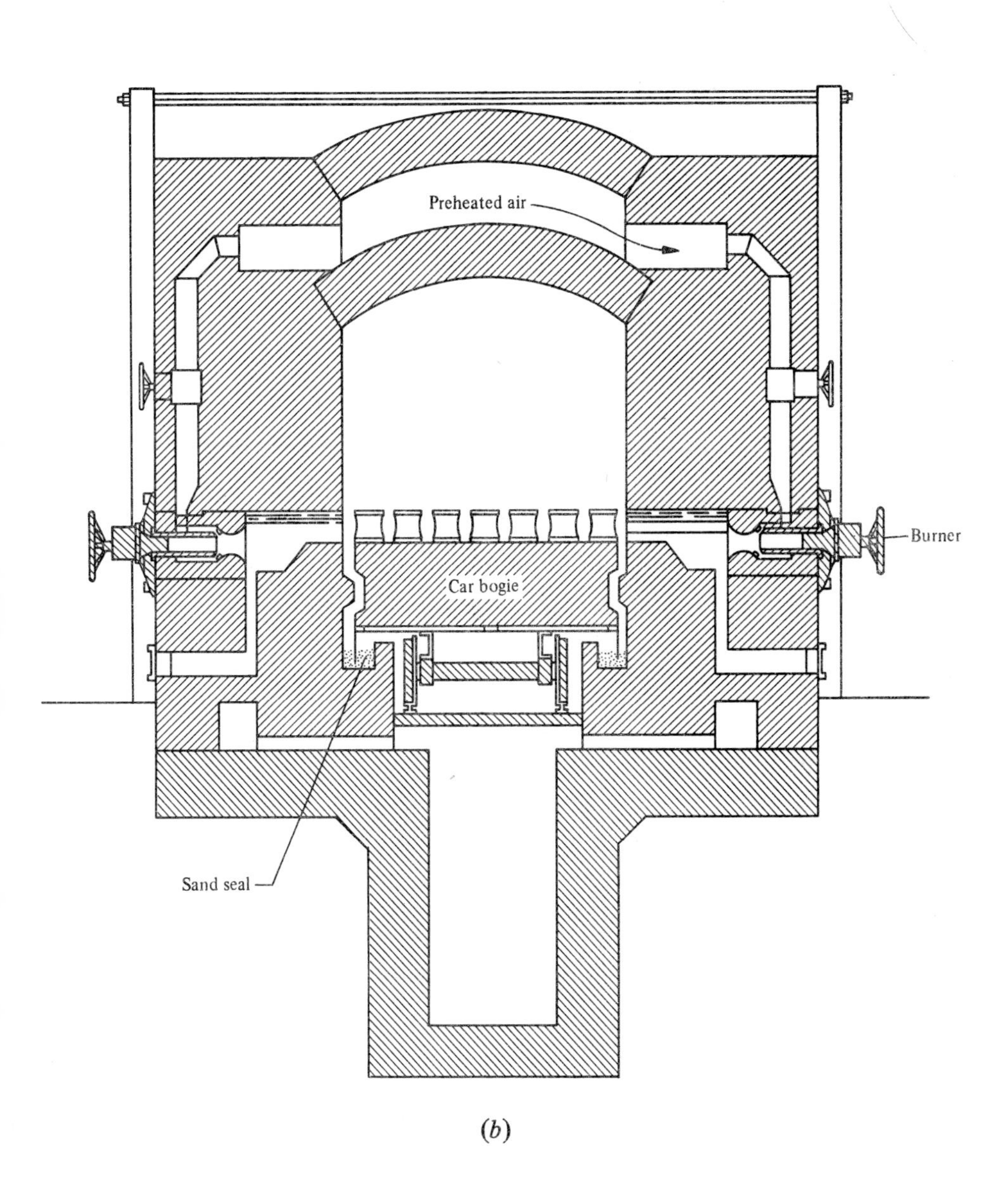

(b)

Figure 13:42 Continuous kilns
(*b*) Car-bogie tunnel kiln: cross-section at burner zone [78]

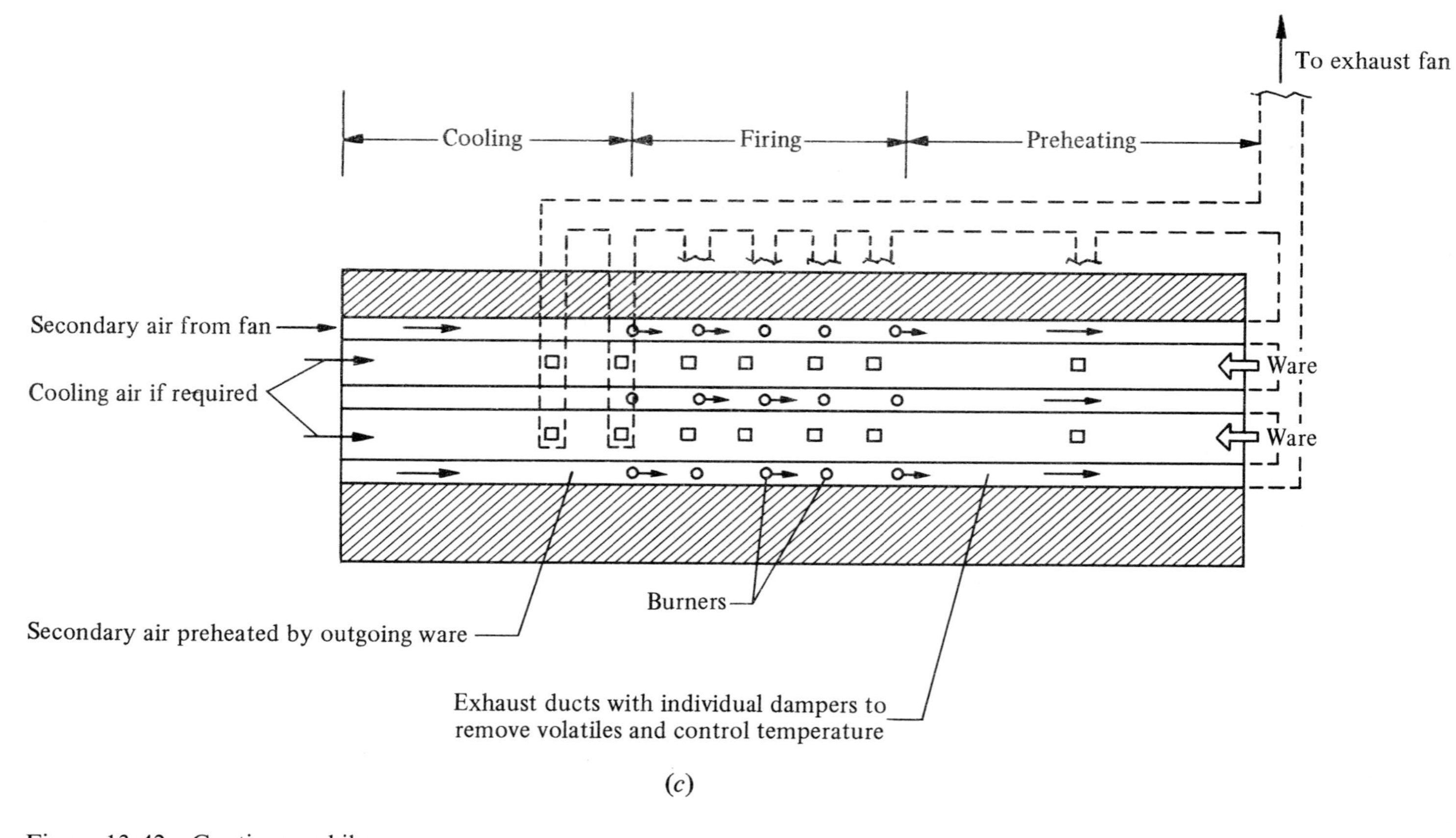

Figure 13:42 Continuous kilns
(*c*) Four-passage multipassage kiln [77]

developed in recent years in which the kiln passages (or single passage) are made much smaller, in many cases only slightly larger than the individual pieces of ware, to minimize lateral temperature differentials and to facilitate much reduced firing cycles. Examples of kilns of this type are multipassage kilns and the Trent kiln which was designed initially for firing sanitary ware.

Figure 13:42(*c*) shows a small multipassage kiln 12 m (40 ft) long. Early electrically heated multipassage kilns had the load moving in opposite directions in alternate passages so that the heat given off by the fired ware in one passage could be used to preheat the incoming ware in adjacent passages. The drawbacks of this method are that the heating and cooling profiles become interdependent, charging and discharging facilities have to be present at both ends of the kiln and uniform loading is required to prevent temperature fluctuations. Gas-fired designs employ unidirectional passages and counter-flow of combustion products and ware. The kiln shown in Figure 13:42(*c*) is a four-passage unit, each passage measuring 380 mm wide by 190 mm high (15 in $\times$ $7\frac{1}{2}$ in). The ware is fed through the kiln on bats, i.e. flat refractory tiles, by a hydraulic pusher. The four passages are heated using six sillimanite combustion chambers fired by low-pressure aerated burners fed with just sufficient primary air to avoid cracking, the remainder of the combustion air being supplied preheated from the cooling zone.

The Trent kiln [79] for sanitary ware was designed on a similar principle except that it employs a single passage, the placing space of which (724 mm $\times$ 673 mm, $28\frac{1}{2}$ in $\times$ $26\frac{1}{2}$ in) is just large enough to take the largest item of ware. Figure 13:43(*a*) shows a longitudinal section and firing curve. A cross-section through the firing zone is shown in (*b*) and through the exhaust zone in (*c*). The ware is loaded on bats which are supported on two rows of bricks, the bricks being pushed through channels in the base of the kiln. By using this method of conveyance a flueway is formed beneath the ware between the bricks avoiding the cold-base problems inherent in conventional tunnel kilns. In addition to the bottom burner there are three side burners on each side of the kiln firing into longitudinal flueways in the side walls. The ware enters the kiln via the exhaust section where the kiln atmosphere and the flue gases are separately exhausted under controlled conditions and then passes through the recirculation zone in which the kiln atmosphere is recirculated to minimize top-to-bottom temperature differentials. In its passage through the first two zones the ware is indirectly preheated by the combustion products from the firing zone passing along flueways to the exhaust system. The ware then passes through the firing zone where it reaches the peak temperature of about 1200 °C. It is then cooled quickly to about 650 °C in the initial rapid cooling zone, then slowly through the β to α quartz modification (575 °C) to eliminate the chance of cracking and finally is cooled quickly to a handleable temperature by the time the kiln exit is reached.

Considerable improvements in firing times have been achieved compared with conventional tunnel kilns, e.g. $11\frac{1}{2}$ hours compared with 36 hours, and the fuel consumption per tonne of finished ware is reduced by about one-half, e.g. 8.3 GJ t^{-1} (80 therm ton^{-1}) compared with 17.6 GJ t^{-1} (170 therm ton^{-1}) for a tunnel kiln in the same factory. The output is about 25 tonnes per week, i.e. one-third of the output from a tunnel kiln. On the other hand the capital cost is less than one-third of a tunnel kiln. Additional advantages of the Trent kiln are the easy loading and unloading compared with the conventional method and the flexibility of operation resulting from low thermal capacity allowing it to be brought up to temperature from cold in three days. Maintenance also is easy since no bogie repairs are required.

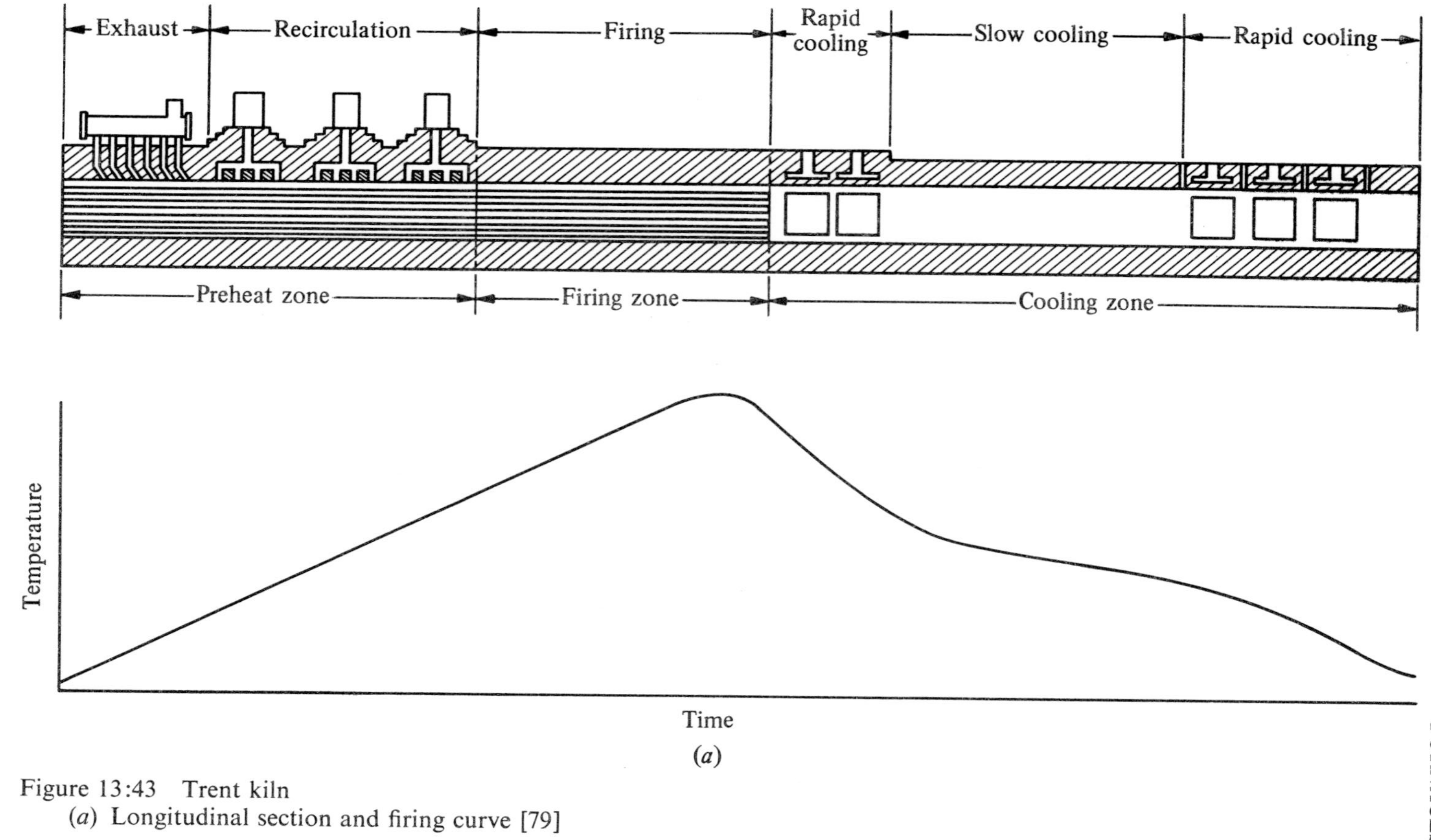

Figure 13:43 Trent kiln
(*a*) Longitudinal section and firing curve [79]

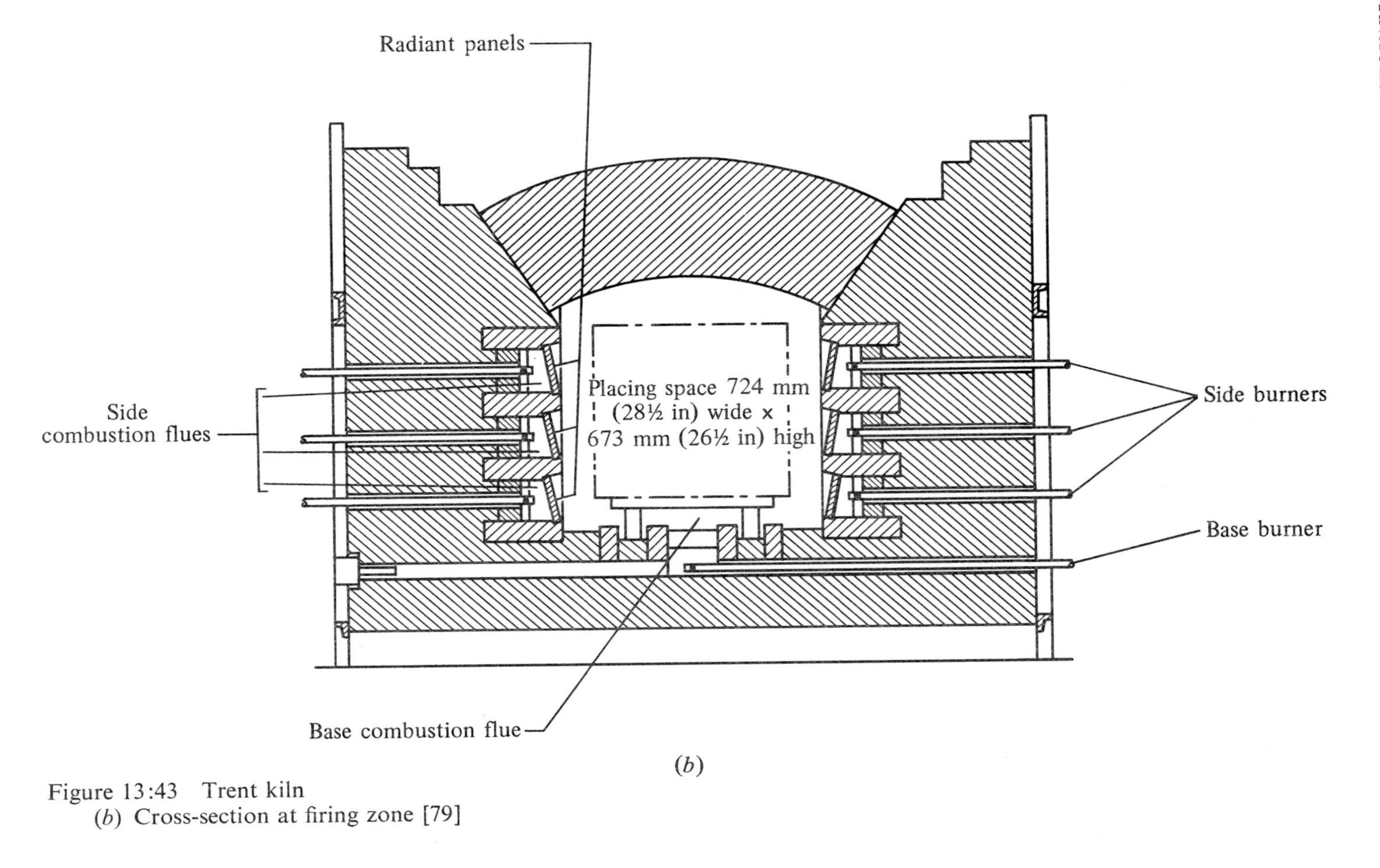

Figure 13:43 Trent kiln
(*b*) Cross-section at firing zone [79]

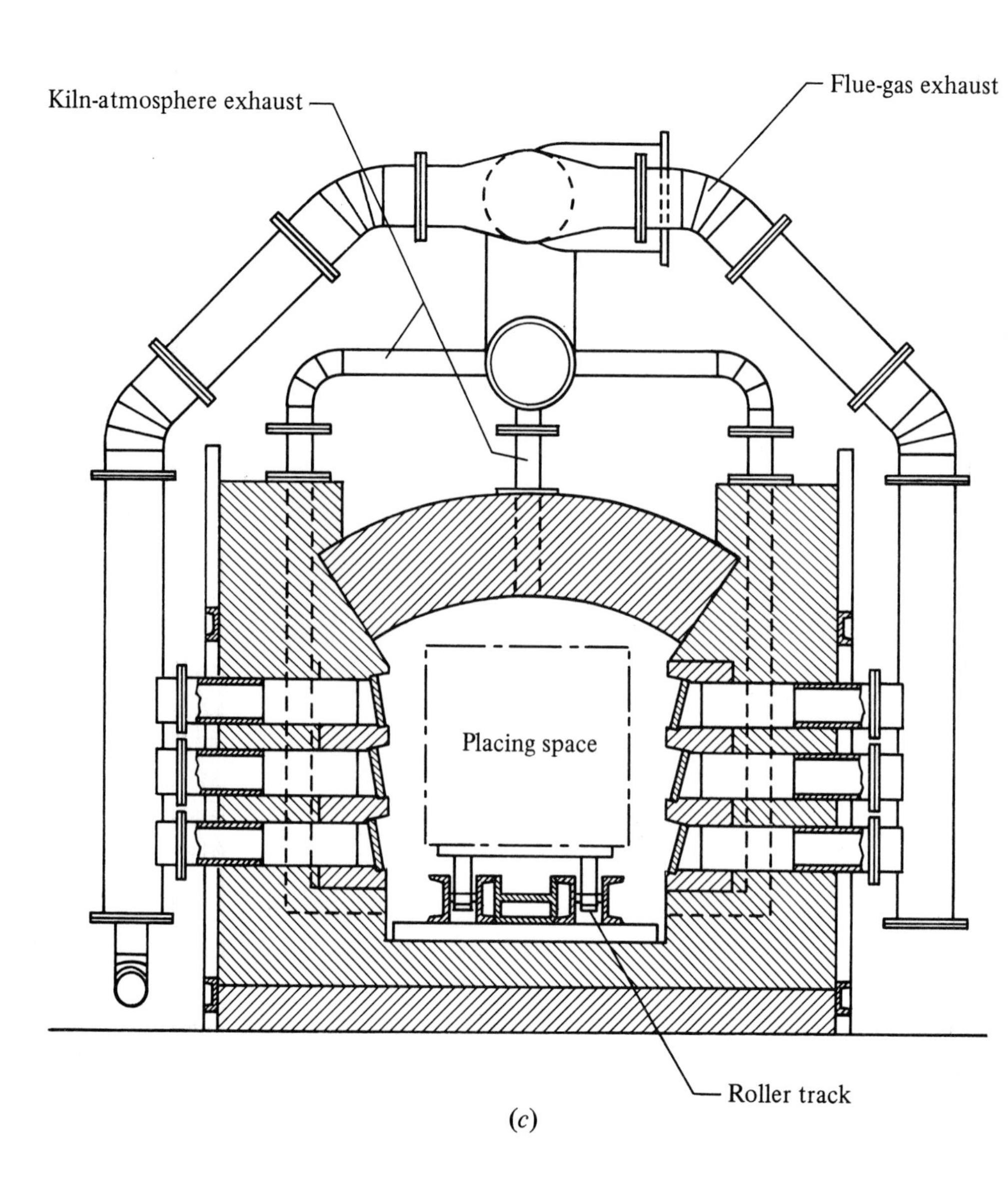

Figure 13:43 Trent kiln
(*c*) Cross-section at exhaust zone [79]

13.6.3 Kilns for heavy-clay products and refractories manufacture

As Johnson [80] has pointed out, the fuel-usage pattern in the British heavy-clay and refractories industries in the last few years shows some parallels with the trends experienced by the pottery industry in the period between the end of the Second World War and the introduction of natural gas, during which coal was very largely replaced by oil, town gas, and even electricity. As indicated in Section 13.6.2, the low sulphur content and good temperature uniformity possible with town gas allowed direct firing as opposed to muffle or saggar firing resulting in town gas being competitive at prices well above those of oil and coal. The situation currently existing in the heavy-clay and refractories market in which natural gas can be sold at prices higher than coal and oil is in some ways similar to that existing in the pottery industry in the 1950s and early 1960s.

There are, however, important differences in the firing of the two types of product:

1 All heavy-clay and refractory goods were fired in directly fired kilns by coal or oil.
2 The fuel cost represents quite a high proportion of the price of the finished heavy-clay product.

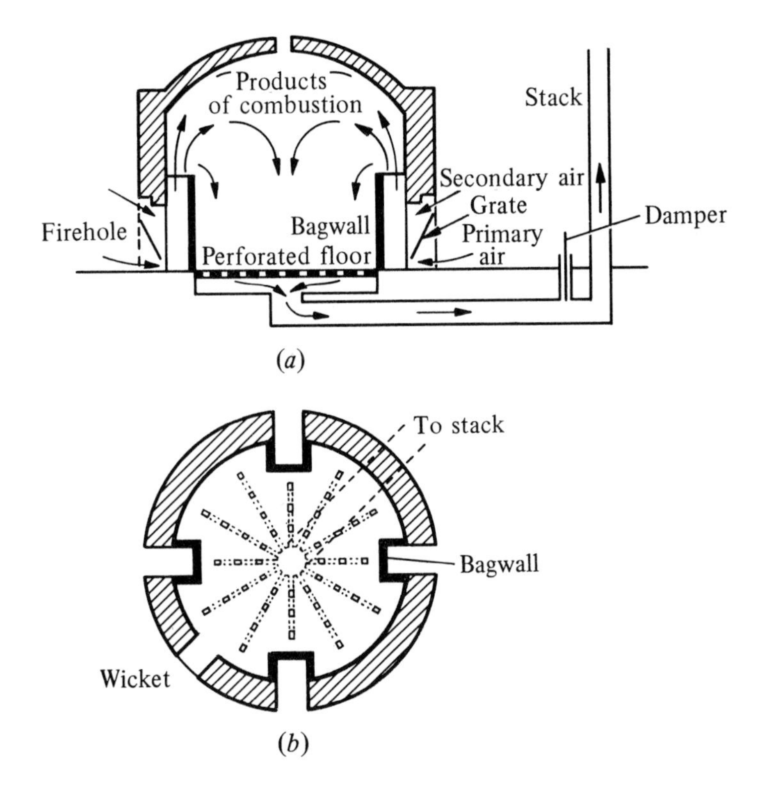

Figure 13:44 Intermittent downdraught kiln [76]
(*a*) Sectional view
(*b*) Plan view showing typical floor flue system

Neither of these factors favoured the competitive position of gas and town gas made few inroads into the market. The market is a very large one; in 1968 the Gas Council reported [76] that the estimated total potential load was around 76.2 PJ (722×10^6 therm) a year excluding power requirements and the fuel requirements of kilns used by the National Coal Board. Individual loads ranged from about 105 GJ up to 845 GJ ($1–8 \times 10^6$ therm) a year. The 1968 report indicated that there had been a substantial increase in usage of heavy fuel oil at the expense of coal and that the majority of building bricks were fired in continuous-chamber (i.e. stationary-hearth) kilns although there was a steady trend towards their replacement by oil-fired, car-bogie tunnel kilns. The report indicated no particular technical problems in changing over all types of kiln to natural-gas firing and it was considered that gas firing should result in large fuel savings over coal firing and smaller but still appreciable savings over oil. Since this time gas has obtained a substantial proportion of the heavy clay and refractories fuel load at prices which take into account the advantages listed in Section 13.6.1.

Intermittent kilns

Intermittent kilns may be circular or rectangular in shape and may be updraught, horizontal-draught or downdraught. Downdraught kilns are the most widely used in Britain being more efficient than the other types. Figure 13:44 shows a typical downdraught intermittent kiln. Downdraught kilns are used for most types of heavy-clay product. Sizes range from 10 000 to 100 000 bricks with a typical capacity in the range 30 000 to 50 000 bricks. Coal is burned at the 'fireholes' around the periphery, the ware being protected from flame impingement by 'bagwalls' which direct the combustion products to the kiln crown from where they are drawn down through the setting to radial flues in the hearth. Some variants of the downdraught kiln exist, including designs with a central interior flueway and others containing flueways in the walls [81].

Continuous kilns

These may be subdivided into:

1 Continuous-chamber or annular kilns, e.g. the Hoffmann kiln (a longitudinal arch, stationary ware kiln) and the Staffordshire kiln (a transverse arch, stationary ware kiln).
2 Car-bogie tunnel kilns.

Figure 13:45 shows a 16-chamber Hoffmann kiln, the operation of which is typical of continuous-chamber kilns. Three chambers are shown firing, the combustion products from which are induced by stack draught or by ID fans through the four preheating chambers to the central exhaust duct thence to the stack. Two drying chambers isolated by paper screens are fed by warm air from the cooling chambers via the hot-air duct. Next to the drying zone is a chamber being 'set' (i.e. loaded), the next one is empty and the remaining chamber is being 'drawn' (i.e. emptied). Air for combustion and drying is induced through the open 'wickets' (doorways) of the last-mentioned three chambers and becomes preheated as it cools the ware in the cooling zone, some of it passing to the drying chambers and the remainder flowing to the firing zones. Each chamber is essentially identical and at regular intervals (say 20 hours) the process is advanced one chamber. The paper division isolating the drying

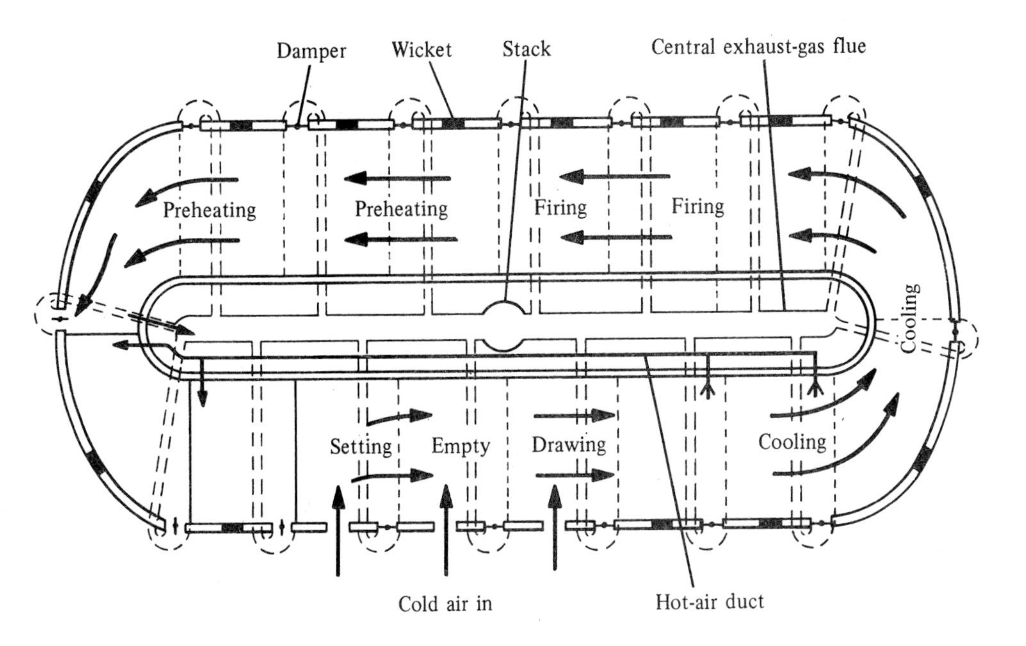

Figure 13:45 Hoffmann continuous-chamber kiln [80]

chamber burns out as the preheating zone advances and this chamber becomes part of the preheating zone. The setting chamber then becomes the second drying chamber and so on. The chamber capacity of Hoffmann kilns is typically 15 000 to 25 000 bricks although capacities up to 45 000 are not unknown.

Transverse-arch kilns operate on a similar principle but consist of arched chambers separated by permanent supporting walls perforated by 'trace-holes'. In general the chamber capacity is larger than the Hoffmann kiln and it is claimed that the transverse-arch kiln provides a more uniform control giving higher efficiencies or higher firing temperatures [81].

Continuous-chamber kilns were originally designed for coal firing and this is done by feeding fine coal, either manually or using automatic stokers, via the fireholes in the arch, the coal falling through gaps left in the setting. Oil firing is done in a similar way, portable oil dispensers inject small slugs of oil into the kiln. The slugs impinge on refractory target bricks on the kiln floor but combustion is under way before the bottom is reached and this assists in obtaining temperature uniformity. Some designs of oil injector incorporate a swinging nozzle head to vary the impingement position. Fuel consumption is affected by the carbon content of the clay. In Fletton bricks, for example, the carbon content is high and the overall fuel consumption is low compared with bricks from noncarbonaceous clays.

Car-bogie tunnel kilns are similar in design to those described for pottery firing. Dimensions are generally much larger, however, for example the kiln described in

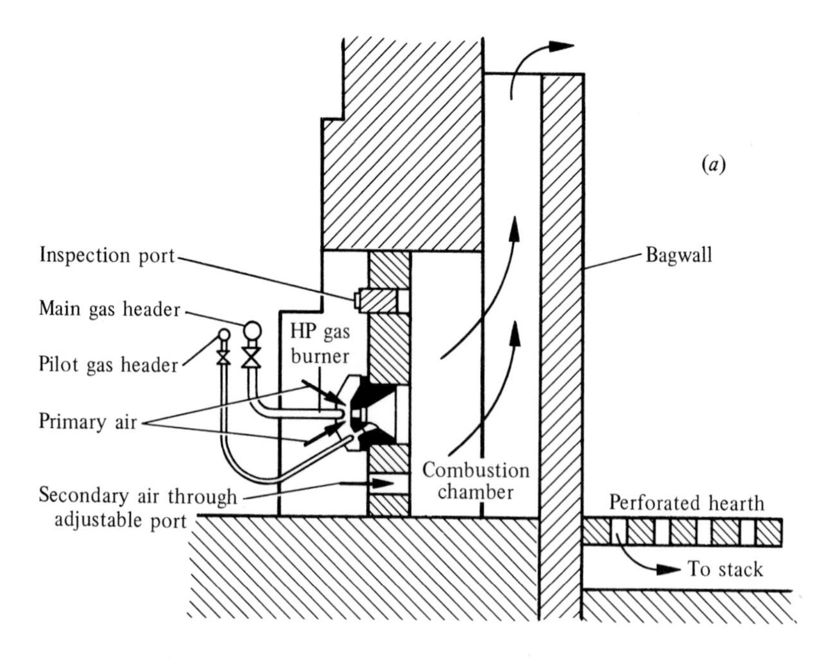

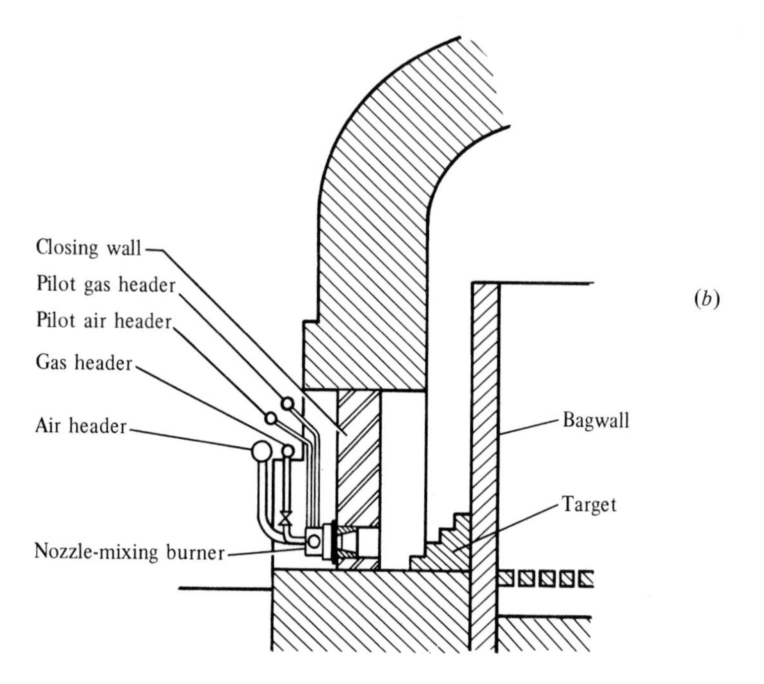

Figure 13:46 Gas firing of intermittent kilns
(*a*) Natural-draught burners [81]
(*b*) Air-blast burners [81]

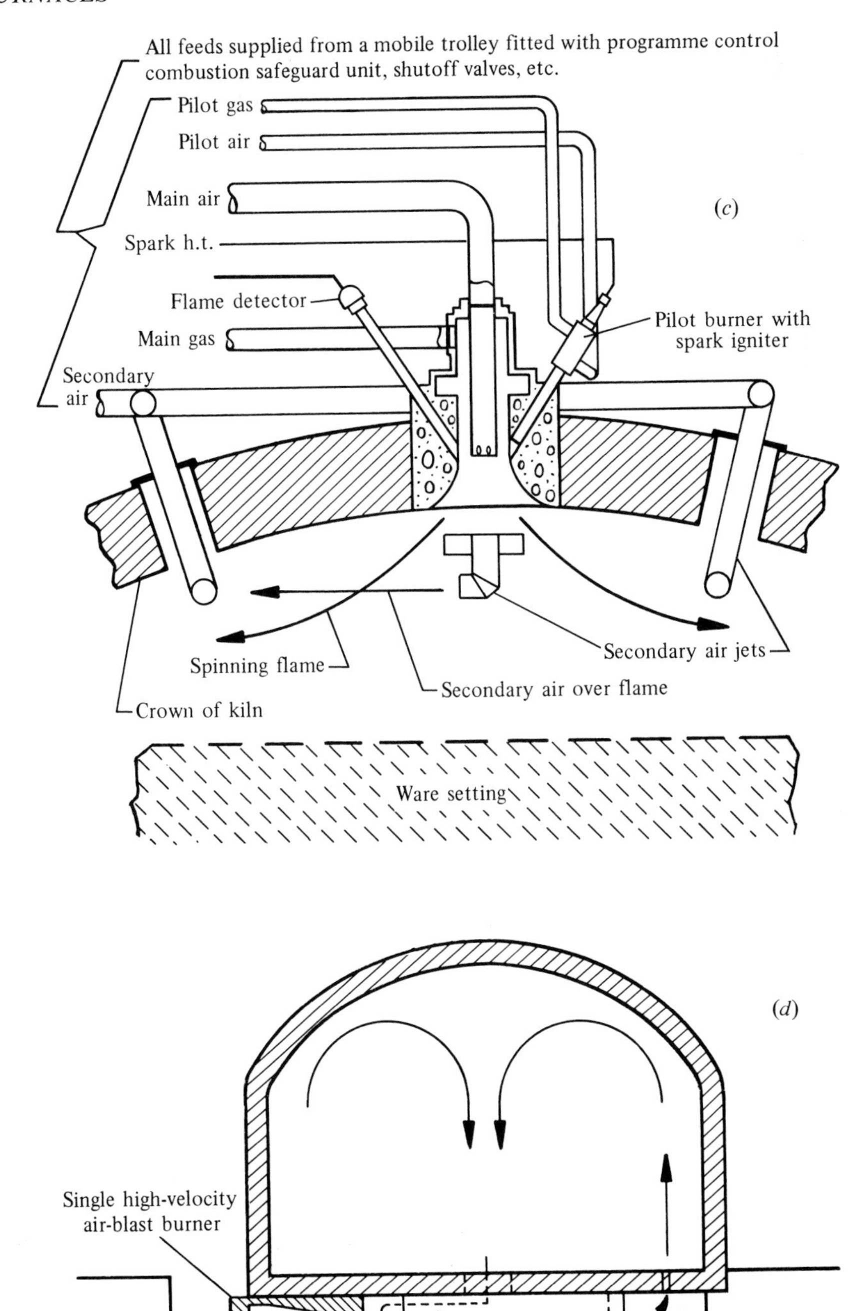

(*c*) Single-burner firing [82]
(*d*) Underhearth firing [83]

[76] measures 131 m long (430 ft). Air from the cooling zones is commonly used in pre-driers.

Changeover of kilns

About 70 per cent of all heavy-clay and refractories products manufactured in Britain are fired in continuous kilns, particularly chamber kilns [80] with a virtually constant hourly fuel demand throughout the year. Additionally, a high load factor is possible in works using intermittent kilns by rotating the firing periods so that one kiln is heating while another is cooling. Interruptible firing is feasible and a number of supply contracts in Britain are on an interruptible basis covering the whole range of oil- and coal-fired intermittent and continuous kilns.

Several alternative systems have been developed for changing over intermittent kilns to gas firing, most of which consist simply of replacing the coal grate (or oil burner) by a gas burner and bricking up the firemouth.

Natural-draught or air-blast burners have both been used, the former using gas at 7 to 14 kPa (70–140 mbar; 1–2 $\mathrm{lbf\,in^{-2}}$) entraining primary air at entry to a divergent refractory quarl (Figure 13:46(*a*)) [81]. Some savings in placing space have been made on changeover using this method by replacing the continuous bagwall by individual combustion chambers.

Figure 13:46(*b*) illustrates the application of tunnel-mixing air-blast burners to intermittent kilns. This method has the advantage of being uninfluenced by wind or changes in stack draught, allows the kiln to operate at positive pressures and facilitates reduction firing if this is required (e.g. blue engineering bricks). Forced-draught portable burners on trolleys have been developed, to be moved from kiln to kiln as required in the same way as the mobile automatic coal stokers they replace [81].

A different approach is shown in Figure 13:46(*c*), which is very much in line with current trends, in which a single automatic flat-flame burner fires from the crown of an intermittent kiln [82]. This system offers two advantages:

Fuel	Coal	Natural gas	Natural gas
Burner system	Hand-fired in fireholes around the kiln	Air-blast burners in fireholes around the kiln	Single air-blast burner in the roof
Setting capacity	51 t (50 ton) of floor tiles	56 t (55 ton) of floor tiles	61 t (60 ton) of floor tiles
Firing time	130 hours	72 hours (45% reduction)	72 hours (45% reduction)
Output per month tonne (ton)	102 (100)	168 (165)	183 (180)
Wastage	15%	0.5%	less than 0.5%
Specific fuel consumption GJ t^{-1} (therm ton^{-1})	8.1 (78)	6.2 (60)	6.2 (60)

Figure 13:47 Comparative results for coal and natural-gas firing of a round down-draught kiln [80]

1 The use of a single burner allows the economic adoption of sophisticated control and safety equipment.
2 The bagwalls may be dispensed with, thereby increasing the kiln capacity.

Figure 13:46(*d*) shows a single high-velocity burner installed below the kiln hearth. A high degree of recirculation is promoted and the method lends itself to the use of a self-recuperative burner [83].

Figure 13:47 shows comparative performance figures for coal firing by hand and gas firing by the methods shown in Figure 13:46(*b*) and (*c*). The specific fuel consumption is reduced in both cases by about 23 per cent and the output is increased by about 9 per cent by top firing.

A major benefit of changeover is the improved ware quality and reduction in wastage which results. All bricks have incurred the full manufacturing costs by the time they leave the kiln and scrap bricks are virtually valueless, consequently even small reductions in scrap rates may result in large financial savings. Carlier [82] has estimated that a reduction by 1 per cent in the scrap rate for a kiln producing 400 000 bricks per week results in an increase of kiln income of about £120 per week (1973 figures).

Johnson [80] has given details of the changeover of a downdraught intermittent kiln using atmospheric burners which showed a specific fuel saving of 36 per cent compared with coal firing by hand (15.2 GJ (144 therm) per 1000 bricks compared with 23.7 GJ (225 therm) per 1000), a 30 per cent increase in annual output and a reduction in the proportion of rejects from 5 to 2 per cent. In addition the kiln required a reduced labour force and the heating and cooling schedule was reduced from 12 days to 8 days.

A major influence in the introduction of natural gas into heavy-clay firing was the question of environmental pollution. Kilns must conform with the requirements of the Clean Air Acts as enforced by the local authority. In addition, however, several types of kiln and firing methods are 'scheduled', i.e. they come under the jurisdiction of the Alkali Inspectorate and these include hand-fired intermittent kilns and blue-brick kilns [82]. When users of these kilns and processes were required to comply with the requirements of the Memorandum on Chimney Heights [85] (which fixes the minimum chimney height based mainly on the sulphur content of the fuel), they were faced with the alternatives of building new nonscheduled kilns, extending chimneys or changing to gas firing. The last option was generally cheapest in capital cost and most of the then scheduled kilns are now gas fired with resulting improvements in performance, paving the way for the changeover of nonscheduled processes [82].

Continuous annular kilns are changed over by inserting burner lances through the top fireholes, the flame developing in gaps left in the brick setting for this purpose (Figure 13:48). The burners and manifolds are portable and are moved along the kiln as the cycle progresses. Where high-pressure gas is available the burner may consist of a single atmospheric injector feeding a heat-resisting steel lance fitted with a flame-retention head. Alternatively, air-blast systems are used to provide the necessary momentum and both premix and concentric-tube nozzle-mixing systems have been employed. The firing method is quite different to the technique used on coal and oil in the respect that the gas burns from the chamber top rather than allowing most of the combustion to take place on the floor. Early doubts that this would be satisfactory have proved groundless. Gibb [81] has described comparative tests between a kiln in

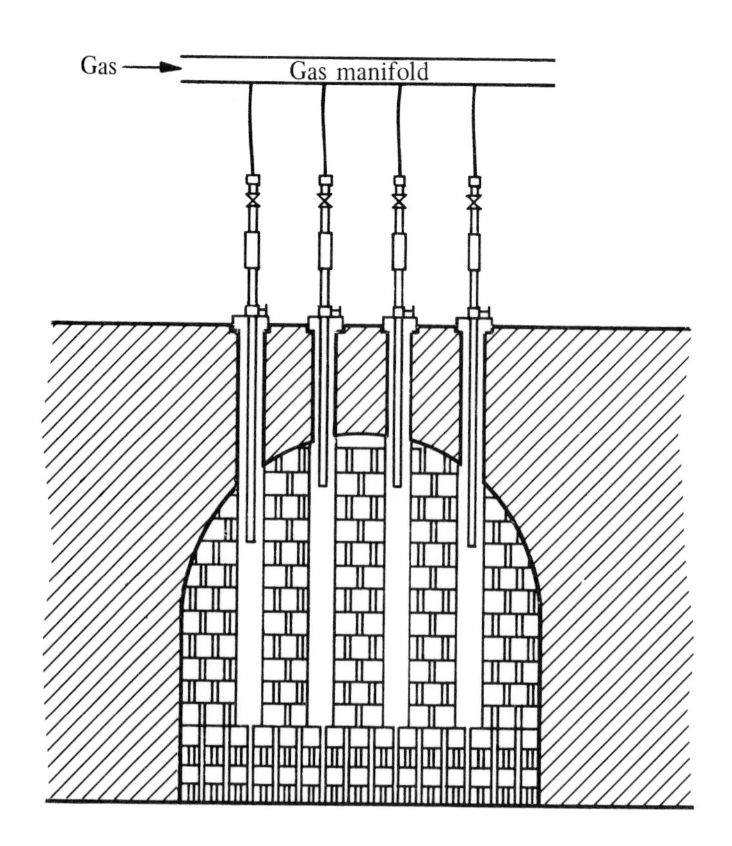

Figure 13:48 Top firing of a Hoffmann kiln [85]

which the burner pressure was pulsed between 14 and 100 kPa (140 mbar and 1 bar, 2 and 15 lbf in^{-2}) at 20 s intervals to produce a flame of alternating length in simulation of the oil-firing method and a kiln in which the pulsation system was omitted. Both systems produced satisfactory temperature distribution and ware of comparable quality.

Trolley-mounted side-firing systems incorporating flame-safeguard equipment have also been applied to chamber-kiln firing [80]. Car-bogie tunnel kilns are relatively modern pieces of equipment and therefore offer less scope for major improvements on changeover than do older kiln designs. Oil-fired tunnel kilns are usually either top fired using oil slug injectors or side fired using air atomization and changeover normally involves firing in the same direction using air-blast or nozzle-mixing burners.

A modern trend is to use intermittent shuttle kilns for increased flexibility of use. These are commonly sited adjacent to the tunnel kilns and use standard kiln cars [82]. They often incorporate a separate preheat chamber to use waste heat from the main chamber during the cooling period. Their efficiency is, however, fairly low and this could be improved by the use of recuperative burners as in the case of pottery shuttle kilns (Figure 13:40). Figure 13:49 illustrates the general layout of a modern car shuttle kiln.

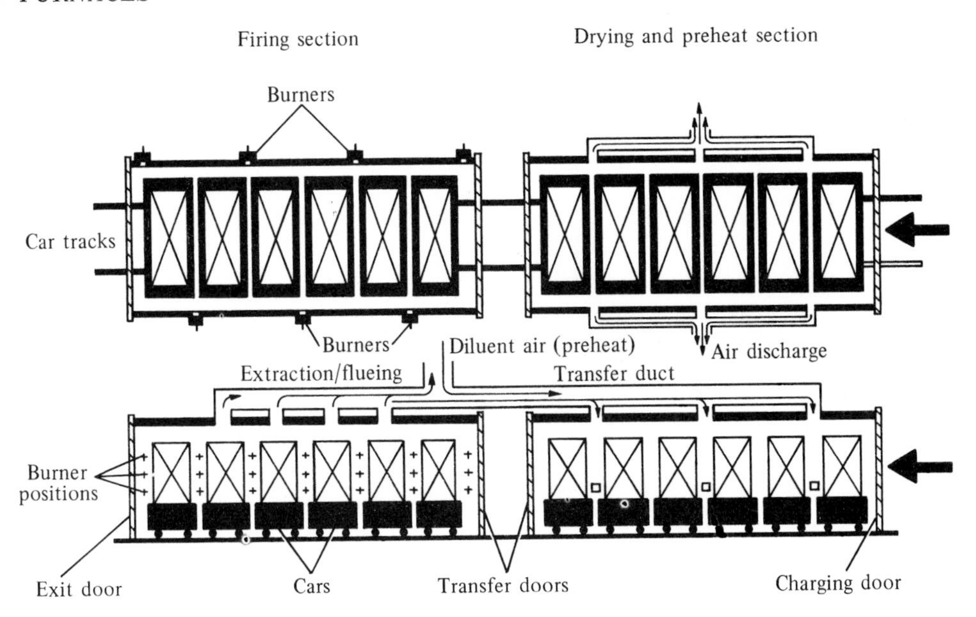

Figure 13:49 Modern car shuttle kiln [80]

Figure 13:50 provides an analysis of a selection of 11 kiln changeovers of various kinds. All of the figures in the gas firing results columns relate to a value of 100 on the original fuel. Thus where the fuel consumption is quoted as 66 in case (1) this indicates that the specific fuel consumption on gas was 66 per cent of that on coal.

As might be expected, the largest savings are those over the crudest previous firing systems. Case (6) shows an increase in fuel consumption on changeover. In fact this kiln uses clay with a carbon content of 12 per cent and the burners can be regarded mainly as igniters, the clay being capable of self-sustaining combustion above 800–850 °C. The fuel consumption is only around 2.1 GJ (20 therm) per 1000 bricks and the increased consumption is mainly due to the increased firing speed. The increased value of the ware justified the changeover [82].

REFERENCES

1 Trinks, W. and Mawhinney, M. H., *Industrial Furnaces*, Vol. 2, 4th edition. New York: Wiley (1967)
2 Gas Council, private communication
3 Francis, W. E., Lawrence, M. N., O'Connor, J. G. and Walker, L. 'The use of high-velocity burners in high-temperature furnaces: tests on a slot-type forge furnace', *Gas Council Res. Commun.* GC82 (1961); *J. Inst. Gas Eng.*, **2** (1962) 468
4 Masters, J., Oeppen, B., Towler, C. J. and Wong, P. F. Y., 'The development and exploitation of rapid heating techniques', *Inst. Gas Eng. Commun.* 849 (1971); *J. Inst. Gas Eng.* **11** (1971) 695
5 Bray, K. F. and Pardoe, T. H., private communication

Case	Type of kiln	Principal product	Firing system		Gas firing: results					
			Original fuel	Gas	Fuel consumption*	Output*	Quality*	Labour*	Repairs	Other comments
1	Down-draught intermittent	Pipes	Coal, hand fed	Natural draught	66	100	100	Reduced	Reduced	Firing time 60% of coal. Smoke eliminated
2	Down-draught intermittent	Tiles	Coal, hand fed	Nozzle mixing	78	120	Waste cut by 5%	66	Much reduced	Water smoking and firing 46% of coal. Smoke eliminated
3	Down-draught intermittent	Tiles	Coal, stoker fed	Nozzle mixing	90	110	Much improved	Reduced	Reduced	Firing time cut by 2 days. Smoke eliminated
4	Down-draught intermittent	Blue bricks	Coal, hand fed	Natural draught	95	—	6% increase in best ware	—	—	Firing time 87% of coal. Smoke eliminated
5	Hoffmann type	Facing bricks	Coal, top fed, mechanical	Top fired, air blast	80	130	13% increase in average value	Reduced	Refractory maintenance up, but now kiln is in good order a reduction is expected	

Case	Type of kiln	Principal product	Firing system		Gas firing: results					
			Original fuel	Gas	Fuel consumption*	Output*	Quality*	Labour*	Repairs	Other comments
6	Hoffmann type	Common and engineering bricks	Coal, top fed, manual	Side fired, natural draught	124	131	15% increase in average value	Reduced	Reduced	Facing bricks can now be fired
7	Belgian type	Refractories	Oil, end fired, fan atomizer	End fired, air blast	85	>100	Improved	Same	—	Smoke eliminated
8	Car tunnel (semi muffle)	Refractories	Oil/air emul., side fired	Side fired, nozzle mixing	83	100	Improved	91	Reduced	Muffle tiles less likely to failure
9	Car tunnel	Engineering brick	Oil slug, top fired	Top fired, air blast	112	111	9.5% increase in average value	Reduced	—	Facing bricks now fired. Smoke eliminated
10	Car tunnel	Facing bricks	Oil-air atom., side fired	Side fired, natural draught	100	110	10% increase in best bricks fired	50	Reduced	—
11	Catenary arch	Special refractories	Oil-air atom., end fired	End fired, nozzle mixing	<100	100	Improved	Same	Same	Kiln can now be cycled to 1750 °C and cooled in 24 hours

*Original fuel = 100

Figure 13.50 Analysis of a selection of kiln changeovers

6 Schnidman, L. (ed.), *Gaseous Fuels*. New York: American Gas Association (1954)
7 Jenkins, I., *Controlled Atmospheres for the Heat Treatment of Metals*. London: Chapman & Hall (1951)
8 Ellingham, H. J. T., *J. Soc. Chem. Ind.*, **63** (1944) 125
9 Warn, J. R. W., *Concise Chemical Thermodynamics in SI Units*. London: Van Nostrand Reinhold (1969)
10 Fleming, P. F., *J. Inst. Fuel*, **47** (1974) 268
11 Ives, D. J. G., *Principles of the Extraction of Metals*. London: Royal Institute of Chemistry (1960)
12 Bray, K. F., private communication
13 Cubbage, P. A. and Ling, K. C., 'The composition of combustion products (with particular reference to the design of exothermic atmosphere generators)', *Gas Council Res. Commun.* GC100 (1963); *J. Inst. Gas Eng.*, **4** (1964) 373
14 Cubbage, P. A. and Whiteside, J. L., 'The generation of heat-treatment atmospheres from town gas: the control of lean exothermic atmospheres', *Gas Council Res. Commun.* GC83 (1961); *J. Inst. Gas Eng.*, **2** (1962) 86
15 Moppett, D. J., 'Latest developments in controlled atmospheres: part 1: stripped rich exothermic atmospheres', Metal Heat Treatment Conference, Bingley Hall, Birmingham; Gas Council Midlands Research Station Ext. Rpt. 64 (1965)
16 Waluszewski, A., 'Natural gas for protective atmospheres', *Ind. Process Heat.*, **7**, No. 12 (December 1967), 22
17 Record, R. G. H., 'Atmosphere monitoring', Heatex 69, paper D6 (1969)
18 Moppet, D. J., 'The generation of atmospheres for carburising', BGC Midlands Research Station External Report No. 52 (1964)
19 Burton, S. and Cubbage, P. A., 'The design of gas-fired vacuum furnaces for copper heat treatment', *Gas Council Res. Commun.* GC125 (1965); *J. Inst. Gas Eng.*, **6** (1966) 280
20 ASM Committee on Production of Aluminium Alloy Castings, *Melting and Casting of Non Ferrous Metals*
21 Fowler, B. J. A., private communication
22 Higgs, A., 'Pot life in gas-fired galvanizing settings', *Inst. Gas Eng. Commun.* 370 (1950); *Trans. Inst. Gas Eng.*, **99** (1949–50) 826
23 Lupton, H. P., *Industrial Gas Engineering*. London: Walter King (1960)
24 Hall, E. M., 'Single burner firing of a galvanising bath', Institution of Gas Engineers Manchester and District Section, Meeting (1972)
25 Bryan, D. J., Masters, J. and Webb, R. J., 'Applications of recuperative burners in gas-fired furnaces', *Inst. Gas Eng. Commun.* 952 (1974)
26 Robertshaw, G. W., 'Gas firing' in *Efficient Use of Energy* (ed. Dryden, I. G. C.). London: IPC Science & Technology Press (1975)
27 Rice, B. J., 'Bulk aluminium melting', London and Southern Junior Gas Association, Meeting (1974)
28 Potter, M. D., 'Changeover of an oil fired reverberatory furnace to natural gas firing', Manchester Junior Gas Association, Meeting (1974)
29 Shelmerdine, A. B., 'Non ferrous metal melting', Gas Council, Midlands Research Station, Publication 83 (1966)
30 Davies, R. M., Lucas, D. M. and Toth, H. E., *Gas Council Res. Commun.* GC184 (1971)
31 Lawrie, W. B. and Ayres, E., 'High speed melting', *The British Foundryman*, **55** (1962) 74

32 Watson, D. and Glen, C. G., 'Thermal aspects of rapid melting of metals — 1', *Gas Council Res. Commun.* GC90 (1962); *J. Inst. Gas Eng.*, **3** (1963) 17
33 Glen, C. G., 'Thermal aspects of rapid melting of metals — 2', *Gas Council Res. Commun.* GC97 (1963); *J. Inst. Gas Eng.*, **4** (1964) 27
34 Lashley, H. and Roberts, S. S., 'Industrial application of a tower-type furnace for bulk aluminium melting', Scottish Section, Institute of Fuel, Meeting, Paisley (1967)
35 Pritchard, R. and Hough, E., 'A tower furnace for baled aluminium scrap recovery', forthcoming
36 Flux, J. H., 'Natural gas in heavy industry', Institution of Gas Engineers, Refresher Course, Pembroke College, Oxford (1969)
37 Wild, R., 'Fuel injection into the blast furnace', *J. Iron and Steel Inst.*, **205** (1967) 245
38 Stift, K. and Radinger, W., 'Practical experience in the use of natural gas in blast furnaces', Institute of Fuel, Conference on the Utilisation of Natural Gas, paper 15 (1968)
39 Blankley, W. E., private communication
40 Allan, J., 'The use of gas to partially replace coke in a cold blast cupola', Industrial Sales Department, Northern Gas, Report (1970)
41 Hirst, C. H. and Ward, T., 'The continuing development of natural gas sales to industry', *Inst. Gas Eng. Commun.* 936 (1974)
42 Arnold, G. D., 'Developments in the use of oxy/gas burners', *J. Inst. Fuel*, **40** (1967) 117
43 Pomfret, K. F., 'Developments in applications of gas for steelmaking', British Gas Corporation Midlands Research Station, Ext. Report 120 (1968)
44 Hansrani, S. P. and Richard, G. P., 'An evaluation of the effect of oxy-fuel techniques on the LD steelmaking process', *J. Inst. Fuel*, **45** (1972) 406
45 Hale, E., 'Application of natural gas in the manufacture of float glass', Manchester District Junior Gas Association, Meeting (April 1974); shortened version in *Gas Eng. & Manage.*, **15** (1975) 59
46 Whiteheart, J. J., Institute of Fuel, Symposium on the Use of Gasification Processes and Gaseous Fuel in Industry, Part I, The Glass Industry, Paper 9, Session 5
47 Cole, S. A., 'Existing and potential use of natural gas in the glass container industry', Yorkshire Junior Gas Association, Meeting (1971); abridged version in *J. Inst. Gas Eng.* **12** (1972), 179
48 Bullens, D. K., *Steel and its Heat Treatment*, Vol. 1, 5th edition. New York: Wiley (1948)
49 Grossmann, M. A. and Bain, E. C., *Principles of Heat Treatment*, 5th edition. Metals Park, Ohio: American Society for Metals (1974)
50 The Welding Institute, *Heat Treatment of Welded Structures* (1969)
51 British Standards Institution, *British Standard 5169 : 1975 Fusion-welded Steel Air Receivers*. London: BSI (1975)
52 British Standards Institution, *British Standard 1515 : 1965 Fusion-welded Pressure Vessels for Use in the Chemical, Petroleum and Allied Industries*. London: BSI (1965)
53 Gadd, E. R., 'Heat treatment of special materials', in *Heat Treatment of Metals* (ed. Institution of Metallurgists). London: Iliffe (1963)
54 'High velocity heating', *Flambeau* (1966)

55 Mitchell, E., 'An engineering concept of heat treatment' in *Heat Treatment of Metals* (ed. Institution of Metallurgists) London: Iliffe (1963)
56 Astley, I. and Merrett, W., 'Annealing and fluidized-bed quenching of Nimonic-alloy sheet', *Sheet Met. Ind.*, **39** (1962) 601
57 Pattison, J. R., *J. Inst. Fuel*, 41 (1968) 345
58 Bosworth, W. T., British Iron and Steel Research Association, Report PE/CONF/25/66 (1966)
59 Dessarts, P. and Fauvel, G., 'Gas burners used for roof firing furnaces for the reheating of steel' (in French), International Gas Union, 12th World Gas Conference, Nice, paper IGU/E 33-73 (1973)
60 Davies, P. A., 'Hot rods—naturally', Manchester Junior Gas Association, Meeting (March 1974); abridged version in *Gas Eng. & Manage.*, **15** (1975) 33
61 Lawrence, M. N. and Spittle, J., 'Forced-convection techniques for gas-fired rapid billet heating', *Gas Council Res. Commun.* GC108 (1964); *J. Inst. Gas Eng.*, **5** (1965) 515
62 Francis, W. E., Moppett, B. E. and Read, G. P., 'Studies of flow patterns and convection in rapid heating furnaces using model techniques', *Gas Council Res. Commun.* GC132 (1966); *J. Inst. Gas Eng.*, **7** (1967) 335
63 Lucas, D. M. and Davies, R. M., 'Mass transfer modelling techniques in the prediction of convective heat transfer coefficients in industrial heating processes', 4th International Heat Transfer Conference, Versailles (1970)
64 Lucas, D. M., Masters, J. and Toth, H. E., 'Prediction of the performance of rapid heating furnaces', *Gas Council Res. Commun.* GC151 (1968); *J. Inst. Gas Eng.*, **9** (1969) 397
65 Gas Council, 'Proceedings of the rapid heating teach-in, Midlands Research Station, April 1970', Midlands Research Station Ext. Report No. 182
66 Masters, J., Oeppen, B. and Towler, C. J., 'Rapid heating—a progress report', *Inst. Gas Eng. Commun.* 914 (1973); abstract in *J. Inst. Gas Eng.*, **13** (1973) 265
67 Knight, S. J. and Proffitt, R., 'Heating in the drop forge. 1. The case for town gas', *Met. Form.*, **33** (1966) 461
68 'Cooled skid rails', Brit. Pat. I 243 626 (1971)
69 Perry, J., 'Modern furnace techniques to reduce decarburisation', in *Decarburisation.* London: Iron & Steel Institute (1970)
70 Johnson, F. J., 'The use of gas in the control of decarburisation', in *Decarburisation.* London: Iron & Steel Institute (1970)
71 Johnson, F. J., 'Reheating furnaces for non-scaling and non-decarburization', Yorkshire Section, Institute of Fuel, Meeting (February 1967); abridged version in *J. Inst. Fuel*, **40** (1967) 468
72 Němeček, J., 'Zunderfreie Erwärmung in direkt beheizten Industrieöfen', *Gas Wärme International*, **16** (1967) 546
73 Delbourg, P. and Bourdil, C., 'Producing non-oxidising atmospheres through air deficient combustion in loose refractory beds', ATG Congress (1966)
74 Rosenthal, E., *Pottery and Ceramics.* London: Penguin (1954)
75 Gibeaut, W. A. and Rice, J. F., 'Effects of kiln atmosphere in ceramic firing', American Gas Association—Institute of Gas Technology, Conference on Natural Gas Research and Technology, Chicago, session 4, paper 3 (1971)
76 Gas Council, 'Gas in the heavy clay and refractories industries', Gas Council Report U221 (1968)
77 Davis, K. and Walker, L., 'Recent developments in the use of gas for ceramic

industry', Institute of Fuel and North Staffs. Fuel Society, Meeting (March 1958)
78 Gas Council, private communication
79 Davis, K. and Manuel, K., 'The development of the West Midlands Gas Board Trent kiln', *Inst. Gas Eng. Commun.* 669 (1964); *J. Inst. Gas Eng.*, **5** (1965) 419
80 Johnson, F. J., 'Natural gas in the heavy clay and refractories industries', International Gas Union, Symposium on Commercial and Industrial Applications of Natural Gas, Brussels (1974)
81 Gibb, J. A., 'Gas firing in the heavy clay industry', Northern Junior Gas Association, Meeting (May 1973)
82 Carlier, R., 'Progress in the use of gas in the heavy clay and refractories industries', Midland Junior Gas Association, Meeting (December 1973)
83 Johnson, F. J., 'Natural gas at Thomas Wragg works', *The British Clayworker* (March 1970); reprinted with title 'Natural gas for clay pipe manufacture', *J. Fuel & Heat Technol.*, **17,** No. 2 (March 1970) 19
84 Bray, K. F., private communication
85 *Memorandum on Chimney Heights.* London: HMSO (1963)
86 Hooper, K. D., 'Town gas in the glass industry in the South Lancs. group of the North Western Gas Board', Manchester Junior Gas Association, Meeting (1958)

14

Boilers and power generation

14.1 BOILERS: INTRODUCTION

Until the arrival of North Sea natural gas the British gas industry's share of the steam-raising market was very small. Although the number of gas-fired steam boilers was substantial they were individually of small output. The steam-raising market was dominated by coal until the 1950s when oil began to replace coal rapidly as the main fuel for industrial boilers. This change was partly due to the introduction of oil-fired packaged boilers which provided a compact unit with all the necessary ancillaries and combining high efficiency and automatic operation.

Coal-fired boilers are appreciably larger than their oil-fired counterparts; the furnaces need to accommodate grates, the heat-release rate is limited by the danger of smoke emission and there is considerable fuel–air stratification necessitating a further increase in furnace volume. An additional limiting factor in the case of shell boilers is the need to avoid high gas temperatures at the entry of the tubes producing ash fusion in the tube ends.

The effect of these factors has been to limit the furnace heat-release rate on coal firing to about 0.52 MW m^{-3} (50 000 Btu ft^{-3} h^{-1}) [1] for shell boilers and even less on water-tube boilers [2].

Oil firing reduces the mixing problems and those due to ash content. Therefore, much higher heat-release rates are practised with oil firing, typically 2.1–3.1 MW m^{-3} (200 000–300 000 Btu ft^{-3} h^{-1}) in the case of shell boilers [1]. Still higher heat-release rates are sometimes possible but are limited by the onset of stack-solids emission [3].

The availability of large quantities of natural gas has allowed the British gas industry to make appreciable inroads into the industrial steam-raising market. At the time of writing (1976) about 2000 industrial shell boilers and 250 water tube boilers have been changed over to natural-gas firing.

Gas firing is on an interruptible basis using dual-fuel burners (see Section 7.12). The vast majority of gas-fired steam boilers were originally designed for another fuel and in many boilers gas is the third fuel to have been used. Over the past few years, however, a growing number of new boilers have been designed specifically for gas–oil dual-fuel firing.

Gaseous fuels are much more easily mixed with air than are oil fuels and their combustion can be completed in a smaller volume. Consequently heat-release rates in shell-boiler furnaces can, in principle, be much higher than 3.1 MW m^{-3} (300 000 Btu ft^{-3} h^{-1}) but sufficient heat transfer must occur in the furnace to avoid excessive temperatures at the beginning of the second pass (see Figure 14:1). Heat transfer and gas and metal temperatures in shell boilers have been the subjects of detailed studies by the gas industry, the main aspects of which are presented in Section 14.3.2.

Section 14.2, which follows, describes in outline the main types of industrial steam boiler at present in use in Britain. In the case of boilers designed for coal or oil firing they are described and illustrated in their original form and their changeover to gas firing is considered in Section 14.4. A very broad division may be made between:

1 Fire-tube or shell boilers in which the hot gases are contained in tubes surrounded by a water space.
2 Water-tube boilers in which the water-filled tubes are surrounded by combustion products.

14.2 BOILER TYPES

14.2.1 Fire-tube or shell boilers

Lancashire boilers (Figure 14:1(*a*))
Two furnace tubes are employed in parallel, the second and third passes taking place around the outside of the shell contained in a brickwork setting. The shell generally has a diameter between about 1.8 m and 3 m (6 and 10 ft) and an overall length of between 5.5 m and 9.8 m (18 and 32 ft) [4, 5]. Steam outputs lie between 1400 and 6400 kg h^{-1} (3000 and 14 000 lb h^{-1}) and working pressures are between 0.83 and 1.24 MPa (8.3 and 12.4 bar, 120 and 180 lbf in^{-2}). The design of the external passes does not lend itself to high efficiency since the flues are sized with bodily access for cleaning as the primary requirement [4]. The brickwork setting imposes a high thermal inertia on the system and may allow hot-gas short-circuiting due to expansion cracking. If superheaters are fitted they are normally located at the entry to the second pass. Variants are the earlier and smaller Cornish boiler which has a single furnace tube and a maximum diameter of about 1.8 m (6 ft) and the 'Super-Lancashire' boiler in which the external flues are dispensed with and replaced by a nest of small-diameter fire tubes below the combustion flues.

Vertical shell boilers (Figure 14:1(*b*))
This rather rudimentary boiler has the advantage of taking up little floorspace. Variants exist incorporating horizontal fire tubes or no fire tubes at all. Outputs rarely exceed 2700 kg h^{-1} (6000 lb h^{-1}).

Economic boilers (Figure 14:1(*c*) and (*d*))
An Economic boiler for a given output occupies only about half the space of the corresponding Lancashire boiler [4]. It possesses one or two furnace tubes forming the first pass. The combustion products then reverse direction in the so-called combustion chamber which is constructed of refractory brickwork in 'dry-back' designs (Figure 14:1(*c*)) and is watercooled in 'wet-back' designs (*d*) and pass through small-

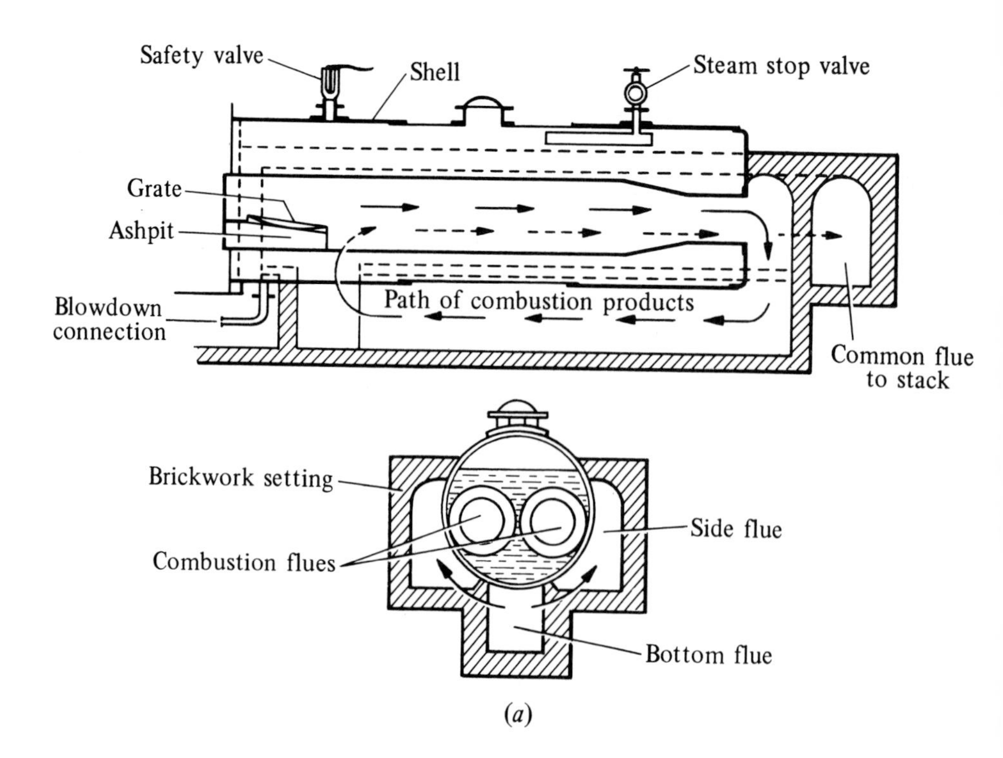

(a)

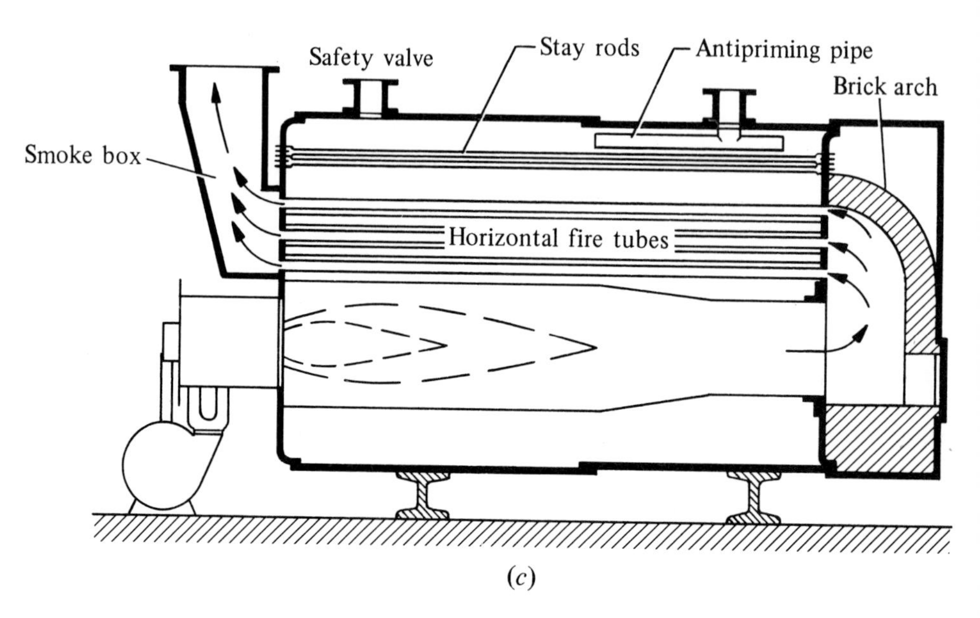

(c)

Figure 14:1

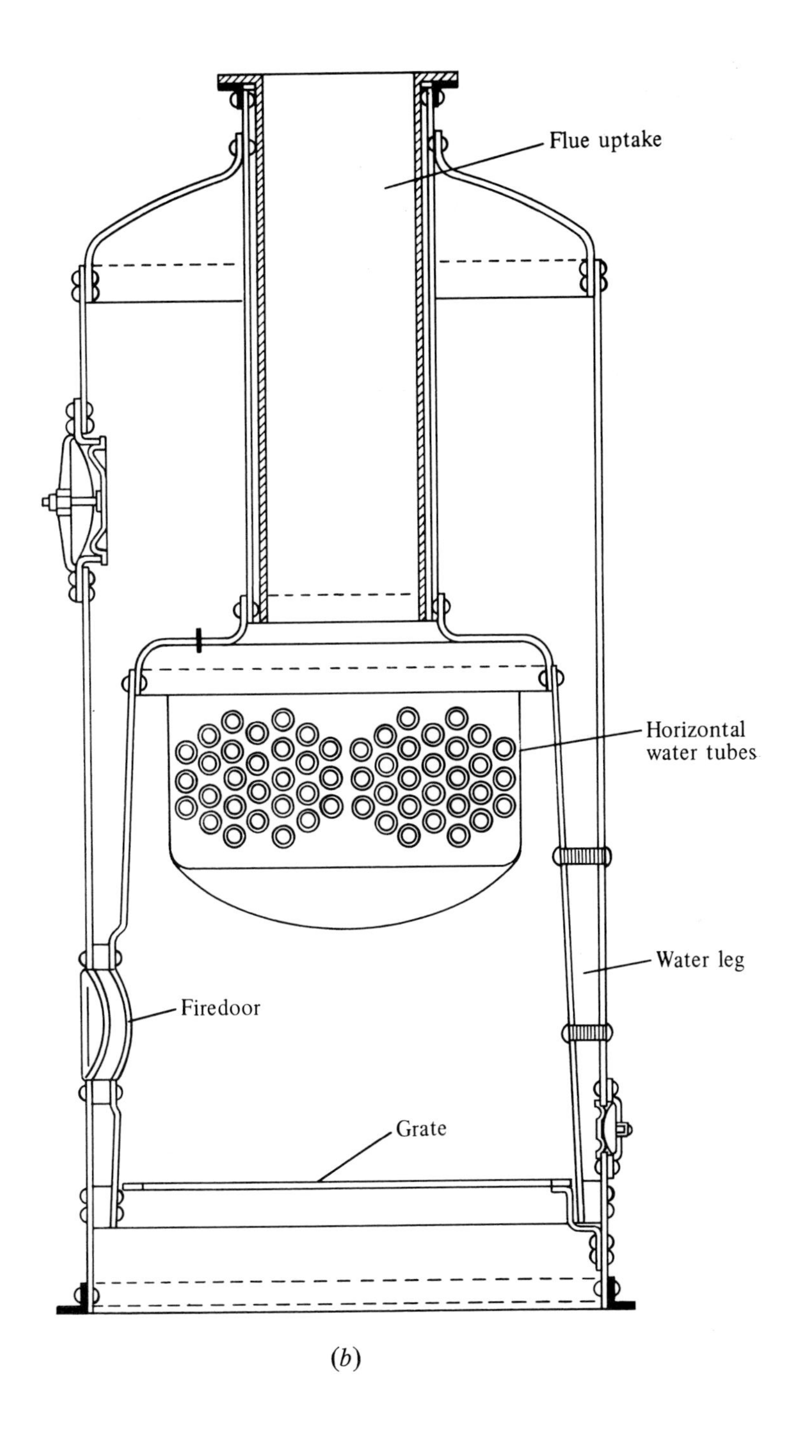
Flue uptake
Horizontal
water tubes
Water leg
Firedoor
Grate

(*b*)

Figure 14:1

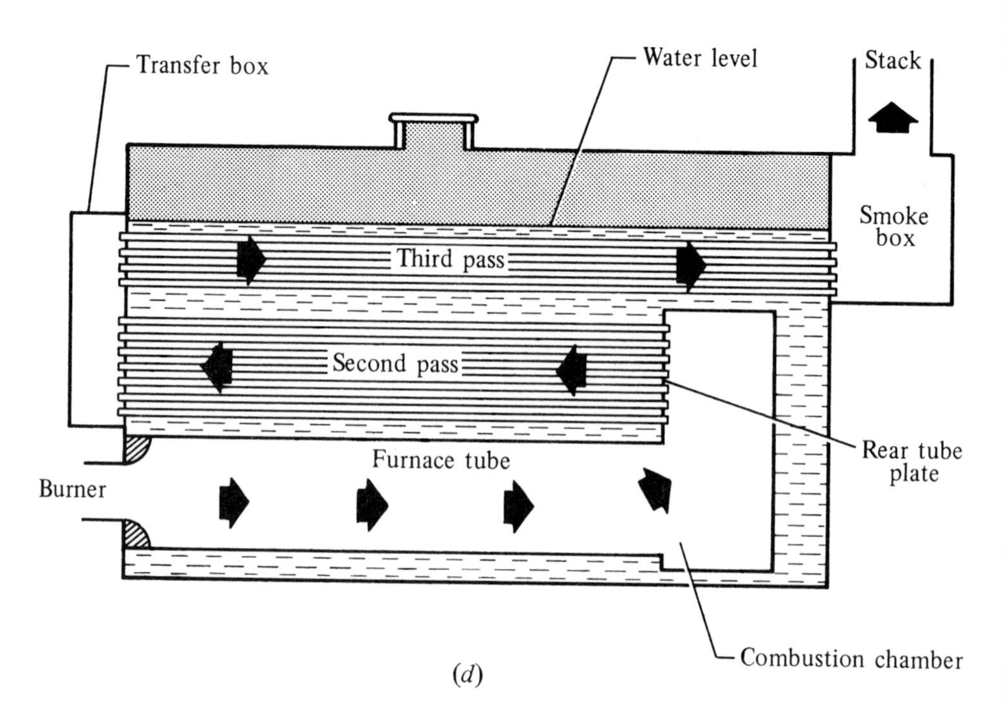

Figure 14:1 Types of fire-tube or shell boilers
(*a*) Lancashire boiler
(*b*) Vertical shell boiler
(*c*) Economic boiler, two-pass dry-back type
(*d*) Economic boiler, three-pass wet-back type

diameter fire tubes forming the second pass. A third pass, if included, similarly consists of a tube bundle. Dimensions typically range from 1.7 m to 3.7 m ($5\frac{1}{2}$ to 12 ft) diameter by 2.1 to 7 m (7 to 23 ft) long [5]. Outputs are between 230 and 23 000 kg h^{-1} (500 and 50 000 lb h^{-1}) at pressures up to 1.7 MPa (17 bar, 250 lbf in^{-2}). Economic boilers were originally designed for coal firing but in more recent years have become almost exclusively oil fired, normally in packaged-boiler form complete with burner, controls, etc., and skid mounted for ease of installation. Thermal efficiencies of over 75 per cent can be achieved without the need for economizers.

Multiple-fire-tube boilers
Small quantities of steam in the range 40 to 400 kg h^{-1} (88 to 880 lb h^{-1}) are commonly supplied by small horizontal or more commonly vertical fire-tube boilers designed specifically for gas firing. The fire tubes, up to a hundred or so in number and up to 50 mm (2 in) diameter are fired by individual burner jets. In vertical models the burner jets are frequently mounted on a swing-out header for ease of lighting. Vertical types are normally natural draught whilst fan-assisted draught may be necessary in the case of boilers with horizontal tubes [6]. The burner controls are of

necessity different to other boiler types due to the multiplicity of small flames. The majority have not been fitted with flame safeguards partly because of the inherent difficulties of doing so and partly because of the small volume and high strength of the individual combustion chambers. Control systems are currently being developed to provide semi- or fully automatic operation as with other boiler types [6].

14.2.2 Water-tube boilers

Industrial water-tube boilers

A shell boiler, because of its construction is normally limited to a maximum working pressure of about 2 MPa (20 bar, 300 lbf in^{-2}) because higher pressures require the use of abnormally thick shell plates. Evaporative capacity is also limited by constructional features to about 23 000 kg h^{-1} (50 000 lb h^{-1}). A water-tube boiler, on the other hand, which consists of a series of drums and headers connected by relatively small-diameter ($<$ 100 mm) water tubes, is capable of much higher working pressures and there is virtually no upper limit to the evaporative capacity.

Pressures up to 12.5 MPa (125 bar, 1800 lbf in^{-2}) are common in industrial practice although higher pressures are used in modern power stations (typically 16 MPa (160 bar, 2300 lbf in^{-2}) but a few supercritical units have been built by CEGB operating at up to 25 MPa (250 bar, 3650 lbf in^{-2}) [2]).

Water-tube boilers are not normally employed below about 20 000 kg h^{-1} (45 000 lb h^{-1}).

Water tubes can be shaped to a wide variety of boiler configurations but essentially water-tube boilers consist of a furnace or combustion chamber occupying about half of the total volume, the other half being termed the convection zone and usually accommodating the superheater. The geometry of the combustion chamber differs totally from that used in shell boilers: it is roughly cuboidal as opposed to long and cylindrical. Figure 14:2 shows a typical industrial water-tube boiler. The furnace chamber may be refractory lined or, more commonly in modern practice, completely surrounded by water tubes, i.e. 'water walls'. Whilst pulverized fuel has dominated water-tube boiler firing for CEGB power generation, oil firing replacing chain-grate stokers has been the pattern with industrial boilers.

Coil boilers

These small-scale boilers, also known as flash boilers or steam generators, may also be classified as water-tube boilers. Their stored-water capacity is very small and steam is available within a few minutes of lighting up. Evaporative capacities are normally in the range 200 to 3000 kg h^{-1} (440 to 6600 lb h^{-1}).

The combustion chamber is formed by the cylindrical space enclosed by the inner coil, the gases then flow multipass through successive layers of coils. The burner normally fires downwards and produces a short bushy flame appropriate to the combustion chamber. Gas-fired and light oil-fired versions are in use. As Robertshaw [6] has pointed out, careful attention is required to proportioning the fuel input to suit the water input and steam pressure. The burner turndown must provide the low input required for low steam flows and to prevent widely fluctuating steam pressures resulting from burner cycling. To provide rapid starting the prepurge air flows are minimized [6] but otherwise the controls are similar to those used on other packaged boiler burners.

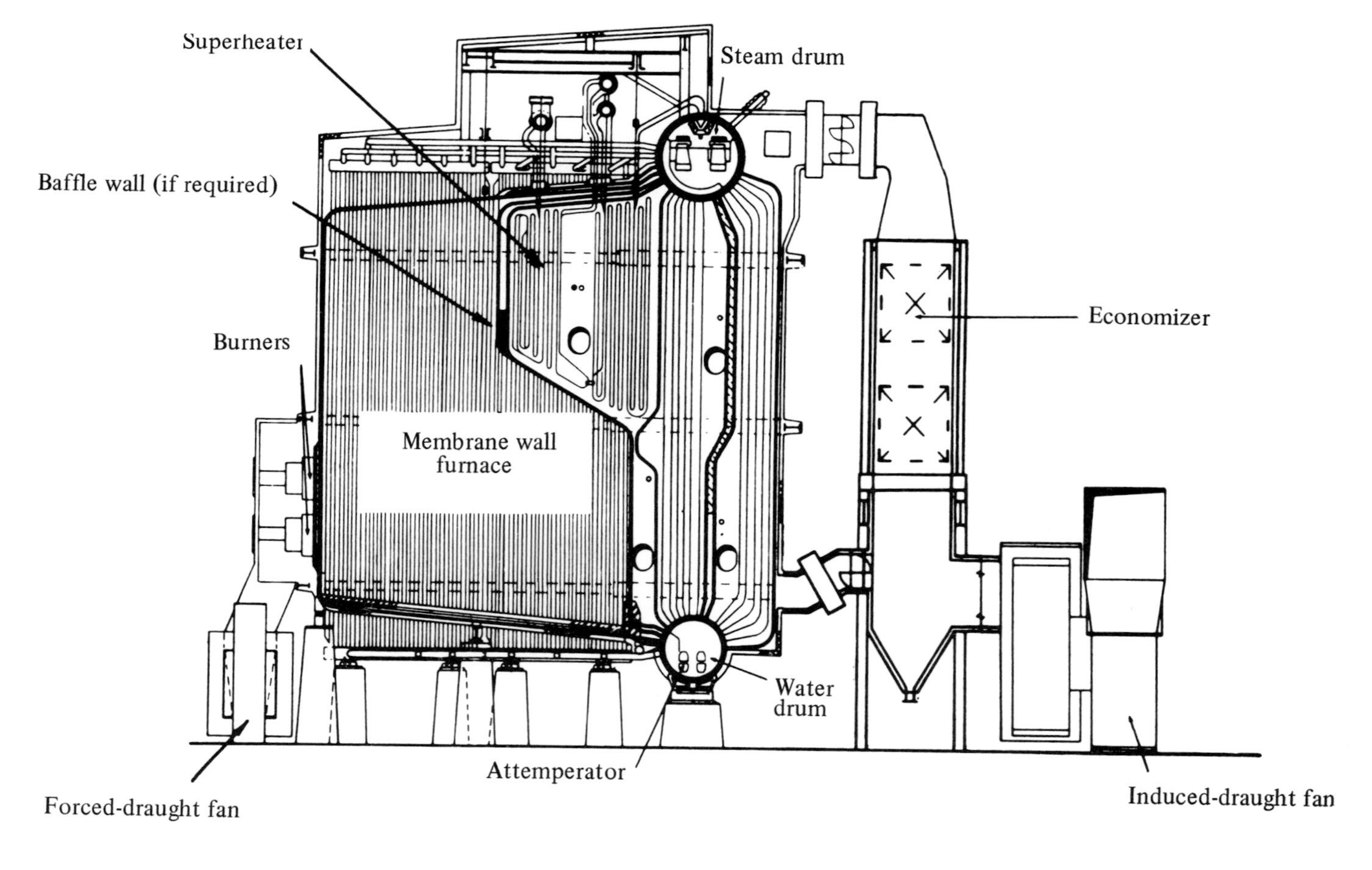

Figure 14:2 Typical industrial water-tube boiler

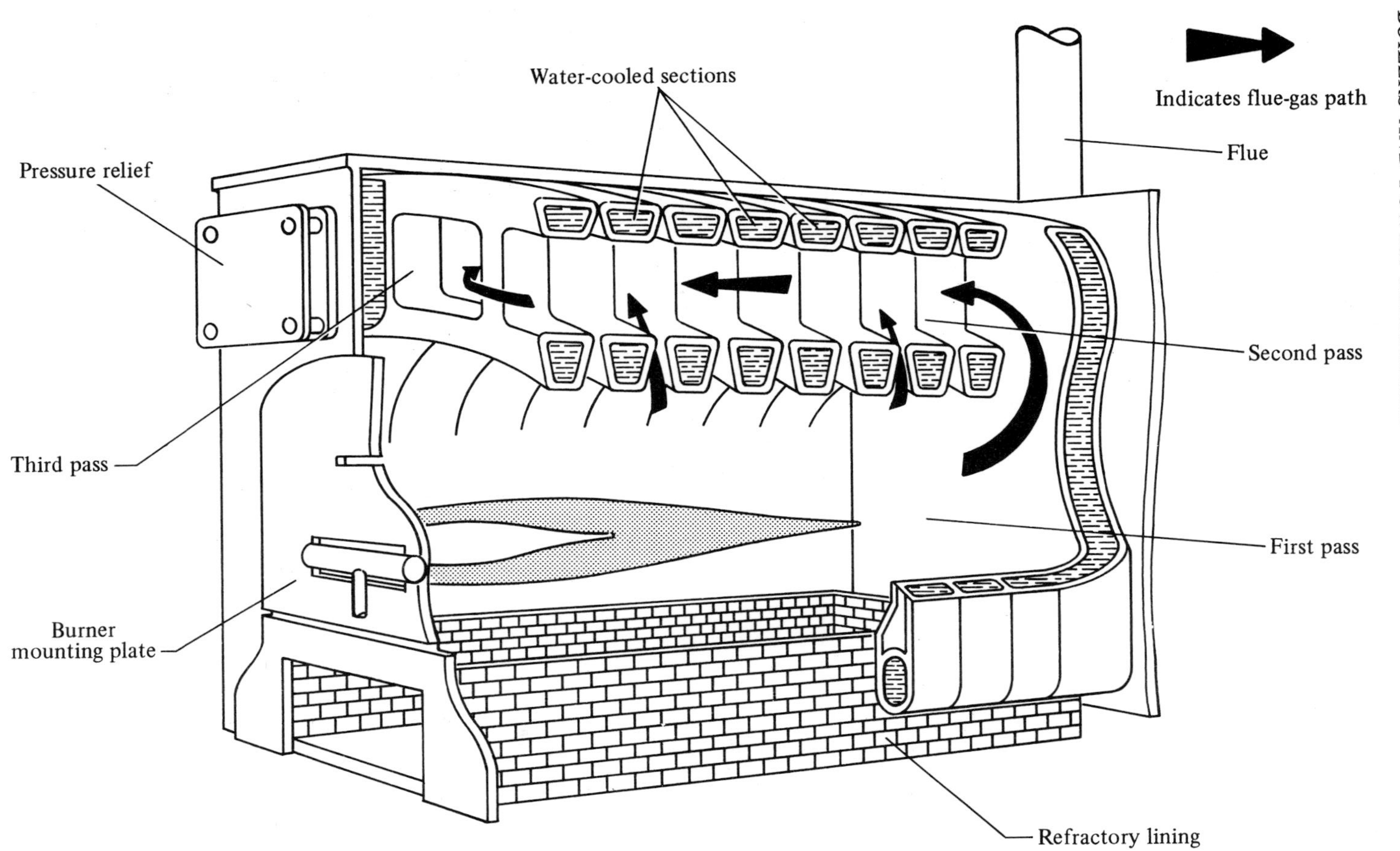

Figure 14:3 Cast-iron sectional boiler

14.2.3 Sectional boilers (Figure 14:3)

Sectional boilers, generally of cast-iron construction and designed for solid-fuel firing, are in wide use in commercial and industrial premises for the provision of hot water and in some cases for the generation of low-pressure steam. Ratings range up to a few thousand kilowatts.

In Britain, a large number of these boilers have been changed over to gas firing, particularly since 1969 when changeover was precipitated by the national shortage of smokeless solid fuel [7].

Their design for solid-fuel firing on natural draught has led to large grate areas, low combustion intensities and over-sized flueways. The changeover and possible uprating of sectional boilers are considered in Section 14.4.3.

Sectional hot-water boilers for central heating and the provision of hot-water services designed specifically for gas firing in the town-gas era were limited to fairly modest outputs and used natural-draught burners.

Current gas-fired designs employ natural-draught systems in the small sizes but make use of forced-draught automatic packaged burners in larger sizes permitting much higher heat-transfer rates and firing intensities. Figures quoted by Cox and Day [7] indicate that an overall heat-transfer rate of about 40 $kW\ m^{-2}$ (12 700 $Btu\ ft^{-2}\ h^{-1}$) would be expected in a modern boiler designed for forced convection compared with about 15 $kW\ m^{-2}$ (4800 $Btu\ ft^{-2}\ h^{-1}$) for a boiler designed for natural draught. The firing intensity may typically be an order of magnitude higher, e.g. 3 $MW\ m^{-3}$ (300 000 $Btu\ ft^{-3}\ h^{-1}$) compared with about 0.3 $MW\ m^{-3}$.

Modular boiler systems in which a number of central-heating boilers, usually of low individual thermal capacity, are controlled to fire in sequence are gaining acceptance as an attractive alternative to the traditional system of one or two large boilers for heating commercial premises.

14.3 CHANGEOVER OF BOILERS TO GAS FIRING

14.3.1 Comparison of boiler fuels

Figure 14:4 presents a comparison of the composition and combustion properties of three major boiler fuels and shows a number of significant differences. Important aspects are the higher carbon content of oil compared with natural gas (line 1) and the fact that the hydrogen content of gas is about twice that of oil and seven times that of coal (line 2). The difference in hydrogen content is reflected in the much higher water-vapour content of the combustion products (line 12), the greater disparity between gross and net calorific values (lines 9 and 10) and the higher dew point of natural gas compared with the other fuels (line 15). These figures indicate an apparent loss in thermal efficiency on changeover from oil firing because of the latent heat content of the additional water vapour in the flue gases. For the same conditions of excess air and flue-gas temperature the extra heat content of the combustion products of natural gas over heavy fuel oil is 4.7 per cent for a typical fuel analysis [8]. In practice this potential loss in efficiency is frequently more than offset by other considerations, e.g. fouling of heat-transfer surfaces, excess-air requirements, etc., as outlined later in this section.

The figures quoted for water dew point (line 15) take no account of the presence of

		Dry steam coal Washed smalls National Coal Board Rank Code No. 201	Medium fuel oil 950 seconds Redwood 1 (234 mm^2 s^{-1})	Natural gas North Sea (typical)
Ultimate fuel analysis by weight (%)				
1 Carbon (C)		77.4	85.98	73.57
2 Hydrogen (H)		3.4	11.08	23.94
3 Oxygen		2.0	0.67	
4 Nitrogen		1.2	0.07	2.49
5 Sulphur		1.0	2.17	negligible
6 Ash		8.0	0.02	—
7 Moisture		7.0	0.01	—
Total		100.00	100.00	100.00
8 $\frac{H}{C}$ by weight		0.0439	0.1288	0.3254
9 Gross calorific value	(Btu/lb)	13 150	18 935	23 074
	(kJ/kg)	30 538	43 972	53 584
10 Net calorific value	(Btu/lb)	12 750	17 885	20 796
	(kJ/kg)	29 609	41 534	48 294
11 Weight of stoichiometric air required for combustion	(lb/gross therm)	76.7	72.8	71.9
	(kg/gross therm)	34.8	33.0	32.6
12 Weight of water vapour in combustion products	(lb/gross therm)	2.86	5.27	9.35
	(kg/gross therm)	1.30	2.39	4.24
13 Stoichiometric CO_2 content of dry flue products	(%)	18.9	15.6	11.7
14 SO_2 emission to atmosphere	(p.p.m. w/w)	1644	2940	negligible
15 Water dew point of flue products	(°F)	95	120	133
	(°C)	35	49	56

Figure 14:4 Boiler fuels: composition and combustion properties [5]

sulphur oxides in the waste gases which elevate the dew point. Whilst the water dew point of a fuel oil may be as low as 50 °C the acid dewpoint is typically 120–140 °C depending on the fuel's sulphur content [5]. In the case of coal the figure is lower because of the lower sulphur content of the coal and the fact that some 10 per cent of the sulphur remains behind in the ash. Clearly the acid dew point governs the final flue-gas temperature since if condensation occurs then corrosion and fouling will result. In addition, acid smut emission may occur. With natural-gas firing, exit temperatures may be 50 °C lower than with oil firing and this strengthens the case for adding economizers to boiler plant for which they have not hitherto been considered worthwhile. It is necessary to incorporate a bypass to prevent the economizer suffering corrosion on oil firing during a period of interruption. Elles [9] has estimated the economic benefits of adding an economizer to a wet-back shell boiler.

14.3.2 Heat release and transfer in boilers

In broad outline the essential difference between the furnace of a water-tube boiler and that of a shell boiler is that the former is roughly cuboidal and the latter a long, relatively narrow cylinder. The flame shape must be well matched to the combustion-chamber geometry and the combustion intensity should be such that the flame volume is somewhat less than the effective volume available. In comparing gas and oil as a boiler fuel, as a general rule oil-firing intensities tend to be limited by the peak heat flux acceptable and with gas the limit is set mainly by the gas exit temperature [10]. The exit temperature is important because of the possibility of excessive tube-end temperatures in shell boilers and differences in superheat conditions in dual-fuel water-tube boilers.

In Lancashire boilers and older Economic boilers the low firing intensity involved (0.3 to 1 MW m^{-3}) has presented no problem with regard to metal temperatures and the same may in general be said for water-tube boilers, most of which have similar firing intensities. In the case of modern packaged shell boilers with firing

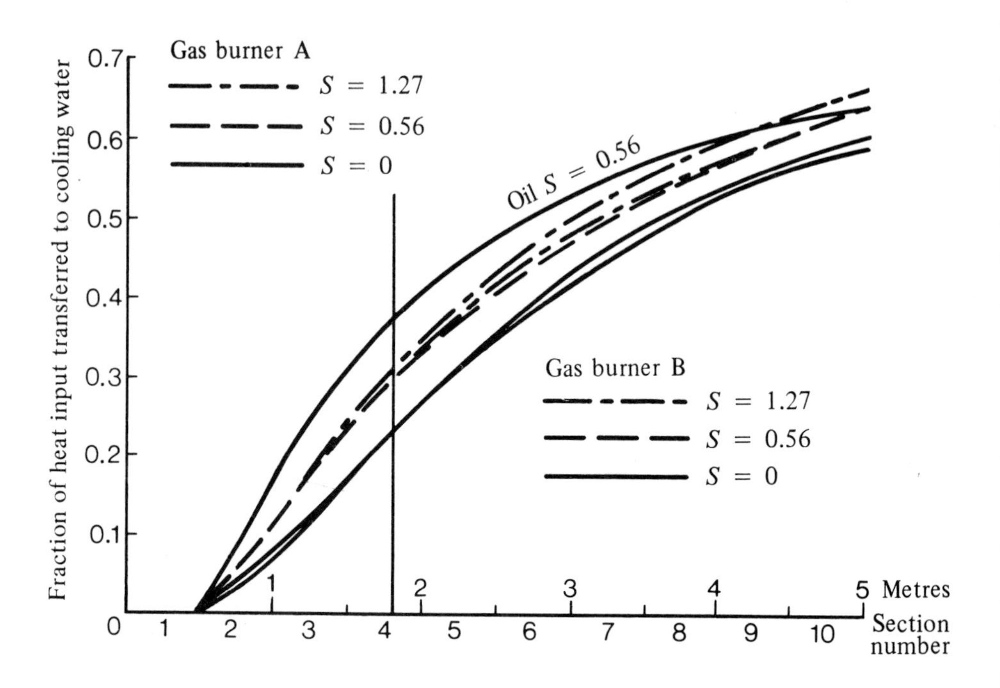

Figure 14:5 Accumulative heat transferred from flames of various fuel and swirl combinations [1, 28]

intensities in the range 1.5 to 3 MW m^3 (150 000 to 300 000 Btu ft^{-3} h^{-1}) despite concern in the early days of changeover that this might lead to excessive back-end temperatures, extensive work by BGC and some boiler manufacturers has shown that, provided a suitable burner is employed, back-end metal temperatures are only marginally greater on gas firing than on oil.

The crucial difference between oil and gas boiler flames is that the former are usually highly luminous whilst gas flames tend to be nonluminous. Thus the heat-transfer pattern in the boiler furnace differs appreciably in the two cases, considerably more heat being transferred (by luminous radiation) from the oil flame. This in principle would be expected to lead to higher gas temperatures at the entrance to the second pass and the possibility of overheating the tube plate and tube ends. A number of boiler manufacturers recommended downrating boilers on changeover by 15 to 40 per cent to prevent this occurring [10].

On the other hand, gas flames have a somewhat higher nonluminous emissivity and the higher water-vapour content increases specific heat capacity and thermal conductivity thereby increasing convective heat transfer. Figure 14:5 [27] shows the progressive heat transfer from oil and gas flames of different swirl numbers (Section 4.8.1) along a water-cooled experimental tube furnace of 0.6 m (2 ft) diameter. Although the experimental furnace differs in a number of respects (in particular length-to-diameter ratio) from practical shell boilers the results serve to illustrate the differences in fire-tube heat flux for gas and oil firing.

A comprehensive theoretical study of shell boilers has been made based essentially on a one-dimensional plug-flow model for the furnace tube and the assumption that the combustion chamber (i.e. the reversal chamber between the first and second passes) can be represented by a single well-stirred zone [10, 11]. Both luminous and nonluminous flames are considered and allowance is made for progressive combustion down the length of the fire tube. (Mathematical modelling of heat transfer in boilers is considered in Chapter 12.) Good agreement has been obtained between theoretical predictions and measurements taken on practical boiler trials. By the end of 1973 a total of 60 boilers had been tested covering 35 combinations of boilers and burners [8]. Field-trials techniques have been refined and the reader is referred to [12] for a detailed description of the measurement of boiler-plant temperature.

Detailed field trials on an 8600 kg h^{-1} (19 000 lb h^{-1}) packaged three-pass wet-back boiler have been reported in detail by Horsler and Lucas [10]. Some results were obtained at 145 per cent MCR (maximum continuous rating) and under these conditions the firing rates are as high as or higher than those known to exist in any current boiler. Figures 14:6(*a*), (*b*), (*c*) and (*d*) show respectively gas temperatures at various points in the boiler on gas and oil firing; tube-end temperatures on gas and oil firing; the effect of excess air on combustion products and metal temperatures for natural-gas firing and the same for oil firing.

It is clear from Figure 14:6 that although the temperature of the combustion products is about 150 °C higher at the end of the firetube on gas firing the corresponding metal temperatures differ by only about 20 °C even at 45 per cent overload. Clearly on this basis there is no necessity for downrating. An increase in excess air level modifies the heat transfer in three ways: the flame temperature is reduced, the flame emissivity is reduced by decreasing the concentration of radiating species and, since the mass flow is increased, the convective heat-transfer coefficient is increased. Figure 14:6(*c*) and (*d*) show the effect on gas and metal temperatures and it is evident that the metal temperatures remain almost constant with both fuels. A few practical

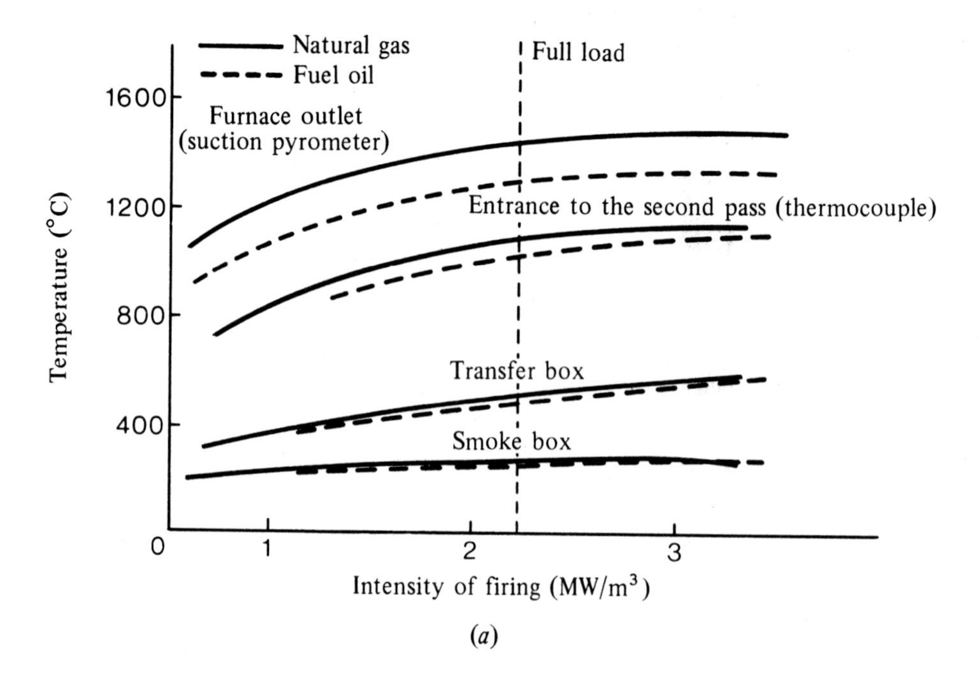

(a)

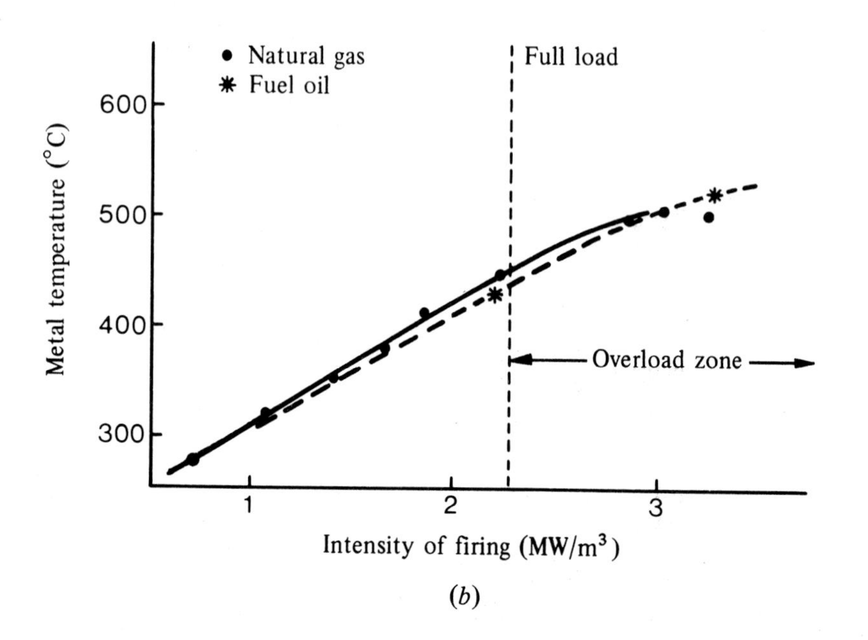

(b)

Figure 14:6

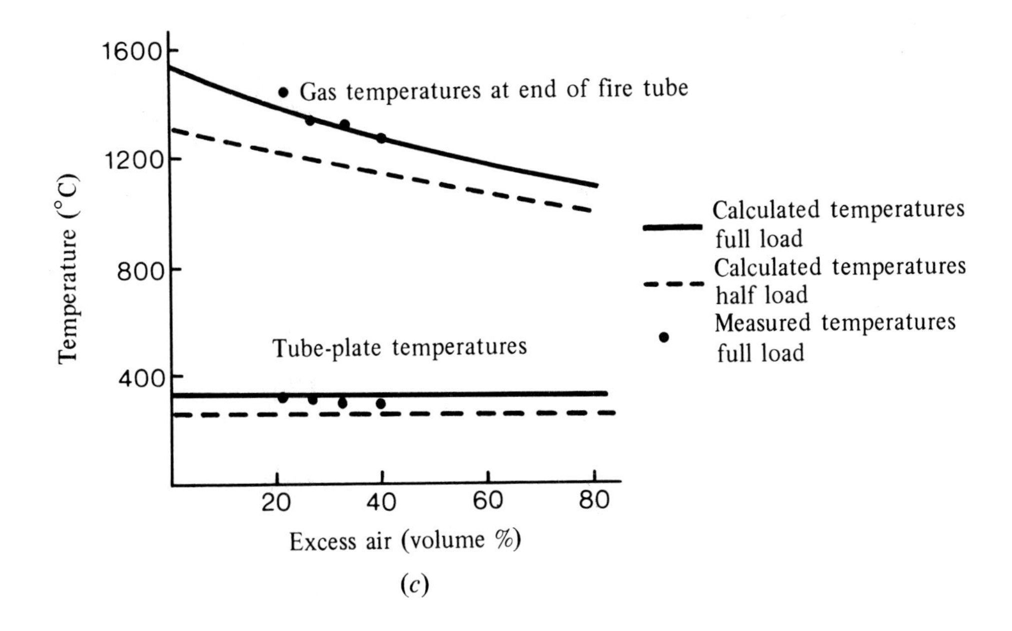

(c)

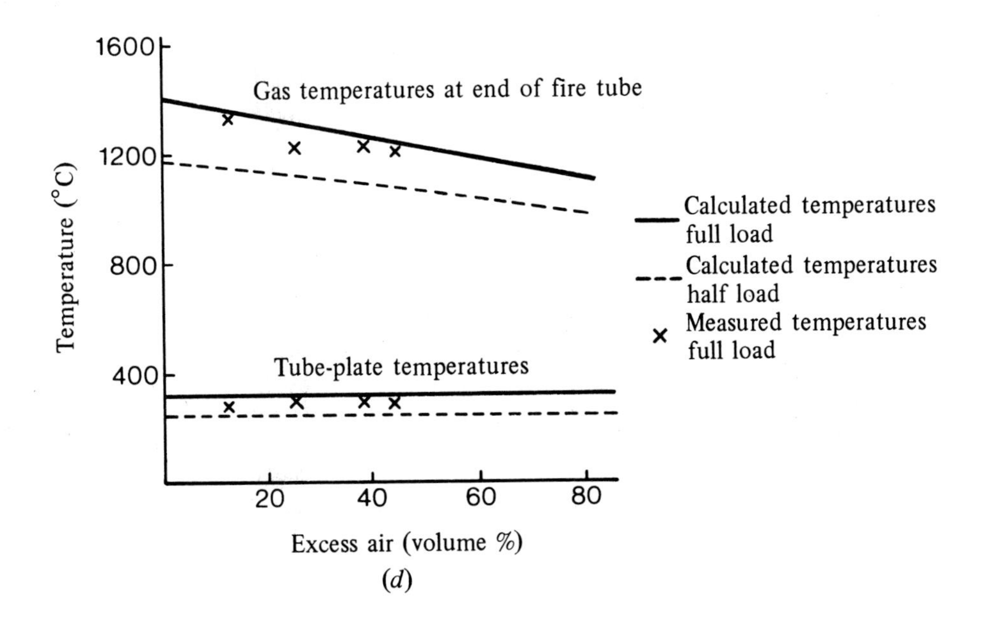

(d)

Figure 14:6

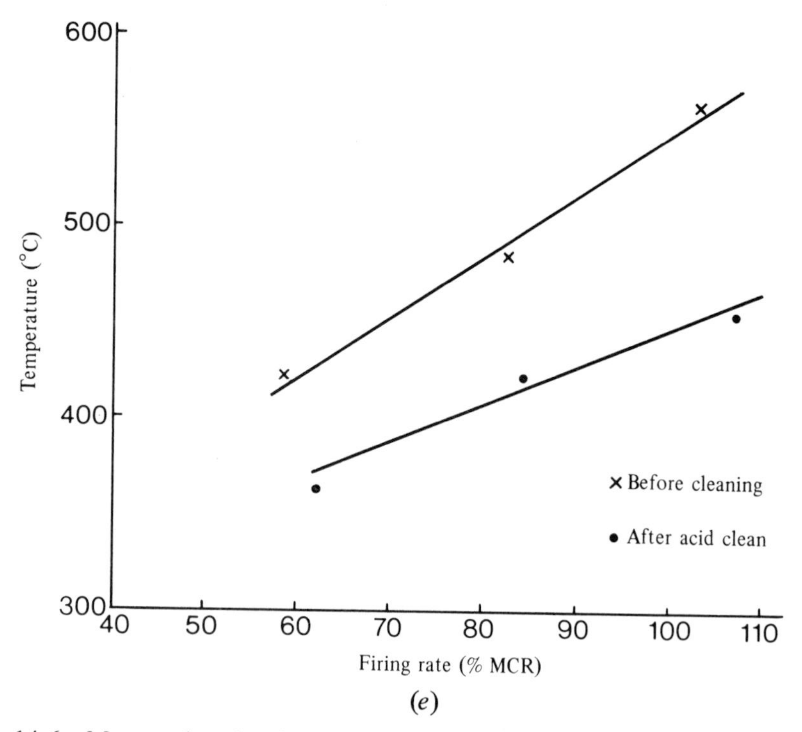

Figure 14:6 Measured and calculated gas and metal temperatures for a packaged boiler of high firing intensity [10]

(*a*) Gas temperatures at various points in the boiler
(*b*) Comparison of tube-end temperatures on natural gas and fuel oil
(*c*) Effect of excess air on gas and metal temperatures for natural-gas firing
(*d*) Effect of excess air on gas and metal temperatures for oil firing
(*e*) Average tube-end metal temperatures before and after cleaning

instances have occurred where the temperature of the furnace-tube exit gases has been substantially higher than calculated. These have been attributed to either the use of an excessively long narrow flame or to poor air distribution across the furnace-tube cross-section. In all cases modification of the burner has led to acceptable back-end temperatures.

In dry-back boilers, reradiation from the refractory back and the reduced area of water-cooled surface tend to increase the heat flux to the tube plate and tube ends. Consequently, firing intensities for such boilers tend to be kept to fairly low levels. Even with highly rated dry-back boilers, however, Horsler and Lucas [10] concluded that changeover presented no cause for concern provided that unsatisfactory burners are avoided. Although the differences between fuels produce differences in metal temperature of only a few degrees, variations in water-side conditions can produce considerable temperature differences. Results have been obtained at 16 per cent excess air for a boiler known to be heavily scaled on the water side before and after an acid clean [8]. The results are shown in Figure 14:6(*e*) which indicates a difference in tube-end temperature before and after cleaning of 94 °C at 100 per cent MCR and

60 °C at 70 per cent MCR. These differences apply regardless of the fuel used and emphasize the importance of paying attention to the water-side conditions.

In the case of water-tube boilers it is more difficult to carry out trials in the same detail as has been possible with shell boilers. In general, however, the temperature of the gas leaving the combustion chamber tends to be higher than on oil firing [8], how much higher and the effect on superheat temperature differing considerably from boiler to boiler and depending on the firing conditions employed. There have been a few isolated cases of increased superheat temperature on changeover. Changeover from coal firing is not likely to give rise to excessive increases in superheat temperature because, although the gas temperature at the outlet of the radiant section is higher, the mass flow through the superheater tubes will be lower and the two effects tend to balance out [8]. Additionally, on gas firing the water-walls and superheaters stay cleaner and the heat-transfer conditions stay constant between annual shutdowns.

The effect on superheat temperature of a changeover from oil depends to a large extent on the amount of excess air used on oil firing. On boilers fitted with attemperators there have been no difficulties. Miller [8] has reported that in only one case (in which de-superheating equipment was not fitted) an excessive increase in superheat temperature occurred but firing a small amount of oil (5–10 per cent of the total input) together with the gas eliminated the difficulty.

14.3.3 Economic aspects of changeover

It is necessary to consider three main cost elements: capital cost, operating costs—e.g. labour and maintenance charges—and fuel price. The comparisons drawn in this section are mostly with oil firing since by far the majority of boiler changeovers have been from oil and, in the few cases of water-tube boilers changed over from coal the business was gained in competition with oil. A consideration of capital and fuel costs is inappropriate here since prices are continually changing. However, some generalized comments regarding operating costs may be made.

Operating costs may be compared on the basis of:

1 Efficiency.
2 Labour.
3 Auxiliary power and steam—for atomization, fuel handling, etc.
4 Maintenance, including boiler cleaning, refractory replacement and mechanical maintenance of burners and fuel-preparation equipment.

A full evaluation of factors (1) to (3) usually indicates an equivalent gain in efficiency which reduces and in many cases outweighs the inherent efficiency deficit referred to earlier.

Efficiency

As indicated in Section 14.3.1 there is a theoretical efficiency deficit for natural-gas firing compared with oil of around 5 per cent for identical firing conditions. However, identical firing conditions do not occur in practice as a result of a number of factors. Since natural-gas combustion products contain no particulate matter there is no fouling of heat-transfer surfaces and no consequent progressive increase in flue-gas temperature. The effects of fouling can be very rapid, a rise of 60 °C in flue-gas exit temperature two weeks after cleaning an oil-fired fire-tube boiler has been recorded [13].

The quantity of excess air used clearly has a significant effect on thermal efficiency. Gas burners can be operated satisfactorily at lower excess-air levels than their oil counterparts. Typical operating levels are quoted in Section 7.12.5. In addition, control of fuel–air ratio is easier to achieve with gas firing and freedom from problems such as nozzle wear and viscosity swings facilitates long-term flow and ratio stability.

Fuel-oil firing leads to the emission of stack solids, i.e. ash together with carbonaceous material, which in addition to being an atmospheric pollutant represents a fuel loss. Stack-solids emission is covered by the Clean Air (Emission of Grit and Dust from Furnaces) Regulations 1971 [3], which permit a maximum of 0.4 per cent of the fuel burned for boilers of MCR up to 40 000 lb h^{-1} (18 000 kg h^{-1}) reducing progressively to just below 0.2 per cent for boilers up to 475 000 lb h^{-1} (215 500 kg h^{-1}).

The practical effect of these factors is that the difference in flue-gas heat content between oil and gas firing reduces to about 2.5 per cent from the theoretical value of 4.7 per cent [8] in the case of shell boilers. For water-tube boilers using similar air–fuel ratio control systems in well operated and maintained systems the difference approaches the theoretical value but 2.5 per cent is a more representative value [8]. In some instances where changeover has been from a poorly controlled oil system direct improvements of up to 6 per cent have been achieved.

Not surprisingly, changeover from coal firing has resulted in significant improvements in efficiency; improvements of around 6 per cent have been obtained on water-tube boilers and even higher for shell boilers.

Labour

Savings in labour costs vary widely with the type and size of boiler plant involved. They range from zero in the case of boilers which were fully automatic and unattended prior to changeover to around 10 per cent equivalent improvement in efficiency for previously coal-fired boilers [8].

Auxiliaries

Auxiliary power and steam requirements are reduced on changeover and this may represent a significant cost saving. The cost items include power (or steam) for oil atomization, fuel preparation and pumping and steam for soot blowing. Compared with oil firing the total saving in fuel costs is normally between 1 and 1.8 per cent of the fuel cost equivalent to an increase in efficiency of around 1 per cent in the general case [8].

Maintenance

Maintenance costs are usually lower on gas firing than on oil and appreciably lower than on coal firing. Figures quoted by Miller [8] indicate equivalent efficiency gains of up to 2 per cent compared with heavy fuel oil and 4 per cent compared with coal.

14.3.4 Burners and boiler modifications

Shell boilers

For shell-boiler firing the burners employed are almost exclusively integral-fan packaged dual-fuel burners (see Section 7.12) rated between 1 MW and 15 MW (3.4×10^6 Btu h^{-1} and 50×10^6 Btu h^{-1}). Normal supply pressures are usually employed but some high-rated burners require boosted gas supplies. The combustion air pressure available needs careful selection to allow for the boiler pressure drop. In the case of Economic boilers the changeover is relatively simple, normally involving a replacement of the existing oil burner without alteration to the boiler itself, although in a few instances spiral turbulators have been installed in the second-pass tubes to increase convection [5]. Consideration should always be given to dressing back the tube ends flush with the tube plate for boilers of high firing intensity and for those known to have a history of poor water treatment and tube-end failure. To compensate for inadequate superheat temperatures in a particular design of three-pass dry-back boiler Greenhalgh and Haslam [14] have described a procedure of reducing the diameter (by a refractory brick ring) of the end of the fire-tubes by 380 mm (15 in) to increase the velocity of the gases impinging on the superheater elements. This resulted in a superheat-temperature increase of about 40 °C on oil and 70 °C on gas both at MCR.

In the case of Lancashire boilers there is more scope for improvement in performance on changeover and a number of different approaches have been adopted in addition to the simple method of firing down each fire tube and using the conventional flue system. Speakman [15] has described modifications in which a series of refractory chequered baffles were built in the fire tube reducing the cross-sectional area by up to 65 per cent or alternatively wedge-shaped piers were used, again increasing the gas velocity and directing the hot gases onto the sides of the fire tubes. These procedures led to an increase of about 8 per cent in the fire-tube heat transfer and $1\frac{1}{2}$–2 per cent in boiler efficiency.

Boiler modifications to permit recirculation of hot gases from the end of the second pass to the inlet end of the fire tubes have been made [5] resulting in an improved thermal efficiency. More extensive modifications are involved in another method outlined in [5] which dispenses with the need to use the external flueways forming the second and third passes. In this system a single burner fires down one fire tube, the front end of the other being blanked off. The combustion products return along the second tube which is divided down the centre to form the second and third passes. The main advantages of this method are that it eliminates the inherent gas-leakage difficulties associated with Lancashire boilers and roughly halves the internal volume to be purged on start-up.

Although dual-fuel operation on gas and oil only has been considered so far there have been a limited number of changeovers from solid-fuel firing in which coal has been retained as the secondary fuel.

Elles [9] has described the changeover of Lancashire and Economic boilers fitted with either chain-grate or low-ram coking stokers. On gas firing the grate is protected by firebricks fed onto the grate at kindling speed until completely covered. On coal firing the gas burner is retracted and coal is fed onto the grate beneath the ignition arch which is already hot from gas firing. The grate travel discharges the firebricks into the bottom of the fire tube, the whole procedure taking about $1\frac{1}{2}$ hours.

Water-tube boilers

Most water-tube boilers are fired by register burners in the range 3 to 20 MW (10×10^6 to 68×10^6 Btu h^{-1}) for industrial purposes rising to about 60 MW (204×10^6 Btu h^{-1}) for power-station boilers [6]. The combustion-air supply ducts (windboxes) are normally incorporated into the boiler walls and are generally supplied from a single large fan rated to provide air at 0.5–2 kPa (5–20 mbar, 2–8 in H_2O) above the combustion-chamber pressure [6]. Register burners are considered in Section 7.12.5.

An important and somewhat neglected aspect of register-burner design is the distribution of combustion air in the windbox. This is particularly so for large watertube boilers in which it is necessary to operate at very low excess-air levels (less than 5 per cent to avoid SO_3 formation on oil firing).

Greenhalgh and Haslam [14] have described in detail the changeover of watertube boilers using coal as the secondary fuel. In this instance the grate was successfully protected on gas firing with 100 mm (4 in) of coal ash, the temperature of the grate being kept down to 300 °C.

Sectional boilers

Solid-fuel and oil-fired cast-iron sectional boilers are normally changed over to gas firing by the addition of forced-draught packaged gas burners of a wide variety of different designs. Although the use of different types of burners does not affect the boiler performance to any great degree some of the designs result in boiler–burner combinations of lower noise level [6]. Since the boilers were designed for solid-fuel firing the gas ways are oversized and the combustion-chamber pressure is essentially atmospheric. The chimney draught needs to be broken either by the use of a draught stabilizer or by admitting diluent air into the flue [6]. To ensure an adequate thermal efficiency the excess-air level should be controlled to less than 25 per cent and this is achieved by sealing the boiler doors or preferably replacing them by a suitable pressure-relief panel (see Section 9.3.3). Details of the recommended design of composite relief panel are given in [16].

Particularly in the smaller sizes natural-draught burner systems are also used in the changeover of sectional boilers. Draught stabilization is particularly necessary in these cases to maintain efficient operation.

Although natural-draught burners have advantages of virtual silence and simplicity of maintenance these are usually outweighed by the advantages of forced-draught systems including:

1 The air flow through the boiler when it is not firing is much less and this can lead to an increase in thermal efficiency of 2 to 5 per cent [16].
2 The performance is affected much less by flue conditions. The excess-air level with atmospheric burners is commonly set at between 30 and 40 per cent whereas forced-draught burners operate satisfactorily with lower levels. The effect of this when running at maximum rating can amount to about 2 per cent difference in thermal efficiency [16].

As indicated earlier the heat flux in sectional boilers designed for solid-fuel firing is very low, around one-third to one-half of that applying in boilers designed for gas firing. This is due primarily to the low combustion-product velocities over the heat-transfer surfaces. Dramatic increases in heat transfer and consequently thermal rating may be obtained on changeover using a forced-draught gas burner by reducing the cross-sections of the second and/or third pass while still retaining the heat-transfer surfaces exposed. Cox and Day [7] have studied the effects of placing blanked-off tubular-steel inserts of both circular and rhomboidal cross-section in the second pass of a number of cast-iron sectional boilers.

The inserts were sized to produce increases in second-pass mean velocities up to 450 per cent and a typical set of results is shown in Figure 14:7. As Figure 14:7 indi-

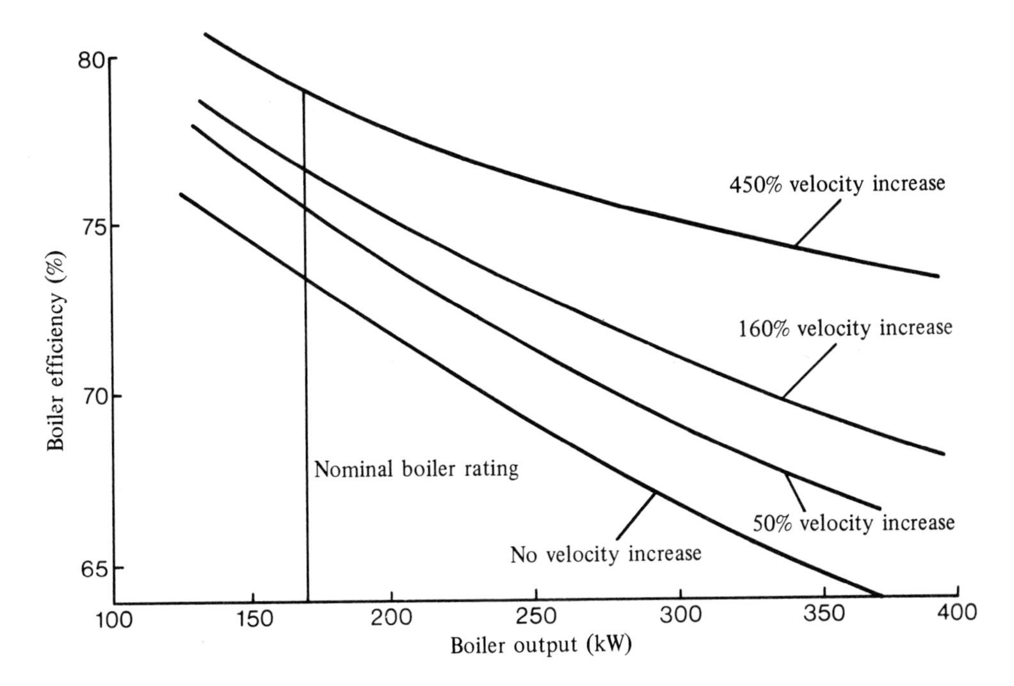

Figure 14:7 Effects of increasing the velocity in the second pass of a sectional boiler (Beeston New Senior boiler using a Langdon Kingsway NG 110 burner) [7]

cates, a 450 per cent increase in velocity increases the efficiency by about 5 per cent at nominal rating or alternatively for the same efficiency the boiler may be operated at about double its nominal rating. Increasing the velocities in this way pressurizes the combustion chamber, for example the pressure drop across the boiler was 88 Pa (0.88 mbar, 0.35 inH_2O) at 160 per cent velocity increase and 378 Pa (3.78 mbar, 1.52 inH_2O) for 450 per cent increase for the same boiler [7]. Increases in third-pass velocity are far less effective due to the lower gas–metal temperature differential. Since uprating leads to pressurization of the boiler gas ways the risk of combustion-product discharge into the boiler room must be guarded against by thorough boiler preparation. In addition it is necessary to protect the back end of the combustion chamber from flame impingement and to avoid hot-gas short-circuiting by sealing the gaps between the combustion chamber and the second and third passes to maximize flue-gas travel.

14.4 LARGE BOILERS: INSTALLATION, SAFETY AND CONTROLS

The principles of safe operation and control are considered in Chapter 9 where the technical standards relating to safety are reviewed. The purpose of this section is to review briefly current practice in their application to the firing of large boilers which by virtue of their high thermal rating and low operating temperature require particular care with regard to safety. Hancock and Spittle [17] reviewed current practice in 1972. The presentation here is largely taken from this paper to which readers are referred for a comprehensive treatment.

14.4.1 Gas supply systems

For large boilers the gas is generally supplied at higher pressures than normal, the usual range being 15 to 150 kPa (150 mbar to 1.5 bar, 2 to 20 lbf in^{-2}). This enables smaller pipes and valves to be used but the minimum pipe size is ultimately limited by high gas velocities leading to noise and possibly pipe erosion. Reference [18a] lays down maximum permitted velocities.

The engineering and installation standards must be stringent. Reference [18a] requires welded pipework and flanged connections for pressures above 100 kPa (1 bar, 14.5 lbf in^{-2}) for all pipework and above 5 kPa (50 mbar, 0.725 lbf in^{-2}) for pipework of 100 mm (4 in BSP) and above. Welding quality is specified in [18a]. There is also a move towards steel-bodied valves because of the possible risk of fracture of cast iron. Pipe and valve layouts should avoid the risk of accidental damage and allow easy access and maintenance. Pipework commissioning should always include cleaning, pressure testing and purging (Section 9.2).

The service-governor house is normally separated from the boiler house in its own enclosure. The service-governor installation [19] includes protection against governor failure giving high outlet pressures normally by fitting an active governor and a monitor governor in series with the monitor set at a slightly higher pressure. To protect against the governor system failing closed, and for maintenance, two or more active and monitor streams may be installed in parallel. For protection against more serious failures over-pressure relief valves or slam-shut valves may be fitted.

Metering is usually by rotary positive displacement meters or by turbine meters (see Sections 8.5.3. and 8.5.4). Due to the very small clearances in some meters these and other control equipment should be protected from dirt by filters on the inlet to the governor installation. The Interim Codes of Practice [18a] specify acceptable gas velocities relative to the degree of filtration employed. The boiler house must be protected by a manual isolating valve and where high line pressures involve the storage of a large volume of gas in the pipework in the boiler house, means must be provided for venting this safely to atmosphere.

14.4.2 Supply safeguards

The performance of safety shutoff valves and systems is considered in Sections 9.4.4 and 9.4.5. Large burner systems require safety shutoff systems incorporating either pressure proving or position checking [18]. Multiple-burner boilers normally have individual safety shutoff systems for each burner, all the systems being checked for closure or leak-tightness before the first burner is lit. A possible alternative involves a gas header supplied from a single valve with individual valves on each burner supply.

An initial header-pressure check is followed by position checking of the individual burner valves [17]. Since pilot rates are limited it is normally adequate to protect these by an unchecked double block and vent system although it is possible to extend the main pressure-checking system to include the pilot valve system. Correct commissioning and routine inspection of safety shutoff systems are greatly facilitated by appropriately placed pressure test points and manual isolating valves and these are specified in [18].

14.4.3 Boiler purging

The importance of an adequate prepurge before an ignition attempt is outlined in Sections 9.3 and 9.8. The presence of dead spaces in the boiler can cause 'slip', that is the purge air bypasses any gas accumulation and neither dilutes it nor sweeps it away. Where dead spaces can exist, consideration must always be given to using greater purge times or air flow rates. In some cases it may be necessary to install additional purge fans to direct air into otherwise inaccessible parts of the boiler. Purge times and rates are specified in [18]. In general for shell boilers having little or no dead space, 30 seconds at the full high fire air rate is normally adequate whilst for water-tube boilers 300 seconds with at least 25 per cent of the high fire combustion air rate is the minimum requirement, these values being subject to at least five volume changes. The purge air flow should be proved either by flow metering or by a combination of monitoring the air pressure and the positions of dampers and burner registers.

14.4.4 Burner ignition

Requirements for establishment of pilot and main flames and permissible energy release during these periods are considered in Sections 9.3.2 and 9.8. On start-up after purging, the air flow is, if necessary, adjusted to a suitable rate for ignition. On single-burner systems this often entails reducing the air flow to around 25 per cent of the full rate. On multi-burner systems where burners are lit sequentially this may also involve adjustment of the register on the burner to be lit, the other registers being left open to maintain a high flow through the boiler.

The best rate for main-flame establishment is debatable. Starting the main burner at full rate is not necessarily unsafe [17] although it is probably safer because of the possibility of a slight ignition delay to keep the rate at a minimum consistent with good cross-lighting and reliable flame establishment.

Automatic control of the complete start-up sequence is becomingly increasingly common, i.e. a 'management system' controls the purge and lighting sequences and continuously monitors during normal running, confirming that the appropriate conditions exist at every stage, e.g. damper positions, boiler-water level, presence or absence of flame, etc.

In the large applications considered in this section, management is normally performed by custom-built equipment using relay or solid-state logic. Some very large water-tube boiler installations are manually operated and here the management system ensures by appropriate interlocks that correct start-up procedures are followed. Control-unit logic is considered in detail in [20].

14.4.5 Operation and supervision

After the start-up sequence has been completed and normal operation commences it is necessary to be able to adjust firing conditions to match the burner throughput to the demand, to optimize the fuel-air ratio and to continue to supervise the safe operation of the boiler. Large packaged burners for shell-boiler firing are nearly always equipped with modulating throughput control and air–gas proportioning is achieved by linked valves. Mechanical linkages present problems of free play and separate valve actuators electrically or pneumatically linked together eliminate some of the difficulties. In the case of water-tube boilers, throughput and ratio control are normally achieved using process-control equipment in which flows are measured and compared with desired values to produce the actuating signal. Boilers of this type are generally multi-burner and it is usual to control the overall air and gas throughputs, the air–gas ratio to each burner being trimmed using adjustable registers. In more sophisticated systems the air–gas ratio may be trimmed automatically in response to flue-gas analysis.

The burner management system supervises the running of the system monitoring all the crucial parameters, e.g. water level, steam pressure, combustion air supply, etc., and taking appropriate action should a potentially unsafe condition occur. Probably the most important interlock is the flame safeguard system (Section 9.5.4). Self-checking systems are necessary for burners firing continuously for periods greater than a week and are desirable in other circumstances [18].

14.4.6 Dual-fuel firing

Since the boilers under consideration in this section are almost exclusively supplied by gas on an interruptible basis it is necessary to make reference to the special requirements of dual-fuel operation. Normally a common gas pilot is used for ignition on both gas and oil firing and the flame-detection system is also common to both fuels. In the case of single-burner systems (and twin-fire-tube installations with two burners) an ultraviolet or flicker-type infrared detector is used and will respond to flames from either fuel. On multi-burner installations where it is necessary to discriminate between the burner flame and those from adjacent burners on either fuel, several approaches have been adopted [17] including the use of a single ultraviolet detector to each burner with switched amplifier gain to allow for the different flame signals from the two fuels and an ultraviolet and cadmium-sulphide detector to each burner, the latter responding to visible radiation from the oil flame and avoiding problems resulting from ultraviolet attenuation by atomized oil.

Switchover from one fuel to the other on single-burner installations is usually done by termination of firing on one fuel followed by a complete restart on the other. Clearly the interrupted fuel must be positively isolated and when this is gas this can be checked by carrying out the normal safety shutoff system checks on start-up on the other fuel. Alternatively an interlocked manual isolating valve may be incorporated in the gas supply which will prevent start-up on the other fuel until the valve is proved closed [21]. This allows the gas safety shutoff system to be serviced and possibly removed while the boiler operates on the other fuel.

On multi-burner operation switchover is generally carried out by shutting down and restarting each burner in turn allowing the plant output to be maintained at a fairly high level.

14.5 POWER GENERATION

14.5.1 Introduction

Any industrial site, mixed building development or complex requires energy in two basic forms, electric power and heat. In addition there may be a demand for cooling in some situations. On an industrial site there is always a demand for power and generally for process heat and heat for providing comfortable working conditions. In commercial buildings such as offices, large stores, supermarkets, hospitals and schools, there are similar requirements but of a different pattern with process demands generally absent.

In the majority of situations in Britain, electrical power is supplied by the area electricity board and the heat energy is provided by either solid fuel, oil or gas. The Central Electricity Generating Board has in the past benefited from the economy of scale that is associated with large turbine plant. It has had access to bulk supplies of relatively cheap fuel, has been able to average out load over a broad area, so saving capital cost for plant needed to meet high maximum demands, and most important has provided a very reliable service. These advantages have enabled the electricity boards to gain and hold the main market for electrical power, consumers preferring to buy a reliable service rather than investing in generating plant of their own with its attendant running and maintenance costs. The only classes of private generation able to survive have been those with a cheap source of energy, usually a by-product of a heat process, or those industries where a large steam demand existed. These combined heat and power systems have been used for many years and the term which is now used to describe them, 'total energy', is therefore not new.

The situation has now arisen in which the electric power in the energy mix has become expensive for a number of reasons:

1 The high cost of generating equipment in the central power stations, transmission and distribution systems.
2 The costs due to the low overall efficiency of generation and transmission.
3 The cost of capital expended to meet future demand, years in advance of receipt of any revenue.
4 The increasing cost of coal and oil.

The high cost of purchased electricity relative to fuel means that the best way to reduce the overall energy costs of an industrial or commercial undertaking is to reduce the electrical cost. Assuming that there is no scope for economy in consumption or improvement in load factor, and that the most advantageous electrical tariff has already been chosen, there is no way in which this can be achieved using purchased electricity.

A total energy scheme may be able to achieve this desired reduction in electrical cost as it depends on economic factors which are different from those which govern the electricity board's costs. In the energy situation which now exists, it is in the national interest that any possibly viable schemes for the application of the total-energy concept should be pursued. In the long term such schemes will help conserve the nation's basic fuel resources. As far as the individual client is concerned, however, the primary purpose is to reduce his overall energy costs. Figure 14:8 demonstrates that from the point of view of fuel efficiency, the total-energy concept has much to commend it. In addition to the cost benefit, however, a client must be satisfied that he will not sacrifice reliability of electricity supply or be inconvenienced in any way at all in comparison to his conventional energy system.

14.5.2 Total energy

Total energy is a title hallowed by use and generally accepted [23], and is defined as

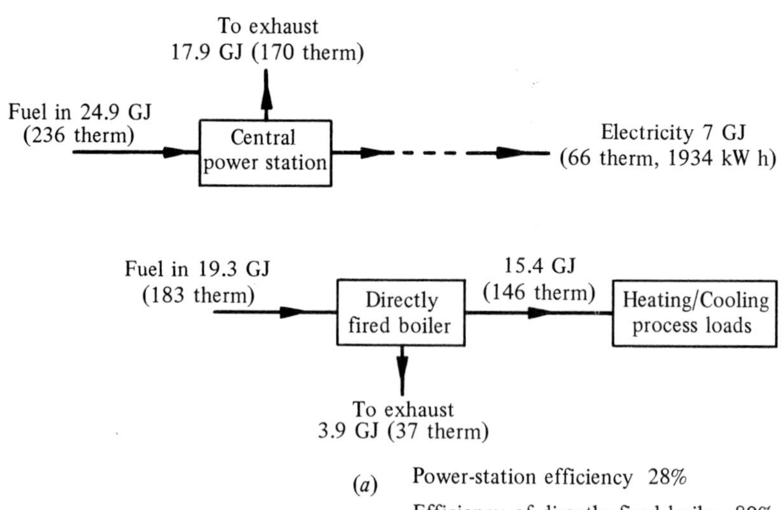

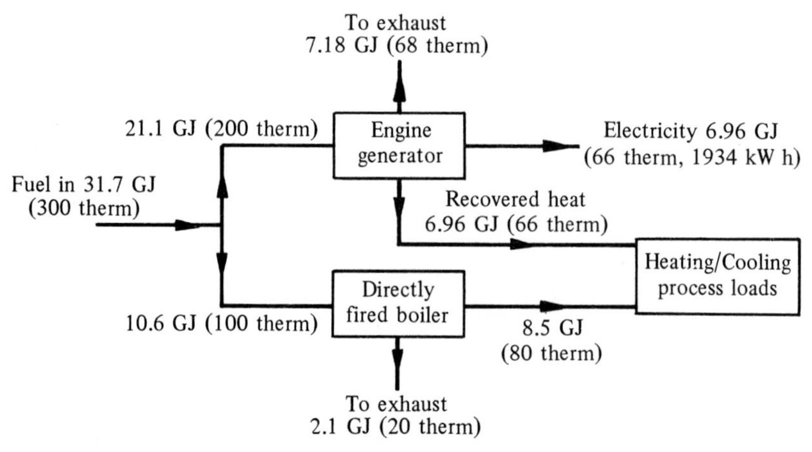

Figure 14:8 Total-energy systems [22]
(*a*) The thermal efficiency of conventional fuel use suffers from the low efficiency of central power generation
(*b*) The overall thermal efficiency of total-energy systems is higher because of the use of waste heat

the supply of all or part of the power and thermal energy required by an industrial or commercial complex by the use of prime movers generating the electrical or mechanical power and fitted with heat-recovery equipment to meet the thermal loads. It is a means of boosting the relatively low thermal efficiencies of standard power-producing equipment by making use of heat that would otherwise be wasted.

Ratio of power and thermal consumption
The heat/power ratios encountered with different products, processes or locations cover an extremely wide range. It is obvious that this ratio is a major factor in determining the feasibility of a total energy system and the type of equipment to be given primary consideration.

The fact that process changes, etc., can result in marked changes in the heat/power ratio during the life of the installed equipment, means that the equipment selected must offer the maximum flexibility. To achieve success the power engineer and the heating engineer must work together to determine whether an integrated approach to the overall problems of energy supply can result in an improved efficiency in the utilization of fuel.

Basic systems for supplying power and heat
The four basic systems available to supply power and heat to an industrial or commercial complex are:

1 Using *purchased electric power* and on-site heat by direct heating and/or indirectly (say by steam) from a purchased fuel.
2 On-site power generation of shaft electrical power using *steam turbines* with heat in the form of steam from a central boiler and power-generating plant.
3 On-site power generation of shaft electrical power from *gas turbines* obtaining heat from the hot waste gases, together with supplementary firing as necessary.
4 On-site power generation of shaft electrical power from *reciprocating engines*, with heat available as low-pressure steam or hot water, together with supplementary firing as necessary.

Systems (2), (3) and (4) are total-energy schemes and become economically viable if the difference between the generation costs and the purchased costs of fuel and power in (1) is sufficient to justify the capital investment in plant and to show a profit on that investment.

This is the simple economic basis of any total-energy feasibility study. Such a study establishes the viability of total energy for a prospective site.

Purchased power. Figure 14:9 shows diagrammatically a conventional public-utility power station supplying electric power to an industrial or commercial complex, the heat requirements of which are supplied indirectly from a steam boiler.

Steam turbines. Steam-turbine systems have been used for many years, as already indicated, in industrial applications with large steam requirements. Their use was based on the availability of comparatively cheap boiler fuels. The chemical, paper and sugar industries have for many years used steam turbines for base-load electricity generation and purchased power to satisfy peak-load demands. Complete on-site generation also has been and is practised, even exporting power to the public supply. The systems possible are:

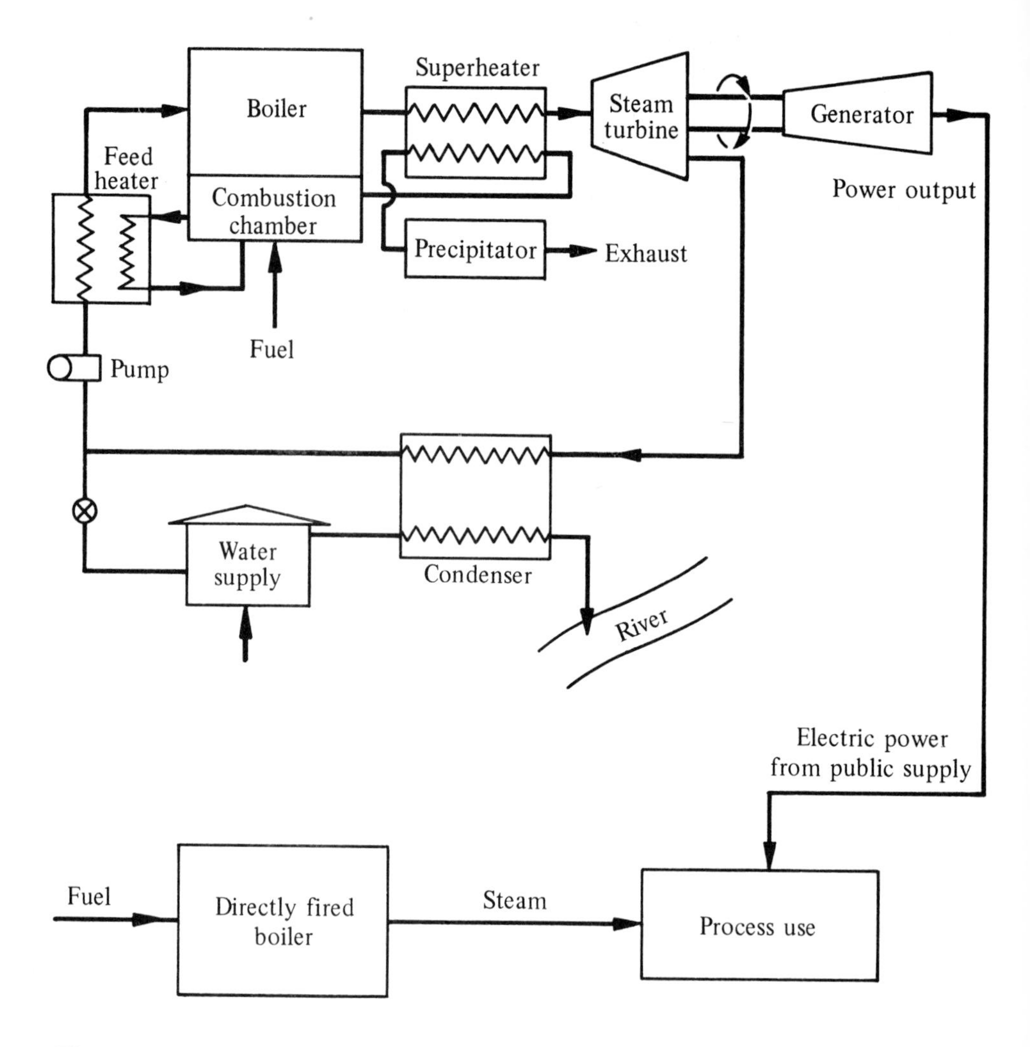

Figure 14:9 Use of electricity purchased from the public supply and steam generated on site

1 *Condensing turbine*. This takes steam at boiler pressure and temperature and expands it through the turbine, exhausting to sub-atmospheric pressure in the condenser. The process steam is supplied either from the same boiler or from a separate boiler. The system provides maximum flexibility for meeting varying process-steam and electrical loads, but it is the least efficient.
2 *Back-pressure turbine*. The turbine takes steam from the boiler and produces power by expanding the steam through the turbine stages before finally exhausting the whole of the steam flow to the heating process. This system provides the maximum economy and the simplest installation. Figure 14:10 shows a traditional industrial back-pressure steam-turbine system. The generation system is most efficient when the power demand corresponds to the output obtainable from the available steam

flow. This is seldom realized in practice. Very often electrical load and steam requirements fluctuate widely and may be out of step with each other.

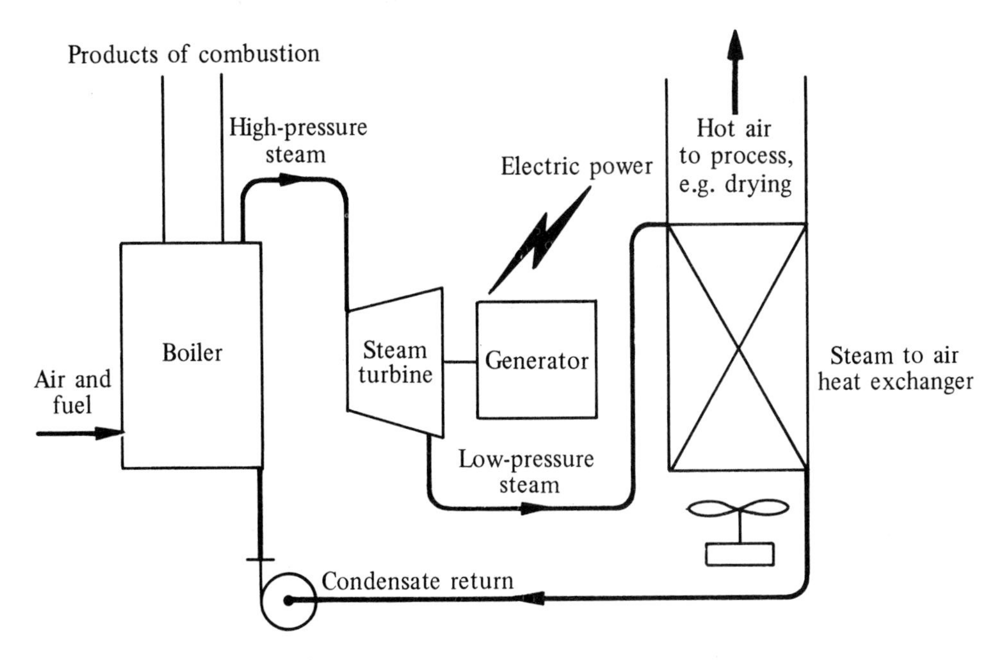

Figure 14:10 Traditional use of a back-pressure steam turbine to supply power and heat

3 *Pass-out condensing turbine.* When the process-steam demand is small in relation to the electrical demand or if it varies widely during the day, a pass-out condensing turbine may provide the best solution. The turbine is a single machine which has a back-pressure turbine section taking steam at boiler pressure and exhausting to the process followed by a condensing turbine section which takes additional steam at the process pressure to make up the required electrical load.

Experience with steam-turbine installations has shown that heat/power ratios of between 0.012 and 0.8 kg steam per MJ are usually required to make them economically attractive. Operation of the equipment at the varying heat/power ratios needed to meet daily and seasonal variations in power and steam consumption results in falling efficiencies which can only be combated by the use of more complex equipment. The maintenance requirements of the prime mover itself are low but this is not always true for the complex support equipment required.

The steam turbine is in itself a very reliable prime mover. The disadvantages of the systems described arise from the complexity of the steam-raising and water-handling equipment, particularly if pressures above 4 MPa (40 bar, 580 lbf in^{-2}) are necessary.

Gas turbines. A gas turbine operates by drawing filtered air from the atmosphere, compressing it to a pressure of 300–600 kPa (3–6 bar, 43.5–87 lbf in^{-2}) depending on the particular turbine, heating the compressed air in a combustion chamber to a

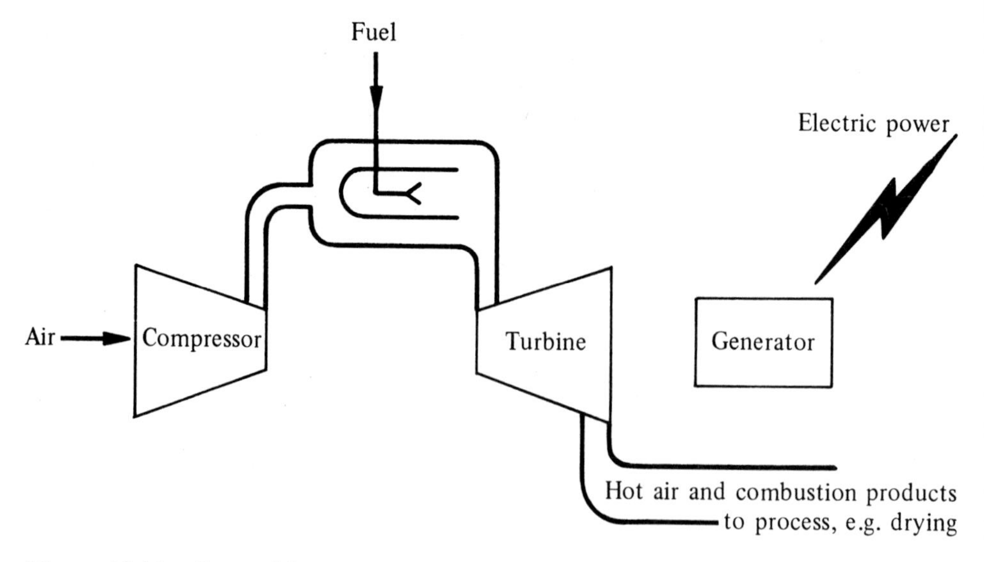

Figure 14:11 Gas-turbine system

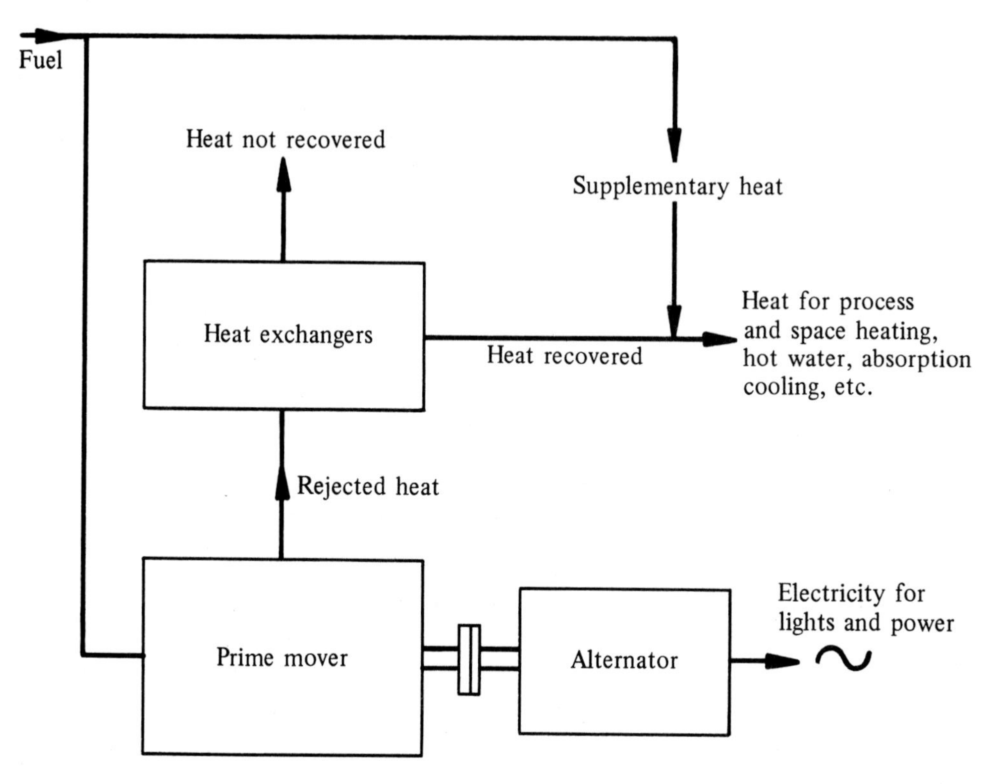

Figure 14:12 Typical total-energy application of a gas-turbine system [9]

controlled temperature and then expanding the hot gases through a power turbine which also drives the air compressor. The shaft speed is high and gearing is included in the installation to reduce the speed to a level suitable for driving an alternator. The high-temperature exhaust gases can then be ducted through a waste-heat boiler to raise steam or, if required, used directly in, say, a drying process. Figure 14:11 shows a gas-turbine system and Figure 14:12 typical total-energy applications.

The full-load overall efficiency of a gas turbine varies between 10 per cent for the small sizes (approximately 0.25 MW) to 25 per cent for the larger sizes (approximately 10 MW). The part-load efficiency tends to be poor with a 20 to 25 per cent drop in efficiency when running at half load.

Gas turbines and steam turbines offer smooth operation with freedom from the vibration and oscillating loads present in the reciprocating engine. Therefore they only require minimal foundations and light support structures. Gas turbines, using air as the working fluid, do not require the complex steam-raising and water-handling equipment needed by steam turbines.

When considering the prime mover only, the gas turbine has the highest capital cost. When the complete installation is considered, total costs become comparable, but comparisons should only be made between similar installations meeting the same specific duties.

The thermal efficiency of all gas turbines is lower than that of the modern reciprocating engine. However, no heat is lost to cooling water and only a very small percentage of the total heat input is lost to lubricating oil and through radiation. This means that a very high proportion of the heat input not converted to power passes to the exhaust and can be recovered by a suitable recovery system. The overall thermal utilization for gas-turbine installations can therefore be very high, ranging from 55 to 90 per cent according to the application. Gas turbines are selective on fuel and normally liquid distillate fuels or gas are specified. The basic fuel cost may be high for some applications, but this can be offset by the higher overall thermal utilization. Gas turbines have a higher exhaust-gas mass flow than other prime movers of equivalent power and the temperature of the gases is approximately 450–500 °C. This enables heat to be readily recovered using simple equipment. Some possible uses for the recovered heat were indicated in Figure 14:12. Available steam is about 4–5 kg per kW of power produced. The high cost of the heat-recovery boilers due to the high gas mass flow is offset by the fact that they only generate steam at process pressure and not the higher pressure demanded by a steam-turbine system.

The fact that gas turbines operate at high overall air–fuel ratios in the order of 60–70 : 1 means that the exhaust gases are only lightly contaminated with combustion products and can be used for many direct drying processes. The dew point is low, so final stack temperatures can be lower than with normally fired equipment operating at low excess-air ratios. The exhaust gases contain about 18 per cent by volume of oxygen and additional fuel can be burned in this gas to increase the quantity of heat available. This is known as auxiliary firing, the maximum additional heat input is determined by the type of heat-recovery equipment used. In terms of steam, the maximum rate will be about 20–30 kg of steam per kW of power produced. When an installation is planned, the extent of the auxiliary firing and the boiler size can be fixed to meet the actual heat/power ratio of the process. During operation the auxiliary firing can be modulated to allow daily and seasonal variations of the process requirements to be met with minimum change in overall efficiency over the range of operation. The additional heat input is converted into steam at 90–95 per cent efficiency since only a relatively small volume of fuel is required to raise the additional steam. Assuming the same exhaust temperature of the waste-heat boiler, the only additional heat lost would be approximately that carried out by the additional volume of exhaust gases.

Gas turbines available for ground duties range from the derated industrialized lightweight aircraft-derivative type to the heavy long-lasting industrial type designed for continuous base-load operation. In this range many types and sizes up to 6 MW are available which enables turbines to be matched closely to the specific duty required.

Advantages of gas turbines include:

1 Low starting torque, allowing a fast start-up.
2 Fast rate of adjustment to load change.
3 Equally suitable for continuous operation over a long period or for intermittent duties.
4 Maintenance costs are low.
5 High power-to-weight ratio.
6 No cooling water required.
7 Pollution-free exhaust.
8 Changeover from gas to standby fuel can be achieved automatically while on load.

Disadvantages of gas turbines include:

1 High initial capital cost.
2 Low overall efficiency.
3 High-frequency noise generated.
4 High-pressure gas required (in excess of 600 kPa (6 bar, 87 $lbf\,in^{-2}$)).

Reciprocating engines (*internal combustion*). The use of reciprocating engines in total-energy schemes is possible in certain applications. These engines have higher shaft thermal efficiencies than gas turbines. In the past, diesel engines have been used to generate electricity on standby duty. The engines suitable for total-energy schemes are dual-fuel types suitable for natural-gas/light-oil firing and spark-ignition types suitable for gas firing. More information on the spark-ignition type is given in Section 14.5.3.

Both the diesel and dual-fuel engines operate on the compression-ignition cycle, the dual-fuel type requiring 5–10 per cent of full-load diesel-fuel input to initiate combustion when nominally firing on gas. The changeover from one fuel to the other can be made easily and automatically while the engine is on load as and when necessary.

There are a number of manufacturers in the UK producing both diesel and dual-fuel engines over the complete range of sizes suitable for total-energy schemes [9]. The overall efficiencies under full-load conditions of the three types of engine are: diesel 38 per cent, dual-fuel (running on gas) 34 per cent and spark ignition 30 per cent.

The remainder of the heat energy is shared approximately equally between the jacket cooling water and the exhaust gases from the engine. The exhaust gases form the only source of high-grade heat. Thus the reciprocating engine is used mainly in cases where the heat/power ratio is low. The heat availability in terms of steam obtained, is approximately 1–2.5 kg steam per kW of power produced [23] and it is only suitable for low-temperature steam heating. It may be possible to recover the low-grade heat from the cooling water for some purpose such as space heating, but in any case it must be removed so the system may require a water–water heat exchanger and cooling tower or fan coolers. Obviously the efficiency is lowered in these cases. Higher thermal requirements can be met by using directly fired boilers to provide the extra steam, but this introduces more complexity into the system to meet the varying power and thermal demands.

Advantages of reciprocating engines include:

1 High overall efficiency.
2 Comparatively low capital cost per kilowatt.

Disadvantages of reciprocating engines include:

1 Necessity for cooling even when the recovered heat is not to be used.
2 Exhaust gases may contain products of incomplete combustion (liquid firing).
3 High starting torque.
4 High maintenance costs.
5 High lubricating-oil consumption.
6 Low power-to-weight ratio, requiring heavy foundations, except for spark-ignition engines of the automotive type.

Feasibility studies [24]
These can be undertaken by the plant manufacturers, consulting engineers, architects, BGC Midlands Research Station or by the BGC regions. The studies entail a great deal of detailed investigation and are costly in terms of man-hours involved, so they are carried out in two stages: an initial study and a detailed study.

Initial study. The first step is to obtain, or estimate in the case of a new site, the hourly heat (steam or direct) and electrical load profiles over as long a period as possible. Consideration must also be given to the probable future trends of these loads and profiles. From the data obtained the fluctuation of the heat/power ratio can be deduced. This ratio enables a good preliminary choice of prime mover to be made.

The gap in the range of heat/power ratios between engines and gas turbines can be filled by producing additional heat in directly fired boilers. From the maximum and minimum electricity demands, the size and number of prime movers can be determined with an allowance for standby equipment. An initial assessment of both the capital and fuel costs can then be calculated, using current information from fuel suppliers and equipment manufacturers.

A first approximation of running costs can be obtained by assuming that the total electricity produced per year is generated at 80 per cent full load for each prime mover considered. This gives the total running time of the generator sets and the fuel input to the prime mover can be costed.

The waste heat recoverable from the installation can now be calculated using the factors quoted for the prime movers and the total running hours. The shortfall, if any, between the heat recoverable and the annual demand can be calculated for gas turbines using an after-firing efficiency figure of 95 per cent and for direct boiler firing using a figure of 80 per cent.

The total fuel costs can now be calculated and, using an estimated maintenance cost, the total running cost is obtained. These costs are then compared with the costs of a conventional purchased-power/directly fired boiler system. A value of the annual saving is obtained that can be related to the net capital cost of the scheme, to provide an indication of payback.

If the initial capital investment required is acceptable and a payback period of less than eight years is obtained, then the scheme warrants a detailed study.

Detailed study. On an existing site, the hourly steam and electrical demands and annual costs are known, so the assessment of loads can be made reasonably accurately, allowing for any future changes. On new sites, the hourly load profiles can only be good estimates.

One other factor which must be considered and which has not been covered yet is air conditioning. This allows a certain amount of manipulation of the overall site heat/power ratio, since both steam and shaft power can be used to drive the chiller. The operating cycle of the refrigeration system governs the mode of drive of the chiller by:

1 Absorption—either direct fuel fired, medium-pressure hot water or steam (2.58 kg steam per kW refrigeration).
2 Compression—electric motor, reciprocating engine or turbine (0.213 kW per kW refrigeration).
3 Combination—such as a steam-turbine absorption system (1.80 kg steam per kW refrigeration).

A complete study can then be made after deciding the type or combination of prime mover, the size and number of units, the type and size of boiler and the fuel. A number of factors influence the size and number of prime movers:

1 Maximum efficiency is usually achieved at or near full load. This is why an operating condition of about 80 per cent of full load is desirable.
2 Standby equipment is necessary to cover failure or maintenance of one generator set. The sizing of equipment is adjusted to allow a practical amount of standby. Gas turbines usually need one unit but engines may require two units.
3 Prime-mover size can also depend on the minimum electrical load, as it is neither economic nor practical to run at low percentage loads.
4 The larger the unit size, the lower the cost per kilowatt and efficiency is also improved.

Waste-heat boilers are usually coupled to individual prime movers but coupling to a pair of gas turbines, for example, may sometimes prove more practical and economic.

Several more factors to be considered which affect the costs and operation of an installation are:

1 Choice of voltage.
2 Choice of manual or automatic operation.
3 Allowable frequency variation.
4 Noise limitations imposed by the surroundings.

The most accurate detailed calculations of capital and running costs, including rates, insurance, lubricating oil, fuel, maintenance and supervision costs, are then compared with the accurate costs of a conventional scheme. The profit per year of the total-energy scheme over the conventional scheme has to repay the extra capital outlay of the total-energy scheme and provide an attractive level of profit on the capital invested in it.

The periods of payback and depreciation of equipment are very important and these financial considerations must be cleared of any doubts early in the study [24].

The market available [24, 25]
The market for total energy lies in the areas of industry and commerce where the operating costs of such an installation show sufficient advantage over the alternative of purchasing power from the area electricity board and producing heat on site to justify the additional capital costs incurred in the scheme. Wherever electricity is needed such a scheme may be feasible, especially when the electrical load factor is in the range of 20 to 70 per cent.

Examples of possible total energy installations include:

1 *Commercial.* Large shopping precincts, office blocks, hotels, hospitals, etc., having a high artificial-lighting load with winter space heating and summer air conditioning.
2 *Industrial.* Processes and industries which satisfy the electrical load factor, such as chemicals, sugar, tobacco, iron and steel, and paper industries.

On a new site, with proper planning and feasibility studies being undertaken at the start of the project a total-energy scheme may well prove to be economically viable.

On existing sites, a number of factors may hinder the development of total energy:

1 Lack of space to house the installation.
2 Difficulty of designing a scheme into the existing site services which may have developed in a random way over a number of years. However, total energy could be considered for a part of a site.
3 Demand by the electricity authority for a standby charge for any electricity supply to the site.
4 Possibility of having to write off capital associated with parts of the existing plant which may be made redundant by the scheme.

Market surveys undertaken by the British Gas Corporation have shown that private power generation in industry is limited to about 300 firms. Until the advent of natural gas the bulk of the fuel used was coal and heavy fuel oil with light fuel oil and gas oil being used to a limited extent for localized power generation and peak lopping. In the last few years large sales of natural gas have been made to industry on either a firm or an interruptible contract basis. Much of this gas is used for direct steam raising which subsequently provides:

1 Electric power by means of steam turbines.
2 Steam for process work.

In the future, the opportunities for natural gas in total-energy/power generation fall into three broad but well defined categories:

1 *Fuel-replacement market.* This is the sector where natural gas will continue to compete with other fuels for bulk sales of firm or interruptible loads, for such purposes as bulk steam raising. The problems of non-premium fuel replacement are well known and marketing strategies are likely to follow the normal BGC region procedures.
2 *Plant-replacement market.* Natural gas here will again be competing with other fossil fuels but where the economic justification for changeover is associated with the replacement of existing heat/power plant. An example is the Singer Manufacturing Company's total-energy installation at Clydebank, where directly fired boilers and pass-out steam turbines were replaced with natural-gas-fuelled gas-turbine sets with exhaust waste-heat boilers and auxiliary firing.
3 *New-installation market.* Here the competition is electrical power from the National Grid. Any feasibility study carried out for a proposed total energy scheme will, as stated earlier, have to be shown to be attractive economically to the client. An example of such a scheme is the natural-gas-fuelled total-energy system installed in the John Player factory in Nottingham.

14.5.3 Selective energy

A selective-energy scheme may provide an alternative method of achieving the efficient use of energy in cases where a total-energy installation is either impractical or too expensive. It is a system using gas prime movers driving alternators which carry the main base electrical load, but with the electricity board supplying standby power and the offpeak requirements when the main sets are not required [26].

Standby generation
The difficulties experienced by the electricity-supply industry in recent years have created an increased demand for standby engine-driven alternator sets for commercial and industrial premises. In the past these have been driven by diesel engines, but now gas spark-ignition engines are available.

The capital and maintenance costs involved have been justified only on the basis of the need for a standby source for essential power requirements. Consideration given to other uses normally leads the user to consider peak-load lopping, i.e. using the plant to supply peak-load power, thereby minimizing the maximum demand metered by the electricity board. Peak-load lopping is necessary only if a maximum-demand tariff is used, under which the charge is in three parts: a maximum-demand charge, a basic unit (kW h) charge, and a fuel-variation charge. Obviously these tariffs are designed to reduce peak loads, which are not attractive to the electricity industry because it has to provide expensive capital plant to meet them although the short duration of the peak-demand periods means there will be little return on the investment. Similar disadvantages would apply to the gas industry if the peak loads were met by a gas-driven prime mover, but if the engines are used for base-load operation the maximum demand on the electricity industry is reduced while the gas load is improved by a more constant demand.

Engines available
Engines used for electrical generation are available from many manufacturers, both British and foreign, in dual-fuel and spark-ignition form. These same engines may also be used in air-conditioning applications (see Chapter 15).

Dual-fuel engines cover the range 180 to 6000 bhp (0.13 to 4.5 MW) and are based upon industrial diesel engines.

Spark-ignition engines cover the range 4 to 6660 bhp (3 kW to 5 MW) and are divided into three main types:

1 *Industrial engines.* Covering the range of output 115 to 6660 bhp (86 kW to 5 MW) these are slower-speed engines derived in the main from their diesel counterparts. They are designed for base-load operation.
2 *Semi-industrial engines.* These are higher-speed engines based upon commercial diesel engines and are available in the range 4 to 1290 bhp (3 to 960 kW). At present, engines of British manufacture are only available in output ratings up to approximately 200 bhp (150 kW) at speeds between 1000 and 3000 rpm [27].
3 *Automotive engines.* These are car petrol engines converted to run on natural gas. By utilizing mass-produced engines the unit cost is comparatively low. The range of power outputs is limited, however, to 10 to 70 bhp (7.5 to 52 kW) over a speed range of 1500 to 3000 rpm but development work is continuing at the British Gas Corporation's Midlands Research Station which has pioneered much of this work.

Selective-energy system

The system is based on the gas-fuelled generator set provided originally for standby power. The set must be time-controlled to start at the beginning and stop at the end of the working day, and is usually connected to the essential load which is normally about 50 to 60 per cent of the total load. The electricity board's supply is connected to the remainder of the load, with automatic switching, so that in the event of an engine failure the essential load is provided by the mains electricity and the non-essential load is switched off. This means that the maximum demand on the electricity supply is not exceeded. If the gas commodity charge per kilowatt-hour is equivalent to the electricity board's unit charge then a system operated for nine hours a day, five days a week, could save up to 25 per cent of the total electricity charges. At 1974 prices, however, for these charges to be equivalent the efficiency of the gas-engine/alternator set would have to be about 35 per cent. The efficiency is usually about 25 per cent and for the example taken the costs of generation and the electricity-authority charge would be about the same. The efficiency of the system can, however, be improved to 36 per cent by making use of the engine cooling water for space heating in the winter.

The result is that in general an engine/alternator set of normal efficiency, running at regular times throughout the year, will show a saving on the electricity authority's charges which will be the difference between the reduction in maximum-demand charge and the Gas Region's standing charge, provided the cooling water is used in the heating season. Base-load lopping will, therefore, give up to a 25 per cent saving on the total charges that would be payable to the electricity authority for the same load.

Market available

Basically, the electricity board's maximum-demand tariff is applicable if the load is 40 kV A or more. Lower loads with other tariffs will not in general prove attractive to a selective-energy scheme. Large plants with electrical demands of 500 kV A will need careful study because the maximum-demand charges reduce and the savings reduce.

The market available is in commercial premises or small industrial establishments with electrical loads within the range quoted.

14.5.4 Novel methods

A number of methods of power generation which may become available in the future and could involve the use of natural gas are outlined below.

Fuel cells
Hydrogen–oxygen cells have been developed for use in space and submarine applications. They are unlikely to be used for large-scale electricity generation because of the cost of producing very pure fuel gases. The use of hydrocarbon fuels such as natural gas also causes difficulty because they are less electrochemically active than hydrogen and produce carbon dioxide in the combustion products. The existing cells are very expensive to make and have running costs of about twice those of a conventional engine.

Thermoelectric
The principle of operation would be to connect a large number of semiconductors in series through an external circuit and then to apply heat to one junction. This would, in effect, 'pump' electrons through the circuit and provide power. The efficiency would probably be only 10 to 20 per cent.

Magnetohydrodynamics (MHD)
In this process a jet of ionized gas flows through a magnetic field. High-voltage

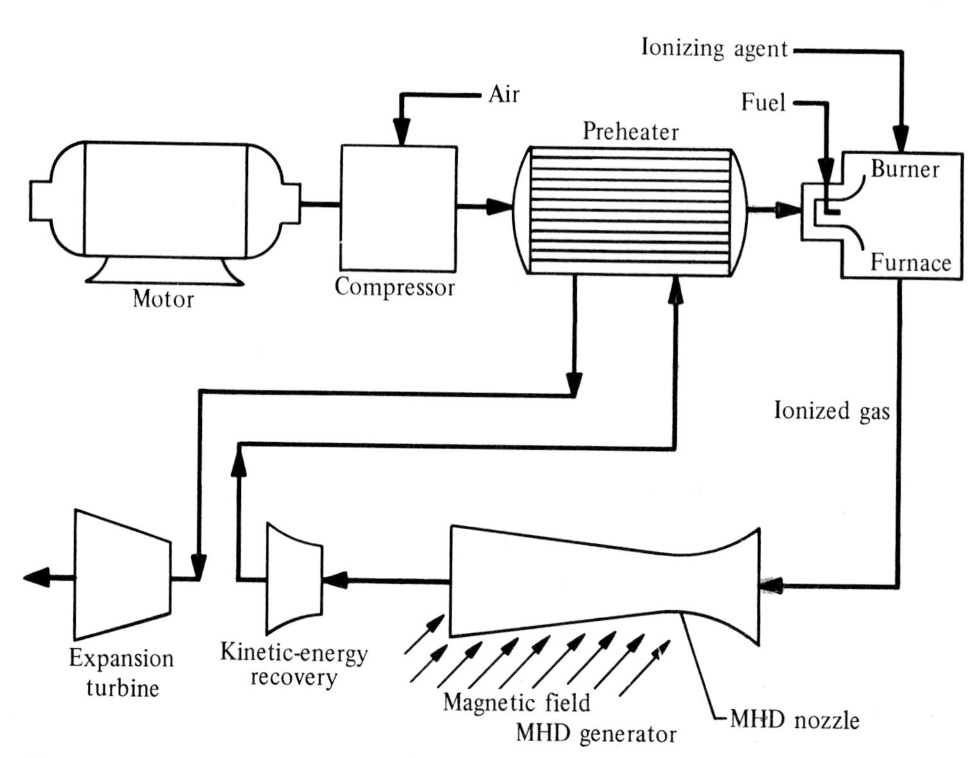

Figure 14:13 Independent magnetohydrodynamic unit

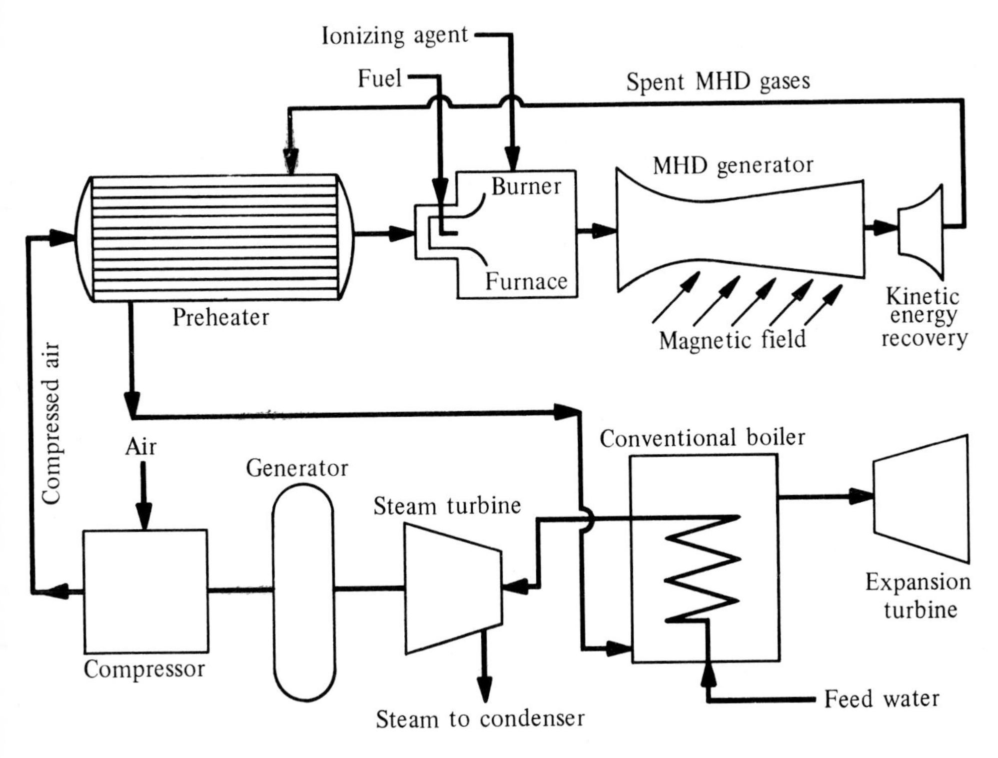

Figure 14:14 MHD-topped conventional plant

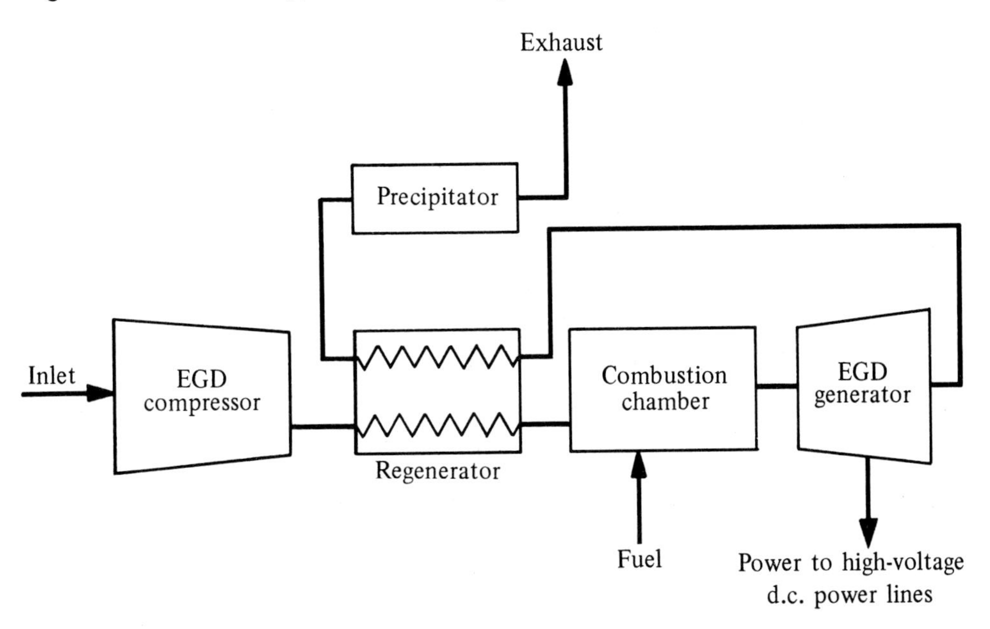

Figure 14:15 Electro-gas dynamics

(> 2000 V) direct current is obtained from electrodes placed in the gas stream. The ionization requires temperatures of 2000 °C or above. Thermal efficiencies of around 55 per cent have been quoted. An independent MHD unit is shown in Figure 14:13 and an MHD-topped conventional plant in Figure 14.14.

Electro-gas dynamics (EGD)
This is another experimental system and is similar to MHD generation but uses an electrostatic field to intercept the gas stream. A possible design of system is shown in Figure 14:15.

Electro-hydrodynamics (EHD)
This system of direct electricity conversion makes use of liquid droplets as the energy-producing medium. The aerosol is forced into a high-velocity gas stream being discharged across a high-intensity electric field. The charged fluid reaches a collector electrode where the aerosol is discharged and the liquid consolidated while the electricity flows through the external circuit.

Figure 14:16 gives the size, temperature range and efficiencies of the various possible generating systems. It is only a guide as performance will vary with load factor, degree of waste-heat utilization, etc.

	Maximum practical efficiency (%)	Temperature range (°C)	Approximate order of optimum size
Thermoelectric	10	150– 500	1 W
Thermionic	20	1000–1600	↓
Fuel cell	70	60–1000	↓
Stirling cycle	50	400– 650	↓
MHD	55	2000–3000	10^7 W
Internal combustion	50	1000–1600	10^5 W
Gas turbine	25	500– 800	↓
Steam turbine	40	400– 650	5×10^8 W

Figure 14:16 Characteristics of various generating systems

REFERENCES

1 Gore, W. A., Gunn, D. C. and Horsler, A. G., 'Natural gas firing of shell boilers', British Gas Corporation Midlands Research Station, Publication No. 203 (1972)
2 How, M. E., 'A review of Institute of Fuel papers since 1950 on coal-fired water tube boilers', *Energy World*, No. 13 (February 1975) 7
3 The Clean Air (Emission of Grit and Dust from Furnaces) Regulations 1971 (Statutory Instrument 1971/162). London: HMSO (1971)
4 Ministry of Power, *Efficient Use of Fuel*, 2nd edition, pp. 273–96. London: HMSO (1958)

5 Elkins, P. C., 'Steam boiler changeover', London and Southern Junior Gas Association, Meeting (1970)
6 Robertshaw, G. W., 'Gas firing' in *Efficient Use of Energy* (ed. Dryden, I. G. C.). London: IPC Science & Technology Press (1975)
7 Cox, R. W. and Day, A. P., 'Uprating sectional boilers', *Gas Eng. & Manage.*, **14** (1974) 231
8 Miller, V. H., 'Steam raising, dual fuel firing and interruptibles', University of Salford and North West Branch of the Institution of Chemical Engineers, Symposium on the Utilization of Natural Gas in the Chemical Process Industries, Salford (April 1974)
9 Elles, H. C., 'Natural gas in the chemical and processing industries', Institution of Chemical Engineers, North West Branch, Meeting (February 1972)
10 Horsler, A. G. and Lucas, D. M., 'The firing of shell boilers', *Gas Council Res. Commun.* GC195 (1972); abstract in *J. Inst. Gas Eng.*, **12** (1972) 276
11 Lucas, D. M. and Toth, H. E., 'The calculation of heat transfer in the fire tubes of shell boilers', *J. Inst. Fuel*, **45** (1972) 521
12 Biswas, D. and Ford, J. D., 'Metal surface temperature measurement within boiler plant', *Steam & Heat. Eng.*, **42,** No. 500 (July 1973) 6
13 Slijpen, J. J., 'Comparative performance of firetube boilers on natural gas and fuel oil', Shell International Gas Ltd, Report S1G/09/8
14 Greenhalgh, A. and Haslam, P., 'Natural gas for boiler firing—some new applications', *J. Inst. Gas Eng.*, **13** (1973) 233
15 Speakman, M. J., 'Aspects of controls and efficiency results for natural gas fired boilers', Institute of Fuel, North Western Section, Meeting (October 1971)
16 Gas Council, *Technical Notes on Changeover to Gas Firing*, 2nd edition. London: Gas Council (1971)
17 Hancock, R. A. and Spittle, P., 'Safety and controls for natural gas firing of industrial boilers', Gas Council Midlands Research Station Publication 202 (1972)
18 Gas Council, 'Interim codes of practice for large gas and dual-fuel burners', Parts 1–4, Reports Nos. 764/70, 766/70, 767/71, 768/71 (1970 and 1971)
18a Amendment No. 1 to [18] (April 1974)
19 IGE TD/5, 'Recommendations for the design of governor installations to consumers' premises', *Inst. Gas Eng. Commun.* 750 (1967)
20 Atkinson, P. G., Grimsey, R. M. and Hancock, R. A., 'Control units for automatic burners', *Gas Council Res. Commun.* GC139 (1967); *J. Inst. Gas Eng.*, **8** (1968) 341
21 Hoggarth, M. L., Pomfret, K. F. and Spittle, P., 'Evaluation of dual-fuel burner performance', *Gas Council Res. Commun.* GC196 (1972); abstract in *J. Inst. Gas Eng.*, **12** (1972) 276
22 Rich, J., 'What price total energy?', *Watson House Bulletin*, **35,** No. 258 (May–June 1971)
23 Stocks, W. J. R., 'Total energy', University of Salford and North West Branch of the Institution of Chemical Engineers, Symposium on the Utilization of Natural Gas in the Chemical Process Industries (April 1974)
24 Pearson, J. and Simpson, P. B., 'Total energy in the gas industry', *Inst. Gas Eng. Commun.* 844 (1971); *J. Inst. Gas Eng.*, **11** (1971) 733
25 Gauthier, C. J., 'Marketing total energy', *Inst. Gas Eng. Commun.* 753 (1967); *J. Inst. Gas Eng.*, **8** (1968) 273

26 Trumble, D. R. and Rattle, J. B., 'Developments in the use of gas engines', Eastern Junior Gas Association, Meeting (December 1974)
27 Wu, H. L., *J. Inst. Fuel*, **42** (1969), 316

15
Other applications

15.1 DRYING

15.1.1 Principles of drying [1]

Drying generally refers to any process in which a liquid is removed from a solid. The type of drying process of interest to industrial gas engineers will involve the application of heat to evaporate moisture, usually water from a solid, and the removal of the resulting vapour.

Section 3.6 dealt with the theory of humidity, psychrometry and drying and showed that drying involves:

1. Heat transfer by radiation, conduction or convection to supply the latent heat necessary to evaporate moisture from a solid material.
2. Mass transfer of liquid or vapour from within the solid to the surface and the removal of vapour from the solid surface.

The removal of water or other solvent from a material sometimes involves processes which are known variously as curing, baking, stoving, polymerizing, etc. A process such as curing gives strength and durability to the finished product and will be discussed as an aspect of drying in this chapter.

The effects of the variables present in the equations governing the drying process on the process may be summarized as follows:

1. Drying-air temperature. The drying rate is directly proportional to the difference between the drying-air temperature and the solid-surface temperature. An increase in air temperature is often the simplest way to increase drying rates and therefore the capacity of the dryer.
2. Solids temperature. Faster and more efficient drying is obtained when the material is fed to the dryer at temperatures above ambient.
3. Air velocity. An increase in air velocity increases the heat and mass transfer coefficients and therefore increases the drying rate. This is due to increased

turbulence at the solid surface reducing the resistance of the laminar film. In applications involving the passage of air through a granular bed of material, the velocity can be increased until the bed becomes fluidized. This increases the area of material available for heat and mass transfer.

4 Final moisture content. An increase in the final desired moisture content decreases the time needed for drying without affecting the drying rate. It is generally uneconomic to dry material below its equilibrium moisture content in atmospheric air.

5 Air distribution. The air must be brought into intimate contact with the whole drying surface to make the best use of its potential for heat- and mass transfer. The air-distribution system must of necessity be very carefully designed in relation to other parameters such as the material being dried, shape and size of drier, and positioning of baffles.

6 Direction of air flow. The air-flow pattern through, over, under or around the material being dried depends on the nature of the material, the means of handling it and the space available.

15.1.2 Convection drying

This is the most common method, the heat transfer occurs by direct contact between the wet solid and hot gases. The resultant vapour is then carried away by the drying medium.

The hot gas can be:

1 Indirectly heated air. Air is blown across a heat exchanger which is heated by combustion products, hot water or steam. The air does not come into direct contact with the heating medium, and therefore cannot be contaminated in any way. The only need for indirectly heated air occurs when the product being dried is very sensitive to contamination or gives off flammable vapour.

2 Directly heated air. The drying air is heated by and mixes intimately with the combustion products from a burner. The fact that only a very few substances cannot tolerate the combustion products of natural gas means that packaged direct gas-fired air heaters located outside the drying system are extensively used. Direct firing gives an increase in efficiency of between 20 and 25 per cent over indirect firing. Temperature control is also improved because by adjustment of heat input, operating temperatures can be reached very quickly and then kept within close limits during operation. This system can also provide a cooling effect, if necessary, by shutting off the gas supply and maintaining the air flow. Two other advantages of this system are cheaper installation costs and adaptability. With a packaged air heater, all the necessary controls are included, which means that replacement costs of standard items, maintenance and labour charges can be kept to a minimum. Existing convection driers can be easily and cheaply converted to direct firing. The faster, more easily controlled drying will give lower costs, increased production and a better product quality.

3 Inert gases. These are only used in very special cases where air oxidizes or otherwise damages the product being dried. In some cases the combustion products of a stoichiometric mixture of fuel and air can be regarded as an inert gas.

4 Steam. Steam is still a common heating medium for use in driers. This is essenti-

ally because the process is often regarded as a medium-temperature process and the steam can provide the correct degree of heat for many drying operations. However, the need for large and expensive ancillary equipment and the limited temperatures are big drawbacks where economy, flexibility and high production rates are required.

In conclusion, air is a very convenient medium for heat and mass transfer. For heat transfer, air has a low specific heat capacity and can therefore be raised in temperature very quickly. For mass transfer, a small increase in the temperature of air can greatly increase the quantity of water it can carry, e.g. a temperature increase from 40 to 44 °C (10 per cent) increases its moisture-carrying capacity by about 35 per cent.

Recirculation
One way in which the thermal efficiency of a drier may be increased and its fuel costs lowered is to reheat some of the used air and pass it back through the drier. Recirculating air also tends to give a more even temperature distribution and therefore a more uniform drying operation than a simple single-stage drier. On the other hand recirculation results in a smaller mean wet-bulb depression. This means that for the same dry-bulb temperature and velocity the drying time will be longer. A limit is reached when the recirculation rate causes the humidity in the drier to rise too quickly and so seriously retard the drying process.

The necessary degree of recirculation can be produced by hot-gas fans or high-velocity jet burners. The choice depends on temperature and economics.

Method of gas–solid contacting
In the process of drying a solid by a hot gas, the heat and mass transfer occur at the surface of the solid. To obtain maximum efficiency there must be:

1 Maximum exposure of the solids surface to the gas phase.
2 Thorough mixing of gas and solids.

A solids bed can exist in a number of ways and the manner in which the gas is made to contact it can also be varied. A number of possible configurations are illustrated below.

Solids condition
1 A static solids bed. An example is a tray drier. There is no relative motion between the particles, but the tray can be moving in any direction.

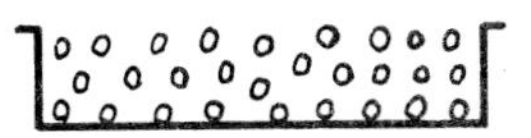

2 A moving solids bed. One example here is a rotary-type drier. In a slightly expanded bed the motion of the solids relative to each other is achieved by mechanical agitation. Another example is a tower drier where solids flow is due to gravity.

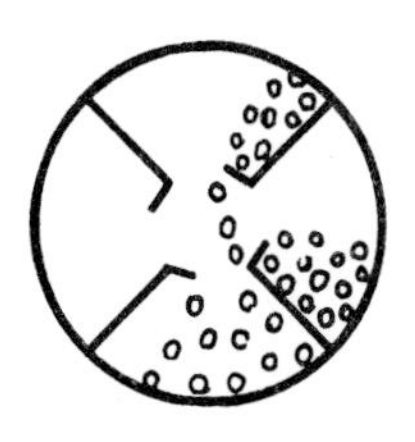

3 A fluidized solids bed. In a fluidized-bed drier, an expanded bed of solids exists where the solids are suspended by the drag forces caused by the gas phase passing at some critical velocity through the voids between the particles. The solids and gas phases intermix, behaving like a boiling fluid.

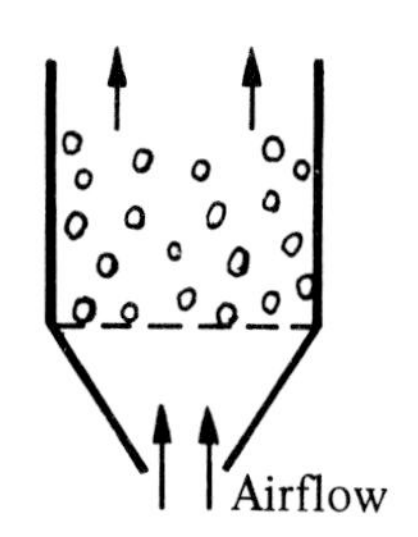

4 A dilute solids bed. This is a fully expanded bed in which the solids are so widely separated that essentially they exert no influence on each other. The density is that of the gas phase alone. One type operating on this principle is the pneumatic conveying drier. Here the gas velocity at all points is greater than the terminal settling velocity of the solids. The particles are therefore lifted and continuously conveyed by the gas.

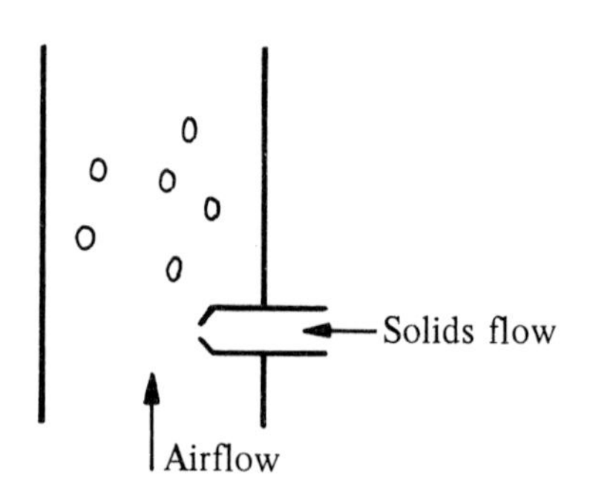

Another type is the spray drier. Here a liquid feed is atomized and projected into a large drying chamber, from which it settles by gravity.

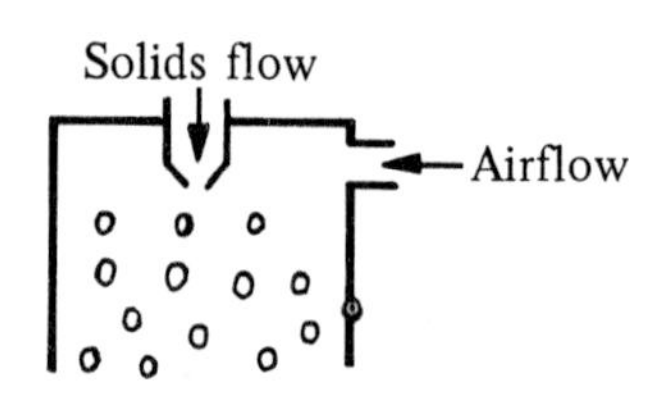

Gas-flow conditions

1 Parallel flow. This type of flow occurs in a tray drier. The gas flows in a direction parallel to the solids surface. Contacting is mainly at the phase interface, but with penetration of gas into the voids between the solids near to the surface.

Airflow

2 Perpendicular flow. The gas flows in a direction normal to the phase interface, e.g. hot air impinges on the surface with some penetration into the voids between the solids being conveyed across the drying chamber.

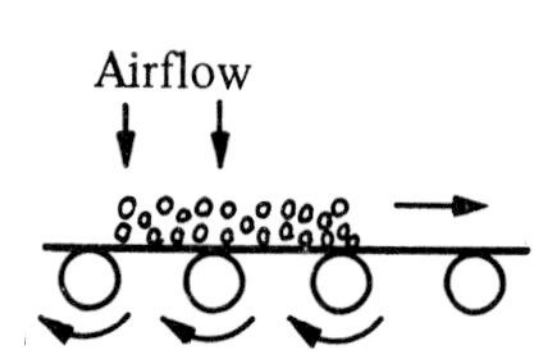

3 Through flow. The gas penetrates and flows through the voids between the solids, circulating more or less freely around individual particles. The solids can be static, moving, fluidized or dilute. The figure shows a perforated tray drier.

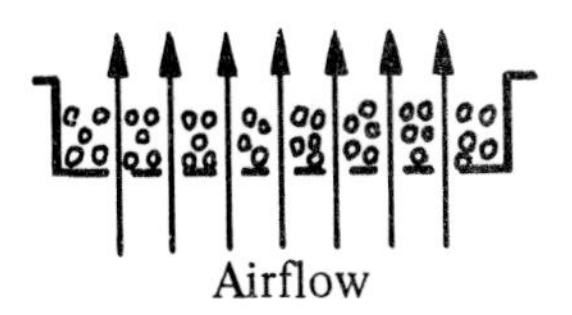

The following flow modes are used to describe continuous drying equipment.

4 Co-current flow. The gas and solids flow in the same direction. Co-current hot air and solids flow upwards in a pneumatic conveying drier.

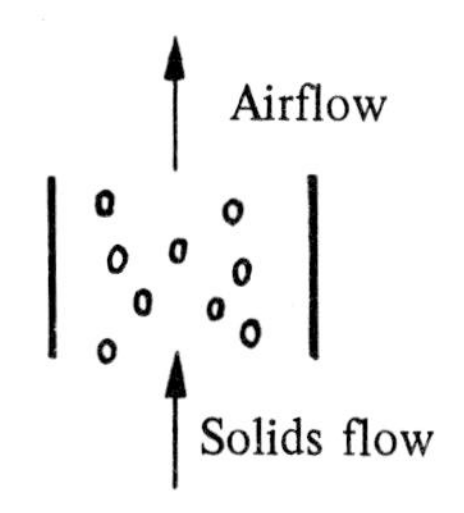

5 Counter-current flow. The direction of gas flow is opposite to the direction of solids movement. The sketch shows granular solids flowing downwards against an upward air flow in a tower drier.

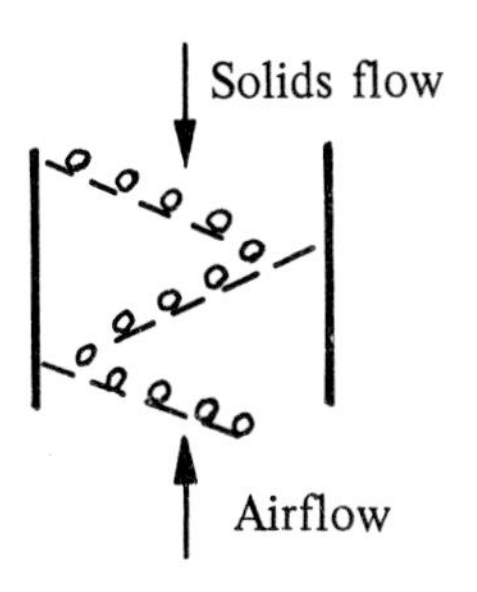

6 Cross-flow. In these the gas flows at right-angles to the direction of solids movement. An example is a cascade-type drier.

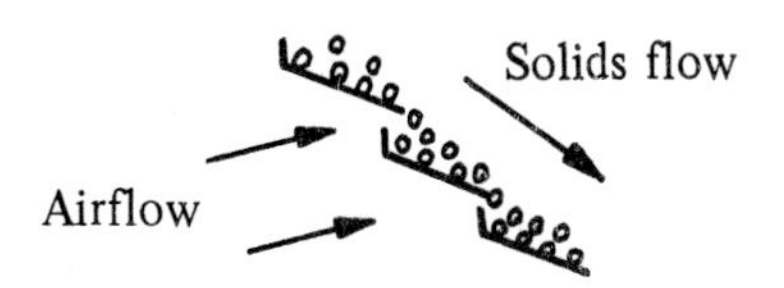

15.1.3 Radiation drying

This process depends on the efficient generation and transmission of infrared radiant energy and its absorption by the article to be dried. The materials being treated must present a flat, smooth plane, normal to the path of the radiation. In practice, by suitably arranging several radiating surfaces, articles with more than one plane surface can be dried successfully. Irregularly shaped articles will suffer from shadowing of parts of the article by the irregularities if radiation is the only drying mode employed. This effect can be lessened by:

1 Reradiation within the irregularities.
2 Conduction through the article if the material of construction is a good conductor.
3 Setting up natural convection currents within the irregularities.

The effect of shadowing as hinted at above is best overcome by supplying forced warm-air convection in addition to the radiating source.

It is apparent from the above that radiation methods of drying are most suitable for treating coated and inked paper and board, textiles, painted metal sheets and other flat products. This range can be extended to cover thin beds of granular, powdered or fibrous products if provision is made to present all the surfaces of the products to the radiation by vibrating or raking the beds to turn the product continuously.

In comparison to the other heat-transfer modes used in drying, radiant heat has a number of advantages:

1 Instantaneous heat. Radiant heaters can reach operating temperatures from cold in a few minutes whereas convection modes usually require 15 to 30 minutes.
2 Simplicity and flexibility. Gas-fired radiant heaters are light, compact and adaptable. They can be installed exactly as required to suit individual products.
3 Cost. Installation charges are lower.
4 Ease of control. The radiant-flux density can be varied quickly to suit a variety of surface conditions.
5 Low losses. Very little of the radiant energy is lost to the surrounding air; units can be open to allow the drying process to be observed.

These factors make radiant dryers particularly suitable for continuous operation. The ability to grade the radiation along the drier means that when the material is wet, strong heat can be applied in the initial stages of drying at the entrance and gentle heat near the exit. Conversely, the product can be heated gently in the initial stages and more strongly in the later stages when heat treatment of the products is required.

Resulting from the nature of the radiant heat there are also a number of disadvantages to this mode of drying as opposed to other methods. Drying rates by radiation alone are adversely affected by:

1 Irregular surfaces. The effect produced is lessened to some extent by reradiation, conduction and the introduction of convection.
2 Different absorption capacities of variously coloured paint and surfaces. Dark, matt surfaces absorb more radiation and therefore dry faster than light polished surfaces. This can be allowed for by adjusting the flux density.
3 Thickness of surface to be dried. This is also affected by the absorptivity of the surface. A highly absorptive surface will give drying in depth, but low absorptivity may result in surface hardening and incomplete drying.

These factors mean that before employing radiation drying, the emissivity of the surface to be dried should be known, indicating the amount of heat which can be absorbed, reflected and transmitted. Then a radiant source is selected with the correct emissivity and wavelength to match those of the material to be dried.

Gas radiant heaters are available with a large range of emissivities which enables them to be used as heat sources in an equally large range of drying applications. High-temperature ceramic plaques attain temperatures of up to 980 °C in about five minutes providing a maximum wavelength intensity of 4.7 μm and a flux density of up to 160 kW m^{-2} (50 720 Btu ft^{-2} h^{-1}). Medium/low-temperature radiant panels are made to attain temperatures of up to 350 °C with a maximum wavelength intensity of 2.7 μm and a flux density of 13 kW m^{-2} (4121 Btu ft^{-2} h^{-1}).

Arrangement of the radiant sources

The radiant heaters can be placed in a variety of ways to suit the particular drying operation. Figure 15:1 shows a number of possible arrangements.

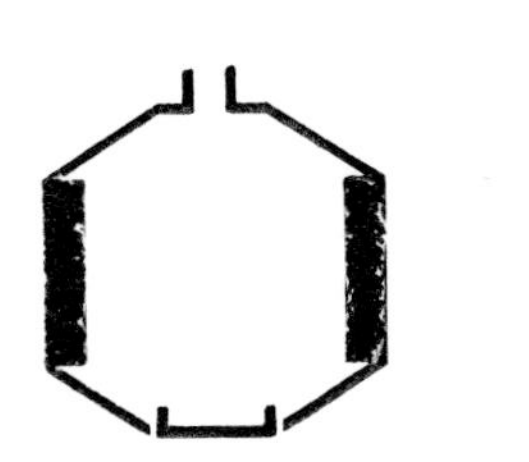
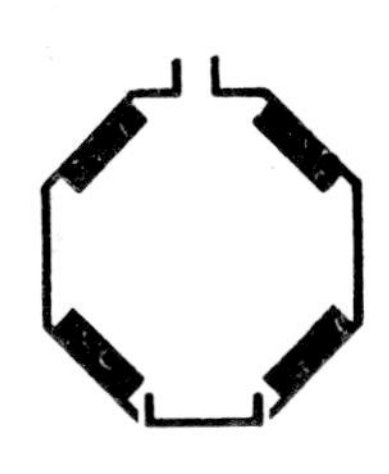
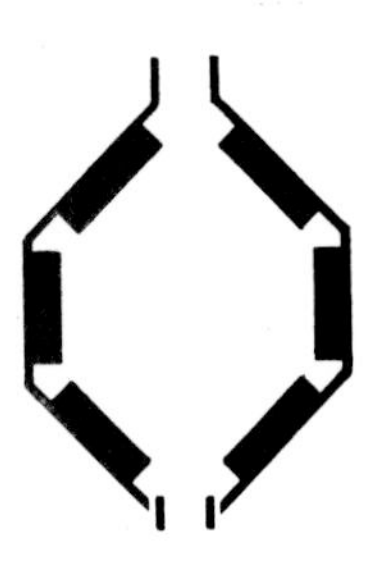

Figure 15:1 Possible arrangements of infrared radiant plaques in a drying tunnel (sectional views)

Usually, when flat articles moving in a horizontal plane are being dried, the radiant sources are placed in the roof of the drier. However, in certain instances the source is placed below the material to achieve some other process advantage, e.g. when drying size coatings in carpet manufacture, the steam from the size rising through the carpet helps to 'burst' the pile.

Radiation onto flat surfaces creates natural convection currents which take away evaporated vapours. This is most effective when the radiation source is mounted vertically or at an angle to the horizontal. When the radiants are mounted horizontally in the roof, the operation is improved if some air circulation is introduced to take away the vapours and combustion products.

The trend in continuous radiation driers is towards high-temperature sources and fast conveyor speeds. Metal surfaces are often suitably placed in the drier to reflect heat from the radiant heaters onto the drying surface to improve efficiency.

15.1.4 Curing

In most drying operations, solvent removal is unconnected with any chemical changes in the product. However, in some instances solvent evaporation is connected with oxidation and hardening of oils and resins. Therefore, in some cases radiation drying may overlap into curing, where the removal of solvent by drying forms a part of the curing process, increasing the strength and durability of the finished product.

15.1.5 Conduction drying

This method of drying involves heat transfer to the wet solid by physical contact with the heated surface of the drying equipment.

The heat source can be:

1. Condensing or superheated steam. This is the most common. It gives heat only in accordance with demand. It is economical in heat consumption, particularly if use is made of low-cost steam from some other process. Efficiencies up to 75 per cent can be obtained. However, efficiency falls as the final desired moisture content increases. The greatest limitation is that high drying temperatures cannot be obtained economically.
2. Hot water or other heat-transfer fluid. These give a faster heat transfer than steam.
3. Combustion gases. These also give a faster heat transfer.
4. Electrically heated elements.

Conduction drying methods are suitable for complete enclosure and therefore adaptable to drying under reduced pressure or in inert atmospheres. Closed systems also enable expensive solvents or gases to be recovered more easily.

15.1.6 Direct-flame drying

This method involves radiation, conduction and convection of heat simultaneously to the article being dried. Control is very important. Regular flame shape and constant combustion characteristics must be guaranteed. The flame must be firm and consistent to avoid burning or scouring the surface of the material being dried.

This method is used for drying printing inks on paper surfaces and in foundries for drying ladles and the blacking on large moulds, but here the use of high-velocity air-blast burners to recirculate combustion products round the ladles obviates the need for direct flame contact.

15.1.7 Dielectric drying

In this method, the material to be dried is placed between two electrodes to which a high-frequency voltage is applied. Heat is then generated rapidly within the mass of material. The great advantage of dielectric drying is its ability to dry the material from the centre—the reverse of all other conventional methods. The risk of overdrying and overheating the surface causing case hardening or other damage is eliminated. The method is expensive, power costs are between 5 and 10 times the fuel costs of conventional methods.

15.2 DRYING PLANT

A survey of drying systems and equipment has been presented by Sloan [2] and the reader is referred to this paper for a comprehensive treatment. The following brief account is drawn largely from Sloan's work and from Wainwright [1].

An important consideration in the choice of a drier for a particular duty is the susceptibility of the material being dried to damage from high temperatures, combustion products or agitation. Materials may be divided into five classes with respect to these effects:

1 The least susceptible materials to these effects are constructional materials such as sands, clays and gravels. Consequently drying is commonly carried out in rotary driers using air and combustion products at high temperatures from a directly fired air heater. The material places no restrictions on the temperature, atmosphere or agitation involved.
2 The second group comprises materials such as fuller's earth and talc which can withstand high temperatures, are unaffected by agitation but may be affected by contaminated atmospheres. Indirectly fired air heaters or radiant heaters are commonly employed but, on natural-gas firing, all but the most susceptible materials should be capable of being dried in directly fired plant.
3 This group, which includes sawdust, lignite, coke, fertilizers and various chemicals, can withstand agitation and contaminated atmospheres but not high drying temperatures (in excess of about 250 °C). Products of this type are usually dried by directly fired methods incorporating recirculation to provide uniform and low temperatures.
4 In this group—which includes many food products such as coffee, milk, starch and eggs, a wide variety of pharmaceuticals, soaps and detergents—agitation is usually acceptable but high temperatures and contaminated atmospheres must be avoided. Drying is usually by indirectly heated air by methods using agitation or spraying.
5 This final group includes metal and ceramic products, large products such as paper and building boards, and products that have undergone a surface treatment. In these cases agitation is either undesirable or impracticable.

Driers may be intermittent or continuous in operation. The latter type is preferable in most cases requiring less labour, fuel and floorspace and producing a more uniform product. On the other hand, batch driers are low in initial cost and maintenance and very versatile in possible applications.

15.2.1 Intermittent driers

Batch drying is generally limited to relatively low product throughputs. Drying times are usually longer than those for continuous driers. Conditions of temperature, humidity and material moisture content are constantly changing which renders precise control difficult. Since the drier structure has to be heated up during each cycle the thermal efficiency is inherently lower than for its continuous counterpart The commonest type of batch drier is the box oven. In fact more than 50 per cent of all low-temperate gas-heated processes are carried out in box ovens [3]. In addition to aqueous drying box ovens are widely used in paint drying and curing. The process of drying involves evaporation of the solvent and in the case of oxidizable paints the process involves the additional stages of oxidation and polymerization. Force drying is the term used to describe paint drying at temperatures up to about 80 °C whilst baking, curing or stoving normally refer to processes above 80 °C. The majority of stoving paints used in industry require oven temperatures around 170 °C. Both directly fired and indirectly fired types of box oven are in use, the former being termed double-case ovens and the latter treble-case ovens (Figure 15:2 (*a*) and (*b*)). The design of pressure reliefs for ovens of these kinds is considered in Section 9.3.3

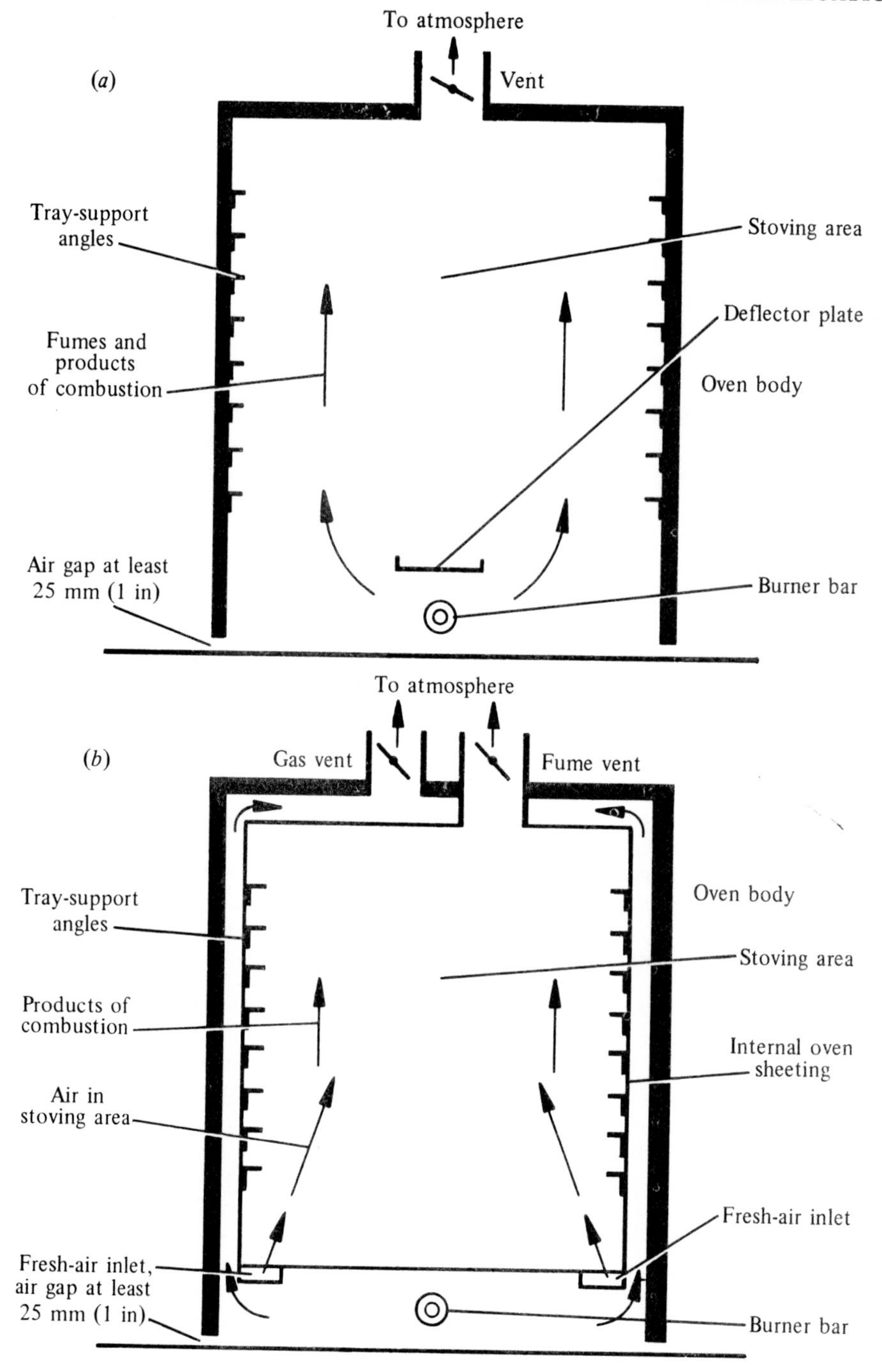

Figure 15:2 Box ovens
(*a*) Double-cased oven, natural convection
(*b*) Treble-cased oven, natural convection

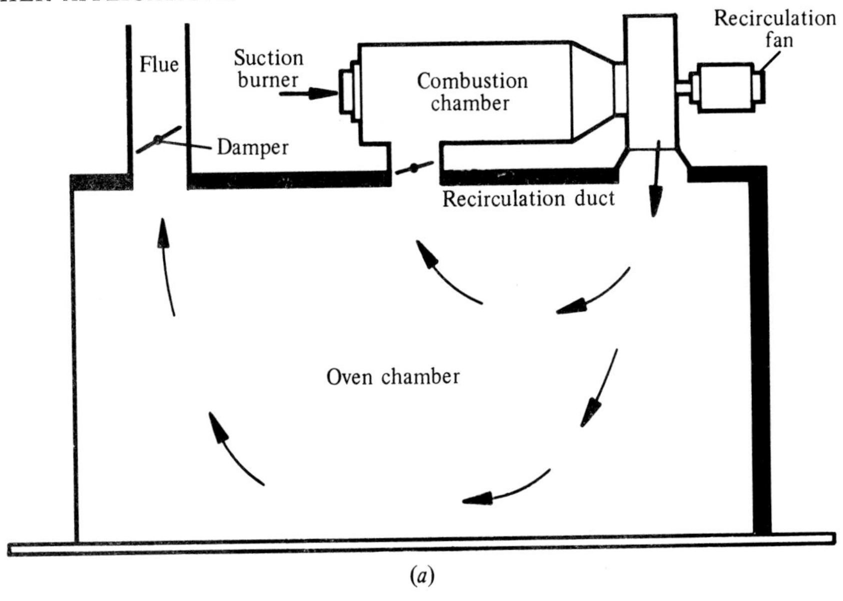

(a)

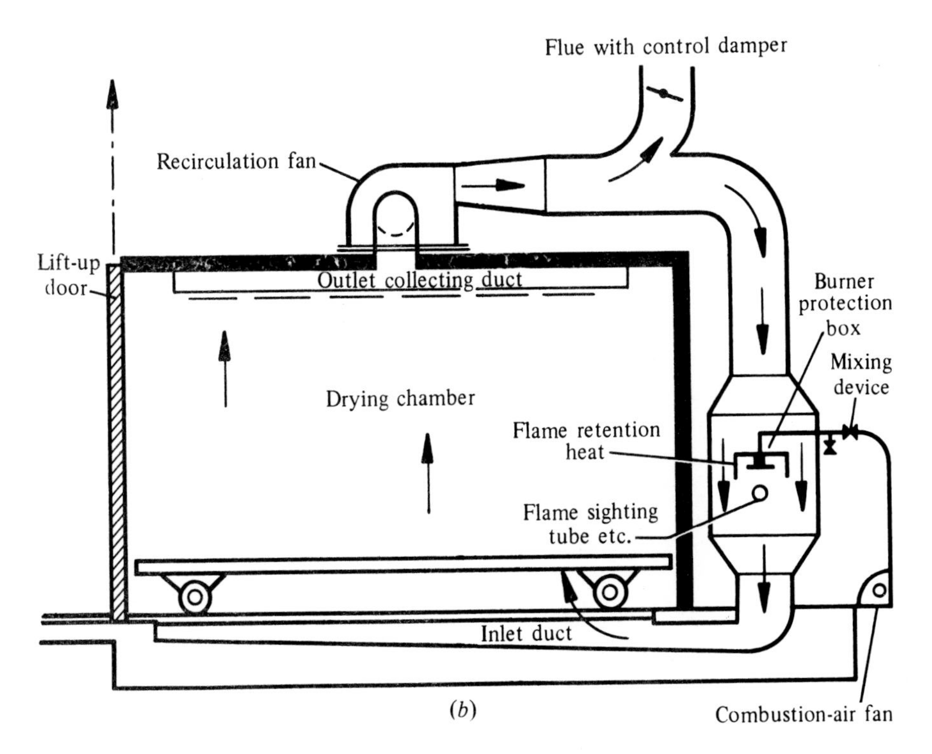

(b)

Figure 15:3 Recirculating drying ovens
(a) Suction-burner type
(b) Forced-draught type

The majority of small box ovens are heated by simple aerated bar burners fitted with thermoelectric combustion safeguard equipment or, increasingly, electronic equipment.

Forced-convection ovens normally make use of what is in effect an integral directly fired air heater, the burner being sited upstream or downstream of the main circulating fan. If it is sited upstream of the fan and the combustion chamber is below atmospheric pressure a suction burner is generally employed (Section 7.3.3) (Figure 15.3). In the alternative system in which the burner is located downstream of the fan, a pressure burner system is required. Forced-draught premix or nozzle-mixing burners

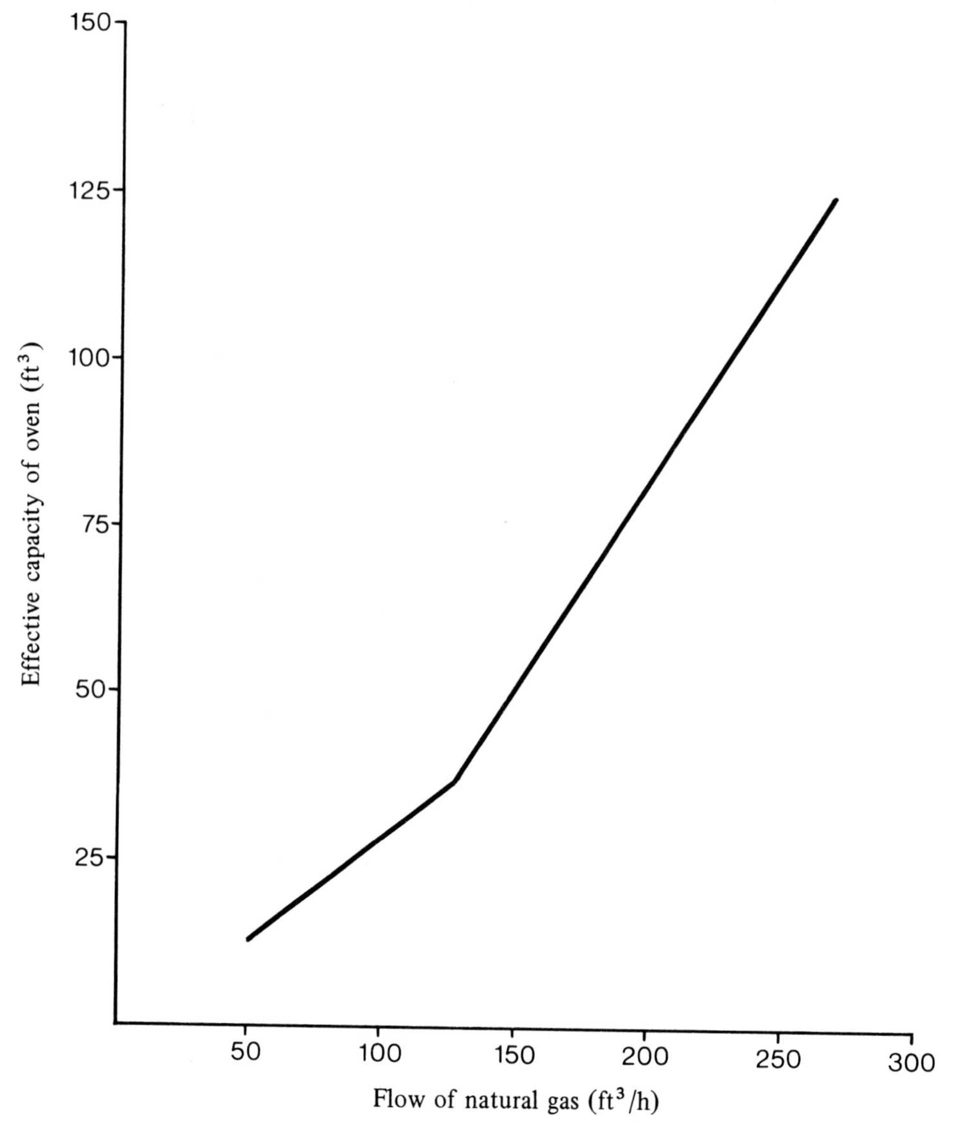

Figure 15:4 Maximum hourly consumption of natural gas by typical box ovens [3]

or, if there is sufficient excess air, neat-gas burners are employed. The use of forced-draught premix burners minimizes the flame size and mixing with the airflow may be enhanced due to the momentum of the combustion products [4]. Both of these factors minimize the space occupied by the combustion chamber. Indirectly fired forced-draught ovens incorporate an integral indirect air heater normally in the form of a single burner firing into a combustion chamber incorporating a tube bank or in some cases using individually fired heat-exchange tubes.

Burner rating for box ovens clearly varies with design and operating conditions. Some guidance to typical maximum gas rates is, however, provided by Figure 15:4 which relates to recirculating types of box ovens.

Drying rooms are sometimes employed for expensive products, such as tobacco, bacon and leather, which are hung or stacked in the room. Very low temperatures and lengthy drying periods are normally involved, recirculation being employed to improve temperature distribution and thermal efficiency.

15.2.2 Continuous driers

These are generally more efficient, faster and more easily-controlled than batch types.

Rotary driers

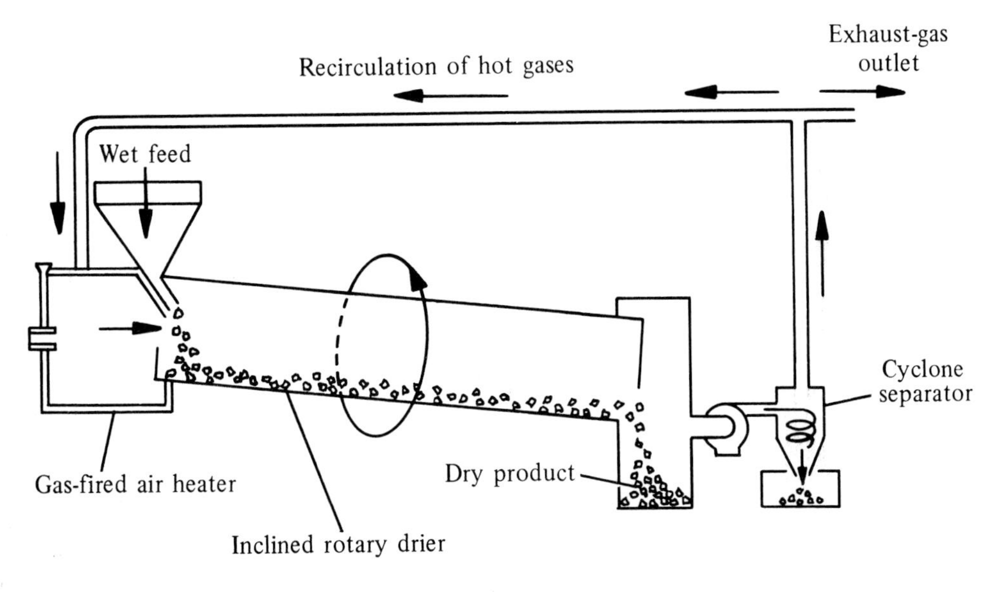

Figure 15:5 Single-shell rotary drier with recirculation

These are used mainly for granular materials like sand and chemicals. In fact far more chemicals are dried by this method than by any other type of drier. Figure 15:5 illustrates a recirculating inclined rotary drier. The inside of the drier is normally equipped with flights running the length of the drier and these lift the material and shower it through the air stream exposing the maximum surface area. Flight geometry affects the drying rate. Normally simple radial flights are used at the feed end for wet materials, whereas flights incorporating a 90° bend are used for free-flowing or nearly dry materials. Airflow may be counter-current or co-current to the material. Counter-current flow is normally used to dry materials to a low final moisture content, the hottest and dryest air contacting the dryest material. Co-current flow is used for heat-sensitive materials. Variants are the double-shell rotary drier and the louvre rotary drier. In the former, hot air passes along a central cylinder then back along the annulus formed between this and the outer shell, the material receiving convected heat as it showers and conducted heat from the inner cylinder. For inlet gas temperatures in the range 530–750 °C evaporation rates are about 32 to 48 kilograms of water per cubic metre of cylinder volume (2 to 3 pounds per cubic foot) for single-shell driers and about 64 to 80 kg m^{-3} (4 to 5 lb ft^{-3}) for double-shell driers.

In the louvre drier the charge is supported by overlapping louvred shelves. No showering is involved and the air flow is through, rather than over, the material. This method produces a minimum of particle breakage.

Double-shell rotary driers are widely used for sand drying. Jet-driven recirculation is often employed, the high temperatures involved precluding the economic use of fan-driven recirculation. The efficiency of a double-shell drier for this process is typically 60 per cent.

Tunnel driers

Tunnel driers include a wide range of conveyance mechanisms, e.g. trucks, conveyors, moving trays, overhead monorails and self-conveying loads.

Tunnel driers are zoned to produce the desired time–temperature profile, air temperature and sometimes humidity being controlled in sections along the tunnel.

Truck or car driers are used widely for ceramic products, often as an adjunct to tunnel kilns and in some cases using heated air from their cooling sections.

Overhead monorail driers are widely used in paint drying and curing. Figure 15:6 illustrates a camel-back continuous stoving oven fired by an integral directly fired air heater. The products suspended from the overhead conveyor enter and leave the oven via the 45° sloped entry and exit openings which provide natural hot-air seals minimizing spillage.

In the case of paper and textiles drying, the material forms its own conveyor being held in tension and moved through the tunnel by rollers. The subject of paper drying is considered in detail in reference [5].

Flash driers or pneumatic driers

In its simplest form a flash drier consists of an air conveyor system using heated air. Drying is generally very fast. Particulate solids are simply dropped into a rapidly moving stream of hot gases or air, gas velocities being typically 18 to 30 m s^{-1} (60 to 100 ft s^{-1}). The solids are usually separated from the gas phase in a cyclone. With materials of very high moisture content the water vapour evolved expands the gas volume to such a degree that it is difficult to obtain even the limited residence time required and in these circumstances systems which recycle part of the product are employed. Size reduction (milling) and drying are sometimes carried out simultaneously.

Spray driers

These may be considered as a variant of the flash drier in which the feed enters the gas stream as a liquid or paste. This method is finding increasing application par-

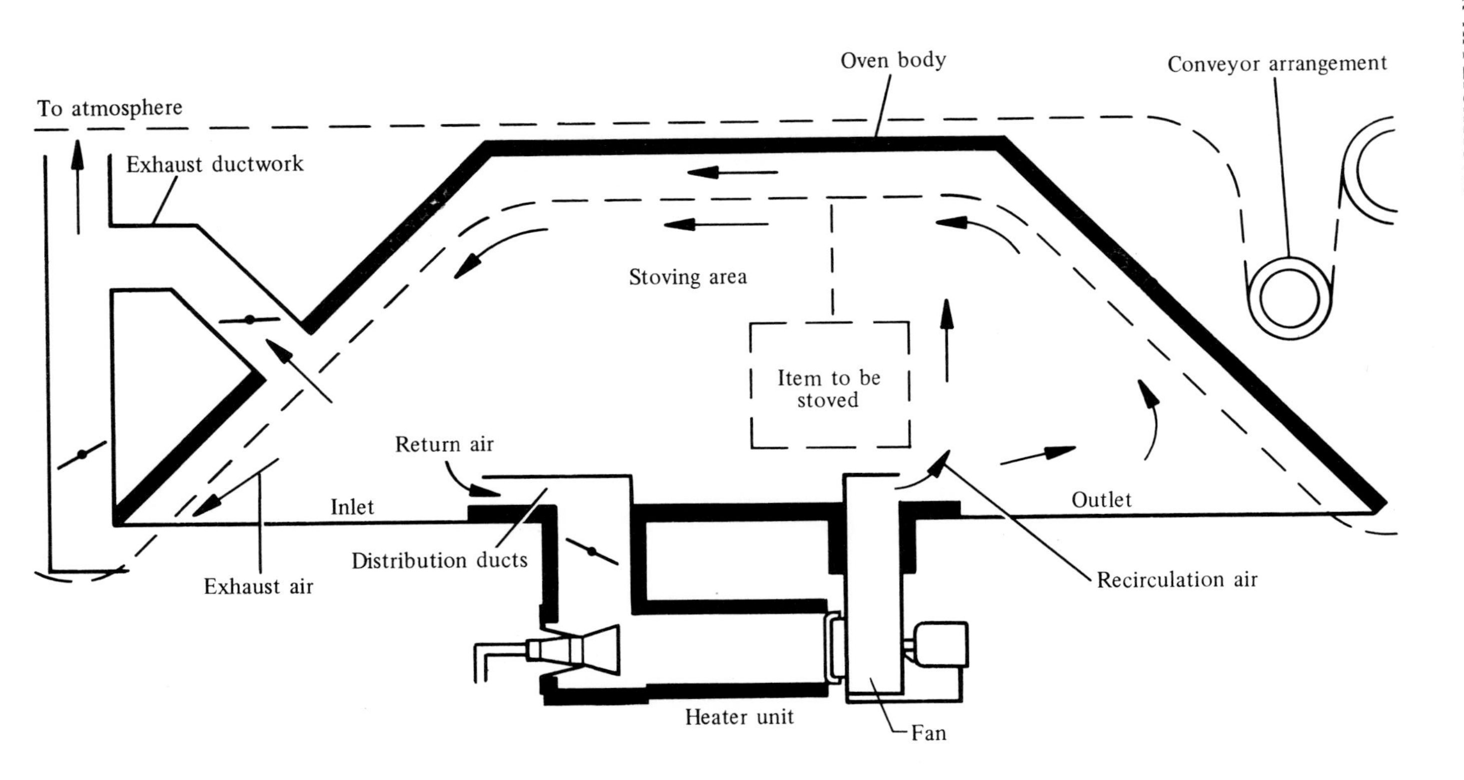

Figure 15:6 Camel-back conveyor oven

ticularly for heat-sensitive materials since although relatively high gas-phase temperatures are employed the drying is so rapid and the residence time so short that the material is not overheated. A wide range of foodstuffs, e.g. milk, coffee and starch, pharmaceuticals and chemicals are dried in spray driers. Figure 15:7 shows the system in outline. The material to be dried enters as a liquid or slurry and is atomized and dried by hot air in a drying chamber. The descent of the particles is delayed by

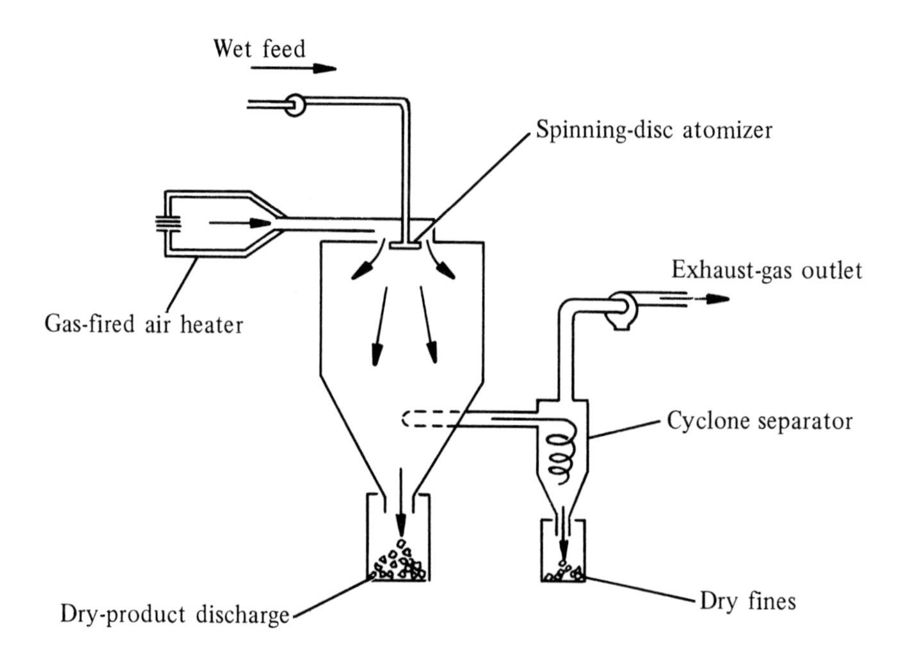

Figure 15:7 Spray drier [1]

the buoyant effects of the escaping gases and a tall tower is not normally necessary. Most of the solids (about 90 per cent) reach the bottom of the chamber and the small proportion of carried-over fines are collected in a cyclone separator. The spray droplets have a very large ratio of surface area to volume and never reach the gas-phase temperature since the air is cooled by evaporation from the droplet and the dry product leaves the system before the hot air can raise its temperature appreciably. The feed may be atomized using a rotating disc, high-pressure nozzle or two-fluid nozzle. A wide variety of spray and chamber arrangements are employed and the reader is referred to reference [2] for a comprehensive treatment.

Spray drying has a number of important advantages: precise spherical particles are produced where size and size distribution can be controlled, maintenance costs are low and a graded marketable product is produced which requires no subsequent treatment.

The air for drying may be indirectly or directly heated. Direct gas-fired air heaters are suitable for almost all materials but direct firing using other, less refined fuels is rarely suitable.

Fluidized-bed driers
These are finding increasing application, primarily in large-scale drying processes, e.g. ores and minerals, chemicals and some food products, e.g. grain, rice and nuts.

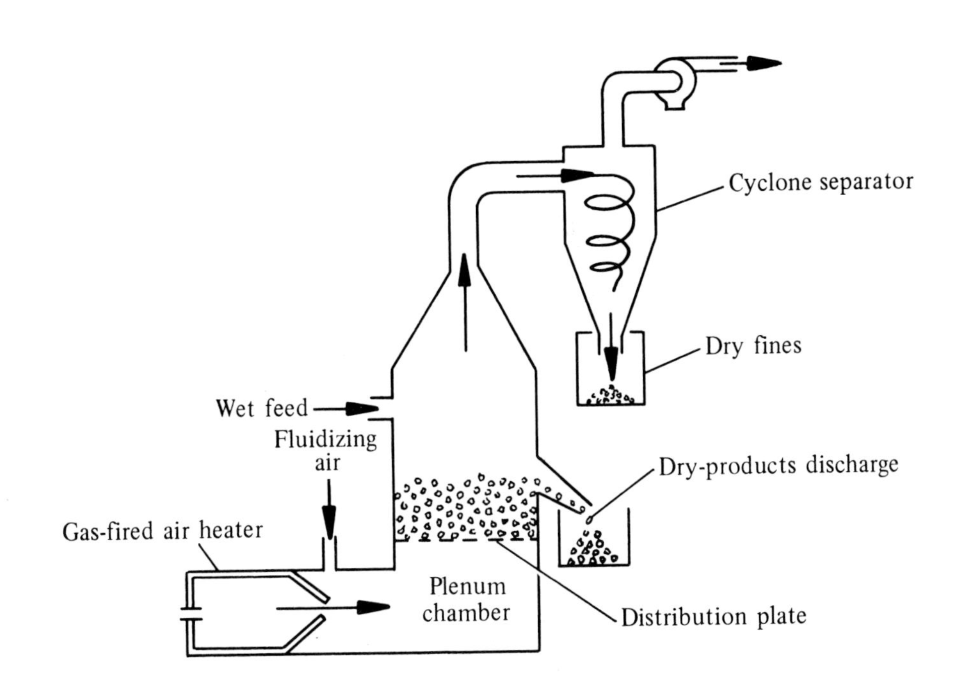

Figure 15:8 Fluidized-bed drier [1]

Figure 15:8 shows the general features of a fluidized-bed drier. Each particle to be dried is surrounded by heated air which supports the particles and the agitation produced provides uniform temperatures in the bed. As with spray drying, direct gas-fired air heaters provide a suitable heating means for almost all materials. Fluidized-bed driers have the following advantages [1]:

1 Heat-transfer rates are very high, leading to compact units of high efficiency. Mass-transfer rates are also high, aided by the large transfer surface available.

2 Particle-breakage rates are low since the particles are cushioned from each other by the surrounding air.

3 Adaptability. Fluidized-bed driers are applicable to most materials and are particularly useful for difficult-to-handle materials. However, a wide particle size range presents problems since the fluidizing velocities needed for large particles will elutriate the smaller particles.

4 Drying and classifying can be done simultaneously, eliminating the need for subsequent screening and dust removal.

15.3 TANK HEATING

15.3.1 Introduction

It was estimated in 1975 that British industry uses some 265 PJ (2500×10^6 therm) of energy a year in heating vats and tanks for metal plating, product washing, dyeing, etc. [6]. In the majority of cases, steam from a central boiler is used either by direct injection into the liquid or indirectly via immersion tubes. Gas firing is a long-standing alternative to these methods and has been applied using equipment ranging from simple open tanks heated by natural-draught bar burners to forced-draught immersion tubes operating at high efficiencies. Although there are large numbers of gas-heated tanks in existence the share of the total market held by gas is relatively small. In general gas has been used where no central steam source is available and this has generally, although by no means exclusively, restricted gas firing to relatively small tanks. Recent developments in the design of forced-draught burners for immersion tubes should result in a much greater market penetration in the future.

15.3.2 External heating

In its simplest form this involves the application of one or more natural-draught bar burners to the tank base. In this rudimentary form, efficiencies are low, heat transfer is restricted to natural convection and there is no control of secondary aeration. Underfired systems can achieve acceptable efficiencies if a correctly designed combustion chamber is employed and the tank is enclosed in a sheet-metal casing to provide ducts for the combustion products along the tank sides and ends, the gases finally discharging through holes along the top edge of the casing. Heat losses from the tank surface and from the bare tank walls may be considerable. A simple metal cover, where applicable, reduces the aqueous surface heat loss by up to 90 per cent [7] and a layer of floating plastic spheres is commonly used as an insulating cover which allows easy entry and withdrawal of treated articles. Heat losses from the tank sides should be minimized by insulation and this, in the form of structural insulating boards may form the outer jacket. Lupton [7] has described a modular system based on ciment-fondu/vermiculite preformed sections. In a number of examples of underfired aqueous degreasing tanks quoted in [7] the efficiencies are about 33 per cent for heating up from cold but no constructional details are quoted. In general a well insulated tank with side ducts will have an efficiency of up to 60 per cent for heating up and around 50 per cent overall. Immersion-tube heating allows higher efficiencies to be achieved since better control over secondary aeration can be obtained, surface insulation is simpler and the temperature differential controlling heat losses is reduced. Additionally in many applications sludge buildup on the tank base may preclude simple bottom firing.

15.3.3 Immersion tubes

Immersion-tube applications cover a wide range of liquid temperatures in addition to aqueous solutions; oil boiling (260–320 °C), salt baths (200–600 °C) and type-metal melting (260–320 °C) are a few examples.

In situations where the nature of the process requires a completely open vessel interior, external heating must be used but in most other cases immersion tubes can be used to advantage. For installations where there are no space restrictions immersion tubes fired by natural-draught burners, the tube providing the flue draught, offer advantages of simplicity. Where space is restricted, pressure-air burners or low-pressure burners with fan extract will be necessary.

The design of natural-draught immersion tubes has been the subject of a number of research investigations [8–10].

The American Gas Association [10] conducted tests on horizontal tubes of 13–150 mm ($\frac{1}{2}$–6 in) diameter and between 1.37 m and 13.7 m ($4\frac{1}{2}$ ft and 45 ft) long using natural gas at a fixed excess-air value of 20 per cent. The results were correlated by the empirical equation:

$$\eta = 20 \log_{10} [(L/\text{ft})^2/(R_h/1000 \text{ Btu h}^{-1})] + 71 \tag{15.1}$$

where η = percentage efficiency (gross)
L = tube length
R_h = rate of heat input

Patrick and Thornton [8] carried out a survey of 63 industrial town-gas immersion-tube-heated systems and obtained a number of regression equations, the one recommended for design purposes being:

$$\eta = 80.9 - 28.3\exp(-0.02L_e/D) \tag{15.2}$$

where D = tube diameter
L_e = effective length of the tube, i.e. centre-line length of immersed tube + 1.1 ft (335 mm) for each 90° elbow or return bend

The absence of a heat-input term from equation 15.2 is taken to indicate that there is no optimum value for natural-draught tubes, the maximum heat input being controlled by combustion requirements.

Tailby and Clutterbuck [9] measured heat-transfer rates and efficiencies using post-aerated and premixed town-gas–air flames in 50 mm (2 in) and 100 mm (4 in) tubes. Again a statistical approach was adopted, the results being correlated by analysis of variance.

The best regression equation for the overall efficiency was:

$$\eta = 72.77 + 79.53 \log_{10}(L/\text{ft}) - 0.4865 (L/\text{ft}) \log_{10}E - 5.683 \log_{10}E - 45.51 \log_{10} (R_h/1000 \text{ Btu h}^{-1}) + 22.62 \log_{10} (D/\text{in}) \tag{15.3}$$

where E is the amount of excess air (per cent).

The application of conventional heat-transfer theory to immersion-tube heating is considered in Section 12.5.2 where it is shown that good agreement with practical measurements may be obtained. Figure 15:9 illustrates typical natural-draught immersion-tube installations. In the Temgas immersion heater (Figure 15:9(*b*)) the tube is replaced by an internally baffled combustion chamber and, since no perforation of the tank side is required, the system is easily applied to existing tanks.

Thornton and Robertshaw [11] have presented a design procedure for immersion tubes fired by low-pressure gas burners operating under natural-draught conditions based on graphical charts derived from equation 15.2. A specific value of heat-input

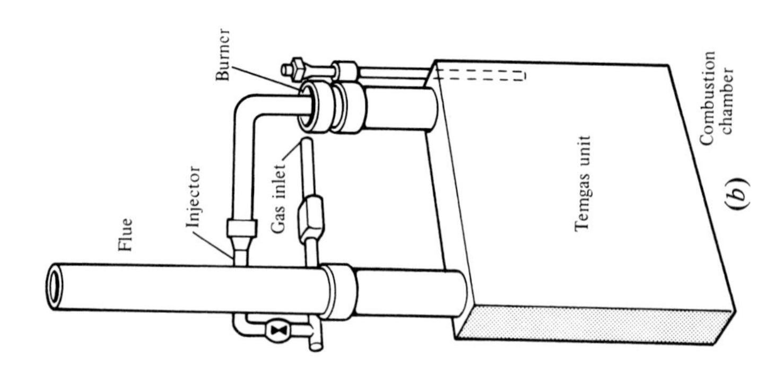

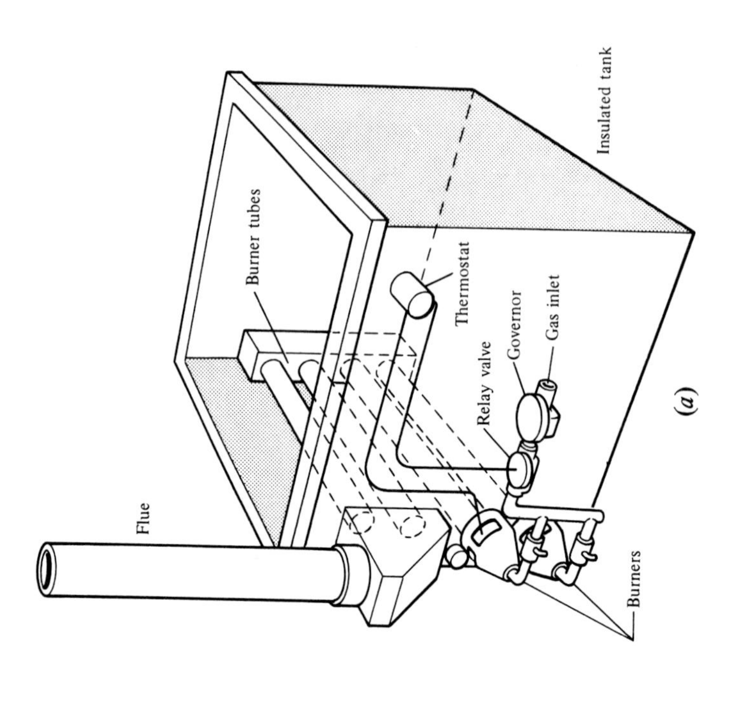

Figure 15:9 Natural-draught immersion-tube systems
(*a*) Simple natural-draught system
(*b*) Temgas immersion heater

rate for each tube size has been selected based on normal conditions of use employing a primary flue height of 1.52 m (5 ft) as shown in Figure 15:10.

Nominal size (in)	2	$2\frac{1}{2}$	3	4	5	6
Nominal size (mm)	50	63	75	100	125	150
Heat-input rate (Btu h^{-1})	18 270	41 700	61 800	113 900	151 000	226 000
Heat-input rate (kW)	5.35	12.2	18.1	33.4	44.2	66.2

Figure 15:10 Selected value of heat input for immersion tubes of various sizes [11]

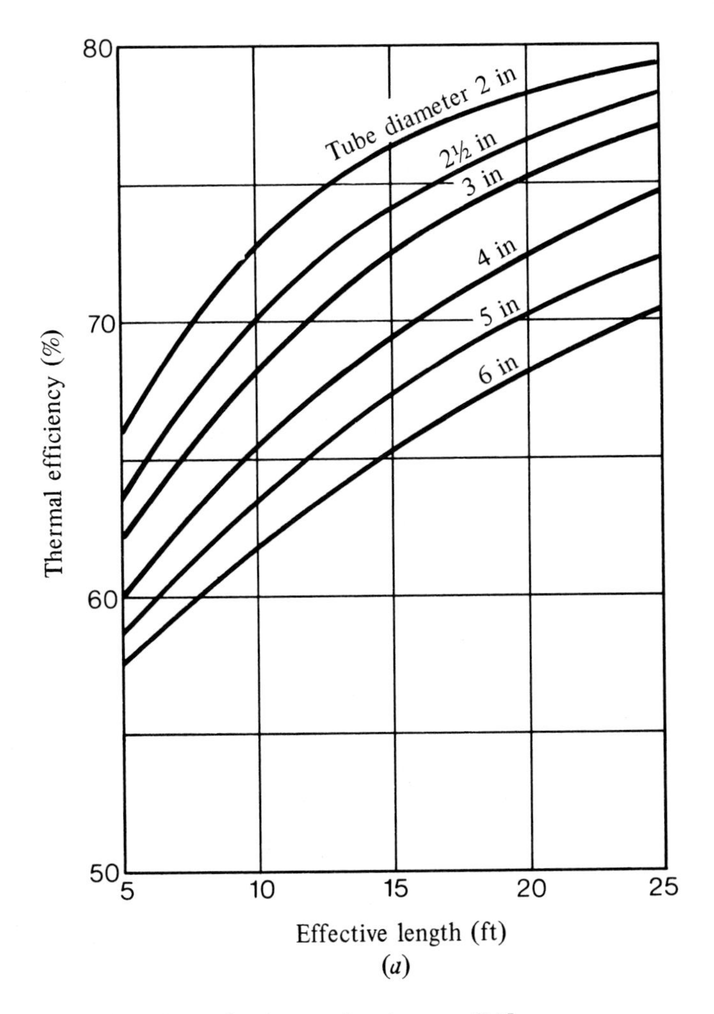

Figure 15:11 Design charts for immersion heaters [11]
(*a*) Variation of thermal efficiency with tube length for tubes of various diameters
(*b*) Rate of heat transfer for tubes of various dimensions

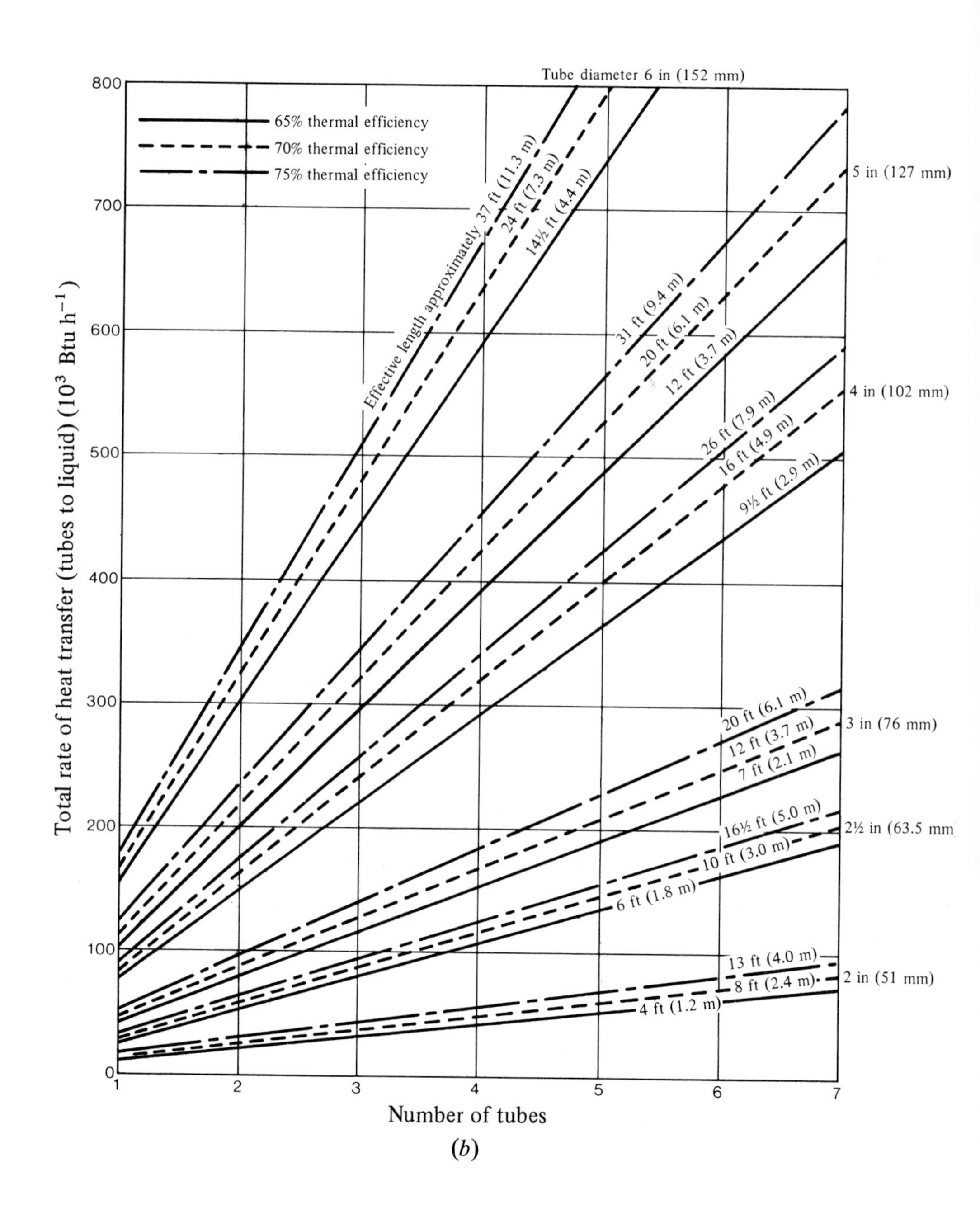

Tube diameter 6 in (152 mm)
65% thermal efficiency
70% thermal efficiency
75% thermal efficiency
Effective length approximately 37 ft (11.3 m)
24 ft (7.3 m)
14½ ft (4.4 m)
5 in (127 mm)
31 ft (9.4 m)
20 ft (6.1 m)
12 ft (3.7 m)
4 in (102 mm)
26 ft (7.9 m)
16 ft (4.9 m)
9½ ft (2.9 m)
20 ft (6.1 m)
12 ft (3.7 m)
7 ft (2.1 m)
3 in (76 mm)
16½ ft (5.0 m)
10 ft (3.0 m)
6 ft (1.8 m)
2½ in (63.5 mm
13 ft (4.0 m)
8 ft (2.4 m)
4 ft (1.2 m)
2 in (51 mm)
Total rate of heat transfer (tubes to liquid) (10^3 Btu h^{-1})
Number of tubes

(*b*)

The design procedure involves the determination of the total heat requirement for heating from cold, making allowances for heat losses from the tank and liquid surfaces, followed by the determination of the number and length of the tubes from Figure 15:11(*b*). This is plotted for three efficiencies, 65, 70 and 75 per cent and the appropriate value of efficiency may be obtained from Figure 15:11(*a*) which relates tube length and efficiency. Finally a suitable tube arrangement is selected bearing in mind the following considerations:

1 The burner, tube and flue positions should be designed to conform with the working requirements.
2 Consideration should be given to the relative economics of using a large number of small tubes or a small number of large tubes and burners.
3 Burners and control systems should be selected to be capable of dealing with low maintenance rates, for example, in baths using effective surface insulation.

Recent work at the British Gas Midlands Research Station has been directed towards the development of high-efficiency forced-draught immersion-tube designs [6, 53]. In this system, which employs a special design of nozzle-mixing burner, heat-release rates comparable with conventional steam tubes can be obtained and the overall efficiency is typically 80 per cent.

In the burner development, particular attention has been paid to avoiding the pressure fluctuations to which closed, high-intensity combustion systems are particularly prone. Flame stabilization is achieved by a combination of reduced port loading and hot-gas recirculation. Gas flowing from an axial pipe is progressively supplied with primary air through small holes in a divergent cone to avoid high air velocities sweeping the flame away. In addition, hot-gas recirculation takes place around the air jets issuing from the holes in the cone. An outer annulus provides the remainder (15–25 per cent) of the air for combustion. The position of the air cone in relation to the entrance to the combustion chamber is crucial in preventing pulsations and incomplete combustion.

For a particular immersion-tube bore, a greatly increased gas throughput is obtained compared with natural-draught systems; for example, a 50 mm (2 in) diameter tube operating at normal gas-supply pressure 2 kPa (20 mbar, 8 inH_2O) has a quoted rating of 110 kW (350 000 Btu h^{-1}) rising to 291 kW (993 000 Btu h^{-1}) at gas and air gauge pressures of 14 kPa (140 mbar, 56 inH_2O) [53]. Comparison with Figure 15:10 illustrates the enormous increase in rating obtained using the forced-draught system. Comparing heat-transfer coefficients, values are typically 100 W m^{-2} °C^{-1} (18 Btu ft^{-2} h^{-1} °F^{-1}) compared with 3–11 W m^{-2} °C^{-1} (0.5–2.0 Btu ft^{-2} h^{-1} °F^{-1}) for natural-draught tubes.

Forced-draught systems incorporating recirculation have been in wide use for many years particularly in the field of low-temperature metal melting. These units are mainly of American origin and, in common with a lot of American equipment, use machine-premix burner systems.

As indicated in Section 7.9, pulsating combustors have been applied to tank heating. For a given pipe diameter the heat input is much greater than with a natural-draught system. The power obtained from the pulsations can be used to propel the combustion products through tubes of high length-to-diameter ratio resulting in high efficiencies and a forced air supply is not required. Figure 15:12 shows an experimental pulsating-combustion immersion-tube unit for tank heating.

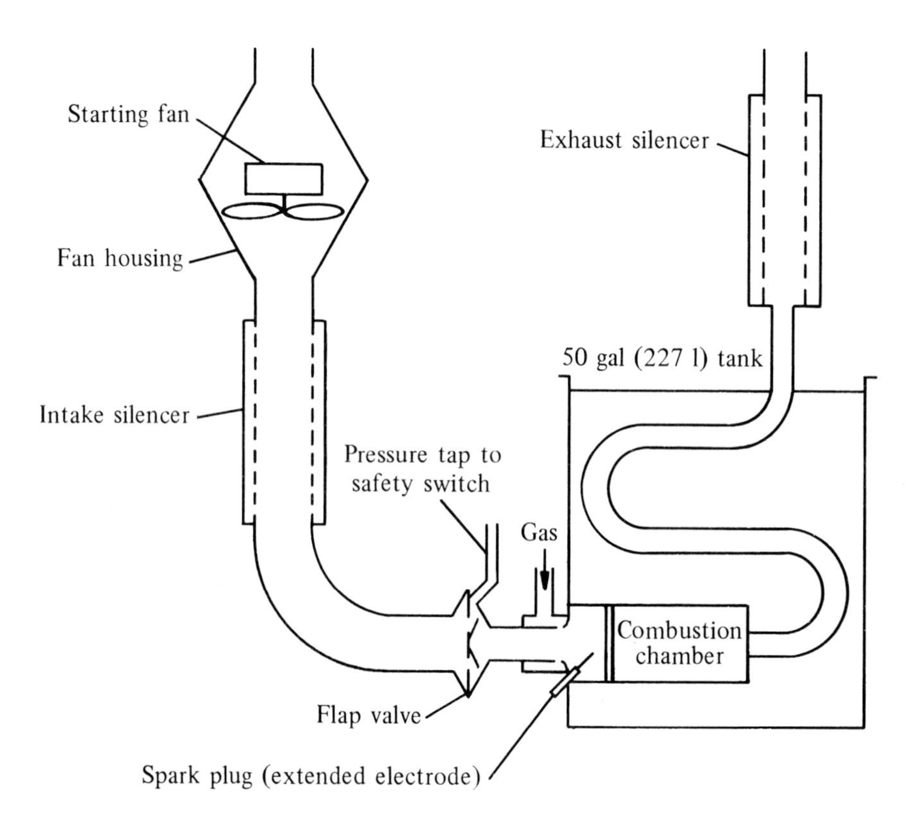

Figure 15:12 Application of a pulsating combustor to tank heating [12]

The unit shown uses a 75 mm/19 mm (3 in/$\frac{3}{4}$ in) Helmholtz unit having a maximum rating of 20 kW (70 000 Btu h^{-1}). A fully automatic start sequence is incorporated, the small fan running only for a few seconds to purge the system of combustion products from previous operating cycles.

15.3.4 Submerged combustion

The tank-heating methods described so far are indirect in the sense that the combustion products are isolated from the tank contents by a metal wall. In submerged-combustion systems the flame is submerged and burns below the liquid surface, the combustion products transferring heat directly to the tank contents.

Some of the heat liberated by the burner is transferred to the burner walls but since the burner is immersed in the tank liquid the heat is quickly transferred to the liquid. Most of the heat liberated by the burner leaves the burner head as sensible heat in the combustion products which take the form of large numbers of very small bubbles. These present a very large total heat-transfer surface so that the combustion products cool quickly to the temperature of the bath liquid, the combustion products and water vapour produced leaving the liquid surface together. Both components contribute individual partial pressures totalling atmospheric pressure in proportion to their

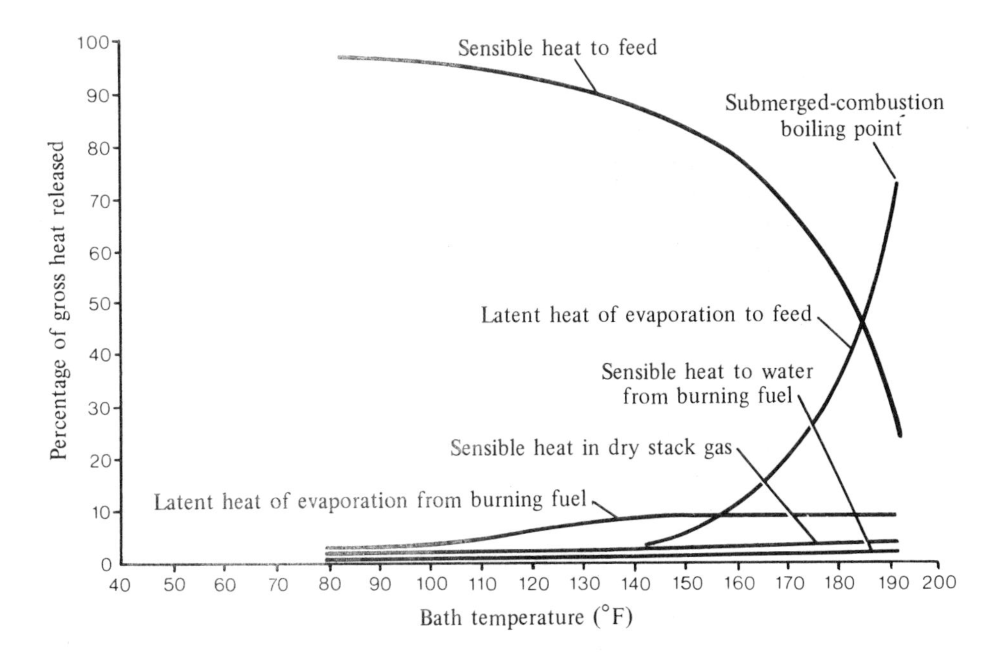

Figure 15:13 Submerged combustion: variation of percentage of gross heat released with bath temperature [13]

relative concentrations. Figure 15:13 shows the effects of water temperature on the heat distribution. The same effects occur in the heating of aqueous solutions and the chart relates to these provided corrections are made to the partial pressure of water because of the presence of the solute (Raoult's law).

As Figure 15:13 shows, as the bath temperature increases the proportion of the heat input appearing as sensible heat in the water reduces and the proportion used to evaporate the water increases. For example, at 50 °C about 1 per cent of the gross heat input appears in the heat content of the dry combustion gases, about 5 per cent as heat of vaporization in the water content of the combustion products and the balance, 94 per cent, appears as sensible heat in the water. At 89 °C, the 'submerged combustion boiling point', no sensible heat is taken by the water, the heat distribution being 2.2 per cent to the dry stack gases, 11 per cent in the water-vapour content of the combustion products and 86.8 per cent in water vapour evaporated from the bath [13].

As a result of these characteristics, submerged combustion has found particular application in:

1 Concentration of dilute solutions and crystallization.
2 Tank heating, particularly of corrosive liquids.
3 Water and effluent treatment.

As Figure 15:13 illustrates, submerged combustion is an effective means of evaporation. Additionally, since the burner-wall area is very small and only a small proportion

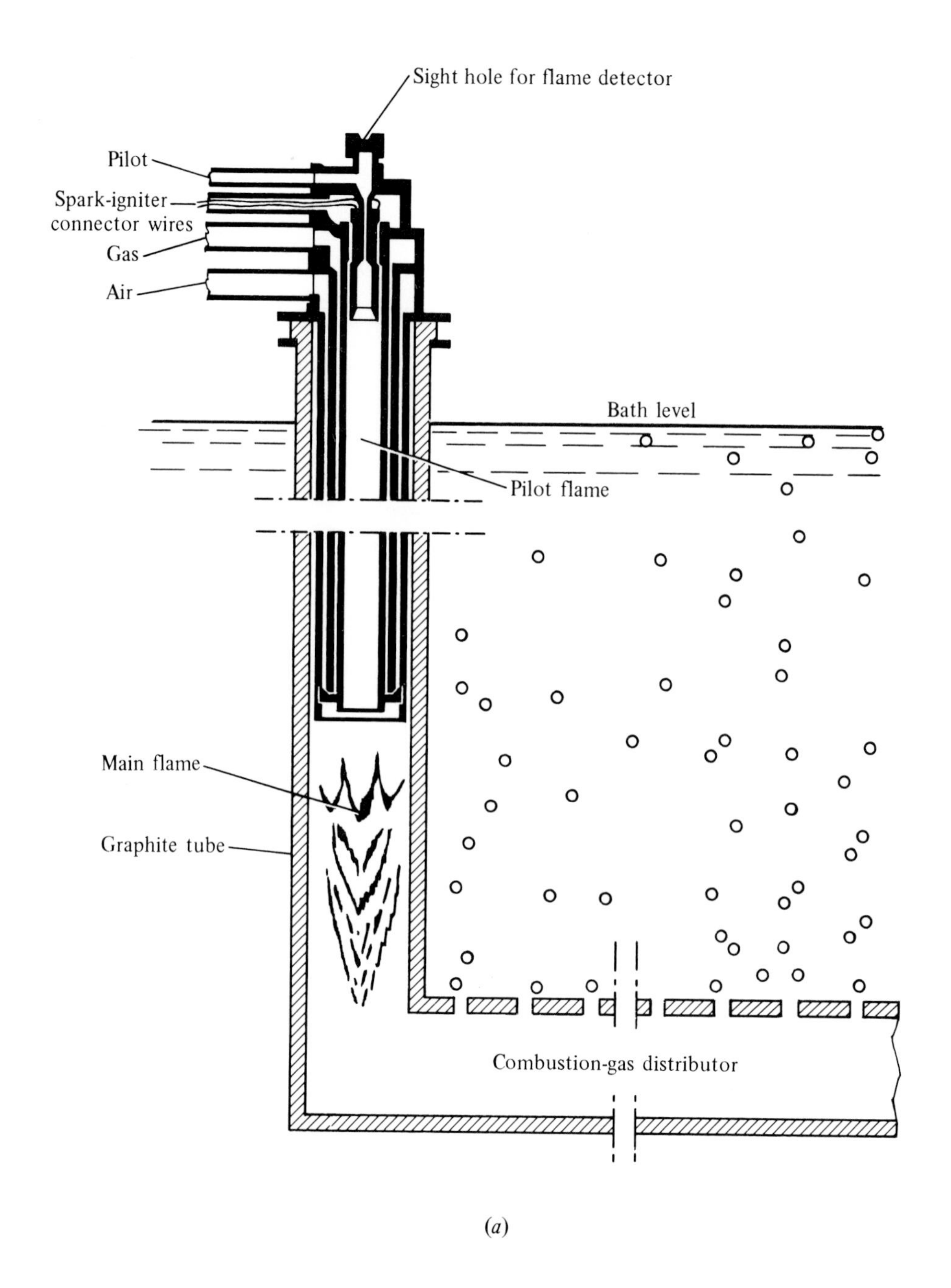

Figure 15:14 Submerged-combustion burners
(*a*) Nozzle-mixing type [3]
(*b*) Machine-premix type [18]

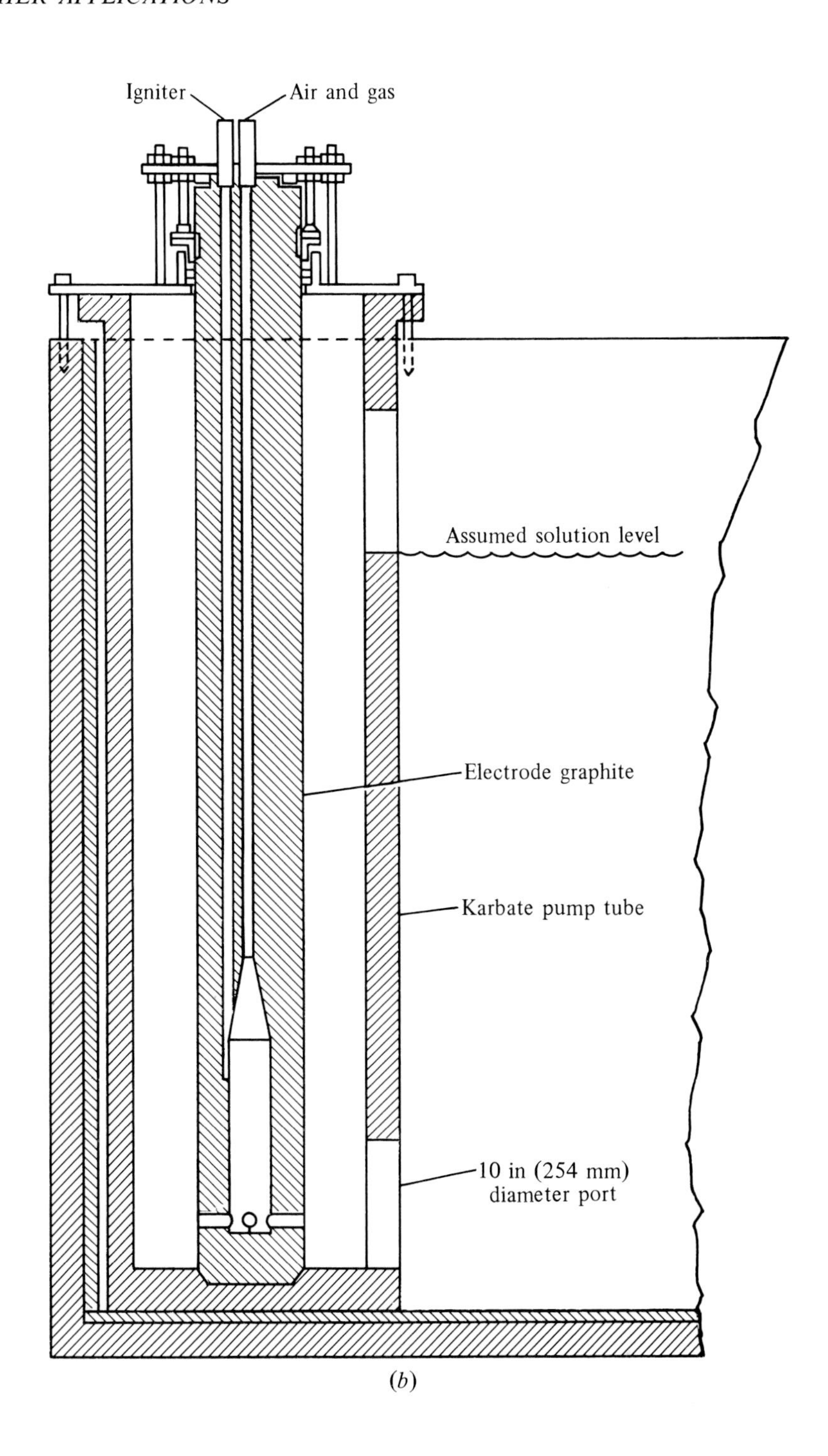
Igniter
Air and gas
Assumed solution level
Electrode graphite
Karbate pump tube
10 in (254 mm)
diameter port

(b)

of the heat transferred passes through it, the difficulties associated with scale formation in conventional systems are virtually eliminated. In addition to concentrating acids and other inorganic solutions, submerged combustion has been used in the concentration of organic solutions such as phthalic anhydride and thiourea. Babich *et al.* [14] have developed a mathematical model to describe the steady-state operation and economic performance of submerged-combustion evaporation equipment for concentrating sulphuric acid. Atkinson [15] and Young and Scott [16] have described the application of submerged combustion in concentrating wet-process phosphoric acid.

Submerged combustion is an efficient method of solution-tank heating up to about 65 °C. It is widely used in the USA for acid pickling. The combustion gases agitate the bath contents giving very small temperature variations. Also it is possible to run the bath at a lower temperature than conventionally heated baths for the same treatment rate. Temperature reductions of up to 11 °C have been reported [17].

The carbon-dioxide-containing combustion products may be used in recarbonation of water, e.g. in the clarification and stabilization of alkaline industrial water and for neutralization of alkaline solutions.

Submerged-combustion burners clearly must ensure complete combustion before the gases leave the burner and the discharged gases must be at a fairly high pressure to overcome hydrostatic pressure and additional system resistances. Nozzle-mixing systems are used but the majority of burners are fed from machine-premix systems.

Figure 15:14(*a*) shows a concentric-tube type fitted with a distributor tube for heating large tanks. The burner shown in Figure 15:14(*b*) was designed for heating sulphuric-acid baths and is fed with mixture at 50 kPa (0.5 bar, 7 $\mathrm{lbf\ in^{-2}}$) from a premix machine via a flame trap. The ignition cycle commences with a 30-second compressed-air purge to clear liquid from the burner and ignition tubes. Photoelectric flame safeguard equipment is incorporated.

15.4 WORKING-FLAME APPLICATIONS

Working flame heating implies direct flame impingement onto the work to provide the required (usually rapid) heating. Working flames are normally used for controlled localized heating in the absence of a furnace enclosure. Applications include heating metals and glass in preparation for a shaping process, local fusion, e.g. oxygas metal cutting, glass piercing and sealing and heating locally to produce metallurgical changes, e.g. surface hardening. Rhodes [19] has presented a comprehensive review of natural-gas working-flame burners and the interested reader is referred to this for a detailed treatment. The brief treatment presented here is largely based on Rhodes's paper. Working-flame burners are either hand-held in the form of the familiar air-gas or oxygas torch or components of automatic machines for glassworking, flame hardening, brazing, etc.

15.4.1 Natural-gas–air burners

Piloted natural-gas–air burners are used widely for working-flame applications and this group constitutes the largest burner range. Open-fire burners are designed to produce a single flame providing outputs of up to 136 $\mathrm{MW\ m^{-2}}$ (300 000 $\mathrm{Btu\ in^{-2}\ h^{-1}}$). The burner head is of the auxiliary flame retention type incorporating a sudden expansion (Section 7.5.2). Burners of this type are used in brazing, soldering and preheating.

Single-port, 'long-focus' (i.e. long cone length) burners incorporate a 'gearwheel' insert to produce auxiliary flames and are commonly used in the glass industry for sealing, shaping and preheating. They may be mounted singly or on curved manifolds to fire around the article being heated. Multiport, short-focus burners or fishtail burners are usually mounted on a manifold to produce a continuous flame for localized hardening of such items as hacksaw blades. Most fishtail burners use an auxiliary flame annulus of crimped ribbon. High heat concentrations are allowed by the use of high pilot-ratio screen burners which operate at a mixture pressure of 2–2.75 kPa (20–27.5 mbar, 8–11 inH_2O). Stabilization is achieved by feeding a relatively high proportion of the mixture to a large area of rolled screen. Uses for these burners include small-scale brazing and the manufacture of light bulbs.

Mixture supplies for gas–air burners are supplied either from air-blast injectors with automatic proportioning by zero governor or from premix machines.

15.4.2 Natural-gas–oxygen burners

Unpiloted natural-gas–oxygen burners are widely used since the high burning velocity of a mixture of oxygen and natural gas does not present the same stability restrictions as occur with natural-gas–air flames. Quite high heat-release rates are possible, e.g. 360–540 MW m^{-2} (0.8–1.2 $\times$ 10^6 Btu in^{-2} h^{-1}). Ports are generally formed by drilling, but fine hypodermic stainless tubing is also used to give ports raised a few millimetres proud of the burner face. Unpiloted burners may also be face-mixing or nozzle-mixing. The characteristics of these burner types are considered in Section 7.13. For very high heat-release rates, for working borosilicate or quartz glass, for example, piloted gas–oxygen burners are necessary. Burner-port loadings up to 2200 MW m^{-2} (4.8 $\times$ 10^6 Btu in^{-2} h^{-1}) are employed at mixture pressures up to 68–73 kPa (680–730 mbar, 11–12 lbf in^{-2}). Relatively small pilot rates are required, 10–15 per cent of the main-flame rate being typical.

The replacement of town gas by natural gas has eliminated two difficulties associated with working-flame burners. First the small but significant sulphur content of town gas caused sulphur 'blooming' on glassware and sulphur removal was frequently practised in glassworking industries. Second the variation in combustion characteristics experienced with manufactured gases caused problems of varying flame length and this led to the development of Wobbe-number or cone-height control systems in which controlled quantities of diluent gas (normally air) were injected into the gas in response to signals from either a Wobbe-number indicator or from a thermocouple located near the tip of a test flame [20].

15.4.3 Flame hardening

Localized intense heating followed by rapid quenching has been used for many years for surface hardening. Flame hardening is easily automated and has the advantage that localized areas may be selected for hardening and, because of high rates of heat transfer to the surface, the core is unaffected by the treatment.

Flame hardening is usually carried out on the finished machined component and it is generally desirable to heat-treat the main body of the component before the final machining operation [7]. Plain carbon steels of carbon content 0.35–0.7 per cent can be employed where more expensive alloy steels could otherwise be required. Rapid and accurate temperature sensing is provided in modern installations by a photoelectric cell comparing the radiation emitted from the workpiece and from a preset filament. In addition the burner and quench speed and/or the movement of the workpiece are also accurately controlled.

The burners used for flame hardening are generally gas–oxygen burners of the slot type having a slot width of about 0.4 mm (0.016 in) for town gas and about 0.6 mm (0.024 in) for natural gas. Flame-stabilization techniques are necessary with natural gas-oxygen burners (burning velocity about 3 m s^{-1} compared with about 7 m s^{-1} for town gas). Auxiliary-flame and sudden-expansion methods are both used. Comparisons between natural-gas and town-gas firing with respect to capacities, feed rates and case-depth limits have been presented by Grönegress and Rossner [21].

15.5 AIR CONDITIONING AND SPACE HEATING

15.5.1 Thermal comfort

Human comfort is a nebulous state of being, philosophically inseparable from a purely subjective interpretation [22]. Nevertheless, a great deal of scientific effort has gone into the pursuit of the identification and quantification of the component parts of comfort. These studies are becoming more important as the complexity of the chemical and physical environment increases and the effects of this complexity on human physical sensitivity and susceptibility are realized.

It is necessary to adopt a broad concept of 'comfort'. A few years ago it was adequate to approach the problem of defining the parameters of comfort simply on the analysis of the thermal balance between a person and his indoor environment. The number of factors which must now be included leads to a definition of comfort as the ecological balance point between a person and his indoor environment. The ideal comfort study should therefore include the complete indoor ecological system. In practice, engineers are concerned with some specific environmental factors which have a bearing on comfort, such as temperature, humidity, air velocity, ventilation rates, odour, dust, air ions and indoor pollutants.

The factors that determine human thermal comfort can be studied on the basis of a heat balance of the body. The body maintains a relatively constant internal temperature. To achieve this, the heat generated through the metabolic life cycle must be dissipated to the environment. This heat generation depends to a large extent on the activity level of the individual as shown in Figure 15:15. It can be dissipated by radiant, convective or evaporative heat transfer, in the extreme case the heat generated would require in excess of 1 ton* of cooling capacity to maintain equilibrium. The body is unable to reject such a quantity of heat or tolerate a significant increase in internal temperature so it regulates itself by collapse. The principal mode of body temperature control is through evaporation. A subject moving from a sleep activity to light factory work increases in metabolic output by 146 W (500 Btu h^{-1}). Below a skin temperature of 36.1 °C (97 °F), convective and radiant heat transfer processes can only

*1 ton of refrigeration (1 TR) is now defined as 288 000 Btu per day. It was originally conceived as the rate of heat extraction provided by melting, in 24 hours, 1 short ton (2000 lb) of ice at 32 °F to make 2000 lb of water at 32 °F. In SI units, 1 TR is 3.517 kW (to 4 significant figures).

dissipate about 59 W (200 Btu h^{-1}) and the balance must be removed by an increase in evaporative cooling. The sensation of thermal comfort is directly related to the degree of evaporative cooling. The comfort quotient decreases as the perspiration rate increases. This phenomenon has been used to quantify human comfort levels by estimating 'evaporative capacity' [22]. Another method is the exposure and subsequent interrogation of subjects to carefully controlled comfort conditions.

Condition	Rate (W)	Rate (Btu h^{-1})
Sleeping	79	270
Sitting at rest	120	400
Office work	130	450
Light factory work	220	750
Slow walk	235	800
Rapid walk	350	1 200
Swimming	590	2 000
Climbing stairs	1300	4 400
Extreme task with exhaustion in 22 seconds	4500	15 500

Figure 15:15 Rate of heat dissipation by a human during various activities

A subject at rest under comfortable conditions will typically be using 10 per cent of the body's maximum evaporative capacity. The feeling of comfort will continue to be associated with 'evaporative capacities' between 10 and 25 per cent. Rates between 25 and 70 per cent will be extremely uncomfortable. Under cold conditions evaporation will be reduced and shivering will commence in order to increase the body's metabolic rate. Although there is no discernible perspiration at 10 per cent evaporative capacity the evaporation should provide a sensation of cooling to a subject.

Mathematical models of thermal comfort have been developed and used to calculate the effect of a number of comfort parameters such as air velocity on human comfort [22].

Commercial buildings [23]

In buildings such as offices and schools, if people are going to work efficiently at visual tasks that can be monotonous, it is important that the working environment should be of a high standard. The most important factor, for the purpose of this section, is the thermal environment.

This is partly controlled by the design and thermal properties of the enclosure of the building and its ability to modify the external climatic conditions.

Investigations into thermal comfort have produced detailed specifications of the physical conditions necessary to achieve a high degree of comfort.

In relation to the clothing worn today in commercial buildings the air temperature at head level should be about 19.5 °C (67.1 °F). Ideally the temperature at foot level should be the same but air distribution problems make this difficult to achieve. It should not, however, be less than 16.6 °C (61.9 °F). The mean radiant temperature of all surfaces forming the enclosure should always be slightly higher than the air temperature, an ideal figure would be 21 °C (69.8 °F). If the floor is heated, thermal discomfort will be felt above a temperature of 24°C (75.2°F), with heated ceilings the limit is related to the height, an average limit would be about 29.4 °C (84.9 °F).

The other factor is the rate of air movement past the human body.

Ideally people prefer a noticeable range of airflow, which means a flow of not less than 0.13 m s^{-1} (0.42 ft s^{-1}). The best range is between 0.2 and 0.3 m s^{-1} (0.67 and 1.0 ft s^{-1}), the upper limit above which thermal discomfort would be experienced is 0.6 m s^{-1}.

The achievement of a satisfactory thermal environment is therefore a complex situation which involves careful consideration of the thermal performance of the building and the thermal performance of the heating and ventilation systems. On new buildings, close collaboration between architects and heating and ventilating engineers is needed, because it is the architect who is deciding on the thermal properties of the building and these will determine to a large extent the mean radiant temperature achieved inside it by the heat input of the ventilation plant.

Traditional buildings with brick or stone load-bearing walls have a high thermal capacity which considerably modifies diurnal conditions to produce minimal effects on the interior. These structures only have about 30 per cent of glazed wall area so the amount of solar radiation admitted does not have a large effect on the internal thermal environment. The space-heating plant does not have to cope with rapid changes in energy demand.

More modern buildings of framed construction have walls of lightweight material and usually include large areas (up to 75 per cent) of glazing. This type of construction considerably reduces the climatic modification characteristic of the building, as the glass admits the diurnal changes in temperature and solar radiation. This can produce a rapid fluctuation of energy demand within the building if thermal comfort conditions are to be achieved. The fact that glass is over 80 per cent transparent to high-frequency solar radiation and almost opaque to reradiated low-frequency heat means that, particularly with double glazing, there will be a buildup of heat due to solar radiation. This can be reduced by external screening; any internal screening will only convert the energy to convection which may again lead to discomfort.

It would appear, therefore, that present-day trends in building construction, which have been determined more by structural and economic factors than environmental factors, have created thermal problems that are difficult to overcome at a time when energy costs, building costs and environmental standards are rising.

Complaints of thermal discomfort in multistorey buildings are usually due to overheating caused by solar gain and inadequate ventilation. The problem can be solved by installing either solar screening or a cooling plant. A cooling plant in such a case will be expensive as it must be sized to cope with the short-duration peak heat gains and it may be difficult to install into the existing structure. Also, buildings have been air-conditioned because their thermal performance has been inadequate to provide satisfactory climatic modification and often the money available is insufficient to provide a plant large enough to give good conditions.

It is essential, therefore, if thermal-comfort standards are to be achieved with reasonable capital and energy consumption for the thermal plant, that the building must have satisfactory climatic-modification characteristics. This can be achieved by reducing the amount of glazing to a figure of about 25 per cent of the outer wall area and employing opaque wall materials with high thermal insulation values. The government and professional institutions must cooperate more fully in the future to ensure that the mistakes of the past are not repeated. The reader is referred to references [24–35] which deal more fully with the factors affecting comfort conditions and the involvement of the gas industry in this subject. Feasibility studies undertaken by design teams incorporating all the professions involved have shown that it should be possible today to design air-conditioned buildings with a high degree of environmental control providing a high standard of working conditions at a cost somewhere between ± 5 per cent of the cost of present-day buildings, designed for daylight with a heating and ventilating plant.

If a building is designed with a high degree of climatic modification, it is then possible, by investigation of the thermal balance, to relate the conductive and ventilation heat losses to the heat gains from the occupants, lighting, office equipment, etc., to establish an external air temperature above which the building requires no heating, but may require cooling. The thermal environment can therefore be carefully controlled if the air coming into the space is always at a temperature lower than the ambient temperature during the occupation period. This means that a high-quality thermal environment can be provided by a moderately sized air-conditioning plant. Any heating required will be limited to periods in mid-winter when the external air temperature is below the balance temperature and preheating will be required when the building is unoccupied.

Returning to the establishment of thermal comfort in existing buildings, they can be divided roughly into two thermal classes. The first class is the daylit buildings with large areas of glass, limited room depths, with usually subdivided accommodation. These require space-heating plant designed to overcome conductive and natural ventilation heat losses in the winter. The second class is the supplementary lit buildings in which the area of glass is somewhere between 40 and 45 per cent of the facade and the rooms are deeper. This feature in multistorey buildings makes mechanical ventilation necessary. These buildings require space-heating plant. Whether a cooling plant is required depends on the density of occupation and the heat gain from the supplementary lighting. The heating and ventilating engineer must look at the building design and decide into which category the building belongs and then, within the limits of the thermal performance of the building itself, attempt to establish a satisfactory thermal environment.

Industrial buildings [36]

The thermal environment is again the most important factor and as with commercial buildings it is affected by the building design. It is also influenced by the nature of the work carried out and the equipment involved. Thermal comfort is usually attained by means of heating and ventilation only. Air conditioning is only provided where the product, equipment or such factors as health and safety make it necessary.

The factors which affect the body will be similar to those in the general description given earlier, making allowances for the industrial situation in the degree of activity, the degree of insulation (protective clothing), ambient air temperatures, surrounding surface temperatures and emissivities, rate of air movement and the moisture content of the air.

The heating and ventilation engineer must work in collaboration with the building designer, the plant engineer and the production engineer responsible for plant layout if a successful scheme is to be devised. Thermal comfort in an industrial building will be affected by four main factors: insulation, building height, roof design, and the plant in the building.

Insulation. The effect of insulation on reducing heat losses and therefore heating requirements is well known. What is perhaps not so well realized is the effect on comfort caused by the lower surface temperatures of poorly insulated buildings. Holt [36] demonstrates using a sensation-of-warmth equation produced by Bedford that a building heated to 21 °C (70 °F) when the outside temperature is −1 °C (30 °F) will have an internal surface temperature of 18 °C (65 °F) if it is insulated to a U value of 1.336 W m^{-2} $°C^{-1}$ (0.2 Btu ft^{-2} $°F^{-1}$ h^{-1}) and will be described as comfortably warm. If the value is 7.95 W m^{-2} $°C^{-1}$ (1.4 Btu ft^{-2} $°F^{-1}$ h^{-1}) then the internal surface temperature will be 2 °C (36 °F) and the building will be described as cool.

The picture changes somewhat if air movement and air moisture content are considered. Humidity is important in buildings used for high-humidity processing such as laundries and dyehouses.

In general, insulation properly applied has the effect of increasing the sensation of warmth.

Building height. This can be summarised by saying that if the problem of providing comfort can be solved by adding heat, then a lower roof is much better and more economical than a high one. This is simply by reduction of surface area through which heat is lost to the outside and by a reduction of the problem of sealing the building, by which heat is lost through in- or ex-filtration. If the problem is one of ventilation, then in general a high building is better than a lower one, because with such problems a temperature gradient will exist in the building and a higher roof-space temperature, which means a lower ventilation rate, will have a less adverse effect on comfort the higher the building. This will be particularly noticeable in a building where solar heat causes the major part of the problem.

Roof design. The construction material is important, for this determines the internal surface temperature under varying internal and external conditions. Poor insulation values mean surface temperatures well below building temperatures in winter. This will necessitate increasing the air temperature or radiant heating. In summer the surface temperatures will be well above air temperatures. The ventilation rate required to offset solar gain is affected by the U value of the material and the height of the roof. The solar gain factor is obviously emphasized if roof glazing is employed.

The location of glazing in the roof design is thus very important. Figure 15:16 shows three typical roof designs and the solar heat input.

Industrial plant. This is usually the greatest modifying factor in thermal comfort in industrial buildings. The problem is usually one of being too hot in summer rather than being too cold in the winter. This problem can arise, however, if there is a large ventilation system associated with the plant or the process, as in paint spraying, welding, etc., where the ventilation rate must be well in excess of that required for normal occupancy. The paper by Holt [36] discusses a number of industrial situations in which overheating problems can occur. The paper also considers the provision of heat in two situations requiring high ventilation rates: (*a*) where ventilation is required because of noxious or toxic fumes or dust emitted by the process which must be captured by local exhaust systems or diluted by general ventilation, and (*b*) where a high ventilation rate is required to absorb moisture emitted by the process which would cause condensation problems if not removed.

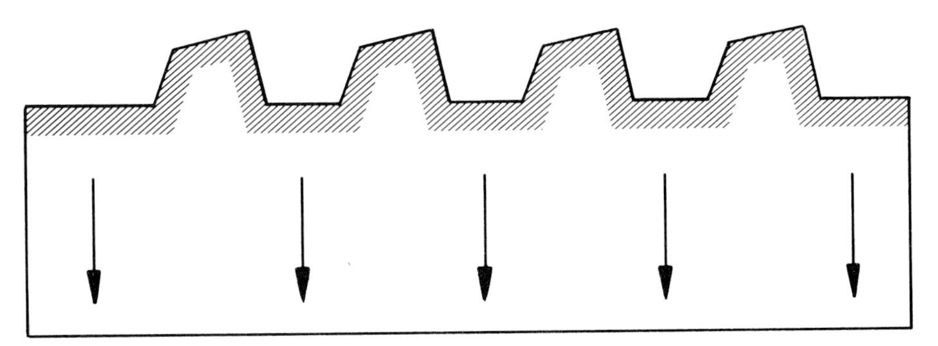

Insulated roof, $U = 1.7$ W/m^2 °C (0.3 Btu/ft^2 h °F), solar-heat input 100 units

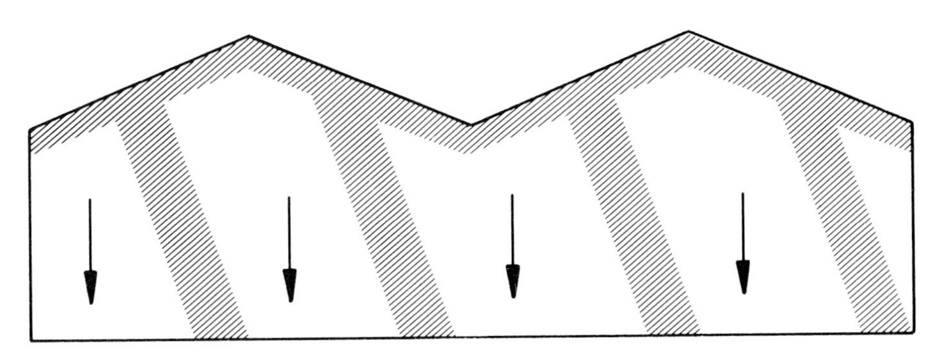

Insulated roof, $U = 1.7$ W/m^2 °C (0.3 Btu/ft^2 h °F), solar-heat input 490 units

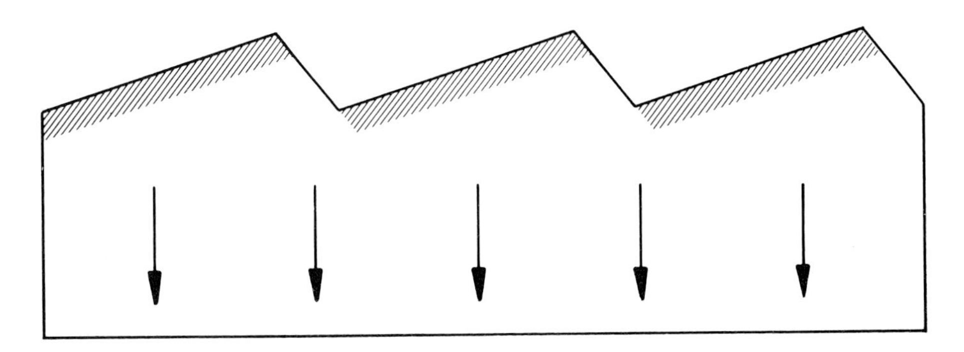

Uninsulated roof, $U = 6.81$ W/m^2 °C (1.2 Btu/ft^2 h °F), solar-heat input 400 units

Figure 15:16 Solar-heat inputs for roofs of various designs [36]

15.5.2 Air conditioning

The reasons why air conditioning is a necessity in many instances and a desirable feature in others have been covered in Section 15.5.1. The market available for gas is covered in Section 15.5.3, where reference is made to the use of gas-fuelled systems in commercial buildings such as offices, department stores and supermarkets, in the public sector for hospitals, civic centres and similar buildings, and finally in the industrial field where the process conditions demand it.

Air conditioning embraces the supply of air to buildings under controlled conditions of temperature, humidity, odour and foreign-body content [37]. In the following sections, descriptions of the different systems, types of equipment and of the role of gas in air conditioning are given.

Absorption system [38, 39]

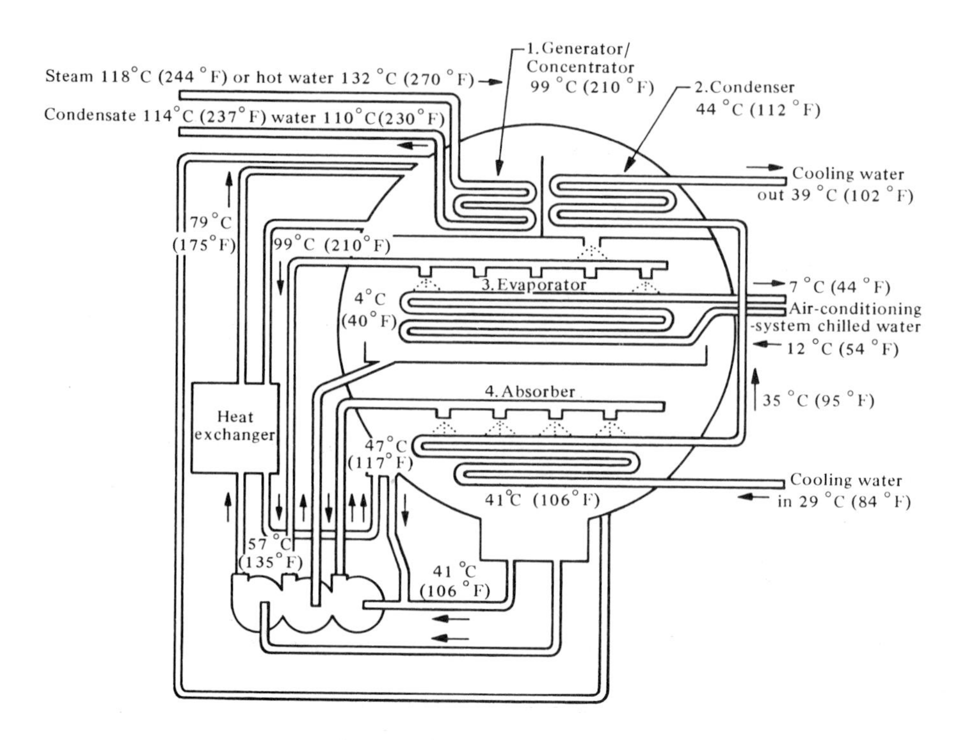

Figure 15:17 Absorption cycle [39]

Absorption cycle. The chilling effect is produced in the absorption cycle illustrated in Figure 15:17 by the evaporation of a fluid, the refrigerant, in the lower-pressure side of a closed system, producing chilled water in the evaporator coil (3). The refrigerant vapour is recovered by the absorbing action of a second fluid, the absorbent, within the absorber (4). This solution of absorbent and refrigerant is pumped into the generator (1) in the higher-pressure side of the system, where heat is applied to drive off the refrigerant as a vapour, which is then condensed to a liquid in the condenser

(2) using external cooling with forced air or recycled water.

Types of equipment. There are two main types of absorption equipment:

1 Directly fired, in which the generator is heated by a gas burner.
2 Indirectly fired, in which heat is supplied as low-pressure steam or medium-pressure hot water passing through coils in the generator.

Directly fired equipment. The three classes of equipment which involve direct fired absorption chilling are: directly fired air-cooled chillers, packaged combination units, and directly fired water-cooled chiller/heaters.

Directly fired air-cooled chillers have capacities in the range 10.5 kW (3 TR) to 35.2 kW (10 TR). These units use ammonia as refrigerant and water as the absorbent. They are always mounted externally, frequently at rooftop level. This is necessitated by the use of ammonia and by the air cooling of the condenser. However, the units are constructed for external mounting and such siting is often advantageous. The use of fanned-air cooling obviates the need for a cooling tower. For larger chilling loads, these units may be mounted in modular arrangements giving the advantage of improved part-load performance and security of the bulk of the load in the event of a breakdown of a single unit.

Packaged combination units employing the air-cooled units described above together with gas-fired warm-air furnaces are capable of supplying cooling or heating in the range 10.5 kW cooling/27.5 kW heating (3 TR/93 750 Btu h^{-1}) to 35.2 kW cooling/71.4 kW heating (10 TR/243 750 Btu h^{-1}). These units must also be externally mounted and are particularly suitable for single-storey buildings of large area, such as supermarkets and warehouses. Units up to 106 kW (30 TR) can be produced if required. An alternative system is to combine the air-cooled units with boilers so that either chilled or hot water can be produced.

Directly fired water-cooled chiller/heaters in the range 52.8 kW cooling/84.5 kW heating (15 TR/288 000 Btu h^{-1}) to 88 kW cooling/141 kW heating (25 TR/480 000 Btu h^{-1}) are also available. These units use the lithium-bromide–water system and can be sited anywhere within a building with complete safety. Figure 15:18(*a*) shows the cooling cycle and (*b*) the heating cycle. In the cooling operation, the condenser and absorber cooling water has to be recycled through a cooling tower. The change-over to heating is carried out by stopping the flow of cooling water to the machine, so that the chilled-water circuit will act as a hot-water circuit.

Indirectly fired equipment. These are available in the range 77.4 kW (22 TR) to 5837 kW (1660 TR). They all use water as refrigerant and lithium bromide as absorbent and operate under a partial vacuum. Cooling is always by water recirculated through a cooling tower. Heat is supplied either by medium-pressure hot water or low-pressure steam, generally in the temperature range of 104 °C to 132 °C. The only moving parts are small pumps, usually integral with the machine, which makes them relatively free from noise and vibration and gives flexibility of siting.

Advantages of absorption equipment

1 The units contain no major moving parts and thus are quiet and free from vibration, except for the fan on air-cooled units. Fanned units, as stated, are always

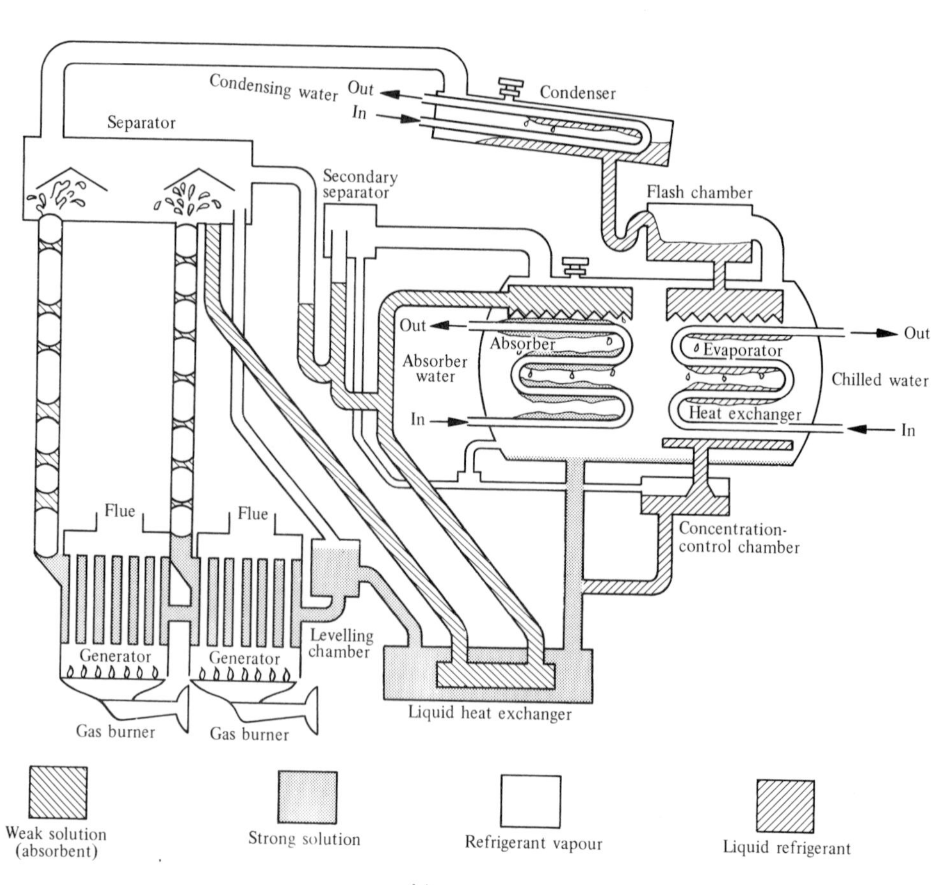
Condensing water
Out
In
Condenser
Separator
Secondary separator
Flash chamber
Out
Absorber water
In
Absorber
Evaporator
Heat exchanger
Out
Chilled water
In
Concentration-control chamber
Flue
Flue
Generator
Generator
Levelling chamber
Gas burner
Gas burner
Liquid heat exchanger
Weak solution (absorbent)
Strong solution
Refrigerant vapour
Liquid refrigerant

(*a*)

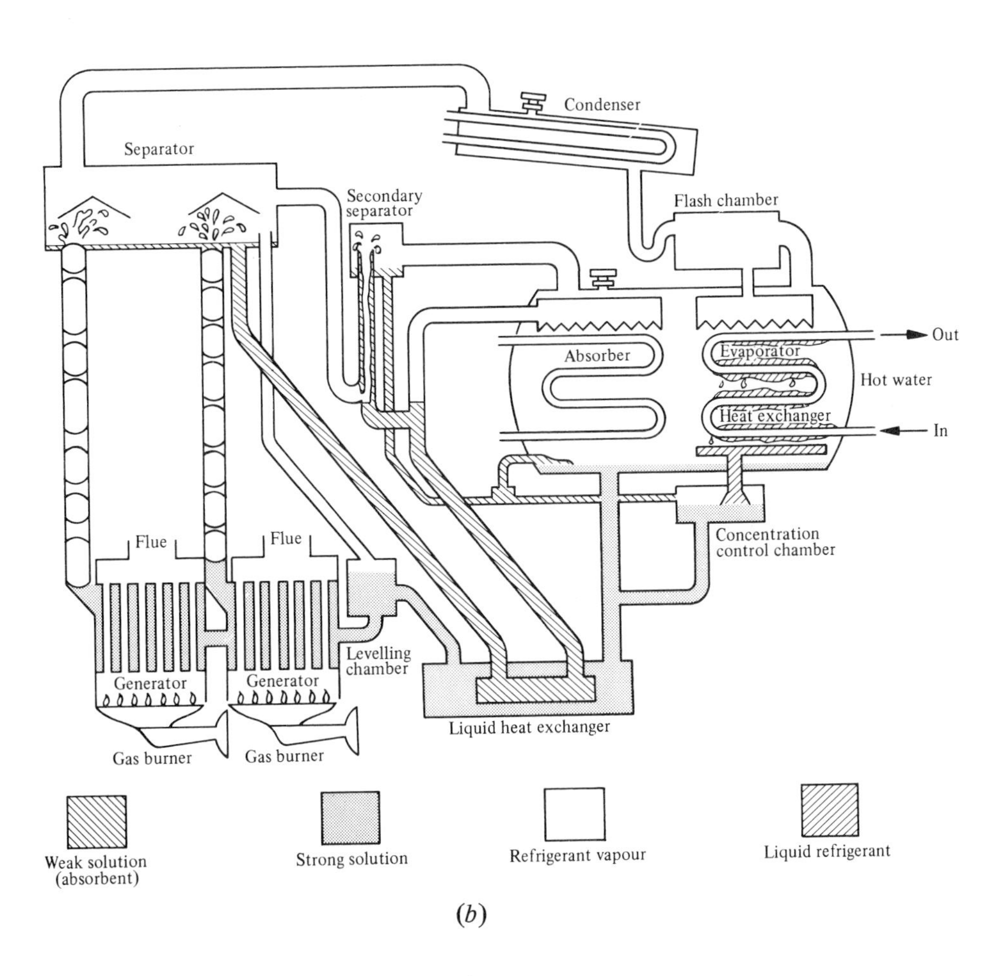

Figure 15:18 Directly fired water-cooled chiller/heater [41]
(*a*) Cooling cycle
(*b*) Heating cycle

positioned externally and in this way low-level noise is not generated in the building. Water-cooled units can be positioned anywhere within the building.

2 Absorption equipment has a long life and low maintenance costs.
3 The units can be operated on part load without sacrifice of efficiency. This part-load capability is important in UK conditions and is illustrated in Figure 15:19.

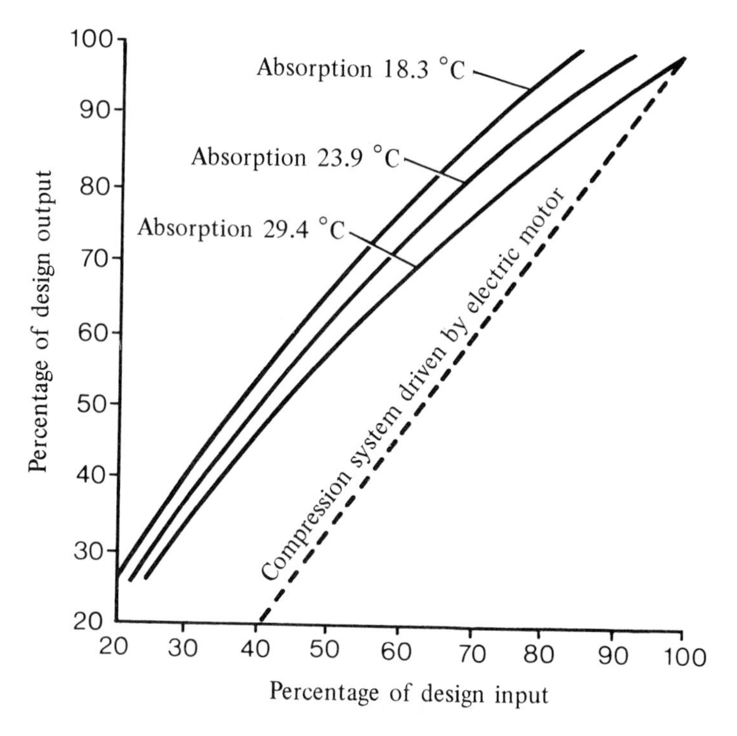

Figure 15:19 Comparison of performance at various loads of absorption systems at various condenser-water temperatures and a typical compression system driven by electric motor [38]

4 Chiller/heaters and packaged combination units require only a single space by eliminating the need for separate boilers.
5 Roof-mounted combination units can provide heating or cooling as demanded to different zones of a building.
6 The ability to satisfy moderate cooling demands by installing the smaller units in modular fashion, gives high efficiency and reduces the amount of standby equipment needed.

Economics. The performance of the heat pump, which is the basis of the chiller, can be expressed as a coefficient of performance [37] defined as

$$\text{coefficient of performance (c.o.p.)} = \frac{\text{heat removed to effect chilling}}{\text{work put in to operate the pump}}$$

An overall c.o.p. can be calculated including the efficiency of the prime mover or source of heat. For large water-cooled vapour-compression machines (see below) the overall c.o.p., including the electric-motor efficiency is about 4.2 while the equivalent indirectly fired absorption machines work at a figure of 0.54, including the boiler efficiency. This means the energy input demands are in a ratio of about 8 to 1. Recently introduced 'double-effect' machines [38] have c.o.p. values of 0.8 to 1.0 and reduce this ratio to about 5 to 1.

Despite these large differences in c.o.p. values absorption refrigeration is competitive in running costs because of the structure of the electricity maximum-demand tariffs (see below and Section 14.5.3) and the difference in cost of electricity and gas (1p/kW h is equivalent to 30p/therm).

Some of the advantages of absorption machines set out above, such as absence of vibration, are not readily quantifiable and improved performance under part-load needs a knowledge of the patterns of load demand exerted by the building in order to give an economic value. However, these must always be considered in any assessment.

The relatively low c.o.p. of absorption-cycle machines involves a corresponding increase in the size of the condenser and cooling tower. This increases relative capital costs but is offset, for large-tonnage machines, by the heavy transformers and expensive switchgear required for electric compression chillers. A comparison between first costs of absorption-cycle chillers and electrically driven machines is given in [38].

The steps in assessing the economic viability of a gas-fired absorption chiller are:

1 From information on the likely usage and other gas consumption within the building, calculate the annual gas cost with and without chilling. Then, by subtraction obtain the marginal chilling cost using the appropriate tariffs.
2 From similar electrical data for the building, calculate the annual electricity cost with and without chilling, The marginal chilling cost is again obtained using the appropriate tariffs.
3 Obtain the annual running costs for the gas and electric equipment by adding the costs of maintenance to the fuel costs, including any other costs where a differential exists, e.g. rates and insurance.
4 Assess the installed capital costs for the two systems. These must include the cost of cooling towers, transformers, cabling, acoustic and vibration treatment, etc., to give true installed costs for each system.
5 As indicated above, there will generally be a higher capital cost for the gas system, particularly in the lower sizes but allowance must be made for the longer life of absorption equipment. The differential capital cost must be compared with the saving in running cost by whatever method, such as simple payback, present worth, discounted cash flow, that is appropriate.

Compression system

Compression cycle [38, 39]. The chilling effect is obtained by utilizing a closed system, shown in Figure 15:20, in which a refrigerant is compressed mechanically in a centrifugal or reciprocating compressor (1) and allowed to evaporate at a lower pressure producing a chilling effect in the water circulating through the evaporator coil (3). The refrigerant vapour is recovered by the mechanical suction applied at the inlet to the compressor and recompressed. Cooling of the superheated refrigerant to a liquid between the compressor and the evaporator takes place in the condenser (2) by means of external cooling, using forced air or recycled water.

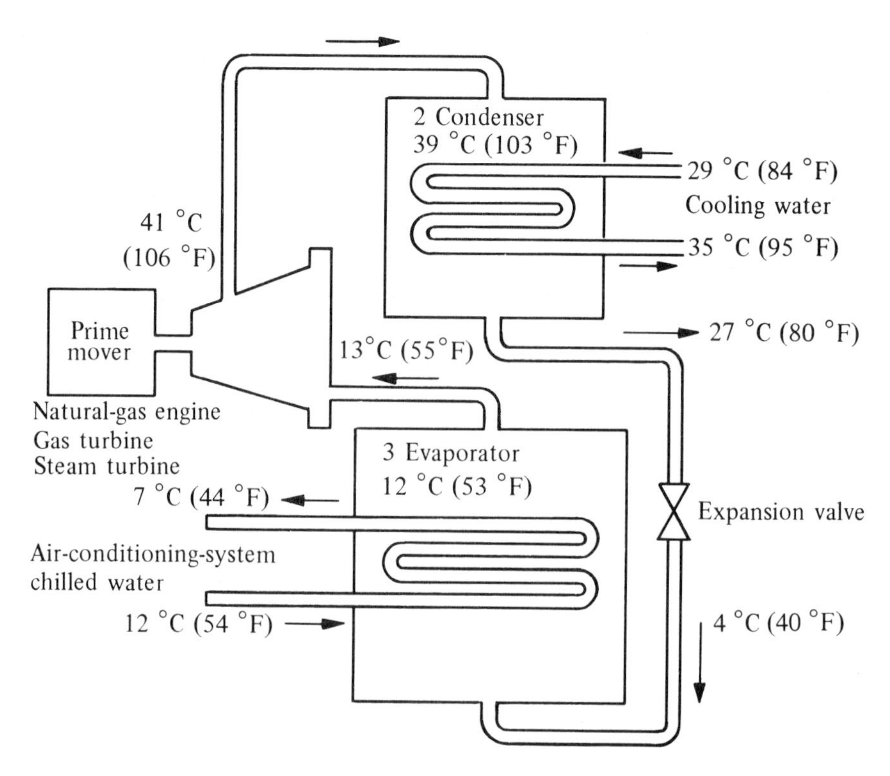

Figure 15:20 Compression cycle [39]

Types of chiller units. The three main types of unit are:

1 *Reciprocating-piston-type compressors.* These are used up to a size of about 600 kW cooling (171 TR). They can rotate at speeds of up to 2800 rpm but are usually run at about 1400–1500 rpm when driven by a four-pole synchronous a.c. electric motor and at 1750 rpm when driven by a gas engine. These are constant-speed machines, which means load variation must be met by unloading compressor cylinders.
2 *Centrifugal compressors.* These rotate at speeds of up to 24 000 rpm and are available in sizes from 350 kW cooling (100 TR) to over 3500 kW (1000 TR) single stage or over 17 500 kW (5000 TR) multistage.
3 *Screw compressors.* These are available in sizes up to 2205 kW (627 TR) rotating at speeds of 2900 rpm.

Driving the compressor. The machines can be driven by three basic types of prime mover:

1 *Electric motors.* As stated above, these are constant-speed devices and suffer from the disadvantage that modulation for part-load has to be achieved by means other than speed reduction. Variable-speed motors can be purchased, but they give only a marginal saving in power consumption at part-load.

2 *Turbines.* Gas turbines range from about 190 kW (250 hp) to 8.9 MW (12 000 hp). Steam turbines range from 1.1 kW (1.5 hp). Steam turbines are normally used with centrifugal compressors, often in a multistage chilling system together with absorption compressors to produce very low temperatures for example as required in the food industry.
3 *Gas engines.* These have already been referred to briefly when discussing total and selective energy schemes. They are available in a great number of sizes from about 19 kW (25 hp) upwards and present a viable alternative to electric motors.

Types of gas engines. The two main types of gas engine are dual-fuel engines and spark-ignition engines.

Dual-fuel engines are compression-ignition engines similar to diesel engines but using gas plus 'pilot' oil as fuel. The size range available is from about 134 kW (180 hp) to 10 MW (13 000 hp). The disadvantages of dual-fuel engines are low power-to-weight ratios, the need for frequent and highly skilled maintenance, high initial capital and installation cost. These disadvantages and the lack of suitable engines in the lower power range make it unlikely that this type of engine will find much application in the commercial market.

Spark-ignition engines are lighter and easier to maintain than dual-fuel engines. They are available in two classes: industrial engines and converted automotive engines.

The industrial engines available cover the range 37 kW to 1.6 MW (50 to 2100 hp). Rotation speeds are from 1200 to 2400 rpm with compression ratios up to 10 : 1. They are cheaper than dual-fuel types but still relatively expensive in both first cost and maintenance.

The ordinary automobile petrol engine is available in a range of sizes and can be simply converted to run on natural gas, by replacing the petrol carburettor with a gas version and advancing the timing. These engines, complete with control gear and associated equipment, are much cheaper per horsepower than their industrial counterparts.

Spark-ignition engines coupled to chilling compressors are available in the range 70.3 kW (20 TR) to 5274 kW (1500 TR).

The siting and installation of the gas engines considered in this section must be carried out with due regard being paid to the engine mounting, noise, supply of ventilation and combustion air, engine cooling, engine starting, exhaust-gas removal, gas supply and speed-control system. Fuller details of these factors are given in [37] and [38].

Advantages of vapour-compression chillers driven by gas engines

1 The main benefit is lower running costs resulting from avoidance of maximum-demand charges for electricity.
2 Most of the compressors available are designed for American 60 Hz electricity and will be downrated by 14 per cent when driven by UK 50 Hz electricity. The gas engine rotating at 1750 rpm restores the output to the design capability.
3 Coefficients of performance are above 1.0 which is about twice that of absorption chillers (except for 'double-effect' machines).

The compressor can be modulated by varying the speed of rotation as well as by unloading cylinders. This gives a considerable increase of efficiency, particularly on part-load. This is illustrated in Figure 15:21.

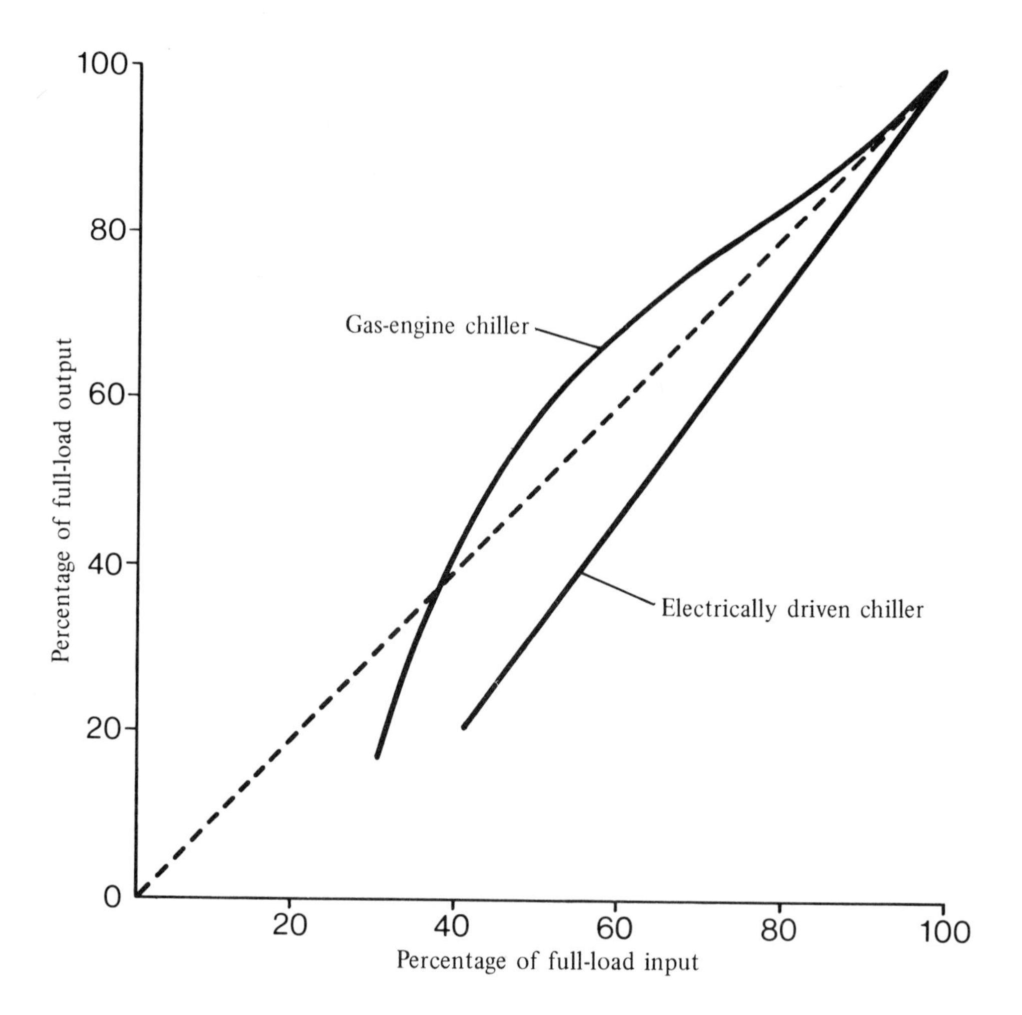

Figure 15:21 Performance at various loads of gas and electric chillers [38]

Economics. The steps in assessing the economic viability of gas-engine drive are similar to those considered for absorption machines:

1 Calculate the marginal chilling cost for gas.
2 Calculate the marginal chilling cost for electricity.
3 Obtain the annual running costs for both the gas and electric equipment.
4 Assess the installed capital costs for both gas-engine drive and electric-motor drive. These must include the cost of cooling towers, transformers, etc., to give true installed costs.
5 The capital cost of the gas-engine-driven compression system will be generally

greater than its electrically driven counterpart. The differential capital cost must be compared with the saving in running cost by an appropriate method.

Dual-service system

It is common in many commercial buildings to produce chilled water by electric-motor-driven vapour-compression equipment and to provide for loss of electric power for lighting and other essential uses by employing a standby diesel generator set. These sets are usually sized so that, in an emergency, trading can continue with much reduced lighting levels and without the ability to produce chilled water for air conditioning.

The 'dual-service' concept is similar to the selective energy concept (Section 14.5.3) and it replaces the electric motor and the diesel engine by a single spark-ignition gas engine. In normal circumstances the gas engine drives a vapour-compression chiller but on failure of the electricity mains supply, the chilling load is dropped and the engine turns an alternator to produce electricity for lighting and other essential uses. The chilling has to be sacrificed during the period of emergency but for most commercial concerns (not for computer suites and other heat-sensitive equipment or processes) loss of chilling can be tolerated for long periods, particularly since the lower lighting level will reduce internal heat gains.

The economics of dual service are very favourable. The running costs of the gas engine are lower than those of the equivalent electric motor and the additional capital cost will be quite low because the cost of the diesel standby, oil storage and electric motor with its controls can be offset against the cost of the gas engine, the larger cooling tower, and the coupling between the engine and the compressor. The use of gas-engine-driven chillers is described in more detail and actual installations studied in [40].

Reliability and maintenance

Water-chilling machines. Absorption units have a particularly good record since there are essentially no moving parts except in machines which include refrigerant and solution pumps and these also have a high degree of reliability. Therefore, these machines operate at full efficiency over many years of service [41].

Maintenance of the absorption system includes the cleaning of heat-transfer surfaces in combustion chambers and possibly cleaning and descaling condenser tubes, steam and hot-water coils. This should not cause problems if proper attention is paid to the water-treatment requirements. A little attention may be required to ensure absolute cleanliness of the burner systems. Manufacturers recommend purging of noncondensible gases occasionally.

Vapour-compression units are also reliable with the same remarks being applicable to the condenser tubes, water treatment and small pumps. The compressor-maintenance requirements differ slightly between reciprocating and centrifugal operation but are normal for the type of equipment. This type of system can, however, be damaged by improper use such as overloading.

Prime mover. Electric motors are reliable prime movers and maintenance requirements are low. They are mechanically simpler than any other type of drive. They suffer, however, by using a very expensive form of energy.

Gas-engine drive is reliable. Time intervals between major overhauls will depend on the particular type of engine but can be from 15 000 to 25 000 hours. Reliability is increased by using breakerless ignition systems, and automatic oil-level and oil-temperature regulators. Maintenance is limited to changing oil and sparking plugs and giving regular attention to the ignition system and general inspection. The level of maintenance is below that required by liquid-fuel engines.

More detailed information on maintenance can be found in [38] as well as the other references quoted in this section.

Market available

Many of the papers already referred to give brief surveys of the possible markets for gas air conditioning. Further details of the prospects of gas in this field can be found in [42] and [43]. The summary below is taken largely from the paper given by Jarvis and Pearson [39].

Demand for air conditioning. The demand is estimated to be increasing at about 10 per cent a year and the trend is expected to continue for at least five years. From a small beginning in the mid-1950s it is calculated that the total value of all air-conditioning systems sold is about £100 million per year. Although slow to develop, the demand has increased as a result of the developments in building design and materials of construction discussed in Section 15.5.1:

1. Deep planned areas involving high internal heat gain from increased lighting.
2. High heat gains from solar radiation.
3. Acoustic sealing of buildings and elimination of opening windows because of high winds in upper storeys of high-rise buildings.

Added to these, there is the need to provide controlled conditions for the increasing number of computers, laboratories and clean rooms used for special processes.

General buildings. In many buildings, such as civic centres, law courts, assembly halls and administrative offices, the cooling requirements create a need for air conditioning which is as important as in the private sector, which has been the major growth area covering offices, shops, supermarkets and department stores.

Hospitals. Here it is not general practice to condition the whole building, but areas that can derive maximum benefit, such as operating theatres and intensive-care units, offer market potential. Single systems can be suitable for various departments as it may be undesirable to mix air in hospitals for medical reasons.

Universities and schools. Again, air conditioning is usually limited to special areas where the activity demands it. These areas are normally isolated from one another in a complex building, so the market is mainly for small systems to serve such areas.

Museums and libraries. These buildings have a general requirement for air conditioning, particularly where the collections housed are sensitive to changes in temperature and relative humidity, or where large numbers of visitors are likely to congregate for any length of time, or again where lighting levels for display purposes are high.

Future demand. This must be estimated in relation to the building programmes in each market sector. In addition to new building, modernization schemes can create demands for air conditioning through the extension of existing systems and by the addition of cooling to existing ventilation and heating systems.

Equipment and installations. The *Technical Guide to Gas Air Conditioning* published by the British Gas Corporation [38], includes a section which illustrates a number of premises in which gas-fired air-conditioning equipment is being used and also sections which give details of the equipment available from manufacturers and a list of equipment suppliers.

15.5.3 Space heating

There are two methods available for raising the internal temperature of industrial and commercial buildings. These are radiant systems and convective systems. A number of methods developed to utilize each of the systems will be described. In many cases the systems can be applied to either commercial or industrial surroundings, due regard having been paid to the specific nature of the duty required.

Radiant-heating systems [44–6]

Overhead radiant heaters. These can provide an efficient heating system in buildings with a roof or ceiling height of not less than 3.4 m (11 ft). The basic unit is a gas-fired ceramic panel and the products of combustion are released directly into the atmosphere being heated. Very high surface-emission rates can be achieved, about 17 W cm^{-2} and units range from 4 to 20 kW. Surface temperatures are in the region of 800–900 °C and the radiant efficiency is about 35–40 per cent.

One of the advantages of this system is that the heaters can be mounted in positions where direct heating of persons, machinery or stock can be achieved without attempting to heat the whole volume of air. However, this concentration of heat on individuals is often limited by factors such as ventilation rate and mobility of persons. The heaters can be controlled thermostatically and/or by time switch either in total or in groups, the latter giving good flexibility.

Overhead-radiant-heater installations are often a little cheaper in capital cost than air-convection systems and running costs may be lower because, as seen earlier, the air temperature can be 3 °C lower than for air-heating systems for similar comfort conditions. However, a high proportion of the heat generated may be carried upwards by the combustion products. Another disadvantage is the possibility of condensation in the roof space, e.g. on steelwork. In general then they are suitable for use in buildings having a fairly high ventilation rate, or for open-air heating, e.g. in loading bays and sports stadia.

Maintenance required is panel cleaning at the start of each heating season. This may be a considerable task on large installations. Panel life is given in [46] as 10 000 h. Figure 7:36(*c*) shows a typical ceramic radiant-heater burner plaque.

Overhead radiant tubes (*radiant pipe heaters*). This is a direct gas-fired radiant emitter. Each unit consists of a 4–7 m length of steel pipe of 65 mm bore, with a combustion chamber at one end and an exhaust fan at the other. They are mounted at a height of 3 to 5 m above the floor. High-temperature gas is drawn along the tube by the fan and rejected to the outside air through a short flue pipe. A feature of one

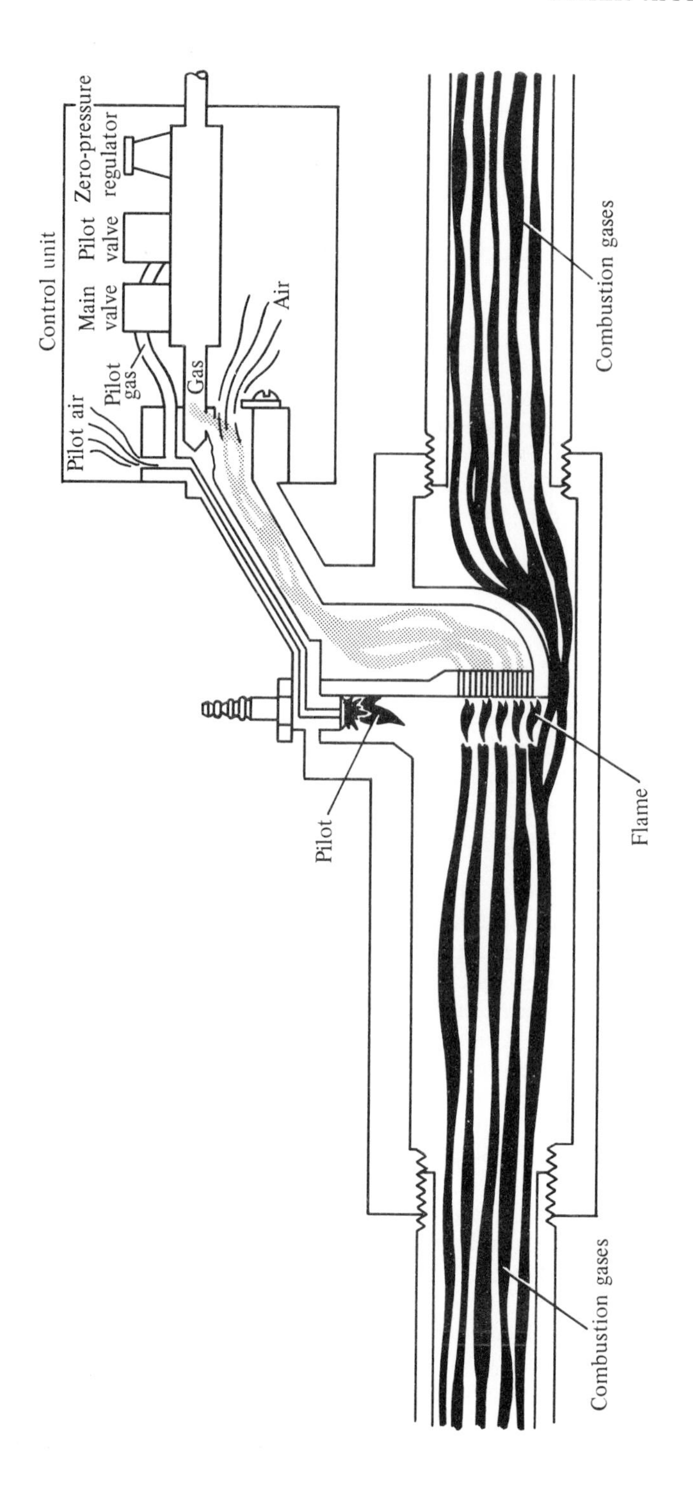

Figure 15:22 Combustion chamber of overhead-radiant-tube space heater [46]

safeguard, as shown in Figure 15:22 is the control. Apart from the normal flame safeguard, purge period, automatic ignition and draught-failure protection, by governing the gas-inlet pressure to zero both gas and air for combustion are induced by fan suction so maintaining a constant air–gas ratio which allows for variations in fan suction. The combustion efficiency is about 90 per cent, radiant efficiency is claimed to be as high as 80 per cent. The tube is screened with a sheet-metal canopy to reduce upward losses. In practice a number of units, up to about 16, can be linked in series with a common exhaust fan. Typical outputs per unit, 4–7 m long, range from 12 to 18 kW.

Capital costs can be from 25 to 70 per cent lower than for traditional piped systems. The high efficiency, plus the savings attributable to lower air temperature requirements, time and zone controlled heating, can reduce fuel cost by 30–50 per cent compared with the traditional system. Figure 15:22 shows a combustion chamber and radiant pipe.

Overhead air-heated radiant tubes. This radiant-tube system uses air as the heat-transfer medium. The radiating surface is made up of banks of circular metal ducts, insulated on the top and protected from cross-draughts at the sides by metal screens.

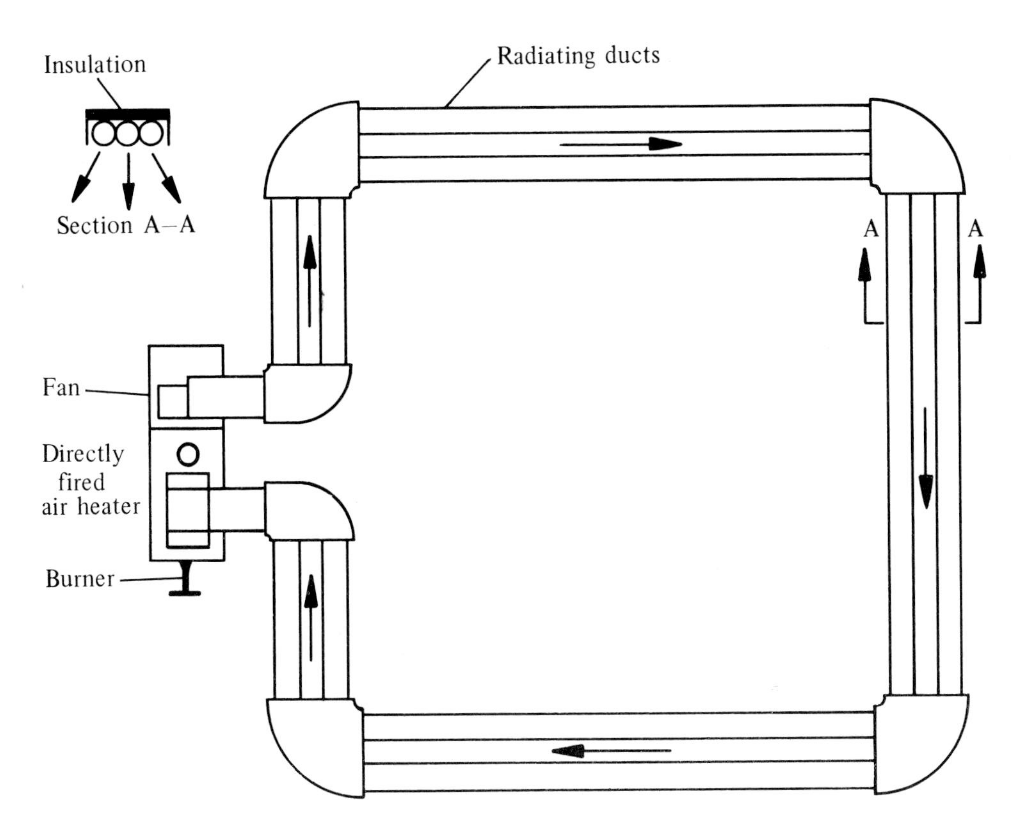

Figure 15:23 Operating principle of overhead air-heated radiant tubes [45]

A directly fired heater supplies air at a temperature of 200 °C and a centrifugal fan drives the air around the closed ducting circuit (Figure 15:23). The average surface temperature of the tube banks is about 150 °C and the radiant efficiency is about 70 per cent. Tube sizes range from 230 to 600 mm and the output per metre run of tube bank ranges from 0.5 to 5.0 kW (0.52 to 5.2 Btu h^{-1} per foot). The load limit per system is about 0.6 MW (2×10^6 Btu h^{-1}). The fan power requirements amount to about 2 per cent of the total energy.

The capital or first cost is less than that of a piped radiant panel system with a central boiler plant, but more than that of a gas-fired unit radiant heater installation. This system can be compared with the traditional piped radiant heating system and has the advantage over directly fired emitters in having a more uniform spread of heat and reduced maintenance because of fewer heat-generating units, but with the usual risk of central plant, that a heater failure means total loss of heat to the building. The heater is flued to the outside air, so there are no condensation problems.

Steam and hot-water radiant systems. The traditional steam and hot-water systems always used unit panels, the main development now is in the use of radiant strip. This consists of continuous lengths of panel, generally mounted horizontally near the roof. Lengths can run to over 100 m. The radiant strip is designed to work on medium- or high-temperature hot water (120–180 °C) or steam. The basic principle is that of a plate heated by contact with a pipe. The radiating efficiency ranges from 50 per cent with the simple pipe grid and air space, to 80 per cent with triangular-section pipes. Emission per unit length depends on the temperature of the medium, the geometry of the strip, the ratio of pipe to plate and the plate material. The commercially available types all represent some compromise between cost and unit emission. Based on average water temperature of 150 °C (302 °F) unit emission for a single-pipe strip can vary from 450 to 600 W m^{-1} (470 to 625 Btu h^{-1} ft^{-1}) with a cost ratio of about 1 : 2. The width of the strip varies with the number of pipes and in practice ranges from 250 to 1500 mm. Commercial applications such as office blocks often use 'sill line' radiators.

Central radiant systems are the most expensive to install but running costs can be among the lowest since the boiler plant can be maintained at high efficiency and heat output can be modulated. In this type of system and the air-heated system alternative fuels to gas are available.

Floor heating. It is not often possible to provide space heating entirely from the floor, unless the building is well insulated and air change is limited. The surface temperature is limited to 24 °C for comfort. The technique is generally to circulate hot water in metal or plastic tubing embedded in the floor. Capital costs can be comparable with central radiant systems.

Forced-convection systems [45–50]

Direct fired blowers. These blowers, which can use gas or light oil, consist of a combustion chamber and secondary heat-transfer matrix. Air is drawn or blown through the heater by a fan and passes directly into the building through adjustable outlets. The products of combustion are vented separately to the outside air. Heater sizes start at a few kilowatts and go up to very large models with outputs in the region of 1 MW (3.5×10^6 Btu h^{-1}) and airflow rate up to 20 m^3 s^{-1} (700 ft^3 s^{-1}).

Temperature distribution needs careful consideration with these units and in general a small number of larger units are less effective than a greater number of small units. The basic design is the subject of British Standard 4256 : Part 2 : 1972, which fixes the limits of operating temperature and airflow rate. Most commercial models work with an air-discharge temperature of 65 °C and the combustion efficiency is about 80–85 per cent on natural gas. They are usually constructed from stainless steel, propeller-type fans are used when discharge is free and blower types when ducting is used, the air supply may be fresh or recirculated. Models are available for mounting overhead or on walls and as free-standing floor units.

High-temperature induction systems. In this system, hot air is generated at a central point and distributed throughout the building by means of ducts, in much the same way as the old type plenum system of combined heating and ventilation. The difference is in the air quantities. Air leaves the generator at about 200 °C with a heat-carrying potential about six times as great as in the traditional warm-air system, so much lower airflow rates are needed. The air-discharge points are designed to induce room air into the jet and so reduce the temperature to a suitable level. They are located at high level at suitable points in the building. The central generator means less maintenance and a higher seasonal efficiency than directly fired blowers. Compared with piped systems such as unit heaters, the high temperature induction system eliminates the boiler house and a secondary distribution medium.

The system is capable of heating large areas from the central generator, the ratings going up to 5 MW (17×10^6 Btu h^{-1}) with flow rates at 200 °C (392 °F) of 35 m^3 s^{-1} (1240 ft^3 s^{-1}). The installation cost of an induction heating system is said to be about two-thirds that of a central piped system with boiler plant.

Unit heaters. The unit heater, which is simply a finned-tube battery and fan, was at one time the only means of heating used with central piped systems. They are still used in many installations, but for new buildings and in replacement the method has been losing ground to directly fired blowers and radiant heating. Units available range from simple battery/propeller fans to high-velocity down-discharge units using centrifugal fans with ratings up to 0.2 MW (0.7×10^6 Btu h^{-1}) and 4 m^3 s^{-1} (140 ft^3 s^{-1}).

For improved temperature distribution, a number of smaller units are preferred to a few larger units. They can be used for replacement air by fitting an air-intake duct to the outside.

The installation costs of unit-heater systems is intermediate between directly fired blowers and central radiant systems. Performance of a battery may suffer from accumulated dust and maintenance of dispersed unit motors, which may be difficult to reach, can be a disadvantage.

Makeup-air heaters. Makeup-air heating generally refers to the heating of air admitted to a building to replace air mechanically extracted at relatively high rates. Makeup-air heating is needed mainly in industrial buildings where special ventilation is required for chemical-process areas, paint-spraying booths, etc.

In the absence of a controlled system for the admittance of air to replace the amount extracted, air would be drawn from ill-fitting doors and windows or even by reverse flow in flues. This would cause discomfort to the occupants by creating cold draughts and lowering the standard of the thermal environment. The incoming air must therefore be heated and distributed to points least likely to affect the occupants adversely. This can be achieved by a fan-assisted air heater and if the fan is rated 5 to 10 per cent greater than the total rating of the extract fans then draughts and flue interference are eliminated.

Such heaters can be of conventional heat-exchanger type indirectly fired by either gas or oil or steam coils placed in the inlet ducting. There is also the direct gas-fired heater which is essentially a gas burner placed in the makeup-air stream. Steam coils can only be economical if the steam used is surplus to the requirements of a much larger process load, so they are rare in commercial premises. Indirect heaters have a higher initial cost because of the heat exchanger and flue system which can be costly in high buildings. The direct gas-fired type has the advantage of a turndown ratio of 35 : 1 compared with the restricted turndown ratio of a typical pressure oil burner. This greater flexibility enables a much closer control of air temperature to be achieved and gives a higher utilization efficiency.

A typical heater is shown diagrammatically in Figure 15:24. There is very little resistance to airflow so outputs can be high and as no flue is required efficiencies of about 90 per cent are obtained. Units available go up to ratings of about 1 MW (3.4×10^6 Btu h^{-1}) and 15 $m^3 s^{-1}$ (530 $ft^3 s^{-1}$). In the USA units of up to 3 MW (10×10^6 Btu h^{-1}) and 50 $m^3 s^{-1}$ (1800 $ft^3 s^{-1}$) are used, the system is widely accepted and has been in use since the mid-1960s. However, these direct gas-fired installations contravene the United Kingdom Building Regulations section M3 [50] although local waivers of these Regulations can be sought and acquired when such a system is proposed.

The objections relate to the effect of the products of combustion on the building occupants. These products are largely carbon dioxide and water vapour, though traces of carbon monoxide, oxides of nitrogen, sulphur dioxide and unburned hydrocarbons may be present. The heater design, however, produces a 50 : 1 dilution of products (i.e. 2 per cent products per volume of fresh air) for a 55 °C rise in temperature. A 30 °C rise, which is more realistic for the UK would give a 100 : 1 dilution with resulting concentrations of 1500 ppm CO_2 and 5 ppm CO.

Air curtains. The air curtain is a means of compensating for the inrush of cold air when large doors are opened. Most air curtains are designed to supply heated air, using a directly fired unit, a steam or hot-water battery. Some use cold air only, in which a bank of fans mounted at the edge of the door deflect the wind vector at the opening.

The warm-air curtain is produced by a fan/heater unit delivering air into a distribution duct either above the door opening or at the side. Nozzles or slots in the duct produce a jet of air across the face of the opening. Air can be partially recycled with the down discharge system through a floor grille. The effectiveness depends on the unit flow rate ($m^3 s^{-1}$ per square metre of opening), the jet velocity and the temperature rise. Current designs range from 0.1 to 1.5 $m^3 s^{-1} m^{-2}$ with a temperature rise of about 80 °C and a velocity of around 10 $m s^{-1}$.

Heat recovery [45,51]. In industrial buildings heated by air-handling systems, combined with mechanical extract ventilation, it is possible to recover a large amount of heat from the exhaust air.

The method is to route the fresh-air inlet and exhaust ducts close together with a heat exchanger between them. One technique is the regenerative wheel in which the

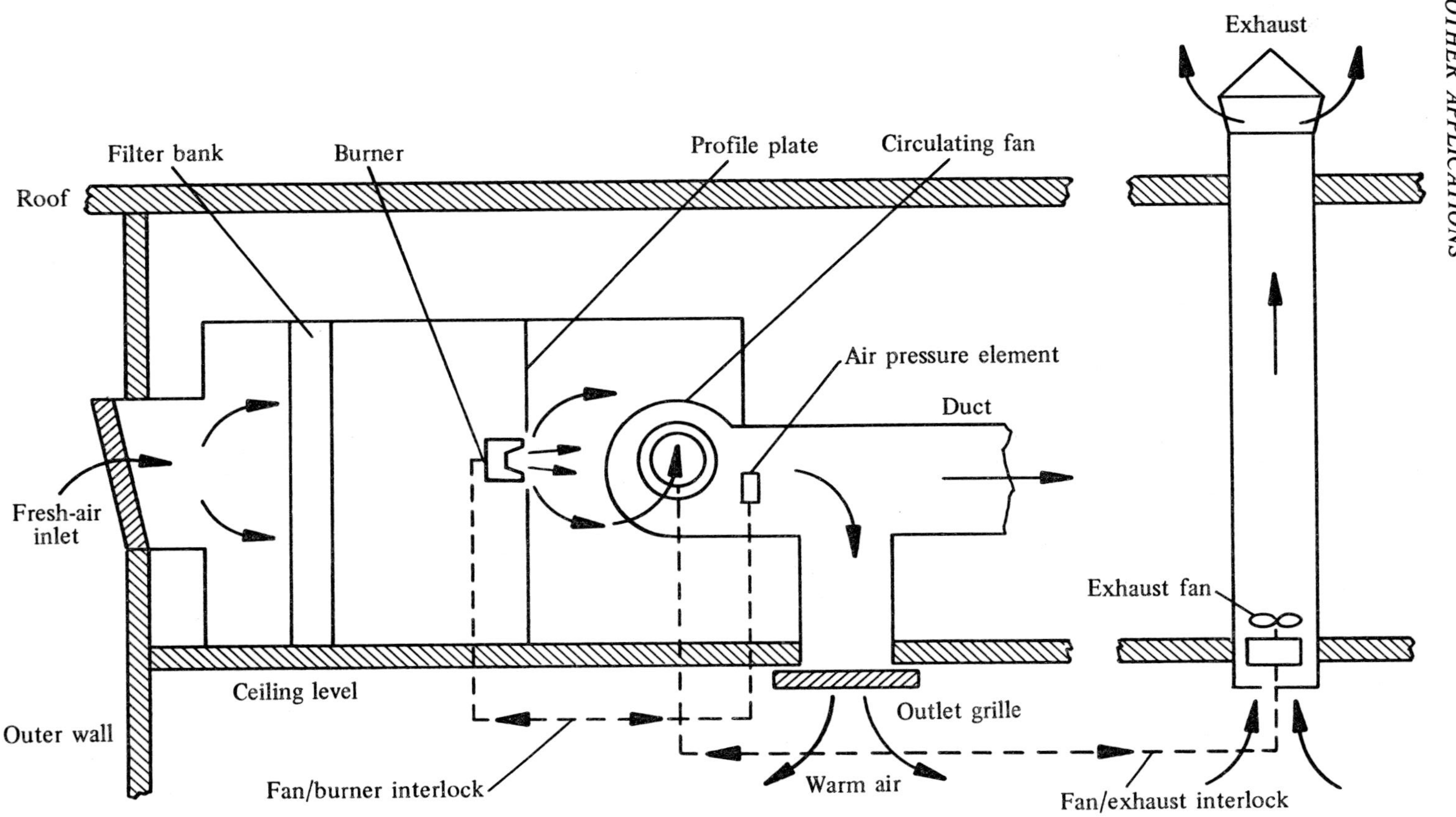

Figure 15:24 Diagram of a typical makeup-air system

metal matrix of the wheel picks up heat from the warm exhaust air and transfers it to the cold incoming air.

Three relatively new methods are the parallel-plate exchanger, the tubular matrix exchanger and the heat pipe. The first consists of a multi-check sandwich of glass plates with air spaces. Incoming air and exhaust air flow through alternate spaces and transfer heat. In the tubular matrix exchanger, one air stream flows through a bank of aluminium or plastic tubes, and the other around the tube bundle. The heat pipe is a sealed tube containing a charge of refrigerant. When one end is heated, refrigerant is evaporated and condenses at the cold end, giving up heat and the liquid is recycled by capillary action through a type of wick. The apparent conductivity is about 500 times that of an equivalent mass of copper. The methods described have a recovery factor of between 50 and 80 per cent.

Market available

In recent years there has been a rapid growth in the use of gas in commercial and industrial space-heating systems and although this is mainly a winter load, when coupled with an air-conditioning role some levelling of seasonal loading is achieved. The future of the space-heating load depends on many factors [50] such as the availability of gas, the desirable 'marketing mix', the price of gas relative to alternative fuels, the level of new building both industrial and commercial and the activity in the equipment-replacement market. The current and future efforts to reduce energy consumption on a national basis must also influence the future space-heating scene.

District heating

Although not strictly a topic of direct interest to an industrial gas engineer, the developments which may take place in the future could possibly be important factors in future commercial and industrial space-heating projects of interest to the gas industry.

The government pressures upon everyone to 'save it' call for an energy balance sheet to be prepared for all new and modernized buildings. This creates the need to understand how individual and communal systems function and how effective and efficient they are in doing a comparable job. In the final analysis, the economics of the alternative systems must be calculated not just in capital terms, but by a full 'cost in use' exercise, as there is little point in saving energy at any price. A review of the subject covering the role of gas, legislation and giving case histories of existing schemes in the UK is given in a paper by Jarvis [52], to which the reader is referred for details.

REFERENCES

1 Wainwright, A., private communication
2 Sloan, C. E., 'Drying systems and equipment', *Chemical Engineering*, **74,** No. 13 (June 1967) 169
3 Pardoe, T. H., private communication
4 Robertshaw, G. W., 'Gas firing' in *Efficient Use of Energy* (ed. Dryden, I. G. C.). London: IPC Science & Technology Press (1975)
5 Margetts, J. and Caine, J. A., 'Natural gas for paper drying', Gas Council Midlands Research Station, Int. Report 308 (1967)
6 'Tank heating method could save fuel imports', *Natural Gas for Industry* (May 1975) 105

7 Lupton, H. P., *Industrial Gas Engineering*, Vol. 3. London: Walter King (1960)
8 Patrick, E. A. K. and Thornton, E., 'Immersion tube heating of aqueous liquids', *Gas Council Res. Commun.* GC47 (1958); *Trans. Inst. Gas Eng.*, **108** (1958–9) 360
9 Tailby, S. R. and Clutterbuck, E. K., 'Heat transfer in horizontal town-gas immersion tubes', *Trans. Inst. Chem. Eng.*, **42** (1964) T64
10 American Gas Association, 'Research in fundamentals of immersion tube heating with gas', *AGA Bull.*, No. 24 (1944)
11 Thornton, E. and Robertshaw, G. W., 'Thermal design of gas-fired tubular immersion heaters', *J. Inst. Fuel*, **38** (1965) 307
12 Francis, W. E., Hoggarth, M. L. and Reay, D., 'A study of gas-fired pulsating combustors for industrial applications', *Gas Council Res. Commun.* GC91 (1962); *J. Inst. Gas Eng.*, **3** (1963) 301
13 Thomas, P. M., 'Submerged combustion', *Industrial Gas* (London), **51** (1965) 176
14 Babich, V. S., Efimov, V. T. and Volov, G. I., 'Determination of optimal conditions for concentration processes in equipment with submerged burners', *Int. Chem. Eng.*, **14** (1974) 226
15 Atkinson, D. R., 'Woodall-Duckham/Nordac submerged combustion evaporator for concentration of phosphoric acid', *Phosphorus and Potassium* (September–October 1967)
16 Young, D. C. and Scott, C. B., 'Wet-process polyphosphoric acid', *Chem. Eng. Prog.*, **59,** No. 12 (December 1963) 80
17 Submerged Combustion (Engineering) Ltd, Telford, Technical Data Sheet TDS 1 (1972)
18 'Submerged combustion developments by C. & W. Walker Ltd', *Flambeau* (1968)
19 Rhodes, K. G., 'Natural gas and the working flame', Manchester and District Junior Gas Association, Meeting (November 1968)
20 Spackman, A. J., 'Cone height control system', *Industrial Gas* (London), **52** (1966) 180
21 Grönegress, H. W. and Rossner, H., 'Erfahrungen mit der Umstellung von Brennhärteanlagen von Leuchtgas-Sauerstoff auf Erdgas-Sauerstoff', *Gas Wärme International*, **16** (1967) 193
22 Kennedy, D. M., 'Environmental control and human factors', *Inst. Gas Eng. Commun.* 818 (1970); *J. Inst. Gas Eng.*, **10** (1970) 645
23 Hardy, A. C., 'Comfort in offices', *Inst. Gas Eng. Commun.* 826 (1970); *J. Inst. Gas Eng.*, **10** (1970) 817
24 The Building Regulations 1972 (Statutory Instrument 1972/317). London: HMSO (1972)
25 The Building Regulations (Second Amendment) 1974. London: HMSO (1974)
26 IHVE Guide (1970)
27 Berry, E. W. and Wright, G. F., 'Heat requirements of buildings', *B.G.C. Watson House Bulletin*, **36,** No. 267 (1972)
28 McNair, H. P., 'Thermal comfort', *B.G.C. Watson House Bulletin*, **36,** No. 263 (1972)
29 Berry, E. W., Bruce, R. D. and Carter, B. C., 'Installation and ventilation', *Gas Council Res. Commun.* GC177 (1970); *J. Inst. Gas Eng.*, **11** (1971) 589
30 Dance, E. W. G., 'Gas in existing buildings', Institution of Gas Engineers, Short Residential Course, Natural Gas and Fuel Conservation, Pembroke College, Oxford (1975)

31 Tipping, J. C., 'Gas in new buildings', Institution of Gas Engineers, Short Residential Course, Natural Gas and Fuel Conservation, Pembroke College, Oxford (1975)
32 National Economic Development Office, *Energy Conservation in the United Kingdom: Achievements, Aims and Options*. London: HMSO (1975)
33 Langdon, F. J., *Modern Offices: A User Survey* (Ministry of Technology Building Research Station, National Building Studies, Research Paper 41). London: HMSO (1966)
34 Ministry of Public Buildings and Works, R. & D., 'Heating installations: the analysis of costs and cost control during design' (1967)
35 Tully, R. E., 'Savings in energy in industrial and commercial buildings', Combustion Engineering Association, Conference, Fuel Conservation or Crisis (1973)
36 Holt, J. E., 'Thermal comfort in factories', *Inst. Gas Eng. Commun.* 827 (1970); *J. Inst. Gas Eng.*, **10** (1970) 828
37 Pattison, J. R. and Pearson, J., 'The role of gas in air conditioning', *Inst. Gas Eng. Commun.* 916 (1973); abstract in *J. Inst. Gas Eng.*, **13** (1973) 265
38 British Gas Corporation, *Technical Guide to Gas Air Conditioning* (1974)
39 Jarvis, D. J. and Pearson, J., 'Gas air conditioning for public buildings', Institution of Gas Engineers, Public Works Congress (1972)
40 Trumble, D. R. and Rattle, J. B., 'The development of uses of gas engines', Eastern Junior Gas Association, Meeting (December 1974)
41 Skerrett, B. and Wright, G. F., 'Air conditioning by gas', Institution of Heating and Ventilating Engineers, Conference (1969)
42 Benson, S. R., 'The prospects for gas in air conditioning', *Refrig. & Air Cond.*, **74,** No. 879 (June 1971) 49
43 Dove, A. A., 'Prospects for gas in air conditioning', *Heat. & Vent. Eng.*, **44** (1970–1) 508
44 Taylor, F. M. H., 'Gas for radiant space heating', *Inst. Gas Eng. Commun.* 859 (1971)
45 Field, A. A., 'Industrial space heating', *Energy World*, No. 12 (January 1975) 9
46 Eldred, D. and Ford, M., 'Commercial load, past, present and future', Manchester and District Junior Gas Association, Meeting (1968)
47 Cox, R. W., 'Direct-fired make-up air heaters', British Gas Corporation Midlands Research Station (1974)
48 Oakley, D., 'Direct-fired make-up air and space heating', Manchester and District Junior Gas Association, Meeting (February 1972)
49 'Make-up air heating', *B.G.C. Watson House Bulletin*, **36,** No. 267 (1972)
50 Ford, M. F., 'Marketing mix—commercial gas', Institution of Gas Engineers, Manchester and District Section, Meeting (December 1974); *Gas Eng. & Manage.*, **15** (1975) 205
51 Manuel, K., 'Fuel economy in industrial process heating', Institution of Gas Engineers, Short Residential Course, Natural Gas and Fuel Conservation, Pembroke College, Oxford (April 1975)
52 Jarvis, D. J., 'Communal gas and communal heating', Institution of Gas Engineers, Short Residential Course, Natural Gas and Fuel Conservation, Pembroke College, Oxford (April 1975)
53 Hoggarth, M. L., Cox, R. W., and Jones, D. A., *Inst. Gas Eng. Commun.* 977 (1975)

Author index

Subject index